TECHNIQUES OF CHEMISTRY

ARNOLD WEISSBERGER, *Editor*

VOLUME I

PHYSICAL METHODS OF CHEMISTRY

PART IIA
Electrochemical Methods

TECHNIQUES OF CHEMISTRY

ARNOLD WEISSBERGER, *Editor*

VOLUME I

PHYSICAL METHODS OF CHEMISTRY, in Five Parts
Incorporating Fourth Completely Revised and Augmented Edition
of Physical Methods of Organic Chemistry
Edited by Arnold Weissberger and Bryant W. Rossiter

VOLUME II

ORGANIC SOLVENTS, Third Edition
John A. Riddick and William S. Bunger

TECHNIQUES OF CHEMISTRY

VOLUME I

PHYSICAL METHODS OF CHEMISTRY

INCORPORATING FOURTH COMPLETELY REVISED AND AUGMENTED
EDITION OF TECHNIQUE OF ORGANIC CHEMISTRY,
VOLUME I, PHYSICAL METHODS OF ORGANIC CHEMISTRY

Edited by
ARNOLD WEISSBERGER
AND
BRYANT W. ROSSITER

Research Laboratories
Eastman Kodak Company
Rochester, New York

PART IIA
Electrochemical Methods

WILEY-INTERSCIENCE

A DIVISION OF JOHN WILEY & SONS, INC.

New York · London · Sydney · Toronto

Library of Congress Catalogue Card Number: 77-114920

ISBN 0 471 92727 9

Printed in the United States of America

10 9 8 7 6 5 4 3 2 1

PLAN FOR
PHYSICAL METHODS OF CHEMISTRY

PART I
Components of Scientific Instruments, Automatic Recording and Control, Computers in Chemical Research

PART II
Electrochemical Methods

PART III
Optical, Spectroscopic, and Radioactivity Methods

PART IV
Determination of Mass, Transport, and Electrical-Magnetic Properties

PART V
Determination of Thermodynamic and Surface Properties

AUTHORS OF PART II

RALPH N. ADAMS,
Department of Chemistry, University of Kansas, Lawrence, Kansas

BERNARD D. BLAUSTEIN,
Pittsburgh Coal Research Center, Bureau of Mines, U.S. Department of the Interior, Pittsburgh, Pennsylvania

ERIC R. BROWN,
Research Laboratories, Eastman Kodak Company, Rochester, New York

RICHARD P. BUCK,
University of North Carolina, Chapel Hill, North Carolina

JACK CHANG,
Research Laboratories, Eastman Kodak Company, Rochester, New York

JOHN L. EISENMANN,
Ionics, Inc., Watertown, Massachusetts

YUAN C. FU,
Pittsburgh Coal Research Center, Bureau of Mines, U.S. Department of the Interior, Pittsburgh, Pennsylvania

DAVID M. HERCULES,
Department of Chemistry, University of Georgia, Athens, Georgia

SEYMOUR L. KIRSCHNER,
Food and Drug Research Laboratories, Inc., New York, New York

ROBERT F. LARGE,
Research Laboratories, Eastman Kodak Company, Rochester, New York
FRANK B. LEITZ,
Ionics, Inc., Watertown, Massachusetts
LOUIS MEITES,
Department of Chemistry, Clarkson College of Technology, Potsdam, New York
OTTO H. MÜLLER,
State University of New York, Upstate Medical Center, Syracuse, New York
ROYCE W. MURRAY,
University of North Carolina, Chapel Hill, North Carolina
RICHARD C. NELSON,
Department of Physics, The Ohio State University, Columbus, Ohio
STANLEY PIEKARSKI,
Plastics Department, E. I. du Pont de Nemours & Company, Wilmington, Delaware
GERHARD POPP,
Research Laboratory, Eastman Kodak Company, Rochester, New York
LEO SHEDLOVSKY,
Consulting Chemist, New York, New York
THEODORE SHEDLOVSKY,
The Rockefeller University, New York, New York
MICHAEL SPIRO,
Imperial College of Science and Technology, London, England
STANLEY WAWZONEK,
University of Iowa, Iowa City, Iowa

NEW BOOKS AND NEW EDITIONS OF BOOKS OF THE TECHNIQUE OF ORGANIC
CHEMISTRY WILL NOW APPEAR IN TECHNIQUES OF CHEMISTRY. A LIST OF
PRESENTLY PUBLISHED VOLUMES IS GIVEN BELOW.

TECHNIQUE OF ORGANIC CHEMISTRY

ARNOLD WEISSBERGER, *Editor*

INTRODUCTION TO THE SERIES

Techniques of Chemistry is the successor to the Technique of Organic Chemistry series and its companion—Technique of Inorganic Chemistry. Because many of the methods are employed in all branches of chemical science, the division into techniques for organic and inorganic chemistry has become increasingly artificial. Accordingly, the new series reflects the wider application of techniques, and the component volumes for the most part provide complete treatments of the methods covered. Volumes in which limited areas of application are discussed can be easily recognized by their titles.

Like its predecessors, the series is devoted to a comprehensive presentation of the respective techniques. The authors give the theoretical background for an understanding of the various methods and operations and describe the techniques and tools, their modifications, their merits and limitations, and their handling. It is hoped that the series will contribute to a better understanding and a more rational and effective application of the respective techniques.

Authors and editors hope that readers will find the volumes in this series useful and will communicate to them any criticisms and suggestions for improvements.

Research Laboratories   ARNOLD WEISSBERGER
Eastman Kodak Company
Rochester, New York

PREFACE

Physical Methods of Chemistry succeeds, and incorporates the material of, three editions of *Physical Methods of Organic Chemistry* (1945, 1949, and 1959). It has been broadened in scope to include physical methods important in the study of all varieties of chemical compounds. Accordingly, it is published as Volume I of the new Techniques of Chemistry series.

Some of the methods described in *Physical Methods of Chemistry* are relatively simple laboratory procedures, such as weighing and the measurement of temperature, or refractive index, and determination of melting and boiling points. Other techniques require very sophisticated apparatus and specialists to make the measurements and to interpret the data; x-ray diffraction, mass spectrometry, and nuclear magnetic resonance are examples of this class. Authors of chapters describing the first class of methods aim to provide all information that is necessary for the successful handling of the respective techniques. Alternatively, the aim of authors treating the more sophisticated methods is to provide the reader with a clear understanding of the basic theory and apparatus involved, together with an appreciation for the value, potential, and limitations of the respective techniques. Representative applications are included to illustrate these points, and liberal references to monographs and other scientific literature providing greater detail are given for readers who want to apply the techniques. Still other methods that are successfully used to solve chemical problems range between these examples in complexity and sophistication and are treated accordingly. All chapters are written by specialists. In many cases authors have acquired a profound knowledge of the respective methods by their own pioneering work in the use of these techniques.

In the earlier editions of *Physical Methods* an attempt was made to arrange the chapters in a logical sequence. In order to make the organization of the treatise lucid and helpful to the reader, a further step has been taken in the new edition—the treatise has been subdivided into technical families:

Part I   Components of Scientific Instruments, Automatic Recording and Control, Computers in Chemical Research
Part II   Electrochemical Methods
Part III   Optical, Spectroscopic, and Radioactivity Methods

Part IV   Determination of Mass, Transport, and Electrical-Magnetic
           Properties
Part V   Determination of Thermodynamic and Surface Properties

This organization into technical families provides more consistent volumes and should make it easier for the reader to obtain from a library or purchase at minimum cost those parts of the treatise in which he is most interested.

The more systematic organization has caused additional labor for the editors and the publisher. We hope that it is worth the effort. We thank the many authors who made it possible by adhering closely to the agreed dates of delivery of their manuscripts and who promptly returned their proofs. To those authors who were meticulous in meeting deadlines we offer our apologies for delays caused by late arrival of other manuscripts, in some cases necessitating rewriting and additions.

The changes in subject matter from the Third Edition are too numerous to list in detail. We thank previous authors for their continuing cooperation and welcome the new authors to the series. New authors of Part II are Ralph N. Adams, Bernard D. Blaustein, Eric R. Brown, Richard P. Buck, Jack Chang, John L. Eisenmann, Yuan C. Fu, David M. Hercules, Seymour L. Kirschner, Robert F. Large, Frank B. Leitz, Royce W. Murray, Richard C. Nelson, Stanley Piekarski, and Gerhard Popp.

We are also grateful to the many colleagues who advised us in the selection of authors and helped in the evaluation of manuscripts. They are for Part II: Dr. Roger C. Baetzold, Dr. Charles J. Battaglia, Dr. Eric R. Brown, Dr. Jack Chang, Dr. Donald L. Fields, Dr. Robert L. Griffith, Dr. Arthur H. Herz, Mrs. Ardelle Kocher, Dr. Robert F. Large, Dr. Louis Meites, Dr. Louis D. Moore, Jr., Dr. Charles W. Reilley, Mrs. Donna S. Roets, Dr. Willard R. Ruby, Mr. Calvin D. Salzberg, Miss Dianne C. Smith, Dr. Donald E. Smith, Dr. Benjamin B. Snavely, Dr. R. Eliot Stauffer, and Dr. John R. Wilt.

The senior editor expresses his gratitude to Bryant W. Rossiter for joining him in the work and taking on the very heavy burden with exceptional devotion and ability.

ARNOLD WEISSBERGER
BRYANT W. ROSSITER

April 1970
Research Laboratories
Eastman Kodak Company
Rochester, New York

CONTENTS

Chapter **VIII**

Chapter **I**

POTENTIOMETRY: OXIDATION-REDUCTION POTENTIALS

Stanley Wawzonek

I THEORETICAL INTRODUCTION

Galvanic Cells

An oxidation-reduction reaction is a reaction involving a transfer of electrons from one chemical species (atom, molecule, or ion) to another. In principle, any such reaction can form the basis of a *galvanic cell*. The only requirement is that the site of oxidation (loss of electrons) be physically separated from the site of reduction (gain of electrons) so that the reaction cannot be completed without the passage of an electric current from one site to the other except through the external circuit. The sites of oxidation and reduction are known as *electrodes*. The electrode at which oxidation occurs is called the *anode*. The site of reduction is called the *cathode*. In *potentiometry* no actual reaction is carried out, that is, no appreciable current is passed. Instead, one measures the tendency for the reaction to occur by measuring the difference in electrical potential between the anode and cathode.

The construction of a galvanic cell is conventionally represented by a line formula, beginning with the anode. If it subsequently turns out that the cell actually operates in the opposite direction, then this can be indicated without rewriting the cell formula by a negative sign for the potential, as discussed below. Thus, for example,

$$\text{Zn}; \text{ZnCl}_2(C_1); \text{AgCl}; \text{Ag} \tag{1.I}$$

represents a cell with a zinc anode, a zinc chloride solution at concentration C_1, and a cathode consisting of silver chloride plated on silver. Each semicolon represents a phase boundary. The electrode reactions of this cell, writing the anode reaction first, are

$$\text{Zn} \rightarrow \text{Zn}^{2+} + 2e^-$$
$$2e^- + 2\text{AgCl} \rightarrow 2\text{Ag} + 2\text{Cl}^-$$

and the overall reaction, obtained on adding these, is

$$\text{Zn} + 2\text{AgCl} \rightarrow 2\text{Ag} + \text{Zn}^{2+} + 2\text{Cl}^-$$

It should be noted that the anode and cathode reactions must be written so as to involve the same number of electrons.

A second example is

$$\text{Zn}; \text{ZnSO}_4(C_1); \text{CuSO}_4(C_2); \text{Cu} \tag{1.II}$$

This cell consists of a zinc anode, a ZnSO_4 solution at concentration C_1, a CuSO_4 solution at concentration C_2, and a copper cathode. The ZnSO_4 and CuSO_4 solutions are in contact but are prevented from mixing with one another. This may be achieved, for example, by a porous ceramic plug.

The electrode reactions of cell (1.II) are

$$Zn \rightarrow Zn^{2+} + 2e^-$$

$$2e^- + Cu^{2+} \rightarrow Cu$$

and the overall reaction is

$$Zn + Cu^{2+} \rightarrow Zn^{2+} + Cu$$

An important difference between cells (1.I) and (1.II) is that whereas cell (1.I) contains phase boundaries only at the two electrodes, cell (1.II) contains an additional boundary between the $ZnSO_4$ and $CuSO_4$ solutions. This boundary is known as a *liquid junction*, and it gives rise to a potential difference, known as a *liquid junction potential*. This potential arises from the fact that the rate of diffusion of Zn^{2+} from left to right across the boundary differs from the rate of diffusion of Cu^{2+} from right to left, so that a potential gradient is established. Unless $C_1 = C_2$, an additional contribution to this potential will arise from a difference between the rate of diffusion of sulfate across the boundary and the net rate of cationic diffusion. In any cell containing a liquid junction, therefore, the overall potential difference across the cell is not purely a measure of the tendency for the occurrence of an oxidation-reduction reaction, but also involves a contribution from the liquid junction potential.

The magnitude of the liquid junction potential may be greatly reduced by the introduction of a *salt bridge*. This is a concentrated solution, or often a saturated solution, of a salt whose cation and anion have close to the same mobility. KCl is the preferred salt, but NH_4NO_3 is often used when the adjacent solutions contain Ag^+ or some other ion that forms a precipitate with chloride ion. The salt bridge introduces two liquid junctions instead of one, that is, if a salt bridge were introduced into cell (1.II), there would be a liquid junction between the bridge and the $ZnSO_4$ solution, and another between the bridge and the $CuSO_4$ solution. The ion concentration in the salt bridge, as we have mentioned, is made very high; the concentrations C_1 and C_2 are usually relatively low. Thus the two liquid junction potentials are largely determined by the relative rates of diffusion of K^+ and Cl^- (or NH_4^+ and NO_3^-) into the adjoining solutions. Since these ions have similar mobilities, the two liquid junction potentials remain relatively small. Moreover, the two potentials have about the same value, but in opposing directions, so that the net liquid junction is kept even smaller. It should be emphasized, however, that it can never be completely eliminated.

The use of a salt bridge is indicated in the line formula for a cell by two vertical lines, for example,

$$Zn; ZnSO_4(C_1)\|CuSO_4(C_2); Cu \qquad (1.III)$$

represents the same cell as cell (1.II) with a salt bridge replacing the direct contact between the $ZnSO_4$ and $CuSO_4$ solutions. It might be noted that junction potentials also arise at all connections outside the cell between dissimilar conductors. These are taken into account by a universal agreement that the connection of the cell to the potentiometer is always by means of a copper wire. The symbol e^- in the electrode reactions thus represents an electron in a copper conductor, and all intervening potentials are automatically included in the cell potential.

Electromotive Force and the Thermodynamics of Cell Reactions

The potential difference across a galvanic cell is in general determined by the relative rates of reaction at the anode and the cathode. These rates in turn are determined by thermodynamic factors, by potential energy barriers at the electrode surface, by rates of diffusion, and so on. They may depend strongly on the current flowing and on electrode design. The results of potential measurement in general can be given no simple interpretation. If, however, the transfer of electrons can be made to occur under electrically *reversible* conditions, then the potential difference depends only on the *thermodynamics* of the cell reaction (including the changes resulting from diffusion at liquid junctions). This is a consequence of the fact that the electrical work done by the reversible transfer of electricity is the *maximum* electrical work, and this, at constant temperature and pressure, is a measure of the loss of free energy occurring in the cell reaction. The potential difference of a galvanic cell, measured under reversible conditions, is called the *electromotive force*, or *emf*, of the cell. It is the only kind of potential measurement that can be simply interpreted.

To obtain the quantitative relation between the emf, E and the free energy change a reaction can be considered which, as written, requires for completion the passage of n faradays of electricity, where $F = 1$ faraday = 1 mole of electrons = 96,493 C. The maximum electrical work that can be done is then given (in joules) by nFE, where E is measured in volts. As has been pointed out, this term represents the decrease in free energy in the overall cell reaction, that is,

$$\Delta G = -nFE. \tag{1.1}$$

It should be noted that if ΔG is desired in calories, then nFE, in joules, must be divided by 4.1840, which represents the number of joules per defined calorie.

The sign of the free energy change determines the direction in which a reaction proceeds spontaneously. A negative sign means that it proceeds spontaneously in the direction in which it is written; a positive sign means it proceeds in the opposite direction. Equation (1.1) defines the corresponding criterion for galvanic cells. A positive value of E means that the cell operates in the direction in which it is written, with the anode at the left of the linear formula. A negative value of E means that it operates in the opposite direction. Experimentally, the anode and cathode are identified by their relative potential; that at the anode must be more negative.

The free energy change in a chemical reaction depends on the activities of the participating substances. In general, for an overall reaction,

$$aA + bB \rightarrow lL + mM$$

$$\Delta G = \Delta G^0 + RT \ln \frac{\alpha_L^l \alpha_M^m}{a_A^a a_B^b}, \tag{1.2}$$

where a_i is the activity of the chemical species i and ΔG^0 is the *standard free energy change* of the reaction, that is, the free energy change that would be observed if all reactants and products remained at unit activity throughout the reaction. From (1.1) we can write down at once the corresponding expression for the emf of a cell in which the same chemical reaction occurs:

$$E = E^0 - \frac{RT}{nF} \ln \frac{a_L^l a_M^m}{a_A^a a_B^b} = E^0 - \frac{2.303RT}{nF} \log \frac{a_L^l a_M^m}{a_A^a a_B^b}, \tag{1.3}$$

where E^0 is the standard emf of the cell, defined as the emf that would be observed if all reactants and products were at unit activity. Equation (1.3) is the basic equation of the galvanic cell. It should be noted that it does not include any liquid junction potentials which may be present, and it is therefore exact only when applied to cells without liquid junction. A list of values of the constant $2.303 \, RT/F$ for various temperatures is given in Table 1.1.

Using (1.3), we can write down, for instance, the emf of cell (1.I) as

$$E = E^0 - \frac{RT}{2F} \ln \frac{a_{Ag}^2 a_{Zn^{2+}} a_{Cl^-}^2}{a_{Zn} a_{AgCl}^2}.$$

Substances present as pure solids have unit activity (by definition) so that this equation reduces to

$$E = E^0 - \frac{RT}{2F} \ln a_{Zn^{2+}} a_{Cl^-}^2. \tag{1.4}$$

Equation (1.4) is exact because cell (1.I) contains no liquid junction. However, the corresponding equation for cell (1.II) or (1.III),

$$E = E^0 - \frac{RT}{2F} \ln \frac{a_{Zn^{2+}}}{a_{Cu^{2+}}}, \tag{1.5}$$

which is only an approximation because it ignores the contribution of the liquid junction potential to E. It is a better approximation for cell (1.III) containing the salt bridge than for cell (1.II), but is nevertheless not exact.

Table 1.1   Values of 2.303 RT/F in Volts for Various Temperatures

Temperature (°C)	2.303 RT/F	Temperature (°C)	2.303 RT/F
0	0.05420	31	0.06035
5	0.05519	32	0.06055
10	0.05618	33	0.06074
12	0.05658	34	0.06094
14	0.05697	35	0.06114
16	0.05737	40	0.06213
18	0.05777	45	0.06312
19	0.05797	50	0.06412
20	0.05816	55	0.06511
21	0.05836	60	0.06610
22	0.05856	65	0.06709
23	0.05876	70	0.06808
24	0.05896	75	0.06908
25	0.05916	80	0.07007
26	0.05935	85	0.07106
27	0.05955	90	0.07205
28	0.05975	95	0.07304
29	0.05995	100	0.07404
30	0.06015		

Single Electrode Potentials

The free energy change for the overall reaction in a galvanic cell can be split into two parts, one for the reaction occurring at the anode and one for the reaction at the cathode. For instance, ΔG for the reaction of cell (1.I) can be written as ΔG for the reaction $Zn \rightarrow Zn^{2+} + 2e^-$ plus ΔG for the reaction $2e^- + 2AgCl \rightarrow 2Ag + 2Cl^-$, that is, as the sum of the free energy change for oxidation at the anode and that for reduction at the cathode. It is more convenient, however, to write all single electrode reactions in the same direction; in America the convention has been to write them in the direction of oxidation, whereas in Europe they are written in the direction of reduction. In 1953 at the 17th IUPAC Conference in Stockholm the Commission on Electrochemistry and the Commission on Physicochemical Symbols and Terminology reached agreement on a "convention concerning the signs of electromotive forces and electrode potentials." This convention essentially conforms to the European system of writing potentials and is used here.

From (1.1) we can write down single electrode potentials corresponding to the free energy changes at each electrode. The total cell potential is the cathode potential minus the anode potential. By using cell (1.I) as an illustration, the anode potential for the reaction $Zn^{2+} + 2e^- \rightarrow Zn$ (written as a reduction according to the IUPAC convention) is

$$E_{Zn^{2+},Zn} = E^0_{Zn^{2+},Zn} - \frac{RT}{2F} \ln \frac{a_{Zn}}{a_{Zn^{2+}}a^2_{e^-}}. \tag{1.6a}$$

We mentioned earlier that the electrons of the electrode reaction are to reside in a copper wire. Since there are no previous conventions regarding a standard state for free electrons we may arbitrarily call this the standard state, so that $\alpha^{e^-} = 1$. The activity of Zn is also unity, so that (1.6a) becomes

$$E_{Zn^{2+},Zn} = E^0_{Zn^{2+},Zn} - \frac{RT}{2F} \ln \frac{1}{a_{Zn^{2+}}}. \tag{1.6b}$$

In the same way the potential for the cathode reaction, $2AgCl + 2e^- \rightarrow 2Ag + 2Cl^-$ also written as a reduction is

$$E_{AgCl,Ag,Cl^-} = E^0_{AgCl,Ag,Cl^-} - \frac{RT}{2F} \ln \frac{a^2_{Ag}a^2_{Cl^-}}{a^2_{AgCl}a^2_{e^-}}. \tag{1.7a}$$

Again solids and electrons are considered to be in standard states at unit activity so (1.7a) simplifies to

$$E_{AgCl,Ag,Cl^-} = E^0_{AgCl,Ag,Cl^-} - \frac{RT}{2F} \ln a^2_{Cl^-}. \tag{1.7b}$$

The overall cell potential is given by

$$E_{cell} = E_{cathode} - E_{anode}$$

$$E = E^0_{AgCl,Ag,Cl^-} - \frac{RT}{2F} \ln a^2_{Cl^-} - E^0_{Zn^{2+},Zn} + \frac{RT}{2F} \ln \frac{1}{a_{Zn^{2+}}} \tag{1.8}$$

or

$$E = E^0_{AgCl,Ag,Cl^-} - E^0_{Zn^{2+},Zn} - \frac{RT}{2F} \ln a^2_{Cl^-}a_{Zn^{2+}}.$$

This result is of course identical to that given by (1.4) with the overall $E^0 = E^0_{AgCl,Ag,Cl^-} - E^0_{Zn^{2+},Zn}$.

Single electrode potentials cannot actually be measured so that standard single electrode potentials such as $E^0_{Zn^{2+},Zn}$ and E^0_{AgCl,Ag,Cl^-} cannot be experimentally determined. Accordingly, we arbitrarily assign a value to one particular single electrode potential, and determine all others relative to it. The universal convention is to set

$$E^0_{H^+,H_2} = 0. \tag{1.9}$$

All other E^0 values for single electrodes can then be determined by measuring E^0 for an overall cell, one electrode of which is either a hydrogen electrode, or another electrode whose E^0 value was previously determined relative to a hydrogen electrode.

In terms of the IUPAC convention, E^0 is negative for electrodes that are better reducing agents than hydrogen and positive for materials that are better oxidizing agents than hydrogen. Other conventions exist but in light of the international agreement they should be abandoned.

Application of Emf Measurements

It can be seen from any of the preceding equations that emf measurements can give us two kinds of results. If we measure the emf of cells with known activities (or concentrations) of the participating substances, then we can determine E^0 for the cell. As we have seen, this is exactly the same as determining the standard free energy of the cell reaction. However, if E^0 is known, then the activities of participating substances can be measured. Only the first application is discussed in the remainder of this chapter.

Activity Coefficients*

The equations for the emf of galvanic cells are expressed in terms of activities and it is often desirable to be able to relate these quantities to concentrations. This presents no problem for gaseous or solid substances. The activity of a gas involved in an electrode reaction may always be replaced by its partial pressure in atmospheres. This assumption becomes inaccurate only at pressures far in excess of 1 atm. The solid substances that participate in electrode reactions are virtually all present as a pure solid phase, and their activity is therefore automatically unity.

The activity of a pure liquid is also unity, and in ordinary dilute solutions, in which the mole fraction of the solvent is, for example, 0.99 or greater, the solvent is so close to being "pure" that its activity can usually be taken as unity without appreciable error. In the case of aqueous solutions a mole fraction of 0.99 corresponds to a concentration of 0.25 M univalent salt. (The water activity is 0.997 in 0.1 M KCl, 0.993 in 0.2 M KCl [2].)

For dissolved ions the activities differ from the concentration and this difference is most important. The relation between these quantities is usually expressed by the equation,

$$a = \gamma C, \tag{1.10}$$

where γ is called the *activity coefficient*. The concentration C may be expressed in terms of *molality*, that is, moles per kilogram of solvent, or in terms of *molarity*, that is, moles per liter of solution. The values of γ depend on the choice of concentration scale. For dilute aqueous solutions, however

---

* For a general discussion see any text on thermodynamics, for example, I. M. Klotz [1].

the difference between molar and molal concentrations is slight, and the activity coefficients on the two scales are almost equal.

In dilute solutions activity coefficients may be estimated by the Debye-Hückel theory, which gives the following relation for the activity coefficient (molar scale) of an ion of charge Z_i

$$-\log \gamma_i = \frac{A Z_i^2 \mu^{1/2}}{1 + B a_i \mu^{1/2}} . \tag{1.11}$$

In this equation A and B are constants depending on the dielectric constant of the solvent and on the temperature:

$$A = \frac{\epsilon^3}{2.303(DkT)^{3/2}} \cdot \frac{(2\pi N)^{1/2}}{(1000)^{3/2}} = \frac{1.8245 \times 10^6}{(DT)^{3/2}} \tag{1.12}$$

$$B = \frac{(8\pi N\epsilon^2)^{1/2}}{(1000 DkT)^{1/2}} = \frac{50.290 \times 10^{-8}}{(DT)^{1/2}}, \tag{1.13}$$

where ϵ is the protonic charge, D the dielectric constant, N Avogadro's number, k Boltzmann's constant, and T the absolute temperature. Values of A and B for aqueous solution are listed in Table 1.2.

Table 1.2 The Constants in the Debye-Hückel Equation for Aqueous Solutions

Temperature (°C)	A Concentration (m)	A Concentration (m)	B Concentration (m)	B Concentration (m)
0	0.4883	0.4883	0.3241	0.3241
5	0.4921	0.4921	0.3249	0.3249
10	0.4960	0.4961	0.3258	0.3258
15	0.5000	0.5002	0.3266	0.3267
20	0.5042	0.5046	0.3273	0.3276
25	0.5085	0.5092	0.3281	0.3286
30	0.5130	0.5141	0.3290	0.3297
35	0.5175	0.5190	0.3297	0.3307
40	0.5221	0.5241	0.3305	0.3318
45	0.5270	0.5296	0.3314	0.3330
50	0.5319	0.5351	0.3321	0.3341
60	0.5425	0.5471	0.3338	0.3366
70	0.5537	0.5599	0.3354	0.3392
80	0.5658	0.5739	0.3372	0.3420
90	0.5788	0.5891	0.3390	0.3450
100	0.5929	0.6056	0.3409	0.3482

The factor μ of (1.11) represents the ionic strength,

$$\mu = \tfrac{1}{2} \sum_i C_i Z_i^2,$$

where the summation extends over all ions present in the solution under consideration. The parameter a_i represents the effective distance between centers of neighboring ions at the distance of closest approach, that is, it represents an effective ionic diameter, including the hydration sphere of the ion. Estimated values usually range between 3 and 8 Å. A tabulation of estimates has been made by Kielland [3] and his values may be used as a basis for calculation; for ions not listed by him, one may use an a_i value for another ion that would be expected to have about the same size.

It should be emphasized that the uncertainty in the choice of a_i introduces an uncertainty into the calculated value of γ_i, which becomes increasingly important as the ionic strength increases. Equation (1.11) is in any event not valid beyond an ionic strength of 0.1 to 0.2 for univalent ions, about 0.05 for divalent ions, and about 0.01 for trivalent ions.

Another important aspect of the Debye-Hückel theory is that it relates the activity of ions to their concentration *as ions*. It cannot be used, for instance, to calculate the activity of an ion from the concentration of an added salt, unless it has been established that the salt in question is 100% dissociated into ions, or unless (as in the case of weak acids and bases) a dissociation constant is introduced as well as an activity coefficient. This restriction is not serious in water solution where most salts are completely dissociated, but it is an important limitation in application to nonaqueous solvents.

Activity coefficients may also be computed from experimental data. Experiments always yield values for the product of activity coefficients of all of the ions of a salt, for example, experiments on KCl yield values for $\gamma_{K^+}\gamma_{Cl^-} = \gamma_{\pm}^2$. To estimate an individual ion activity coefficient, one makes use of one of a number of possible assumptions. The best is probably that of MacInnes, which states that in KCl solutions $\gamma_{K^+} = \gamma_{Cl^-}$ (K^+ and Cl^- have about the same hydrated radii in water solution) and that γ_{K^+} and γ_{Cl^-} in other solutions have the same values they would have in KCl at the same ionic strength. The activity coefficients of other ions can then be computed from the mean ion activity coefficients of their chlorides or potassium salts.

Almost as good an assumption is that of Guggenheim, which states that for all univalent electrolytes $\gamma+ = \gamma-$. For extension to higher valence types and for tabulation of experimental values of mean ion activity coefficients of various salts, the reader is referred to standard texts on electrochemistry [4].

To illustrate the magnitude of activity coefficients and the differences involved in the use of the various assumptions listed above, some calculated values are shown in Fig. 1.1. It can be seen that for univalent ions the various estimates differ from one another by less than 0.02 in $\log \gamma$ up to an ionic

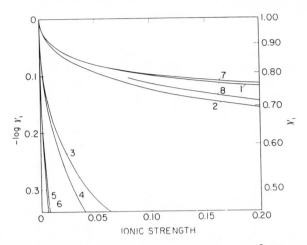

Fig. 1.1 Activity coefficients of ions in aqueous solution at 25°. Curves 1 to 6 are calculated by equation (11), as follows: 1, univalent ion with $a_i = 6A$; 2, univalent ion with $a_i = 3A$; 3, divalent ion with $a_i = 6A$; 4, divalent ion with $a_i = 3A$; 5, tervalent ion with $a_i = 6A$; 6, tervalent ion with $a_i = 3A$. Curves 7 and 8 are experimental values for Cl^- in HCl, curve 7 being based on the Guggenheim assumption and curve 8 on the MacInnes assumption.

strength of 0.1. For ions of higher valence types, however, the uncertainty is greater.

Activity coefficients of *uncharged solutes* (organic compounds) may be taken as unity at ionic strengths below 0.1, that is, in the region in which the activity coefficients of ionic solutes can be estimated with accuracy.

2 POTENTIOMETERS

It was mentioned earlier that we are interested in the potential of a galvanic cell only when it can be measured under reversible conditions. Therefore, a simple voltmeter cannot be used,* for it draws a small current from the cell, that is, it measures the potential when the cell reaction is proceeding spontaneously at a small rate. The device designed for reversible measurements is known as a potentiometer, and is schematically illustrated by Fig. 1.2. The potential of the cell is opposed by a known variable potential. If this potential is greater than that of the cell to be measured, it will cause current to flow through a galvanometer in one direction; if it is smaller the current will flow in the opposite direction. At a single intermediate point no current flows at all, and at this point the variable potential is equal to the cell potential. Provided that the same balance point is approached from either direction,

---

* Commercial high input impedance voltmeters based on operational amplifiers draw negligible current from the cell and can measure cell emf accurately.

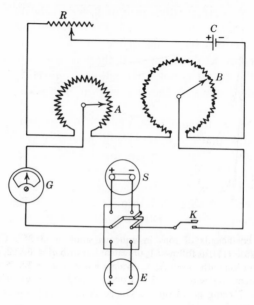

Fig. 1.2 Circuit diagram for a potentiometer.

the measurement is a reversible one, and the measured potential is equal to the emf of the cell.

In Fig. 1.2, C is the battery that is the source of the variable potential. It should be a high-capacity, low-discharge storage cell, which should undergo no appreciable change in voltage during the time of measurement. In series with this battery are a variable resistance R, one or more resistance boxes B, and a slide-wire A. By adjusting the contacts A and B, the resistance between them can be given any desired value. Since the potential between A and B is the product of this resistance and the current flowing through the potentiometer circuit, $E = IR$, the potential can also be given any desired value. In commercial potentiometers the resistance boxes and slide-wire are already calibrated in volts; they therefore require that the current through the potentiometer circuit have always the same fixed value. This value can be obtained by adjustment of the resistance R. To indicate when the adjustment has been correctly made, we use a Weston standard cell. The reaction of this cell is $Cd + Hg_2SO_4 \rightarrow CdSO_4 + Hg$. The operation of the cell is reversible, with an emf of close to 1.0184 V at 25°C, with only a small temperature coefficient. Weston standard cells are commercially available, and, if calibrated against Bureau of Standards' cells, are provided with calibration certificates. They should be checked against a new standard cell about once a year. To make the correct adjustment of R, we set the contacts A and B so that they read the correct emf of the cell. The standard cell S is then put into the galvanometer circuit by means of the double-pole double-throw switch of Fig. 1.2. The key K is depressed. If there is a galvanometer deflection, indicating that the circuit is not balanced, R is adjusted. This procedure is repeated until depression of the key no longer leads to a galvanometer deflection.

In measurement of the unknown emf E, the double-pole double-throw switch is reversed so that E is in the circuit instead of S. The contacts A and B are adjusted until there is no galvanometer deflection on depression of the key. At the end of the measurement the emf of the standard cell is again checked. It should give the correct value without adjustment of R. As mentioned earlier, the measured balance point must be the same when approached from either direction. If it is not, then the cell being measured is not behaving reversibly.

Commercial potentiometers vary in precision. A pH meter may be used as a potentiometer. Most models have a precision of about 1 mV, and this is sufficient for many purposes, such as potentiometric titrations. The most precise pH meters have a precision of about 0.1 mV. The familiar "student potentiometer" has a precision of about 0.01 mV, and the "type K" is readable to 0.001 mV or better. The "type K" potentiometer also has convenient built-in features. A separate slide-wire is provided for the standard cell so that the resistance R may be adjusted frequently without disturbing the settings of contacts A and B. There are also three separate keys with protective resistances. The key with the largest resistance is used when one is far from the balance point and prevents a violent deflection of the galvanometer under these circumstances. The key with the lowest resistance is used only when one is very close to the point of balance. Commercial potentiometers more sensitive than the "type K" are never required for the measurement of the emf of galvanic cells.

The galvanometer required depends on the internal resistance of the cell being used. A more sensitive type is required for solutions of low ion concentration. Enclosed lamp and scale galvanometers suffice for most purposes. Their sensitivity is of the order of 0.04 μA/mm. The galvanometers should always be critically damped so that after the key is released and the circuit opened the pointer swings back to zero without oscillation.

3 OXIDATION-REDUCTION POTENTIALS: THEORY

Introduction

Reversible electrodes formed by immersing an "indifferent" metal such as gold or bright platinum into a solution containing both the oxidized and reduced form of a reversible oxidation-reduction system, for example, Sn^{4+} and Sn^{2+} or quinone and hydroquinone, are known as oxidation-reduction electrodes. The metal in this system acts merely as a conductor for making electrical contact just as in the case of a gas electrode such as a hydrogen or oxygen electrode. There is no fundamental difference between an oxidation-reduction electrode and a system consisting of a metal and its ions. The term "oxidation-reduction" which appears in current usage is treated separately only as a matter of convenience, since it is restricted to systems in which both components are soluble in the solution in contact with the "inert" electrode.

Most oxidations and reductions of organic compounds are irreversible and such reactions cannot be studied by thermodynamic methods. A limited class of organic substances, quinones, hydroquinones, and related compounds, however, undergo reversible oxidations and reductions and their reactions may be followed by potentiometric methods. These compounds may also be studied polarographically and the results obtained in a number of examples are comparable to those obtained potentiometrically but are not as accurate. The rapidity of the polarographic method has made it more popular than the potentiometric method in the past decade. The polarographic results, however, may be complicated by the adsorption of the compound on the mercury electrode (e.g., methylene blue) and by hydration of one of the compounds involved in the electrode reaction (e.g., ascorbic acid).

Oxidation-reduction potentials are of interest in organic chemistry since (1) they provide a method of characterizing the oxidizing power of systems, for example, containing a quinone or, conversely, the reducing strength, for example, of a hydroquinone, (2) they can be used in structure determination, for example, when the quinone can exist in tautomeric forms, and (3) from the changes in potential involved with pH they can be used to study the ionization relations of compounds and to estimate their ionization constants.

Determination of Standard Oxidation-Reduction Potentials

In principle, the determination of the standard potential of an oxidation-reduction system involves setting up electrodes containing the oxidized and reduced state at known activities and measuring the potential E by combination with a suitable reference electrode and inserting this value into the following equation:

$$E = E^0 - \frac{RT}{nF} \ln \frac{a_L{}^l a_M{}^m}{a_A{}^a a_B{}^b}, \qquad (1.14)$$

where $aA + bB \rightleftharpoons lL + mM$ represents the cell reaction. For the quinhydrone (an equimolecular compound of quinone and hydroquinone) electrode, which has been studied extensively, the cell is:

$$\text{Au; Hydroquinone, Quinone, HCl}(m)\|\text{HCl}(m); \text{H}_2(\text{Pt})} \qquad \text{(1.IV)}$$

and the electrode reaction is:

$$\text{Quinone} + 2\text{H}^+ + 2e^- \rightleftharpoons \text{Hydroquinone}$$

with the potential referred to the normal hydrogen electrode being expressed by the following equation if the change in liquid potential is ignored

$$E = E^0 - \frac{RT}{2F} \ln \frac{\gamma_{\text{QH}_2}}{\gamma_\text{Q}} + \frac{RT}{F} \ln a_{\text{H}^+}. \qquad (1.15)$$

Since Q and QH_2 are uncharged, the ratio of activity coefficients can be assumed to be one. The potential determined in this manner at a given pH is then commonly reported as $E^{0\prime}$ and is the formal potential of the electrode at a particular pH. This value ($E^{0\prime}$) is related to E^0, the standard oxidation potential, by the following equation

$$E^{0\prime} = E^0 - 2.303 \frac{RT}{F} \text{pH}. \qquad (1.16)$$

When the hydrogen and quinhydrone electrodes are immersed in the same solution the cell can be formulated by

$$\text{Au; Quinhydrone, H}^+(a); \text{H}_2(\text{Pt})} \qquad \text{(1.V)[5]}$$

with no liquid junction potential being involved, and the expression for the corresponding potential is:

$$E = E_\text{Q} - E_\text{H} = -\frac{RT}{F} \ln a_{\text{H}^+} + E^0 - \frac{RT}{2F} \ln \frac{\gamma_{\text{QH}_2}}{\gamma_\text{Q}} + \frac{RT}{F} \ln a_{\text{H}^+}$$

$$= E^0 - \frac{RT}{2F} \ln \frac{\gamma_{\text{QH}_2}}{\gamma_\text{Q}}. \qquad (1.17)$$

The use of quinhydrone assures equal concentrations of quinone and hydroquinone, but the activity and activity coefficient ratios are altered by the presence of other dissolved substances. The resulting change in potential often termed the "salt error" can be evaluated by determining the potential for various concentrations of acid or electrolyte. A plot of the potential against the total concentration and an extrapolation to zero concentration gives the standard oxidation potential E^0 of the quinhydrone electrode or of the system quinone-hydroquinone. This value was determined to be $+0.69938 \pm 00003$ V at $25°\text{C}$.

Determination of Approximate Standard Potentials

Difficulties involved in obtaining the quinhydrone in other quinone-hydroquinone systems, or in determining the concentrations existing at equilibrium, since the reduced form may be oxidized rather easily by the oxygen of the air before a measurement can be made, and the slowness in establishing the equilibrium with the "inert" metal of the electrode, which is inherent in the method of mixtures, can be avoided by the potentiometric titration method of Clark [6]. This titration can be either oxidative or reductive and is carried out in the absence of oxygen since contact of this gas with the inert electrode develops a potential (gas electrode). The pure oxidized form of the system such as a quinone is dissolved in a buffer, known amounts of a reducing solution are added in the absence of air, and the solution is kept agitated by means of a current of pure nitrogen. The buffer is required since hydrogen ions are involved in most organic systems and the pH must be kept constant. The potential of an inert electrode, for example, platinum or gold, immersed in the reacting solution is measured after each addition of the titrant by combination with a reference electrode. The liquid junction potential involved in such measurements usually is large enough to mask any salt error introduced, that is, the variation in the ln of γ_Q/γ_{QH_2} from 1. For all practical purposes, therefore, the activities may be taken as equal to the molar concentrations.

It might be thought that the presence of the inorganic oxidizing or reducing agent can interfere with the procedure, that is, that the electrode potential might reflect the inorganic redox equilibrium rather than the organic one. It is easily shown, however, that this does not occur. Thus in the oxidation of hydroquinone with ceric ions, the reaction occurs as follows:

$$QH_2 + 2Ce^{4+} \rightleftharpoons Q + 2Ce^{3+} + 2H^+$$

The equilibrium constant for this reaction is

$$K = \frac{a_Q a_{Ce^{3+}}^2 a_{H^+}^2}{a_{QH_2} a_{Ce^{4+}}^2}, \tag{1.18}$$

and since

$$\ln K = \frac{nF}{RT} E^0 \tag{1.19}$$

K for the above reaction can be written as

$$\ln K = \frac{2F}{RT} (E^0_{Ce^{4+}, Ce^{3+}} - E^0_{Q, QH_2}) \tag{1.20}$$

or

$$\ln \frac{a_Q a_{Ce^{+3}}^2 a_{H^+}^2}{a_{QH_2} a_{Ce^{4+}}^2} = \frac{2F}{RT} (E^0_{Ce^{4+}, Ce^{3+}} - E^0_{Q, QH_2}). \tag{1.21}$$

By rearranging this expression,

$$\frac{RT}{2F} \ln \frac{a_Q a_{H^+}^2}{a_{QH_2}} + \frac{RT}{2F} \ln \frac{a_{Ce^{3+}}^2}{a_{Ce^{4+}}^2} = E^0_{Ce^{4+},Ce^{3+}} - E^0_{Q,QH_2} \tag{1.22}$$

$$E^0_{Q,QH_2} - \frac{RT}{2F} \ln \frac{a_{QH_2}}{a_Q a_{H^+}^2} = E^0_{Ce^{4+},Ce^{3+}} - \frac{RT}{2F} \ln \frac{a_{Ce^{3+}}^2}{a_{Ce^{4+}}^2}, \tag{1.23}$$

that is, the potential of the inorganic system is automatically the same as that of the organic system. A platinum electrode inserted into such a solution therefore gives a potential characteristic of the organic system.

In order for the oxidation or reduction of a system to be complete within the limits of accuracy of ordinary volumetric analysis, it is necessary that the concentration of one form at the end point be at least 10^3 times that of the other, or that the oxidation or reduction be complete within 0.1% or better. The equilibrium constant (equation (1.18)) should therefore be smaller than 10^{-6} if n is the same for both systems interacting, or 10^{-9} if n is one for one system and two for the other.

By making use of an equation similar to (1.20), it can be shown that in order to fulfill these conditions the standard potentials should differ by at least 0.34 V if n is 1 for both systems, 0.26 V if n is 1 for one system and 2 for the other, or 0.18 V if n is 2 for both. If the inorganic oxidizing or reducing agent represents an irreversible system, it will have no or little effect upon the potential of the inert electrode.

The results obtained for the titration of a buffered (pH 6.98) solution of 1-naphthol-2-sulfonate indophenol at 30°C are plotted as ordinates against the volumes of added reagent as abscissas (Fig. 1.3). The point at which the

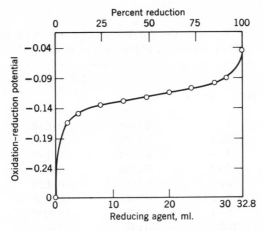

Fig. 1.3 Reduction of 1-naphthol-2-sulfonate indophenol (Clark).

potential undergoes a rapid change is that corresponding to complete reduction, and the quantity of the reducing solution then added is equivalent to the whole of the oxidized organic compound originally present. From the amounts of reducing agent added at various stages, the corresponding ratios of the oxidized [Ox] to the reduced form [Red] may be calculated without any knowledge of the initial amount of the former or of the concentration of the titrating agent. If t_c is the volume of titrant added when the sudden change of potential occurs, that is, when the reduction is complete, and t is the amount of titrant added at any point in the titration, then at any point [Ox] is equivalent to $t_c - t$, and [Red] is equivalent to t, provided the titrant employed is a powerful reducing agent. Accordingly, in the equation

$$E = E^{0'} - \frac{RT}{nF} \ln \frac{a_{\mathrm{Ox}}}{a_{\mathrm{Red}}} \qquad (1.24)$$

Replacing the ratio of the activities by the ratio of concentrations gives the following equation:

$$E = E^{0'} - \frac{RT}{nF} \ln \frac{[\mathrm{Ox}]}{[\mathrm{Red}]} = E^{0'} - \frac{RT}{nF} \ln \frac{t_c - t}{t}, \qquad (1.25)$$

where $E^{0'}$ is the formal potential of the system for the pH employed in the titration. Values of $E^{0'}$ can thus be obtained for a series of points on the titration curve, and if the system is behaving in a satisfactory manner these values should be approximately constant. The results obtained by applying the above equation to the data in Fig. 1.3 are given in Table 1.3.

The experiment described above can also be carried out by starting with the reduced form of the system and titrating it with an oxidizing agent, for example, potassium dichromate. The formal potentials obtained in this

Table 1.3 Evaluation of the Approximate Formal Potential of 1-Naphthol-2-Sulfonate Indophenol at 30°C at pH 6.98 [7]

t	Reduction (%)	E_{obs}	$\dfrac{RT}{2F} \ln \dfrac{t_c - t}{t}$	$E^{0'}$
4.0	12.2	−0.1479	−0.0258	−0.1221
8.0	24.4	−0.1368	−0.0148	−0.1220
12.0	36.6	−0.1292	−0.0072	−0.1220
16.0	48.8	−0.1224	−0.0006	−0.1218
20.0	61.0	−0.1159	+0.0058	−0.1217
24.0	73.2	−0.1085	+0.0131	−0.1216
28.0	85.4	−0.0985	+0.0230	−0.1215
32.8 (t_c)	100.	−0.036	—	—

manner agree with those derived from the titration of the oxidized form with a reducing agent and also with the potential measured in mixtures made up from known amounts of oxidized and reduced forms.

These data can be also used to determine n, the number of electrons involved in the oxidation-reduction system. This value can be obtained from the slope of the flat portion of the titration curve shown in Fig. 1.3. The flatter the curve the larger is n. The value of n is obtained more exactly by plotting the measured potential E against log [Ox]/[Red] or its equivalent log $(t_c - t)/t$. This plot, according to the following equation

$$E = E^{0'} - \frac{2.303RT}{nF} \log \frac{t_c - t}{t},$$ (1.26)

should be a straight line of slope $-2.303 \, RT/nF$ (cf. Table 1.1). The results derived from Fig. 1.3 are plotted in Fig. 1.4 and give a slope of -0.030 at

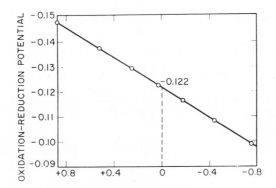

Fig. 1.4 Determination of n and $E^{0'}$.

30°C so that n is equal to 2. The formal potential is given by the point at which the ratio [Ox]/[Red] is 1 or log [Ox]/[Red] = 0.

The slope is also often reported as the difference in millivolts between the 50 and 25% points and is known under these circumstances as the index potential E_i. When E_i is 28.6 (±0.2) mV at 30°C, the redox system is a univalent one. If E_i is 14.3 (±0.2), the redox system is a bivalent one.

Determination of Acid Dissociation Constants

In the treatment of the quinone-hydroquinone systems thus far the possibility of the hydroquinone ionizing as an acid has not been considered.

Generalization of the equation for the electrode reaction involving a quinone-hydroquinone system can be represented by

$$R + 2H^+ + 2e^- \rightleftarrows H_2R$$

The hydroquinone is a weak dibasic acid for which two ionization equilibria are possible,

$$H_2R \rightleftharpoons H^+ + HR^-$$
$$HR^- \rightleftarrows H^+ + R^{2-}$$

for which the ionization constants K_1 and K_2 can be expressed as

$$K_1 = \frac{[H^+][HR^-]}{[H_2R]} \qquad K_2 = \frac{[H^+][R^{2-}]}{[HR^-]} . \qquad (1.27)$$

The potential of the electrode is given by

$$E = E^0 + \frac{RT}{nF} \ln \frac{[R][H^+]^2}{[H_2R]} . \qquad (1.28)$$

Since the reduced substance can be present in three states, H_2R, HR^- and R^{2-}, the total stoichiometric concentration $[H_2R]_s$ is equal to the sum of the concentration of these three species:

$$[H_2R]_s = [H_2R] + [HR^-] + [R^{2-}] \qquad (1.29)$$

Assuming that the activities and the concentrations of these substances are the same and inserting for $[HR^-]$ and $[R^{2-}]$ the values derived from the expressions for K_1 and K_2:

$$[H_2R]_s = [H_2R] + \frac{[H_2R]}{[H^+]} K_1 + \frac{[H_2R]}{[H^+]^2} K_2K_1 \qquad (1.30)$$

$$[H_2R] = \frac{[H^+]^2[H_2R]_s}{[H^+]^2 + K_1[H^+] + K_2K_1} , \qquad (1.31)$$

which when substituted in the electrode reaction gives

$$E = E^0 + \frac{RT}{2F} \ln \frac{[R]}{[H_2R]_s} + \frac{RT}{2F} \ln ([H^+]^2 + K_1[H^+] + K_1K_2). \quad (1.32)$$

If K_1 and K_2 are very small, as they are for hydroquinone, the above equation becomes,

$$E = E^0 + \frac{RT}{2F} \ln \frac{[R]}{[H_2R]_s} + \frac{RT}{F} \ln [H^+] \qquad (1.33)$$

or the equation used in previous discussions. This equation indicates how the oxidation-reduction potential varies with hydrogen ion concentration in regions where the ionization of the hydroquinone is not important. A change in pH of 1 unit shifts the potential by 2.303 RT Volts/F (cf. Table 1.1). This is the basis for the use of the quinhydrone electrode in determining pH. A critical examination of the kinetics of the quinhydrone electrode is given by Vetter [8].

If the ionization constants are appreciable, the more complicated equation [see (1.32)] must be used, which affords a method in certain examples for estimating K_1 and K_2. Measurements are made with the ratio [R]/[H$_2$R] equal to unity and the pH of the solution is altered by the addition of acid or alkali. The results for the β-anthraquinone sulfonate system which is stable in alkaline medium, are shown in Fig. 1.5.

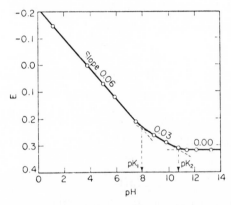

Fig. 1.5 Variation of the potential of β-anthraquinone sulfonate with pH.

It will be seen that the plot consists of three straight lines, one with a slope of 0.06, another with a slope of 0.03, and the third with a slope of zero. At low pH values the term $[H^+]^2$ is large in comparison with $K_1[H^+]$ and K_1K_2. Therefore, as long as K_1 is negligible in comparison with $[H^+]$, the plot of E^0 against pH will be a straight line with a slope of 2.303 RT/F or at 30°C, 0.060. The first line corresponds to the condition $[H^+] \gg K_1$.

If $[H^+]^2$ is negligible in comparison with $K_1[H^+]$ but still larger than K_2, the variation of E^0 with pH should be

$$E = E^0 + \frac{RT}{2F} \ln K_1[H^+] \qquad (1.34)$$

or a straight line with a slope of 2.303 $RT/2F$ or 0.030 at 30°C. The second line corresponds to the condition $[H^+] \gg K_2$. Finally, if $K_1K_2 \gg [H^+]$ then,

$$E = E^0 + \frac{RT}{2F} \ln K_1K_2 \qquad (1.35)$$

and the slope is not affected by pH and is zero or horizontal. This condition is illustrated in the third line. A rough estimate of pK_1 and pK_2 can be made graphically as shown by extending the straight lines until they cross and reading the corresponding pH values. There is no independent evidence that ionization constants obtained in this way are correct, but they are probably of the right order of magnitude.

The results discussed above are immediately evident if the equation used is differentiated with respect to pH:

$$\frac{dE}{dpH} = -2.303 \frac{RT}{2F} \frac{2[H^+]^2 + K_1[H^+]}{[H^+]^2 + K_1[H^+] + K_1K_2}. \tag{1.36}$$

For $[H^+] \gg K_1 \gg K_2$, $K_1[H^+]$ and K_1K_2 can be neglected and

$$\frac{dE}{dpH} = -2.303 \frac{RT}{F}. \tag{1.37}$$

For $K_1 \gg [H^+] \gg K_2$, $2[H^+]^2$, $[H^+]^2 + K_1K_2$ are insignificant and

$$\frac{dE}{dpH} = -2.303 \frac{RT}{2F}. \tag{1.38}$$

For $K_1 > K_2 \gg [H^+]$, $2[H^+]^2 + K_1[H^+]$ can be neglected and

$$\frac{dE}{dpH} = 0. \tag{1.39}$$

If $K_1 = [H^+]$ and K_1 is much greater than K_2, then

$$\frac{dE}{dpH} = -2.303 \frac{RT}{2F} \frac{3}{2} = -0.045. \tag{1.40}$$

Such a slope is intermediate between 0.060 and 0.030 and corresponds to a point on the first bend of the curve or the point of interaction of the two lines of slope 0.060 and 0.030 as shown in Fig. 1.5. For $[H^+] = K_2$,

$$\frac{dE}{dpH} = -2.303 \frac{RT}{2F} \frac{2[H^+]^2 + K_1[H^+]}{[H^+]^2 + K_1[H^+] + K_1[H^+]}. \tag{1.41}$$

Since $K_1 \gg [H^+]$,

$$\frac{dE}{dpH} = -2.303 \frac{RT}{2F} \frac{K_1[H^+]}{2K_1[H^+]} = -0.015 \text{ at } 30°C. \tag{1.42}$$

This value corresponds to the point of intersection of the two lines of slope 0.030 and zero as shown in Fig. 1.5. The pK values for β-anthrahydroquinone sulfonate are 7.9 and 10.6 and correspond to ionization constants of 1.23×10^{-8} and 2.51×10^{-11}, respectively. The pHs for these values represent the points at which $[RH_2]$ is equal to $[RH^-]$ and $[RH^-]$ is equal to $[R^{2-}]$, respectively.

The results obtained at various pH values for quinone-hydroquinone are given in Fig. 1.6 to illustrate the behavior of a substance (quinone) susceptible to irreversible changes (decomposition). Examination of the upper graph in the figure shows that the change at the pK of the first ionization occurs normally but that at about the point at which the pK of the second ionization should be, the slope reverts to the original 0.058. Furthermore the slope

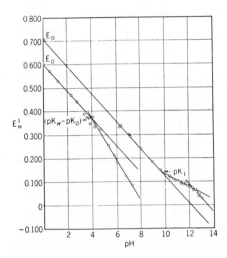

Fig. 1.6 Formal potentials of the hydroquinone system at different pH values.

between pH 9.8 and 12.3 is not the theoretical 0.029 but 0.0261. The irregular slope indicates that in this region the potential is determined by the quinone-hydroquinone and hydroxyquinone-hydroxyhydroquinone systems. The latter is formed by the action of alkali on quinone and gives the behavior indicated by the lower line in Fig. 1.6 [9].

Complex behavior of this type is frequently encountered. The reduced form may have several stages of acidic dissociations; in addition, the oxidized form may exhibit one or more acidic dissociations. Basic dissociation, if it occurs, is treated as an acidic ionization of the conjugate acid. The method of treatment given above can be applied to any case no matter how complex, and the following general rules have been derived which facilitate the analysis of pH-potential curves for oxidation-reduction systems of constant stoichiometric composition [7, 10].

1. Each bend in the curve may be related to an acidic dissociation constant; if the curve becomes steeper with increasing pH, the dissociation has occurred

in the oxidized form (cf. Fig. 1.6), but if it becomes flatter it has occurred in the reduced form (cf. Fig. 1.5).

2. The intersection of the extensions of adjacent linear parts of the curve occurs at the pH equal to pK for the particular dissociation group responsible for the bend.

3. As each dissociation becomes involved the slope of the plot changes by 2.303 RT/nF Volt per pH unit where n is the number of electrons involved in the change between the oxidized and reduced state.

Two-Stage Oxidation-Reductions

The overall reduction of a quinone to a hydroquinone involves two electrons

$$Q + 2e + 2H^+ \rightleftharpoons QH_2$$

The actual mechanism of this change as to whether the electrons are added stepwise or not is important. Potentiometric studies were the first to demonstrate the stepwise addition of electrons in certain examples. More recently the polarographic method has shown that this stepwise addition is general for simple quinones using dimethylformamide or acetonitrile as a solvent.

The addition of one electron to quinone produces an intermediate known as a semiquinone,

$$Q + e^- \rightleftharpoons Q^-$$

This substance is an anion radical which is stabilized by resonance similar to the triarylmethyl free radicals.

The semiquinone upon further reduction gives the hydroquinone dianion

$$Q^- + e^- \rightleftharpoons Q^{2-}$$

Superimposed on these reactions is the reaction of Q^- and Q^{2-} with hydrogen ions in aqueous media to form QH and QH_2. The free radical QH is much less stabilized by resonance and immediately picks up another electron and forms QH^- or QH_2 in the presence of hydrogen ions. The overall reaction for the first stage in the presence of hydrogen ions is then

$$Q + H^+ + e^- \rightleftharpoons QH$$

The intermediate QH, in addition to this reaction, may disproportionate or may dimerize to the meriquinone which is regarded as a molecular compound ($QH_2 \cdot Q$).

$$2QH \rightleftharpoons Q + QH_2$$

The demonstration that semiquinones are stable in strong alkali or in aprotic solvents (CH_3CN) (polarographic study) is in agreement with the above scheme.

p-Phenylenediamines in strong acid medium form a similarly stabilized radical ion by the loss of an electron,

The product is known as a Würster salt and on further oxidation gives the diimine,

which is analogous to a quinone in structure.

The stability of these semiquinone intermediates depends upon the structure of the compounds and the solvent composition involved. Figure 1.7, for example, shows the variation in the shape of the titration curve with change of acidity for the dye hydroxythiazine.

The treatment of these systems in aqueous solution is now discussed.

For meriquinone formation the stages of oxidation-reduction may be written:

$$QH_2 \cdot Q(2QH) + 2H^+ + 2e^- \rightleftarrows 2QH_2$$
$$2Q + 2H^+ + 2e^- \rightleftarrows QH_2 \cdot Q$$

so that if E_1' is the formal potential of the first stage at a definite hydrogen ion concentration,

$$E = E_1' + \frac{RT}{2F} \ln \frac{[QH_2 \cdot Q]}{[QH_2]^2} . \qquad (1.43)$$

If the original amount of the reduced form $[QH_2]$ in a given solution is a, and x equivalents of a sufficiently strong oxidizing agent are added, x moles of QH are formed and these dimerize to form $\frac{1}{2}x$ moles of meriquinone, $QH_2 \cdot Q$; an amount of $a - x$ moles of QH_2 remains unchanged. It follows that in a solution of volume v

$$E = E_1' + \frac{RT}{2F} \ln \frac{\frac{1}{2}x/v}{(a-x)^2/v^2} = E_1' + \frac{RT}{2F} \ln \frac{x}{(a-x)^2} + \frac{RT}{F} \ln \frac{v}{2} \quad (1.44)$$

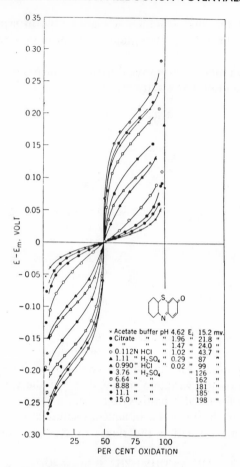

Fig. 1.7 Titration of hydroxythiazine at different pH values (S. Granick, L. Michaelis, and M. P. Schubert, *J. Am. Chem. Soc.*, **62**, 1802 (1940)).

or the potential depends upon the volume of the solution and the position of the potentiometric titration curve since the oxidation of QH_2 varies with the concentration of the solution. At constant volume the above equation becomes

$$E = E_1' + \frac{RT}{2F} \ln x - \frac{RT}{F} \ln (a - x) \qquad (1.45)$$

so that in the earlier stages of oxidation, that is, when x is small, the last term on the right-hand side can be regarded as constant and the slope of the titration curve corresponds to a process in which two electrons are involved, that is, n is 2. In the later stages, however, the change of potential is determined mainly by the last term and the slope of the curve changes to that of a one-electron system, that is, n is effectively unity.

When a true semiquinone is formed, for example, in strong alkali, the two stages of oxidation-reduction are

$$Q^- + e^- \rightleftarrows Q^{2-}$$
$$Q + e^- \rightleftarrows Q^-$$

so that for a definite hydrogen ion concentration,

$$E = E_1' + \frac{RT}{F} \ln \frac{[Q^-]}{[Q^{2-}]} = E_1 + \frac{RT}{F} \ln \frac{(x)}{(a-x)}, \qquad (1.46)$$

where x and a represent similar quantities to those mentioned for the meriquinone. From this expression it is evident that the titration curves are independent of volume and have the same slope, $n = 1$, throughout the curve.

The assumption has been made in the above that the two stages of oxidation are fairly distinct but in most studies in aqueous media in the presence of H^+ the two stages usually overlap. The treatment becomes complicated and is only mentioned briefly here. The reactions under these conditions can be represented by

$$QH + H^+ + e^- \rightleftarrows QH_2$$
$$Q + H^+ + e^- \rightleftarrows QH$$

and

$$E = E_1' + \frac{RT}{F} \ln \frac{[QH]}{[QH_2]} \qquad (1.47)$$

$$E = E_2' + \frac{RT}{F} \ln \frac{[Q]}{[QH]} \qquad (1.48)$$

are the potentials for these stages at a definite pH where E_1' and E_2' are the formal potentials for these stages. The potential can also be formulated for the overall reaction,

$$Q + 2H^+ + 2e^- \rightleftarrows QH_2$$

$$E = E_m' + \frac{RT}{2F} \ln \frac{[Q]}{[QH_2]}, \qquad (1.49)$$

where E_m' is the usual formal potential for the overall reaction at a definite hydrogen ion concentration.

$$E_m' + \frac{RT}{2F} \ln \frac{[Q]}{[QH_2]} = E_1' + \frac{RT}{F} \ln \frac{[QH]}{[QH_2]} = E_2' + \frac{RT}{F} \ln \frac{[Q]}{[QH_2]} \qquad (1.50)$$

or

$$2E'_m + \frac{RT}{F} \ln \frac{[Q]}{[QH_2]} = E'_1 + E'_2 + \frac{RT}{F} \ln \frac{[Q]}{[QH_2]} \qquad (1.51)$$

and

$$E'_m = \frac{E'_1 + E'_2}{2}, \qquad (1.52)$$

in which E'_1 and E'_2 are the centers of the first part and second part of the titration curve and E'_m is the potential of the middle of the curve.

An equation for the potential E observed during an oxidation-reduction titration which includes a consideration of the formation of semiquinones can be obtained as follows. Using the relationship

$$Q + QH_2 \rightleftharpoons 2QH$$

an expression using the law of mass action can be obtained in which

$$K = \frac{[QH]^2}{[Q][QH_2]}, \qquad (1.53)$$

where K is the semiquinone formation constant.

If a is the initial amount of QH_2 being titrated and x is the number of moles of oxidant added in a titration, at any point of the titration,

$$[QH] + 2[Q] = x \qquad (1.54)$$

since the formation of Q represents two steps of oxidation and QH only one. By making use of the above relationships and letting $x/a = X$, the following equation can be derived. [11]

$$E = E'_m + \frac{RT}{2F} \ln \frac{X}{2 - X} + \frac{RT}{2F} \ln \frac{X - 1 \pm \sqrt{(X-1)^2 + 4X(2-X)K}}{1 - X \pm \sqrt{(X-1)^2 + 4X(2-X)K}}$$

$$(1.55)$$

and gives the variation of $E - E'_m$ with x/a in terms of the semiquinone formation constant.

Some of the results obtained for various K values are shown in Fig 1.8. As long as K is small, the titration curve throughout has the shape of a normal two-electron oxidation-reduction system since the above equation approaches

$$E = E'_m + \frac{RT}{2F} \ln \frac{X}{2 - X} \qquad (1.56)$$

and no break occurs at $X = 1$. This equation corresponds to the line labeled $K = 0$ in Fig. 1.8. For K between 4 and 16, the slope corresponds to that of a one-electron process but there is no break at the midpoint. The presence of

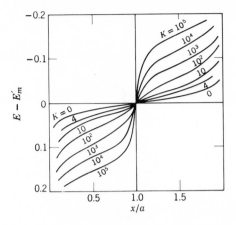

Fig. 1.8 Titration curves for semiquinone formation.

a semiquinone is often indicated in these cases, however, by the appearance of a color that differs from the quinone or hydroquinone. When the semi-quinone formation constant exceeds 16, a break appears at the midpoint and the extent of this break becomes larger as K increases since the curve approaches the form

$$E = E'_m + \frac{RT}{F} \ln \frac{[\text{Ox}]}{[\text{Red}]} \tag{1.57}$$

in two separate steps involving one electron each. The results of such a titration are shown in Fig. 1.9 for the oxidation of reduced α-hydroxyphena-zine with benzoquinone.

The values of E'_1, E'_2, and E'_m depend upon the pH of the solution, and since QH_2, QH, and Q may possess acidic or basic functions, the slope of the curves of these formal potentials against pH may change direction at various points and crossings may occur. In Fig. 1.10 such plots for pyocyanine (N-methyl-α-hydroxyphenazine) give a crossing point of the three curves at pH = 5. Here the semiquinone constant K is 1 and $E'_m = E'_1 = E'_2 = 1$, and the maximum ratio of semiquinone to total dye, $(QH/a)_{\max}$ is 0.333. To the left of this crossing point, at higher acidities, the curves diverge, $E'_2 > E'_m > E'$, and the semiquinone formation constant is large. To the right of the crossing point there is the reverse order, $E'_2 < E'_m < E'_1$ and the semiquinone formation is very small. In this region all calculations depend on very slight variations of E_i and have relatively large limits of error which are shown in the graph on the assumption that the index potential may be uncertain to ±0.2 mV. Such errors in data can be reduced by using an

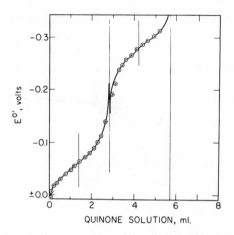

Fig. 1.9 Potentiometric titration of α-hydroxyphenazine with quinone.

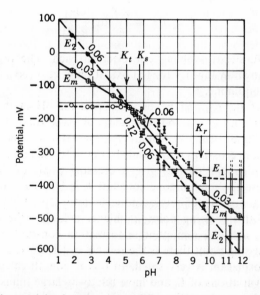

Fig. 1.10 Formal potentials of pyocyanine at different pH values (L. Michaelis, E. S. Hill, and M. P. Schubert, *Biochem. Z.*, **255**, 66 (1932).)

iterative method for analysis of the potentiometric curves employing computers [12]. The bends of the curves correspond to the acidic ionization constants of the reduced, semioxidized, and totally oxidized forms K_{QH_2}, K_{QH}, and K_Q, respectively.

4 OXIDATION-REDUCTION POTENTIALS: EXPERIMENTAL

Reductive Titration: Reducing Agents

Usually the Ox form of a reversible redox system is stable in air, for example, methylene blue. In this case, a solution of Ox is made up in a buffer, the oxygen of the solution is thoroughly removed by bubbling with oxygen-free nitrogen, a reducing agent is added from a buret in successive portions, and at each step the potential is measured. During the titration the solution is guarded against the access of any trace of oxygen.

The buffer should have an ionic strength μ not smaller, or at least not much smaller, than 0.1, and the concentration of the dye should be in the neighborhood of 10^{-3} M. If the concentration is too low ($\ll 10^{-4}$ M), the poise of the system may not be sufficient to establish reproducible potentials. If it is too high (10^{-2} or 10^{-1} M), the capacity of the buffer may not be sufficient to keep the pH constant during the titration if the Ox and Red forms differ with respect to their acidic properties. The reducing agent must be in homogeneous solution and may be the Red form of another reversible redox system of a much more negative normal potential than that of the system to be measured. Or, the reducing agent may be some irreversible system which performs the reduction practically instantaneously, in such a manner that its oxidation product does not interfere with the measurement because of any secondary chemical reaction. The concentration of the reducing agent depends on the purpose of the titration. This may be: (1) the determination of the amount of some oxidizable (or reducible) substance in the solution. Here, the titer of the agent used for titration must be exactly known, and the only aim of the titration is to ascertain the end point. Not all of the reagents mentioned below, especially among the reducing agents, are suitable for making up a standard solution of exactly known titer. In such examples the amount of reagent can be determined by titrating the same volume of agent used with a standard oxidizing agent. Or it may be: (2) the determination of the formal potential and of some other characteristic properties of a redox system. Here, the titer of the oxidizing or reducing agent need be known only approximately. The end point of titration manifests itself by the jump of the potential, and it is the shape of the titration curve from which the wanted data are to be derived. The abscissa, originally plotted in terms of milliliters of the added reagent, is readjusted in terms of percentage of oxidation or reduction.

Out of the multitude of reducing or oxidizing agents, a few are described that have proved to be especially convenient. Some other reagents may be found in the review by Furman [13].

Sodium Dithionite, $Na_2S_2O_4$ (*Sodium Hydrosulfite*). Application of this compound requires detailed directions. The solid substance, when dry, is perfectly stable in air. In solution, however, it is autoxidizable to an extremely high degree. It is not possible to keep a stock solution even approximately at its original titer. The solution must be prepared for immediate use. Even when kept protected from oxygen, it may gradually, sooner or later, undergo an irreversible decomposition, recognizable by turbidity caused by colloidal sulfur. This decomposition does not seem to occur in a slightly alkaline solution.

Dithionite is a very valuable and efficient reducing agent, yet its use is restricted. (1) It can be used only at pH > 4. (2) One must consider that, while acting as a reducing agent, it is oxidized to the bisulfite ion, HSO_3^-. This ion undergoes reactions with many compounds, for example, it combines with quinones, and so on. Furthermore, bisulfite ion, or rather its nonionized form, namely, sulfurous acid, is itself a reducing agent. Although it is much less powerful than dithionite and acts at a much slower rate, it may cause the potentials to drift in time. Dithionite can be used to titrate only dyes of such a potential range that sulfurous acid or bisulfite ion does not reduce them. Dyes with a formal potential decidedly less positive than methylene blue can be titrated, provided they form no addition compounds with bisulfite. Methylene blue itself cannot be titrated with dithionite (unless the titration is carried out within a few minutes) because it is reduced at a slow rate by sulfurous acid. Examples of dyes that can be successfully titrated with dithionite are riboflavin, rosindone sulfonate, and most of the phenazine derivatives, including the safranines.

Since the stability of sodium dithionite is improved by slightly alkaline reaction, its solution may be prepared in about 0.001 M sodium hydroxide. The buffer capacity of such a dilute sodium hydroxide solution is very small. Adding it to a well-buffered solution of great buffer capacity does not change the pH of the buffer to any appreciable extent. The following directions for making up a dithionite solution are therefore derived: A flask containing 20 to 30 ml of water, after adding 1 drop of 0.05 M sodium hydroxide, is closed with a stopper with two holes, one holding a glass tube as an inlet of pure nitrogen, the other as an outlet for the gases. A stream of nitrogen is bubbled for 15 min; then, while the gas flows, the stopper is lifted and a weighed amount of sodium dithionite is added; the stopper is reset, and the bubbling is continued for 5 min more. The glass tube, which has served as a gas outlet, is then pushed farther in so that the lower end is deep below the surface of the solution which, by the application of nitrogen pressure, is pushed into the buret previously washed with nitrogen.

In estimating the amount of solid dithionite, it should be kept in mind that the commercial preparation is about 90% pure, that a small part may be lost by oxidation while the solution is prepared, and that one molecule of dithionite delivers two electrons (or two hydrogen atoms). The titer of the solution is known only to a rough approximation. For most purposes this does not matter, as can be seen later, and one can dispense with a determination of the titer.

Titanous Chloride. The commercially obtainable solution of titanous chloride is made up in 15% hydrochloric acid and is rather stable in this condition. When diluted with water, so as to establish a less acid solution, it is easily oxidized in air. Therefore, any dilute solution must be immediately filled into the buret and protected from oxygen. Titanous chloride is a most useful reagent for reductive titration under the condition that the redox system is dissolved in a very acid solution, for example, $\lesssim 0.1$ N hydrochloric or sulfuric acid; and the hydrochloric acid content of the dilute titanous chloride is so low that its addition to the redox solution does not change the pH to any appreciable extent. Because of its potential range, titanous chloride is a useful and convenient reducing agent when applied to acid solutions. If one wishes to carry out a titration in extremely acid solution, for example, 5 M sulfuric acid, the stock solution of titanous chloride may be diluted, not with pure water, but with the same acid in such a concentration that the final concentration of the acid approximately matches that of the redox solution. In such a case, special attention must be paid to establishing constant liquid junction potentials (described under Measurements in Extremely Acid Solutions p. 44).

If one wants to use titanous chloride in less acid or even neutral solutions of the redox system, laborious procedures must be applied. Clark prepared solid titanous chloride, almost free of excessive hydrochloric acid, by evaporating the stock solution *in vacuo* and, protected from air, dissolving it in any desired weakly acid solution, or even in a citrate or tartrate buffer in which complex compounds are formed that are soluble even at rather high pH. The redox system can be dissolved in a similar buffer. In general, one may restrict the use of titanous chloride to very acid solutions, for which it is the most convenient reducing agent for titration.

Chromous Chloride. Chromous chloride is prepared, according to Conant and Cutter [14] by reduction of chromic chloride, using amalgamated zinc and hydrochloric acid. It is either kept over amalgamated zinc or is filtered through glass wool in an atmosphere of carbon dioxide. If the solution is filtered from suspended material and the acidity is not too high, it will keep for months without evolution of hydrogen. If working with a 1 to 10 N sulfuric acid (but not $>10\ N$) solution, chromous sulfate can be prepared by dissolving powdered chromium metal directly in the acid in a stream of hydrogen, with some heating if necessary. This solution can be directly pushed into the buret by nitrogen.

An approximately neutral, very poorly buffered, and, for this reason, convenient solution of a chromous salt can be made as follows. Chromoacetate is prepared according to Clark and Perkins [15] by reducing an aqueous solution of chromic chloride in hydrochloric acid with zinc dust. From the filtrate, crystalline, red chromoacetate is precipitated by addition of sodium acetate. The crystals are washed with water and dissolved in a large volume of air-free water. All manipulations are carried out in a stream of carbon dioxide. The final solution is washed with hydrogen and kept under hydrogen. The titer of such a solution is constant over weeks, for example, 0.0037 N.

Vanadous Chloride and Sulfate.   These compounds have been prepared by Conant and Cutter [14] as follows: Vanadium pentoxide is reduced by amalgamated zinc and acid; although very much zinc is needed, the procedure is convenient and yields a solution containing a large amount of vanadous salt. As regards the range of usefulness of chromous and vanadous salts in their acid solutions, the same considerations hold as for titanous chloride.

Leuco Dyes.   These dyes can be used as reducing agents provided their formal potential is decidedly less positive than that of the redox system to be titrated, so as to ascertain a distinct jump, or least inflection, of the potential curve at the end point of titration. For a number of cases, according to the formal potential of the redox system to be titrated, sodium indigo disulfonate in its leuco form may be used. Rosindone sulfonate (Rosinduline GG) in its leuco form is by far the most convenient "dye" to be used as a titrating agent. It can be used over a wide pH range and is stable even in very alkaline solution, a property shared by few other dyes of sufficiently negative potential and otherwise suitable properties as titrating agents. Rosindone sulfonate is not commercially available. Since it is almost indispensable for systematic redox studies, its preparation is described in the following section.

Preparation of Rosindone Monosulfonate (Rosinduline GG).   A solution of 5 gm of rosindone trisulfonate (rosinduline 2B, National Aniline and Dye Co.) in 60 ml of water is heated in a sealed tube at 180°C for 24 hr. Orange rosettes of free rosindone monosulfonic acid appear on cooling. The solution is acidified with 10 ml of 2 N hydrochloric acid and the precipitate collected on the filter. If the starting material is good, the filtrate is practically colorless.

The yield is 1.9 gm (70%). To convert the acid into the well-crystallized sodium salt it is suspended in 200 ml of water and neutralized with sodium carbonate. Then 2 gm of sodium chloride are added and the solution heated to boiling. After 24 hr at room temperature, the product is filtered and washed with a little 1% sodium chloride solution. The compound should show at pH 4.64 the formal potential of +0.130 V, when the leuco dye is titrated with potassium ferricyanide, and an index potential of 15.1 mV.

The solution of the leuco dyes, either of rosindone sulfonate or of any other suitable dye, is made up as follows: A solution of the dye in adequate concentration (about 10 to 30 times that of the redox solution to be titrated) is made up, either in water or, if solubility properties permit, in the same buffer as is used as solvent for the redox system. However, since the volume change during the titration is small, preferably not more than one-tenth, the pH of the buffered redox system is not changed beyond the limits of error even if the dye is dissolved in pure water. The latter is often preferable because most dyes are more soluble in pure water than in salt solutions. If the concentration of the redox system is $5 \times 10^{-4} M$, the concentration of the titrating leuco dye should be about $10^{-3} M$. The solubility of sodium rosindone sulfonate is about $2 \times 10^{-2} M$, except in very acid solution, in its yellow form, which is much less soluble. Therefore the restrictions imposed on the use of this dye because of its solubility properties are not serious.

A bottle is filled with the dye solution and closed with a two-hole stopper providing an inlet and outlet for gases. Then a small amount of colloidal palladium solution is added and hydrogen is bubbled through the solution until complete reduction of the dye is accomplished. For rosindone sulfonate, the color, originally red, turns on complete reduction to pale yellow or yellowish brown, according to the concentration. The amount of palladium solution depends on the specimen of colloidal palladium used. It is found, for instance by adding one drop of a 1% aqueous solution to each 10 ml of the dye solution. If after 10 min of bubbling the reduction does not proceed visibly, more should be added. The reduction may take, according to the concentration of the dye, from 10 to 30 min. It often starts but does not go to completion, the catalyst becoming inactive. If this happens, another drop should be added. The fact that the catalyst becomes inactive after a while is advantageous because it makes unnecessary the removal of the palladium before using the leuco dye for the titration. In any case the total amount of palladium should be so small that its gray color is not at all, or only slightly, noticeable in the leuco dye solution. When the reduction is complete, the hydrogen is displaced by nitrogen bubbled through the solution for a sufficient time, about 10 min, with 5 bubbles/sec. The tube which served as a gas outlet is pushed deep below the surface of the solution, and the solution is transferred by pressure of nitrogen into the buret previously washed with nitrogen.

A modification of the method, used especially by Clark, is the addition of small pieces of platinum asbestos instead of colloidal palladium solution, and interposing of a glass wool filter between the solution and the buret in order to filter out the platinum asbestos. Even so, microscopic particles of colloidal platinum may be carried over. This is not important because the platinum catalyst also is usually inactivated at the time the solution is used. The only disadvantage in the use of a leuco dye as a reducing agent lies in the fact that the color changes in the redox system during the titration are obscured. The formal potential of rosindone sulfonate is so negative that only rarely are occasions encountered when, in reversible redox systems, no recognizable end point of titration is established.

Ascorbic Acid. Sometimes, at pH < 6, a freshly prepared solution of ascorbic acid may be successfully used as a reducing agent for systems with a fairly positive formal potential.

Sodium Biphenyl. This reagent is used in the aprotic solvents tetrahydrofuran or hexamethylphosphoramide and is prepared by an overnight reaction at room temperature of a 0.1 *M* solution of biphenyl in the solvent with a sodium mirror [16]. The reaction proceeds to about 16% conversion and at this stage the system seems to reach its equilibrium. The concentration of sodium biphenyl is determined by titrating aliquots of the solution with hydrochloric acid or with methyl iodide.

Oxidative Titration: Oxidizing Agents

A compound in the reduced form can in many cases be titrated with an oxidizing agent. If a compound present in the oxidized form is to be titrated with an oxidizing agent, it is first reduced by a small amount of colloidal palladium and hydrogen gas. The completeness of reduction can be recognized by the change in color. It may be mentioned that when bubbled long enough with hydrogen in the presence of colloidal palladium a bright platinum electrode reaches a constant potential corresponding to that of the hydrogen electrode and that this potential can be used for the measurement of pH at the beginning of the titration. Then the hydrogen is driven out with pure nitrogen as thoroughly as possible. A check as to the completeness of the removal of hydrogen is the following. When the hydrogen potential is established, the bubbling with nitrogen is started, the potential gradually shifts toward the positive direction. The bubbling is kept on until the potential (which is not well reproduced at two bright platinum electrodes) is about 100 mV more negative than the formal potential of the system to be titrated. If the formal potential of the redox system is so positive that this condition cannot be fulfilled, one must use good judgment as to when to start the titration. During the titration, the behavior of the potential rise shows how well the starting time for the titration was chosen, and proper correction for the true zero point may be applied. The following discussion illustrates the manner of handling the situation.

Suppose the removal of hydrogen was practically complete before the first portion of the oxidizing agent was added. Then, even after the addition of the first few drops of the oxidizing agent, a constant and reproducible potential (at least within 1 mV or so) will be established within about half a minute after the addition of the oxidizing agent. Further addition causes the potential to rise and to be firmly established within a very short time, reproducible at different electrodes in the vessel to within 0.1 mV, and further on, even better. Toward the end of the titration, the speed at which the potential is being established, and its reproducibility, may be lessened. This may happen when the oxidation is, for example, 95% complete. In the very neighborhood of the end point, both the stability and the reproducibility of the potential are much less satisfactory. However, on plotting the titration curve, one will see that this uncertainty has no significance as to the determination of the end point.

If the removal of hydrogen was not perfect, however, the potential at the beginning of the titration may not rise sufficiently, or even drift back in time. Then, before proceeding with the titration, the bubbling with nitrogen must be continued to remove the last traces of hydrogen, which may be only slowly released if a somewhat larger amount of colloidal palladium is present. On proceeding with the titration, the potentials behave regularly, and from the plot of the titration curve the true zero point may be extrapolated. Although such an experience is not desirable, the experiment need not be discarded provided the extrapolation is small.

Bromine and Chlorine. Solutions at any pH can be titrated with bromine and freshly prepared solutions can be easily standardized as to titer, which is maintained over a sufficiently long time. Similarly, a freshly prepared aqueous solution of chlorine may be used.

Potassium Ferricyanide. A freshly prepared solution of this reagent can be used over a wide pH range, excluding only extremely acid solutions. A similar reagent of still higher potential is $K_3Mo(CN)_8$, potassium molybdicyanide, described by Buchnall and Wardlaw [17].

Potassium Dichromate. This compound can be used for acid solutions of pH < 4. At higher pH values, the oxidation is sluggish and may cause the potentials to be established with undue slowness.

Ceric Sulfate or Thallic Sulfate. Ceric sulfate or thallic sulfate dissolved in sulfuric acid of suitable concentration can be used as oxidizing agents if work must be done in strongly acid solutions.

Benzoquinone.   Water is boiled to remove dissolved oxygen and, after cooling down to about 80°C, a proper amount of finely ground quinone is dissolved in it; the solution, still rather warm, is sucked into the buret. It is stable for a sufficiently long time, but gradually decomposes. It is not recommended for the titration of alkaline solutions because quinone is very unstable in alkali and may decompose before the added amount is thoroughly mixed with the solution. For acid solutions it is a fairly convenient oxidizing agent. It has the advantage that no overoxidation takes place in such cases in which the Ox form of the redox system is likely to be irreversibly overoxidized with any appreciable speed. However, the various possible reactions which quinone may undergo make necessary good judgment in the use of this otherwise very useful reagent.

Dyes.   For redox systems of sufficiently positive potential range, the Ox forms of stable dyes of very positive range can be successfully used as oxidizing agents. The use of phenolindophenol is especially recommended and the compound is easily prepared [18] but not commercially available. Other commercially available dyes of the indophenol class should be properly tested on standard substances because the available specimens are rarely pure compounds.

Lead Tetraacetate.   For solutions in water-free glacial acetic acid, a solution of lead tetraacetate in glacial acetic acid can be used as an oxidizing agent [19, 20].

Copper(II) Perchlorate.   The anhydrous salt is prepared by dissolving tetrakis(acetonitrile)copper(II) perchlorate prepared from nitrosyl perchlorate and copper metal in anhydrous acetonitrile [21].

Gold(III) Chloride.   The anhydrous salt is prepared by heating $HAuCl_4 \cdot 4H_2O$ slowly in a stream of chlorine to 200°C, maintaining this temperature for 0.5 hr, cooling under chlorine, and then flushing with nitrogen [22].

Apparatus for Redox Titrations

For the measurement of potentials, a potentiometer or a pH meter can be used.

The electrode is constructed by spot-welding a small piece of platinum foil to a short piece of no. 26 platinum wire, which is then sealed into a narrow soft-glass tube. Pyrex glass cannot be used because its coefficient of thermal expansion is very different from that of platinum, with the result that a perfect seal cannot be obtained.

Whether the electrode should consist of bright platinum or gold has been the subject of much discussion. However, the difference in behavior is slight and there is no reason why the blank platinum electrode should not be used in routine work. It should be frequently cleaned in cleaning solution and in

10% sodium hydroxide, and thoroughly washed in water thereafter. A gold electrode is most easily prepared by coating bright platinum with gold. For this purpose, a platinum electrode is used as cathode, with another platinum electrode as anode, in a solution of 1% gold chloride and excess of potassium cyanide. Current is supplied from a 4-V battery with proper resistance to allow a very slow deposit of gold. The coated electrode is then washed in nitric acid and water. The gold electrode is supposed to establish potential more quickly and more reliably than the platinum electrode in relatively weakly poised redox systems, but sometimes the opposite may be true, and when the system is well poised there should be no difference between the two types. Even mercury can be used as an indifferent electrode when the redox potential of the system is such that all traces of mercury ions, which may be formed at the surface of the mercury electrode [23] on immersing it in the solution, are reduced. This condition is fulfilled for all redox systems with a potential range less positive than that of the indophenol dyes. For phenolindophenol, for instance, mercury cannot be used as an electrode, but it can be used for indigo sulfonates. Since mercury does not absorb any hydrogen gas, the electrode vessel may even be bubbled with hydrogen instead of nitrogen. However, a mercury electrode is much less convenient to handle than a platinum electrode and is therefore not often be used.

It is necessary always to have at least two platinum electrodes in the titration vessel and to read the potentials of each for comparison. When the poise of the system is sufficient, that is, if the redox system is perfectly reversible and the ratio of the oxidized and the reduced forms of the dye is within the range of 10:1 to 1:10, the potentials should agree to at least 0.1 mV; usually, they are better. In the neighborhood of the starting point and of the end point, the agreement may be less perfect. This may also happen in the neighborhood of the 50% point with a bivalent redox system in which the two successive steps are extremely widely separated and, consequently, the titration curve around the 50% point of titration is very steep. However, if the slope of the titration curve is extremely steep an uncertainty of the potential of a few millivolts has in general no bearing on the results obtained from the titration curve.

The electrode vessel, should in general be cylindrical, about 4 to 5 cm in diameter and about 12 cm high (Fig. 1.11). If feasible, 20 to 40 ml of the solution to be titrated should be used, or at least enough to cover the electrode metals entirely. For special cases all dimensions can be increased or decreased. The upper end of the vessel is closed with a well-fitting rubber stopper with a number of holes through which glass tubes are tightly inserted. These tubes provide for two platinum electrodes (or even three), gas inlet, a gas outlet, a potassium chloride–agar bridge, a buret, and a reserve hole which may be closed or used for inserting a thermometer, or for the addition of a reagent, or for the third platinum electrode.

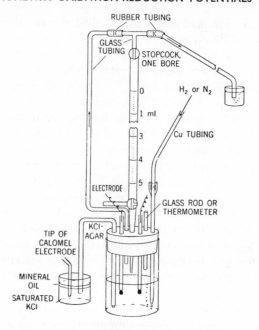

Fig. 1.11   Assembly for oxidimetric titration.

The gases used are pure nitrogen and, alternately, hydrogen. The latter is used when the solution, in the presence of palladium, is being reduced. Nitrogen is used to bubble out the hydrogen, or the oxygen of dissolved air, as the need may be, and is also used throughout the titration. The speed of bubbling nitrogen should be great when oxygen or hydrogen is being removed (about 10 bubbles per second), and slow during the titration proper (about 3 bubbles/sec).

Hydrogen and nitrogen are best supplied from cylinders with reduction valves. The gas train is shown in Fig. 1.12. From the cylinders, heavy-walled vacuum rubber tubes, which have been soaked for a day in about 5% sodium hydroxide solution and thoroughly washed with water, lead the gas to the purifying system in such a way that alternatively the one or the other gas can be used. The gases pass first to a wash bottle with 10% sodium hydroxide, and then to a glass tube filled with copper filings, and are heated in the furnace to about 400 to 450°C. If care is taken that the temperature never exceeds 500°, Pyrex glass can be used. If no temperature control is applied, combustion-tube glass is necessary; however, it is more difficult to build up all glass parts with it. Temperature control can be secured in the use of Pyrex glass with a thermocouple of alumel-chromel, the soldered contact of the two metals being placed right above the Pyrex glass tube and covered by

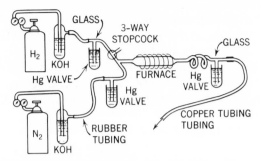

Fig. 1.12 Train for the purification of hydrogen and nitrogen. The glass tube after leaving the furnace should be slightly bent downward to avoid reflux of water of condensation into the furnace.

the lid of the furnace. The tube is filled either entirely with copper filings, freed from grease, or by alternate sections, each 10 cm in length, of copper filings and tightly rolled copper gauze, which keeps the filings more firmly in place. After leaving the furnace, the gases should never be exposed to any rubber tubing which may give off volatile impurities and are never reliable with respect to impermeability to traces to oxygen. The end of the glass tube as it protrudes from the furnace is bent into a spiral for greater elasticity. The end is sealed to copper tubing of the flexible, elastic type, with De-Khotinsky, or a similar, cement. This copper tubing, of about 2-mm i.d.,

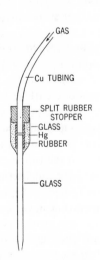

Fig. 1.13 Joint between copper and glass tubing.

leads to the gas inlet of the electrode. An airtight connection with the inlet tube may be obtained as follows. The upper end of the inlet tube is equipped with a glass bell (Fig. 1.13) surrounding the open end. The protruding end has slipped over it a piece of rubber tubing, not too thin-walled, exhibiting some rigidity and prepared as mentioned above, of such a length as to protrude from the opening of the glass tube, for example, 2 cm, into the bell. The end of the copper tubing is then pushed into this rubber tubing so as to be in contact with the end of the glass tubing, and the beaker is filled with mercury. The mercury is prevented from spattering by closing the beaker with a rubber stopper after first splitting the stopper into halves which can be properly put in place. The lower end of the gas inlet tube is drawn out to a fine, but not too fine, opening, and reaches almost to the bottom of the titration vessel.

The high efficiency of this method of purification of nitrogen has recently been corroborated by Liebhafsky and Winslow [24]. They recommend especially the admixture of some hydrogen to the nitrogen to be purified by the heated copper so that the oxygen is converted to water rather than to copper oxide. It is, of course, not permissible to take the chance of deliberately adding hydrogen to nitrogen to be used in potentiometric work. However, the conditions recommended by these authors are approximately fulfilled here since the same copper tube is used alternately for the purification of hydrogen and nitrogen.

Test for Purity of Nitrogen.   In order to test the completeness of the purification of the nitrogen gas from traces of oxygen, the following procedure is easy to apply and highly recommendable. The gas is led through a vessel similar to the electrode vessel but containing only inlet and outlet tubes for the gas. A piece of yellow phosphorus (carefully and freshly cut from a large piece stored under kerosene) is placed in this vessel and the flow of gas started. As the oxygen pressure decreases, a halo (white cloud) will be visible which entirely disappears on continued bubbling. The gas may then be said, for all practical purposes, to be entirely free from oxygen [25].

Although this simple procedure for purification of nitrogen combined with the possibility of the alternate use of nitrogen and hydrogen is satisfactory in every respect, an alternate method used by Cameron [9] may be described here. Cameron purified and tested the nitrogen as follows:

"Gas from the cylinder was passed through two towers of solid ammonium carbonate, and then through a tower of soda lime, leaving a definite partial pressure of ammonia in the gas. The mixture was passed through a quartz tube packed with platinized asbestos free from sulfur and heated to 450 to 500°C. Ammonia remaining after the reaction of ammonia and oxygen was removed with wash towers of 30 per cent sulfuric acid colored with bromophenol blue to indicate exhaustion, and acid spray and carbon dioxide were removed with a wash of 60 per cent potassium hydroxide. A bottle of a 0.1 per cent solution of reduced indigodisulfonate in the line indicated complete absence of oxygen, even at high rates of flow. It is important to insure saturation of the gas stream with ammonia, however. Exhaustion of any reagent was readily noted, and the line could be operated for long periods of time without attention."

The gas outlet is a glass tube secured with a liquid seal. The agar bridge is the same as that described previously. The end closed with a ground stopper is submerged in the solution, while the other end is dipped into a beaker containing a saturated solution of potassium chloride covered with a layer of white oil. Care should be taken that the end of the agar level is not, by accident, prevented by a drop of white oil from making contact with the solution. The tip of the calomel electrode is submerged in the same beaker of potassium chloride solution.

Figure 1.11 shows the type of buret recommended. The mixing of the reagent withdrawn from the buret may be accomplished by the bubbling gas, the speed of which may be temporarily increased for the purpose. Good judgment will show whether this procedure is sufficient. Agitation of the vessel briefly (1 sec) after each addition of reagent from the buret is preferred. Although this can be done by hand, a magnetic stirrer is more convenient. This device should be grounded or turned off during the measurement of the potential.

As regards temperature control, there is no objection to working at room temperature, provided the temperature is kept the same, within 0.5°C, in the calomel electrode and in the titration vessel. It is preferable, however, always to carry out the measurement at one temperature. In the Michaelis laboratory a compartment big enough for all manipulations was built in the laboratory as an air thermostat at 30°C. Even so, the temperature both in the calomel electrode and in the titration vessel was checked by a thermometer permanently enclosed in and protruding from the vessel. The potentiometer was outside the constant-temperature compartment. A water bath for temperature control can be used only with utmost precaution (see below), for the water film on the electrode walls is likely to cause stray currents. However, in Clark's laboratory, a thermostat with white oil, in which the electrodes are immersed, is successfully used.

A somewhat different apparatus used by Cameron [9] for the discontinuous titration of several organic developing agents is shown in its essential parts in Fig. 1.14. A double-walled beaker, the top of which is flattened and ground plane, is fitted with a flanged glass cap, also ground plane, carrying a water seal for the stirrer, a glass electrode, and a neck in which can be inserted a stopper carrying the electrodes. Lead-in tubes for the salt bridge and for the gas supply to a porous alundum bubble head are sealed through the cap. Three small necks on the cap permit insertion of buret tips with pieces of small rubber tubing for gasketing. A thermometer mounted in this same cap

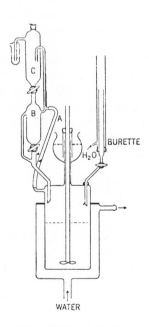

Fig. 1.14 Cameron assembly for oxidimetric titration.

permits determination of the reaction mixture temperature. Water is pumped through the jackets of the reaction vessel, the half-cell, and the burets from a 35-gal thermostat operating at 20°C ± 0.01°. The addition tube in Fig. 1.14 makes it possible to add the oxidizing solution rapidly and flush the vessel with buffer solution without admitting atmospheric oxygen. With the pinch clamp on rubber tubing *A*, nitrogen from the reaction vessel sweeps out compartment *B* containing the oxidizing solution, and bubbles through the buffer solution in compartment *C*, before escaping to the air through the bubbler. If the clamp is removed from the tubing, nitrogen from the reaction vessel can replace the solutions when the lower and upper stopcocks are opened in succession. The stopcocks are both of large enough bore to permit 10 ml of solution to be added in about 2 sec.

Measurements in Extremely Acid Solution

When a potentiometric titration is to be made on a solution that is much more acid than about 1 *N* hydrochloric acid, the following considerations may be helpful. First of all, the necessity of more or less continuous bubbling with gas makes it unfeasible to use hydrochloric acid as an acid because of the evaporation of gaseous hydrogen chloride. However, sulfuric acid can be used up to a concentration of 10 *M* or more. The determination of pH with the hydrogen electrode is impossible, for not only is the liquid junction potential very large, of incalculable magnitude, and very changeable with time, but the very definition of pH breaks down. However, a useful titration curve for a reversible redox system can be obtained [26] as follows: The closed tip of the potassium chloride–agar bridge is immersed before use for many hours in an acid solution of the same composition as that of the experiment proper, while the other end is kept under a saturated potassium chloride solution. The acid is renewed several times. A quasi-stationary concentration gradient of the acid is gradually established within the agar tube. When the agar tube is now used for the titration, the liquid junction potential is relatively stable. Although its magnitude is unknown and even not reproducible, it is constant within approximately 1 mV during the time necessary for the titration. The absolute values of the potentials have no significance, but the shape of the potential curve can be utilized for conclusions, as discussed below. It may be mentioned that the titration curves obtained with certain bivalent redox systems in solutions of such high acidity can be used to make up a rational extension of the acidity scale, which may be regarded as an extension of the pH scale [27].

Measurement of pH in Redox Titration

All redox titrations must be carried out in well-buffered solutions which must fulfill two conditions. (1) They must possess a buffer capacity sufficient to keep the pH constant during the titration; and (2) the pH must be known either by using one of the standardized buffer solutions or, better, by direct pH measurement before or after the titration, or, when suitable, both before and after. Most of the usual buffers can be used. A few remarks suffice.

a. Borate buffer may cause erroneous results because of its inclination to form complex compounds, especially but perhaps not exclusively, with molecules containing two adjacent hydroxyl groups. Indigo sulfonates, for example, seem to yield erroneous results when titrated in borate buffer.

b. Glycine buffer cannot be used with quinones and probably with some other compounds because of chemical interaction. N,N-Dimethylglycine buffer [28] can be safely used instead. Its pH range is about 0.5 pH units higher than that of glycine, the latter being around 9.5.

Buffers that may be recommended are those composed of a mixture of the following acids and their sodium or potassium salts: phosphoric acid either as a mixture of primary with secondary phosphates for the pH range from 5.5 to 7.5, or as a mixture of secondary and tertiary phosphates for pH from 10.5 to 11.3 or so, prepared by mixing $N/15$ disodium phosphate with sodium hydroxide in amounts so as not to convert the secondary entirely into tertiary phosphate. Because of the large variation of pH in these buffers with change of ionic strength, the pH should be measured in the final mixture and tabulated values not relied upon. Acetate buffer can be safely used at pH from 5.8 to 3.4, a buffer of lactic acid and sodium lactate at pH from 4.5 to 3, and citrate buffer at pH from 3 to 1.5. For the pH range from 7.5 to 9.8, dimethyl barbiturate buffer [29], prepared from sodium dimethyl barbiturate (sodium Veronal) and hydrochloric acid is almost indispensable because of the deficiencies mentioned of other buffers in this pH range. For pH in the neighborhood of 1, hydrochloric acid can be used, for example, a solution of 0.1 N hydrochloric acid plus 0.5 N potassium chloride. For extremely acid solutions outside the well-defined pH range, solutions of sulfuric acid are used, with the special precautions with respect to liquid junction described in the Section on Measurements in Extremely Acid Solution (p. 44). The pH during the titration is measured with a glass electrode.

Titration of Unstable Systems

If a component of a redox system is labile so as to undergo spontaneously some reaction other than that involved in the establishment of the redox equilibrium, no steady potential is obtained. If such a secondary reaction is sufficiently slow, the polarographic method is advised for the determination of the formal potential since measurements can be made more rapidly than by the potentiometric method. In the latter technique the drifting potential is extrapolated to a value corresponding to the initial state as it exists or is imagined to exist before the decomposition takes place. Continuous titration of such a system is impossible, but a discontinuous titration can sometimes be helpful. For instance, to the solution of the stable reductant an amount of oxidizing agent is added that is estimated to produce a certain amount of the oxidant, which is unstable. The ratio of oxidant to reductant is fixed at least momentarily. The potential is read as quickly as possible and its drift in time is pursued. Extrapolation gives an approximate value of the original potential. Then another portion of the reductant is mixed with a larger portion of the oxidizing agent and the procedure repeated. When all the experiments are grouped together, a composite titration curve can be constructed and utilized as a regular titration curve, sometimes with good success, depending upon the lability of the system. Methods of "quick" titration, with or without extrapolation to zero time, have been used by Clark et al. [7], by Biilmann and Blum [30], Fieser [31], and others.

The same principle has been used by Ball and Chen [32] in a method based on a principle devised for other purposes by Hartridge and Roughton [33]. A solution of the reductant (for the details of this delicate procedure, the original literature must be consulted) under investigation and a solution of the oxidizing agent flow through a mixing chamber and then through a tube which at certain intervals is provided with platinum electrodes. If the mixing rate is high in comparison with the oxidation rate, the potentials along the tube change according to the rate at which the oxidation takes place. Each electrode shows a steady potential. With a stable system, such as Fe^{3+}-Fe^{2+}, the potentials along the tube are alike. With a deteriorating system, the change in potential resulting from the deterioration is superimposed on the change caused by the oxidation process. The potential then can be extrapolated to zero time, and the rate of change gives the rate of the deteriorating reaction.

For another method of discontinuous titration with the apparatus (Fig. 1.14), Cameron [9] may be quoted:

"The experimental procedure followed consisted in adding 100 ml of buffer solution, 0.2 M in buffering ion and with a sodium ion concentration fixed by addition of sodium sulfate, as necessary, to the reaction vessel, and determining the pH with two hydrogen electrodes. The glass electrode was then read, and the hydrogen displaced with pure nitrogen with the substitution of blank platinum and gold electrodes for the platinized ones. The addition tube was set in place with the measured amount of oxidizing solution and buffer solution for flushing, and half an hour allowed for the sweeping out of hydrogen and air. Ten milliliters of reducing agent was then added from one of the jacketed burets. The lower stopcock of the addition tube was opened, and at the same time an electric timer was started which gave a single stroke on a bell at 15-sec intervals. The upper stopcock was then opened and the lower chamber flushed into the reaction vessel."

"Readings were taken at 15-sec intervals for 3 min and at 1-min intervals for 7 min. The glass electrode was then read again, and the pH of the solution as originally determined with the hydrogen electrode was corrected for any change in pH which had occurred. Readings were taken with a student-type potentiometer. The potential of the glass electrode was read with the same potentiometer, using a ballistic galvanometer and a 1 microfarad condenser as null point indicator".

"Addition of the solutions raised the temperature of the reaction mixture-slightly, but this effect could be disregarded, since in most cases the extrapolation to zero time was made from a graph which was a straight line up to 10 min."

Interpretation of Titration Curves

In the interpretation of the titration, first of all, the symmetry of the curve around its midpoint (at 50% oxidation) should be checked. The potentials from 0 to 5% and those from 95 to 100% are not very reliable in this respect. In more difficult cases, that is, at very low concentration of the redox system or when the substance to be titrated is contaminated with some other oxidizable or reducible material, the potential ranges of which overlap somewhat with the redox system proper, the unreliable range may extend to 10% and from 90%, respectively. The best method for first orientation is to check whether the difference in potential, the so-called index potential E_i is, for example, within 0.2 mV, the same, no matter whether it is taken as the difference at the 50 and 25% points of the titration, or at the 75 and 50% points. For instance, in Fig. 1.15 E_i is 22 mV.

If the titration curve is symmetrical:

1. When E_i is 28.6 mV (the limit of error may usually be regarded as ±0.2 mV), with corresponding values chosen for other temperatures in proportion to the values of Table 1.1), the redox system is a univalent one. In exceptional cases, recognizable by the fact that during the titration a twofold change in color takes place, it is a bivalent redox system with a semiquinone formation constant $= 4$ (see below). Such cases never causes any doubt because this event can occur, for any one redox system, only at one particular pH. The same dye at any other pH shows an E_i different from 28.6 mV.

2. When E_i is 14.3 (±0.2) mV at 30°C, the redox system is a bivalent one, without the formation to any measurable extent of any intermediate step by univalent oxidation.

3. When the index potential E_i in a redox system previously known to be bivalent is >14.3 (±0.2) mV, there exists an intermediate univalent step of oxidation. If, on varying the initial concentration of the redox system (for example, from 5×10^{-4} to 3×10^{-3} M) in the same buffer, E_i is unchanged,

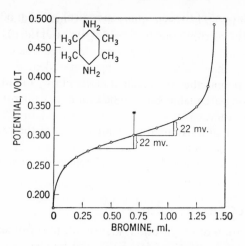

Fig. 1.15 Titration of the *p*-diaminodurene system.

the intermediate step of oxidation is a free semiquinone radical. However, if E_i depends on the initial concentration, the intermediate step is not only represented by this free radical but also by a dimerization product of the radical with which it is in equilibrium.

4. When E_i is $\lessgtr 40$ mV, no other point of inflection should appear than that at 50% oxidation. Whenever E_i is >40 mV, in addition to the point at 50% oxidation, two lateral points of inflection appear. This demonstrates even at first inspection that the oxidation proceeds in two successive steps. When E_i is extremely large, for example, $\gg 200$ mV, the two lateral points of inflection lie at 25 and at 75% oxidation. In such a case, the first half of the curve (from 0 to 50%) may be considered as representing a univalent oxidation which, considered separately, has $E_i = 28.6$ mV; and the same holds for the second half, from 50 to 100%.

5. The semiquinone formation constant K has already been defined as:

$$K = \frac{[QH]^2}{[Q][QH_2]},$$

where [QH] is the concentration of the semiquinone radical, [QH$_2$] that of the reduced form, and [Q] that of the (totally) oxidized form of the redox system. One derives K from E_i as follows

$$\sqrt{K} = 10^{E_i/0.06} - 3 \times 10^{-E_i/0.06},$$

where E_i, the absolute value of the index potential in volts, is always taken as a positive number. (The factor 0.06 holds for 30°C. Compare Table 1.1.)

The maximum amount of the semiquinone radical exists at 50% of oxidation, and the ratio of the semiquinone to the total dye, $[QH/a]_{max}$, at that point is

$$[QH/a]_{max} = \sqrt{K}/(2 + \sqrt{K}). \tag{1.58}$$

6. The formal potential of the bivalent system QH_2, Q or the mean formal potential E'_m is the potential at 50% oxidation.

7. The formal potential of the lower step of oxidation E'_1 or the formal potential of the QH_2, QH system is

$$E'_1 = E'_m - 0.03 \log K, \tag{1.59}$$

where K is the semiquinone formation constant, and E'_m the potential at 50% of the oxidation. The formal potential of the higher step of oxidation E'_2, or the formal potential of the QH, Q system is

$$E'_2 = E'_m + 0.03 \log K \tag{1.60}$$

and we always have

$$E'_m = (E'_1 + E'_2)/2.$$

E'_m may be interpreted as the mean formal potential, or the formal potential of the system considered as a bivalent redox system. E'_1 is the formal potential of the lower (univalent) step of the oxidation, E'_2 that of the higher (likewise univalent) step.

If the titration curve is not symmetrical about the point of 50% oxidation, and/or the shape of the curve is not independent of the concentration of the redox system:

1. The curve is symmetric but its slope varies with the concentration. The intermediate step of oxidation is not represented by a free radical QH, alone, of the same molecular size as Q or QH_2, but by a dimerization product of the radical in equilibrium with the free radical. This is true, for instance, for phenanthraquinone-3-sulfonate in alkaline solution. On lowering the concentration, one may reach a concentration range in which the titration curve no longer depends on the concentration. In this range of concentration, the dissociation of the dimer into free radicals is practically complete, and then the mathematic analysis of the limiting curve for extremely high dilution is the same as before. From the deviation at higher concentration, the constant of dimerization can be calculated. As to the details of such calculations, the original papers may be consulted [35].

A single case has been encountered thus far, namely, phenanthraquinone sulfonate in acid solution, in which in a homogeneous solution an intermediate step of oxidation is formed consisting practically entirely of the dimerized form of the radical. Here, at very low concentration, no intermediate step is formed in a distinctly measurable amount, the curve having an index potential of 14.3 mV within the limits of experimental error; but at higher concentration the slope of the titration curve becomes increasingly dependent on the concentration, and the formation of an intermediate step of oxidation can be recognized also by the twofold color change during the titration.

2. The titration curve is not symmetrical around its 50% point. In this case, the molecular size of the Q form differs from that of the QH_2 form. In many dyes, the Q form, at not too low a concentration, has the tendency to form polymers which, at not too high a concentration, are usually dimers.* The QH_2 forms (leuco dyes) are not known to have a comparable tendency to polymerize in solution [37].

In the concentration range from 10^{-2} to 10^{-4} M, this phenomenon has no conspicuous effect on the shape of the titration curve. It is just noticeable, for instance, with methylene blue. In higher concentrations the phenomenon can be studied by spectrophotometric methods but not very well potentiometrically because, as stated above, it is hardly possible to keep the pH constant during a titration unless the concentration of the dye is very much smaller than that of the buffer.

A special case of a titration curve which is neither symmetrical about its midpoint nor independent of the initial concentration of the substance to be titrated is encountered if the chemical process takes place according to

$$T + 2e^- \rightleftharpoons 2R$$
$$T + 2H^+ + 2e^- \rightleftharpoons 2R$$

Two examples are

$$Br_2 + 2e^- \rightleftharpoons 2Br^-$$
$$RSSR + 2H^+ + 2e^- \rightleftharpoons 2RSH$$

where RSH is any sulfhydryl compound such as thioglycolic acid, and RSSR is the corresponding disulfide. Most of the sulfhydryl compounds, however, do not form truly reversible redox systems at the platinum electrode. The reason for this is probably the poisoning of the surface similar to that observed in the catalytic reduction of organic compounds using a platinum catalyst. An apparent exception to this generalization is the reversible oxidation of thiourea to formamidine disulfide in acid medium [38].

$$
2\underset{\underset{NH_2}{|}}{\overset{\overset{NH_2^+}{\|}}{C}}{-}S^- \rightleftharpoons \underset{\underset{NH_2}{|}}{\overset{\overset{NH_2^+}{\|}}{C}}{-}S{-}S{-}\underset{\underset{NH_2}{|}}{\overset{\overset{NH_2^+}{\|}}{C}} + 2e^-
$$

---

* It may be noted that, according to G. N. Lewis and J. Bigeleisen [36], leuco methylene blue is dimeric at the temperature of liquid air in a rigid solvent. There is no evidence of dimerization at room temperature.

The reaction involved is indicated above and is independent of pH. The pK of thiourea from the data given must be less than 0.

The electrode equation is the following

$$E = E^{0'} + 0.03 \log \frac{[\text{RSSR}]}{[\text{RSH}]^2},\qquad (1.61)$$

where the constant $E^{0'}$ may be considered as the "formal potential," namely, the potential under the condition that the logarithmic term vanishes. Obviously, this does not happen either when the ratio of the concentration of the oxidized and the reduced forms is $1:1$, or at any other fixed value of this ratio, but rather this formal potential depends on the absolute concentration. The titration curve is not symmetrical about its midpoint at 50%; the point of inflection lies at that point where the ratio of the molar concentration of the reduced form to the (equivalent) concentration of the oxidized form equals $2 - \sqrt{2}$, or at 41.4% oxidation. This asymmetry is not very striking on inspection of the graph. Much more striking is the fact that the shape of curve, and also the formal potentials, depends on the initial concentration of the substance. The dependence of potentials on the initial concentration is very much more sensitive criterion for the existence of this type of redox system than the asymmetry of the titration curve. It should be mentioned that Preisler [39] has also shown the existence of truly reversible titratable redox systems of a sulfhydryl compound with a symmetrical titration curve, such as dithiobiuret-dihydrodithiobiuret.

The selenoureaformamidine-formamidine diselenide system behaves similarly in acid medium [40].

The cystine-cysteine system apparently behaves in a truly reversible manner at a mercury electrode. Potentials measured potentiometrically at a cathodically pretreated mercury electrode [41] are in good agreement with the values obtained from combustion data [42], and from spectrophotometric studies of the ferric complex of cysteine [43].

It should be added that the theory of reversible two-step oxidation can be extended, with a slight modification, to another type of system. In the preceding section the reversibility was restricted to a bivalent oxidation, which could be separated into two successive univalent steps. Usually, the reversibility does not extend over a wider range than the two-step oxidation. However, some cases have been described in which a reversible bivalent oxidation can be extended to a further reversible bivalent oxidation. The first case to fulfill this condition at least approximately concerned a naturally occurring vegetable pigment from the plant *Mercurialis perenuis* [44]. A carefully investigated case [45] is the oxidation of hexahydroxybenzene to tetrahydroxyquinone, and further to rhodizonic acid. Each of these two steps is reversible and is a bivalent oxidation which, under the conditions studied, does not give rise to any appreciable amount of semiquinone. Such a case can be dealt with in the same way as before with the exception that each time where the factor RT/F was formerly used in the underlying theory the factor $RT/2F$ must be applied. Here also, one can distinguish the formal potential of the lower step, that of the higher step, and the mean formal potential of the whole system. The plot of such formal potentials resembles in principle that of the example shown in Fig. 1.9.

Nonaqueous Titrations

Glacial acetic acid as a titration medium is unsatisfactory. Platinum indicating electrodes are found to be sluggish and insensitive in this medium, making precise measurements difficult. In the vicinity of the equivalence point, the potentials of the electrodes reach a maximum and then decrease when both Cr(VI) and Ce(IV) are used as oxidants. Addition of water improves the sensitivity of the titration. Solvent mixtures of 75% acetic acid–25% dilute aqueous perchloric acid (by volume) were found to be suitable for studies on arylferrocenes [46].

Mixtures of 1 part by volume of glacial acetic acid and 1 part of 0.5 N sulfuric acid were suitable for investigation of hydroquinones [47].

The use of a 90% dioxane–10% water mixture is necessary for the study of electron-exchange polymer. The oxidation-reduction potentials obtained are less accurate since the activity effects are less known and the reactions are slow in this system [48].

Acetonitrile, which has seen wide use in polarography, is a suitable solvent for potentiometric titration. Studies in anhydrous acetonitrile allow an assessment of hydrogen ion effects in the system. Cupric chloride has been used as an oxidizing agent for substituted benzidines and arylamines [21].

The reference electrode was Ag/0.01 M AgNO$_3$ in CH$_3$CN.

Gold chloride (AuCl$_3$) has been used in anhydrous acetonitrile saturated with lithium chloride for the determination of redox potentials of arylated phenols. The formal potentials of the stable phenoxyl radicals were in good agreement with polarographic data [22].

The reference electrode was Ag-AgCl.

Hexamethylphosphoramide is a suitable solvent for the determination of redox potentials of aryl substituted olefins and aromatic polynuclear hydro-carbons using 10^{-2} M sodium biphenyl as the titrant. In this solvent the alkali salts of organic anions are completely dissociated in contrast to tetra-hydrofuran in which ion pairing occurs [16, 49].

Reversible Organic Redox Systems

Organic redox systems that have so far been shown to be thermodynami-cally reversible in aqueous medium are given in Table 1.4. The sulfhydryl

Table 1.4 Reversible Organic Redox Systems

Oxidant	*Reductant*
Quinones	Hydroquinones
Quinone-imines	Aminophenols
Quinone-dimines	Phenylenediamine
Azobenzenes	Hydrazobenzenes
Nitrosobenzenes	Phenylhydroxylamines
Indophenol dyes	Leuco compounds
Indamine dyes	Leuco compounds
Azine dyes	Leuco compounds
Oxazine dyes	Leuco compounds
Thiazine dyes	Leuco compounds
Indigoid dyes	Leuco compounds
Flavin dyes	Leuco compounds
Arylarsonic acids	Arylarsenoxides
Nitroguanidine	Nitrosoguanidine
Ascorbic acid	Dehydroascorbic acid
Tetrazolium salt	Formazan*

* B. Jambor, *J. Chem. Soc.*, **1958**, 1604.

compounds (mercaptans) such as thioglycolic acid or cysteine probably belong to this class even though considerable confusion exists in this field. As has been mentioned in the last section, the platinum electrode is not satisfactory for a study of this system but specially pretreated electrodes give values that are in agreement with reports obtained by other methods.

The use of aprotic solvents such as acetonitrile and hexamethylphosphoramide has indicated that alkylated benzidines [21]

and arylated phenols [22]

$$ArOH - e^- \rightleftharpoons ArO\cdot + H^+$$

are reversibly oxidized in the former and that arylated olefins and poly-nuclear hydrocarbons

$$ArH + e^- \rightleftharpoons ArH^{\dot{-}}$$

are reduced reversibly in the latter [49].

Indicators for Biological Systems

Many cases have been discovered in biochemistry in which two molecular species, differing only in the level of oxidation but not behaving as reversible redox systems, become reversible systems upon addition of an enzyme. Nevertheless, no metal electrode responds to such an enzyme-activated system so as to establish a redox potential. These systems, however, can be brought into equilibrium with a regular redox system in the solution. The latter then impresses its potential on the electrode and so permits an indirect measurement of the state of the other system in terms of a redox potential. The principle of those measurements is the same in all cases; the first case discovered, by Thunberg [50] and Quastel [51], may therefore be sufficient to illustrate the experimental procedure.

Fumaric acid and succinic acid differ only by two hydrogen atoms. A mixture of the two, by the addition of an enzyme, can be said to form a reversible redox system. The substance used as an enzyme is succinodehydrogenase, extracted from muscle tissue. On adding a small amount of methylene blue to the substrate-enzyme system, methylene blue is partially reduced to such an extent that an equilibrium between the substrate and the dye system is established. Provided the concentration of the dye is very much smaller than that of the organic acids, the redox state of the dye is just an indicator for that of the organic acids. The reaction between the organic acid system and the dye is slow, and the establishment of the definite potential requires considerable time, several minutes to an hour.

The dye used as indicator must have a formal potential not too different from that of the inert system. Only if in the final state of equilibrium the concentration of the oxidized and the reduced form of the indicator dye is of the same order of magnitude is the poise of the dye system sufficient to establish a reproducible potential of thermodynamic significance. In order to test the reversibility of the system, one has to carry out measurements with varied ratios of the two acids to show that the variation of the potential is that expected for a reversible system.

This method is based on the assumption that the following reaction is reversible:

$$Red + Dye_{Ox} \rightleftharpoons Ox + Dye_{Red}$$

At equilibrium the potentials of both oxidation-reduction systems are the

Table 1.5 $E^{0'}$ Values for Oxidation-Reduction Indicators at 30°C as a Function of pH

Indicator	$E^{0'}$										
	pH 5.0	pH 5.4	pH 5.8	pH 6.2	pH 6.6	pH 7.0	pH 7.4	pH 7.8	pH 8.2	pH 8.6	pH 9.0
Phenol-m-sulfonate indo-2,6-dibromophenol	+0.390	+0.366	+0.342	+0.319	+0.295	+0.273	+0.251	+0.229	+0.207	+0.187	+0.168
o-Chlorophenol indophenol	—	—	—	+0.288	+0.262	+0.233	+0.203	+0.170	+0.139	+0.108	+0.082
Phenol indophenol	—	—	—	+0.276	+0.254	+0.227	+0.200	+0.170	+0.139	+0.110	+0.083
o-Bromophenol indophenol	—	—	+0.308	+0.284	+0.259	+0.230	+0.200	+0.167	+0.137	+0.105	+0.079
2,6-Dichlorophenol indophenol	+0.366	+0.339	+0.310	+0.279	+0.247	+0.217	+0.189	+0.162	+0.137	+0.113	+0.089
Phenol indo-2,6-dibromophenol	—	—	—	—	—	+0.218	+0.190	+0.163	+0.138	+0.114	+0.090
m-Cresol indophenol	—	—	—	+0.259	+0.233	+0.208	+0.185	+0.160	+0.134	+0.105	+0.076
o-Cresol indophenol	—	—	—	+0.243	+0.217	+0.191	+0.168	+0.143	+0.116	+0.068	+0.057
2,6-Dichlorophenol indo-o-cresol	+0.335	+0.307	+0.277	+0.245	+0.212	+0.181	+0.152	+0.125	+0.099	+0.075	+0.051
Toluylene blue	+0.221	+0.196	+0.173	+0.151	+0.132	+0.115	+0.101	+0.088	+0.075	+0.063	+0.051
Thymol indophenol	—	—	+0.244	+0.222	+0.198	+0.174	+0.148	+0.123	+0.097	+0.069	+0.041
1-Naphthol-2-SO$_3$H-indo-3,5-dichlorophenol	+0.262	+0.236	+0.210	+0.181	+0.150	+0.119	+0.088	+0.060	+0.034	+0.010	−0.012
Thionine	+0.138	+0.112	+0.100	+0.087	+0.074	+0.062	+0.050	+0.037	+0.025	+0.014	+0.001
Methylene blue	+0.101	+0.077	+0.056	+0.039	+0.024	+0.011	−0.002	−0.014	−0.026	−0.038	−0.050
Indigo tetrasulfonate	+0.065	+0.041	+0.017	−0.006	−0.027	−0.046	−0.062	−0.077	−0.090	−0.102	−0.114
Indigo trisulfonate	+0.032	+0.008	−0.016	−0.039	−0.061	−0.081	−0.099	−0.114	−0.127	−0.140	−0.152
Indigo disulfonate	−0.010	−0.034	−0.057	−0.081	−0.104	−0.125	−0.143	−0.160	−0.174	−0.187	−0.199

same and at a constant pH and for a two-electron change are expressed as

$$E^{0'}[\text{Dye}] - 0.03 \log \frac{[\text{Dye}_{\text{Ox}}]}{[\text{Dye}_{\text{Red}}]} = E^{0'} - 0.03 \log \frac{[\text{Ox}]}{[\text{Red}]} \qquad (1.62)$$

or

$$\log \frac{[\text{Dye}_{\text{Red}}]}{[\text{Dye}_{\text{Ox}}]} \frac{[\text{Ox}]}{[\text{Red}]} = \frac{E^{0'} - E^{0'}[\text{Dye}]}{0.03}. \qquad (1.63)$$

In calculating $E^{0'}$, the potential of the system is measured potentiometrically or the ratio $\text{Dye}_{\text{Ox}}/\text{Dye}_{\text{Red}}$ is determined spectrophotometrically. For some purposes it may be sufficient to take a number of samples of the solution concerned and to add to each a different indicator and to observe which are reduced (or oxidized). If one indicator is decolorized and the other is not, the potential must lie between the standard potentials of these two indicators at the hydrogen ion concentration (pH) of the solution. A series of oxidation-reduction indicators used in biological work together with their formal potentials ($E^{0'}$) at various pH values is given in Table 1.5.

This method has the same limitation as that mentioned in the use of indicators in pH determinations. Interaction between the systems may occur other than oxidation-reduction and give an erroneous value. The cysteine-cystine system, for example, gives a value that deviates by 170 mV from the value obtained by other methods [52].

Quasi-reversible and Irreversible Systems

As mentioned above, it may happen that in a redox system one of the forms is not a stable molecular species but gradually undergoes an irreversible change. Sometimes it is the reduced form, Red, and sometimes the Ox form that is labile. Systems have never been encountered with stable or even moderately stable Red and Ox forms in which the semiquinone radical is conspicuously labile so as to undergo an irreversible decomposition more rapidly than the Red or Ox forms. According to the velocity of such a secondary irreversible reaction, a potentiometric oxidative or reductive titration can be executed with fair accuracy provided the whole titration is performed within a short time. Samples of redox systems with a labile Red form are the indigo sulfonates in alkaline solution (pH 11 to 12). Examples with a labile Ox form are the aromatic p-diamines, of which the Ox form is the diimine, which are so labile in aqueous solution that no reliable end point can be obtained. In some of these cases [53], namely, if the two steps of the bivalent oxidation are sufficiently separated so as not to overlap too much, the end point of the first step of oxidation can be obtained more or less satisfactorily. Any irreversible decomposition of one of the forms of the system causes the potential to drift. If the drift is small, say 0.2 mV/min, and the titration can be finished within 5 min, a rather useful titration curve can be obtained and the required data can be derived from it, such as the formal potential, the semiquinone formation constant, and so on. The more rapid the decomposition, the less satisfactory the titration curve and, in extreme cases, which only too often occur, it may be quite useless. Such systems are transitions to irreversible systems. Special techniques for labile systems were mentioned in the section on Titration of Unstable Systems.

Among the irreversible systems, two types can be distinguished, although transitional cases may occur. In the first class, oxidation-reduction can be accomplished in a reversible way, speaking purely chemically but not thermodynamically. For instance, the di- and triphenylmethane dyes can be completely reduced in a slightly acid solution, for example, by zinc dust, and reoxidized with 100% yield, for example, by oxygen with a trace of a ferric salt as catalyst. Reduction is possible only with a very powerful reducing agent, and reoxidation only by a very powerful oxidizing agent, probably because of the fact that the semiquinone formation constant is extremely small, so that the concentration of the semiquinone, which is necessary for the progress of the bivalent overall oxidation, is, under all conditions, vanishingly small. A transitional case is that of duroquinone (dissolved in 20% pyridine + 80% water, for better solubility). This system can be titrated perfectly well from pH 14 to 6, but at pH <6 the establishment of the potentials becomes sluggish and the results are not strictly reproducible.

In the other type of irreversible system, either the oxidation or the reduction can be brought about only by strong agents, often only in a roundabout way, and the yield on reversing the process, if any, is small. Such a system is encountered when an azo dye is reduced in acid solution, where the first stable reduction product results from the irreversible splitting of the $N{=}N$ group into two NH_2 groups. Here, there is no meaning in a "potential" in the thermodynamic sense.

Conant and Fieser [54] have introduced the terms "apparent oxidation potential" and "apparent reduction potential." The apparent potential is the potential of a reversible redox system which is just able to bring about the oxidation (or reduction) of the irreversible system at a noticeable rate. The speed of the reaction can be observed by watching the drift of the potential of the reversible agent when acting upon the irreversible one. Nitro compounds belong to this group of irreversibly reducible substances.

References

1. I. M. Klotz, *Chemical Thermodynamics*, Prentice-Hall, Englewood Cliffs, New Jersey, 1950.
2. R. A. Robinson and R. H. Stokes, *Electrolyte Solutions*, Academic Press, New York, 1955.

3. J. Kielland, *J. Am. Chem. Soc.*, **59**, 1675 (1937).

4. Robinson and Stokes, *Electrolyte Solutions*, Academic Press, New York, 1955; H. S. Harned and B. B. Owens, *The Physical Chemistry of Electrolyte Solutions*, 2nd ed., Van Nostrand Reinhold, New York, 1950.

5. F. Hovorka and W. C. Dearing, *J. Am. Chem. Soc.*, **57**, 446 (1935).

6. W. M. Clark, *J. Wash. Acad. Sci.*, **10**, 255 (1920).

7. W. M. Clark et al., "Studies on Oxidation-Reduction," *U.S. Pub. Health Serv.*, *Hyg. Lab. Bull.*, No. 151 (1928).

8. K. J. Vetter, *Z. Elektrochem.*, **56**, 799, 1952.

9. A. E. Cameron, *J. Phys. Chem.*, **42**, 1217 (1938).

10. L. Michaelis, *Oxidation-Reduction Potentials*, Lippincott, Philadelphia, 1930.

11. L. Michaelis, *Chem. Rev.*, **16**, 243 (1935).

12. R. D. Draper and L. L. Ingraham, *Arch. Biochem. Biophys.*, **125**, 802 (1968).

13. N. H. Furman, *Ind. Eng. Chem.*, *Anal. Ed.*, **14**, 367 (1942).

14. J. B. Conant and H. B. Cutter, *J. Am. Chem. Soc.*, **48**, 1016 (1926).

15. W. M. Clark and M. E. Perkins, *J. Am. Chem. Soc.*, **54**, 1230 (1932).

16. J. Jagur-Grodzinski, M. Feld, S. L. Yang, and M. Szwarc, *J. Phys. Chem.*, **69**, 628 (1965).

17. W. R. Buchnall and W. Wardlaw, *J. Chem. Soc.*, **1927**, 2981.

18. W. M. Clark, B. Cohen, M. X. Sullivan, H. D. Gibbs, and R. K. Canan, *U. S. Pub. Health Serv.*, *Hyg. Lab. Bull.*, No. 151 (1928).

19. L. Michaelis and S. Granick, *J. Am. Chem. Soc.*, **62**, 2241 (1940).

20. A. Berka and J, Zyka, *Collection Czech. Chem. Commun.*, **24**, 105 (1959).

21. B. Kratochvil, D. A. Zatko, and R. Markuszewski, *Anal. Chem.*, **38**, 770 (1966).

22. K. Dimroth and K. S. Kraft, *Chem. Ber.*, **99**, 264 (1966).

23. L. Michaelis and H. Eagle, *J. Biol. Chem.*, **87**, 713 (1930).

24. H. A. Liebhafsky and E. H. Winslow, *J. Am. Chem. Soc.*, **68**, 2734 (1946).

25. H. von Wartenberg, *Z. Elektrochem.*, **36**, 295 (1930).

26. L. Michaelis, M. P. Schubert, and S. Granick, *J. Am. Chem. Soc.*, **62**, 204 (1940).

27. L. Michaelis and S. Granick, *J. Am. Chem. Soc.*, **64**, 1861 (1942).

28. L. Michaelis and M. P. Schubert, *J. Biol. Chem.*, **115**, 221 (1936).

29. L. Michaelis, *J. Biol. Chem.*, **87**, 33 (1930).

30. E. Biilmann and J. Blum, *J. Chem. Soc.*, **125**, 1719 (1930).

31. L. F. Fieser, *J. Am. Chem. Soc.*, **52**, 4915 (1930).

32. G. Ball and T. T. Chen, *J. Biol. Chem.*, **102**, 691 (1933); R. N. J. Saal, *Rec. Trav. chim.*, **47**, 73 (1928); H. Schmid, *Z. Physik. Chem.*, **A141**, 51 (1929); *ibid.*, **148**, 321 (1930).

33. A. H. Hartridge and F. J. W. Roughton, *Proc. Roy. Soc. (London)*, **B104**, 376 (1923). See Chapter X, "Rapid Reactions," by F. J. W. Roughton and Britton Chance in Vol. VIII of this series.

34. L. Michaelis, M. P. Schubert, and S. Granick, *J. Am. Chem. Soc.*, **61**, 1981 (1939).

35. L. Michaelis and M. P. Schubert, *Chem. Rev.*, **22**, 437 (1938); L. Michaelis and E. S. Fetcher, *J. Am. Chem. Soc.*, **59**, 2460 (1937). L. Michaelis and G. Schwarzenbach, *J. Biol. Chem.*, **123**, 527 (1938).

36. G. N. Lewis and J. Bigeleisen, *J. Am. Chem. Soc.*, **65**, 2419 (1943).

37. E. Rabinowitch and L. F. Epstein, *J. Am. Chem. Soc.*, **63**, 69 (1941).

38. P. W. Preisler and L. Berger, *J. Am. Chem. Soc.*, **69**, 322 (1947).

39. P. W. Preisler and M. M. Bateman, *J. Am. Chem. Soc.*, **69**, 2632 (1947).

40. P. W. Preisler and T. N. Scortia, *J. Am. Chem. Soc.*, **80**, 2309 (1958).

41. J. C. Ghosh, S. N. Raychandhuri, and S. C. Ganguli, *J. Indian Chem. Soc.*, **9**, 43 (1932); J. C. Ghosh and S. C. Ganguli, *Biochem. J.*, **28**, 381 (1934); D. E. Green, *ibid.*, **27**, 678 (1933).

42. H. Borsook, E. L. Ellis, and H. M. Huffman, *J. Biol. Chem.*, **117**, 281 (1937); G. Becker and W. A. Roth, *Z. Physik. Chem.*, **169**, 287 (1934).

43. N. Tanaka, I. M. Kolthoff, and W. Stricks, *J. Am. Chem. Soc.*, **77**, 2004 (1955).

44. R. K. Cannon, *Biochem. J.*, **20**, 927 (1926).

45. P. W. Preisler, L. Berger, and E. S. Hill, *J. Am. Chem. Soc.*, **69**, 326 (1947).

46. J. G. Mason and M. Rosenblum, *J. Am. Chem. Soc.*, **82**, 4206 (1960).

47. V. C. Giza, K. A. Kun, and H. G. Cassidy, *J. Org. Chem.*, **27**, 679 (1962).

48. L. Luttinger and H. C. Cassidy, *J. Polymer Sci.*, **22**, 271 (1956).

49. A. Cserhegyi, J. Jagur-Grodzinski, and M. Szwarc, *J. Am. Chem. Soc.*, **91**, 1892 (1969).

50. T. Thunberg, *Skand. Arch. Physiol.*, **46**, 339 (1925).

51. J. H. Quastel and M. D. Whetham, *Biochem. J.*, **18**, 519 (1924).

52. G. S. Fruton and H. T. Clarke, *J. Biol. Chem.*, **106**, 667 (1934).

53. L. Michaelis, M. P. Schubert, and S. Granick, *J. Am. Chem. Soc.*, **61**, 1981 (1939).

54. J. B. Conant and L. F. Fieser, *J. Am. Chem. Soc.*, **45**, 2194 (1923).

General

Branch, G. E. K., and M. Calvin, *Theory of Organic Chemistry*, Prentice-Hall, Englewood Cliffs, New Jersey, 1941.

Clark, W. M., B. Cohen, M. X. Sullivan, H. D. Gibbs, and R. K. Canan. "Studies on Oxidation-Reduction," *U.S. Pub. Health Serv., Hyg. Lab. Bull.*, No. 15,1 (1928).

Clark, W. M., *Oxidation Reduction Potentials of Organic Systems*, Williams and Wilkins, Baltimore, 1960.

Furman, H. N., "Potentiometric Titrations," *Ind. Eng. Chem., Anal. Ed.*, **14,** 367 (1942); *Anal. Chem.*, **22,** 33 (1950); *ibid.*, **23,** 21 (1951); *ibid.*, **26,** 84 (1954).

Glasstone, S., *Introduction to Electrochemistry*, Van Nostrand, Reinhold, New York, 1942.

Kolthoff, I. M., and Furman, N. H., *Potentiometric Titration*, Wiley, New York, 1931.

MacInnes, D. A., *The Principles of Electrochemistry*, Van Nostrand, Reinhold, New York, 1939.

Meites, L., *Handbook of Analytical Chemistry*, McGraw-Hill, New York, 1963.

Michaelis, L., *Oxidation-Reduction Potentials* (translated by L. B. Flexaer) Lippincott, Philadelphia, 1930.

Michaelis, L., "Occurrence and Significance of Semiquinone Radicals," *Ann. N.Y. Acad. Sci.*, **40,** 39 (1940).

Murray, R. W., and Reilley, C. N., *Anal. Chem.*, **34,** 313R (1962).

Murray, R. W., and Reilley, C. N., *Anal. Chem.*, **36,** 370R (1964).

Reilley, C. N., *Anal. Chem.*, **28,** 671 (1956).

Reilley, C. N., *Anal. Chem.*, **30,** 765 (1958).

Reilley, C. N., *Anal. Chem.*, **32,** 185R (1960).

Remick, A. E., *Electronic Interpretation of Organic Chemistry*, Wiley, New York, 1943.

Roe, D. D., *Anal. Chem.*, **38,** 461R (1966).

Toren, Jr., E. C., *Anal. Chem.*, **40,** 402R (1968).

Chapter **II**

POTENTIOMETRY:
pH MEASUREMENTS AND ION
SELECTIVE ELECTRODES

Richard P. Buck *

The author expresses gratitude to Professor Charles Tanford, Department of Biochemistry, Duke University, for allowing direct reproduction of some material relating to the measurement of pH from his chapter "Potentiometry," in *Physical Methods of Organic Chemistry*, Arnold Weissberger, Ed., Part IV, 3rd ed., Interscience, New York, 1960.

1 INTRODUCTION TO ION-SELECTIVE ELECTRODES

Recent commercial availability of a variety of electrodes whose zero current potentiometric response is, under restricted conditions, attributable primarily to the activity of a single cation or anion is the result of a confluence of several events. Need for concentration measurements of ions in mixtures has always existed. However, requirements for continuous monitoring or production of an analytical signal in a form suitable for recording, computer data handling, or control have made classic batch analysis undesirable. Through the efforts of physical chemists and electrochemists, ion-exchange membrane processes at the microscopic level have become very well understood since World War II, including both theory and experimental techniques. Simultaneously, biochemists and physiologists have measured parameters of transport in more complex membranes. With a new grasp of interfacial equilibria and bulk transport processes, attention has turned both backward to a reinterpretation of the classic glass electrodes used for pH measurement and forward to the search for materials with properties necessary for construction of electrodes responsive to alkali metals. The latter are now available and serve both as an analytical tool and as a model for alkali-permeable, cell membranes.

In recent years most of the significant electrode discoveries have involved materials for electrodes of the membrane type, that is, electrodes whose potential originates at two interfaces and in the intervening membrane bulk. At present, ion-selective electrodes, including the venerable pH glass electrodes, imply membrane electrodes. However, current interest in these various membranes should not obscure the well-known electrodes of the zeroth, first, and second kinds (Section 6) which are selective for many uni-, di-, and trivalent cations and a variety of anions. Were it not for anion-selective electrodes of the second kind used as reference probes, membrane cells could not function. A survey of these metallic interface electrodes is included in a later section.

Ion-Exchange Membranes as Selective Electrodes

Materials suitable for membrane electrode applications are broadly classified as ion exchangers. They are generally insoluble in water although they may absorb water. The extent of internal hydration is a balance between the osmotic pressure-driven uptake and the strength of the bonds that constrain the membrane to a fixed volume. The broad classification of ion exchanger includes materials that exchange ions of the same kind as their constituents. Thus an AgCl wafer is an ion exchanger for Ag^+, as can be demonstrated by exposing the wafer to radioactive Ag^+ and counting the incorporated radiosilver after different lengths of exposure. Solid, homogeneous membranes include glasses based on alkali silicate compositions, synthetic polymer membranes with acidic or basic functional groups, inorganic crystals or cast and pressed pellets, some naturally occurring clays and minerals, and certain treated natural polymers, for example, oxidized collodion. Liquid homogeneous membranes include hydrophobic organic acids and bases such as long-chain dialkyl esters of phosphoric acid, long-chain amines and quaternary ammonium salts, and certain long-chain-substituted phenanthroline complex salts, either alone or in solvents [1–4]. An intriguing further example of a liquid membrane is a liquid hydrocarbon containing electrically neutral complex-forming compounds. For the construction of alkali selective membranes, these additives include depsipeptides such as valinomycin, macrotetrolides such as nonactin, monactin, and others in the actin series, and various enniatins and cyclic polyethers.

However, many of the same homogeneous solid exchangers can be used in heterogeneous membrane configurations in which an ion exchanger is embedded in an inert solvent-rejecting solid matrix such as paraffin or silicone rubber. Materials not readily cast into membrane shape can be used in this mode: natural minerals of the zeolite and montmorillonite types, synthetic molecular sieves, and cation exchangers prepared by combining group-IV coxides with group-V and -VI oxides, for example, zirconium phosphate.

Nomenclature of Ion-Selective Electrodes

Terms used to describe synthetic ion exchangers have carried over into the literature and theory of selective electrodes. In cation exchangers cations are mobile and can be reversibly exchanged for other cations. While synthetic cation exchangers can be completely converted from one cation to another because of the loose structure and high ionic mobility, other solid membranes, typically glasses, do not reach bulk ion-exchange equilibrium but do equilibrate their surface layers—an important step in determining the interfacial potential. In anion exchangers the anions are mobile. Solid synthetic ion exchangers and glasses contain fixed "sites", that is, ionic groups such as $-SO_3^-$, $-SiO^-$, $-NR_3^+$, and so on, which are not free to diffuse because of the cross-linked structure of the solid. Liquid ion exchangers contain relatively more mobile sites which can be distributed under combined diffusive and electric forces.

Ion-exchange equilibrium at the surface can be viewed as an extraction process. A synthetic resin or glass membrane in the sodium form is considered as Na^+R^-, where R^- is a high-molecular-weight polyelectrolyte. An acid H^+X^- being extracted into this phase releases (exchanges) H^+ for Na^+ but, in addition, brings X^- into the resin phase as well. Extraction equilibrium demands that $a_{H^+} a_{X^-}$(aqueous) $= a_{H^+} a_{X^-}$(membrane). When the activities of H^+ and X^- (aqueous) are small compared with the concentration of R^-, a reasonable assumption, then H^+ is accumulated in the membrane to a much higher activity than in solution, while the activity of X^- in the membrane is correspondingly much smaller. For very large concentrations of R^-, H^+ (membrane) $\approx R^-$, while in the limit $R^- \rightarrow \infty$, X^- (membrane) ≈ 0. In this limit, it is proper to consider the extraction process merely as an ion exchange of one cation for another quite independent of the nature of the anion in solution. This phenomenon of ion exchange described for a sodium resin and an acidic environment can be expressed quantitatively and is an example of a Donnan equilibrium. The interfacial potential associated with the extraction is called the *Donnan potential*. Anion rejection by a cation exchanger is an example of *co-ion exclusion* where a co-ion is any ion in solution of the same sign as the membrane sites. The mobile, exchanging species are called *counterions*. When co-ions are well excluded from a membrane and counterions are the only mobile species, the membrane is considered "permselective" [5].

A Brief Historical Survey

Membrane electrodes selective for a specific ion are not new. The most successful and widely used membranes for hydrogen ion made from ion-exchanging glass were studied by Cremer [6] and documented and interpreted by Haber and Klemensiewicz [7]. Nernst was interested in electrical properties of liquid membranes as early as 1902 [8]. Use of glass electrodes was limited for many years to low-resistance glasses; the best known was Corning 015 (approximately 22% Na_2O, 6% CaO, 72% SiO_2). Upon development of electronic voltmeters with high input impedance by Howard Cary and Arnold Beckman, potentials developed across membranes with resistances up to 1000 MΩ could be measured. From 1937 to the present, improved glasses for pH determinations appeared with excellent mechanical strength, hydrolytic stability, and freedom from response to alkali metal ions in high pH solutions but not without a compromise on resistivity. The goal of strength and freedom from errors obscured the possibility that low resistance, high alkali error glasses might possess a unique property of their own: selectivity toward alkali metal ions. Lengyl and Blum [9] had already investigated enhanced sodium responses of alumino- and borosilicates. Almost simultaneously, Leonard and Arthur of Beckman Instruments, and Eisenman and co-workers investigated glasses suitable for selective response to sodium and potassium. The Beckman pNa^+ electrode appeared about 1959 and the so-called "cation"-sensitive electrode a short time later.

Transport of ions through solid salts and crystals constitutes an important test of the theories of the defect solid state. Measurements of transport parameters primarily occupied physicists during the 1920s and 1930s. Jost has summarized this work [10]. Silver chloride, for example, studied by Koch and Wagner [11], shows transport of current almost exclusively by silver ions as interstitials. At the same time Kolthoff and Sanders [12] demonstrated that silver chloride and bromide cast pellets used as membranes between solutions with widely differing silver activities responded as though two conventional cells were back to back. Thus the membrane cell

$$\text{Ag; AgCl; } K^+Cl^- a_{Ag^+} = \frac{K_{sp}^0}{a_{Cl^-}} \text{ ; AgCl; } Ag^+NO_3^- \mid K^+NO_3^- \mid KCl; Hg_2Cl_2; Hg \quad (2.I)$$
$$\text{(sat'd)}$$

| ref. 1 | | Soln 1 | |mem-| | soln 2| | salt | |ref. 2 |
| | | | | brane | | | | bridge | | |

was equivalent to

$$\text{Ag; AgCl; } K^+Cl^-; \text{ AgCl; Ag; AgCl; Ag}^+NO_3^- \mid K^+NO_3^- \mid \text{KCl; Hg}_2Cl_2; \text{Hg} \quad (2.II)$$
(sat'd)

$$|\!\longleftarrow\!\!-\!\!-\!\!\text{cell 1}\!\!-\!\!-\!\!\longrightarrow\!|$$
$$|\!\longleftarrow\!\!\!-\!\!\!-\!\!\!-\!\!\!\text{cell 2}\!\!-\!\!\searrow\!\!\longrightarrow\!|$$

which is two reversible cells with a common Ag terminal. The emf (right terminal–left terminal) is given by

$$\text{emf} = E(\text{Ref. 2}) - E(\text{Ref. 1}) + \frac{RT}{F} \ln \frac{a_{Ag^+}(\text{soln 1})}{a_{Ag^+}(\text{soln 2})} + \sum E_J.$$

Terms in this equation are defined and analyzed later in this section. The advantage of the membrane configuration was that solutions 1 and 2 could contain oxidants such as $KMnO_4$ without distortion of the measured potential. These results were apparently overlooked by analytical chemists and other practitioners of potentiometry with the result that the discovery of a crystalline material as a membrane electrode for fluoride by Frant and Ross [13] was a surprise to virtually everyone. Various crystals, cast wafers, and pressed pellets used as selective electrodes have been investigated since 1966.

During the past 40 years a variety of other solid homogeneous and heterogeneous membrane electrode systems have been proposed, prepared, and tested. An extensive literature on sintered clays as selective electrodes for metal ions exists through the pioneering work of Marshall [14]. Collodion membranes, parchment membranes containing precipitates formed in place, and precipitates or other insoluble materials in inert binders are by no means of recent origin. Good recent reviews of the status of heterogeneous membranes including the commercially available Pungor electrodes together with earlier review and original source references are found in articles by Covington [15] and Sollner [4].

Of perhaps greater potential versatility are electrodes made from liquid ion exchangers. While potentials were known to develop across liquid ion exchangers [16], the first practical electrode making use of the effect for calcium activity was described by Ross [17]. Quite rapidly thereafter, selective electrodes were constructed for other divalent cations and monovalent anions. Liquid membranes are particularly important because of the flexibility that exists in their construction. Numerous solvents and additives influence their behavior in addition to the wide variety of ion exchangers that can be used. Whereas in solids high-charge cations and anions have markedly less mobility than monovalent ions, mobilities in the liquid phase are comparable. Thus compositions of solid ion exchangers, especially glasses, which are responsive selectively to divalent ions will probably never be found.

In the foregoing short historical survey, those types of membranes that have reached commercial realization have been emphasized. Various other membranes have been studied. Extensive measurements on synthetic ion-exchanger electrodes have been made as tests of the Teorell-Meyers-Sievers membrane potential theory [18]. Although the possibility of practical applications doubtless occurred to researchers, membrane potential responses of sulfonate and amine resin membranes are highly nonselective; all ions of constant charge behave sufficiently similarly to prevent use as selective electrodes. The same holds for collodion and protamine-collodion membranes.

Electrode Availability in the United States

Selective ion electrodes including pH glass electrodes are available from a variety of sources in the United States. pH electrodes with documented characteristics are available from Beckman Instruments, the Coleman Division of Perkin Elmer, Corning Glass Works, Leeds and Northup, Sargent Co. and A. H. Thomas, to mention the main suppliers. Quite excellent European electrodes, Radiometer for example are also available. Glass electrodes responsive to sodium, potassium, silver, ammonium, and perhaps other monovalent ions are made by Beckman Instruments, Corning Glass Works and Orion Research. Solid-state electrodes of the single-crystal, cast disk, or pressed pellet types are available from Beckman Instruments, the Coleman Division of Perkin Elmer, and Orion Research. Liquid ion-exchanger electrodes are available from Corning Glass Works and Orion Research. Heterogeneous membranes of the Pungor type are available from Radelkis Electrochemical Instruments, Budapest, through National Instrument Laboratories, Rockville, Maryland. Compensations of commercially available electrodes are summarized in Tables 2.1A through 2.1D.

2 ORIGIN OF pH- AND ION-SELECTIVE ELECTRODE RESPONSE

Although electrochemical cells are described in many textbooks and reference works, the emphasis is almost always placed on aqueous cells using metallic electrodes. Such cells are indeed important for both physical and analytical measurements and they provide excellent examples of closed systems for the description and discussion of thermodynamic relations. The equilibrium entire cell potential or emf at zero current is a measure of thermodynamic activity of species involved in potential-determining processes. Alternatively, the potential is related to the Gibbs free energy for the cell reaction, if it were allowed to proceed in a reversible way.

The latter view of electrochemical cell emfs is unfortunately restrictive. In potentiometry a net electrochemical cell reaction does not proceed at zero current. It is therefore more important to recognize the existence of

Table 2.1A   Glass Electrodes*

Ion Detected in the Principal Applications	Composition	Interferences
H^+	$Li_2O-BaO-La_2O_3-SiO_2$	$Li^+, Na^+ > K^+$
Na^+ ⎱	⎧ $Li_2O-Al_2O_3-SiO_2$	$Ag^+ > H^+$
Ag^+ ⎰	⎨ $Na_2O-Al_2O_3-SiO_2$	$H^+ > Na^+$
K^+		H^+, Na^+, NH_4^+
NH_4^+		H^+, Na^+, K^+
Li^+		
Tl^+		
Cu^+	{ $Na_2O-Al_2O_3-SiO_2$	H^+, Na^+, K^+
R_4N^+		
Also Rb^+ and Cs^+		

* Compiled from Beckman Instruments, Inc. bulletins and G. A. Rechnitz, "Ion Selective Electrodes," *Chem. Eng. News*, June 12, 1967, p. 146.

Table 2.1B   Solid-State Electrodes*

Ion Detected	Useful Measurement Range pM or pA	Composition	Interferences
F^-	0–6	LaF_3	OH^-
Cl^-	0–4.3	$AgCl-Ag_2S$	$NH_3, Br^-, I^-, CN^-, S_2O_3^{2-}$, S^{2-}
Br^-	0–5.3	$AgBr-Ag_2S$	$NH_3, I^-, CN^-, S_2O_3^{2-}$, S^{2-}
I^- ⎱	0–7.3	$AgI-Ag_2S$	⎰ $S^{2-}, CN^-, S_2O_3^{2-}$
CN^- ⎰	2–6		⎱ $S^{2-}, S_2O_3^{2-}$
S^{2-}	0–17	Ag_2S	Hg^{2+}
SCN^-	0–5	$AgSCN-Ag_2S$	Br^-, I^-, S^{2-}, NH_3, $CN^-, S_2O_3^{2-}$
Ag^+	0–17	Any AgX or Ag_2S	Hg^{2+}
Cu^{2+}	0–8	$CuS-Ag_2S$	Hg^{2+}, Ag^+
Pb^{2+}	1–7	$PbS-Ag_2S$	Hg^{2+}, Ag^+, Cu^{2+}
Cd^{2+}	1–7	$CdS-Ag_2S$	Hg^{2+}, Ag^+, Cu^{2+}

* Compiled from Ross [95] and Covington [99].

Table 2.1C Liquid Ion Exchangers*

Ion Detected	Useful Measurement Range pM or pA	Composition	pH Range	Interferences
Ca^{2+}	0–5	$(RO)_2PO_2^-$	5.5–11	$H^+ > Zn^{2+} > Fe^{2+} > Pb^{2+} > Cu^{2+} > Ni^{2+} > Sr^{2+} > Mg^{2+} > Ba^{2+}$
Ca^{2+}/Mg^{2+} (divalent ion)	0–8	$(RO)_2PO_2^-$	5.5–11	$H^+ > Zn^{2+} > Fe^{2+} > Cu^{2+} > Ni^{2+}$
Cu^{2+}	1–5	$R-S-CH_2COO^-$	4–7	$Fe^{2+} > H^+ > Zn^{2+} > Ni^{2+}$
Pb^{2+}	2–5	$R-S-CH_2COO^-$	3.5–7.5	$Cu^{2+} > Fe^{2+}$
Cl^-	1–5	NR_4^+	2–11	$ClO_4^- > I^- > NO_3^- > Br^- > OH^-$
ClO_4^-	1–5	$FeL_3(ClO_4)_2$†	4–11	$OH^- > I^-$
BF_4^-	1–5	$NiL_3(BF_4)_2$	2–12	$I^- > NO_3^- > Br^-$
NO_3^-	1–5	$NiL_3(NO_3)_2$	2–12	$ClO_4^- > I^- > ClO_3^- > Br^- > S^{2-}$

* Compiled from Ross [95].
† L = substituted *o*-phenanthroline.

Table 2.1D Heterogeneous, Pungor-Type Electrodes*

Ion Detected	Composition in Silicone Matrix
F^-	LaF_3, CaF_2
Cl^-	AgCl
Br^-	AgBr
I^-	AgI
S^{2-}	Ag_2S
Ag^+	Ag_2S and all AgX
SO_4^{2-}	$BaSO_4$†
PO_4^{3-}	$Mn(III)PO_4$† or $BiPO_4$

* From A. K. Covington, "Ion-Selective Electrodes," *Chem. Brit.*, **5**, 388 (1969).
† There is some doubt that these reported electrodes are operable.

interfacial charge exchange processes and the occurrence of homogeneous charge transport which are the sources of cell potential in the zero current, quiescent state. The total cell emf is made up of a series of interfacial potentials, that is, potential differences between phases which contain one or more charged species in rapid, reversible exchange equilibrium. While these single interface potentials are not precisely thermodynamically defined, they are meaningful electrostatically and may be combined to yield thermodynamically defined properties. The key concept is the electrochemical potential by which the entire field of interfacial potentials may be unified. Recognition of the electrochemical potential makes possible an understanding of promising new ion-selective electrodes which do not involve metal-metal ion half cells for ion activity measurement in the classic way.

Relevant features of electrochemical cells for activity measurement are electrostatic. These electrostatic phenomena occur at interfaces between charged or uncharged conducting phases: solid-solid, solid-liquid and liquid-liquid. Within the conducting phase there are mobile, charged species: ions and electrons. In metallic conductors these are electrons; in ionic solids such as glass, single crystals such as LaF_3, and pressed pellets such as AgI and Ag_2S, charge carriers are ions with some, usually small, contribution made by electrons; in conducting liquids charge carriers are ions. Mobilities of ions and electrons within phases and across interfaces are intimately related to the development of electrochemical cell potentials or emfs.

The Electrochemical Potential Concept

An ionic salt dissolved in a solvent at constant temperature and pressure possesses a defined chemical potential which is the partial molar Gibbs free energy. This quantity $\mu(salt)$ is a function of the salt activity level according to

$$\mu(salt) = \mu^0(salt) + RT \ln a_+ a_- \tag{2.2}$$

for a uni-univalent salt. μ^0 is a reference point standard free energy and is a function of temperature. In any real process or chemical reaction of this salt, $\mu(salt)$ figures in the expressions for the Gibbs free energy of the reaction and in the equilibrium constant. For every chemical process or reaction, both cation and anion are involved in reaching the new equilibrium state. For this reason individual ion activities are not experimentally accessible. This does not mean that overall processes cannot be subdivided hypothetically into steps involving only ions of one kind or one sign. In the context of selective electrodes, one such process is the transfer of ions of one sign from a metal or a glass to an adjacent solution. Obviously, as cations pass from metal or glass into solution as the result of an electrolytic current, anions are simultaneously delivered to the solution from a reference electrode to maintain electroneutrality of the phase. At the site of the cation transfer, it is reasonable to ask how to describe the process in thermodynamic terms, realizing that the anion may be entering the electrolyte at a remote point. Perhaps more significant of the need to consider processes involving ions of one sign is the mobile cation exchange occurring at the metal or glass interface in the absence of net current flow and the absence of net anion influx into the solvent medium.

Guggenheim [19] has suggested that the partial molar free energy of a single *charged* species be considered and called an electrochemical potential defined by

$$\tilde{\mu}_i = \mu_i + z_i F \varphi = \mu_i^0 + RT \ln a_i + z_i F \varphi \qquad (2.3)$$

where μ_i is the chemical potential the ion would contribute to the salt chemical potential if it were uncharged, and $z_i F \varphi$ accounts for the energy of the particle in the medium of local electrostatic potential φ, the so-called "inner" potential of a phase. The chemical potential of the salt is the sum of the electrochemical potentials of the component ions as is verified by substituting (2.3) into (2.2), where the potential terms disappear and

$$\mu^0(\text{salt}) = \mu^0(\text{cations}) + \mu^0(\text{anions}). \qquad (2.4)$$

In the same way that the chemical potential of a salt is equal in all phases containing this salt at equilibrium, the assertion is made that for any process involving the rapid, reversible distribution of an ion of one sign between phases at equilibrium, the electrochemical potential of that ion is constant in each phase. This assertion then defines the "inner" potential difference between phases undergoing a rapid ion-exchange or electron-exchange process.

Interfacial Potential Differences for Metals, Glasses, and Insoluble Salts

Many properties of the metallic state can be explained in terms of a model in which the metal is an array of positive ions surrounded by an electron cloud. A mole of zinc metal is conceived of as 2 moles of electrons constrained by a periodic lattice of Zn^{2+} ions. The inner electrons within the zinc ion need not be specifically considered. Semiconductors can be treated in a similar way after accounting for the forbidden energy levels. The chemical potential of the metal per mole is given by

$$\mu_{Zn} = \tilde{\mu}\,(Zn^{2+}) + 2\tilde{\mu}\,(\text{electrons}), \qquad (2.5)$$

where $\tilde{\mu}$ (electrons) is the Fermi level energy or the electron chemical potential and the chemical part is taken to be zero. When two metals, for example, Zn and Cu, are contacted, electrons transfer until the electron chemical potentials are equal. Copper becomes negative and zinc positive. A further description can be made in terms of measurable quantities, but this is beyond the needs of this chapter.

When a metal is in contact with a solution of its ions and the exchange of ions reaches equilibrium, then for zinc in a zinc ion-containing solution, the "inner" potential difference is defined by

$$\varphi(\text{metal}) - \varphi(\text{solution}) = \frac{RT}{2F}\left[\frac{\mu^0_{Zn^{2+}}(\text{soln}) - \mu^0_{Zn}}{RT} + \ln \frac{a_{Zn^{2+}}(\text{soln})}{a_{Zn^{2+}}(\text{metal})}\right]. \quad (2.6)$$

Since the parameters of this equation are not known, interfacial potentials are determined by reference to a complete cell containing a hydrogen electrode at unit activity and fugacity. For the hydrogen electrode,

$$\varphi(\text{metal}) - \varphi(\text{solution}) = \frac{RT}{F}\left[\frac{\mu^0_{H^+}(\text{soln}) - \mu^0_{H_2}/2}{RT} + \ln \frac{a_{H^+}(\text{soln})}{(P_{H_2})^{1/2}}\right]. \quad (2.7)$$

The quantity $(\mu^0_{H^+}(\text{soln}) - \mu^0_{H_2}/2)$ is assumed to be zero by international convention. As long as the measuring circuit uses lead wires of a common metal, for example, copper, all metal contact potentials cancel and the measured emf for unit activity zinc ion is called $E^0_{Zn^{2+}/Zn}$ so that

$$E^0_{Zn^{2+}/Zn} = \varphi(\text{zinc}) - \varphi(\text{soln } a_{Zn^{2+}} = 1)$$

$$= \frac{RT}{2F}\left[\frac{\mu^0_{Zn^{2+}}(\text{soln}) - \mu^0_{Zn}}{RT} - \ln a_{Zn^{2+}}(\text{metal})\right]. \quad (2.8)$$

For other activities of zinc ion,

$$E_{Zn^{2+}/Zn} = \varphi(\text{zinc}) - \varphi(\text{soln})$$

$$= E^0_{Zn^{2+}/Zn} + \frac{RT}{2F}\ln a_{Zn^{2+}}. \quad (2.9)$$

Obviously, the interfacial potential difference defined by the measured and measurable quantities in (2.9) differs from the unknowable true value by the additive constant in (2.7) taken to be zero. All that can be said concerning the interfacial potential magnitude is that increases in zinc ion activity make the electrode "inner" potential more positive relative to the solution.

In contrast to the metal-solution interface where interfacial potential can be closely related to E for the half cell, there are no conventions for interfacial potentials at other ionic interfaces. However, this omission is dictated only by practical circumstances, that is, there are no convenient references such as the hydrogen electrode which can be inserted into ionic solids, glasses, and ion-conducting crystals. Glass membranes and other selective ion barriers are always used in double cells in which both sides of the membrane are exposed to ionic contacts, either liquid or solid. If attention is focused on each of the interfaces, the electrochemical potential concept can be applied to each in the same way it was applied above to a single metal–metal ion interface.

Typical high-quality pH-sensitive glass membranes are chiefly lithium silicates with lanthanum and barium ions as lattice "tighteners" added to retard silicate hydrolysis and lessen alkali metal, chiefly sodium ion, mobility. Lithium ions are the bulk mobile charge carriers under an applied electric field. After the membrane is soaked in water, the surface layer is depleted of Li^+ which is replaced by H^+. Virtually all surface silicate anion sites, "fixed" sites, are neutralized by H^+ ions. Content of H^+ decreases in a complex way with increasing distance into the membrane, while Li^+ content increases in such a way that the sum of positive ions (charge carriers and other cations) balance the presumed uniform fixed-site concentration which is about 20 M for typical pH-sensitive glasses. This idealized model can be amended to allow for osmotic pressure-driven uptake of water with consequent hydrolysis of the surface silicate chains to yield perhaps a lower fixed-site density at the surface than exists in the bulk. The model must also be corrected for the existence of space charge within a few Debye lengths inside the surface. However, since the electrochemical potential concept is best applied in electrically neutral phases, we can refer interfacial potential difference to points just beyond the space charge regions in each phase. These arguments apply equally to membranes of sulfonated polystyrene resins or other synthetic membranes containing high concentrations of fixed-site ionic groups homogeneously distributed.

The interfacial potential at one side of a pH-sensitive glass membrane is given by

$$\varphi(\text{membrane}) - \varphi(\text{soln})$$

$$= \frac{RT}{F}\left[\frac{\mu^0_{H^+}(\text{soln}) - \mu^0_{H^+}(\text{membrane})}{RT} + \ln\frac{a_{H^+}(\text{soln})}{a_{H^+}(\text{membrane})}\right]. \quad (2.10)$$

If in addition to rapid reversible equilibrium of hydrogen ions, the solution and the membrane surface contain other univalent ions, for example, Na^+ at equilibrium, then an expression equivalent to (2.10) with subscripts Na^+ applies also. Since the interfacial potential is unique, the chemical potential terms are related through the ion-exchange constant for the reaction

$$Na^+(\text{soln}) + H^+(\text{membrane}) \rightleftharpoons Na^+(\text{membrane}) + H^+(\text{soln}) \quad (2.11)$$

$$K_{H^+/Na^+} = \exp\left[\frac{1}{F}(\mu^0_{Na^+} - \mu^0_{H^+})\text{soln} - \frac{1}{F}(\mu^0_{Na^+} - \mu^0_{H^+})\text{membrane}\right].$$

$$(2.12)$$

By mass balance on concentration,

$$(C_{Na^+} + C_{H^+})\ \text{membrane} = \text{fixed sites} = \omega(\text{concentration units}). \quad (2.13)$$

Then the interfacial potential can be rewritten

$$\varphi(\text{membrane}) - \varphi(\text{soln}) = \frac{RT}{F}\left[\frac{\mu_{H^+}^0(\text{soln}) - \mu_{H^+}^0(\text{membrane})}{RT}\right.$$

$$\left. + \ln\frac{\bar{\gamma}_{Na^+}}{\bar{\gamma}_{H^+}\omega}\left(a_{H^+} + K_{H^+/Na^+}\frac{\bar{\gamma}_{H^+}}{\bar{\gamma}_{Na^+}}a_{Na^+}\right)\right], \quad (2.14)$$

where $\bar{\gamma}$ parameters correspond to membrane activity coefficients. This expression is an embryonic form of the Nicolsky equation for multiionic membrane responses. It shows that the interfacial potential reflects changes in solution activities for species existing in equilibrium in both phases.

Ionic crystals such as lanthanum fluoride and the silver halides used in selective ion electrodes contain point, line, and area defects. Although selective ion electrodes do not require high degrees of lattice perfection (in fact pressed microcrystalline pellets are usually suitable), this short discussion emphasizes perfect crystals where point defects predominate. Line and area defects may profoundly affect measured parameters such as conductivity but do not change the origin of potentials that relate to the vacant lattice sites, interstitial ions, and impurity atoms. Point defects in ionic crystals seem to occur as one of two types—although other possibilities have been considered [20]. The first type, Frenkel disorder, can be represented as

$$\text{occupied lattice site} \rightleftharpoons \text{vacancy} + \text{interstitial}$$
$$Ag^+ \rightleftharpoons \boxed{-} + Ag^{+*} \quad (2.15)$$
$$F^- \rightleftharpoons LaF_2^+ + F^{-*}$$

for silver salts and for lanthanum fluoride. Experimental evidence is needed to demonstrate which ions in a lattice are subject to Frenkel disorder. Transport measurements have demonstrated the high mobility of Ag^+ and F^-, from which the conclusion is drawn that they are the defect species. Another low energy–barrier defect is the Schottky disorder in which ions of both signs are transferred to the surface to extend the lattice, leaving behind vacancies of both signs. Electroneutrality in the bulk requires that equal numbers of vacancies of each kind exist. They may occur side by side as neutral vacancy pairs, or they may be separated from each other. The bulk conductivity receives contributions from both defects although the higher-mobility species may predominate.

The concentrations of point defects are strong functions of temperature. For many crystals the concentrations and mobilities are so low at room temperature that the crystals are insulators. For reasons which are not understood, a mere handful of materials show striking intrinsic ionic conductivity at room temperature. If these materials are insoluble, they are possible candidates for construction of selective electrodes. For crystals with conductivities too low to be useful at dimensions necessary for physical strength, impurities may be added which greatly increase either interstitial or vacancy concentrations. For instance, vacancies in silver chloride may be increased by addition of $PbCl_2$, while interstitials are created upon addition of Ag_2S. Europium(II) fluoride in LaF_3 increases conductivity presumably by vacancy introduction. As little as a few parts per million impurity need be homogeneously incorporated to cause a large effect. For intrinsic ionic conductors containing $\sim 10^{-11}$ moles of vacancy-interstitial pairs as in AgCl, 1 ppm of impurity obviously has a marked effect.

In applying the electrochemical potential concept to ionic-conducting crystals, the problem arises on the mechanistic level as to the nature of the ion-exchange process. Two expressions can be written depending on whether exchange occurs between aqueous ions and vacancies or aqueous ions and interstitials. At equilibrium the number of vacancies and interstitial species are proportional to the square root of the number of ions in the crystal and the number of possible interstitial sites. Without further detail we can write

$$\varphi(\text{crystal}) - \varphi(\text{soln})$$

$$= \text{constant (dependent on temperature)} + \ln \frac{a_{Ag^+}(\text{soln})}{a_{Ag^+}(\text{crystal})}. \quad (2.16)$$

For the fluoride membrane electrode, the logarithmic term contains fluoride activities and is preceded by a negative sign.

Diffusion Potentials

Another very important source of potential difference within a cell occurs whenever the composition of a phase containing electrolyte is *not* at uniform activity. In contrast to interfacial potentials, these diffusion potentials arise spontaneously within a single phase when nonuniform distributions of electrolytes seek to become uniform by diffusing from regions of high activities to low. Diffusion potentials in the liquid phase between two merging electrolytes are called *junction potentials*; they may introduce a time dependence and an ultimate limit on the reproducibility of zero-current potentiometric measurements. They also create an uncertainty in the interpretation of single ion activities. Perhaps more important in the context of selective ion electrodes is that all the membranes in use as selective electrodes can be considered examples of constrained liquid junctions. That is, the total membrane potential between points in solution outside the membrane on either side consists of the two interfacial potentials *plus* the internal diffusion potential of mobile charge carriers within the membrane itself. The latter quantity can on occasion be zero, but it must be considered in the most general treatment of ion-selective electrode responses.

In this section liquid junctions are briefly considered together with the diffusion potential for permselective membranes. Diffusion potentials are a result of nonequilibrium conditions and processes. They are not subject to strict thermodynamic treatment except insofar as the adjacent solutions contain the same salts at concentrations differing by an incremental amount. Then, for a junction consisting of i ions at concentration C_i, charge z_i (including sign), and bulk transference numbers t_i adjacent to a solution with the same ions at $C_i + dC_i$

| uniform concentration C_i potential φ_1 | junction region | uniform concentration $C_i + dC_i$ potential φ_2 | (2.III) |

the junction potential is given on thermodynamic grounds by

$$dE_J = d(\varphi_2 - \varphi_1) = -\frac{RT}{F} \sum_i \frac{t_i}{z_i} d \ln a_i \qquad (2.17)$$

and for a real junction considered to be made up of a series of thin sections

$$E_J = \varphi_2 - \varphi_1 = -\frac{RT}{F} \int_1^2 \sum_i \frac{t_i}{z_i} d \ln a_i \qquad (2.18)$$

The transference numbers for a single salt with cation of charge z_c and anion of charge z_a are

$$t_c = \frac{z_c u_c}{z_c u_c - z_a u_a} \qquad \text{and} \qquad t_a = 1 - t_c \qquad (2.19)$$

where u_i is an ion mobility related to the ion diffusion coefficient by $D_i = u_i RT$ or $D_i = u_i RT/F$, depending on units chosen for u_i. If all ions in a junction situation have the same mobilities, the junction potential is always zero. This leads to a simple rule: the sign of a junction potential is quickly found by noting which side has the higher ionic activity. Then note which ion is the more mobile. Cations and anions attempt to diffuse separately from high to low activities. The more mobile ion moves ahead at short times until a field is created which drags the slower moving species. The faster ion is slowed and the slower speeded until the salt moves in an electrically neutral way with an overall mean diffusion coefficient, the salt diffusion coefficient, or the equivalent salt mobility,

$$D_{\text{salt}} = \frac{D_c D_a (z_c - z_a)}{z_c D_c - z_a D_a}$$

$$u_{\text{salt}} = \frac{u_c u_a (z_c - z_a)}{z_c u_c - z_a u_a}. \qquad (2.20)$$

The faster ion takes its charge into the lower concentration side and determines the sign of the dilute side. At short times the concentration profiles are nonlinear, creating a very, very small space charge and Warburg-type impedance. In the steady state concentration profiles are linear. The magnitude of the junction potential requires integration of (2.18), which in turn requires knowledge of single ion activities and the exact concentration-distance-time profiles. Ordinarily, integrations are possible by assuming constant activity coefficients. Various equations by Henderson for the mixture junction, Sargent and Lewis for two salts with a common ion, and Planck for the constrained junction in the steady state are given in texts such as MacInnes, *The Principles of Electrochemistry* [21].

Real liquid junctions are free flowing and often involve solutions on the two sides with no ions in common. There are no equations for accurate calculation of these complicated junctions. Furthermore, junction potentials cannot be measured for all possible systems. The cases that can be measured rely on the difference in emf of electrochemical cells that can be physically realized with and without junctions. Despite this lack of precise knowledge, junction potentials are believed to be small, less than 10 mV unless an acid or base is present; or, the difference in concentration between the two sides is very large. A junction between 0.1 M HCl and 0.01 M HCl has a potential of 40 mV with the dilute side positive. For KCl at the same concentrations, the junction potential is 1.0 mV, with the dilute side negative since the mobility of K^+ is slightly less than Cl^-.

For the mixed case KCl (0.1 M) and HCl (0.1 M), the migration of Cl^- is virtually zero and the HCl side is negative with respect to KCl by 28 mV. Values of junction potentials are given in Table 2.2. Use of a salt bridge of a

Table 2.2   Liquid Junction Potentials at 25°C*

Boundary	E_J (mV)†	Boundary	E_J (mV)†
LiCl(0.1 F) | KCl(0.1 F)	− 8.9	KCl(0.1 F) | KCl(3.5 F)	+ 0.6
NaCl(0.1 F) | KCl(0.1 F)	− 6.4	NaCl(0.1 F) | KCl(3.5 F)	− 0.2
NH$_4$Cl(0.1 F) | KCl(0.1 F)	+ 2.2	NaCl(1 F) | KCl(3.5 F)	− 1.9
NaOH(0.1 F) | KCl(0.1 F)	−18.9	NaOH(0.1 F) | KCl(3.5 F)	− 2.1
NaOH(1 F) | KCl(0.1 F)	−45	NaOH(1 F) | KCl(3.5 F)	−10.5
KOH(1 F) | KCl(0.1 F)	−34	KOH(1 F) | KCl(3.5 F)	− 8.6
HCl(0.1 F) | KCl(0.1 F)	+27	HCl(0.1 F) | KCl(3.5 F)	+ 3.1
H$_2$SO$_4$(0.05 F) | KCl(0.1 F)	+25	H$_2$SO$_4$(0.05 F) | KCl(3.5 F)	+ 4

* The data are taken from G. Milazzo, *Elektrochemie*, Springer-Verlag, Vienna, 1952; cf. R. G. Bates, *Electrometric pH Determinations*, Wiley, New York, 1954, p. 41.

† $E_J = \varphi_{right} - \varphi_{left}$.

high concentration of KCl minimizes but does not totally eliminate junction potentials. Bridges are discussed elsewhere in this chapter.

The mechanistic view of ionic transport offers an alternate and perhaps more satisfying description of diffusion potentials. This description based on the force-velocity equations of Nernst and Planck leads to the same results as the thermodynamic treatment. The flux of ions in moles per square centimeter per second through a unit cross section is determined by the diffusive and electric forces. The sum of these forces is the gradient of the electrochemical potential discussed above. Flux J_i of ion i is given by

$$J_i = -C_i u_i \frac{\partial}{\partial x} [\mu_i^{\,0} + RT \ln a_i + z_i F \varphi] \tag{2.21}$$

for the one dimensional case. If activity coefficients are independent of distance, then

$$J_i = -RT u_i \left(\frac{\partial C_i}{\partial x} + \frac{z_i F C_i}{RT} \frac{\partial \varphi}{\partial x} \right). \tag{2.22}$$

For cations and anions moving together across a junction, the current is zero:

$$I = F \sum_{i=1}^{n} z_i J_i = 0, \tag{2.23}$$

and the fluxes of charge (electroneutrality of diffusion) are equal. For a single salt, fluxes are related according to

$$z_c J_c = -z_a J_a, \tag{2.24}$$

and the gradient of the diffusion potential is easily found to be

$$\frac{\partial \varphi}{\partial x} = -\frac{RT}{F} \left[\frac{z_c u_c \dfrac{\partial C_c}{\partial x} + z_a u_a \dfrac{\partial C_a}{\partial x}}{z_c^{\,2} u_c C_c + z_a^{\,2} u_a C_a} \right]. \tag{2.25}$$

Since for any salt

$$C_c = -\frac{z_a}{z_c} C_a = C, \tag{2.26}$$

where C = formal or stoichiometric concentration, then

$$\frac{\partial \varphi}{\partial x} = -\frac{RT}{F} \left[\frac{u_c - u_a}{z_c u_c - z_a u_a} \right] \frac{\partial \ln C}{\partial x}. \tag{2.27}$$

Note that this expression is the space derivative of the thermodynamic expression. Integration of this equation between two regions of constant activity yields a *time-independent* junction potential. This desirable behavior can only be expected for junctions involving salts or mixtures of salts of the same kind on both sides of the junction. The above equations can be generalized for mixtures.

Within a solid phase, such as glass containing fixed negative sites ($-SiO^-$ groups), and perhaps several mobile univalent charge carriers, Li^+, Na^+, and H^+ for example, diffusion drives the mobile species slowly toward uniformity of activity. The more mobile ions establish the sign of the diffusion potential as they move in response to their activity gradient while at the same time maintaining electroneutrality with respect to the fixed sites at all points except near the interfaces where space charge exists. By combining flux equations subject to the zero-current condition, the diffusion potential for a glass membrane has the form:

$$\varphi(\delta) - \varphi(0) = \frac{RT}{F} \ln \left[\frac{u_{H^+}C_{H^+} + u_{Na^+}C_{Na^+} + u_{Li^+}C_{Li^+} \text{ at side } 0}{u_{H^+}C_{H^+} + u_{Na^+}C_{Na^+} + u_{Li^+}C_{Li^+} \text{ at side } \delta} \right]. \quad (2.28)$$

The concentrations are to be evaluated just inside the space charge region at side 0 and at side δ for a glass of thickness δ. The sum·of the concentrations equals the fixed site concentration and the exact distribution of sites among cations depends upon their external activities through their respective ion-exchange equilibrium constants. Note that the internal diffusion potential within a glass membrane depends only on concentrations just inside the surface and is independent of the slow adjustment of bulk concentrations with time. The diffusion potential is established as soon as ion-exchange equilibrium is reached. This very important fact permits the exceedingly rapid response of glass electrodes to changing external activities. Some limitations on response rate are discussed later. If a steady-state distribution of ions throughout a glass were required for measurement of a steady potential, then glass membranes would be useless as analytical devices. The diffusion potential gradient (the internal electric field) is constant for membranes with uniformly distributed sites.

Liquid ion-exchange membranes in the steady state have an internal diffusion potential of the same form as (2.27) even though the sites are mobile. The sites are distributed linearly across the membrane and, electroneutrality exists in the bulk but not at the interfaces. The potential gradient is not uniform in general. Permselectivity is achieved despite the fact that sites are mobile because the net flux of sites is restricted to zero inasmuch as they must remain inside the membrane phase.

Synthetic ion-exchange membranes with high fixed-site content obey the same conditions as stated above for glasses. When the fixed-site concentration is low, that is, comparable to ionic concentration in the external solutions, co-ion rejection is not achieved and ions of both sign can transport. This breakdown of permselectivity leads to internal diffusion potentials with slopes less than $2303RT/zF$ mV per decade concentration change, roughly proportional to the difference between the transference numbers of the permeating ions. Details of the diffusion potentials for this general case are summarized by Helfferich [22].

The internal diffusion potential for solid-state pressed-pellet and crystal electrodes can be treated on the same basis as glass electrodes. A detailed analysis requires consideration of the rates of formation and disappearance of charge carriers in the bulk in addition to the usual forces in the flux equation.

Membrane Potentials and the Selectivities of Electrodes

A membrane electrode is a portion of a membrane cell and consists of a membrane, a solution called the inner or filling solution and an internal, reversible half-cell, variously called a reference electrode or reference probe. The membrane cell contains the membrane electrode dipped into a second solution, usually the solution under investigation, and a second external half-cell or reference probe. The organization of ion exchanging membrane cells have been described in detail in recent books by Helfferich [22] and by Marinsky [23]. Both junction cells and junction-free cells are important. These can be written, for example,

Junction Cell:
Ag; AgX; inner or filling soln; membrane; test soln | $K^+NO_3^-$ | K^+Cl^- dilute or sat'd;
$$Hg_2Cl_2; Hg \quad (2.IV)$$
Junctionless Cell:
Ag; AgCl; inner soln with Cl^-; membrane; test soln containing Cl^-; AgCl; Ag   (2.V)
General Form:
$$\text{membrane (ion-selective electrode); test soln } | \text{ or } \| \text{ ext. ref.} \quad (2.VI)$$

The semicolon indicates an interface, while the single vertical line is a junction. Both are regions of a potential difference which contribute to the overall cell potential. A double vertical line is a conventional symbol of a junction at which the potential difference is negligible through the use of a salt bridge.

The measured cell potential at zero current is the sum of all of the interfacial and junction potentials. These were analyzed in the foregoing sections. The measured cell potential using any of the commercial pH meters is the potential difference between the electrons in the wire lead from the electronic voltmeter to the inner reference probe minus that in the wire lead of the same metal to the external reference electrode. Consequently,

$$E_{\text{measured}} = \varphi(\text{inner soln}) - \varphi(\text{test soln}) + \sum E_J + E(\text{int. ref.}) - E(\text{ext. ref.}),$$
$$(2.29)$$

where φ (inner soln) $- \varphi(\text{test soln}) = \Delta E$ membrane.   (2.30)

Note that measured E_{cell} is the negative of the conventional emf. For cell (2.V) junction potentials are absent but the reference electrodes contribute a value to the measured emf depending on the chloride activity, since

$$E(\text{int. ref.}) - E(\text{ext. ref.}) = + \frac{RT}{F} \ln \frac{a_{Cl^-}(\text{ext. ref.})}{a_{Cl^-}(\text{int. ref.})} \qquad (2.31)$$

If a cation-selective membrane is considered in cell (2.V), and the composition of the test solution is a chloride salt of the cation, then apparently super-Nernstian response will occur. A change in cation activity, measured at the membrane, and the simultaneous change in chloride activity measured at the external reference probe add together to give a response to the mean activity of the salt.

Junction cells avoid this complication and are more convenient to use despite the unknown junction potentials. However, there is the ever-present possibility that reference half-cell components may diffuse into the test solution on long standing. The more usual problem that must be guarded against is clogging of the junction between the external reference and the test solution. This problem is well known to those who use calomel or silver chloride reference electrodes made up in nonaqueous media such as dimethylformamide. Saturation of the reference with an alkali halide and calomel or silver halide results in a high concentration of soluble mercury and silver species. Not only does chloride diffuse out of the junction into the test solution, but the soluble heavy metal complexes diffuse as well. The junction is easily clogged if the test solution contains a precipitant such as sulfide or hydroxide. This problem is severe even in aqueous solutions containing sulfide. Double junctions are required of the form:

$$\text{Hg; Hg}_2\text{Cl}_2\text{; KCl(satd)} \mid \text{KNO}_3 \mid \text{test soln} \qquad (2.VII)$$
$$\text{or Ag; AgCl; NaCl(0.1 } M) \mid \text{NaClO}_4 \mid \text{test soln} \qquad (2.VIII)$$

Although they are not commercially available and must be constructed by the user, unsaturated reference electrodes using dilute halides, $\sim 0.01\ M$ to $0.1\ M$, to match the approximate ionic strength of the test solution are preferred for membrane electrode cells.

In (2.29) the term φ (inner soln) $- \varphi$ (test soln) is the membrane potential of the selective ion electrode. Since negligible current flows in a potentiometric measurement, there is no iR drop in the solutions on either side of the membrane. The inner potentials φ of these phases are constant provided the concentrations of salts are uniform as a result of stirring. The membrane potential is found by summing the two interfacial potentials with the internal diffusion potential according to

$$\Delta E_{\text{membrane}} = \underbrace{[\varphi(s)' - \varphi(m)']}_{\text{Ref. } \Delta\varphi} + \underbrace{[\varphi(m)' - \varphi(m)]}_{\text{Diff. } \Delta\varphi} + \underbrace{[\varphi(m) - \varphi(s)]}_{\text{Test } \Delta\varphi} \qquad (2.32)$$

where the m and s designations stand for membrane and solution. For a pH glass electrode exposed to hydrogen and sodium ions in the test solution while the filling solution contains a buffer

$$\Delta E_{\text{membrane}} = \frac{RT}{F} \ln \left[\frac{u_{\text{H}^+}a_{\text{H}^+}/\bar{\gamma}_{\text{H}^+} + K_{\text{H}^+/\text{Na}}u_{\text{Na}^+}a_{\text{Na}^+}/\bar{\gamma}_{\text{Na}}}{u_{\text{H}^+}a_{\text{H}^+}/\bar{\gamma}_{\text{H}^+}(\text{inner soln})} \right] \quad (2.33)$$

This equation is usually written in a form given by Nicolsky, although it was empirically found earlier,

$$\Delta E_{\text{membrane}} = \frac{RT}{F} \ln \left[\frac{a_{\text{H}^+} + K_{\text{H}^+/\text{Na}^+}^{\text{pot}}a_{\text{Na}^+}(\text{test soln})}{a_{\text{H}^+}(\text{inner soln})} \right] \quad (2.34)$$

$K_{\text{H}^+/\text{Na}^+}^{\text{pot}}$ is a combination of parameters: the ion-exchange equilibrium constant $K_{\text{H}^+/\text{Na}^+}$ defined in (2.12), mobilities u, and activity coefficients $\bar{\gamma}$ within the membrane phase. The general expression for K^{pot} is

$$K_{\text{H}^+/i}^{\text{pot}} = K_{\text{H}^+/i} \left(\frac{u_i\bar{\gamma}_{\text{H}^+}}{u_{\text{H}^+}\bar{\gamma}_i} \right) \quad (2.35)$$

and it is also called the selectivity coefficient for ion i relative to hydrogen ion. Expressions of the same form as (2.34) for the membrane potential of glass electrodes seem to hold for liquid, solid-state, and heterogeneous membranes of the Pungor type.

$$\Delta E_{\text{membrane}} = \frac{RT}{zF} \ln \left[a_1 + \sum_i K_{1/i}^{\text{pot}}a_i \right] + \text{constant} \quad (2.36)$$

The expression for $\Delta E_{\text{membrane}}$ must be combined with the junction potentials and reference electrode contributions according to (2.29). The constant term in (2.36) contains the contribution to the cell potential from the constant inner solution activities.

The quantity $K_{1/i}^{\text{pot}}$ is usually defined in such a way as to be less than unity. That is, for many electrodes the response to a particular ion is dominant, for example, H^+ in the case of a glass electrode. At a given activity level in the test solution, this ion gives the most positive measured potential of all other monovalent positive ions. Consequently, this ion is designated a_1 and the selectivity for other monovalent ions referred to 1 is less than unity. Beware of the nomenclature for selectivity coefficients. Some authors prefer $K_{i/1}^{\text{pot}}$ as the coefficient for the activity of the ith species. Quite often the reciprocal of $K_{i/1}^{\text{pot}}$ is used as a sensitivity coefficient to describe how much more sensitive the electrode is for 1 relative to i. For example, a typical general-purpose glass electrode has a selectivity coefficient $K_{\text{H}^+/\text{Na}^+}^{\text{pot}} \approx 10^{-11}$. The electrode is therefore 10^{11} times more sensitive to H^+ than to Na^+. The potential response of this glass electrode at $a_{\text{H}^+} = 1$ requires $a_{\text{Na}^+} = 10^{11}$ to raise the measured potential by $(RT/F) \ln 2$ or 18 mV. A third interpretation is that an activity $a_{H^+} = 10^{-11}$ makes the same contribution to the potential as $a_{Na^+} = 1$. In connection with liquid ion-exchange membranes, K^{pot} examples are perhaps clearer since the dominant ion and interference are independent of each other. A calcium-selective electrode shows a barium interference to the extent $K^{pot}_{Ca^{2+}/Ba^{2+}} = 0.01$. The electrode potential measured for a junction cell omitting reference and junction potentials is

$$E_{\text{measured}} = \frac{RT}{2F} \ln (a_{Ca^{2+}} + 0.01 a_{Ba^{2+}}) + \text{constant}. \tag{2.37}$$

The selectivity constant clearly means that the response attributable to Ca^{2+} alone at $a_{Ca^{2+}}$ would be matched in the absence of calcium by 100 times larger activity of barium.

In many instances ions can be found that are far more responsive at an electrode than the ion for which the membrane was intended. The calcium electrode, for example, responds more positively for a given activity of Zn^{2+} than for the same activity of Ca^{2+}. Since $K^{pot}_{Ca^{2+}/Zn^{2+}} = 3.2$; $K^{pot}_{Ca^{2+}/Fe^{2+}} = 0.8$; $K^{pot}_{Ca^{2+}/H^+} = 10^7$, $a_{Zn^{2+}}$ must be less than $1/3.2$ times the lowest $a_{Ca^{2+}}$ to be measured, $a_{Fe^{2+}} < \dfrac{1}{0.8}$ times, and $a_{H^+} < 10^{-7}$ times the lowest value of $a_{Ca^{2+}}$. Selectivity coefficients are defined for a monovalent ion electrode and a divalent interference by

$$E_{\text{measured}} = \frac{RT}{F} \ln (a_{M^+} + K^{pot}_{M^+/M^{2+}} a_{M^{2+}}^{1/2}) + \text{constant}. \tag{2.38}$$

For the opposite situation,

$$E_{\text{measured}} = \frac{RT}{2F} \ln (a_{M^{2+}} + K^{pot}_{M^{2+}/M^+} a_{M^+}^{2}) + \text{constant}. \tag{2.39}$$

An example of the latter is the sodium interference on the Cu^{2+} electrode where $K^{pot}_{Cu^{2+}/Na^+} = 5 \times 10^{-4}$.

A plot of E measured versus $\log a_{Ca^{2+}}$ computed as $C_{Ca^{2+}} \gamma_{\pm}$ at 25°C yields a straight line, a "Nernst curve" of slope about 29 mV/decade $a_{Ca^{2+}}$. If barium were present at constant activity $a_{Ba^{2+}}$ for the series of calcium solutions, the measured potential would level off as shown in Fig. 2.1. By extrapolation of the constant potential back to the Nernst curve, a corresponding $a_{Ca^{2+}}$ is read off. A simple calculation gives

$$K^{pot}_{Ca^{2+}/Ba^{2+}} = \frac{a_{Ca^{2+}}(\text{intercept})}{a_{Ba^{2+}}(\text{added})}. \tag{2.40}$$

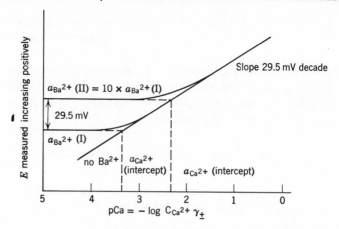

Fig. 2.1 An example of ion interference—the effect of constant barium levels on a calcium response curve. A theoretical "Nernstian" response according to (2.37) is a straight line $a_{Ba^{2+}}(I) \approx 5 \times 10^{-2}$; $a_{Ba^{2+}}(II) \approx 5 \times 10^{-1}$.

Data can be used more effectively by "best fitting" trial values of K^{pot} in the region of curvature. Observed values of K^{pot} can be obtained for a range of $a_{Ba^{2+}}$ levels to determine whether or not the selectivity is independent of the interference activity level. If the selectivity coefficient is not constant, then deviations tend to be in the direction of higher values of K^{pot} (i.e., more interference) when determined at low interference levels. This effect probably relates to slow approach to equilibrium at the lower activity levels.

A further type of measurement clarifies this effect. A constant selectivity implies that E versus $\log a_{Ca^{2+}}$ and E versus $\log a_{Ba^{2+}}$ using pure solutions would be straight-line Nernst curves displaced on the potential scale by a constant value as shown in Fig. 2.2. This is the basis of the two-solution measurement of electrode selectivity.

$$\Delta E_{measured} = E(a_{Ca^{2+}} = 0.1) - E(a_{Ba^{2+}} = 0.1) = -\frac{RT}{F} \ln K^{pot}_{Ca^{2+}/Ba^{2+}}.$$

$$(2.41)$$

$\Delta E_{measured}$ could be determined by comparing any pair of constant-activity solutions. Actually, at low activities the pure Nernst curves are not obeyed. If the spread between curves increases or decreases, $K^{pot}_{Ca^{2+}/Ba^{2+}}$ will apparently decrease or increase, respectively. If the two-point method involving pure solutions is used (and it is not recommended for careful work), activity should be high, ~ 0.1 M, to achieve equilibrium rapidly. In any case selectivity values are not high-precision quantities and ought not be used to compute unknown activities. They primarily serve as an index indicating approximately what

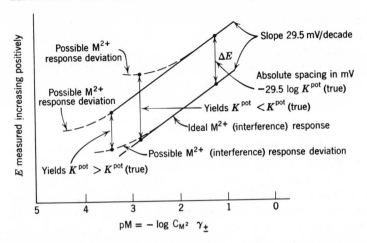

Fig. 2.2 Illustration of K^{pot} determination by the two-point method using ΔE measured for equal activities of M^{2+} and M^{2+} (interference).

conditions of solution purity are required to obtain a negligible level of interference.

3 ELECTRODES AND ELECTRODE VESSELS FOR pH DETERMINATION

Hydrogen Electrode

The hydrogen electrode consists of a catalytically active surface immersed in a solution saturated with hydrogen gas. The electrode is constructed by spot-welding a small piece of platinum foil to a short piece of no. 26 platinum wire, which is then sealed into a narrow soft-glass tube. Pyrex glass cannot be used because its coefficient of thermal expansion is very different from that of platinum, with the result that a perfect seal cannot be obtained. The exposed platinum is plated with platinum black, which is the catalytic surface. This is done by making the electrode the cathode in an electrolytic cell containing a 1 to 3% solution of chloroplatinic acid to which a little lead acetate has been added. A current of 200 to 400 mA is passed through the cell for 1 to 3 min. After plating, the electrode is washed with distilled water (preferably overnight in running water) and stored under water. The glass tube is then filled with mercury and electrical contact is made by dipping a wire in the mercury.

The electrode is placed in any convenient glass vessel and a stream of hydrogen is bubbled through the adjacent solution. The hydrogen should be purified, that is, oxygen should be removed. This is most conveniently accomplished by means of platinum catalysts, which are available commercially in cartridges which fit directly onto hydrogen cylinders. Typical half-cells containing hydrogen electrodes are shown in Figs. 2.3 and 2.4.

The reaction at the hydrogen electrode $H^+ + e^- \rightarrow \frac{1}{2}H_2$, and the emf of the half-cell is

$$E = +\frac{RT}{F} \ln \frac{a_{H^+}}{P_{H_2}^{1/2}}, \tag{2.42}$$

where a_{H^+} is the activity of hydrogen ions in the solution and P_{H_2} the pressure of the hydrogen gas with which the solution is equilibrated. There is no E_0 term in (2.42) because E_0 for the hydrogen electrode is zero in aqueous media.

For surface-active solutions, for example, for proteins and other colloidal solutions, the bubbling of hydrogen through the solution is not possible since foaming results. For such solutions a closed vessel containing hydrogen gas and the desired liquid solution is used. The latter is saturated with the gas by rocking the electrode vessel back and forth for 30 to 45 min. This type of half-cell, developed by Clark [24], is shown in Fig. 2.5.

The hydrogen electrode can be employed only for a limited variety of solutions. Cyanide, H_2S, and many other sulfur compounds, arsenic, the

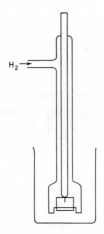

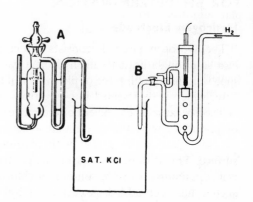

Fig. 2.3 Hydrogen electrode (Hildebrand type). The electrode is dipped into the solution to be measured.

Fig. 2.4 Apparatus for pH determination using a hydrogen electrode, a calomel electrode, and a beaker of saturated KCl as salt bridge. Stopcock B is a "conducting" stopcock (Fig. 2.10). Stopcocks A and B are briefly opened before each emf reading to permit a drop of solution to enter the beaker, forming a broad, diffuse liquid junction.

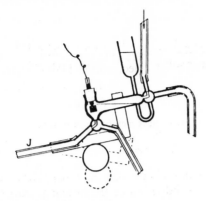

Fig. 2.5 Hydrogen electrode vessel for surface-active solutions (Clark type). The tube J is filled with saturated KCl solution.

cations of metals more noble than hydrogen (e.g., Cu, Ag, Hg), all poison the catalytic surface. Organic compounds which may be reduced at a platinum surface also interfere. The poisoning action of mercury should be especially noted. Since mercury is used to make electrical contact with the platinum wire, great care must be taken that none is spilled and that the platinum-glass seal is perfect.

A catalytic surface of reduced activity can be prepared by plating the platinum disk with palladium black rather than platinum black. This is done by substituting a chloropalladous acid for the chloroplatinic acid in the plating procedure [25]. This electrode is satisfactory for solutions of some organic compounds that interfere in the use of platinized electrodes. Particularly important is the use of this electrode with potassium hydrogen phthalate, which is one of the primary standards in pH measurement. Palladized electrodes attain equilibrium much more slowly than platinized electrodes and should be used only when the latter cannot be employed.

Glass Electrodes

A pH-sensitive glass membrane between two solutions is the source of a membrane potential which is linearly related to the negative logarithm of the hydrogen ion activities in contact with the membrane. Contributions to the membrane potential include the interfacial and the internal diffusion potential as discussed in Section 2. The overall cell potential was described in Section 2 also. For glass electrodes commercially available, one has the option of calomel or silver–silver chloride internal reference electrodes. Saturated calomel external references are normally but not universally used. With the internal reference attached to the positive input of a pH meter, the measured cell potential is

$$E_{cell} = \frac{RT}{F} \ln \frac{a_{H^+}(\text{test soln})}{a_{H^+}(\text{inner soln})} - \frac{RT}{F} \ln \frac{a_{Cl^-}(\text{inner soln})}{a_{Cl^-}(\text{ext. ref.})}$$

$$+ E^0_{AgCl/Ag} - E^0 \text{calomel} + E_J \quad (2.43)$$

for cells of the type

$$\text{Ag; AgCl; HCl; glass membrane; test soln} \parallel \text{satd calomel;} \qquad (2.IX)$$
$$\text{or KCl + buffer} \qquad\qquad\qquad\qquad\quad \text{ext. ref.}$$
$$\text{(inner soln)}$$

or an equivalent cell containing a calomel inner reference. At a given temperature and pressure, subdivision of the potential into E_{glass} and $E_{ext. ref.}$ according to

$$E_{cell} = E_{glass} - E_{ext. ref.} + E_J = E^{0'}_{glass} + \frac{RT}{F} \ln a_{H^+}(\text{test soln}) + E_J,$$

$$(2.44)$$

yields

$$E_{glass} = E^0_{glass} + \frac{RT}{F} \ln a_{H^+}(\text{test soln}). \qquad (2.45)$$

E^0_{glass} includes E^0 for the inner reference and terms including the activities of H^+ and Cl^- within the glass electrode filling (inner) solution, while $E^{0'}$ also includes the external reference potential. It is important to note that E^0_{glass} and $E^{0'}_{glass}$ are functions of temperature. Except for the magnitude of the constant term, the glass electrode potential behaves thermodynamically in identical fashion to a hydrogen electrode but has none of the disadvantages of the latter. It comes to equilibrium rapidly and has no catalytic surface which can be poisoned. There is no oxidation-reduction process with which metals or other reducible substances can interfere.

However, the glass electrode has disadvantages of its own. Alkali silicate glasses are intrinsically unstable with respect to hydrolysis. Fortunately, even in acidic and basic media, the rate is slow because of the presence of multivalent ions in the glass network which act empirically as "tighteners." Particularly, the presence of potassium in glasses weakens the lattice leading to microcracking at the surface and enhanced hydrolysis. Sodium glasses which are the bases of older pH glasses are also prone to hydrolysis with the development of surface films of partially hydrated silica gel. On prolonged use especially in alkaline solutions, the films become visible and the electrode response becomes both slow and sub-Nernstian. Modern lithium-based glasses do not show this degradation in response and serve as the basis of high-quality, wide-pH-range glass electrodes. There is considerable literature which asserts that good pH glasses, that is, those indicating Nernstian response show a correlation between response quality and hygroscopic character. It appears now that these results apply mainly to sodium glasses, and that uptake of water is necessary to facilitate migration of protons through the hydrolyzed layer to the fixed sites. Glasses that hydrolyze extensively must not be allowed to dry out without loss of response— although a hydrogen fluoride etch restores response in most cases. Lithium glasses which are less susceptible to hydrolysis are also less sensitive to the need for hydroscopicity.

Equation (2.45), indicating pure pH response of glass electrodes, is limited in validity to pH values roughly from 1 to 12 depending on the glass composition. The more general form of (2.45) is

$$E_{\text{glass}} = E^0_{\text{glass}} + \frac{RT}{F} \ln [a_{\text{H}^+} + K^{\text{pot}}_{\text{H}^+/i} a_i](\text{test soln}), \qquad (2.46)$$

where $i = \text{Na}^+$ and Li^+ and K^{pot} is the selectivity coefficient described in detail in Section 2. Values of selectivity coefficients for pH glasses are not well documented. Error measurements attributable to sodium at high pH values have been tabulated by Bates [26] for many commercial electrodes. Assuming (2.46) holds, $K^{\text{pot}}_{\text{H}^+/\text{Na}^+}$ could be calculated. Since pH glasses are prepared from commercial sources of materials whose impurity levels vary, it is not clear that selectivity coefficients remain constant and meaningful from time to time within a given manufacturer's lines. Also, there is evidence that alkali ion errors obey a more complicated form inasmuch as the activity coefficients within the glasses are concentration dependent. Eisenman and Karreman have proposed a two-parameter equation [27]:

$$E_{\text{glass}} = E^0_{\text{glass}} + \frac{nRT}{F} \ln [a_{\text{H}^+}^{1/n} + (K^{\text{pot}}_{\text{H}^+/i} a_i)^{1/n}] \qquad (2.47)$$

where $n > 1$ and has higher values in relation to the hardness of the glass.

In strong acids glass electrodes slow negative errors. It is not clear whether this effect is the result of a breakdown in co-ion exclusion or the loss or change in character of fixed sites. The dependence of glass electrode properties on composition has been discussed in encyclopedic fashion by Isard [28]. The origin of the Nicolsky equation and a further elaboration of it to take account of a range of ion-exchange constants corresponding to different strengths of fixed sites has been given and extensively documented by Nicolsky et al. [29]. Eisenman has summarized his theory of the relation between glass composition and potentiometric selectivity [30].

Glass electrodes are available in many forms. Literally hundreds of designs can be purchased from various manufacturers. In addition to variation in glasses, internal reference electrodes, and filling solutions, electrodes come

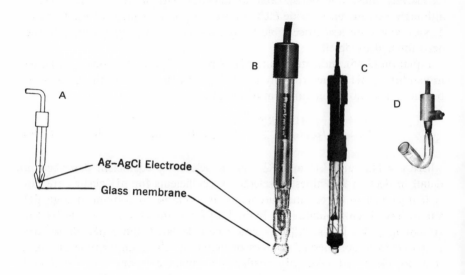

Fig. 2.6 Four commercial glass electrodes. Types *A* and *B* dip into the solution to be measured. Type *A* has a small hemispherical dome of pH-sensitive glass. This style is rugged and shock insensitive. Type *B* has a large, relatively fragile (compared with *A*) blown bulb of high-resistance, low-error pH glass. Type *C* is a combination electrode with the outer concentric cylinder as the reference electrode. Type *D* is a 1-drop electrode with pH glass built in.

in various sizes and shapes, suitable for large and microsize samples and capable of sterilization and use *in vitro* and *in vivo*. Examples of bulb types, a combination glass-reference electrode and an electrode for 1 drop of sample are shown in Fig. 2.6.

Glass electrodes are constructed from cylinders of very high resistance, usually a lead glass which is coated internally with a hydrophobic material to reduce conductivity up the wall. The pH-sensitive glass hemisphere or bulb is blown or pulled across the lower extremity. The internal electrode and filling solution are then added and sealed to complete the electrodes. Only low-resistance glasses are usually used in the hemisphere configuration. High-resistance, lower-error glasses must be used in high-area, thin-film bulbs to reduce resistance as much as possible. Resistance values range from 10 to 500 MΩ. Electrode leads must be carefully shielded to avoid spontaneous electrostatic and ac pickup. For careful work the entire electrode and cell should be enclosed in a grounded copper wire screen called a Faraday cage.

Glass Microelectrodes

The need to measure pH in 50 μl of solution or in multiple-phase biological samples of similar size or less has led to the development of microelectrodes for pH, pNa, and pK, among others, and correspondingly miniature reference electrodes. These electrodes, some of which are illustrated in Figs. 2.7 and 2.8, can be used for both *in vitro* and *in vivo* measurements including flowing systems. Friedman's recent review should be consulted for the latter [31]. The problems involved in constructing and using microelectrodes should not be underestimated. Electrodes whose tips have a diameter of a few microns are fragile and difficult to fill with electrolyte. Resistivities from 10^9 to 10^{11} Ω require excellent shielding and a quality electrometer for valid measurements.

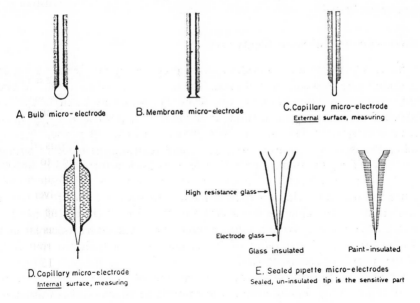

A. Bulb micro-electrode   B. Membrane micro-electrode   C. Capillary micro-electrode
                                                                  External surface, measuring

High resistance glass→

Electrode glass→

Glass insulated   Paint-insulated

D. Capillary micro-electrode   E. Sealed pipette micro-electrodes
Internal surface, measuring   Sealed, un-insulated tip is the sensitive part

Fig. 2.7 General types of glass microelectrodes [32].

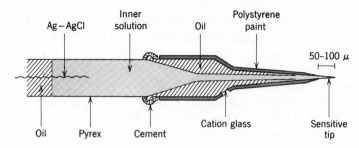

Fig. 2.8   Concentric cation glass microelectrode with external insulation [32].

Liquid junction problems with blood and other biological fluids are complicated by the presence of proteins. These problems have been extensively documented, and ways of avoiding and minimizing junction effects, such as the use of a pNa electrode as reference, are described by Khuri [32]. The tip potential problem is interesting in itself and is an example of a diffusion potential. Micro open tips of glass contain fixed sites. These retard the transfer of anions and cause a potassium chloride-filled tip junction to be positive outside and negative inside. This effect can be used to advantage for pH measurement simply by using a small open-ended, solution-filled capillary as a pH electrode. The potential response is not necessarily Nernstian but can be calibrated.

Silver–Silver Chloride Electrodes

Silver–silver chloride electrodes may be prepared by first sealing a no. 26 platinum wire into soft or flint glass. The wire is cleaned in nitric acid and then covered by silver in one of two ways. The wire may be either electroplated by being made the cathode in a solution of purified $KAg(CN)_2$, or the platinum wire (in the form of helix) may be covered with porous silver by filling the helix with a silver oxide paste and reducing the oxide to porous silver by heating for 10 min at 500°C. The silver oxide is formed from $AgNO_3$ by the addition of NaOH. Silver oxalate may be used in place of silver oxide. After the platinum wire has been covered with silver, part of the silver is converted to silver chloride by electrolyzing the silver electrodes, as anodes, in a 1 M solution of twice-distilled hydrochloric acid, using a current of a few milliamperes. The duration of the current should be such that 15 to 20% of the silver is converted to AgCl.

Silver–silver chloride electrodes may require 24 hr or more to come to equilibrium. They should be stored in chloride solutions of about the same concentration as that found in the solution to be measured. This procedure reduces the time required for equilibrium. The electrode reaction is $AgCl + e^- \rightarrow Ag + Cl^-$, and the half-cell potential is

$$E = E^0_{\text{AgCl/Ag}} - \frac{RT}{F} \ln a_{\text{Cl}^-}, \qquad (2.48)$$

where a_{Cl^-} is the activity of chloride ion in the solution with which the electrode is in contact. Values of the standard potential E^0 at several temperatures are given in Table 2.3. For a review of recent work on this electrode, see Covington [33].

Table 2.3 Standard Potential of the Silver–Silver Chloride Electrode

Temperature (°C)	$E^0_{\text{AgCl/Ag}}$	Temperature (°C)	$E^0_{\text{AgCl/Ag}}$
0	+0.23655	40	+0.21208
5	+0.23413	45	+0.20835
10	+0.23142	50	+0.20449
15	+0.22857	55	+0.20056
20	+0.22557	60	+0.19466
25	+0.22234	70	+0.18782
30	+0.21904	80	+0.1787
35	+0.21565	90	+0.1695
		95	+0.1651

Calomel Electrodes

The commonest of all electrodes used is the calomel electrode. It is almost universally employed as the reference electrode in conjunction with other electrodes. Its construction is illustrated in Fig. 2.9. A paste is made of Hg_2Cl_2 and Hg and this is placed over a mercury pool, which makes contact by means of a platinum wire with an external tube filled with mercury into which a wire leading to the potentiometer can be dipped. A solution of KCl is placed over the $Hg-Hg_2Cl_2$ paste (this solution is previously saturated with Hg_2Cl_2) and a side arm makes contact with a salt bridge. Calomel electrodes are ordinarily used only in cells with liquid junctions.

The reaction at the calomel electrode in the direction of reduction is

$$\tfrac{1}{2}Hg_2Cl_2 + e^- \rightarrow Hg + Cl^-$$

and the half-cell potential is given by

$$E = E^0_{\text{Hg}_2\text{Cl}_2/\text{Hg}} - \frac{RT}{F} \ln a_{\text{Cl}^-}. \qquad (2.49)$$

The standard potential at 25°C is +0.2680 V.

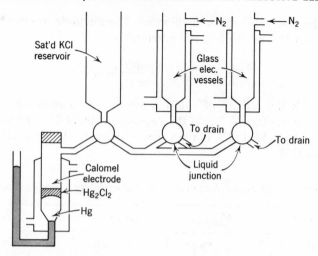

Fig. 2.9 Apparatus for pH measurement with dipping glass electrodes, illustrating the use of three-way stopcocks to make the liquid junction.

In view of the fact that the calomel electrode is usually used as a reference electrode and is separated by a salt bridge from a second electrode, which is the one of real interest in the determination of potential, the concentration of KCl can be fixed at any desired value. Customarily the solution contains 0.1 N, 1 N, or saturated KCl. For these concentrations the values of E given by (2.49) have been estimated at several temperatures, as listed in Table 2.4. E for this cell is called E_C in later sections. A recent review of this type of electrode has been given by Covington [33].

Salt Bridges

The most reliable salt bridges are those that contain fresh saturated KCl solution, or NH_4NO_3 if KCl cannot be used (for instance, if the test solution

Table 2.4 Potential of Calomel Half-Cell

Temperature (°C)	0.1 N KCl	1 N KCl	Satd KCl
12	+0.3362	—	+0.2528
15	+0.3360	—	+0.2508
20	+0.3358	—	+0.2476
25	+0.3356	+0.2802	+0.2444
30	+0.3354	—	+0.2417
35	+0.3353	—	+0.2391
38	+0.3352	—	+0.2375

contains Ag^+ or other metal ions which precipitate in contact with Cl^-). Such a saturated solution is generally denser than the solutions in the electrode vessels, and it is therefore placed below these, to avoid mixing by convection. A very simple way of making a junction between two half-cells, using a saturated solution, is to dip the outlets from the half-cells into a beaker containing saturated KCl, as shown in Fig. 2.4. If this procedure is used, both half-cells must be sealed (to prevent liquid flow) or they must be equipped with stopcocks which can conduct electricity though closed. Ordinary stopcocks with an ungreased strip about the middle are generally used for this purpose. This strip fills with solution, so that continuity of liquid is maintained when the stopcock is closed, but leakage of liquid through the stopcock is extremely slow (see Fig. 2.10).

A more suitable procedure for making contact is shown in Fig. 2.9. This technique uses three-way stopcocks. Each half-cell can be rinsed and filled without disturbing the rest of the apparatus. Fresh saturated KCl solution is then run through the stopcock at the bottom of each half-cell, while the solutions in the half-cells remain undisturbed. Finally, the stopcock at the bottom of the

THIN LAYERS OF GREASE

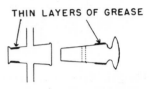

Fig. 2.10 Construction of a stopcock which conducts electricity when closed.

KCl reservoir is turned so as to join the two half-cells; both of these are then connected to the bridge. Only one of the half-cells needs to be sealed. The other may be open to the atmosphere, as is necessary, for instance, if it is a hydrogen half-cell.

If half-cell 1 of Fig. 2.9 is a saturated calomel electrode, its solution is identical with that of the salt bridge and it can then be connected directly to the salt bridge without a stopcock. In this case the two remaining stopcocks may be replaced by a single stopcock, as shown in Fig. 2.11, which shows the MacInnes-Belcher cell for pH measurement with a glass and saturated calomel electrode.

For approximate measurements (accuracy 2 mV) it is sufficient to use a salt bridge immobilized in a gel. It is formed by dissolving about 3 gm of agar-agar in 100 ml of saturated salt solution by heating close to the boiling point. The resulting hot solution is poured or sucked into a U-tube and gels on cooling. The tips of the U-tube should be tapered down to a small diameter to prevent the gel from falling out.

Such a bridge can be used to connect two half-cells as shown in Fig. 2.12. The same bridge should not be used for too many measurements because the solutions into which it is dipped gradually diffuse into it, and will diffuse back into the surrounding solution during subsequent measurements. Use

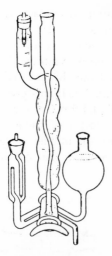

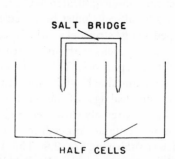

Fig. 2.11 Classic apparatus for pH measurement using the Mac-Innes-Belcher glass electrode and a calomel electrode.

Fig. 2.12 Use of an agar gel saturated with KCl as a salt bridge.

of a bridge of this type introduces difficulties if one wishes to make measurements on solutions of low ionic strength because the diffusion of KCl from the bridge into the surrounding solution increases the density of the solution. The net result is that the surrounding liquid flows toward the bottom of the electrode vessel. In effect the solution in the electrode vessel is stirred and causes an increase in the rate of diffusion. (It should be noted that salt bridges in which liquid saturated KCl is used are always connected to the *bottom* of the electrode vessels so that this stirring process is avoided. The rate of diffusion from agar-supported KCl solution is nearly as rapid as that from liquid KCl solution.)

Mention should be made finally of the commercial saturated calomel electrodes that are equipped with a small hole through which an asbestos fiber runs. This fiber is saturated with the electrode solution, that is, saturated KCl, and it forms the salt bridge to the solution into which the electrode is dipped. This arrangement is subject to the same objections as an agar-supported salt bridge and should be avoided for solutions of low ionic strength. The fiber junction also becomes easily clogged by various substances (e.g., proteins). Other saturated calomel reference electrodes have small cracks or annular regions of contact between inner and test solutions. These controlled leaks are made by inserting a glass ball or rod or platinum wire into the outer glass jacket during manufacture. Mismatched coefficients of expansion allow the leak to form while the wire or other glass remains firmly in place. Porous frit junctions made of corundum, sintered glass, or leached glass are also available in reference junctions. Examples are illustrated in Fig. 2.13.

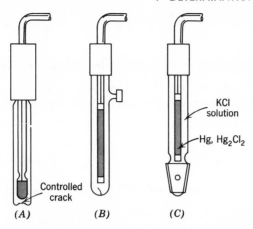

KCl
solution

Hg, Hg$_2$Cl$_2$

Controlled
crack

(A)   *(B)*   *(C)*

Fig. 2.13 Three forms of calomel reference electrodes. (*A*) Leeds and Northrup, (*B*) Beckman fiber type, (*C*) Beckman sleeve type.

4 DETERMINATION OF pH

It is well known that it is impossible to measure the activity of a single ion species in solution. The reason for this is that activity is defined in terms of chemical potential, which in turn is defined as the free energy increment per mole of substance added, $\partial G/\partial n$, at constant temperature and pressure, the amounts of all other substances in the solution being held constant. It is clearly impossible to perform this operation for a single ion species, for one cannot add an ion to a solution without, at the same time, adding an ion of opposite charge.

Nevertheless, as we shall see, it is possible to measure a quantity which is closely related to the activity of a single ion. In particular, considerable effort has been invested in trying to determine a quantity closely related to the activity of hydrogen ions, and it is this value that we are concerned with when we measure pH. This quantity may be described in general as,

$$pH \simeq -\log a_{H^+}. \qquad (2.50)$$

There are two approaches to the problem. One is an attempt to determine $-\log a_{H^+}$ itself by means of an appropriate nonthermodynamic assumption. One can often do this with a certainty of about 0.005. The method is relatively laborious and is virtually never employed in routine determinations. Its main importance lies in its use as a source of primary pH standards. The second method is to define a practical pH scale without undue concern as to how closely it may measure a suitably defined "$-\log a_{H^+}$." This method is the one employed for the vast majority of measurements. Its use is especially convenient in determining a pH value relative to a primary pH standard. The pH values obtained by this method rarely differ by more than 0.02 from values obtained by the more exacting procedure. Both approaches are described below.

Absolute Determination of a Quantity Close to $-\log a_{H^+}$

For this method one uses a cell without liquid junction, containing hydrogen and silver–silver chloride electrodes. The solution to be measured must contain chloride ions at a known concentration C. The cell is described by the linear formula,

$$\text{Pt; } H_2(\text{pressure } p); \text{ soln } X(C_{Cl^-} = C, a_{H^+} \text{ unknown}); \text{AgCl; Ag} \qquad (2.X)$$

By a combination of (2.42) and (2.48), the conventional emf of the cell becomes

$$E(\text{Ag} - \text{Pt}) = E_{AgCl/Ag} - E_{H^+/H_2} = E^0 - \frac{RT}{F} \ln \frac{a_{H^+} a_{Cl^-}}{P_{H_2}^{1/2}}, \qquad (2.51)$$

where E^0 is just the standard potential of the silver–silver chloride electrode. The standard hydrogen electrode potential in water is zero by definition.

In place of the activity of chloride ion, we may write the product of concentration and activity coefficient, $a = C\gamma$, so that

$$E(\text{Ag} - \text{Pt}) = E^0 - \frac{2.303RT}{F} \log \frac{C_{Cl^-}\gamma_{Cl^-}}{P_{H_2}^{1/2}} - \frac{2.303RT}{F} \log a_{H^+}. \qquad (2.52)$$

All the quantities in this equation are experimentally determinable except γ_{Cl^-} and $-\log a_{H^+}$. Within an uncertainty of the order of 0.01 it is, however, possible to calculate $\log \gamma_{Cl^-}$ for dilute solutions. This calculation is now standardized by agreement using an extended form of the Debye-Hückel equation known as the Bates-Guggenheim convention [34]. Thus $-\log a_{H^+}$ may be determined within the limit of this uncertainty.

The procedure here outlined can be used only for solutions that do not interfere with the operation of the hydrogen electrode, have a low ionic strength, and contain a known concentration of chloride ion. It is a simple matter, however, to obtain measurements for solutions that do not contain chloride ion by determining $-\log a_{H^+}$ by (2.52) in the presence of varying amounts of chloride ion, and then extrapolating to zero concentration of chloride ion.

Practical pH Measurements and the Defined pH Scale

The practical pH scale is defined by means of the emf of the following cell containing a liquid junction:

$$\text{Pt; } H_2(\text{pressure } p); \text{ soln } X \mid KCl(0.1 \ N, 1 \ N, \text{ or satd}); Hg_2Cl_2; Hg \qquad (2.XI)$$

The emf of this cell is merely the difference between the emf of the hydrogen half-cell and the reduction potential of the calomel cell, i.e.,

$$E = E(\text{Hg} - \text{Pt}) = E_C - \frac{RT}{F} \ln \frac{a_{\text{H}^+}}{P_{\text{H}_2}^{1/2}} + E_J \qquad (2.53)$$

where E_C is the potential of the calomel half-cell, given by (2.49) or by the numerical values of Table 2.4, and E_J is the junction potential at the salt bridge-solution X interface.

The practical pH scale is defined by (2.53). Thus

$$\text{pH} = \frac{2.303F}{RT} [E - (E^{0'} + E_J)] \qquad (2.54)$$

where $E^{0'}$ is equal to $-E_C$ if $P_{\text{H}_2} = 1$ atm; otherwise it is equal to $-E_C + (RT/2F) \ln P_{\text{H}_2}$. It should be noted that $E^{0'}$ is negative since E_C is positive.

Cell (2.XI) cannot be used for absolute pH measurement because of the inherent uncertainty in E_J. At the time of this writing, the quantity $E^{0'} + E_J$ lies in the range -0.3353 to -0.3358 V for 0.1 N calomel reference and -0.2441 to -0.2446 V for saturated calomel reference when dilute aqueous solutions of intermediate pH values 3 to 11 are used. Thus $E^{0'} + E_J$ can be considered constant though not precisely known [35]. At high acidities and alkalinities of solution X, the change in E_J causes the sum to vary outside the ranges stated here. Inasmuch as the quantity $pa_{\text{H}^+} = -\log a_{\text{H}^+}$ can be determined quite accurately for dilute buffers by the method described above, some of the uncertainty in the junction potential can be avoided. There are seven certified standard buffer mixtures for which pa_{H^+}, called pH_S, is known to better than ± 0.005 units. Their values are internally consistent to better than ± 0.004 units. When one standard is used in any practical cell of the kinds described, $E^{0'} + E_J$ for this cell is established since

$$E^{0'} + E_J = E - \frac{2.303F}{RT} \text{pH}_S \qquad (2.55)$$

This statement is also true if the measurement is made with a pH meter rather than a potentiometer. The "standardize" control automatically introduces a biasing voltage corresponding to $E^{0'} + E_J$ such that the pH value read is pH_S. This definition of pH is conventional and is accepted by all American scientists with access to the defined standards. Further discussion of relative pH measurements appears later.

pH Standards

The two methods just given find their principal use in the creation of a series of *pH standards*. These are solutions that possess stable pH values that need be carefully measured only once. Depending on the assumptions used in connection with the methods of absolute pH determination, somewhat different pH values are assigned to the various standard solutions by the British Standards Institution and the Japanese Standards Association. In general the disagreement is rarely as high as 0.02 pH units. We shall use here the pH values assigned by the United States National Bureau of Standards, which are based on the emf of cells without liquid junctions. The primary standard solutions, together with their pH values, are shown in Table 2.5. The values are taken from Bates [36] and Staples and Bates [37].

As indicated in Table 2.5, some of these standard substances can be purchased (quite inexpensively) from the United States National Bureau of Standards. The material is provided in powder form, and instructions for the preparation of solutions are included. It is most important that these instructions be followed exactly. In particular, the use of water free from acidic and basic substances is essential. Distilled water should be boiled to remove carbon dioxide and should be redistilled if it contains copper or other

Table 2.5 pH_S of NBS Primary Standards from 0 to 95°C*

Temperature (°C)	0.05 M KH$_2$ Citrate	0.025 M Each NaHCO$_3$, Na$_2$CO$_3$	KH Tartrate (satd at 25°C)	0.05 M KH Phthalate	0.025 M KH$_2$PO$_4$, 0.025 M Na$_2$HPO$_4$	0.008695 M KH$_2$PO$_4$, 0.03043 M Na$_2$HPO$_4$	0.01 M Borax
0	3.863	10.317	—	4.003	6.984	7.534	9.464
5	3.840	10.245	—	3.999	6.951	7.500	9.395
10	3.820	10.179	—	3.998	6.923	7.472	9.332
15	3.802	10.118	—	3.999	6.900	7.448	9.276
20	3.788	10.062	—	4.002	6.881	7.429	9.225
25	3.776	10.012	3.557	4.008	6.865	7.413	9.180
30	3.766	9.966	3.552	4.015	6.853	7.400	9.139
35	3.759	9.925	3.549	4.024	6.844	7.389	9.102
38	—	—	3.548	4.030	6.840	7.384	9.081
40	3.753	9.889	3.547	4.035	6.838	7.380	9.068
45	3.750	9.856	3.547	4.047	6.834	7.373	9.038
50	3.749	9.828	3.549	4.060	6.833	7.367	9.011
55	—	—	3.554	4.075	6.834	—	8.985
60	—	—	3.560	4.091	6.836	—	8.962
70	—	—	3.580	4.126	6.845	—	8.921
80	—	—	3.609	4.164	6.859	—	8.885
90	—	—	3.650	4.205	6.877	—	8.850
95	—	—	3.674	4.227	6.886	—	8.833

* These reagents can be purchased in powdered form from the Bureau of Standards.

Table 2.6 Compositions of Primary Standard Buffer Solutions*

(*Weight of buffer substance, in air near sea level,
per liter of buffer solution at 25°*)

Solution	Buffer Substance	Weight in Air (gm)
Tartrate, about 0.034 M	$KHC_4H_4O_6$	Satd at 25°C
Phthalate, 0.05 M	$KHC_8H_4O_4$	10.12
Phosphate, 1:1		
0.025 M	KH_2PO_4	3.388
0.025 M	Na_2HPO_4	3.533
Phosphate, 1:3.5		
0.008695 M	KH_2PO_4	1.179
0.03043 M	Na_2HPO_4	4.302
Borax, 0.01 M	$Na_2B_4O_7 \cdot 10H_2O$	3.80

* From Bates [36].

contaminants. The same pH standards, as well as those not available from the Bureau of Standards, can be prepared in the laboratory from recrystallized reagent grade chemicals using the weights indicated in Table 2.6. Again the use of good distilled water is essential. In view of the low cost of the Bureau of Standards samples, occasional checking of homemade standards against these should be carried out. Standard solutions can also be purchased from most supply houses.

In addition to the seven primary standards for the operational pH scale, three supplementary standards of pa_{H^+} have been established. One of these, 0.05 M potassium tetroxalate, is highly acidic ($pa_{H^+} = 1.68$ at 25°C), and the other, a solution of calcium hydroxide saturated at 25°C, is highly alkaline ($pa_{H^+} = 12.45$ at 25°C). Assigned values are given in Table 2.7.

Some other useful solutions of documented pH are given in Table 2.8.

These pH values in general may be considered to be correct within ± 0.02.

Determination of Relative pH

The availability of pH standards makes it possible to determine the pH of solutions relative to the pH of a standard solution. Using cell (2.XI), for instance, one can determine the emf given by (2.54) with a standard solution of assigned pH. The same cell can then be used to measure the emf with the unknown solution in place of the standard solution. The pH is then given by (2.56), where the subscript S refers to the standard solution

$$E - E_S = \frac{RT}{2.303F}(pH - pH_S) + (E_J - E_{JS}) \qquad (2.56)$$

Table 2.7 pH_S of NBS Secondary Standards

Temperature ($^\circ$C)	pa_{H^+} *Values of* 0.05 M $KH_3(C_2O_4)_2 \cdot H_2O$*	pa_{H^+} *Values of* $Ca(OH)_2$ *Satd at 25°C**
0	1.666	13.423
5	1.668	13.207
10	1.670	13.003
15	1.672	12.810
20	1.675	12.627
25	1.679	12.454
30	1.683	12.289
35	1.688	12.133
38	1.691	12.043
40	1.694	11.984
45	1.700	11.841
50	1.707	11.705
55	1.715	11.574
60	1.723	11.449
70	1.743	—
80	1.766	—
90	1.792	—
95	1.806	—

Temperature ($^\circ$C)	pa_{H^+} *Values of* 0.02 M *Piperazine Phosphate†*	pa_{H^+} *Values of* 0.05 M *Piperazine Phosphate†*
0	6.580	6.589
5	6.515	6.525
10	6.453	6.463
15	6.394	6.404
20	6.338	6.348
25	6.284	6.294
30	6.234	6.243
35	6.185	6.195
40	6.140	6.149
45	6.097	6.106
50	6.058	6.066

* From Bates [36].

† From A. B. Hetzer, R. A. Robinson and R. G. Bates, *Anal. Chem.*, **40**, 634 (1968).

Table 2.8   Additional pH Standards with pH Values at 25°C

Solution	pH
0.1 M HCl	1.10
0.01 M HCl, 0.09 M KCl	2.07
0.1 M acetic acid, 0.1 M Na acetate	4.65
0.01 M acetic acid, 0.01 M Na acetate	4.71
0.025 M NaH succinate, 0.025 M Na$_2$ succinate	5.40
0.01 M Na$_3$PO$_4$	11.72
0.05 M NaOH	12.62

and $E_J - E_{JS}$ is defined to be zero for dilute solutions. Measurement with respect to a standard and calculation of pH by this equation are valid at any constant temperature for which pH$_S$ is given. Limits are imposed by any changes in the junction potential value at the external reference junction when the test solution is substituted for the standard solution. For dilute solutions similar to the standards, the difference between the junctions is very small and is negligible. Consequently, primary standards measured against each other agree to better than 0.004 pH units. When a high-ionic-strength solution or an electrolyte in a nonaqueous solvent is used, the "residual" junction difference (and the differences in $E^{0\prime}$ for nonaqueous media) have to be considered. The residual junction $\bar{E}_J$, in pH units, is given by:

$$\bar{E}_J = \frac{F}{2.303RT}(E_J - E_{JS}) \qquad (2.57)$$

The majority of pH determinations in modern laboratories are relative determinations. However, the glass electrode is universally used in preference to the hydrogen electrode because of its greater convenience and because it can be used for a much greater variety of solutions. The cell most often employed is

$$\text{glass electrode; soln } X \parallel \text{satd calomel electrode} \qquad (2.\text{XII})$$

By (2.44), and with the same assumption as before, the measured potential E of this cell becomes

$$E = E^0_{\text{glass}} - E_C + E_J = E^{0\prime}_{\text{glass}} + E_J - \frac{RT}{2.303F}\text{pH} \qquad (2.58)$$

The constant $E^{0\prime}_{\text{glass}}$ of (2.58) has a different value for each electrode used. Its dependence on reference potentials and composition of the filling solution was described in (2.44) and (2.45). Note that the convention chosen here to report membrane cell emfs is the reverse of that used for galvanic cells.

If we now measure the emf E_S of a standard solution with pH = pH_S, and then, in the same apparatus, the emf of a solution of unknown pH, then, by (2.58), we obtain again (2.56) for the pH of the unknown solution relative to that of the standard.

The scales of most pH meters are calibrated directly in terms of pH rather than volts, that is, they are in fact made to read $(2.303F/RT)E$, rather than E itself. Resistors in series with the slide-wire may be adjusted so that the pH reading with the standard solution is exactly pH_S. The pH of an unknown solution can then be read directly. The validity of this measurement as discussed just above again depends on the assumption that the "residual" junction is zero and that $E^{0\prime}_{\text{glass}}$ for the overall cell remains constant. Measurements of the pH using pH meters are easily extended beyond the scale provided. For instance, the scale of a Beckman pH meter (Model G), which extends only to pH 13, can be extended to pH 14 by setting the scale for the standard solution being used to 1 pH unit less than its actual value, for example, if borax of pH 9.18 is being used, the scale is set to pH 8.18. A scale reading of pH 13 will then correspond actually to pH 14. This comment is included because so many Beckman Model G meters are still preferred. Ac line-operated meters have largely replaced the older battery-operated models. These instruments have also passed through tube-type models to all-solid-state versions. The highly developed "Research" pH meter of Beckman, the highly stable Leeds and Northrup models and those of Corning and Orion are recommended for precise work. Expanded scale models in which 1 pH unit (or less) can be expanded full scale are useful for detecting small changes in pH.

Most pH meters provide temperature adjustment by means of a variable resistor, that is, they provide electrical means for adjusting the factor $2.303RT/F$ to any temperature within a limited range. Use of this convenience requires that the electrode be temperature compensated. The electrode and its reference electrode must possess an "isothermal" point at 0 V. This is accomplished by making pH 7.00 correspond to 0 V at all temperatures through providing the glass electrode internal solution with a buffer whose pH change with temperature exactly compensates the temperature changes of the internal and external reference electrodes. The Nernst curves then pass through the isothermal point and only their slopes need correction as provided by the variable resistor. To be sure, it is best to calibrate the instrument by making it read the standardized value of the standard warmed or cooled to the desired temperature as well as setting the proper slope adjustment. However, it is equally convenient to leave this resistor always set at 25°C and to use the multiplying factor $T/298.16$ to correct one's readings. At any temperature T (absolute), with the temperature resistor set at 25°C, the scale reading is $(T/298.16)$pH; that is, it is necessary to set the scale with the standard solution at the pH_S value stated for $T°$(Abs.), and to multiply the scale reading observed for the unknown solution by $(298.16/T)$.

Any of the electrodes and electrode vessels described in Section 3 can be used for relative pH measurement. For most purposes it is sufficient to use commercial glass and calomel electrodes which dip into the solution whose pH is to be determined. Provided that the temperature is measured and provided that the temperature variation of the pH of standard solutions is taken into account in the standardization procedure, this technique should be accurate to within ±0.02 at ionic strengths of about 0.01 and higher. At lower ionic strengths the use of dipping calomel electrodes is not recommended, for the reasons given in Section 3.

It should be noted that temperature control is usually desirable. The pH of most solutions varies with temperature, and even though the precision of the measurement may be good, the variation with temperature may introduce a considerable uncertainty if temperature is merely measured rather than controlled. It is also important to note that atmospheric carbon dioxide is readily absorbed by alkaline solutions with a resulting decrease in pH owing to the reaction $CO_2 + OH^- \rightarrow HCO_3^-$. This is not a source of error when hydrogen electrodes are used because the air in the hydrogen electrode vessels has been replaced by hydrogen. When glass electrodes are used, it is necessary that the air be replaced by another gas, which is usually nitrogen. The nitrogen can be bubbled through or passed over the solution being measured.

Use of the Quinhydrone Electrode

In a later chapter the oxidation-reduction potential of solutions containing quinone and hydroquinone will be discussed. This potential will be seen to be pH dependent and can thus be used as a measure of pH. However, the method is subject to numerous errors and has no advantages over the use of glass electrodes. It was frequently used before glass electrodes were commercially available but is rarely employed today.

Use of Indicators

Indicators are substances that can exist in acidic and basic forms, the two forms having clearly distinguishable colors. Since the equilibrium between these forms depends on pH, the color of the indicator can be used as a measure of pH over a limited range. Mixtures of different indicators or a series of mixtures can be made up to cover the entire accessible pH range. Paper strips impregnated with indicator are commercially available. While the use of indicators has been displaced by the glass electrode for the determination of pH in aqueous media, indicators retain an important position in acid-base equilibrium measurements in nonaqueous solutions. Indicators are still useful in acid-base titrimetry and especially accurate results can be achieved by the photometric titration procedures. These are not limited to indicators but can be used for acids and bases which themselves absorb radiation in the visible or ultraviolet regions of the spectrum. Recommendations for indicators continue to appear in the literature. An enormous list presently exists as cataloged by Kolthoff [38]. More recent recommendations are given by Bates [39] and by Kolthoff et al. [40]. A detailed account of indicator equilibria, salt effects, and solvent effects in the context of pH determination using photometry and colorimetry is given by Bates [39].

The ability of such indicators to measure pH is greatly impaired by the fact that the substances that act as indicators also tend to combine with a large number of other organic substances (e.g., proteins). In the presence of these substances, the acid-base equilibrium constants of the indicators may therefore be greatly changed, and the colors of the two forms may also be altered. The same difficulties arise if nonaqueous or mixed solvents are used. The indicator method of measuring pH is thus a hazardous one; its use is not recommended without a careful study of the system. In any event, the indicator method has a maximum accuracy of about 0.2 pH unit.

For very strongly acidic solutions, however, potentiometric methods cannot be used. This happens, for instance, when one wishes to determine the base strength of very weak bases, which may require measurements in concentrated sulfuric acid or other solution in which the hydrogen ion activity may exceed unity by several powers of 10. The pH may reach values of -9. Although the meaning of these measurements is clear on thermodynamic grounds, practical measurements are not reliable when using glass electrodes. Under these conditions an indicator method is the only one available, and has been developed by Hammett and others [41].

Since pH is defined in terms of an emf measurement, Hammett has defined for the indicator method a closely related "acidity function," H_0, such that

$$H_0 = -\log \frac{(a_{H^+}\gamma_B)}{(\gamma_{BH^+})}, \tag{2.59}$$

where B and BH^+ represent any organic base and its conjugate acid. A similar function H_- for indicators that are negative in the basic form is defined

$$H_- = -\log \frac{(a_{H^+}\gamma_{B^-})}{(\gamma_{BH})}. \tag{2.60}$$

These functions, which are important in highly acidic or basic solutions of high dielectric constant, have been tabulated and are discussed in a recent review by Boyd [42].

Determination of Hydrogen or Hydroxyl Ion Concentration

It will be seen subsequently that it is often desirable, for the purpose of making a material balance, to determine from potentiometric pH measurements the actual *concentration* of *free* hydrogen ions in a solution, rather than the hydrogen ion activity. As an approximation, to estimate order of magnitude, this concentration may be considered to be the same as the activity, that is, one may set pH $\simeq -\log C_{H^+}$. Thus at pH 1, $C_{H^+} \simeq 0.1\ M$; at pH 2, $C_{H^+} \simeq 0.01\ M$, and so on. At pH 3 or 4 and above, the hydrogen ion concentration is thus nearly always negligibly small in comparison with the concentrations of other substances in a solution.

In a similar way the concentration of free hydroxyl ions can be estimated. In Section 5 the dissociation constant of water, $K_w = a_{H^+}a_{OH^-}$, is discussed, and tabulated values are shown in Table 2.12. These values may be used to compute a_{OH^-}. Thus at 25°, at pH 13, $C_{OH^-} \simeq 0.1$; at pH 12, $C_{OH^-} \simeq 0.01$, and so on. At pH 10 or 11 or below, the concentration of hydroxyl ions is nearly always negligible in comparison with the concentrations of other substances.

From pH 3 or 4 to pH 10 or 11, both C_{H^+} and C_{OH^-} are negligibly small and values of these concentrations are normally not required. Below pH 3 or 4 and above pH 10 or 11, however, it is often necessary to evaluate these concentrations. In so doing, the assumption that activities are equal to concentrations becomes inadequate. A better approximation is to assume that pH $= -\log a_{H^+}$ and to calculate C_{H^+} by using activity coefficients, that is,

$$-\log C_{H^+} = pH + \log \gamma_{H^+} \tag{2.61}$$

or, using K_w,

$$-\log C_{OH^-} = pK_w - pH + \log \gamma_{OH^-}, \tag{2.62}$$

where $pK_w = -\log K_w$.

Since pH is in fact not equal to $-\log a_{H^+}$ the best procedure is to estimate *empirically* the relation between pH and C_{H^+} *or* C_{OH^-}. One prepares solutions of known concentrations of hydrogen and hydroxyl ions (e.g., solutions of $HClO_4$, HCl, KOH, and so on) as similar as possible to the solutions of which the pH is being measured. The ionic strength especially should be the same, an objective achieved by the addition of NaCl, KCl, or other convenient salt. The solutions must not, of course, contain any weak acids or bases, or salts thereof which could contribute hydrogen or hydroxyl ions to the solution. The measurement of pH on these solutions can then be used to provide a relation between pH and C_{H^+} or C_{OH^-} which can reasonably be expected to be applicable to the unknown solution. This may be done in tabular form, as illustrated by Table 2.9 or graphically as shown in Fig. 2.17 (blank titration curve).

Table 2.9 Typical Data for the Evaluation of Hydrogen and
Hydroxyl Ion Concentrations at 25°C, $\mu = 0.15$*

C_{H+} (HCl) (moles/kg H_2O)	$-\log C_{H+}$	pH	Difference [pH $-$ ($-\log C_{H+}$)]
0.02331	1.632	1.710	0.078
0.01170	1.932	2.009	0.077
0.004667	2.331	2.405	0.074

C_{OH^-}(KOH) (moles/kg H_2O)	$-\log C_{OH^-}$	pH	pK_w $-$ pH	Difference [(pK_w $-$ pH) $-$ ($-\log C_{OH^-}$)]
0.03056	1.515	12.308	1.689	0.174
0.01524	1.817	12.010	1.987	0.170
0.002972	2.527	11.299	2.698	0.171

* The ionic strength was provided by addition of KCl. The data here given were used to estimate C_{H+} and C_{OH^-} in protein solutions at the same ionic strength.

pH Buffer Solutions

Mixtures of weak acids and their salts or weak bases and their salts have the property of establishing the pH of an aqueous solution and are called buffers. Even though the pH may be high, the rapid equilibrium between the two forms (proton donor-acceptor pair) maintains the hydrogen ion or hydroxyl ion activities. Dissociation and recombination of water is also rapid so that a_{H+} or a_{OH^-} become fixed when the other is established. Buffers have the important property that addition of strong acid or strong base in amounts less than the buffer concentration have only a mild effect on the pH. Pure water or a fully ionized salt solution such as KNO_3 or $NaClO_4$ has a pH of about 7 in the absence of dissolved carbon dioxide. Addition of 1 ml of 0.1 N NaOH or 1 ml of 0.1 N HCl to 1 liter of solution changes the pH to 10 or to 4—changes in H^+ by a factor of 1000 in each case. If the liter of solution contains 0.05 M K_2HPO_4 and 0.05 M KH_2PO_4 whose pH is also close to 7, then addition of the same quantity of a strong acid or base has only a minor effect on the pH, ± 0.002 pH units, respectively.

Buffers make possible the regulation of pH corresponding to small activity levels of H^+ and OH^- which could not be done as well by the dilution of solutions of strong acids and bases. Buffer action is a consequence of the equilibrium between weak acids and bases and water. The common forms of equilibria are

$$H_xA^{z+1} + H_2O \rightleftharpoons H_{x-1}A^z + H_3O^+ \tag{2.63}$$

for example:

$$HOAc + H_2O \rightleftharpoons OAc^- + H_3O^+ \tag{2.64}$$
$$H_2PO_4^- + H_2O \rightleftharpoons HPO_4^{2-} + H_3O^+ \tag{2.65}$$
$$RNH_3^+ + H_2O \rightleftharpoons RNH_2 + H_3O^+ \tag{2.66}$$

In the Brønsted-Lowry notation:

$$Acid_1 + Base_2 \rightleftharpoons Base_1 + Acid_2 \tag{2.67}$$

where H_2O—H_3O^+ is the acid-base pair of the solvent. Strong base buffers of the form

$$B^z + H_2O \rightleftharpoons BH^{z+1} + OH^- \tag{2.68}$$

also involve proton exchange, but the solvent acid-base pair is H_2O—OH^-. Ammonia and the more basic amines fit this category. The effect of adding strong base or acid to a mixture of weak acid and its basic form is to convert the forms of the buffer without producing a significant upset in free H_3O^+ or OH^-. The equilibrium ionization properties of the first group of buffers are described by equilibrium constants in the form:

$$K_a = \frac{a_{H^+}a_{A^-}}{a_{HA}} = \frac{C_{H^+}C_{A^-}}{C_{HA}} \cdot \frac{\gamma_{H^+}\gamma_{A^-}}{\gamma_{HA}} \tag{2.69}$$

and for the second group by

$$K_b = \frac{a_{OH^-}a_{BH^+}}{a_B} = \frac{K_w a_{BH^+}}{a_{H^+}a_B} = \frac{K_w C_{BH^+}}{C_{H^+}C_B} \cdot \frac{\gamma_{BH^+}}{\gamma_{H^+}\gamma_B}, \tag{2.70}$$

where K_w is the dissociation constant of water. For buffers whose component activities are greater than a_{H^+} (which is always the case for any practical application), then the activity of hydrogen ions is determined by component activities according to

$$a_{H^+} = K_a \frac{a_{HA}}{a_{A^-}}, \tag{2.71}$$

where a_{HA} and a_{A^-} are fixed by the quantity of buffer components added. If these are equal, then

$$pa_{H^+} = pK_a. \tag{2.72}$$

Activity coefficients must be considered so that equimolar mixtures of HA and A^- do not yield pH values exactly equal to pK_a. Ignoring activities for the moment

$$pH \simeq pK_a + \log \frac{C_{A^-}V}{C_{HA}V}$$
$$\simeq pK_a + \log \frac{m^*_{A^-}}{m^*_{HA}}, \tag{2.73}$$

Table 2.10   pH Ranges of Buffer Solutions*

Acidic Component	Basic Component	pH Range	Ref.†
HCl	Glycine	1.0–3.7	a
HCl	Na_2H citrate	1.0–5.0	a
p-Toluenesulfonic acid	Na p-toluenesulfonate	1.1–3.3	d
KH sulfosalicylate	NaOH	2.0–4.0	
HCl	KH phthalate	2.2–4.0	b
Citric acid	NaOH	2.2–6.5	e
Citric acid	Na_2HPO_4	2.2–8.0	e
Furoic acid	Na furoate	2.8–4.4	d
Formic acid	NaOH	2.8–4.6	
Succinic acid	Borax	3.0–5.8	e
Phenylacetic acid	Na phenylacetate	3.4–5.1	d
Acetic acid	Na acetate	3.7–5.6	e
KH phthalate	NaOH	4.0–6.2	b
NaH succinate	Na_2 succinate	4.8–6.3	d
Na_2H citrate	NaOH	5.0–6.3	a
NaH maleate	NaOH	5.2–6.8	
KH_2PO_4	NaOH	5.8–8.0	b
KH_2PO_4	Borax	5.8–9.2	c
NaH_2PO_4	Na_2HPO_4	5.9–8.0	a
HCl	Triethanolamine	6.7–8.7	
HCl	Na diethylbarbiturate	7.0–9.0	e
Diethylbarbituric acid	Na diethylbarbiturate	7.0–9.0	
H_3BO_3 or HCl	Borax	7.0–9.2	a, e
HCl	Tris(hydroxymethyl)-aminomethane	7.2–9.0	
H_3BO_3	NaOH	8.0–10.0	b
K p-phenolsulfonate	NaOH	8.2–9.8	
Glycine	NaOH	8.2–10.1	a
NH_4Cl	NH_4OH	8.3–9.2	
Glycine, Na_2HPO_4	NaOH	8.3–11.9	
HCl	Ethanolamine	8.6–10.4	
Borax	NaOH	9.2–11.0	a
$NaHCO_3$ or HCl	Na_2CO_3	9.2–11.0	c, e
Borax	Na_2CO_3	9.2–11.0	c
Na_2HPO_4	NaOH	11.0–12.0	e

* From Bates [43].

† Key to references:

^{a} S. P. L. Sørensen, *Biochem. Z.*, **21**, 131 (1909); **22**, 352 (1909); *Ergeb. Physiol.*, **19**, 393 (1912).

^{b} W. M. Clark and H. A. Lubs, *J. Biol. Chem.*, **25**, 479 (1916); W. M. Clark, *The Determination of Hydrogen Ions*, 3rd ed. Williams and Wilkins, Baltimore, 1928, Chapter 9.

^{c} I. M. Kolthoff and J. J. Vleeschhouwer, *Biochem. Z.*, **189**, 191 (1927).

^{d} W. L. German and A. I. Vogel, *Analyst*, **62**, 271 (1937).

^{e} H. T. S. Britton, *Hydrogen Ions*, 4th ed., Vol. I, Van Nostrand, Reinhold, New York, 1956, Chapter 17.

where V is the solution volume so that $m^*_{A^- \text{ or } HA}$ is a total number of moles. Addition of m^* total moles of strong acid changes the pH by converting $m^*_{A^-}$ to $m^*_{A^-} - m^*$ and m^*_{HA} to $m^*_{HA} + m^*$. Thus

$$\text{pH} \simeq pK_a + \log \frac{m^*_{A^-} - m^*}{m^*_{HA} + m^*}. \tag{2.74}$$

Addition of strong base simply changes the signs in front of m^*. As long as $m^*_{A^-}$ and m^*_{HA} are large relative to m^*, the pH of a buffer is a slowly varying function of m^*. The ability of a buffer solution to resist changes in pH is measured quantitatively by β, the buffer index where

$$\beta = \frac{dm^*}{d\text{pH}}. \tag{2.75}$$

The pH values of buffer solutions are subject to change when the buffer is diluted or an inert salt is added. These are primary salt effects readily predicted from the concentration dependences of the activity coefficients. For example, since

$$\text{pH} = pK_a + \log \frac{C_{A^-}}{C_{HA}} + \log \frac{\gamma_{A^-}}{\gamma_{HA}} \tag{2.76}$$

in the first class of buffers, dilution raises the value of γ_{A^-}, while an inert salt decreases it. Therefore dilution increases pH while inert salt decreases it. These effects can be predicted for buffer systems of other charge types by writing out the equilibrium expression and recalling that activity coefficients of ions decrease with ionic strength independent of charge sign as the square of the charge. These effects, temperature effects, and a general description of buffers are found in Bates [43]. Some useful buffer systems are given in Table 2.10.

The Measurement of "pH", pa_{H^+}, and $pa^*_{H^+}$ in Nonaqueous or Mixed Solvents

The preceding discussion has concerned itself entirely with pH determination in aqueous media. It is often desirable to be able to determine a quantity related to hydrogen ion activity in other solvents. In principle, there is no reason why equally valid measurements cannot be made in the majority of such solvents, but the actual number of such studies has been small. As a consequence, there does not exist for nonaqueous solvents an agreed set of assumptions that lead to the measurement of a quantity at least closely related to the hydrogen ion activity.

The method of Section 4 using cell (2.X), a cell without liquid junction, is applicable to any solvent that meets the following requirements: that the hydrogen electrode operate reversibly in the solvent, that NaCl or some other halide dissociates completely in the solvent while the corresponding silver halide does not, and that the silver–silver halide electrode operate reversibly in the solvent. Even then, new E^0 (Ag − Pt) values have to be determined in each solvent. These conditions have been met for mixtures of water with dioxane, methanol, and other alcohols, as well as for pure methanol. The E^0 (Ag − Pt) values in these solvents have been tabulated by Robinson and Stokes [44]. At 25°C, for instance, E^0 (Ag − Pt) for cell (2.X) is 0.2223 V in water, 0.2030 V in 20% dioxane, 0.0640 V in 80% dioxane, −0.0415 V in 82% dioxane, and −0.3113 V in pure methanol. The activity coefficient of chloride ion can be calculated provided that the chloride is dilute and completely dissociated. The use of this method is limited as for aqueous solutions by the sensitivity of the hydrogen electrode to numerous substances that can act as reducing or oxidizing agents.

The spontaneous chemical reaction occurring in the aqueous cell (2.X) as would occur if the leads were shorted through a large resistor is

$$\tfrac{1}{2}H_2(P) + AgCl \rightarrow Ag + H^+ + Cl^- \quad \Delta G^0 = F(E^0_{H^+/H_2} - E^0_{AgCl/Ag}) \quad (2.77)$$
$$= F[E^0(Pt - Ag)]$$

and ΔG^0 is negative. Since H_2, AgCl, and Ag are unit activity materials, ΔG is a measure of the stability of H^+ and Cl^- in the respective solvents. Comparing two cells, one using water corresponding to $\Delta_w G^0$ with another using one of the solvents listed corresponding to $\Delta_s G^0$, it is clear that the transfer reaction

$$H^+ + Cl^- \text{ (both in solvents } s) \rightarrow H^+ + Cl^- \text{ (both in water)} \quad (2.78)$$

is accompanied by a negative free energy change

$$\Delta G^0 \text{ (transfer)} = \Delta_w G^0 - \Delta_s G^0. \quad (2.79)$$

The Debye-Hückel theory allows one to calculate $_w\gamma_i$ for H^+ and Cl^- based on a standard state of infinite dilution in water and likewise $_s\gamma_i$ for a standard state based on infinite dilution in solvent s. Obviously the activity of HCl in solvent s does not approach unity relative to water, but some number greater than unity, ~400 in ethanol, ~100 in MeOH, and values between 100 and unity for alcohol–water mixtures. This factor is the thermodynamically defined "medium effect" for the neutral species HCl.

$$\Delta G^0 \text{ (transfer)} = -2RT \ln {_m\gamma_\pm} = -RT \ln {_m\gamma_{H^+}} {_m\gamma_{Cl^-}}, \quad (2.80)$$

where $_m\gamma_{H^+}$ and $_m\gamma_{Cl^-}$ are the unknown single-ion medium effects for H^+ and Cl^-. The free energies of H^+ in water and in solvent are

$$_w G_{H^+} = {_w G^0_{H^+}} - RT \ln C_{H^+} {_w\gamma_{H^+}} \quad (2.81)$$

$$_s G_{H^+} = {_s G^0_{H^+}} - RT \ln C_{H^+} {_s\gamma_{H^+}}, \quad (2.82)$$

where

$$_sG^0_{\mathrm{H}^+} = {_w}G^0_{\mathrm{H}^+} + RT \ln {_m}\gamma_{\mathrm{H}^+} \tag{2.83}$$

and consequently

$$E^0_{\mathrm{H}^+(\text{solvent})/\mathrm{H}_2} = \frac{RT}{F} \ln {_m}\gamma_{\mathrm{H}^+}. \tag{2.84}$$

When cell (2.XI) is used with solution X partially aqueous, but a totally aqueous external reference, then

$$E = E(\mathrm{Hg} - \mathrm{Pt}) = E_C - \frac{RT}{F} \ln \frac{_sa_{\mathrm{H}^+} \, _m\gamma_{\mathrm{H}^+}}{P^{1/2}_{\mathrm{H}_2}} + E_J \tag{2.85}$$

where $_sa_{\mathrm{H}^+} = C_{\mathrm{H}^+} \, _s\gamma_{\mathrm{H}^+}$. Similarly, for the glass electrode cell [see (2.XII)],

$$E_{\text{glass}} - E_{\text{Hg}} = E^0_{\text{glass}} - E_C + \frac{RT}{F} \ln {_s}a_{\mathrm{H}^+} \, _m\gamma_{\mathrm{H}^+} + E_J. \tag{2.86}$$

At the present time values of $_m\gamma_{\mathrm{H}^+}$ are not available, although extensive programs are underway to define this quantity by convention. Perhaps the most promising route involves the equipartitioning of the extraction coefficient for triisoamylbutylammonium tetraphenylborate proposed by Popovych [45]. This assumption seems to be superior to the assumption that reference couples may exist that have equal ionic medium effects in all solvents.

When a pH glass electrode is standardized in water and then used to measure conventional relative pH in a solvent, the measurement yields

pH(solvent) − pH$_S$(aqueous)

$$= -\log {_s}a_{\mathrm{H}^+} - \log {_m}\gamma_{\mathrm{H}^+} + \frac{F}{2.303RT} (E_J - E_{JS}). \tag{2.87}$$

It is now very clear that there is a dilemma with regard to the meaning of pH (solvent) by this relative measurement. Writing a definition

$$pa^*_{\mathrm{H}^+} (\text{solvent}) = -\log C_{\mathrm{H}^+} \, _s\gamma_{\mathrm{H}^+} = -\log {_s}a_{\mathrm{H}^+} \tag{2.88}$$

$$\text{pH (solvent)} - \text{pH}_S (\text{aqueous}) = pa^*_{\mathrm{H}^+} + \delta, \tag{2.89}$$

where δ is defined in the current literature as

$$\delta = -\log {_m}\gamma_{\mathrm{H}^+} + \bar{E}_J, \tag{2.90}$$

$pa^*_{\mathrm{H}^+}$ is a quantity that can be measured colorimetrically or by junctionless cells of the type (2.X) where a nonaqueous solvent is used throughout. Unfortunately, measurements against aqueous standards contain the unknown term δ. If nonaqueous $pa^*_{\mathrm{H}^+}$ standards were available, $_m\gamma_{\mathrm{H}^+}$ would not figure and relative $pa^*_{\mathrm{H}^+}$ values could be easily measured. To achieve this goal, $pa^*_{\mathrm{H}^+}$ standards for all solvents would have to be determined. Where studies have been made, in the cases of methanol–water and ethanol–water, δ values are known so that $pa_{H^+}^*$ and pH can be interconverted. On the other hand $p(a_{H^+}^*\,_m\gamma_{H^+})$ is interesting in its own right since it represents the activity in a solvent relative to water. The ability to know this value is limited by our inability to determine $\bar{E}_J$.

Relative pH measurements can also be made, of course, in the same way as for aqueous solutions, in a cell with liquid junction such as cell (2.XI). Cells of this and similar type, employing glass electrodes, have been used in a variety of nonaqueous solvents including, for instance, glacial acetic acid, acetonitrile, nitrobenzene, nitromethane, chlorobenzene, dioxane, xylene, chloroform, pyridine, dimethylsulfoxide, dimethylformamide, and so on. In the solvents of low dielectric constant it has been usual to saturate the glass membrane with water before use and at frequent intervals thereafter (this is done by standing the electrode in water and then wiping it dry).

The principal difficulty in this type of measurement is to find a suitable pH standard. Such a standard must clearly be in the same solvent as the unknown solutions. If either HCl or $HClO_4$ is completely dissociated in a solvent, then solutions of these acids can be used as standards, with the assumption that $pH = -\log a_{H^+}^* = -\log C_{H^+}\,_s\gamma_{H^+}$ being calculated. An alternative method is to use a mixture of an acid and its conjugate base, for example, acetic acid and acetate. The pH of such an acid can be computed if the dissociation constant is known. Needless to say, the number of known dissociation constants in nonaqueous media is very small. Examples and means for computing pH from equilibrium measurements in glacial acetic acid are given by Kolthoff and Bruckenstein [46]. In many so-called pH measurements in nonaqueous solvents, an attempt at valid standardization is not made at all. The numbers obtained may then be useful because of their reproducibility and they can be used as end-point indicators in acid-base titrations, but they bear little relation to hydrogen ion activity, and they cannot be used to compute acid dissociation constants.

A discussion of solvent acid-base character including tables of known dissociation constants in nonaqueous media occurs in Section 5. Several source books are now available on the properties of solvents [47–50]. In addition to these, acid-base properties are discussed in relation to pH determination and titrimetry in books by Kucharsky and Safarik [51] and Huber [52].

pH Determination in Unstable Solutions

It is sometimes necessary to determine the pH of solutions that are unstable and undergo changes in pH with time. Since it may take several minutes even for glass electrodes to come to equilibrium, it is necessary to measure pH in such unstable systems by a flow technique in which the electrode is placed in a flowing stream of solution which is continually being freshly prepared by mixing of stable components. An example of the application of this technique to the measurement of pH in acid solutions of hemoglobin is given by Steinhardt and Zaiser [53].

5 TITRATION CURVES AND ACID DISSOCIATION CONSTANTS

Acidimetric Titrations

Acidimetric titration consists of adding from a buret, stepwise, small portions of standardized strong acid, such as HCl, or a strong base, such as NaOH, and measuring the pH at each step. One objective of such a titration is to measure the total amount of base or acid in the solution being titrated. This is done by determining the "end point" of the titration, that is, the point at which the base or acid present has been exactly neutralized. This point is recognized by a rapid change in pH with addition of titrant.

The total amount of acid or base in the solution, once determined, can be used to indicate the purity of an organic substance by comparison of the amount of acid or base present with the amount expected on the basis of the number of grams of the substance in the solution. If the purity of a substance is reasonably well established, the total amount of base or acid present can be used to determine the molecular weight (neutralization equivalent), since this must be the weight of substances corresponding to an integral number of moles of acid or base.

Typical titration curves of a strong, weak, and very weak acid are shown in Fig. 2.14. As Fig. 2.14 shows, it is not really necessary to measure the pH as such in order to determine the end point of a titration curve. The emf of the cell can be used directly. For this reason acidimetric titrations can be carried out in nonaqueous solvents as readily as in aqueous solution, even though the relation between pH and emf in such solvent may not have been established (see preceding section). Examples of this type of titration, using cell (2.XII), are shown in Fig. 2.15. Fritz [54] has used this technique to titrate organic bases with $HClO_4$ in acetic acid and to titrate organic acids with C_2H_5ONa in alcohols, benzene, or toluene.

Titrations of this kind can be carried out automatically by using *automatic titrimeters*. These devices are commercially available. They are equipped with recorders which make a direct plot of the titration curves. Titration curves may be measured on very small amounts of materials by using special glass electrodes and ultramicroburets [55].

Titration may be carried out on a mixture of organic acids as well as on individual acids. If the dissociation constants of the acids in the mixture are sufficiently different, $\Delta pK \simeq 5$, then the titration curve of each individual

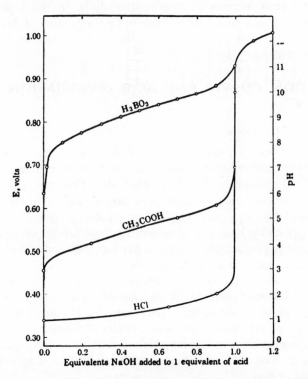

Fig. 2.14 Titration curves of a strong, weak, and very weak acid [53].

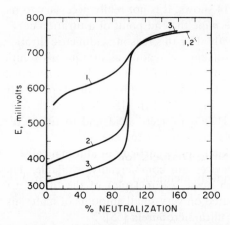

Fig. 2.15 Titration curves of three organic bases in 1:1 acetonitrile acetic acid using 0.1 N HClO$_4$ as titrant [55]. Curve 1, p-nitroaniline; curve 2, p-bromoaniline; curve 3, N-methylaniline.

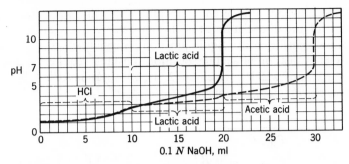

Fig. 2.16 Titration curves of mixtures of organic acids: (solid line) hydrochloric and lactic acids; (dashed line) hydrochloric, lactic, and acetic acids.

acid is clearly distinguishable, as shown in the example of Fig. 2.16. The amount of base required for each step of the titration is a measure of the amount of each acid present in the mixture. In the example in Fig. 2.16, each of the acids is present in equal amount.

Dissociation Constants of Acids and Bases

A very important use of titration curves or pH measurements is the determination of acid dissociation constants, that is, the equilibrium constants for reactions such as,

$$H_2O + CH_3COOH \rightleftharpoons H_3O^+ + CH_3COO^-$$
$$H_2O + NH_4^+ \rightleftharpoons H_3O^+ + NH_3$$
$$H_2O + HSO_4^- \rightleftharpoons H_3O^+ + SO_4^{2-} \cdot$$

(2.91)

We are using here the Brønsted definition of an acid as a substance capable of dissociating a proton. The product of the reaction is known as the *conjugate base*. It always bears a charge more negative by 1 unit than the charge on the acid.

Acid dissociation can be represented by the general equation (2.63) or the formal expression (2.67). The general form of the thermodynamic dissociation constant corresponding to this reaction is given in (2.69). Formulations for basic dissociation equilibria are given in (2.70).

Exact Thermodynamic Dissociation Constants

Although, as previously mentioned, the activity of hydrogen ions is not a measurable quantity, the dissociation constant given by (2.69) and (2.70) can be exactly determined. The procedure is quite laborious and for most purposes is not necessary. It is required, however, if one wishes to use the dissociation constants to measure thermodynamic quantities such as ΔH^0 and ΔS^0 for the dissociation, for these depend on the difference between dissociation constants at different temperatures and thus necessitate a precise knowledge of the dissociation at each temperature free from assumptions about activity coefficients or liquid junction potentials.

The exact evaluation of acid dissociation constants requires the use of cell (2.XI), the emf of which is given by (2.51). Substituting for a_{H^+} by equation (2.69) and rearranging, we obtain

$$E + \frac{2.303RT}{F} \log \frac{C_{Cl^-} C_{HA^z}}{P_{H_2}^{1/2} \, C_{A^{z-1}}}$$

$$= E^0 - \frac{2.303RT}{F} \log K_a - \frac{2.303RT}{F} \log \frac{\gamma_{Cl^-} \gamma_{HA^z}}{\gamma_{A^{z-1}}} . \quad (2.92)$$

By measuring E for various mixtures of HA^z, A^{z-1}, and Cl^-, one can determine the left-hand side of equation (2.92). Its value varies from solution to solution owing to variation in the term containing the activity coefficients. This term becomes zero, however, as the concentrations of all solutes tend to zero. Thus $E^0 - (2.303RT/F) \log K_a$ can be evaluated by plotting the left-hand side of equation (2.92) as a function of ionic strength and extrapolating to zero ionic strength [56]. Where either HA^z or A^{z-1} bears zero charge its concentration does not contribute to the ionic strength; it has a relatively small effect on the activity coefficients. For exactness, however, the concentration of such a substance should be decreased along with those that do contribute to the ionic strength, so that the total solute concentration becomes zero at the same time as the ionic strength.

Some typical data obtained by this method are shown in Table 2.11. The value of such data lies, of course, in the detailed insight they give into the basic principles underlying the ionization of organic compounds. The entropy values, for instance, provide a measure of the extent of interaction with the solvent. The products of dissociation of carboxylic acid (H^+ and $R-COO^-$) are much more highly solvated than the parent acid so that there is a large negative ΔS^0. This negative ΔS^0 increases in the dioxane–water mixtures, indicating that only water serves as solvation agent: its greater scarcity in the mixed solvents leads to the greater entropy change. In the dissociation of NH_4^+, however, the products (H^+ and NH_3) appear to be no more hydrated than is NH_4^+, since $\Delta S^0 \simeq 0$.

One of the most important dissociation constants determined exactly by the method here given or by similar exact methods is that of H_2O,

$$K_w = a_{H^+} a_{OH^-} / a_{H_2O}, \quad (2.93)$$

values for which are listed in Table 2.12. For solutions that contain a total concentration of all dissolved species (ions and molecules) less than about 0.1, it is usually sufficiently accurate to assume that $a_{H_2O} = 1$, as discussed

Table 2.11 Exact Thermodynamic Dissociation Data at 25°C [57]

	$pK_a = -\log K_a$	ΔH^0 (cal/mole)	ΔS^0 (cal/deg.-mole)
In Water			
HCOOH	3.752	−25.6	−27.253
CH_3COOH	4.756	−94.7	−22.079
C_6H_5COOH	4.201	96.6	−18.902
Citric acid, K_1	3.128	996.9	−10.969
Citric acid, K_2	4.761	582.6	−19.827
Citric acid, K_3	6.396	−803.5	−31.959
o-Phthalic acid, K_1	2.950	−637.0	−15.633
o-Phthalic acid, K_2	5.408	−495.9	−26.406
NH_4^+	9.245	12,477.3	−0.450
$CH_3NH_3^+$	10.624	13,088.1	−4.718
$(CH_3)_2NH_2^+$	10.774	11,848.7	−9.558
$(CH_3)_3NH^+$	9.800	8,792.6	−15.343
Acetic Acid in Mixed Solvents			
20% Dioxane	5.292	−50.5	−24.383
45% Dioxane	6.307	−445.3	−30.355
70% Dioxane	8.321	−613.3	−40.132
82% Dioxane	10.140	−1,345.7	−50.902
50% Glycerol	5.271	244.1	−23.306

earlier, so that

$$K_w = a_{H^+}a_{OH^-}. \tag{2.94}$$

Accurate values of K_w are necessary for a number of calculations. For instance, they may be used to compute the base dissociation constant of a base from the dissociation constant of its conjugate acid. For example, for NH_4OH (which is merely hydrated NH_3, that is, $a_{NH_4OH} \equiv a_{NH_3}$),

$$K_b = \frac{a_{NH_4^+}a_{OH^-}}{a_{NH_3}} = \frac{a_{NH_4^+}a_{OH^-}a_{H^+}}{a_{NH_3}a_{H^+}} = \frac{K_w}{K_a}. \tag{2.95}$$

It should be noted that in older literature the dissociation of bases such as NH_4OH was always described in terms of K_b. This is no longer the practice. The dissociation of bases is now nearly always described in terms of the K_a of the conjugate acid. The older literature also contains frequent references to "hydrolysis constants." For instance, the hydrolysis constant of acetate ion is the constant for the reaction $CH_3COO^- + H_2O \rightleftharpoons CH_3COOH + OH^-$

Table 2.12   The Ionization Constant of Water

Temperature (°C)	$K_w \times 10^{14}$	$pK_w = -\log K_w$
0	0.1139	14.943
5	0.1846	14.734
10	0.2920	14.535
15	0.4505	14.346
20	0.6809	14.167
25	1.008	13.996
30	1.469	13.833
35	2.089	13.680
40	2.919	13.535
45	4.018	13.396
50	5.474	13.262
55	7.297	13.137
60	9.614	13.017

This constant is clearly equivalent to K_b for the acetate ion:

$$K_b = \frac{a_{CH_3COOH}a_{OH^-}}{a_{CH_3COO^-}a_{H_2O}} = \frac{a_{CH_3COOH}a_{OH^-}a_{H^+}}{a_{CH_3COO^-}a_{H_2O}a_{H^+}} = \frac{K_w}{K_a}. \qquad (2.96)$$

Approximate Dissociation Constants

While the method of the preceding section must be used if it is desired to obtain precise thermodynamic data, it is for most purposes sufficient to determine acid dissociation constants by an approximate method, using the practical pH scale. The approximate constant for the general case differs from that defined by (2.69) in two ways: (1) the activity coefficients are absorbed into the constant, and (2) the approximation is made that pH = $-\log a_{H^+}$. Thus taking logarithms of (2.69),

$$pK' = -\log K' = pH - \log \frac{C_{A^{z-1}}}{C_{HA^z}} \qquad (2.97)$$

Because the activity coefficients $\gamma_{A^{z-1}}$ and γ_{HA^z} are incorporated in pK', its value depends on ionic strength. As long as the ionic strength is sufficiently low (0.1 for univalent ions, about 0.05 for bivalent ions, about 0.01 for trivalent ions) correction for these activity coefficients can be made by the Debye-Hückel theory. The pK' values so corrected should lie within a few hundredths of pK_a, the residual error arising from the uncertainty in the activity coefficient correction and in the assumption pH = $-\log a_{H^+}$. Uncorrected pK' values may vary from pK_a by 0.1 or 0.2, especially if HA^z or A^{z-1} is an ion of charge greater than unity.

The general procedure in determining approximate dissociation constants is to start with a given concentration of the acid form of an acid-base pair and to add increments of a strong base, or alternatively, to start with the basic form and to add increments of a strong acid. At each addition the pH is measured, usually using the method for relative pH measurements. It remains only to compute $C_{A^{z-1}}/C_{HA^z}$ in order to determine pK' by (2.97). This ratio is calculated from the amount of strong acid or base added. If the pH measured is in the range from pH 3 or 4 to pH 10 or 11, then the concentration of free H^+ or OH^- at the end of the addition is usually negligible. All of the titrant added must have been used in the conversion of HA^z to A^{z-1}, or vice versa. Thus C_{HA^z} and $C_{A^{z-1}}$ can be directly calculated from the amounts of reagents added and pK' is obtained at once from (2.97). This procedure is illustrated by Table 2.13. If, however, the pH of a solution is below pH 3 to 4 or above pH 10 to 11, then some of the added titrant has been used to create a concentration of free H^+ or OH^- in the solution, and a correction must be made before C_{HA^z} and $C_{A^{z-1}}$ can be calculated. This procedure is illustrated by Table 2.14. The corrections were made by using the factors of Table 2.15, relating pH to the concentration of free acid.

The method here described can be used to determine pK' values only between the limits of $pK' = 2$ to $pK' = 12$; however, if one is able to use concentrations of the organic acid in the vicinity of $1\ M$, the range can be extended to $pK' = 1$ and $pK' = 13$. The reason for this is that the free H^+ or OH^- concentration must always be smaller in the range of titration than the concentration of H^+ or OH^- used in the acid-base reaction. Otherwise, the major portion of the added titrant remains in the free form, and the correction for this amount cannot then be made with sufficient accuracy to give a reliable pK' value.

A plot of the data of Table 2.14 is shown in Fig. 2.17. The pK' value could have been obtained as the midpoint of this curve since $pK' = pH$ where $C_{HA^z} = C_{A^{z-1}}$.

In most instances curves such as that of Fig. 2.17 can be conveniently obtained directly from titration curves such as those of Fig. 2.14. This procedure has the advantage of requiring only a small amount of material. At the same time, the exact concentration of the substances being titrated need not be known since the end point of the titration can be used as a

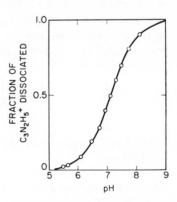

Fig. 2.17 The hydrogen ion dissociation curve of imidazolium ion. The pK' value obtained from the pH at the midpoint is 7.12.

Table 2.13 The Acid Dissociation Constant of Imidazole
at 25°C, Ionic Strength 0.15

$$\begin{array}{ccc} CH{=\!\!=}CH & & CH{=\!\!=}CH \\ | \quad\quad | & & | \quad\quad | \\ HN \quad\quad NH^+ & \rightleftharpoons & HN \quad\quad N \quad + H^+ \\ \diagdown C \diagup & & \diagdown C \diagup \\ | & & | \\ H & & H \end{array}$$

Equivalents KOH *Added* Per Equivalent $C_3N_2H_5^+NO_3^-$	*Relative Concentrations*			$pK' = pH - \log\dfrac{(C_3N_2H_4)}{(C_3N_2H_5^+)}$
	$C_3N_2H_5^+$	$C_3N_2H_4$	pH	
0.087	0.913	0.087	6.09	7.11
0.190	0.810	0.190	6.48	7.11
0.395	0.605	0.395	6.92	7.11
0.494	0.506	0.494	7.10	7.11
0.594	0.406	0.594	7.30	7.13
0.694	0.306	0.694	7.48	7.12
0.804	0.196	0.804	7.76	7.15

measure of this concentration. The correction for free acid or base can be determined directly by a blank titration, which records the amount of acid or base required to bring the solvent alone to any pH value.

For best accuracy the ionic strength should change little during the titration. This goal may be attained by suitable adjustment of the concentrations of the titrant and of the solution being titrated. Often salt is added to titrant or solution or both. The blank titration should be performed on pure solvent adjusted to the same ionic strength as prevails during the titration itself. If it is intended to obtain pK' values with an accuracy of only 0.1 to 0.2, then

Table 2.14 The Acid Dissociation Constant of Glycylglycine at
25°C, Ionic Strength 0.15

$$^+H_3N{-}R{-}COOH \rightleftharpoons H^+ + {}^+H_3N{-}R{-}COO^-$$

Initial conc. $^+H_3N{-}R{-}COO^-$	0.1000	0.1000	0.1000
Added conc. HNO_3	0.0286	0.0476	0.0762
pH	3.60	3.22	2.74
Corresponding free conc. HNO_3	0.0003	0.0007	0.0020
"Bound" HNO_3	0.02830	0.0469	0.0742
Final conc. $^+H_3N{-}R{-}COO^-$	0.0717	0.0531	0.0258
Final conc. $^+H_3N{-}R{-}COOH$	0.0283	0.0469	0.0742
pK'	3.20	3.17	3.20

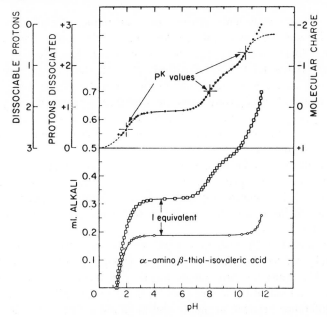

Fig. 2.18 The titration curve of α-amino-β-thiolisovaleric acid [58]. Sample titration: ○, blank titration to correct for base added in neutralizing free H⁺ (hydrogen ion dissociation curve); □, experimental titration curve; ●, corrected titration curve obtained by subtracting the blank curve from the experimental curve.

one need ordinarily not be concerned with the ionic strength. An example of the determination of pK' from titration curves is illustrated in Fig. 2.18 by the titration curve for α-amino-β-thiolisovaleric acid, which is a tribasic curve and thus a succession of three titration curves of the simple type so far discussed. The upper curve in this figure is the true hydrogen ion dissociation curve, resulting from subtracting the blank titration from the experimental titration curve. It should be noted in looking at Fig. 2.18 that there are numerous equivalent ways of labeling the ordinate, depending upon the reference point chosen.

Dissociation Constants of Organic Acids in Relation to Their Structure

The pK values of organic acids vary in a systematic manner with their structure. The pK values of numerous acids in water solution are well known, and a selection is given in Table 2.15. They may be used as an aid in qualitative organic analysis (in the following discussion the numbers in parentheses refer to Table 2.15).

Aliphatic saturated carboxylic acids (e.g., nos. 1 and 2) have a pK at 25°C near 4.8. Polar substituents in the α-position reduce the pK to somewhere between 2.5 and 4.0, depending on the polarity of the substituent (nos. 3, 5, 6, and 7). Polar substituents in the β-position are less effective: they may reduce pK to 4.0–4.5 (no. 4). Positively charged substituents reduce pK to about 2.0 if in the α-position (no. 8) and by a diminished amount as the distance of the charged group from the dissociating proton decreases (no. 9). Conversely, negatively charged substituents raise the pK because they attract rather than repel protons (no. 10). Unsaturation lowers the pK, but only by a few tenths (no. 11).

Similar effects can be observed in the aliphatic amines (nos. 12–18). It should be noted that primary, secondary, and tertiary amines show little difference and that neighboring carbonyl groups show an especially enhanced effect (no. 15) because of the possibility of forming resonance structures of the type $^-O-C=NH_2^+$.

Aromatic carboxylic acids have a pK near 4.0 (no. 19). Substituents on the ring usually reduce the pK by an amount depending not only on the nature of the substituent, but also on its position in the ring, because the latter affects the number of resonance possibilities (nos. 20–25).

The difference between aromatic and aliphatic amines (nos. 26 and 14) is much more pronounced than that between the corresponding carboxylic acids. In the case of alcohols, the difference is even greater: the acid dissociation constant of phenol (no. 28) exceeds that of methanol (no. 27) by about 2×10^7, presumably because of resonance stabilization of the aromatic RO^- ion. Electron-withdrawing substituents tend to increase even further the acidity of the aromatic hydroxyl group (nos. 29–35). It should be noted that *ortho* substitution is usually much more effective than either *meta* or *para* substitution. The nitrophenols (nos. 33–35) are an exception, because of stabilization of *o*-nitrophenol (but not the nitrophenolate ion) by hydrogen bonding.

Sulfhydryl groups are generally more acidic than the corresponding alcohols (nos. 36 and 37).

The acidity of heterocyclic substances is as yet incompletely understood [58]. The dissociation constants depend strongly on the degree of unsaturation, on the number of noncarbon atoms in the ring, and on their relative positions (nos. 38–42). Substituents have strong, specific effects, highly sensitive to the position of substitution (nos. 42–46).

Special mention should be made of guanidine (no. 47), the acid ion of which is strongly stabilized by resonance.

A full discussion of the theory of these and other effects of structure on acidity, or a complete listing of experimental pK' values is beyond the scope of this work. Two extensive source books on organic acid and base dissociation constants in aqueous media have been prepared through the International Union of Pure and Applied Chemistry [59, 60]. Relations between

Table 2.15 Selected pK Values in Aqueous Solution at 25°C*

Compound	pK Value	Compound	pK Value
1. CH_3COOH	4.76	24. $m\text{-HO—C}_6H_4\text{—COOH}$	4.08
2. $CH_3(CH_2)_2COOH$	4.82	25. $p\text{-HO—C}_6H_4\text{—COOH}$	4.54
3. $CH_3\text{—CH}_2Br\text{—COOH}$	3.0	26. $C_6H_5NH_3^+$	4.6
4. $CH_2Br\text{—CH}_2\text{—COOH}$	4.0	27. CH_3OH	17
5. $CH_2OH\text{—COOH}$	3.83	28. C_6H_5OH	9.78
6. $HOOC\text{—CH}_2\text{—COOH}$	2.86	29. $o\text{-Cl—C}_6H_4OH$	8.49
7. $CH_3\text{—CO—COOH}$	2.5	30. $p\text{-Cl—C}_6H_4OH$	9.18
8. $^+H_3N\text{—CH}_2\text{—COOH}$	2.35	31. $2,3\text{-Cl}_2\text{—C}_6H_3OH$	7.45
9. $^+H_3N\text{—(CH}_2)_3\text{—COOH}$	4.03	32. $2,4,6\text{-Cl}_3\text{—C}_6H_2OH$	6.42
10. $^-OOC\text{—(CH}_2)_2\text{—COOH}$	5.64	33. $o\text{-NO}_2\text{—C}_6H_4OH$	7.25
11. $C_2H_5\text{—CH}{=}\text{CH—COOH}$	4.7	34. $m\text{-NO}_2\text{—C}_6H_4OH$	8.28
12. $CH_3\text{—CH}_2\text{—NH}_3^+$	10.7	35. $p\text{-NO}_2\text{—C}_6H_4OH$	7.21
13. $(CH_3CH_2)_2NH_2^+$	11.1	36. CH_3SH	10.7
14. $(CH_3CH_2)_3NH^+$	10.8	37. C_6H_5SH	7.2
15. $CH_3\text{—CO—NH}_3^+$	−1	38. Pyrazole (—NH$^+$)	0.4
16. $^+H_3N\text{—(CH}_2)_2\text{—NH}_3^+$	7.0	39. Pyrazole (—NH$^+$)	2.5
17. $^-OOC\text{—CH}_2\text{—NH}_3^+$	9.78	40. Imidazole (—NH$^+$)	7.0
18. $^-OOC\text{—CH}_2\text{—NH—CO—CH}_2\text{—NH}_3$	8.25	41. Triazole (—NH$^+$)	2.6
19. $C_6H_5\text{—COOH}$	4.20	42. Quinoline (—NH$^+$)	4.9
20. $o\text{-Cl—C}_6H_4\text{—COOH}$	2.94	43. 2-OH-quinoline (—NH$^+$)	−0.3
21. $m\text{-Cl—C}_6H_4\text{—COOH}$	3.83	44. 3-OH-quinoline (—NH$^+$)	4.3
22. $p\text{-Cl—C}_6H_4\text{—COOH}$	3.98	45. 4-OH-quinoline (—NH$^+$)	2.3
23. $o\text{-HO—C}_6H_4\text{—COOH}$	2.98	46. 5-OH-quinoline (—NH$^+$)	5.2
		47. Guanidine ($={}$NH$_2^+$)	13.6

* The dissociation constants are those for the final proton in each formula.

acid-base properties and structure are discussed in chapters in books by Ingold [61], Braude and Nachod [62], Gould [63], Hine [64], Wiberg [65], and most recently by Liberles [66]. A review of the properties of weak bases by Arnett [67] has appeared. Profound discussions of structure-reactivity relations including acid-base properties can be found in works by Leffler and Grunwald [68] and Ritchie and Sagen [69].

In biochemistry and macromolecular chemistry acid-base equilibria with appropriate examples are considered by Edsall and Wyman [70], Tanford [71], and Morawetz [72].

Polybasic Acids

We have shown in Fig. 2.18 the titration curve of a polybasic acid in which the dissociating groups are so different that each successive step in the dissociation is well separated. The titration curves of such acids are essentially a succession of titration curves for monobasic acids. The situation, however, is not always as simple. Figure 2.19, for instance, shows the titration curve of

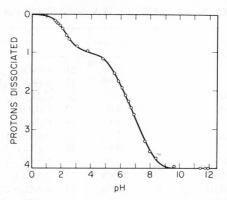

Fig. 2.19 The hydrogen ion dissociation curve of histidyl-histidine. [J. P. Greenstein, *J. Biol. Chem.*, **93**, 479 (1931).]

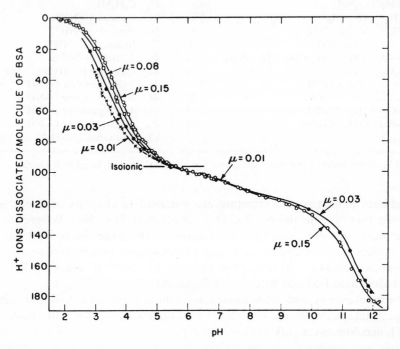

Fig. 2.20 The hydrogen ion dissociation curve of bovine serum albumin at several ionic strengths. [C. Tanford, S. A. Swanson, and W. S. Shore, *J. Am. Chem. Soc.*, **77**, 6414 (1955).]

histidyl-histidine, which has four dissociable groups. Three of these (two imidazole groups and an α-amino group) lie so close together in pK that they are titrated in the same pH range. Even more complex are titration curves of proteins, an example of which is shown in Fig. 2.20. The analysis of such titration curves to obtain meaningful pK values may become quite complicated [73].

It is worth mentioning that in these more complicated cases spectroscopy is often an invaluable aid to potentiometry. Many acidic groups have a visible or ultraviolet spectrum differing for the acidic and basic forms. The proton dissociation of such a group may thus be followed independently of other groups which are indistinguishable in a potentiometric titration. In principle, this procedure can be used for all types of acidic groups, employing infrared absorption. However, quantitative measure of absorption in the infrared is much less readily accomplished than in the visible or ultraviolet range of the spectrum.

Dissociation Constants in Nonaqueous or Mixed Solvents

Many organic substances are insoluble or only sparingly soluble in water. They can therefore be titrated only in mixed solvents or in pure organic solvents. This does not introduce any additional experimental difficulty in the determination of exact thermodynamic dissociation constants by hydrogen electrodes in cells without liquid junctions (provided that Cl^- is soluble in the solvent used); some examples of pK_a values in such solvents are cited in Table 2.11. The determination of approximate pK' values, using relative pH measurements with the glass electrode, does introduce new problems, as discussed under pH measurements in Section 4. If the pH can be measured, however, there is no further difficulty in determining the pK' [74]. This measurement, however, overlooks the detailed equilibria, that actually exist between an un-ionized form and a fully dissociated species in solvents of low dielectric constant. Whereas in methanol and ethanol the dissociation equilibrium of an acid is merely shifted in favor of the undissociated form by 5 pH units relative to water, in media of dielectric constant less than 13, extensive ionized but undissociated species exist. Ion pairing plays a very important role in describing the equilibrium state in solvents such as acetic acid, ethylenediamine, and pyridine. Below dielectric constants of about 5, ionization still occurs but only through formation of scarcely dissociated triple and quadruple ion aggregates. In acetic acid, for example, even a strong acid such as perchloric ionizes and dissociates according to the reaction:

$$HClO_4 + HOAc \underset{\text{ionization}}{\overset{}{\rightleftharpoons}} \underset{\text{ion pair}}{H_2OAc^+ ClO_4^-} \underset{\text{dissociation}}{\overset{}{\rightleftharpoons}} H_2OAc^+ + ClO_4^-$$

The overall dissociation constant is about 10^{-5}; however, ion pair formation is quite extensive. Considerable dissociation equilibrium data now exist for glacial acetic acid [46] and acetonitrile [75].

A nonaqueous solvent is classified as amphiprotic if it can act as a Brønsted acid-base pair. Amphiprotic solvents are rated relative to water in their proton-donating and -accepting properties. Amphiprotic solvents that are much stronger acids than water are called protogenic. In this class are hydrofluoric, sulfuric, formic, acetic, and trifluoroacetic acids. More basic solvents than water are protophilic, for example, liquid ammonia, butylamine and ethylenediamine. Hydroxylic solvents, water, methanol, ethanol, and other alcohols are designated as such. Their acid-base properties lie between protogenic-protophilic extremes.

Aprotic solvents are those with negligible acid properties and only slight basic character compared with water. These include nitriles, ketones, ethers, esters, amides, dimethyl sulfoxide, and pyridine. Benzene, hexane and other hydrocarbons, chloroform, carbon tetrachloride, and halogenated benzenes are also in this category.

The acid-base character of a solvent stems from its self-ionizing property described in its autoprotolysis constant

$$2HS \rightleftharpoons H_2S^+ + S^- \qquad (2.98)$$

$$K_s = a_{H_2S^+} a_{S^-} \qquad (2.99)$$

for unit activity solvent. A recent tabulation is included in Table 2.16. A

Table 2.16 Solvents for Acid-Base Reactions*

Substance	pK_S	Dielectric Constant
Amphiprotic, comparable to water		
Ethanol	19.1	24.3
Methanol	16.7	32.6
Water	14.0	78.5
Amphiprotic, protogenic		
Acetic acid	14.45	6.13
Formic acid	6.2	58.5
Hydrogen fluoride	>11.7 (0°C)	84 (0°C)
Sulfuric acid	3.6	100
Amphiprotic, protophilic		
Ammonia	27, 37.7 (−50°C)	16.9, 27 (−50°C)
Ethylenediamine	~13 ± 1	12.9
Aprotic		
Dimethyl sulfoxide	<20	46
Acetonitrile	>28	36.7
Methyl isobutyl ketone	>30	13.1

* Data refer to 25°C unless stated otherwise.

comprehensive discussion of aprotic solvents is given by Davis [76]. Some dissociation constants for acids and bases in glacial acetic acid, methanol, and ethanol are given in Tables 2.17 and 2.18. Other sources of nonaqueous equilibrium constant data and discussion are found in chapters by Ritchie [77], Prue [78], and Drago and Purcell [79].

The real problem in nonaqueous solvents lies in the fact that very few data are available in the literature. The pK values in solvents other than water are very different from those of Table 2.15, but a systematic study of the variations is lacking. In general, nonaqueous solvents of dielectric constant less than that of water shift the dissociation equilibria in favor of the *uncharged* forms. Thus the pK values of carboxyl groups are increased (Table 2.11); those of organic amines, however, are decreased. The lack of quantitative information on these changes means that pK values in nonaqueous solvents can be used for qualitative analysis at the present time only if parallel studies are made on model compounds.

Determination of Stability Constants of Metal Complexes

Organic compounds that combine with metal ions in solution are generally basic compounds which also combine with hydrogen ions. Examples are imidazole, glycinate ion, o-phenanthroline:

Each of these can combine with a hydrogen ion to form the conjugate acid, or with a metal ion to form a metal complex. Combination with metal is thus clearly a competition between metal and hydrogen ions for the same site, so that the extent of combination can be evaluated from the effect of the metal ion on the titration curve of the organic base. This effect is always to reduce the pH of a mixture of an acid HA^z and its conjugate base A^{z-1} below the value it would have in the absence of metal. The metal withdraws base from participation in the acid-base equilibrium so that dissociation of acid, with release of hydrogen ion, occurs until equilibrium, according to (2.91), is again established.

Usually a metal ion can combine with several molecules of the organic base so that for different initial compositions the average number $\bar{n}$ of complexed organic molecules or ions per metal ion may vary from zero to a maximum number N equal to the effective coordination number of the metal ion for the particular ligand being used. To evaluate $\bar{n}$ from a pH measurement requires only a simple material balance. Since the acid HA^z does not combine with metal ordinarily, its concentration is determined at once from the added

Table 2.17 Overall Dissociation Constants
in Glacial Acetic Acid at 25°C*

Compound	pK
Base	
Tribenzylamine	5.4
Pyridine	6.10
Dimethylaminoazobenzene	6.32
Ammonia	6.40
Potassium acetate	6.14
Sodium acetate	6.68
Lithium acetate	6.79
2,5-Dichloroaniline	9.48
Urea	10.24
Water	12.53
Acid	
Perchloric	4.87
Sulfuric	7.24
p-Toluenesulfonic	8.44
Hydrochloric	8.55
Salt	
Sodium perchlorate	5.48
Potassium chloride	6.88
Urea hydrochloride	6.96
Lithium chloride	7.08
pK_S = 14.45	

* From I. M. Kolthoff, E. B. Sandell, E. J.
Meehan, and S. Bruckenstein *Quantitative
Chemical Analysis*, Macmillan, New York, 1969.

Table 2.18 Dissociation Constants in Water, Methanol, and Ethanol*

Acid	pK_{H_2O}	pK_{CH_3OH}	p$K_{C_2H_5OH}$
Benzoic	4.18	10.4	9.7
o-Nitrobenzoic	2.21	8.4	7.4
m-Nitrobenzoic	3.47	9.2	8.3
p-Nitrobenzoic	3.40	8.9	8.1
Salicylic	2.97	8.7	7.9
Picric	0.80	3.8	3.8

* From I. M. Kolthoff, J. J. Lingane, and W. O. Larson, *J. Am. Chem. Soc.* **60,**
2512 (1938).

concentration corrected for the amount dissociated into A^{z-1}. This correction is determined from the free hydrogen ion concentration, and is most often negligibly small. Equation 2.91 now gives at once the ratio $C_{A^{z-1}}/C_{HA^z}$, hence the concentration of *free* A^{z-1}. All the organic base originally added and not accounted for as either free A^{z-1} or HA^z must be complexed. This amount divided by the metal ion concentration is, by definition, equal to $\bar{n}$.

The successive stability constants are defined as

$$k_1 = [MA]/[M]c$$
$$k_2 = [MA_2]/[MA]c \qquad\qquad (2.100)$$
$$\cdots$$
$$k_N = [MA_N]/[MA_{N-1}]c$$

where c is the concentration of free A^{z-1}, and $\bar{n}$ is related to them by the equation,

$$\bar{n} = \frac{k_1 c + 2k_1 k_2 c^2 + \cdots + N k_1 k_2 \cdots\cdots\cdots k_N c^N}{1 + k_1 c + k_1 k_2 c^2 + \cdots + k_1 k_2 \cdots\cdots k_N c^N} \qquad (2.101)$$

If $\bar{n}$ is determined over a sufficiently wide range of c (i.e., of pH, for c increases with increasing pH), then all of the stability constants can be evaluated by suitable mathematical procedures.

A detailed account of theory and measurements of stability constants can be found, for instance, in two classic books by Bjerrum and by Martell and Calvin [80, 81] and in more recent expositions [82, 83, 84].

There are two exhaustive sources of equilibrium (formation) constants for complex species. These are by Yatsimirskii and Vasilev [85] and by Sillen and Martell [86].

6 ELECTRODES FOR CATION AND ANION ACTIVITY (pM AND pA) MEASUREMENTS

Electrodes Based on Metal–Metal Ion Half Cells

Metallic electrodes whose half-cell potentials depend upon ionic activities are called "indicator electrodes." Their ability to function requires at least that they be part of an entire electrochemical cell, for example,

$$M;\ M^{n+}|\ \text{or}\ \|\text{ref.} \qquad\qquad (2.XIII)$$

where the measured emf is given by

$$\text{emf} = E(\text{ref}) - E(M) = -E^0_{M^{n+}/M} - \frac{RT}{nF}\ln a_{M^{n+}} + E_{\text{ref}} - E_{J(\text{ref}-M^{n+})}$$
$$(2.102)$$

The reference electrode could be saturated calomel, silver–silver chloride in normal chloride, or several other reliable references. In a junction cell configuration, an uncertainty is introduced in the measurement of $a_{M^{n+}}$ because of the junction potential. The latter can be minimized but not nullified by use of a salt bridge. The most reliable measurements are obtained from junctionless cells. However, this type of cell is restricted to measurements of solutions with an anion in common with the reference electrode. For example, a halide solution of M^{n+} can be measured in cells of the form

$$M; M^{n+}, nX^- \text{ or } n/2X^=; AgX \text{ or } Ag_2X; Ag \qquad (2.XIV)$$

where X is a halide, thiocyanate, sulfide, or any of the insoluble carboxylic acid salts. Another junctionless cell which can be used for metal ion couples in solutions of constant pH, constant sodium, or potassium activity employs a pH, pNa or pK-sensitive glass electrode as reference. Presumably, other specific electrodes could be used for references as well. The pH must not only be constant but adjusted to a value such that $E_{M^{n+}/M}$ is positive with respect to the hydrogen electrode at the given pH. This condition establishes thermodynamic stability of the metal electrode with respect to spontaneous chemical oxidation by H^+. It need not be obeyed if the electrode corrosion process is slow. However, the cell emf may not even then be stable or reflect accurately the activity of M^{n+}. One further type of cell is the "concentration" cell

$$M; M^{n+} \text{ at } C_1 \| M^{n+} \text{ at } C_2; M. \qquad (2.XV)$$

Here the emf (right-left) is

$$E = \frac{RT}{nF} \ln \frac{C_2 \gamma_2}{C_1 \gamma_1}. \qquad (2.103)$$

In each of these cells, the measured emf is related to an ill-defined ionic activity, not a concentration. Furthermore, there is the clear disadvantage compared with a potentiometric titration that the measurement is logarithmically related to activity rather than linearly related to concentration. Consequently, great precision is required to measure an apparent activity to better than 1 %. For the junctionless cells,

$$a_{M^{n+}} = C\gamma_{\pm} \qquad (2.104)$$

gives a relation between measured activity and concentration of a pure salt in terms of the tabulated mean activity coefficients which can be obtained by a variety of methods of which the "isopiestic" technique is preferred [87]. For the junction cells,

$$a_{M^{n+}} = C\gamma_? \qquad (2.105)$$

using Guggenheim's notation [88] where $\gamma_?$ requires a knowledge of both ionic interactions and mobilities. Mixtures with inert salts, which are generally present in any practical system, further complicate the relationship.

The problem of relating cell measurements to concentrations of responsive ions is virtually unsolvable in any reasonable theoretical model. Of course this view has been held for a long time by scholars in the field of hydrogen ion measurement (see Section 4). Measurement of metal ion activities, as with hydrogen ion activities, requires some standard procedure so that the results can have defined significance regardless of whether the desired transition from activity to concentration via (2.104) or (2.105) can be made. At the present time it seems most reasonable to establish pM^{n+} scales analogous to the NBS operational pH scale. Even if standards are defined (and progress has been made—see below) [89], researchers then must learn to think in terms of defined ionic activities without demanding a method to relate activities back to concentrations. The ionic activity, not the concentration, is crucial in chemical reactions.

Short of the goal of operational pM scales and development of pM standards, a useful technique employed extensively by Scandinavian solution chemists is activity calibration at high ionic strength. Solutions of known M^{n+} concentrations are prepared at high levels of an inert salt, typically 2 M KNO_3 or $NaClO_4$ and the potential-concentration plot prepared. The assumption is then made that the activity coefficient is chiefly determined by the inert salt and is therefore constant for all mixtures in the range of concentrations of M^{n+} that are much less than the inert salt. A Nernst slope passes through the points, at least at lower M^{n+} activities, and the concentration axis can be considered an activity axis with an arbitrary zero. Unknowns are measured using direct potentiometry after adding inert salt at the same level as the standards. This technique lends itself to the use of concentration cells. A single known M^{n+} solution in high inert salt is placed in one compartment. The unknown is treated with inert salt at the same level and the cell potential measured. The concentration is calculated directly from (2.103) because the activity coefficients are the same for M^{n+} in both solutions. Furthermore, the junction potential is virtually zero and can be ignored.

Some metallic electrodes are commercially available from Beckman Instruments and other manufacturers of glass and reference electrodes. Metallic electrodes are conveniently prepared from a $\frac{1}{4}$-in. diameter rod which is waxed into an oversized glass tube somewhat shorter than the rod. The lower end is cut off flush with the glass so that a flat surface is exposed to the solution. The upper end protrudes through the glass tube and is used to connect the electrode to its lead wire by means of an alligator clip, for example.

Classification of Metal–Metal Ion-Based Electrodes

The metal-metal ion electrodes designated as M/M^{n+} and used as examples in the foregoing section are called *electrodes of the first kind* because one interface is involved. Of course inert electrodes such as $Pt/Fe(CN)_6^{4-}$, $Fe(CN)_6^{3-}$ that involve one interface are legitimately electrodes of the first kind. Perhaps they should be considered electrodes of the zeroth kind since electrons but not ions cross the interface. Many metal–metal ion electrodes which might be suitable as selective electrodes are virtually eliminated from consideration for use in acidic air-saturated aqueous solutions. Since the metal must be thermodynamically stable with respect to air oxidation especially at low ion activities, E for the metal–metal ion couple over its range of usefulness should be greater than E for the O_2/H_2O couple. In neutral solutions, suitable electrodes are restricted to Hg_2^{2+}/Hg and Ag^+/Ag. If oxygen is removed from a cell by deaeration with nitrogen or helium, other electrodes become feasible: Cu^{2+}/Cu, Bi^{3+}/Bi, Pb^{2+}/Pb, Cd^{2+}/Cd, Sn^{2+}/Sn, Tl^+/Tl, and Zn^{2+}/Zn. Hydrolysis and complex formation may occur in a given solution. Each electrode in a proposed application must be considered in the light of possible interferences. In acidic solutions, Zn^{2+}/Zn, Mg^{2+}/Mg, Na^+/Na, and so on cannot be used. However, dilute amalgam electrodes that respond to Ca^{2+}, Na^+, and K^+ have been used in neutral and basic solutions for thermodynamic measurements. They are not to be considered practical electrodes.

Other metal–metal ion electrodes which may be readily identified by consulting Latimer's table of oxidation potentials [90], that is, Ni^{2+}/Ni, Co^{2+}/Co, Fe^{3+}/Fe, and elements of groups IV and V, are not useful as electrodes of the first kind. These electrodes suffer from irreversibility as a result of slow electron transfer, ion hydrolysis with formation of blocking surface films, or slow homogeneous chemical steps coupled with electron transfer. As long as the surface remains unblocked, even electrodes involving slow electron transfers respond thermodynamically, that is, they register the reversible potential provided two conditions are met: (1) the current passed by the measuring circuit is small compared with the exchange current and (2) sufficient time is allowed to achieve chemical equilibrium among all chemical steps involving production and dissipation of the simple ion.

Since electrodes M/M^{n+} that are reversible can be used to indicate free ion activities, it follows that they cannot indicate total "activity" of species M in the $n+$ state if some complexing agent (ligand) is present that consumes M^{n+}. This fact is both a blessing and a curse. For analytical purposes a total ionic activity can be measured only in a medium free from a complexing agent such that all metal ions are free, unhydrolyzed, solvated ions. However, a means for measuring free ion activity has allowed the potentiometric method to be widely applied to identification of complex species and determination of complex formation constants. These broad applications require a monograph for complete exposition; an example is Ref. 82. An excellent recent introduction to the chemistry of complex ion equilibria is Chapter 6 in Ref. 40. Electrodes of the first kind are subject to interference by oxidants which attack the metal electrode. In principle, all oxidants whose half-cells have more positive E^0 values than the electrode are potentially corrosive and sources of potential distortion.

Electrodes of the second kind involve two interfaces and are usually, but not exclusively, responsive to anion activities. By finding a suitable insoluble salt of a given anion, an electrode of the first kind can be converted to an electrode of the second kind. Examples include

$$\text{Ag; AgI; I}^- \tag{2.XVI}$$

$$E = E^0_{\text{Ag}^+/\text{Ag}} + \frac{RT}{F} \ln K^0_{\text{sp}} - \frac{RT}{F} \ln a_{\text{I}^-}$$

$$= E^0_{\text{AgI}/\text{Ag}} - \frac{RT}{F} \ln a_{\text{I}^-} \tag{2.106}$$

and

$$\text{Hg; Hg}_2\text{Cl}_2\text{; Cl}^- \tag{2.XVII}$$

$$E = E^0_{\text{Hg}_2^{2+}/\text{Hg}} + \frac{RT}{2F} \ln K^0_{\text{sp}} - \frac{RT}{F} \ln a_{\text{Cl}^-}$$

$$= E^0_{\text{Hg}_2\text{Cl}_2/\text{Hg}} - \frac{RT}{F} \ln a_{\text{Cl}^-}. \tag{2.107}$$

K^0_{sp} is the thermodynamic solubility product obtained by extrapolation to infinite dilution. Thus

$$E^0_{\text{AgI}/\text{Ag}} = 0.7994 + 0.05914 \log 10^{-16.08} \qquad \text{at } 25°C$$

$$= -0.168 \text{ V} \tag{2.108}$$

and

$$E^0_{\text{Hg}_2\text{Cl}_2/\text{Hg}} = 0.268 \text{ V} \qquad \text{at } 25°C. \tag{2.109}$$

Then for the iodide couple,

$$E = -0.168 - 0.0591 \log a_{\text{I}^-} \qquad \text{at } 25°C, \tag{2.110}$$

and for the calomel couple

$$E = 0.268 - 0.0591 \log a_{\text{Cl}^-} \qquad \text{at } 25°C. \tag{2.111}$$

Electrodes of the second kind exist for many anions, both inorganic and organic. The underlying electrode must be reversible, and consequently, mercurous, silver, lead, cadmium, and bismuth electrodes are used in conjunction with corresponding insoluble salts. A sketch for an $\text{Ag}/\text{Ag}_2\text{S}$ electrode is shown in Fig. 2.21. The insoluble salt should be in contact with

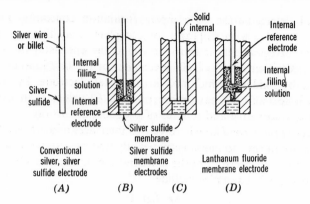

Fig. 2.21   Examples of the construction of conventional (*A*) and membrane-style selective electrodes (*B*, *C*, and *D*). *B* and *C* show the relatively thick pressed pellet membranes in a cylindrical inert plastic body. *D* shows the preferred high seal area with relatively small internal solution contact area configuration for thin crystal electrodes. From Light [104].

the metal so that the electrode metal can sense the correct free cation activity in equilibrium with the much higher bulk concentration of anion. A porous salt coating is prepared over an electrode wire or billet by anodizing the electrode in a solution of a soluble salt containing the precipitant anion. (See Section 3 and Refs. 33 and 91.) Here the assumption is made that the electrode operates, in fact, as an electrode of the first kind. This assumption must be true for electrodes coated with insoluble salts which are nonconductors of ions. Otherwise, the electrode would be simply out of contact with the solution, separated by an electrical insulator. However, it is known for silver salts that silver ions are mobile and, consequently, a silver wire with a uniform coating of AgCl (for example, one made by dipping the wire in molten salt) probably functions as a true electrode of the second kind. Note that there is a close relation here to membrane electrodes using silver halide films or pellets as discussed below.

In view of the assumption regarding mechanism, it is not surprising that anion activity measurements can be made using a metal electrode and the powdered insoluble salt simply added to the solution. For example, chloride activities can be measured by saturating a solution with AgCl powder using a silver wire as indicator and a suitable reference electrode separated by a halide-free salt bridge to avoid contamination. The critical feature of this short-cut method is that the solution must be vigorously stirred to insure that the region around the silver wire is truly saturated and representative of correct free silver ion activity.

Limitations on electrodes of the second kind are severe. First, they can be used only in a range of activity levels of anion such that the solution remains saturated. As a rule of thumb, the linear potential versus log activity plot according to (2.110) or (2.111) holds for anion activities greater than $\sim100(K_{sp}^0)^{1/2}$. Negative deviations from linearity appear at lower activities and the potential ultimately becomes independent of added a_{X^-} since the saturating salt produces a level of X^- exceeding that to be measured. As with electrodes of the first kind, solutions containing oxidants may attack the electrode metal or salt to create an unstable or unusable electrode potential. Interferences from other anions can occur if they too form an insoluble salt with the cation of the underlying electrode. Consider the reaction

$$Br^- + AgCl \rightleftarrows AgBr + Cl^- \tag{2.112}$$

$$K = \frac{a_{Cl^-}}{a_{Br^-}} = \frac{K_{sp}^0(AgCl)}{K_{sp}^0(AgBr)} = \frac{10^{-9.75}}{10^{-12.31}} = 10^{2.56}. \tag{2.113}$$

In using an Ag/AgCl electrode for a_{Cl^-}, bromide must be absent or less than $10^{-2.56}$ times the lowest anticipated a_{Cl^-} level to be measured. Otherwise, when $a_{Br^-} > 10^{-2.56}a_{Cl^-}$ the precipitate of AgCl metathesizes into AgBr and the electrode potential becomes independent of a_{Cl^-} at all lower activity levels. This problem is quite general and care must be taken to work with solutions in which potential interferences are absent or removed to a safe low-activity level.

Electrodes of the second kind, while representing a class of selective electrodes, are better known as reference electrodes. Preparation and operation of these electrodes in this context are covered extensively by Ives and Janz [91] and, more recently, by Covington [33].

A final class of electrodes, *electrodes of the third kind*, involving three interfaces have been known since the work of LeBlanc and Harnapp [92]. Electrodes responsive to Tl^+, Cu^{2+}, and Ca^{2+} can be made from the following combinations:

TlI/AgI/Ag	(2.XVIIIA)
CuS/HgS/Hg	(2.XVIIIB)
CaSO₄/PbSO₄/Pb or	
CaWO₄/Hg₂WO₄/Hg or	(2.XVIIIC)
CaC₂O₄/MC₂O₄/M	

where M = Zn, Cu, Pb, and Hg.

This author duplicated the results on the calcium-selective electrodes, but found them sluggish at best and virtually useless except in deaerated solutions.

The principle of exploiting an electrode of known reversibility to measure activities of ions for which no first kind electrode exists is a good one. To convert from one cation response to another requires three interfaces and two insoluble salts and leads to sluggish response. In two papers Reilley and co-workers [93] showed how to use the same principle with an electrode of the first kind:

$$Hg/HgY^{2-}, CdY^{2-}, Cd^{2+}$$

or

$$Hg/HgY^{2-}, MY^{+n-4}, M^{n+}$$ (2.XIX)

where $M^{n+} = Cd^{2+}, Sn^{2+}, Ba^{2+}, Mg^{2+}, Zn^{2+}, Cd^{2+}, Cu^{2+}, Pb^{2+}, Mn^{2+}, Co^{2+}, Ni^{2+}, VO^{2+}, Al^{3+}, Ga^{3+}, In^{3+}, Tl^{3+}, Cr^{3+}, Bi^{3+}, Sc^{3+}, Y^{3+}, La^{3+}, Ce^{3+}, Pr^{3+} Nd^{3+}, Sm^{3+}, Zr^{4+}$ and Hf^{4+} and H_4Y is ethylenediaminetetraacetic acid.

Solid-State Crystal and Pressed Pellet Electrodes

The commercial adaptation of low-resistance, permselective cast disk and pressed pellet membranes made from Ag_2S, $AgCl$, $AgBr$ (or $AgCl$-Ag_2S, $AgBr$-Ag_2S), AgI-Ag_2S, $AgSCN$-Ag_2S, Ag_2S-CuS, PbS, and CdS has provided chemists with new electrodes for the measurement of halide, thiocyanate, sulfide, cyanide, Ag^+, Cu^{2+}, Cd^{2+}, and Pb^{2+} activities. These electrodes are constructed in cylindric form from nonconducting plastic with the solid membrane sealed at the bottom as shown in Fig. 2.21b and d. The membrane must provide the only electrical continuity between the inner filling solution and the outer "test" solution. The seal is critical and while a small leak may be tolerable if it remains permselective, the time constant for response of the electrode is lengthened. Temperature cycling of the electrodes is permissible, but care should be taken to avoid thermal shocks.

These electrodes contain a filling solution or a gel which is initially saturated with the membrane material and may contain either added $AgNO_3$, a soluble salt of the membrane anion, or an inert salt whose anion neither precipitates Ag^+ nor forms a less soluble silver salt than the membrane. Reversible equilibrium of Ag^+ at the inner surface determines the internal interface potential. The metal wire (inner reference) is silver and may be coated with the membrane material although this step is not necessary. If the membrane has some electronic conductivity, the inner filling solution and/or inner reference may be omitted. Although the details are not yet clear, it appears that a silver salt pellet can function as a membrane if the silver wire is connected directly to the inside surface of the membrane (Fig. 2.21c). It is known that glass membranes function when contact to the inside of the bulb is made through mercury or through AgCl solidified in place from a melt. Perhaps only ionic equilibrium at these solid-solid interfaces is required. Definitive experiments on these unusual interfaces have not been made. Although AgCl and AgBr form pressed pellets from powders, AgI tends to crumble when the pressure on the pellet is released. Since soft Ag_2S is more insoluble than any of the other membrane materials, it may be incorporated and serves as a binder. Mixed pellets CuS, CdS and PbS with Ag_2S produce electrodes with response to activities of Cu^{2+}, Cd^{2+}, and Pb^{2+} ions through the typical equilibrium

$$Cd^{2+} + Ag_2S \rightleftharpoons CdS + 2Ag^+ \tag{2.114}$$

Since equilibrium is rapidly achieved with these reversible sulfides, the electrode response to Ag^+ is a measure of the divalent activities.

Solid-state electrodes are handled in the same way and with the same care as glass electrodes. They are lowered into a stirred solution to be measured and the circuit completed with an external reference electrode. While the saturated calomel electrode in its own salt bridge is commercially available and is useful for glass electrodes and LaF_3 electrodes, halide-free or double-junction references are preferred for the silver salt membranes. Several half-cells including mercury–mercurous sulfate or mercurous carboxylate salts are suitable [33]. The oxide or hydroxide-containing references are not useful in neutral or acid solutions. The so-called double-junction references which are commercially available are generally useful although they compromise freedom from halide with an added junction potential uncertainty. When a saturated calomel is used, KNO_3 solution fills the outer bridge so that only NO_3^- contaminates the sample. Unsaturated references such as Ag; AgCl; 0.1 M NaCl $|$ NaClO$_4$ $|$ permit only ClO_4^- contamination.

When using sulfide, iodide, and fluoride electrodes, one need not worry about dissolving away the membrane material in neutral or mildly acidic solutions. For chloride and bromide electrodes, it is recommended that powdered AgCl or AgBr be added to the test solution to make sure it is saturated with the membrane material prior to inserting the selective electrode.

When the potentiometer, pH meter, or other voltmeter is attached to the membrane cell, the selective electrode should be connected to the positive terminal and the external reference electrode to the negative. The potential measured is that of the lead to the inner reference minus that of the lead to the external reference electrode. The measured potentials for several electrodes can be expressed:

$$E \text{ (measured)} = \Delta E \text{ (membrane)} + E \text{ (inner ref.)} - E \text{ (ext. ref.)} + \sum E_J, \tag{2.115}$$

where the sense of the junction potentials E_J is the solution potential nearest the inner terminal less the adjoining solution. Since

$$\Delta E_{\text{membrane}} = \frac{RT}{F} \ln \frac{a_{Ag^+}(\text{test soln})}{a_{Ag^+}(\text{inner soln})}, \tag{2.116}$$

E (saturated calomel electrode) $= +0.242$ V, and

$$E_{Ag^+/Ag} = +0.799 + 0.059 \log a_{Ag^+}, \tag{2.117}$$

potentials for the following cell examples can be readily computed.

Example I:

$$\left|\begin{array}{cc} \text{salt} & \text{test} \\ \text{bridge} & \text{soln} \end{array}\right| \quad \longleftarrow\text{selective electrode}\longrightarrow$$

$$\text{saturated calomel} \mid KNO_3 \mid Ag^+NO_3^-; Ag_2S; \text{inner soln}; Ag_2S; Ag \quad (2.XX)$$
$$\text{ext. ref.} \qquad 1 \qquad 2 \qquad\qquad\qquad \text{int. ref.}$$

$$E \text{ (measured } 25°) = 0.0591 \log a_{Ag^+} + 0.799 - 0.242 + E_{J1} + E_{J2} \text{ V.} \tag{2.118}$$

Example II: Same cell with test solution containing sulfide at $a_{S^{2-}}$

$$E(\text{measured } 25°) = -\frac{0.0591}{2} \log a_{S^{2-}}$$

$$+ \frac{0.0591}{2} \log K_{sp}^0(Ag_2S) + 0.557 + E_{J1} + E_{J2} \text{ V.} \tag{2.119}$$

The term $0.0591/2 \log K_{sp}^0$ has the value -1.45 V since $-\log K_{sp}^0$ for Ag_2S is 49.2. Experimental tests of (2.118) and (2.119) have been made by Durst [94] with the result shown in Fig. 2.22.

Example III:

$$\text{satd calomel} \mid KNO_3 \mid \text{halide soln}; AgX; \text{inner soln}; AgX; Ag \quad (2.XXI)$$
$$\text{ext. ref.} \qquad 1 \qquad 2 \quad a_{X^-} \qquad\qquad\qquad \text{int. ref.}$$

$$E(\text{measured } 25°) = -0.0591 \log a_{X^-} + 0.0591 \log K_{sp}^0(AgX)$$

$$+ 0.557 + E_{J1} + E_{J2} \text{ V.} \quad (2.120)$$

where $-\log K_{sp}^0$ has the value 9.75, 12.31, and 16.08 for AgCl, AgBr and AgI, respectively.

Example IV:

$$\begin{array}{cc} \text{satd} & \mid KNO_3 \mid \text{cadmium}; Ag_2S\text{-CdS}; \text{inner soln}; Ag_2S; Ag \quad (2.XXII) \\ \text{calomel ref.} & \text{soln } a_{Cd^{2+}} \end{array}$$
$$\qquad 1 \qquad\qquad 2$$

$$E(\text{measured at } 25°) = \frac{0.0591}{2} \log a_{Cd^{2+}} + \frac{0.0591}{2} \log \frac{K_{sp}^0(Ag_2S)}{K_{sp}^0(CdS)}$$

$$+ 0.557 + E_{J1} + E_{J2} \text{ V.} \quad (2.121)$$

The K_{sp}^0 term has the value -0.68 V using $K_{sp\ CdS}^0 = 10^{-26.1}$. These equations contain unknown parameters and merely serve as guides for directional changes in potential. Electrodes must be standardized against solutions of ionic strength comparable with unknowns. E_{J1} and E_{J2} apply at junctions 1 and 2.

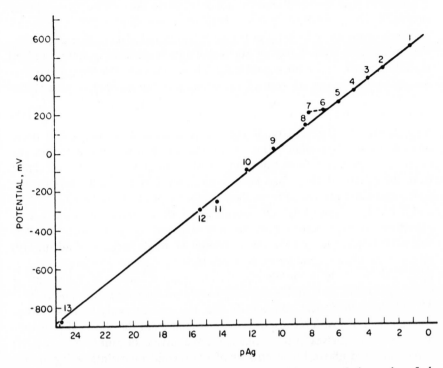

Fig. 2.22 Silver sulfide electrode response for silver activity test solution vol. $= 5 \mu l$.

(Solid line: 59.2 mV/pAg)

Point	Solution Composition	~pAg (calc)	E(mV)
1	$10^{-1}\,M\,AgNO_3$	1.1	+550
2	$10^{-3}\,M\,AgNO_3$	3	+438
3	$10^{-4}\,M\,AgNO_3$	4	+385
4	$10^{-5}\,M\,AgNO_3$	5	+323
5	$10^{-6}\,M\,AgNO_3$	6	+260
6	$10^{-7}\,M\,AgNO_3$	7	+225
7	$10^{-8}\,M\,AgNO_3$	8	+213
8	Sat'd AgI	8.2	+150
9	Sat'd AgI + $10^{-6}\,M$ KI	10.3	+21
10	Sat'd AgI + $10^{-4}\,M$ KI	12.3	−91
11	Sat'd AgCl + $1\,M$ $Na_2S_2O_3$	14.2	−256
12	Sat'd AgCl + $0.1\,M$ KI	15.5	−298
13	$0.1\,M\,Na_2S$ + $1\,M$ NaOH	24.9	−872

From Durst [3].

Limitations on the conditions for use of the silver salt solid-state membrane electrodes are generally less severe than when the same salt is used as an electrode of the second kind. For the silver halide and silver sulfide electrodes, the limiting low level of Ag^+ or anion that can be detected is determined by the solubility of the membrane material. These limits are determined approximately by taking the νth root of K_{sp}^0 where ν is the number of ions in the salt. The limits are $\sim 10^{-5}\,M\,Ag^+$ or Cl^-, $10^{-6}\,M\,Ag^+$ or Br^-, $10^{-8}\,M\,Ag^+$ or I^-, $10^{-17}\,M\,Ag^+$ or S^{2-} for AgCl, AgBr, AgI, and Ag_2S electrodes, respectively. As the activity of the ion is decreased to about 100 times limiting value, Nernstian points are obtained. Below these activities the potential responses curve away from the theoretical Nernst curve calculated from the equations above and become constant. Limiting activities of Pb^{2+}, Cu^{2+}, and Cd^{2+} are found from their solubility products to be 10^{-13}, 10^{-17}, and $10^{-13}\,M$, respectively. Of course all of these lower limits on cation activities are hypothetical since reliable standard solutions with activities below $10^{-7}\,M$ for testing the theory cannot be made because of adsorption of the ions on all conducting surfaces including glass. However, tests of limits can be made in mobile equilibrium systems such as acidified sulfide solutions. In this case sulfide activities can be produced down to about $10^{-19}\,M$ and the sulfide electrode does in fact respond correctly at these levels. Another test of low-level responses is provided from the shape of potentiometric titration curves in the vicinity of the equivalence point.

As discussed above for electrodes of the second kind, interfering anions are those that metathesize the membrane material into a less soluble salt. AgCl membranes are therefore sensitive to interference by a larger number of anions than AgBr, and so on. As a general rule, uninegative interfering anions with a solubility product K_{sp}^0 (interference) less than that of the membrane material must be removed to activity levels below the lowest activity of the ion being determined multiplied by K_{sp}^0 (interferences)/K_{sp}^0 (membrane). For example iodide and bromide must be reduced to levels less than $\frac{1}{2} \times 10^6$ and $\frac{1}{400}$ times the lowest chloride activity anticipated when using an AgCl membrane. The maximum tolerable sulfide is given by a slightly different function

$$a_{S^{2-}} = 10^{-49.2}(a_{X^-})^2/(K_{sp}^0)^2 \qquad \text{for AgX} \qquad (2.122)$$

where a_{X^-} is the lowest level anticipated in the experiment. Complex ions CN^- and $S_2O_3^{2-}$ also interfere with halide electrodes because of dissolution of the membrane with the formation of soluble silver complexes. Solutions of strong reducing agents known to attack silver halides destroy the effectiveness of the membrane electrodes. These are typically photographic developers in basic media. Strong oxidants in acidic media such as MnO_4^-, $Cr_2O_7^{2-}$, and Ce(IV) may slowly attack the anions of the membrane material and distort the equilibrium potential. There is little documentation on this point and users should study their systems before relying on the measurements. Other details are discussed by Ross [95].

By far the most welcome of the solid-state electrodes is that selective to fluoride ion constructed by Frant [13]. The membrane is a single crystal of LaF_3 which may be used in pure form or doped with various divalent ions. The crystal is sealed to the bottom of a plastic barrel of cylindrical shape (Fig. 2.21). The seal is critical and any partial cracks either in the crystal itself or in the seal lead to decreased speed of response. Other trivalent rare earth fluorides behave similarly. LaF_3 is an anionic conductor with a specific conductivity of $\sim 10^{-7}$ ohms^{-1} cm^{-1}. Typical membranes have a dc resistance using F^- contacts of 10^5 to 10^6 Ω. Fluoride mobility has been demonstrated [96] by temperature dependence of X-ray-determined lattice parameters, narrowing of the F^{19} NMR line over a range of temperatures and inferred from conductivity vs temperature measurements using blocked electrodes.

One commercially available electrode uses an inner filling solution 0.1 M NaF and 0.1 M NaCl with a Ag/AgCl inner reference. For a junction cell with a saturated calomel external reference of the form

$$\text{satd calomel} \mid \text{soln } a_{F^-}; \underset{\substack{\text{test} \\ \text{membrane}}}{\text{LaF}_3} \quad \underset{0.1 \, M \, \text{NaCl; AgCl; Ag}}{0.1 \, M \, \text{NaF}} \qquad \text{(2.XXIII)}$$

$$E(\text{measured at } 25°C) = -0.0591 \log \frac{a_{F^-}}{0.1\gamma_{F^-}} + 0.222$$

$$- 0.0591 \log (0.1\gamma_{Cl^-}) - 0.242 + E_{J1}. \quad \text{(2.123)}$$

Since there are quantities in this equation that cannot in general be calculated, the electrode must be calibrated. Experimentally, a Nernstian response of this form is observed from saturated fluoride down to about 10^{-6} M. Response times at the lowest levels are 1 to 3 min but are much shorter at the higher activities. Some short-term drift has been observed. Butler's review [97] should be consulted.

The quality of response at low-activity levels of fluoride makes this electrode superior to electrodes of the second kind based on aluminum or bismuth proposed earlier. The apparent solubility (activity) product of LaF_3 implied by the potentiometric response is about 5 powers of 10 smaller than that computed from the titration [98] curve of fluoride with lanthanum, $K_{sp}^0 = 3 \times 10^{-19}$. This anomaly may reflect a thermodynamically sound particle size effect since LaF_3 appears to be a solid with a high surface tension. However, the effect should disappear as the particles are allowed to "ripen." Covington [99] thinks the precipitate may be a hydrated form of LaF_3 with a unique solubility product. The electrode is remarkably free from interferences. Phosphate, acetate, halides, nitrate, sulfate, and bicarbonate are

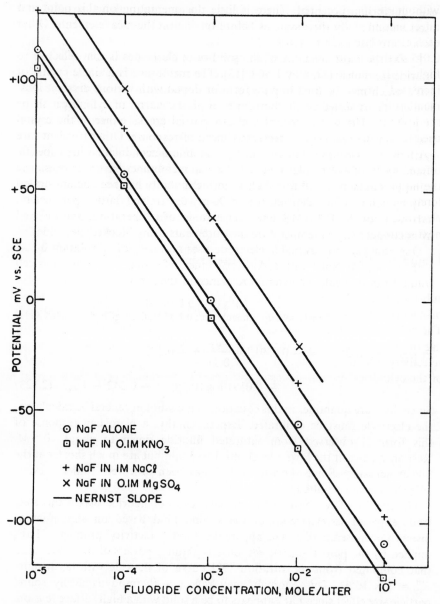

Fig. 2.23 Potential of LaF_3 membrane electrode as a function of fluoride concentration in various supporting electrolytes [97].

without specific effect. These ions influence the ionic strength and so indirectly affect measured E through γ_{F^-}. Calibration of the electrode must be done in each electrolyte as shown in Fig. 2.23. Hydroxide shows a $K^{pot}_{F^-/OH^-}$ of about 0.015 so that at $10^{-6}\ M\ F^-$, the pH must not be above 8 and correspondingly higher pH values (12 at 0.1 $M\ F^-$) are tolerable as the fluoride level is increased. Strong complexing agents for lanthanum, which dissolve the membrane, are expected to interfere by shortening the range of response at the low activities of fluoride. One example is citrate. The electrode may be used in nonaqueous media, but see Butler's comments [97].

A new and convenient solid electrode responsive to calcium activities is offered by Beckman Instruments. The response characteristics are similar to the liquid electrode offered by Orion with the exception that sodium interference on the solid electrode is greater while the pH range of the solid electrode is greater than for the liquid electrode. The latter is discussed in the next section.

Liquid Ion-Exchange Membrane Electrodes

Liquid ion-exchange systems for cations are salts of acidic, hydrophobic substances, mixed with solvent and held in the form of a membrane. This problem has been solved in two ways by Orion Research as shown in Fig. 2.24. The bulk or pool of liquid ion exchanger is made in a chemical form containing the ion for which the electrode is intended with whatever diluents and selectivity modifiers are chosen. This pool, contained within the outer part of the cylindrical barrel, saturates a support membrane held in place across

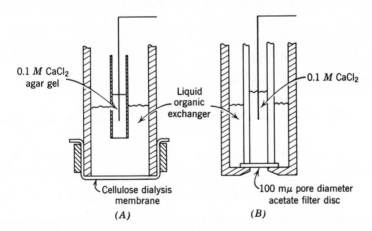

Fig. 2.24 Liquid membrane electrodes. (*A*) A simple configuration suitable for work where high resistance and long time response are not important. (*B*) A thin-layer configuration with low resistance and fast time response [95].

the lower opening. Suitable inert membrane materials include Millipore filter (cellulose acetate), dialysis membrane, or even porous glass. Surface tension, wick action, keeps the liquid in the membrane. Filling the inner part of the barrel and exposed to the internal face of the membrane is a solution with Ca^{2+}, in this case common to the membrane and anion, Cl^-, reversible at the $Ag/AgCl$ inner reference.

Liquid membrane electrodes are more sensitive to their solution environment than are solid-state electrodes. The temperature must be kept within limits specified by the manufacturer so that neither water permeates the membrane nor membrane liquids bleed into the aqueous solution. Mixed aqueous-nonaqueous media that do not dissolve the membrane can be acceptable systems for measurement. However, it seems best to restrict measurements to aqueous media. Otherwise, liquid membranes are handled much as glass or solid-state electrodes are. They should be carefully placed in the solution to be studied and the solution stirred. The external reference electrode is not critical for the cation-sensitive liquid membrane electrodes and any of the commerical examples are suitable. The anion-selective electrodes for Cl^-, ClO_4^-, NO_3^-, and BF_4^- activities must be used with double-junction external reference electrodes so that neither interfering ions nor ions of the kind being determined come in contact with the test solution. For the intermediate bridge, an inert salt solution containing ClO_4^- is preferred for all save the perchlorate electrode. Chloride is not a serious interference for the ClO_4^-, NO_3^-, and BF_4^- electrodes.

The calcium membrane electrode in a test solution containing calcium at $a_{Ca^{2+}}$ when completed with a saturated calomel reference has the form

$$| \longleftarrow\!\!\!-\!\!\!-\!\!\!-\text{Ca electrode}\longrightarrow |$$
$$\text{Hg; Hg}_2\text{Cl}_2\text{; KCl} \,|\, \text{test soln} \,|\, \text{liquid membrane} \,|\, \text{CaCl}_2 \;0.1\; M\text{; AgCl; Ag}$$
$$\text{satd} \qquad a_{Ca^{2+}} \qquad\qquad\qquad\qquad\qquad\qquad\qquad (\text{2.XXIV})$$

The calcium electrode potential is

$$E_{Ca^{2+}} = \frac{RT}{2F} \ln a_{Ca^{2+}}(\text{test}) - \frac{RT}{2F} \ln (0.1)(0.2)^2 \gamma_{\pm}^{3} + .222 \quad (2.124)$$

and the cell potential with the calcium electrode connected to the positive terminal is

$$E_{cell} = E_{Ca^{2+}} - E_C + E_J = E_{Ca^{2+}} - 0.242 + E_J. \quad (2.125)$$

A perchlorate-sensitive electrode formulated in a cell according to

$$| \longleftarrow\!\!\!-\text{perchlorate electrode}\longrightarrow |$$
$$\text{Hg; Hg}_2\text{Cl}_2\text{; KCl}_{satd} \,|\, \text{test soln} \,|\, \text{liquid membrane} \,|\, \text{AgClO}_4\text{; Ag} \quad (\text{2.XXV})$$
$$a_{ClO_4^-}$$

has the potential

$$E_{ClO_4^-} = -\frac{RT}{F} \ln a_{ClO_4^-}(\text{test}) + \frac{RT}{F} \ln a_{Ag^+}a_{ClO_4^-}(\text{inner}) + 0.799 \text{ V.}$$

(2.126)

The cell potential is

$$E_{cell} = E_{ClO_4^-} - E_C + E_J.$$

(2.127)

Limitations on the usefulness of these electrodes is imposed by the solubilities of the membrane components. As discussed in connection with solid state electrodes, use of more and more dilute solutions of the ion to which the electrode is selective produces a series of Nernstian points until the activity of the solution is within a factor of 100 of the solubility of the membrane. Positive or negative deviations occur until at the limit an inert salt solution saturated with the membrane material yields a constant potential reflecting the solubility of the membrane salt. This solubility effect is illustrated in Figs. 2.25 and 2.26.

The theory of response of liquid membranes is formally similar to the glass electrode or other fixed site cases. In the steady state the flux of mobile sites is zero so that the distribution in space becomes fixed in terms of external

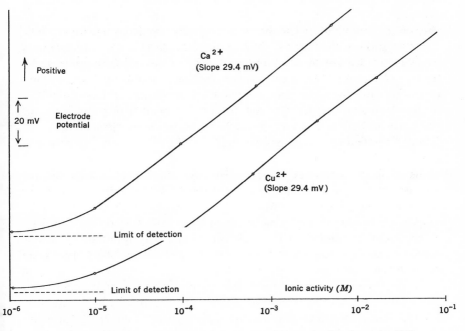

Fig. 2.25 Calibration curves for cupric and calcium ion-selective electrodes [95].

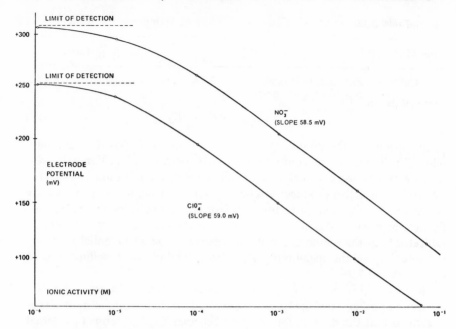

Fig. 2.26   Calibration curves for nitrate and perchlorate ion-selective electrodes [95].

activities of counter ions. The theory rests chiefly on charge balance and ion-exchange surface equilibrium. Other ions such as H^+, Na^+, and various divalent ions are bound to show interference in the cation-sensitive membranes. The extent of interference is related to their stabilities in the membrane as reflected by their ion-exchange equilibrium constants and their mobilities in the membrane. The form of the membrane potential response equation was given in Section 2. Approximate values of the selectivities are given in Table 2.19 and an actual demonstration of the effect of Na^+ on Ca^{2+} response is illustrated in Fig. 2.27.

One of the consequences of these selectivity data is that the electrodes are occasionally more sensitive to the interference than to the ion for which electrode was intended. Thus by soaking a calcium electrode in Zn^{2+} or Fe^{2+} solution, or better, by preparing the membrane with Zn^{2+} or Fe^{2+} in place of Ca^{2+}, an electrode for Zn^{2+} or Fe^{2+} is obtained. This procedure probably works for other divalent ions such as Ba^{2+}. However, a Nernstian response to Ba^{2+} could not be expected in excess of Ca^{2+}.

The time response of liquid membranes is about 30 sec in pure solutions or longer in mixtures. This is not surprising, but the precise explanation is lacking. The steady-state zero-current theory assumes ion-exchange equilibrium at the "surface." However, since the membrane already contains the

Table 2.19 Selectivity Constants for Liquid Membrane Electrodes*

Ion Measured	Selectivity Constants	Ion Measured	Selectivity Constants
Ca^{2+}	Zn^{2+} 3.2, Fe^{2+} 0.80, Pb^{2+} 0.63, Cu^{2+} 0.27, Ni^{2+} 0.08, Sr^{2+} 0.02, Mg^{2+} 0.01, Ba^{2+} 0.01, H^+ 10^7, Na^+ 0.0016	NO_3^-	Cl^- 6×10^{-3}, OAc^- 6×10^{-3}, $S_2O_3^{2-}$ 6×10^{-3}, SO_3^{2-} 6×10^{-3}, F^- 9×10^{-4}, SO_4^{2-} 6×10^{-4}, $H_2PO_4^-$ 3×10^{-4}, PO_4^{3-} 3×10^{-4}, HPO_4^{2-} 8×10^{-5}
Ca^{2+} and Mg^{2+}	Zn^{2+} 3.5, Fe^{2+} 3.5, Cu^{2+} 3.1, Ni^{2+} 1.35, (Ca^{2+} 1.0), (Mg^{2+} 1.0), Ba^{2+} 0.94, Sr^{2+} 0.54, Na^+ 0.01	ClO_4^-	OH^- 1.0, I^- 1.2×10^{-2}, F^- 2.5×10^{-4}, NO_3^- 1.5×10^{-3}, Cl^- 2.2×10^{-4}, Br^- 5.6×10^{-4}, SO_4^{2-} 1.6×10^{-4}, OAc^- 5.1×10^{-4}
Cu^{2+}	Na^+ 5×10^{-4}, K^+ 5×10^{-4}, Mg^{2+} 1×10^{-3}, Ca^{2+} 2×10^{-3}, Sr^{2+} 1×10^{-3}, Ba^{2+} 1×10^{-2}, Zn^{2+} 3×10^{-2}, Ni^{2+} 1×10^{-2}, H^+ 1×10^1, Fe^{2+} 1.4×10^2	Cl^-	ClO_4^- 32, I^- 17, NO_3^- 4.2, Br^- 1.6, OH^- 1.0, OAc^- 0.32, HCO_3^- 0.19, SO_4^{2-} 0.14, F^- 0.10
Pb^{2+}	Cu^{2+} 2.6, Fe^{2+} 0.08, Zn^{2+} 0.003, Ca^{2+} 0.005, Ni^{2+} 0.007, Mg^{2+} 0.008	BF_4^-	OH^- 10^{-3}, I^- 20, NO_3^- 0.1, Br^- 4×10^{-2}, OAc^- 4×10^{-3}, HCO_3^- 4×10^{-3}, F^- 10^{-3}, Cl^- 10^{-3}, SO_4^{2-} 10^{-3}
NO_3^-	ClO_4^- 10^3, I^- 20, ClO_3^- 2, Br^- 0.9, S^{2-} 0.57, NO_2^- 6×10^{-2}, CN^- 2×10^{-2}, HCO_3^- 2×10^{-2}		

* Ref 95

same ion as that being measured, the rate of establishment of the interfacial potential is limited only by solution mixing. The response time might be expected to be very short as observed with solid electrodes. However, Nernstian potential establishment *also* requires that the internal membrane diffusion potential reach its steady-state value via adjustment in the liquid ion-exchanger concentration profile. This diffusion-controlled process probably accounts for the relatively slow response compared with fixed-site membranes where steady-state concentration profiles need not be established. In the presence of interfering ions, the ion-exchange process must establish

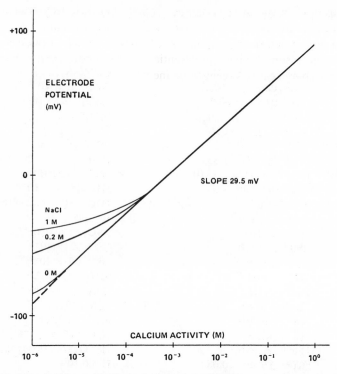

Fig. 2.27   Effect of Na^+ in calcium ion activity measurements. $CaCl_2$ varied in the presence of the indicated background molarity of NaCl [95].

new ion activities in the "surface." This process is not necessarily instantaneous inasmuch as the volume of the surface region surely is finite and some time must be required for equilibration.

The knotty problems in this field are the prediction of selectivities for a given ion exchanger-additive-solvent matrix or, conversely, the prediction of a suitable ion exchanger for creation of a selective electrode for a specific ion. On the average, an arbitrary ion exchanger is nonselective except on the basis of charge type. Liquid membranes are known that show general responses to most organic anions, for example [100]. Creation of specificity in a liquid ion-exchange system is probably more feasible than in a solid synthetic resin membrane system because of the flexibility in additives and solvents. A key concept is the formation of weakly bonded but stabilized structures which are preferred in the membrane phase.

Neutral extraction membranes, described briefly in Section 1, are low dielectric liquids, typically decane containing 10^{-4} to 10^{-7} M extractant: valinomycin, nonactin, or monactin. Some slightly polar solvents can be used, for example, octanol or a polar additive may be included in the mixture. Three electrodes available commercially are made by Beckman Instruments (solid and liquid membrane versions) and Orion Research. These electrodes respond to potassium activity preferentially over sodium and other alkali ions. Higher-charge cations seem to be inert. Valinomycin membranes show great potassium sensitivity, $K^{pot}_{K^+/Na^+} \sim 10^{-3}$, while the actin-based membranes show $K^{pot}_{K^+/Na^+} \sim 10^{-2}$. Both of these values are remarkably better than has been observed with potassium-sensitive glass electrode compositions. The actin electrodes are about four times more responsive to NH_4^+ than to K^+ so that the ammonium level should be reduced where possible by pH adjustment, using $Ba(OH)_2$ and boiling. Hydrogen ions seem to interfere and their concentration should be kept a factor of 10 less than the lowest potassium level to be measured, for example, pH 5. Potassium response is Nernstian from saturation to about 10^{-4} M for cells of the form

$$\text{Hg; } Hg_2Cl_2\text{; } \underset{\text{satd}}{KCl} \mid \text{test soln; membrane; } \underset{KHCO_3}{KCl} \text{ ; AgCl; Ag} \quad (2.\text{XXVI})$$
$$\underset{a_{K^+}}{}$$

where

$$E(\text{Ag} - \text{Hg}) = \frac{RT}{F} \ln a_{K^+} + \text{constant.} \quad (2.128)$$

Responses at temperatures from 5 to 50°C have been studied. Response times and range of linearity improve with increasing temperature. Response times are less than 1 min for step increases in potassium activity and are longer but less than 3 min at room temperature on decreasing concentration steps. Apparently, some diffusional processes are involved although almost nothing is known with certainty on the origin or distribution of the potential across these unusual membranes.

pM-Responsive Glass Membrane Electrodes

Alkali metal silicates containing Na_2O, Li_2O, and to some extent K_2O have been studied extensively as membrane materials for alkali metal-selective electrodes [28]. These formulations contain, in addition, trivalent oxides, typically Al_2O_3, which seem to moderate the electrical character of the negative fixed sites [30]. Virtually every reasonable divalent, trivalent, tetravalent, and even some of the pentavalent oxides have been incorporated into glass compositions with the hope of achieving selectivity for one or another of the alkali ions in the presence of other mono- and divalent cations.

Eisenman's oft-quoted systematic study of the Na_2O-Al_2O_3-SiO_2 glass membrane system shows that even a three-oxide system can show many surprising differences in membrane selectivity. Glasses with high enough SiO_2/Na_2O ratios for hydrolytic stability and low Al_2O_3 are essentially pH glasses, while at 27% Na_2O–5% Al_2O_3–68% SiO_2 the response becomes selective for potassium over sodium at depressed hydrogen ion levels. The actual selectivity is $H^+ > K^+ > Na^+ > Li^+$. Note that a composition close to this is thought to be used in the well-known Beckman "cation" electrode. As the alumina content is raised to $11\% Na_2O$–$18\% Al_2O_3$–$71\% SiO_2$, the selectivity at neutral pH is $Na^+ > Li^+ > K^+$; hydrogen ion response still dominates at low pH. Higher alumina glasses are quite difficult to make because of their high melting points.

pM-responsive glass electrodes are historically of three types. The best known is the pH glass described at length in Section 3. The pNa electrodes, which have relatively low resistance compared to pH glasses, are an off-shoot of the pH glass developments. In the third category fall cation electrodes, which are those aimed chiefly at the determination of pK^+. Both the sodium-sensitive and the cation-sensitive electrodes are slower responding and less stable over a long term than the pH glass. The sodium electrodes have selectivities over potassium from 10^3 to 10^5, depending on the source of the electrode. The cation-sensitive electrodes favor potassium over sodium by about 7 to 10, although the selectivity improves slightly on prolonged soaking [101].

Both electrodes respond to their dominant ion from saturation to less than $10^{-4} M$ in a Nernstian fashion. The cation electrodes measure sodium activities in the absence of potassium about as well as the sodium electrode does. Hydrogen ion concentration should be suppressed by use of calcium or barium hydroxides to anywhere from 10^{-2} to 10^{-4} times the lowest anticipated sodium or potassium concentration.

As with pH electrodes, pM electrodes tend to hydrolyze at the surface with the formation of a hydrated silica layer of composition different from the bulk glass. This statement has no direct proof (as has been obtained for films on pH glasses), but there are many effects that are best explained in terms of a hydrated layer. The causes of hydrolysis were discussed earlier in connection with pH electrodes. These glasses are intrinsically unstable with respect to hydrolysis and cannot be exposed for long times to alkaline solutions without a resultant shortening in the life of the electrode. Impedance measurements, dynamic response to step change in activity, long-term drift of potential, and uptake of radioactive sodium suggest at least a partially diffusive transport layer on these glasses.

An expected feature of these electrodes is response to other monovalent ions. The sodium-sensitive electrode is quite sensitive to silver and yields a suitable Nernstian response such that it can be used in argentometric titrations. The cation-sensitive electrode, in the absence of sodium and potassium, responds to NH_4^+, Li^+, and others given in Table 2.1A. The cation electrode responds well enough to Rb^+, Cs^+, and Ag^+ to serve as a sensor in potentiometric titrations of these ions. This electrode has also been used in activity measurements in ethanol, acetone, ethylene glycol, and dimethyl-formamide mixtures with water. A wealth of material [2] has been compiled on the applications of these electrodes.

Practical cells, reference electrodes, and precautions in handling are the same as for glass electrodes for pH measurement described in Section 3.

Heterogeneous Membrane Electrodes

The most promising of the large variety of heterogeneous membranes are those using a silicone binder described as the Pungor-type [15]. A number of these were mentioned in Section 1 because of their availability. In each of these an ion-exchanging insoluble powder is fixed in a hydrophobic binder. The powders selected (note examples in Table 2.1D) are usually among those used in the solid-state crystal and pressed pellet membranes. One guiding hypothesis in the construction of new heterogeneous electrodes is to use ion-conducting solids as powders which would serve as single crystal electrodes if they could be obtained in thin form with low resistance. The Pungor-type heterogeneous membranes allow fabrication of these materials in thin sections while permitting a flexible and mechanically strong form which would be difficult or impossible to achieve otherwise. Of the possible binder materials, including polyethylene, polystyrene, polyvinyl chloride, and paraffin wax, the most suitable material is silicone rubber because of its flexibility and resistance to cracking and swelling. Like the other binders, it is water repellant. The liquid silicone rubber precursor is mixed about 50-50 by weight with the selected fine powder, ground to 5 to 15-μ average particle size, and laid out in thin form for curing. The dried film must be checked for freedom from holes.

The high concentration of suspended solid presumably is needed to insure contact necessary for ion conductivity through the membrane. The implication of this statement is that Pungor-type membranes operate on the same principle as solid-state membranes. However, the existence of sulfate and phosphate electrodes suggests that an alternate mechanism may prevail. The diffusion potential may not occur through the bulk of the ion-conducting, touching particles, but rather through thin layers of solution continuous from one side of the membrane to the other. These layers are visualized as wetting the surfaces of the particles but as being sufficiently thin that surface charges of adsorbed ions determine the extent of "along the surface" diffusion. This type of controlled diffusion leads to a diffusion potential which is the same as the "tip" potential occurring in a narrow glass tube with fixed surface sites. The author has found that $BaSO_4$ precipitated in excess sulfate gives a cation-dependent membrane response, while $BaSO_4$ precipitated in excess $BaCl_2$ gives a membrane selective to sulfate and other anions. This result is consistent with the thin film hypothesis whereby the adsorbed ion, barium in the second case, serves as a fixed-site restricting co-ion mobility and permitting counterion transport so that a Nernst response might occur in the limit of ideal permselectivity.

Many of the heterogeneous electrodes have solid homogeneous electrode counterparts, and there are no particular advantages to recommend the heterogeneous versions. However, practical electrodes for sulfate, phosphate, and perhaps other ions constructed by the Pungor approach have been reported [15]. These examples are important because solid or liquid electrodes cannot or have not been successfully formed because of high resistance, fragility, or other reasons. Pungor-style electrodes incorporating graphite, carbides, borides, and other electronic conductors have been demonstrated as inert electrodes comparable with graphite rods, and so on. Heterogeneous membranes with synthetic ion exchangers including chelating resins and several salts other than those in Table 2.1D have been reported [15].

The electrodes that are commercially available are constructed in forms quite similar to solid-state or glass electrodes. Instead of blowing a bulb across the lower opening of a high-resistance glass cylinder, the silicone membrane is sealed across the working end. An inner reference electrode and inner filling solution containing ions common to the membrane and the internal reference electrode are used inside the tube to complete the electrode. To outward appearances the completed electrodes resemble any other probe-style sensor.

Most of the literature prior to 1969 on the heterogeneous electrodes developed by Pungor is in Hungarian journals. A complete historical, theoretical, and practical survey of these electrodes by Covington has recently appeared [15].

pM Standards

The technique for establishment of $\log a_{H^+}$ of primary standard buffers using junctionless cells was described in Section 4. From these values conventional pH values of unknowns can be determined in junction cells with an uncertainty attributed only to the residual junction potential difference.

Three standard pM(S) materials are now available whose activities at four different concentrations in water are precisely known. These values are given in Table 2.20. These solutions are suitable for standardizing glass, solid-state, and liquid ion-exchange membrane electrodes responsive to sodium, calcium, chloride, and fluoride activities.

Some Potentiometric Applications

Ion-selective glass, solid-state crystal or pressed pellet, and liquid exchanger electrodes have clearly extended the field of direct potentiometry. The remarkable growth in this field in one decade, as a topic of extensive research

6 ELECTRODES FOR CATION AND ANION ACTIVITY MEASUREMENTS 155

Table 2.20 Suggested Reference Standard Values, pM(S), at 25°C

Material	*Molality* (mole kg^{-1})	pNa(S)	pCa(S)	pCl(S)	pF(S)
NaCl	0.001	3.015	—	3.015	—
	0.01	2.044	—	2.044	—
	0.1	1.108	—	1.110	—
	1.0	0.160	—	0.204	—
NaF	0.001	3.015	—	—	3.015
	0.01	2.044	—	—	2.048
	0.1	1.108	—	—	1.124
CaCl$_2$	0.000333	—	3.537	3.191	—
	0.00333	—	2.653	2.220	—
	0.0333	—	1.887	1.286	—
	0.333	—	1.105	0.381	—

and as a field for commercialization, attests to the importance of electrodes based on membrane potentials. Obviously, practical electrode for activities of fluoride, perchlorate, nitrate, and calcium could not have been achieved following along classic lines involving metal–metal ion or metal-salt systems.

The new electrodes described in this chapter have found immediate acceptance and wide applicability. It is not possible to survey these here; merely the areas of application can be indicated.

In titrimetry potentiometric end-point detection with the fluoride electrode has displaced classic colorimetric methods. Good results can be achieved using either lanthanum or thorium as titrant in unbuffered aqueous or aqueous ethanol media [98]. A titration curve is illustrated in Fig. 2.28. The technique of null point potentiometry, which has been exploited for low-level silver and halide analyses using conventional electrodes, can be performed with these newer electrodes. The procedure involves two solutions in concentration cell configuration. The sample unknown activity is matched by titration into a second solution of comparable ionic strength until a "null" potential is observed. Durst and Taylor [102] determined 380 pg of fluoride in 10 μl with an uncertainty of only 2 pg. A special sample cell using an inverted LaF$_3$ electrode was built for this purpose. Closely related methods are the "spiking" techniques in which an unknown activity of fluoride in a known volume V_x is increased by addition of a small volume V_s of a more concentrated solution C_s. The new activity should be about 10 times the unknown and the total volume not significantly increased. If S is the slope of the response curve, $\pm \dfrac{RT}{zF}$ for ideally responding ions of absolute charge z, the

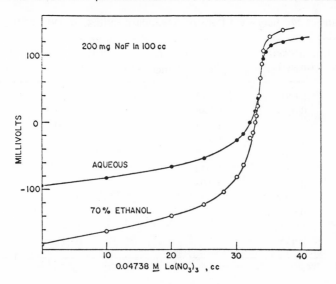

Fig. 2.28 Influence of ethanol on the titration curve of fluoride ion with lanthanum ion in neutral, unbuffered medium. In both cases 200.0 mg of NaF in an original volume of 100 cc was titrated [98].

unknown concentration may be computed from

$$C_{\text{unknown}} = C_s \left(\frac{V_s}{V_x + V_s} \right) \left[10^{\Delta E / S} - \left(\frac{V_x + V_s}{V_x} \right) \right]^{-1} \quad (2.129)$$

where ΔE is the potential change upon "spiking." The reverse process whereby a quantity of an unknown is used to dilute a standard has been applied also. Titration, matching potential, and standard addition techniques are not limited to fluoride analysis and could be useful in metal ion analyses using the appropriate selective electrodes.

New thermodynamic and kinetic measurements have become possible through the availability of selective electrodes. Mean activity coefficients can, in principle, be measured using junctionless cells composed of one cation-selective and one anion-selective electrode chosen from the group of selective ion electrodes. In the range of concentrations where selectivity is ideal, mean activities of mixtures could be studied as well. A number of complex ion stability studies have been made on systems involving alkali metal ions and fluoride ions which were inaccessible to study previously. Rechnitz's extensive contributions in this field have been reviewed [103]. Rechnitz's group has been responsible for a number of kinetic studies using selective electrodes. His original papers given in Ref. 103 should be consulted for practical details.

Continuous monitoring of water hardness, sulfide in industrial streams, and fluoride in natural waters (after addition of an ionic strength control and buffer solution) are a few examples in the industrial analysis field. The review by Light [104] should be examined.

References

1. F. Helfferich, *Ion Exchange*, McGraw-Hill, New York, 1962.
2. G. Eisenman, Ed., *Glass Electrodes for Hydrogen and Other Cations*, Marcel Dekker, New York, 1967.
3. R. A. Durst, Ed., *Ion Selective Electrodes*, National Bureau of Standards Special Publ. No. 314, U.S. Govt Printing Office, Washington, D.C., 1969.
4. K. Sollner, *Ann. N. Y. Acad. Sci.*, **148**, 154 (1968).
5. Ref. 1, Chapter 1, for a more complete discussion.
6. M. Cremer, *Z. Biol.* **47**, 562 (1906).
7. F. Haber and Z. Klemensiewicz, *Z. Phys. Chem.*, **67**, 385 (1909).
8. W. Nernst and E. H. Riesenfeld, *Ann. Phys.*, **8**(4), 600 (1902).
9. B. Lengyl and E. Blum, *Trans. Faraday Soc.*, **30**, 461 (1934).
10. W. Jost, *Diffusion in Solids, Liquids and Gases*, 3rd ed., Academic Press, New York, 1960, Chapters 2, 3, and 4.
11. E. Koch and C. Wagner, *Z. Phys. Chem.* (B), **38**, 295 (1937).
12. I. M. Kolthoff and H. L. Sanders, *J. Am. Chem. Soc.*, **59**, 416 (1937).
13. M. S. Frant and J. W. Ross, *Science*, **154**, 1553 (1966).
14. C. E. Marshall, *J. Phys. Chem.*, **48**, 67 (1944) and earlier references therein. Two later works on divalent ion electrodes are in *J. Am. Chem. Soc.*, **70**, 1297, 1302 (1948).
15. A. K. Covington in *Ion Selective Electrodes* National Bureau of Standards Special Publ. No. 314, U.S. Govt. Printing Office, Washington, D.C., Chapter 3.
16. Early work is analyzed by L. Michaelis, *Hydrogen Ion Concentration*, Williams and Wilkins, Baltimore, 1926.
17. J. W. Ross, *Science* **156**, 1378 (1967).
18. Ref. 1, Chapter 8.
19. E. A. Guggenheim, *Thermodynamics*, 3rd ed., Interscience, New York, 1957, Chapter 10; *J. Phys. Chem.*, **33**, 842 (1929).
20. N. F. Mott and R. W. Gurney, *Electronic Processes in Ionic Crystals*, 2nd ed., Dover, New York, 1964, Chapter 2.
21. D. A. MacInnes, *The Principles of Electrochemistry*, Dover, New York, 1961, Chapter 13.
22. Ref. 1, Chapter 8.
23. J. A. Marinsky, Ed., *Ion Exchange*, Vol. 1, Marcel Dekker, New York, 1966, Chapter 1.
24. W. M. Clark, *The Determination of Hydrogen Ions*, 3rd ed., Williams and Wilkins, Baltimore, 1928.

25. W. J. Hamer and S. F. Acree, *J. Res. Natl. Bur. Std.*, **33**, 87 (1944).
26. R. G. Bates, *Determination of pH, Theory and Practice*, Wiley, New York, 1964, Chapter 10.
27. Ref. 2, Chapter 2.
28. J. O. Isard, in *Glass Electrodes for Hydrogen and Other Cations*, G. Eisenman, Ed., Marcel Dekker, New York, 1967, Chapter 3.
29. B. P. Nicolsky, M. M. Shultz, A. A. Belijustin, and A. A. Lev, in *Glass Electrodes for Hydrogen and Other Cations*, G. Eisenman, Ed., Marcel Dekker, New York, 1967, Chapter 6.
30. G. Eisenman, in *Glass Electrodes for Hydrogen and Other Cations*, G. Eisenman, Ed., Marcel, New York, 1967, Chapter 7.
31. S. M. Friedman in *Glass Electrodes for Hydrogen and Other Cations*, G. Eisenman, Ed., Marcel Dekker, New York, 1967, Chapter 16.
32. R. N. Khuri, in *Glass Electrodes for Hydrogen and Other Cations*, G. Eisenman, Ed., Marcel Dekker, New York, 1967, Chapter 18; in *Glass Microelectrodes*, M. Lavallée, O. F. Schanne and N. C. Hébert, Eds., Wiley, New York, 1969 Ch. 13.
33. A. K. Covington, in *Ion Selective Electrodes*, R. A. Durst, Ed., National Bureau of Standards Special Publ. No. 314, U.S. Govt. Printing Office, Washington D.C., Chapter 4.
34. Ref. 26, p. 57.
35. Ref. 26, p. 30.
36. Ref. 26, Chapter 4.
37. B. R. Staples and R. G. Bates, *J. Res. Natl. Bur. Standards*, **73A**, 37 (1969).
38. I. M. Kolthoff (C. Rosenblum, transl.) *Acid-Base Indicators*, Macmillan, New York, 1937, pp. 103ff; O. Tomiček, *Chemical Indicators* (A. R. Weir, transl.), Butterworths, London, 1951.
39. Ref. 26, Chapter 6.
40. I. M. Kolthoff, E. B. Sandell, E. J. Meehan, and S. Bruckenstein, *Quantitative Chemical Analysis*, 4th ed., Macmillan, London, 1969, pp. 691–696.
41. L. P. Hammett, *Physical Organic Chemistry*, McGraw-Hill, New York, 1940.
42. R. H. Boyd, *Solvent-Solute Interactions*, J. F. Coetzee and C. D. Ritchie, Eds., Interscience, New York, 1969, Chapter 3.
43. Ref. 26, Chapter 5.
44. R. A. Robinson and R. H. Stokes, *Electrolyte Solutions*, 2nd ed., Academic Press, New York, 1959.
45. O. Popovych, *Anal. Chem.* **38**, 558, 1966. See also O. Popovych and A. J. Dill, *ibid.*, **41**, 456 (1969).
46. I. M. Kolthoff and S. Bruckenstein, *Treatise on Analytical Chemistry*, I. M. Kolthoff and P. J. Elving, Eds., Part I, Section B, Interscience, New York, 1959, Chapter 13.
47. H. H. Sisler, *Chemistry in Non-Aqueous Solvents*, Van Nostrand Reinhold, New York, 1961.
48. T. C. Waddington, Ed., *Non-Aqueous Solvent Systems*, Academic Press, New York, 1965.

49. J. Lagowski, Ed., *The Chemistry of Non-Aqueous Solvents*, 2 vols., Academic Press, New York, 1962.
50. J. F. Coetzee and C. D. Ritchie, Eds., *Solvent-Solute Interactions*, Wiley, New York, 1969.
51. J. Kucharsky and L. Safarik, *Titrations in Non-Aqueous Solvents*, Elsevier, New York, 1965.
52. W. Huber, *Titrations in Nonaqueous Solvents*, Academic Press, New York, 1967.
53. J. Steinhardt and E. M. Zaiser, *J. Biol. Chem.*, **190**, 197 (1951).
54. J. S. Fritz, *Acid-Base Titrations in Non-Aqueous Solvents*, G. F. Smith Chem. Co., P.O. Box 1611, Columbus, Ohio.
55. Emil Greiner and Co., 2026 N. Moore St., New York, New York.
56. Ref. 44, Chapter 12, for the extrapolation procedures.
57. Calculated from data given by R. A. Robinson and R. H. Stokes, *Electrolyte Solutions*, Academic Press, New York, 1955. This text should be consulted for more extensive compilations.
58. See A. Albert et al., *J. Chem. Soc.*, **1948**, 2240; *ibid.*, **1956**, 1294.
59. G. Kortum, W. Vogel, and K. Andrussow, *Dissociation Constants of Organic Acids in Aqueous Solutions*, Butterworths, London, 1961.
60. D. D. Perrin, *Dissociation Constants of Organic Bases in Aqueous Solutions*, Butterworths, London, 1965.
61. C. K. Ingold, *Structure and Mechanism in Organic Chemistry*, Cornell Univ. Press, Ithaca, New York, 1953.
62. E. A. Braude and F. C. Nachod, Eds., *Determination of Organic Structures by Physical Methods*, Academic Press, New York, 1955, Chapter 14.
63. E. S. Gould, *Mechanism and Structure in Organic Chemistry*, Holt, New York, 1959, Chapter 4.
64. J. Hine, *Physical Organic Chemistry*, 2nd ed., McGraw-Hill, 1962, Chapter 2.
65. K. B. Wiberg, *Physical Organic Chemistry*, Wiley, New York, 1964, Chapter 9.
66. A. Liberles, *Introductions to Theoretical Organic Chemistry*, Macmillan, N.Y. 1968, Chapter 3.
67. E. M. Arnett, "Quantitative Comparisons of Weak Organic Bases," in *Progress in Physical Organic Chemistry*, Vol. 1, Interscience, New York, 1963.
68. J. E. Leffler and E. Grunwald, *Rates and Equilibria in Organic Reactions*, Wiley, New York, 1963.
69. C. D. Ritchie and W. F. Sagen, "An Examination of Structure-Reactivity Relationships," in *Progress in Physical Organic Chemistry*, Vol. 2, Interscience, New York, 1964.
70. J. T. Edsall and J. Wyman, *Biophysical Chemistry*, 2 vols., Academic Press, 1958.
71. C. Tanford, *Physical Chemistry of Macromolecules*, Wiley, New York, 1961, Chapter 8.
72. H. Morawetz, *Macromolecules in Solution*, Interscience, New York, 1965, Chapter 7.

73. See, for instance, the discussion of cysteine by G. Gorin, *J. Am. Chem. Soc.*, **78**, 767 (1956) and by R. E. Benesch and R. Benesch, *ibid.*, **77**, 5877 (1955). For proteins see C. Tanford, in *Electrochemistry in Biology and Medicine*, Wiley, New York, 1955.

74. See, for instance, A. L. Bacrella, E. Grunwald, H. P. Marshall, and E. L. Purlee, *J. Org. Chem.*, **20**, 747 (1955).

75. J. F. Coetzee, "Ionic Reactions in Acetonitrile," in *Progress in Physical Organic Chemistry*, Vol. 4, Interscience, New York, 1967.

76. M. M. Davis *Acid-Base Behavior in Aprotic Organic Solvents*, *Natl. Bur. Standards Monograph* 105, Washington, D.C., 1968.

77. Ref. 50, Chapter 4.

78. J. E. Prue, in *The International Encyclopedia of Physical Chemistry and Chemical Physics*, Topic 15, Vol. B *Ionic Equilibria*, 1966, Chapter 9.

79. R. S. Drago and K. F. Purcell, "The Coordination Model for Non-Aqueous Solvent Behavior," in *Progress in Inorganic Chemistry*, Vol. 6, Interscience, New York, 1964.

80. J. Bjerrum, *Metal Ammine Formation in Aqueous Solution*, 2nd ed., P. Haase and Son, Copenhagen, 1957.

81. A. E. Martell and M. Calvin, *Chemistry of the Metal Chelate Compounds*, Prentice-Hall, Englewood Cliffs, New Jersey, 1952.

82. F. J. C. Rossotti and H. Rossotti, *The Determination of Stability Constants*, McGraw-Hill, New York, 1961.

83. A. Ringbom, *Complexation in Analytical Chemistry*, Interscience, New York, 1963.

84. F. P. Dwyer and D. P. Mellor, Eds., *Chelating Agents and Metal Chelates*, Academic Press, New York, 1964.

85. K. B. Yatsimirskii and V. P. Vasilev, *Instability Constants of Complex Compounds* (D. A. Paterson, trans.) Pergamon Press, Oxford, 1960.

86. L. G. Sillen and A. E. Martell, *Stability Constants of Metal-Ion Complexes*, The Chemical Society, London, 1964.

87. Ref. 44, Chapter 8.

88. E. A. Guggenheim, *J. Phys. Chem.*, **34**, 1758 (1930).

89. Ref. 3, Chapter 6.

90. W. M. Latimer, *The Oxidation States of the Elements and Their Potentials in Aqueous Solution*, 2nd ed., Prentice-Hall, Englewood Cliffs, New Jersey, 1952.

91. D. J. G. Ives and G. J. Janz, *Reference Electrodes, Theory and Practice*, Academic Press, New York, 1961.

92. M. LeBlanc and D. Harnapp., *Z. Phys. Chem.* (*Leipzig*), **166**, 321 (1933).

93. C. N. Reilley and R. W. Schmidt, *Anal. Chem.*, **30**, 947 (1958); C. N. Reilley, R. W. Schmidt, and I. W. Lamson, *ibid.*, **30**, 953 (1958).

94. Ref. 3, Chapter 11.

95. Ref. 3, Chapter 2.

96. A. Sher, R. Solomon, K. Lee, and M. W. Muller, *Phys. Rev.*, **144**, 593 (1966).

97. Ref. 3, Chapter 5.

98. J. J. Lingane, *Anal. Chem.*, **39**, 881 (1967); **40**, 935 (1968).
99. A. K. Covington, *Chem. Brit.*, **5**, 388 (1969).
100. C. J. Coetzee and H. Freiser, *Anal. Chem.*, **40**, 2071 (1968); **41**, 1128 (1969).
101. E. W. Moore and J. W. Ros, *Science*, **148**, 71 (1965).
102. R. A. Durst and J. K. Taylor, *Anal. Chem.*, **39**, 1374, 1483 (1967); R. A. Durst, *ibid.*, **40**, 931 (1968).
103. Ref. 3, Chapter 9.
104. Ref. 3, Chapter 10.

General

PART 1

Helfferich, F., *Ion Exchange*, McGraw-Hill, New York, 1962.

PART 2

Bates, R. G., "Electrode Potentials," in *Treatise on Analytical Chemistry*, Part I, Vol. I, I. M. Kolthoff and P. J. Elving, Eds., Interscience, New York, 1959, Chapter 9.

Latimer, W. M., *The Oxidation States of the Elements*, 2nd ed., Prentice-Hall, Englewood Cliffs, New Jersey, 1952.

MacInnes, D. A., *The Principles of Electrochemistry*, Dover, New York, 1961, Chapter 13.

PART 3

Eisenman, G., Ed., *Glass Electrodes for Hydrogen and Other Cations*, Marcel Dekker, New York, 1967.

Eisenman, G., R. Bates, G. Mattock, and S. M. Friedman, *The Glass Electrode*, Interscience, New York, 1965.

Lavallée, M., Schanne, O. F., Hébert, N. C., Eds., *Glass Microelectrodes*, Wiley, New York, 1969.

Lingane, J. J., *Electroanalytical Chemistry*, 2nd ed., Interscience, New York, 1958, Chapters 4 and 5.

Phillips, J. P., *Automatic Titrators*, Academic Press, New York, 1959.

PART 4

Bates, R. G., *Determination of pH, Theory and Practice*, Wiley, New York, 1964.

Bates, R. G., "Concept and Determination of pH," in *Treatise on Analytical Chemistry*, Part I, Vol. I, I. M. Kolthoff and P. J. Elving, Eds., Interscience, 1959, Chapter 10.

Bell, R. P., *The Proton in Chemistry*, Cornell Univ. Press, Ithaca, New York, 1959.

Britton, H. T. S., *Hydrogen Ions, Their Determination and Importance in Pure and Industrial Chemistry*, 2 vols., Van Nostrand Reinhold, New York, 1956.

Gyenes, I., *Titration in Nonaqueous Media*, Van Nostrand Reinhold, New York, 1967.

Kolthoff, I. M., "Concepts of Acids and Bases," in *Treatise on Analytical Chemistry*, Part I, Vol. I, I. M. Kolthoff and P. J. Elving, Eds., Interscience, New York, 1959, Chapter 11.

Kolthoff, I. M., and S. Bruckenstein, "Acid-Base Equilibria in Nonaqueous Solutions," in *Treatise on Analytical Chemistry*, Part I, Vol. I, I. M. Kolthoff and P. J. Elving, Eds., Interscience, New York, 1959, Chapter 13 and Refs. 51 and 52.

Van der Werf, C. A., *Acids, Bases and the Chemistry of the Covalent Bond*, Van Nostrand Reinhold, New York, 1961.

PART 5

Bruckenstein, S., and I. M. Kolthoff, "Acid-Base Strength and Protolysis Curves in Water," in *Treatise on Analytical Chemistry*, Part I, Vol. I, I. M. Kolthoff and P. J. Elving, Eds., Interscience, New York, 1959, Chapter 12.

Butler, J. N., *Ionic Equilibrium*, Addison-Wesley, Reading, Massachusetts, 1964.

Greenstein, J. P., and M. Winitz, *The Chemistry of the Amino Acids*, Vol. I, Wiley, New York, 1961, Chapter 4.

Kolthoff, I. M., and V. A. Stenger, *Volumetric Analysis*, 2nd ed., Vol. II, Interscience, New York, 1947.

Monk, C. B., *Electrolytic Dissociation*, Academic Press, New York, 1961.

Ringbom, A., "Complexation Reactions," in *Treatise on Analytical Chemistry*, Part I, Vol. I, I. M. Kolthoff and P. J. Elving, Eds., Interscience, New York, 1959, Chapter 14.

Robinson R. A., and R. H. Stokes, *Electrolyte Solutions*, 2nd ed., Academic Press, New York, 1959, revised ed., 1965.

Sillen, L. G., "Graphical Presentation of Equilibrium Data," in *Treatise on Analytical Chemistry*, Part I, Vol. I, I. M. Kolthoff and P. J. Elving, Eds., Interscience, New York, 1959, Chapter 8 and Refs. 82, 83, and 84.

PART 6

Durst, R. A., Ed., *Ion Selective Electrodes*, National Bureau of Standards Special Publ. No. 314, U.S. Govt. Printing Office, Washington, D.C., 1969.

Chapter **III**

CONDUCTOMETRY

Theodore Shedlovsky and Leo Shedlovsky

---

1 INTRODUCTION

The transport of electricity through matter is effected by the movement of electric charges. These may be electrons, as in metallic conductors, positive and negative ions in electrolytic conductors, or charged particles, as well as ions, in colloidal conductors. Most conducting solutions, fused salts, and many solid ones are electrolytic conductors.

The extent to which electric current flows through an electrolytic solution depends on the applied voltage and on the number, charge, and mobility of the ions or other charged particles present. The number present depends on the concentration and, with weak electrolytes, on the extent of ionization, which also depends on concentration, temperature, pressure, and the nature of the solvent. The mobility depends on the geometric nature of the ions, which may be solvated, the viscosity, and the temperature.

Electric current can be led into and out of an electrolytic solution through solid electrodes, other electrolytic systems, or gaseous conductors. The cell as a whole, however, generally contains two metallic electrodes, an anode and a cathode, at which chemical reactions take place. Negative ions, or anions, move towards the anode where oxidation occurs; positive ions, or cations, move toward the cathode where reduction occurs. The nature of these chemical changes, or electrode reactions, is not discussed in any detail in this chapter. They concern us only insofar as they may introduce difficulties into conductivity determinations.

2 ELECTROLYTIC CONDUCTANCE

Definitions

According to Ohm's law the *current*, I, flowing in an electric conductor is directly proportional to the *electromotive force*, E, and inversely proportional to the *resistance*, R, in the conductor. The respective units, ampere, volt, and ohm, for these quantities are defined so that

$$I = V/R \tag{3.1}$$

The *resistance*, $R = V/I$, of a homogeneous material under given conditions depends not only on the intrinsic property but also on the quantity and shape of the sample. If the material is of uniform cross section with an area of A square centimeters and is l centimeters in length, its longitudinal resistance is given by the equation

$$R = \rho \cdot l/A \tag{3.2}$$

in which ρ is the intrinsic property *specific resistance*, that is, the resistance per centimeter with a uniform cross-sectional area of 1 cm². The reciprocal of resistance is termed conductance, and the reciprocal of specific resistance ρ is termed *specific conductance, κ*.

The specific conductance of an electrolytic solution depends only on the ions present, and therefore it varies with their concentration. Our attention is focused on the ionic nature of the solution by the use of the quantity *equivalent conductance*, Λ. It refers to a particular electrolyte and may be defined as the conductance, attributable to that electrolyte, of a layer of solution 1 cm long having a uniform cross section and containing 1 equiv. of solute. It is related to the specific conductance κ by the equation

$$\Lambda = 1000\kappa/c, \tag{3.3}$$

in which c is the concentration in gram equivalents per liter of solution.

The analogous quantity, *molar conductance*, Λ_m, is given by

$$\Lambda_m = 1000\kappa/c_m, \tag{3.4}$$

where c_m is the concentration in moles per liter. The ratio of Λ_m to Λ is thus the number of positive (or negative) ionic charges obtainable from one molecule of the substance.

The equivalent conductance of an electrolyte, such as a salt, acid, or base, varies with concentration in a given solvent at a specific temperature and pressure. This may be attributable both to a change in the number of conducting ions as well as to a change in the ionic mobilities.

The ionic mobility is defined as the net ionic velocity in the direction of the electric potential per unit field (1 V per centimeter). In aqueous solutions, as well as in many organic solvents, the equivalent conductance of electrolytes decreases with increasing concentration from the values that correspond to infinite dilution. This phenomenon is explained by a consideration of several theoretical points.

Theory of Electrolytic Conductance

General

Arrhenius proposed the classic ionic or electrolytic dissociation theory in 1883. He assumed that the decrease in equivalent conductance of an electrolyte with increasing concentration was attributable to a decrease in the proportion of conducting ions, but that their mobilities remained unaltered. Thus he expressed the "degree of ionization" x as the ratio of the equivalent conductance Λ at a given concentration to the value extrapolated to infinite dilution Λ_0:

$$x = \Lambda/\Lambda_0. \tag{3.5}$$

This theory appeared to hold fairly well for weak electrolytes such as acetic acid in water, but not for strong ones such as most salts, strong acids, and bases. The difficulty lay in the assumption that the ionic mobilities are independent of concentration.

Milner [1] attempted a quantitative mathematical calculation for the interionic, electrostatic force effects of the ions on themselves, but his theoretical treatment led to serious mathematical difficulties. It remained for Debye and Hückel [2] to formulate an interionic attraction theory in a more useful form. A further improvement in the theory, as applied to electrical conductance, was presented by Onsager [3].

In the interionic attraction theory of electrolytes, the properties of the solutions are considered to be attributable to the interplay of electrostatic forces and thermal motion. The first of these tends to give the ions a definite arrangement, and the second tends to produce a random distribution. With these ideas it is possible to account for the influence of concentration on ionic mobilities and on thermodynamic properties, and also satisfactorily to explain variations in conductance of dilute solutions of strong as well as weak electrolytes.

As a result of electrostatic (coulomb) forces and thermal motion, any selected ion in a solution, a positive one for instance, has on a time average more negative ions near it than if the distribution were purely random. This is known as the "ion atmosphere" of the selected ion. From a thermodynamic point of view, the presence of the ion atmosphere reduces the activity coefficients of the ions. It influences the electrolytic conductance by the so-called "electrophoretic effect" and "time of relaxation effect" both of which tend to decrease the ionic mobilities with increasing ionic concentration.

When an electric potential is applied, the ion atmosphere, which has a net charge equal but opposite in sign to that of the central ion, tends to move with its associated solvent molecules in a direction opposite to that of this ion. This retards the central ion, which in effect moves against a countercurrent of solvent amounting to an increase of viscous drag (Fig. 3.1). Because of its analogy to the retarding effect of the solvent on the motion of colloidal particles in an electric field, this phenomenon has been called the *electrophoretic effect*.

When no external electric field is applied on a solution, the ion atmosphere is spherically symmetrical with its electrical center of gravity at the central ion. If, however, the ion is made to move by an external field, the ion atmosphere must also move to adjust for this change. This takes place rapidly (during the short time of relaxation) but not instantaneously, so that the ion atmosphere is distorted, its electrical center of gravity is shifted to a point behind the central ion, and its motion is slowed up by an electrostatic retarding force (Fig. 3.2). This phenomenon is called the *time of relaxation effect*.

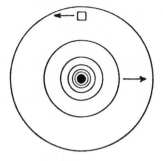

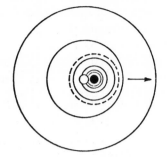

Fig. 3.1 Electrophoretic effect.   **Fig. 3.2** Time of relaxation effect.

In very dilute solutions the "thickness" of the ion atmosphere, $1/\kappa$, decreases with the square root of the ionic concentration and with the first power of ionic charge or, more precisely, with the square root of the ionic strength. The retarding effect on ionic mobility attributable to both the electrophoretic and time of relaxation effects turns out to be inversely proportional to this thickness, which is usually of the order of ångströms. Moreover, the former is proportional to the ionic mobility at infinite dilution, and the latter to the fluidity of the solvent.

The quantitative conductance equation for very dilute electrolytic solutions derived by Onsager [3] based on these considerations is

$$\Lambda = \Lambda_0 - (\alpha\Lambda_0 + \beta)\sqrt{c}, \qquad (3.6a)$$

where c is the ion concentration and the constants α and β depend on the nature of the solvent and the temperature. For uni-univalent electrolytes

$$\alpha = \frac{8.20 \times 10^5}{(DT)^{3/2}} \qquad \text{and} \qquad \beta = \frac{82.5}{(DT)^{1/2}\eta}$$

where D is the dielectric constant at the absolute temperature T, and η is the viscosity.

The Onsager equation is the correct *limiting* expression for the conductance of *strong* electrolytes [3, 4]. This is shown graphically in Fig. 3.3 for the equivalent conductance of several salts in water at 25°C, in which the limiting slopes of the curves are quantitatively in excellent accord with those calculated from the theory. However, positive deviations from the theoretical conductance (broken line) appear as the concentration is increased. These can be accounted for by an empirical extension of the Onsager equation [5], namely,

$$\Lambda_0 = \Lambda_0' - Ac - Bc \log c + Dc^2, \qquad (3.6b)$$

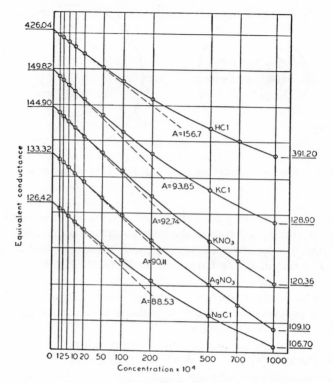

Fig. 3.3 Equivalent conductance of typical uni-univalent electrolytes in water at 25°C.

in which $\Lambda_0' = (\Lambda + \beta\sqrt{c}\,(1 - \alpha\sqrt{c}))$ from (3.6a). In many cases, such as for most univalent salts and strong acids in water up to a concentration of about 0.1 N, the last two terms in (3.6b) can be omitted.

If the Onsager equation had been derived for equivalent resistance instead of equivalent conductance, it would have taken the form:

$$\frac{1}{\Lambda} = \frac{1}{\Lambda_0} + \frac{(\alpha\Lambda_0 + \beta)\sqrt{c}}{\Lambda_0^2}. \tag{3.6c}$$

This equation is in closer accord with measurements on strong electrolytes up to higher concentrations than (3.6a) and is more convenient in analyzing conductance data on weak electrolytes [6].

The original Debye-Hückel-Onsager treatments of the conductance problem considered point charges in the ionic model. Introduction of boundary conditions for the ionic radius $r = a$ immediately led to terms of order c. Onsager and Fuoss derived a general equation of continuity for mixtures of electrolytes [7]. Before integration they dropped terms of higher order than linear in the charge of the reference ion e_j. These neglected terms gave rise to terms of order c and also to transcendental terms in the conductance equation. Therefore considering the ions as spheres instead of point charges requires retention of the higher terms in the differential equation. In their approach to the problem, Fuoss and Onsager used the following procedure. First, they adjusted the solution of the equation of continuity, correct to terms of order e_j, to the boundary conditions for $r = a$ rather than for $r = 0$. These conditions correspond to a nonsingular potential field at infinite values of r, continuity of field at $r = a$, and impenetrability of two ions, since the normal component of relative velocity must vanish at $r = a$. Second, they inserted the first-order solution in the higher-order (inhomogeneous) terms of the equation of continuity, which they then resolved as a perturbation problem. The result is an explicit form of the relaxation term in the conductance equation, valid up to terms of order $c^{3/2}$, which combined with the 1932 calculation of the electrophoresis term, gives a conductance equation which can be written in compressed form as

$$\Lambda = \Lambda_0 - (\alpha'\Lambda_0 + \beta')\sqrt{c}, \tag{3.6d}$$

where α' and β' are now functions of concentration of the general pattern:

$$\alpha' = \alpha[1 + P(\kappa a)Tr(\kappa a)],$$

where $P(\kappa a)$ indicates polynomial expressions and $Tr(\kappa a)$ transcendentals.
   In its limiting form (3.6d) leads to

$$\Lambda = \Lambda_0 - (\alpha\Lambda_0 + \beta)\sqrt{c} + Dc \log c + (J_1 c - J_2 c^{3/2})(1 - \alpha\sqrt{c}) \tag{3.6e}$$

in which the constants J_1 and J_2 are explicit functions of ion size, a, Λ_0, and of the properties of the solvent; the constant D is, however, independent of a. The last two terms in (3.6e) can be approximated by $cB(\kappa a)$ up to concentrations of about 0.01 N. The authors [8] give numerical tables for computing $B(\kappa a)$ and describe a method for obtaining the two empirical parameters Λ_0 and a from conductance data.
   The new conductance equation contains only two arbitrary constants, Λ_0, which has definite physical significance, and the ion size, a. The latter is a required parameter of molecular dimensions to include all the mathematical and physical approximations necessary to solve the conductance problem in closed form. The equation has been tested against experimental data for the cases of lithium, sodium, and potassium halides in water; the ion sizes found are in the range of values obtained in the computation of activity coefficients. The limiting equivalent conductances agree within a few hundredths of a unit with the values obtained by the empirical Λ_0' extrapolation method [see

(3.6b)]. Fuoss and Onsager used their equation [see (3.6e)] over a concentration range below 0.1 N and imply that it should not be used at higher concentrations. Since their result depends on explicit use of the Debye–Hückel electrical field potential,

$$\psi_j^0 = e^{\kappa a - \kappa r} \epsilon_j / r\epsilon(1 + \kappa a),$$

and since the use of κ the reciprocal of the ion atmosphere in turn implies a continuous distribution of charge, this upper limit of concentration seems reasonable. For aqueous solutions of ordinary 1:1 salts, κa is numerically about the same as $c^{1/2}$; hence $c = 0.1$ corresponds to κa of about 0.3, which corresponds to approximately three ionic diameters. The concept of an ion atmosphere certainly loses physical significance when average interionic distances reach this order of dimension.

The new conductance equation (3.6d) provides a theoretical explanation of the long-known experimental fact that the conductance curve for normal strong electrolytes in water approaches its limiting tangent from above (Fig. 3.3); long-range interionic forces can give only positive deviations and any observed negative deviations are explained by short-range effects such as ion association. Although it makes relatively little difference whether one uses the limiting Onsager tangent or the new equation, which shows positive deviations from this tangent, as a criterion for analyzing conductance curves of very weak electrolytes, this is not the case for "intermediate" electrolytes. Here the new equation should be used in combination with the mass action law. Fuoss [9] has done so and has shown how the values for Λ_0, a, and the ionic association constant A (the reciprocal of the ionization constant K) can be obtained by graphical means from conductance data on low concentrations of moderately weak electrolytes, even of those approaching $A = 1$. He points out, however, that a viscosity correction is required if one of the ionic species is large compared to the solvent molecules. Fuoss and Kraus [10] applied this treatment of conductance data to several salts in dioxane–water mixtures with gratifying results. For example, they found that the value of a (4 Å) remained constant within a few percent for sodium bromate solutions as the composition of the solvent was varied from pure water to 55% dioxane with dielectric constant of 31.5.

Effect of High Field Strength

It was observed by Wien [11] that the equivalent conductance of electrolytes appeared to increase with the use of high electrical potential gradients for the measurements. This efiect is in accord with the interionic attraction theory. When the ionic velocities attributable to high field strengths become sufficiently great (of the order of 10–100 kV/cm), the ion atmospheres hardly have time to form, with the result that both the electrophoretic and time of relaxation effects become correspondingly diminished. In accordance with the observations of the Wien effect, the theory predicts an asymptotic approach of the curves of Λ versus field strength to the Λ_0 values. Also, as would be expected, the relative increase in conductance is most marked in the case of relatively concentrated solution of high valence electrolytes in which the interionic effects are most marked.

An unexpected experimental result was the large Wien effect shown by weak acids and bases. It appears, therefore, that for such weak electrolytes high field strengths tend to produce increased ionization.

Effect of High Frequency

In order to avoid complications arising from polarizing effects at the electrodes, it is customary to carry out conductance measurements with alternating current. It was shown by Debye and Falkenhagen [12] that when sufficiently high frequencies of current (over 5 Mc) are used the equivalent conductance increases and approaches a limiting value somewhat lower than Λ_0. The theoretical explanation of this phenomenon lies in the fact that at high frequencies the ion atmosphere does not fail to form, as in the Wien effect, but as mentioned earlier, it has insufficient time to distort fully.* Consequently, although the electrophoretic effect retains its full influence, the time of relaxation effect decreases. Thus, as frequency is increased, equivalent conductance approaches $\Lambda_0 - \beta\sqrt{c}$ rather than Λ_0 [see (3.6a)].

3 MEASUREMENT OF ELECTROLYTIC CONDUCTANCE

Kohlrausch Method

The passage of current from an electrode to a solution causes a chemical reaction which may produce changes in the composition of the solution in that region. Moreover, this may also create potentials at the electrodes and introduce serious errors in the measurements unless such "polarization" effects can be made negligible.

Alternating currents are generally used in making conductance measurements so that polarization is constantly reversed. Platinum electrodes with a light coating of "platinum black," deposited electrolytically, are also used.

---

* When an ion moves under a low-frequency applied electric field, its ion atmosphere must continually tend to build up in front of and decay behind this ion. The time required for this readjustment, which is not zero, is the so-called time of relaxation. A distortion of the atmosphere thus results in the sense that its electrical center of gravity is behind the moving ion and exerts a retarding force on the ion's migration. The effect is of course greater with faster moving ions. However, with the use of high frequency alternating currents, whose frequency is not negligible compared to the time of relaxation, the ion moves relatively slightly before it reverses its direction of motion, so that its ion atmosphere becomes correspondingly less distorted.

The "platinization" (platinum black is an adherent coating of finely divided platinum) greatly increases the surface and materially reduces polarization. The measurements of resistance, from which the conductance is obtained, are usually made with the help of a Wheatstone bridge. This method was first used by Kohlrausch (1875) and is still almost universally employed.

A schematic arrangement of the Kohlrausch equipment is shown in Fig. 3.4. The conductivity cell forms one arm of a Wheatstone bridge ab, a standard variable resistance box R forms another arm ac, and a calibrated slide-wire resistance bc provides the third and fourth arms bd and cd. Alternating

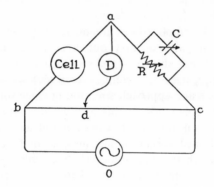

Fig. 3.4 Kohrausch's bridge circuit.

current is supplied to the bridge by a source of ac power O across bc, and a detector telephone D is connected across ad. Since with alternating current, reactances, that is, capacities and inductances, as well as resistances are important in balancing the bridge, a variable condenser C is provided for the purpose. When the position of contact d is adjusted for zero or minimum sound in telephone detector D, the potentials of points a and d are equal, or very nearly so. This gives the following condition:

$$R_{cell}/bd = R/dc, \tag{3.7}$$

from which the resistance of the cell R_{cell} can be computed since the three other factors are known.

When the bridge is accurately balanced, the reactances X, as well as the resistances must be adjusted, so that

$$R_{ab}/R_{bd} = R_{ac}/R_{dc} \tag{3.8}$$

and

$$X_{ab}/X_{bd} = X_{ac}/X_{dc}. \tag{3.9}$$

In most of the recent work, the slide-wire has been replaced by fixed ratio arms, which usually consist of two closely adjusted, nonreactive resistances, that is, resistances with low distributed capacity and negligible inductance. The source of ac power is commonly some form of electronic oscillator, and an electronic amplifier is used to obtain greater sensitivity in the measurements.

The theory and design of ac bridges for measuring electrolytic solutions have been discussed by Jones and Josephs [13a] and by Shedlovsky [13b]. When a direct current flows through a conductor, leakage can take place only if insulation is inadequate. With alternating currents the case is more complicated because the current may also leak away through electrostatic and electromagnetic linkings, commonly known as "stray couplings." Such stray couplings may be electrostatic capacities between the parts of the circuit and capacities between these parts and the ground. There may also be mutual inductive effects attributable to linkages with stray magnetic fields. The effects produced by these couplings are greater and more troublesome at higher frequencies.

A bridge supplied with alternating current in which the position of some of the capacitative couplings are shown appears in Fig. 3.5. The capacity

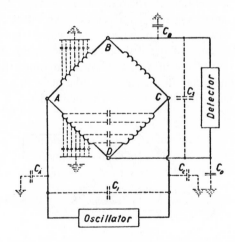

Fig. 3.5 Capacity couplings in an ac bridge [13b].

paths from one lead of the oscillator to the other, either directly (C_1) or through ground (C_A and C_C in series), merely shunt the current input and do not affect the bridge balance. Similarly, the coupling between the leads to the detector does not influence the bridge reading but merely shunts the detector. However, couplings are possible that may disturb the balance of the bridge.

For example, current that leaks from branch AD, through CB to the terminal B, and back to the oscillator through branch BC would have an effect on the bridge balance. (It should be pointed out that bridge terminals A, B, C, and D include the leads appended to them. If they are not too long, their resistance is usually negligible.) Although some of these numerous possible paths tend to neutralize each other, the effect of these paths must be controlled; otherwise, the bridge readings depend on the relative positions of the units in the circuit and on the surroundings.

Precise measurements are not possible unless troublesome couplings are neutralized by making them symmetrical with respect to the terminals of the detector. It is not possible to avoid electrical asymmetry entirely in the construction of the apparatus, particularly if the bridge is used for measurements with various frequencies of current, requiring readjustments in the oscillator, or perhaps in different oscillators. It is not practicable to reduce these couplings so as to make them negligible, especially at higher frequencies, and in the measurements of dilute solutions of high resistance. For these reasons it is necessary to provide means for controlling and balancing these disturbing effects. This is accomplished by the "Wagner earthing device."

It can be shown [13b] that the influence of stray capacity currents from a pure resistance to a shield makes the resistance behave as if it were somewhat larger and had a small appended inductance. An effect also appears that amounts to the presence of impedances, such as resistance capacity combinations, from the ends of the resistance to the shield. Three conductors join at each terminal of the bridge, as shown diagrammatically in Fig. 3.6, in which it is assumed that all the leakages take place to a common point E, which is usually "earthed," and that interbranch leakages are negligible. It is

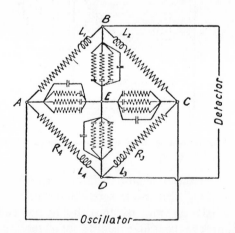

Fig. 3.6 Effect of capacity couplings in an ac bridge [13b].

at once apparent from this figure that paths AEB and CEB, for instance, act as shunts on arms AB and BC, respectively. Therefore errors result unless these paths happen to be adjusted in the same proportion as AB and BC.

The Wagner earthing device [13b] eliminates the difficulty by providing adjustable impedances (resistance and capacity) across AE and CE. The procedure is first to balance the bridge itself with the detector across BD, then to balance to Wagner arms with the detector across BE or DE, and finally to readjust the main bridge balance with the detector in the original position. Thus points B, D, and E are all brought to the same potential and no shunting errors can result.

Figure 3.7 represents a Wheatstone bridge circuit [13b] for electrolytic conductance work of high precision. The bridge is direct reading; that is,

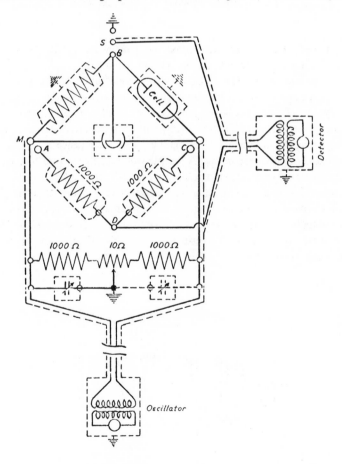

Fig. 3.7   Ac bridge circuit for conductance measurements [13b].

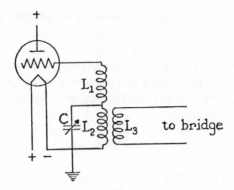

Fig. 3.8   Simple vacuum tube audiooscillator.

the ratio arms AD and CD are equal, so that when both resistance and condenser balances are achieved for the main bridge and Wagner earth circuits the resistance of MB is equal to the resistance of the conductivity cell. A schematic circuit for a simple vacuum-tube audiooscillator is shown in Fig. 3.8. L_1 and L_2 are two fixed inductances, and C is a condenser whose capacity may be varied if different frequencies of current are desired. The alternating current is carried to the bridge from inductance L_3, which picks it up electromagnetically from L_2. A simple one-stage amplifier for use as the bridge detector is shown schematically in Fig. 3.9. It may be coupled to the bridge through a transformer, as shown. Condenser C serves to "tune" the circuit to the frequency of current used in the bridge, and T represents the telephone detector. However, if greater compactness of assembly is desired, the transformer is replaced by resistance coupling of the amplifier to the bridge through a condenser.

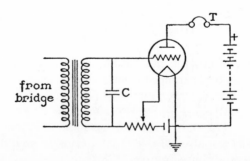

Fig. 3.9   Simple vacuum tube amplifier.

More recently, a high-precision conductance bridge, assembled from commercially available components, has been described by Janz and McIntyre [14]. While fundamentally similar in design to the bridge described above [13b], the ratio arms of their bridge consist of very closely matched inductances of low resistance. No Wagner ground circuit is required; the entire assembly, with oscillator of several frequencies and visual detector, is compact and the bridge is simple to operate.

Direct Current Method

Electrolytic conductance has sometimes been measured with dc and nonpolarizable electrodes [15], such as silver–silver chloride electrodes in chloride solutions, mercury–mercurous sulfate in sulfate solutions, and hydrogen electrodes in acids. The current may be introduced through these electrodes into the cell, whose design is not critical and whose resistance is measured on an ordinary dc Wheatstone bridge. Another alternative is to introduce the current through two electrodes whose nature is not important as long as the electrode reaction products are not allowed to enter the portion of the cell containing the reversible electrodes. The potential difference between these electrodes is measured, as well as the current flowing through the cell, and the resistance between them is computed from Ohm's law. However, the dc method suffers from the limitation that it is applicable only to those electrolytes for which nonpolarizable electrodes can be obtained.

Conductivity Cells

For measuring electrolytic conductance, cells are used which are constructed of relatively insoluble glass, such as Jena 16III or Pyrex, and firmly sealed platinum electrodes. For work of very high precision with extremely dilute aqueous solutions, quartz cells have been used. The resistance R of the solution in the cell is measured, as described above, and the conductance κ is computed from the equation

$$\kappa = \frac{\text{cell constant}}{R}. \tag{3.10}$$

If the cell is a tube of uniform cross section with the electrodes at the ends, the cell constant can be determined from geometric measurements, of length (centimeters) per area (square centimeters) [see (3.2)]. However, in practice, the cell constant is obtained from (3.10) after measuring the resistance of a solution having a known specific conductance. Solutions of potassium chloride are usually employed (Table (3.1)) for this purpose since the specific conductance of several concentrations of this salt at several temperatures has been determined with a very high degree of accuracy [16]. The values given in this table do not include the conductance attributable to the water, which must be added and which should be less than $\kappa_{H_2O} = 10^{-6}$ in work with dilute solutions.

It is good practice to choose cells with appropriate cell constants, so that the resistance does not fall far below 1000 Ω nor above 10,000 to 30,000 Ω. In the former case one avoids complications resulting from excessive polarization and too high current densities; in the latter case one avoids errors

Table 3.1   Specific Conductance, κ, of Potassium Chloride
Solutions for Cell Constant Determinations*

Temp., °C	Grams KCl *per* 1000 *grams water*		
	76.627	7.4789	0.74625
0	0.065176	0.0071379	0.00077364
18	0.097838	0.0111667	0.00122052
25	0.111342	0.0128560	0.00140877

* G. Jones and B. C. Bradshaw, *J. Am. Chem. Soc.*, 55, 1780 (1933).

attributable to insulation leakage. If, however, very high resistances must be measured, it is advisable to shunt the cell with a calibrated high resistance, 10,000 Ω, for instance, and then compute the cell resistance from the parallel circuit equation

$$\frac{1}{R_b} = \frac{1}{R_c} + \frac{1}{R_s},$$ (3.11)

where R_b is the resistance of the parallel combination measured in the bridge and R_s is the resistance of the shunt.

It has been shown that many cells that had been commonly used in careful conductivity work suffer from a source of error inherent in the cell design [17]. Thus, in cells with filling tubes relatively close to the electrode leads, disturbing parasitic currents can flow through capacity-resistance paths and may produce apparent variations in cell constants. Such variations of cell constants at high resistances had been observed by Parker [18], who wrongly attributed the effect to adsorption. Jones and Bollinger [17] showed that this so called "Parker effect" disappeared if the filling tubes and electrode leads in the cells were spaced in a manner that avoids appreciable stray currents. A properly designed cell is shown in Fig. 3.10. Contact is made to the sealed platinum electrodes through the mercury pools, shown as solid areas.

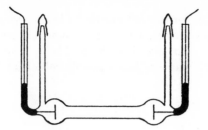

Fig. 3.10   Conductivity cell (after Jones and Bollinger).

With very dilute solutions it is desirable to use a cell of relatively large volume so that increasing concentrations can be built up and successively measured without risk of contamination from atmospheric or other impurities. Some workers have used cells consisting essentially of a flask with a pair of dipping electrodes, as is represented diagrammatically in Fig. 3.11a. However, dipping electrodes may also lead to errors in conductivity measurements if high precision is important. With the cell indicated in Fig. 3.11a, even if polarization is negligible, the total current measured in the bridge consists not only of that flowing directly between the electrodes through the solution and a capacity current between the electrodes (including the leads), but also of a parasitic current which flows from the electrode leads by electrostatic capacity to the solution and then through it. This latter current is neither constant nor directly proportional to the resistance of the solution, and thus tends to produce apparent variations in the cell constant.

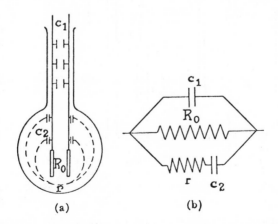

(a)   (b)

Fig. 3.11   "Parasitic" currents in cell with dipping electrodes.

The electrical circuit for the cell in Fig. 3.11a is shown in sufficient detail in Fig. 3.11b. Since this circuit, which also includes electrode polarization not shown in Fig. 3.11b, is balanced in the bridge by a simple resistance and capacity in parallel, it can be shown theoretically that when the bridge is balanced

$$\frac{1}{R_b} = \frac{1}{R_0} + \frac{1}{r} - \frac{1}{\omega^2 C_2^2 r^2 R_0^2}, \tag{3.12}$$

in which ω is the angular frequency, $2\pi f$. The last term depends both on the frequency of current used in the bridge and on the resistance of the solution. To avoid this type of error, the electrode leads should be removed from within the flask, as has been done in the cell shown in Fig. 3.12. Here, the electrodes consist of truncated cones of platinum foil the outer surfaces of which are sealed to the glass.

This type of electrode construction is useful, not only because it assures the rigidity of the position of the electrodes, but also because it avoids the trapping of air bubbles in the cell, which should of course be completely filled with solution. For this reason the cell shown in Fig. 3.13 is particularly useful in work with protein or detergent solutions, which have a tendency to form bubbles.

The "platinizing" of electrodes is carried out electrolytically after filling the cell, which is first cleaned with chromic sulfuric acid mixture or aqua

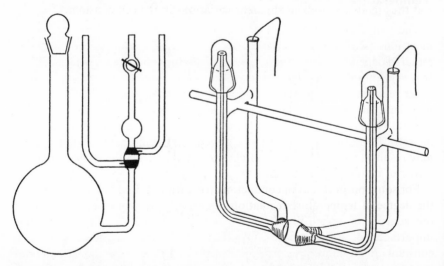

Fig. 3.12 Flask-type conductance cell (Shedlovsky).

Fig. 3.13 Conductance cell for protein and other solutions.

regia, with a solution containing about 3% chloroplatinic acid and a trace (about 0.02%) of lead acetate. The lead tends to favor the formation of a fine, adherent deposit. A current of several milliamperes is sufficient, and it is best to stop the electrolysis when the color of the electrode just changes from light brown to black. The current is then reversed and the procedure repeated for the other electrode. The cell is then thoroughly washed with good distilled water, with which it should be filled for storage.

If the substances to be investigated tend to be decomposed by platinum black, the electrodes may first be "platinized" and washed, then heated to redness until the deposit assumes a gray appearance on cooling. If smooth electrodes are used, definite polarization effects will occur, and it is then necessary to extrapolate measurements of resistance at several frequencies to infinite frequency. This can be done by extrapolation of resistance to $1/f^2 = 0$, where $f =$ current frequency.

It is obvious that the resistance of the leads connecting the cell to the bridge should be negligible or of known magnitude. If long leads are used, they should be shielded wire with the metallic sheath grounded.

To minimize temperature variations resulting from heat exchange between the electrodes of the thermostated cell and the leads, it is convenient to immerse two test tubes containing mercury in the thermostat. Short copper wires provide contact between the electrodes and the mercury pools from which leads are then carried to the bridge.

Temperature Control

The temperature coefficient of conductance is about 2% per degree for most ions (see Appendix). Obviously, the degree of temperature regulation required depends on the accuracy necessary in the conductance determination. It is very important to have accurate temperature control within the narrow range required for high precision conductance measurements. The thermostat may have any appropriate and convenient design, but it is better to use mineral oil rather than water in the bath. This is because the relatively high dielectric constant and conductance of water, as compared with mineral oil, may result in sufficient leakage of electrical energy from the cell to introduce errors of disturbing magnitude.

For some purposes, as in conductometric titrations and other applications, the degree of temperature regulation need not be very precise, and many of the sources of error discussed above may also be correspondingly less important. For other purposes, as in the accurate determination of ionization constants or of concentrations, the control of temperature should be correspondingly greater. If an accuracy of one or two hundredths of a percent is desired, the thermostat should regulate well within a hundredth of a degree.

Purity of Solvent and Solvent Correction

Distilled water always contains some carbon dioxide, and sometimes ammonia and other traces of impurity. The conductance attributable to these substances is a most serious limitation in work with dilute solutions that have a low conductivity.

For most general purposes, adequately pure "conductivity" water can be prepared by redistilling good distilled water to which a small quantity of alkaline permanganate or Nessler's solution has been added for the removal of ammonia. Precautions should be taken to avoid the carrying over of any spray. A quartz, resistance glass, or pure tin condenser should be used. About one-fourth of the distillate is rejected and the rest collected in a thoroughly clean, steamed container with provision for avoiding contamination by carbonic acid or ammonia in the air. Such water should have a specific conductance of less than 1.0×10^{-6} mhos cm^{-1}. For still purer water, special distilling equipment must be used [19]. In recent years, however, ion-exchange resins have come into common use for obtaining low conductance water with ease and little cost, but it may contain traces of organic matter.

If organic solvents are used, they should be both carefully purified by distillation and protected from contamination by moisture which often has a marked influence on the conductance.

The solvent conductance, which is usually very low, is determined in a cell of low cell constant, with a shunt if necessary, as discussed above. The cell is refilled with fresh portions of solvent until a reproducible value is obtained. This procedure is also followed in all conductance determinations to assure that no contamination occurs from material retained on the electrodes or glass walls from solutions previously present in the cell.

In general, the conductance of the solvent can be subtracted from the apparent conductance of the solution. This is true of most salt solutions. However, in certain cases the solvent correction is a complicated matter, which depends on the substance being measured and on the nature of the impurity. Of course, the concentration of the solution is an important factor since it determines the magnitude of the conductance and therefore the relative importance of the much smaller part played by the solvent. For example, in very dilute acetic acid solutions, the conductance attributable to a trace of carbonic acid in the original solvent is repressed, but not that attributable to traces of salt. In acetic acid solutions of higher concentrations, the exact form of the solvent correction becomes trivial because of its inconsequential magnitude in comparison with the total conductance. Since conductance is a readily measurable property of solvents, it can often be used to help determine the purity if conducting contaminants are suspected, and it can indicate the progress of removal of electrolytes. As purification proceeds, the solvent attains a constant, minimum conductance. The purity of "conductivity" water and other solvents, such as alcohol or anhydrous acetic acid, can be estimated in this way.

It is obvious, however, that the conductance method is of little use if the presence of nonconducting impurities are in question.

4 APPLICATIONS

Determination of Ionization Constants

In solvents of high dielectric constant such as water, strong electrolytes appear to be completely dissociated, whereas in solvents of low dielectric constant this is not the case.

It is generally assumed that only a portion of a weak electrolyte is in the form of "free" ions while the rest is undissociated, and that an equilibrium exists between the two forms of the solute in accordance with the law of mass action. However, the undissociated solute need not consist of stable molecules bound by chemical or quantized forces, but may be present as ion pairs. The electrostatic force between a pair of ions of opposite charge that have approached each other closely tends to hold them together as a dipole. If the average kinetic energy of the solvent molecule is greater than the potential energy of the ion pair, it will soon be disrupted. This is probably the case for strong electrolytes in solvents of high dielectric constants. If, however, the average kinetic energy of the solvent molecules is less than the mutual potential energy of the ion pair, it will exist until it is struck by an exceptionally fast solvent molecule. This viewpoint predicts that there should be relatively more ion pairs when the dielectric constant is low, when the ions are small, when the charges are large, and when the temperature is low.

According to the Arrhenius dissociation theory, weak electrolytes were found to follow the Ostwald dilution law:

$$\Lambda = \Lambda_0 - \frac{c\Lambda^2}{K_c\Lambda_0}, \qquad (3.13)$$

an expression derived from the law of mass action in terms of concentrations and the assumptions that the ions are perfect solutes, and that the ratio Λ/Λ_0 represents the degree of dissociation. However, the first assumption is true only as a very rough approximation in dilute solutions; and the second assumption can be valid only if the mobilities of the ions do not change with the concentration, which is contrary to fact.

The ionic equilibrium in a solution of a simple weak electrolyte is represented by the equation

$$A^+ + B^- = AB$$

According to the law of mass action,

$$K = \frac{[A^+][B^-]}{[AB]} = \frac{c^2 x^2 \gamma^2}{c(1 - x)\gamma_u}, \tag{3.14}$$

in which K is the thermodynamic ionization constant, the brackets refer to activities, c is the concentration, x is the degree of ionization, and γ and γ_u are the activity coefficients of the ions and "undissociated molecules" or ion pairs, respectively.

In dilute solutions the ion pairs may be considered as normal solutes with $\gamma_u = 1$. The activity coefficient f for the "free" ions can be computed from the Debye-Hückel equation:

$$-\log \gamma = a\sqrt{cx}, \tag{3.15}$$

in which cx is the ion concentration and the constant $a = 0.509$ in water at 25°C for univalent electrolytes. It increases linearly with decreasing dielectric constant and absolute temperature.

The degree of ionization x can be obtained from the ratio of the actual equivalent conductance Λ to the equivalent conductance Λ_e of the completely ionized solute at the same ion concentration, that is,

$$x = \Lambda/\Lambda_e. \tag{3.16}$$

If, for example, the weak electrolyte is an acid, HA, the values of Λ_e at different ion concentrations can be computed from three sets of conductance data on strong electrolytes [4, 20]. For example,

$$\Lambda_e = \Lambda_{H^+Cl^-} + \Lambda_{Na^+A^-} - \Lambda_{Na^+Cl^-} = \Lambda_{0(HA)} - A\sqrt{cx} + Bcx + \cdots \tag{3.17}$$

$$x = \frac{\Lambda}{\Lambda_{0(HA)} - A\sqrt{cx} + Bcx + \cdots}. \tag{3.18}$$

The details of the computations for x involve a short series of approximations since the ion concentration contains x as a factor.

If the ion concentration is not too high, the degree of ionization can be obtained from the conductance data on the weak electrolyte itself [6], using a modified form of the Onsager equation [see (3.6b)]. Thus

$$x = \frac{\Lambda}{\Lambda_0} F, \tag{3.19}$$

in which

$$F = [(Z/2) + \sqrt{1 + (Z/2)^2}]^2$$

or, approximately, $F = 1 + Z$. The function Z is defined as

$$Z = \frac{\alpha\Lambda_0 + \beta}{\Lambda_0^{3/2}} \sqrt{C\Lambda}.$$

Combining this expression for x with the mass action law results in

$$\Lambda F = \Lambda_0 - \frac{F^2 f^2 \Lambda^2 C}{K\Lambda_0}, \tag{3.20}$$

analogous to the older Ostwald dilution equation, or

$$\frac{1}{\Lambda F} = \frac{1}{\Lambda_0} + \frac{f^2 F\Lambda C}{K\Lambda_0^2}, \tag{3.21}$$

similarly analogous to an equation that had been used by Kraus and Bray.

The activity coefficient f can be computed from the Debye-Hückel equation:

$$-\log f = a\sqrt{cx} = a\sqrt{C\Lambda F/\Lambda_0}.$$

The computations are first carried through with a provisional value of Λ_0 and plots are prepared of (3.21) or (3.20). These plots, which should be linear, are extrapolated to a more nearly correct value of Λ_0. As a rule, two or three series of computations and plots yield a constant value of Λ_0. The value of the ionization constant K can thus be found readily from the slope of the last plot. Intermediate electrolytes, that is, those that are only slightly "weak," have already been discussed in Section 2.

Measurements on the conductance of acetic acid in mixtures of water and methanol [21a], water and ethanol [21b], and water and 1-propanol [21c] have been made over the entire range of solvent compositions. The acetic acid ionization constants have been obtained for these mixed solvents at several temperatures except in the case of methanol which is given only for 25°C.

Determination of Concentration and Solubility

Conductance is a sufficiently sensitive function of concentration to be useful in its quantitative determination. Since the temperature coefficient of conductance is appreciable, about 2% per degree for most ions in water, good temperature regulation is essential for accurate work.

The procedure is obvious. A calibration conductance curve, or analytical function, is first established for the pure substance in the pure solvent for a number of known concentrations at a definite temperature. A solvent conductance correction should be used in work of high precision, so that the values refer only to the conductance of the solute under consideration. The solution whose concentration is to be found is then measured conducto-metrically at the same temperature, and the value sought is read from the curve or computed analytically. This method for determining the concentration of pure electrolytes is capable of great accuracy and has been frequently employed.

The solubility of slightly soluble salts can be determined from conductance measurements of the saturated solutions if the ionization is of a simple form, complete, and not complicated by ion complexes [22]. The computation is made with the equation that defines equivalent conductance:

$$\Lambda = 1000\kappa/C.$$

The specific conductance κ is measured, and if the solubility is very low it is sufficient to use for Λ the Λ_0 value computed from the limiting ionic conductances. The solubility c is obtained more accurately from the equation

$$\Lambda = \Lambda_0 - (\alpha\Lambda_0 + \beta)\sqrt{c} = 1000\kappa/c \qquad (3.22)$$

[see (3.6)], which can be solved by a method of approximation, or from

$$\frac{c}{1000\kappa} = \frac{1}{\Lambda_0} + \frac{(\alpha\Lambda_0 + \beta)}{\Lambda_0^2}\sqrt{c} \qquad (3.23)$$

which is a simple quadratic equation in $\sqrt{c}$.

Determination of Extent of Hydrolysis

The hydrolysis of many salts results in relatively large changes in the conductance of the solutions. For example, aniline hydrochloride hydrolyzes to form nonconducting aniline and an equivalent very mobile hydrogen ion. The net change in conductance obtained from this hydrolysis is attributable to the replacement of hydrogen ions by phenylammonium ions.

If h is the extent of hydrolysis, the specific conductance κ of the hydrolyzed solution is

$$\kappa = (1 - h)\kappa' + h\kappa_{HCl}, \qquad (3.24)$$

in which κ' is the corresponding conductance of the unhydrolyzed solution and κ_{HCl} that of hydrochloric acid at the *same* concentration. κ' can be measured by adding enough aniline base to the salt to repress the hydrolysis. Its own ionization is repressed by the phenylammonium ions present.

The extent of hydrolysis is then found from

$$h = \frac{\kappa - \kappa'}{\kappa_{HCl} - \kappa'}. \qquad (3.25)$$

MacInnes [4] thus computed the hydrolysis constant for aqueous solutions of aniline hydrochloride at 25°C from Bredig's [23] data:

$$K_h = \frac{h^2 C}{(1 - h)} = 2.3 \times 10^{-4}.$$

Conductometric Titrations

In the course of many ionic reactions accompanied by neutralizations, complexations, oxidations, reductions, and precipitations, there are changes in conductance which can be used to determine the end points as well as the progress of the reactions. The conductometric method is particularly useful for very dilute solutions and for ionic reactions to which potentiometric methods are not readily applicable. Accuracy of conductometric titrations depends on ionic conductances, strengths of the acids and bases, hydrolysis, stabilities of complexes, and the solubilities of precipitates involved in the process.

The conductance of a solution depends on the sum of the conductances contributed by all the ionic species present. In the course of some titrations, the change in conductance takes place by displacement of highly mobile ions such as hydrogen and hydroxyl ions by the much less mobile ions of salt solutions, and also by a decrease in ionic concentrations by neutralization. Consequently, the end points can be defined in titrations that involve acids and bases. Frequently, oxidation-reduction reactions also involve changes in hydrion concentrations.

Experimental Procedures

The cell design and the precision of the conductance measurement required depends upon the magnitude of the change in conductance. In many titrations changes in conductance are large and the design of the cell is not critical. A convenient form is shown in Fig. 3.14. The conductance is measured after each successive addition of reagent from an appropriate buret has been thoroughly mixed in the cell.

An appreciable change in volume of solution can be avoided by titrating with a solution that is 10 or 20 times the concentration of the initial solution. In most work the small dilution during the course of titrations can be neglected. If dilution is appreciable, the measured values of reciprocal resistance or specific conductance can be corrected by the fractional increase in volume. If the cell is not

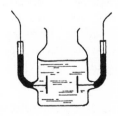

Fig. 3.14 Conductance cell for titrations.

thermostated, it can be provided with a thermometer so that corrections can be made, if necessary, from a knowledge of the appropriate temperature coefficients of conductance for the ions present (see Appendix).

From a graph of the conductance against the volume of titrant added, the end point is located by the intersection of two linear portions. A curvature near the end point which may be attributable to hydrolysis, dissociation, or solubility of the reaction product should often be disregarded in determining the end point.

The conductance can be measured with an ac Wheatstone bridge, although most of the refinements discussed in an earlier section of this chapter can usually be omitted. Several 60- to 1000-cycle bridges which are operated on an ac line are commercially available. These assemblies usually use a galvanometer or "magic eye" tube as a visual detector of the null point. However, a high-sensitivity conductometric titrator makes it possible to titrate successfully in the presence of neutral salts, when the change in conductance during the titration is very small.

Several recording instruments are available for conductivity measurements such as are used in automatic titrations [24]. In such titrations the reagent is added at a fixed rate. In addition, compensating control mechanisms, including cooling units, maintenance of inert atmospheres, and so on, have also been used [25]. Conductometric titrations are occasionally made without placing the metal electrodes in direct contact with the solution. Determinations are usually carried out at high frequencies, in the megacycle range. In this method, which has been referred to as oscillometry, the measurement of conductance is complicated by the fact that the electrodes are not immersed in the solution. Conductometric analyses of this type, have been made at radiofrequencies [26]. An apparatus with which electrodes are avoided entirely makes use of a transformer bridge at audiofrequencies [27].

Separating electrodes from the solution avoids the effect on conductivity of a precipitate on the electrodes and prevents interactions of concentrated acids with them. It also provides a means of monitoring the appearance of bands in chromatographic columns in which electrode contact is made with the outside of the columns. A method using high-frequency alternating current with electrodes that are not in direct contact with solution had been employed for acid-base, precipitation, soluble complex, and oxidation-reduction titrations [28].

The literature [29] should be consulted for details of oscillometric methods and other applications.

Examples of Conductometric Titrations

Some typical conductometric titration curves are shown in Fig. 3.15. The measurements were obtained with cells whose electrodes were immersed in the solutions.

Titration of a strong acid with sodium hydroxide is represented in Fig. 3.15a. At first, the large decrease in conductance is attributable to a replacement of hydrogen ion, with high mobility, by sodium ion, with much lower mobility. The equivalence point is easily obtained as the intersection of two straight lines. After neutralization is complete, the rising linear portion of the graph is attributable to an excess of sodium hydroxide. A plot similar to that shown in Fig. 3.15a would result if a strong base were titrated with hydrochloric acid.

The course of a titration of acetic acid ($K_a = 1.75 \times 10^{-5}$) with a strong base is shown in Fig. 3.15b. When base is added, at first the initial low conductance is decreased because the lower conductance resulting from neutralization of part of the acid is not compensated by the added conductance attributable to the small rise in salt concentration, but later, the conductance increases with the growth in salt concentration. These opposing

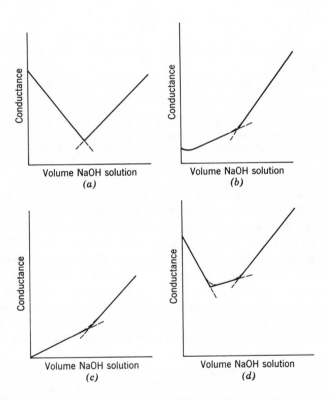

Volume NaOH solution
(a)

Volume NaOH solution
(b)

Volume NaOH solution
(c)

Volume NaOH solution
(d)

Fig. 3.15(a) Strong acid plus strong base: HCl + NaOH. (b) Weak acid plus strong base: HOAc + NaOH. (c) Very weak acid plus strong base: H_3BO_3 + NaOH. (d) Mixture of strong and weak acid plus strong base (HCl + HOAc) + NaOH.

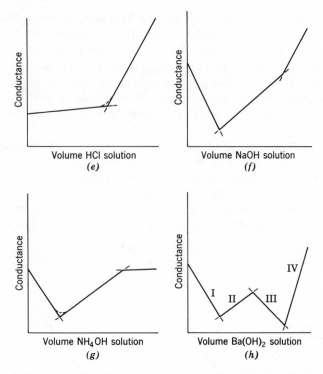

Fig 3.15 (*contd.*)   (*e*) Salt of weak acid plus strong base: NaOAc + HCl. (*f*) Phosphoric acid plus strong base: H_3PO_4 + NaOH. (*g*) Phosphoric acid plus weak base: H_3PO_4 + NH_4OH. (*h*) Phosphoric acid plus barium hydroxide: H_3PO_4 plus $Ba(OH)_2$.

effects form a minimum in the curve. The curvature near the end point results from the hydrolysis of sodium acetate which increases hydroxyl ion concentration. The last part of the curve is attributable to a steady rise in concentration of sodium hydroxide. The first drop in the curve is disregarded and the end point is taken at the intersection of the two linear sections.

The titration of weak acids is not feasible when the extent of hydrolysis is too great and the region of the curve near the end point is not linear because of an increase in hydroxyl ion concentration and in conductance. Obviously, this hydrolysis effect becomes more pronounced the weaker the acid. If the dissociation constant is less than 10^{-11}, a break in the titration curve cannot be defined.

However, good titrations are obtained for weak acids such as boric acid ($K_a = 6.4 \times 10^{-10}$) and phenol ($K_a = 10^{-10}$) for 0.1 N solutions but not for 0.01 N solutions. The plot in Fig. 3.15c is for boric acid titrated with sodium hydroxide. Initially, the conductance of the solution is extremely low. As the titration proceeds, the rise in conductance is the result of salt formation. After the equivalence point, an excess of hydroxide increases the slope of the curve. Boric acid or phenol cannot be satisfactorily titrated with ammonia because the hydrolysis error extends through most of the titration.

For the analysis of mixtures of acids or bases, conductometric titrations are often preferred to the usual potentiometric methods. The titration plot of a mixture of hydrochloric and acetic acids with sodium hydroxide is shown in Fig. 3.15d. The first break in the curve corresponds to the end point for the strong acid and the second break for the weak acid. If hydrolysis effects, shown by the curved portion near the second end point, extend too far to establish clearly the middle linear branch, a sharper break can be obtained by adding ethanol to the solution of mixed acids.

The plot in Fig. 3.15e represents the titration of sodium acetate, a salt of a weak acid, with hydrochloric acid, a strong acid. Here again, addition of ethanol to the solution decreases the ionization of the weak acid, and in turn minimizes the curvature near the end point. The first portion of the plot shows a small rise in conductance that is attributable primarily to the replacement of acetate ion by the chloride ion. The liberated weak acetic acid hardly affects the conductance. After the end point, the conductance rises sharply as the concentration of hydrion resulting from excess acid is increased.

The titration of two weak acids with sodium hydroxide will give two well-defined end points if the dissociation constants differ by at least a factor of 10. In a titration phosphoric acid behaves as a mixture of a relatively strong acid ($K_1 = 7.5 \times 10^{-3}$) and two weak acids ($K_2 = 6.2 \times 10^{-8}$ and $K_3 = 4.2 \times 10^{-13}$). A titration of phosphoric acid with sodium hydroxide is shown in Fig. 3.15f and in Fig. 3.15g where the titrant is aqueous ammonia. In each case the first break in the curve corresponds to one-third of the acid concentration and the second break to the rest of the acid. This second end point is more sharply defined when the titrant is aqueous ammonia.

In the neutralization of phosphoric acid with barium hydroxide [30], three end points in the titration are shown in Fig. 3.15h. Each break in the curve corresponds to the interaction of one-third of the acid. In the third branch of the curve, the neutralization is accompanied by the precipitation of barium phosphate. Satisfactory titrations are obtained if the concentration of the phosphoric acid is less than 0.02 N and if the solution is stirred vigorously for 3 to 5 min after each addition of titrant.

By using a high sensitivity conductometric apparatus, titrations can be made in the presence of excess of indifferent electrolyte [25]. Satisfactory end points have been obtained in the following titrations: (1) HCl + KCl titrated with NaOH (KCl concentration 10,000 times HCl concentration), (2) H_3PO_4 + KCl titrated with HCl (KCl concentration 450 times H_3PO_4

concentration), (3) ferrous salt + HCl titrated with $K_2Cr_2O_7$ (HCl concentration 300 times ferrous salt concentration), (4) disodium ethylenediaminetetraacetate (Na_2EDTA) + NaOH titrated with $CaCl_2$ (NaOH concentration 100 times Na_2EDTA).

Other examples and details are given in Ref. 25.

A change in solvent is sometimes necessary to analyze acidic or basic compounds by conductometric titrations. It was noted before that sharper end points are obtained in an ethanol–water solvent than in aqueous solution when a mixture of hydrochloric and acetic acids is titrated with sodium hydroxide and when sodium acetate is titrated with hydrochloric acid.

Satisfactory conductometric titrations of very weak bases can be made in a slightly acidic solvent such as 50% aqueous ethanol containing acetic acid. In such a solvent the dissociation of the base is increased so that hydrolysis of the salt is reduced. For example, well-defined end points are obtained with p-phenylenediamine, p-toluidine, or glycine titrated with trichloracetic acid. Two end points are determined with mixtures of p-toluidine and glycine [31, 32].

Conductometric titrations, in which a reaction product is a precipitate, are limited by effects of solubility, change in solvent, rate of precipitation, adsorption, occlusion, coprecipitation, and poorly defined precipitates, all of which often interfere with the determinations of the end points.

Titrations in which only precipitation takes place involve ion replacement. For example, silver in silver nitrate solution is well determined by titration with lithium chloride. Here the net decrease in conductance is attributable to the replacement of Ag^+ by Li^+ and the conductances are in the ratio of about 3 to 2.

Sulfate may be quantitatively determined by conductometric titration in 50% aqueous methanol of an approximately 10^{-4} M solution with 0.1 M barium acetate. A high-sensitivity recording titrator with smooth platinum electrodes to avoid undesirable adsorption effects has been used for this purpose [25]. As noted before, the titrant should be added at a slower rate than that of the precipitation. The net decrease in conductance results from the replacement of SO_4^{2-} by $2CH_3COO^-$.

Conductometric titrations can be used to determine the composition of complexes, but in general the values obtained are only qualitative.

A typical example is the titration of potassium aspartate and alaninate, two amino acid ions with cupric salt [33]. From the two breaks in the curves shown in Fig. 3.16, for each amino acid the formation of two coordination complexes, CuA and CuA_2 is indicated, with proportions of 1 mole of amino acid ion to each mole of cupric ion when the latter is in excess, and when the amino acid ion is in excess, 2 moles are coordinated to each mole of cupric ion.

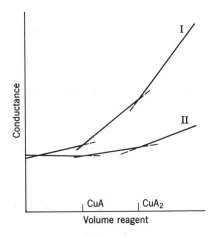

Fig. 3.16 (I) Cupric nitrate plus potassium aspartate. (II) cupric nitrate plus potassium alaninate.

Oxidation-reduction reactions sometimes show large changes in conductance and the completion of the reactions can be determined by conductometric titration. For example, in the oxidation of arsenious acid by iodine to form arsenic acid and hydriodic acid, the initial low conductance is attributable to weak arsenious acid ($K = 6 \times 10^{-10}$). The conductance increases during the titration since arsenic acid and hydriodic acid are much stronger electrolytes and very mobile hydrions are formed. After the arsenious acid is completely oxidized, there is no additional change in conductance. The end point is well defined by the intersection of two linear portions of the curve. The net ionic reaction is as follows.

$$AsO_3^{3-} + I_2 + 3H_2O \rightleftharpoons AsO_4^{3-} + 2I^- + 2H_3O^+.$$

In the conductometric titration of ferrous salt in acidified solution oxidized with potassium dichromate solution, the shape of the curve is mostly influenced by the removal of hydrion. The best results are obtained by titrating under an atmosphere of nitrogen and by acidifying the solution with phosphoric acid. The net ionic reaction is as follows.

$$6Fe^{2+} + Cr_2O_7^{2-} + 14H_3O^+ \rightleftharpoons 6Fe^{3+} + 2Cr^{3+} + 21H_2O$$

There are many other analyses that can be made by conductometric titrations. Hydroxybenzenes, phenols, amino acids, organic phosphates, polypeptides, proteins, vanillin, aniline, pyridine, sulfophthaleins, and alkaloids such as codeine and atropine are other examples among organic compounds. The literature should be consulted for specific applications [24, 25, 30–34].

Determination of Critical Micelle Concentration of Colloidal Electrolytes

The electrolytic conductance of a dilute aqueous solution of a colloidal electrolyte is similar to that of salts such as sodium chloride. Above a certain concentration, there is a sharp change in the slope of the curve of the specific or equivalent conductivity against concentration. This is represented in curve C Fig. 3.17 and in Fig. 3.18 for aqueous sodium n-dodecyl sulfate which

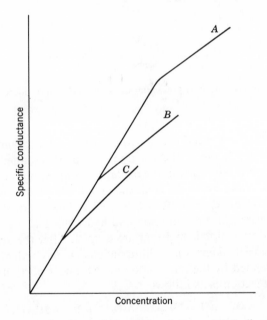

Fig. 3.17 Conductance of sodium alkyl sulfate solutions. (A) Sodium n-decyl sulfate; (B) mixture of sodium n-decyl sulfate and sodium n-dodecyl sulfate; (C) sodium n-dodecyl sulfate.

is a typical anionic colloidal electrolyte [35]. A break in the curves corresponds to the critical micelle concentration (c.m.c.) at which there is a pronounced change in the characteristics of the solution, such as an increase in solvent properties attributable to the formation of micelles that are molecular aggregates of the solute. Sodium n-dodecyl sulfate has an aggregation number of about 70 and a micellar weight of about 20×10^3. For longer alkyl chains and in the presence of added salts, the c.m.c. is lower and the micellar molecular weight is greater. Usually, the aggregation numbers cover a range from about 50 to 150 [36].

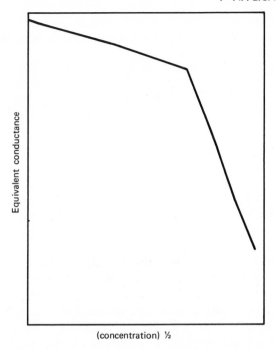

(concentration) ½

Fig. 3.18

Micelles have a net charge much lower than the number of nonmicellar anions in the aggregate. This is attributed largely to interaction with counterions. For sodium alkyl sulfate, the net charge is negative, whereas for a cationic colloidal electrolyte such as a long-chain tetralkylammonium salt, the net charge is positive.

Under any given conditions of temperature and concentration, micelles of a particular average size are formed if the solute consists of only one electrolyte. If the solute consists of more than one component, the conductivity of the solution does not show a sharp break at the c.m.c., and both the size and composition of the micelles are not constant. For example, the sharp break in the plot of specific conductance against concentration of sodium dodecyl sulfonate alone becomes less pronounced when a nonionic surface active agent, dodecyl hexaoxyethylene glycol monoether is present. For example, when there is twice as much nonionic as anionic constituent, the curve is smooth and a break is not well defined [37].

Micellar solutions have solubilization properties which are not shown at nonmicellar concentrations of colloidal electrolytes. This behavior is also reflected in a rapid rise in solubility of a colloidal electrolyte above the Krafft temperature, the temperature at which the solubility is equal to the c.m.c. The Krafft temperature can be determined by the measurement of the conductivity of the solution of a colloidal electrolyte in the presence of undissolved solid phase at various temperatures.

Determination of Rates of Reactions

Very rapid as well as slow reaction rates in solution have been measured by conductometry. Flow methods are restricted by the rate of mixing of the reactants which is not less than about 1 msec. For the study of much faster reactions with a half-time as short as 10^{-9} sec, Eigen and his associates [38, 39, 40] have introduced and developed useful methods which depend on the measurement of relaxation effects. A mixture of the reactants at equilibrium is perturbed by altering an external parameter such as the electric field density, the pressure, or the temperature. The kinetics of the reaction is obtained from the rate of change for the equilibrium concentrations. An excellent treatment of relaxation methods is given in Ref. 39. These methods are particularly suitable for the study of ionic reactions in solution.

The conductivity of weak electrolytes in solution increases at high field strengths because of an increase in ionic dissociation. Changes in the electric field density are produced by discharging a condenser through a reaction mixture. Heating effects are kept at a minimum by using solutions of high resistance. Conductometry furnishes a sensitive detector of the change in ionic concentrations in the solutions. Customary bridge arrangements together with oscillographic measurements are used [39]. Impulses of 10^{-7}-sec duration on an oscillograph may be registered photographically [41].

In the so-called pressure-jump method, a change in static pressure is applied to small volumes of electrolytes, and the minute changes in chemical equilibria are detected by precise conductometry. High frequency bridge methods are well adapted for the measurement of such very small differences in concentration.

As an interesting example of the application of the pressure jump method the relaxation time of the ionic dissociation of beryllium sulfate may be cited [42]. The concentration changes are detected by conductance. More extensive kinetic studies by the same method were made of reactions of metal complexes containing cations such as Fe^{3+} with Cl^-, ClO_4^-, NO_3^-, and SO_4^{2-} [43]. Relaxation effects in solutions of cobalt(II)-malonate were studied with the use of shock waves followed by conductance measurements [44].

In the temperature-jump method, the process of dissociation or association in ionic reactions is accompanied by a change in conductivity which may be followed conveniently by oscillographic observation of alteration in the bridge balance [39]. Flow methods can be used for somewhat slower reactions. A precision conductance apparatus for investigating ionic reactions in solution has been described and used to study the reaction:

$$Be^{2+} + HF = BeF^+ + H^+$$

by a stopped-flow method in which the conductance changes during the reaction by about 1% [45]. With this procedure the reaction rate can be studied at the relatively high ionic concentration of 0.2 M. Rates of substitution reactions of organometallic halides, especially those of silicon and germanium, with half-times from milliseconds to hours have been measured with a stopped-flow apparatus in which electrical and optical procedures were used [46]. Here, a flow trace, a reaction trace, and a timing trace were observed on an oscilloscope. The rate of the reaction $CO_2 + OH^- = HCO_3^-$ was studied by another stopped-flow method in which changes in concentration were detected by the application of conductometry with rapid recording [47]. Slower reactions such as the solvolysis of alkyl chloride [48], in which the half-time is about 2 hr, and slow heterogeneous processes such as the solution in water of fatty acid soaps can easily be measured by conductance [49].

Many other specific applications of conductometry will occur to the reader. The aim of this chapter has been, however, to discuss methods, along with some essential theoretical background, and to present a limited number of typical examples of applications.

In the Appendix some useful conductance data are given, and the general references that follow cite literature of interest that has not been directly referred to in the text.

5 APPENDIX*

Limiting Ion Conductances in Water at 25°C and
Approximate Temperature Coefficients

$$\lambda_2^{\,\circ} = \lambda_{25}^{\,\circ} [1 + a(t - 25)]$$

Cation	$\lambda_{25}^{\,\circ}$	a	Anion	$\lambda_{25}^{\,\circ}$	a
K^+	73.52	0.0189	Cl^-	76.34	0.0188
Na^+	50.11	0.0209	Br^-	78.4	0.0187
H^+	349.82	0.0142	I^-	76.8_5	0.0186
Ag^+	61.92	0.0197	NO_3^-	71.44	0.0180
Li^+	38.69	0.0226	HCO_3^-	44.48	—
NH_4^+	73.4	0.0192	OH^-	198	0.0160
Tl^+	74.7	0.0187	$CH_3CO_2^-$	40.9	—
$\frac{1}{2}Ca^{2+}$	59.50	0.0211	$CH_2ClCO_2^-$	39.7	—
$\frac{1}{2}Ba^{2+}$	63.64	0.0206	$CH_3CH_2CO_2^-$	35.81	—
$\frac{1}{2}Sr^{2+}$	59.46	0.0211	$CH_3(CH_2)_2CO_2^-$	32.59	—
$\frac{1}{2}Mg^{2+}$	53.06	0.0218	ClO_4^-	68.0	—
$\frac{1}{3}La^{3+}$	69.6	—	$C_6H_5CO_2^-$	32.3	—
$\frac{1}{3}Co(NH_3)_6^{3+}$	102.3	—	$\frac{1}{2}SO_4^{2-}$	79.8	0.0196
			$\frac{1}{3}Fe(CN)_6^{3-}$	101.0	—
			$\frac{1}{4}Fe(CN)_6^{2-}$	110.5	—

Limiting Equivalent Conductances of Electrolytes in Methanol at 25°C

Salt	Λ_0	Salt	Λ_0
LiCNS	101.8	CsCl	113.6
NaCNS	107.0	$LiNO_3$	100.2
KCNS	114.5	$NaNO_3$	106.4
RbCNS	118.2	KNO_3	114.5
CsCNS	123.2	$RbNO_3$	118.1
NH_4CNS	118.7	$CsNO_3$	122.9
LiCl	90.9	$Mg(CNS)_3$	120
NaCl	96.9	$Ca(CNS)_2$	122
KCl	105.0	$Sr(CNS)_2$	122
RbCl	108.6	$Ba(CNS)_2$	125

* Data for the four tables in the Appendix are taken in part from D. A. MacInnes, *Principles of Electrochemistry*, Reinhold, 1939, pp. 342, 357–358, 339, and 349, respectively.

Equivalent Conductances of Electrolytes in Water at 25°C

Electrolyte	Concentration, (equiv per liter)								Ref.
	0.0000	0.0005	0.001	0.005	0.01	0.02	0.05	0.10	
NaCl	126.45	124.50	123.74	120.65	118.51	115.76	111.06	106.74	a
KCl	149.86	147.81	146.95	143.55	141.27	138.34	133.37	128.96	a
LiCl	115.03	113.15	112.40	109.40	107.32	104.65	100.11	95.86	b, c
HCl	426.16	422.74	421.36	415.80	412.00	407.24	399.09	391.32	b
NH_4Cl	149.7	—	—	—	141.28	138.33	133.29	128.75	d, e
KBr	151.9	—	—	146.09	143.43	140.48	135.68	131.39	f, g
KI	150.3_8	—	—	144.37	142.18	139.45	134.97	131.11	d
NaI	126.94	125.36	124.25	121.25	119.24	116.70	112.79	108.78	g
NaO_2CCH_3	91.0	89.2	88.5	85.72	83.76	81.24	76.92	72.80	h, d
$NaO_2CCH_2CH_3$	85.92	84.24	83.54	80.90	79.05	76.63	—	—	i
$NaO_2C(CH_2)_2CH_3$	82.70	81.04	80.31	77.58	75.76	73.39	69.32	65.27	i
KNO_3	144.96	142.77	141.84	138.48	132.82	132.41	126.31	120.40	b
$KHCO_3$	118.00	116.10	115.34	112.24	110.08	107.22	—	—	j
$AgNO_3$	133.36	131.36	130.51	127.20	124.76	121.41	115.24	109.14	b
NaOH	248	246	245	240	237	233	227	221	k
$\frac{1}{2}CaCl_2$	135.84	131.93	130.36	124.25	120.36	115.65	108.47	102.46	l
$\frac{1}{2}BaCl_2$	139.98	135.96	134.34	128.02	123.94	119.09	111.48	105.19	l
$\frac{1}{2}SrCl_2$	135.80	131.90	130.33	124.24	120.29	115.54	108.25	102.19	l
$\frac{1}{2}MgCl_2$	129.40	125.61	124.11	118.31	114.55	110.04	103.08	97.10	l
$\frac{1}{2}Ca(OH)_2$	258	—	—	233	226	214	—	—	m
$\frac{1}{2}Na_2SO_4$	129.9	125.74	124.15	117.15	112.44	106.78	97.75	89.98	n, o
$\frac{1}{2}NiSO_4$	—	118.7	113.1	93.2	82.7	72.3	59.2	50.8	p
$\frac{1}{3}LaCl_3$	145.9	139.6	137.0	127.5	121.8	115.3	106.2	99.1	q
$\frac{1}{4}K_4Fe(CN)_6$	184	—	167.24	146.09	134.83	122.82	107.70	97.87	r

[a] T. Shedlovsky, A. S. Brown, and D. A. MacInnes, *Trans. Electrochem. Soc.*, **66**, 165 (1934).
[b] T. Shedlovsky, *J. Am. Chem. Soc.*, **54**, 1411 (1932).
[c] D. A. MacInnes, T. Shedlovsky, and L. G. Longsworth, *J. Am. Chem. Soc.*, **54**, 2758 (1932).
[d] L. G. Longsworth, private communication.
[e] L. G. Longsworth, *J. Am. Chem. Soc.*, **57**, 1185 (1935).
[f] L. G. Longsworth, private communication.
[g] R. A. Lasselle and J. G. Aston, *J. Am. Chem. Soc.*, **55**, 3067 (1933).
[h] D. A. MacInnes and T. Shedlovsky, *J. Am. Chem. Soc.*, **54**, 1429 (1932).
[i] D. Belcher and T. Shedlovsky, private communication.
[j] T. Shedlovsky and D. A. MacInnes, *J. Am. Chem. Soc.*, **57**, 1705 (1935).
[k] G. H. Jeffery and A. I. Vogel, *Phil. Mag.*, **17**, 582 (1934); M. Randall and C. C. Scalione, *J. Am. Chem. Soc.*, **49**, 1486 (1927); J. Goworecka and M. Hlasko, *Roczniki Chem.*, **12**, 403 (1932).
[l] T. Shedlovsky and A. S. Brown, *J. Am. Chem. Soc.*, **56**, 1066 (1934).
[m] T. Noda and A. Miyoshi, *J. Soc. Chem. Ind. Japan*, **35**, Suppl. Bldg., 317 (1932); F. M. Lea and G. E. Bessey, *J. Chem. Soc.*, **1937**, 1612.
[n] From sum of ion mobilities.
[o] T. Shedlovsky, private communication.
[p] K. Murata, *Bull. Chem. Soc. Japan*, **3**, 47 (1928).
[q] G. Jones and C. E. Bickford, *J. Am. Chem. Soc.*, **56**, 602 (1934).
[r] G. Jones and F. C. Jelen, *J. Am. Chem. Soc.*, **58**, 2561 (1936).

Ionization Constants of Acids at 25°C from Conductance Measurements

Acid	$K \times 10^5$	Ref.	Acid	$K_1 \times 10^5$	Ref.
Acetic	1.753	a	Carbonic	0.0431	k
Monochloroacetic	139.6	b	Malonic	139.7	l
Propionic	1.343	c	Succinic	6.63	l
n-Butyric	1.506	c	Glutaric	4.54	l
Benzoic	6.30	d–f	Adipic	3.72	l
o-Chlorobenzoic	119.7	e	Pimelic	3.10	l
m-Chlorobenzoic	15.06	e	Suberic	2.99	l
p-Chlorobenzoic	10.4	e	Methylmalonic	8.47	m
o-Bromobenzoic	140	g	Ethylmalonic	10.9	m
p-Bromobenzoic	10.7	g	n-Propylmalonic	10.3	m
p-Fluorobenzoic	7.22	g	Dimethylmalonic	7.06	m
Phenylacetic	4.88	i	Methylethylmalonic	15.4	m
o-Chlorophenylacetic	8.60	f	Diethylmalonic	70.8	m
m-Chlorophenylacetic	7.24	f	Ethyl-n-propylmalonic	78.4	m
p-Chlorophenylacetic	6.45	h	Di-n-propylmalonic	92.0	m
o-Bromophenylacetic	8.84	f	Phenylmalonic	277	n
p-Bromophenylacetic	6.49	h	Cyclopropane-1,1-di-carboxylic	150	o
p-Iodophenylacetic	6.64	h	Cyclobutane-1,1-di-carboxylic	7.55	o
p-Methoxyphenylacetic	4.36	f			
Acrylic	5.50	i	Cyclopentane-1,1-di-carboxylic	5.96	o
Lactic	13.87	j	Cyclohexane-1,1-di-carboxylic	3.54	o

^{a} D. A. MacInnes and T. Shedlovsky, *J. Am. Chem. Soc.*, **54**, 1429 (1932).
^{b} B. Saxton and T. W. Langer, *J. Am. Chem. Soc.*, **55**, 3638 (1933); T. Shedlovsky, A. S. Brown, and D. A. MacInnes, *Trans. Electrochem. Soc.*, **66**, 165 (1934).
^{c} D. Belcher, *J. Am. Chem. Soc.*, **60**, 2744 (1938).
^{d} F. G. Brockman and M. Kilpatrick, *J. Am. Chem. Soc.*, **56**, 1483 (1934).
^{e} B. Saxton and H. F. Meier, *J. Am. Chem. Soc.*, **56**, 1918 (1934); see also J. F. C. Dippy, F. R. Williams, and R. H. Lewis, *J. Chem. Soc.*, **1935**, 343.
^{f} J. F. C. Dippy and F. R. Williams, *J. Am. Chem. Soc.*, **1934**, 1888.
^{g} J. F. C. Dippy, F. R. Williams, and R. H. Lewis, *J. Am. Chem. Soc.*, **1935**, 343.
^{h} J. F. C. Dippy and F. R. Williams, *J. Am. Chem. Soc.*, **1934**, 161.
^{i} W. I. German, G. H. Jeffery, and A. I. Vogel, *J. Am. Chem. Soc.*, **1937**, 1604.
^{j} A. W. Martin and H. V. Tartar, *J. Am. Chem. Soc.*, **59**, 2672 (1937).
^{k} T. Shedlovsky and D. A. MacInnes, *J. Am. Chem. Soc.*, **57**, 1705 (1935).
^{l} G. H. Jeffery and A. I. Vogel, *J. Chem. Soc.*, **1935**, 21.
^{m} G. H. Jeffery and A. I. Vogel, *J. Chem. Soc.*, **1936**, 1756.
^{n} S. Basterfield and J. W. Tomecko, *Can. J. Res.* **8**, 447 (1933).
^{o} W. L. German, G. H. Jeffery, and A. I. Vogel, *J. Am. Chem. Soc.*, **1935**, 1624.

References

1. S. R. Milner, *Phil. Mag.*, **23**, 551 (1912); **25**, 743 (1913).
2. P. Debye and E. Hückel, *Physik. Z.*, **24**, 305 (1923).
3. L. Onsager, *Physik. Z.*, **27**, 388 (1926); **28**, 277 (1927).
4. D. A. MacInnes, *The Principles of Electrochemistry*, Van Nostrand Reinhold, New York, 1939.
5. T. Shedlovsky, *J. Am. Chem. Soc.*, **54**, 1405 (1932); T. Shedlovsky and A. S. Brown, *ibid.*, **56**, 1066 (1934).
6. T. Shedlovsky, *J. Franklin Inst.*, **225**, 739 (1938). R. M. Fuoss and T. Shedlovsky, *J. Am. Chem. Soc.*, **71**, 1496 (1949).
7. L. Onsager and R. M. Fuoss, *J. Phys. Chem.*, **36**, 2689 (1932).
8. R. M. Fuoss and L. Onsager, *J. Phys. Chem.*, **61**, 668 (1957), *Proc. Natl. Acad. Sci. U.S.*, **45**, 807 (1959).
9. R. M. Fuoss, *J. Am. Chem. Soc.*, **79**, 3301 (1957).
10. R. M. Fuoss and C. A. Kraus, *J. Am. Chem. Soc.*, **79**, 3304 (1957); A. S. Brown, *ibid.*, **56**, 1066 (1934).
11. M. Wien, *Ann. Phys.*, **83**, 327 (1927); **85**, 795 (1928); **1**, 400 (1929).
12. P. Debye and H. Falkenhagen, *Physik. Z.*, **29**, 121, 401 (1928); O. M. Arnold and J. W. Williams, *J. Am. Chem. Soc.*, **58**, 2613, 2616 (1936).
13a. G. Jones and R. C. Josephs, *J. Am. Chem. Soc.*, **50**, 1049 (1928).
13b. T. Shedlovsky, *J. Am. Chem. Soc.*, **52**, 1793 (1930).
14. G. J. Janz and D. E. McIntyre, *J. Electrochem. Soc.*, **108**, 272 (1961).
15. E. D. Eastman, *J. Am. Chem. Soc.*, **42**, 1648 (1920); L. V. Andrews and W. E. Martin, *ibid.*, **60**, 871 (1938); H. E. Gunning and A. R. Gordon, *J. Chem. Phys.*, **10**, 126 (1942); **11**, 18 (1943); R. E. Jervis, D. R. Muir, J. P. Butler, and A. R. Gordon, *J. Am. Chem. Soc.*, **75**, 2855 (1953).
16. G. Jones and B. C. Bradshaw, *J. Am. Chem. Soc.*, **55**, 1780 (1933); G. Jones and L. T. Prendergast, *J. Am. Chem. Soc.*, **59**, 731 (1937); R. W. Brenner and T. G. Thompson, *J. Am. Chem. Soc.*, **59**, 2372 (1937); C. W. Davies, *J. Chem. Soc.*, **1937**, 432, 1326; J. E. Lind, Jr., J. J. Zwolenik, and R. M. Fuoss, *J. Am. Chem. Soc.*, **81**, 1557 (1959).
17. G. Jones and G. M. Bollinger, *J. Am. Chem. Soc.*, **53**, 411 (1931).
18. H. C. Parker, *J. Am. Chem. Soc.*, **45**, 1366, 2017 (1923).
19. C. A. Kraus and W. B. Dexter, *J. Am. Chem. Soc.*, **44**, 2468 (1922); I. Bencowitz and H. T. Hotchkiss, Jr., *J. Phys. Chem.*, **29**, 705 (1925); A. I. Vogel and G. H. Jeffery, *J. Chem. Soc.*, **1931**, 1201.
20. C. W. Davies, *J. Phys. Chem.*, **29**, 977 (1925); M. Sherrill and A. A. Noyes, *J. Am. Chem. Soc.*, **48**, 1861 (1926); D. A. MacInnes, *ibid.*, **48**, 2068 (1926).
21a. T. Shedlovsky and R. L. Kay, *J. Phys. Chem.*, **60**, 151 (1956).
21b. H. O. Spivey and T. Shedlovsky, *J. Phys. Chem.*, **71**, 2171 (1967).
21c. M. Goffredi and T. Shedlovsky, *J. Phys. Chem.*, **71**, 4436 (1967).
22. F. Kohlrausch, *Z. Physik. Chem.*, **44**, 197 (1903); **64**, 129 (1908); J. W. McBain and C. R. Peaker, *Proc. Roy. Soc. (London)*, **A125**, 394 (1929); *J. Phys. Chem.*, **34**, 1033 (1930).
23. G. Bredig, *Z. Physik. Chem.*, **13**, 289 (1894).

24. L. E. Ashman, R. S. Cohen, J. A. Glass, H. R. Stillwell, and H. E. Jones, *Instruments*, **24**, 710 (1951); R. D. Goodwin, *Anal. Chem.*, **25**, 263 (1953); L. J. Anderson and R. R. Revelle, *Anal. Chem.*, **19**, 264 (1947); W. Boardman, *Chem. Ind.* **1963**, April 6 (No. 14), p. 565; D. W. Colwin and R. C. Propst, *Anal. Chem.*, **32**, 1858 (1960); G. W. Ewing, *Instrumental Methods of Chemical Analysis*, 2nd ed., McGraw-Hill, New York, 1960, p. 390; R. L. Garman, *Ind. Eng. Chem. Anal. Ed.*, **8**, 146 (1936).

25. T. R. Mueller, R. W. Stelzner, D. J. Fisher, and H. C. Jones, *Anal. Chem.*, **37**, 13 (1965).

26. G. G. Blake, *Conductometric Analysis at Radio Frequency*, Chemical Publishing Co., New York, 1953.

27. S. R. Gupta and G. J. Hills, *J. Sci. Instr.*, **33**, 313 (1956).

28. W. J. Blaedel and H. V. Malmstadt, *Anal. Chem.*, **22**, 734 (1950); W. J. Blaedel and H. T. Knight, *Anal. Chem.*, **26**, 741 (1954).

29. C. R. Cornelius, J. A. Griffiths, and D. I. Stock, *J. Phys. Chem.*, **62**, 47 (1958); K. Cruse, *Z. Chem.*, **5**(1) 1 (1965); C. N. Reilley, "High-Frequency Methods," in *New Instrumental Methods in Electrochemistry*, P. Delahay, Ed., Interscience, New York, 1954, pp. 319–345; G. Szabo and S. B. Nagy, *J. Sci. Instr.*, **39**, 414 (1962).

30. Z. P. Yakusheva, *J. Appl. Chem.* (*USSR*), **35**(10) 2108 (1962); translated from *Zh. Prikl. Khim.*, **35**(10), 2195 (1962).

31. J. Rose, *Advanced Physical-Chemical Experiments*, Wiley, New York, 1964, pp. 17–20.

32. F. Gaslini and L. Z. Nahum, *Anal. Chem.*, **32**, 1027 (1960).

33. N. C. Li and E. Doody, *J. Am. Chem. Soc.*, **72**, 1891 (1950); **74**, 4184 (1952).

34. J. W. Loveland, in *Treatise on Analytical Chemistry*, I. M. Kolthoff, P. J. Elving and E. B. Sandell, Eds., Part 1, Vol. 4, Interscience, New York, 1963, Chapter 51.

35. E. D. Goddard and G. C. Benson, *Canadian J. Chem.*, **35**(9), 986 (1957); E. K. Mysels and K. J. Mysels, *J. Colloid. Science*, **20**, 315 (1965); K. J. Mysels and R. J. Otter, *J. Colloid. Sci.*, **16**, 462 (1961).

36. K. Shinoda, E. Hutchinson, and P. Van Rysselberghe, eds., *Colloidal Surfactants*, Academic Press, New York, 1963, pp. 20–23.

37. J. M. Corkill, J. F. Goodman, and J. R. Tate, *Trans. Faraday Soc.*, **60**, 986 (1964).

38. M. Eigen, *Discussions Faraday Soc.*, **17**, 194 (1954).

39. M. Eigen and L. de Maeyers, in *Technique of Organic Chemistry*, S. L. Friess, E. S. Lewis, and A. Weissberger, Eds., Vol. 8, Part 2, Interscience, New York, 1963, Chapter 18, pp. 895–1054.

40. B. Havsteen, in *Biochemical Society Symposium*, T. W. Goodwin, Ed., Academic Press, New York, 1966, pp. 53–67.

41. M. Eigen and J. Schoen, *Z. Elektrochem.*, **59**, 483 (1955).

42. H. Strehlow and M. Becker, *Z. Elektrochem.*, **63**, 457 (1959).

43. H. Strehlow and H. Wendt, *Z. Elektrochem.*, **64**, 131 (1960).

44. H. Hoffmann and K. Pauli, *Naturwissenschaften*, **53**(4), 358 (1966).

45. P. A. Tregloan and G. S. Laurence, *J. Sci. Instr.*, **42**(12), 869 (1965).

46. R. H. Prince, *Z. Elektrochem.*, **64**(1), 13 (1960).
47. J. A. Sirs, *Trans. Faraday Soc.*, **54**, 201 (1958).
48. B. L. Murr, Jr., and V. J. Shiner, Jr., *J. Am. Chem. Soc.*, **84**, 4672 (1962).
49. L. Shedlovsky, G. D. Miles, and G. V. Scott. *J. Phys. Colloid Chem.*, **51**, 391 (1947).

General

CONDUCTOMERTIC TITRATIONS

Epstein, Y. A., *Biokhimia*, **30**, 964 (1965) (Acid alkaline function of RNA and assay of ribonuclease activity by conductometric titration).

Ersepke, Z., *Chemie*, **8**, 48 (1952) (A review of differential methods in conductometric titrations).

Hall, N. F., and W. F. Spengeman, *Trans. Wisconsin Acad. Sci.*, **30**, 51 (1937) (Conductometric titration of organic bases in glacial acetic acid).

Khudyakova, T. A., *Pharm. Zentralhalle*, **105**(4), 54 (1966) (Chrono conductometric analysis of mixtures of amino acids with other acids).

Kolthoff, I. M., *J. Am. Chem. Soc.*, **87**(5), 1004 (1965) (Effect of acid-base dissociation of salts on conductometric-titration curves in acetonitrile).

Lydersen, D. L., *Z. Anal. Chem.*, **218**(3), 161 (1966) (Conductometric-titration using only one experimental value of first branch of titration curve).

Patel, K. C., and R. P. Patel, *J. Inorg. Nucl. Chem.*, **28**(8) 1752 (1966) (Spectrophotometric and conductometric studies in chelate formation in copper–tetra benzyl ethylene diamine system).

Wojtszak, Z., *J. Polymer Sci.*, Pt. A, 3(10), 3613 (1965) (Viscometric-conductometric titrations of alkaline earth metal cations and poly [methacrylic acid]).

TEXTS

Britton, H. T. S., "Conductometric-Analysis," in *Physical Methods in Chemical Analysis*, W. G. Berl, Ed., Academic Press, New York, 1951, pp. 51–104.

Fuoss, R. M., and P. Accascina, *Electrolytic Conductance*, Interscience, New York, 1959.

Glasstone, S., *An Introduction to Electrochemistry*, Van Nostrand Reinhold, New York, 1942.

Hague, B., *Alternating Current Bridge Methods*, Pitman, London, 1930.

Hamer, W. J., and R. E. Wood, in *Handbook of Physics*, E. U. Condon, Ed., McGraw-Hill, New York, 1958, Chapter 9, pp. 4-138–4-158 (Electrolytic conductivity and electrode processes).

Loveland, J. W., "Conductometry and Oscillometry," in *Treatise on Analytical Chemistry*, I. M. Kolthoff, P. J. Elving, and E. B. Sandell, Eds., Pt. 1, Theory and Practice, Vol. 4, Interscience, New York, 1963, pp. 2569–2627.

MacInnes, D. A., *The Principles of Electrochemistry*, Van Nostrand Reinhold, New York, 1939.

Robinson, R. A., and R. H. Stokes, *Electrolyte Solutions*, Academic Press, New York, 1955.

MISCELLANEOUS

Anderson, C. B., and S. E. Wood, *J. Chem. Ed.*, **42**, 658 (1965) (60 Cycle conductance bridge for general chemistry laboratory).

Bourrelly, P., *J. Chim. Phys.*, **62**(6), 673 (1965) (Differential conductometric study of kinetics of enzymatic hydrolysis of urea).

Bruning, W., and A. Holtzer, *J. Am. Chem. Soc.*, **83**, 4865 (1961) (Effect of urea on hydrophobic bonds in dodecyl-trimethyl ammonium bromide from conductivity measurements).

Feltham, R. D., and R. G. Hayter, *J. Chem. Soc.*, **1964**, 4587 (Method for the determination of electrolyte type in aqueous and non-aqueous solvents).

Gavis, J., *J. Chem. Phys.*, **41**(12) 3787 (1964) (Electrical conductivity of low dielectric constant liquids by d.c. measurement).

Greenke, R. A., and H. B. Mark, *Anal. Chem.*, **38**, 340 (1966) (Conductometric determination of carbonyl compounds).

Liler, M., *J. Chem. Soc.*, **1965**, 4300 (Conductometric determination of basic dissociation constants of weak bases in sulfuric acid).

Michaels, A. S., and L. Mir, *J. Phys. Chem.*, **69**(5), 1447 (1965) (Conductometric study of polycation polyanion reactions in dilute aqueous solutions).

Muir, Jr., B. L., and V. J. Shiner, Jr., *J. Am. Chem. Soc.*, **84**, 4672 (1962) (Precise conductance measurements and determination of rate data for solvolysis of alkyl chloride).

Packter, A., *J. Polymer Sci.*, Pt. A, **1**(10) 3021 (1963) (Conductivity and viscosity measurements on poly (sodium acrylates) of different molecular weights. Walden rule not applicable to such solutions).

Popovych, O., *Anal. Chem.*, **38**, 117 (1966) (Conductometric determination of solubility of Ag salts in solvents of medium and low dielectric constant).

Savedoff, L. G., *J. Am. Chem. Soc.*, **88**, 664 (1966) (Conductance of electrolytes in anhydrous acetone).

Schmidt, K., *Rev. Sci. Instr.*, **37**(5), 671 (1966) (Simple transistorized conductance bridge with phase sensitive detector).

Wolff, H. H., *Rev. Sci. Instr.*, **30**, 1116 (1959) (Wagner-earth and other null instrument capacity neutralizing circuits).

Chapter **IV**

TRANSFERENCE NUMBERS
Michael Spiro

I INTRODUCTION AND BASIC THEORY

The electrical conductance of an electrolyte solution is a measure of the extent to which *all* the ions present in the solution move in the direction of an applied electric field and so carry the resulting electric current. The relative extent to which a given type of ion i participates in this process, and thus the proportion of the current it carries, is expressed by its ionic or electric transport number t_i. Unfortunately, in the majority of cases (whenever, in fact, there are complex ions present) we simply cannot measure t_i. What we can and do always measure, however, is the relative transfer on the passage of current, not of a particular type of ion, but of a particular ion-constituent R. This we call the transference number T_R [1–3]. All these quantities are now to be related to the properties of the ions in the solutions, both to clarify the picture and because the main reason for determining transference numbers at all is to obtain information about the species in the solution. The equations are illustrated with reference to a simple electrolyte, aqueous oxalic acid ($H_2 Ox$). For readers not interested in the detailed derivations, the definition on p. 209 and equation (4.10) are the most essential parts of the following discussion.

Conductance

Let us consider a solution containing ions $1, 2, \ldots i$. Ion i has an algebraic charge number z_i (e.g., $+1$ for H^+, -2 for Ox^{2-}) and an arithmetic charge number $|z_i|$ (e.g., $+1$ for H^+, $+2$ for Ox^{2-}). If it is present to the extent of M_i mole liter^{-1}, its normality C_i in equiv liter^{-1} is

$$C_i = M_i |z_i|. \tag{4.1}$$

(In pure SI terminology, liter is replaced by cubic decimeters.) Under the influence of an applied electric field X volt cm^{-1}, any ion such as i moves in the direction of the field with a constant velocity v_i cm sec^{-1} that is proportional to the field, that is,

$$v_i = u_i X, \tag{4.2}$$

where u_i is the mobility of i in cm^2 sec^{-1} volt^{-1}. It is related to the equivalent conductance of the ion λ_i in cm^2 ohm^{-1} equiv^{-1}, by

$$\lambda_i = u_i F, \tag{4.3}$$

where F is the Faraday constant (96,487 coulomb equiv^{-1}).

Consider an imaginary reference plane (Fig. 4.1) of area A in the solution and perpendicular to the applied field. In time t seconds there passes across the plane all the ions of type i contained in the volume $v_i t$(centimeters) $\times$ A(square centimeters), or $v_i t A M_i/1000$ mole of i. These carry an electric charge of $|z_i| Fv_i t A M_i/1000$ coulomb. The total flux of charge transferred across the plane in both directions by all the anions and cations is therefore

$$\sum_i z_i Fv_i t A M_i/1000.$$

The current I in amperes is obtained on dividing by t, and the specific conductivity κ (ohm^{-1} cm^{-1}) is, from its definition and Ohm's law,

$$\kappa = I/AX = \sum_i |z_i| Fu_i M_i/1000 = \sum_i C_i \lambda_i/1000. \tag{4.4}$$

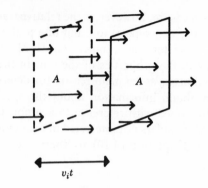

Fig. 4.1   Schematic flow diagram.

If there is only one electrolyte present (such as oxalic acid) of normality C equiv litre^{-1}, its equivalent conductance is given by

$$\Lambda C = 1000\kappa = \sum_i |z_i| \, Fu_i M_i = \sum_i C_i \lambda_i. \tag{4.5}$$

Ionic or Electric Transport Number

It was shown above that the charge carried across the reference plane in Fig. 4.1 by ions of type i is $|z_i| \, Fv_i tAM_i/1000$ coulomb or $|z_i| \, v_i tAM_i/1000$ faraday. Now the number of faradays carried by ions of type i across the reference plane (which is taken as fixed relative to the solvent) when 1 faraday of electricity passes across that plane is the ionic or electric transport number, and thus

$$t_i = \frac{|z_i| \, v_i AM_i/1000}{\sum_i |z_i| \, v_i AM_i/1000} = \frac{|z_i| \, u_i M_i}{\sum_i |z_i| \, u_i M_i} = \frac{|z_i| \, \lambda_i M_i}{\sum_i |z_i| \, \lambda_i M_i} = \frac{\lambda_i C_i}{\sum_i \lambda_i C_i} = \frac{\lambda_i C_i}{\Lambda C}. \tag{4.6}$$

It is clear that t_i is also the fraction of the total current carried by the ionic species i, and although this is the usual definition, it is of little use in practice since we possess no devices for measuring current fractions.

It follows that t_i is dimensionless, either positive or zero, and that

$$\sum_i t_i = 1. \tag{4.7}$$

For example, in an aqueous solution of oxalic acid there are present as solutes the (hydrated) ionic species H^+, HOx^-, and Ox^{2-}, and the molecular species

H_2Ox. From (4.6)

$$t_{H^+} = \lambda_{H^+} M_{H^+} / \Lambda C$$

$$t_{HOx^-} = \lambda_{HOx^-} M_{HOx^-} / \Lambda C$$

$$t_{Ox^{2-}} = 2\lambda_{Ox^{2-}} M_{Ox^{2-}} / \Lambda C$$

$$t_{H_2Ox} = 0$$

$$\Lambda C = \lambda_{H^+} M_{H^+} + \lambda_{HOx^-} M_{HOx^-} + 2\lambda_{Ox^{2-}} M_{Ox^{2-}}.$$

The basic principle of measurement is evident from the definition. The number of moles of i passing some reference plane (such as an etch mark on a tube) per faraday is determined, and with the charge number of species i known the number of faradays carried by these ions follows immediately. This is sufficient when the electrolyte dissociates into two ions only, as in the case of aqueous HCl. However, a serious difficulty arises when we consider a solution in which complex ions exist; aqueous oxalic acid is a typical example. Here, because of the rapid dynamic equilibrium between the ionic and molecular species in the solution, it is impossible to measure experimentally the number of moles of H^+ or of HOx^- or of Ox^{2-} *ions* that pass a reference plane. Hydrogen is being carried toward the cathode as H^+ ions and toward the anode as part of HOx^- ions, and the difference between the two flows, the *net* number of moles of hydrogen being transferred in one direction, is the only quantity we can determine experimentally (by titration with alkali, for example). Similarly, we can obtain the net transfer of oxalate in the form of both HOx^- and Ox^{2-} ions but not the transfer of each ionic species separately. Clearly, we must introduce another kind of transference number to define the quantity that we actually measure in the laboratory.

Ion-Constituent Transference Number

It is convenient to begin with the concept of "ion-constituent" or radical. Thus in oxalic acid solution the ion-constituent H^+ exists in the forms of the ions H^+, the ions HOx^- and the molecules H_2Ox, while the ion-constituent Ox^{2-} exists in the form of the species HOx^-, Ox^{2-}, and H_2Ox. It is the concentration of H^+ ion-constituent, not that of H^+ ions, that is determined by chemical analysis such as titration with alkali, and likewise it is the transfer of an ion-constituent, and not that of an individual ion, that can be measured when an electric current flows. This brings us to the formal definition: "The transference number, T_R, of a cation- or anion-constituent R is the net number of faradays carried by that constituent in the direction of the cathode or anode, respectively, across a reference plane fixed with respect to the solvent, when one faraday of electricity passes across the plane." Clearly, T_R is dimensionless and

$$\sum_R T_R = 1. \tag{4.8}$$

Numerically, T_R is equal to the net number of equivalents of R traversing the plane per faraday, or the net number of moles of R passing for every $|z_R|$ faradays. It should be noted that T_R can be not only positive or zero but also negative. A well-known example is a concentrated aqueous solution of cadmium iodide in which the presence of many mobile CdI_3^- and CdI_4^{2-} ions results in a *net* transfer of cadmium *ion-constituent* toward the anode. (The ionic or electric transport number of cadmium *ions* is of course still positive though, as explained above, it cannot be measured.)

A quantitative relation between T_R and the properties of the ions in solution is easily arrived at by supposing that each mole of ion i contains $N_{R/i}$ moles of ion-constituent R. For example, in 1 mole of CdI_3^- ions there are 3 moles of iodide ion-constituent and 1 mole of cadmium ion-constituent. The electrical charge on any ion is then the algebraic sum of the charges of its component ion-constituents, that is,

$$\sum_R z_R N_{R/i} = z_i. \tag{4.9}$$

It was shown earlier that $v_i t A M_i / 1000$ mole of ion i cross the reference plane in Fig. 4.1 in time t. The number of moles of ion-constituent R crossing the plane as part of the ionic species i is therefore $N_{R/i} v_i t A M_i / 1000$. Before summing over all species to find the total transfer of R we must take into account two directional effects. First, cations and anions carry a given ion-constituent in opposite directions and so there is need of a factor $z_i/|z_i|$ or $|z_i|/z_i$ that is $+1$ for any cation and -1 for any anion. Second, a factor $z_R/|z_R|$ is required as well because the definition of transference number distinguishes between cation and anion constituents. Thus the net number of moles of R crossing the reference plane is

$$\sum_i \frac{z_R}{|z_R|} \cdot \frac{|z_i|}{z_i} \cdot \frac{N_{R/i} v_i t A M_i}{1000}.$$

To obtain T_R we must multiply by $|z_R|$ to obtain the net number of faradays transferred by R and divide by the total flux of charge, in faradays, transferred across the plane in both directions. Cancelling common factors and utilizing (4.2) and (4.3), we finally arrive at

$$T_R = \frac{\sum_i z_R \left(\frac{|z_i|}{z_i} \right) N_{R/i} u_i M_i}{\sum_i |z_i| u_i M_i} = \frac{\sum_i z_R \left(\frac{|z_i|}{z_i} \right) N_{R/i} \lambda_i M_i}{\sum_i |z_i| \lambda_i M_i}$$

$$= \frac{\sum_i \left(\frac{z_R}{z_i} \right) N_{R/i} \lambda_i C_i}{\sum_i \lambda_i C_i}. \tag{4.10}$$

The relation between T_R and t_i is therefore

$$T_R = \sum_i \left(\frac{z_R}{z_i}\right) N_{R/i} t_i. \qquad (4.11)$$

The two transference numbers are equal only for an electrolyte that dissociates into not more than two ionic species, for then ion-constituent R and ion i become synonymous.

A quick check proves that (4.10) and (4.11) do indeed fulfill condition (4.8):

$$\sum_R T_R = \sum_R \sum_i t_i N_{R/i} z_R / z_i = \sum_i \sum_R t_i N_{R/i} z_R / z_i$$

$$= \sum_i t_i \sum_R N_{R/i} z_R / z_i = \sum_i t_i \qquad \text{[by (4.9)]}$$

$$= 1 \qquad \text{[by (4.7)]}.$$

Let us now apply these equations to find the transference numbers of the hydrogen and oxalate ion-constituents in an aqueous oxalic acid solution:

$$T_{H^+} = \frac{\lambda_{H^+} M_{H^+} - \lambda_{HOx^-} M_{HOx^-}}{\lambda_{H^+} M_{H^+} + \lambda_{HOx^-} M_{HOx^-} + 2\lambda_{Ox^{2-}} M_{Ox^{2-}}} \qquad (4.12)$$

$$T_{Ox^{2-}} = \frac{2\lambda_{HOx^-} M_{HOx^-} + 2\lambda_{Ox^{2-}} M_{Ox^{2-}}}{\lambda_{H^+} M_{H^+} + \lambda_{HOx^-} M_{HOx^-} + 2\lambda_{Ox^{2-}} M_{Ox^{2-}}}. \qquad (4.13)$$

The factor of 2 arises in each case because $z_{Ox^{2-}} = -2$. A further small term ought in each case to be added or subtracted to allow for any net migration of H_2Ox molecules carried along with the ions, but this is normally omitted [4].

Should the concentration of the oxalic acid be so high that the relative concentration of Ox^{2-} is negligible, it would be possible to choose as ion-constituents either H^+ and Ox^{2-} as above or else H^+ and HOx^-. In the latter case:

$$T_{H^+} = 1 - T_{HOx^-} = \frac{\lambda_{H^+} M_{H^+}}{\lambda_{H^+} M_{H^+} + \lambda_{HOx^-} M_{HOx^-}} = \frac{\lambda_{H^+}}{\lambda_{H^+} + \lambda_{HOx^-}}. \qquad (4.14)$$

This is a typical equation for a simple 1:1 electrolyte. The only restriction in the choice of ions as constituents is that they must not dissociate into smaller entities under the given experimental conditions and no ambiguity arises as long as a given choice is clearly stated and thereafter consistently adhered to. Where the solute reacts appreciably with the solvent, solvonium and solvate ion-constituents (e.g., H^+ and OH^- in water) must be included. Solutions of amino acids are examples of this [2].

Further applications of (4.10) to other inorganic and organic electrolytes may be found elsewhere [1–3]; of particular interest is its use in a recent review of transport properties of polyelectrolyte solutions [5]. It must be emphasized once again that what is experimentally measured is always the transference number of an ion-constituent, and only for solutions containing no more than two ionic species (e.g., strong electrolyte solutions) is this property also numerically the same as the ionic or electric transport number. Most elementary textbooks mention only the latter, and it is for this reason that in the present chapter the ion-constituent transference number has been treated in some detail.

A comment is necessary about the frame of reference used in the definition of transference numbers in solution: an imaginary plane fixed with respect to the solvent as a whole. There is no doubt that this is an operational concept suitable for macroscopic experiments, and yet it leaves something to be desired from the microscopic viewpoint. Molecules of the solvent may be crudely classified into those that are free and those that are solvating and traveling with the ions; could one but set up a plane stationary with respect to that free part of the solvent that does not move on the passage of current, one would be able to determine "true" or "absolute" transference numbers [6]. Unfortunately, no one has yet been able to do so. Much effort has been expended on this subject in the past, but both of the ingenious techniques employed have been shown to be invalid. In the first a small amount of a so-called inert reference substance such as sucrose was added and the movement of solvent measured relative to it, until it was found that the results depended on the substance chosen and on its concentration [7]. The reason was simple—the reference substance, too, was polar, and so to some extent also solvated and traveled with the ions. In the second approach the electrolyte solution was divided into two (or more) sections by a diaphragm or membrane stretched across the cell, and the amount of solvent carried through during electrolysis was measured volumetrically or gravimetrically. Apart from the problem of electroosmosis, the main obstacle to success was that the type of diaphragm influenced the results, and indeed by using ion-exchange membranes one can obtain at will transference numbers ranging from 0 to 1. Glass frits are the most innocuous, but to the best of the author's knowledge no systematic series of such experiments with glass frits of graded porosity has yet been carried out. We are therefore forced to return to using the solvent as a whole to fix the position of the reference plane. In moving boundary experiments one measures the transference number relative to the cell and then an appropriate volume correction must be applied; this is discussed further under the appropriate technique.

Pure molten salts, for which the solvent is also the electrolyte itself, form a special case and have been discussed elsewhere [8].

Ion-Constituent Conductance and Mobility

The equivalent conductance of an ion-constituent R is given by

$$\bar{\lambda}_R = T_R \Lambda = \sum_i N_{R/i} \lambda_i \left(\frac{z_R}{z_i}\right)\left(\frac{C_i}{C}\right) \tag{4.15}$$

and its mobility by [cf. (4.3)]

$$\bar{u}_R = \frac{\bar{\lambda}_R}{F} = \sum_i N_{R/i} u_i \left(\frac{z_R}{z_i}\right)\left(\frac{C_i}{C}\right). \tag{4.16}$$

For a simple weak electrolyte such as acetic acid in water, we have

$$\bar{u}_{OAc^-} = u_{OAc^-} C_{OAc^-}/C = \alpha u_{OAc^-}, \tag{4.17}$$

where α is the degree of dissociation of the acid. If the electrolyte is completely dissociated, α is unity so that for HCl in water, for example,

$$\bar{u}_{Cl^-} = u_{Cl^-} \tag{4.18}$$

These equations are employed later to determine the stability or otherwise of possible moving boundary systems.

2 VARIATION WITH PHYSICAL FACTORS

Variation with Concentration

At infinite dilution (or more correctly, at infinitesimal ionic strength [9]) ion association is absent. Transference numbers extrapolated to this state, called limiting transference numbers, are then very simply related to the limiting equivalent conductances of the constituent ions and of the electrolyte as a whole by

$$T_\pm^0 = \frac{\lambda_\pm^0}{\lambda_\pm^0 + \lambda_\mp^0} = \frac{\lambda_\pm^0}{\Lambda^0} \tag{4.19}$$

whatever the valence type of the ions. (This depends on choosing as the ion-constituents the two ions that the electrolyte actually dissociates into in this state.) Thus (4.19) shows that combination of limiting transference number and limiting equivalent conductance produces the limiting ionic conductance, virtually the only property of an individual ion that can be unambiguously determined. Once a table of single ionic conductances has been drawn up for a given solvent [10], (4.19) allows us to calculate the limiting transference numbers for many electrolytes.

Transference numbers are essentially ratios of conductances and so usually vary less with concentration than do conductances themselves. The concentration dependence of the transference numbers of most *completely dissociated symmetrical electrolytes* (symmetrical in terms of charge) follows a standard pattern in dilute solutions (Fig. 4.2). As the concentration increases, the transference numbers increase when they exceed 0.5, decrease when they are less than 0.5, and change hardly at all when they are close to 0.5. Moreover, the rate of change of transference number with concentration is greater the larger is the value of $(T - 0.5)$. The explanation is given by the

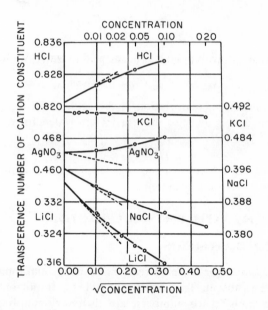

Fig. 4.2   Variation with concentration of the transference numbers of some uni-univalent electrolytes in water at 25°C [11]. The dotted lines are the theoretical limiting slopes.

now classic picture of an ionic atmosphere around each ion treated by the Debye-Hückel-Onsager (D-H-O) theory and its various modifications [12a]. Briefly, when an ion moves in an electric field, its conductance λ_i is reduced both by an electrophoretic term λ_e (resulting from the counterflow of solvent in the ion's atmosphere) and by a relaxation or asymmetry effect caused by the electrostatic drag as the ion and its atmosphere move in opposite directions. Together,

$$\lambda_i = (\lambda_i^0 - \lambda_e)\left(1 - \frac{\Delta X}{X}\right), \qquad (4.20)$$

where $\Delta X/X$ is the fractional change in the applied field X attributable to the relaxation effect [13, 14]. This equation does not take into account certain interactions between the electrophoretic and relaxation effects [15]. For electrolytes that are symmetrical in terms of charge, λ_e and $\Delta X/X$ are the same for both ions. Thus

$$T_{\pm} = \frac{\lambda_{\pm}^{0} - \lambda_e}{\Lambda^0 - 2\lambda_e} \qquad (4.21)$$

since the relaxation effect cancels out. Introduction of (4.19) and rearrangement yields Kay and Dye's general expression [14]

$$T_{\pm} = T_{\pm}^{0} + \frac{2\lambda_e(T_{\pm} - 0.5)}{\Lambda^0}. \qquad (4.22)$$

According to the Fuoss-Onsager extension of the D-H-O theory [12a–14],

$$\lambda_e = \beta |z| \sqrt{I}/(1 + Ba\sqrt{I}) \qquad (4.23)$$

for a $z:z$ valent electrolyte, where

$$\beta = 41.25/\eta(\epsilon T)^{1/2}$$

$$B = 50.29 \times 10^8/(\epsilon T)^{1/2}$$

$$I = \tfrac{1}{2} \sum_i M_i z_i^2 = M z^2.$$

Here I is the ionic strength, η the viscosity of the solvent in poise, ϵ the dielectric constant of the solvent, T the absolute temperature, and a the distance of closest approach of cation and anion in centimeters. In water at 25°C and 1 atm, $\beta = 30.3$ and $B = 0.329 \times 10^8$ [12b].

The D-H-O theory is based on a model of hard-sphere ions in a solvent continuum, and it is not therefore surprising that (4.22) holds only up to about 0.03 N in aqueous solutions. In solvents of smaller dielectric constant it is theoretically applicable only in still lower concentration ranges, and a recent analysis [16] has demonstrated that even here the fit of (4.22) to the best non-aqueous experimental data is far from satisfactory and may require physically unreasonable values of a. For practical extrapolation purposes it is recommended that a sensible a value, say 3–5 Å, be inserted into (4.23) and the $T_{\pm}^{0}$ values calculated from the resulting (4.22) be plotted against concentration. Such a graph will normally be linear, as indeed the closely related semiempirical Longsworth plots normally are [17].

In very dilute solutions in any solvent, $Ba\sqrt{I} \ll 1$, and (4.22) tends to the limiting form

$$T_{\pm} = T_{\pm}^{0} + 2\beta |z| (T_{\pm}^{0} - 0.5)\sqrt{I}/\Lambda^0 \qquad (4.24)$$

for symmetrical electrolytes. This is the equation of the dotted lines in Fig. 4.2, and it can be seen that these serve well as limiting tangents to most of the experimental T_+ versus $\sqrt{C}$ plots. The limiting theoretical equation for *completely dissociated unsymmetrical electrolytes* is

$$T_\pm = T_\pm{}^0 + \{(|z_+| + |z_-|)T_\pm{}^0 - |z_\pm|\}\beta\sqrt{I}/\Lambda^0, \qquad (4.25)$$

but this does not describe the experimental behavior as successfully [12a, 18, 19].

Symmetrical electrolytes showing ion association, either in the form of molecules (e.g., CH_3COOH) or ion pairs (e.g., $Zn^{2+}SO_4{}^{2-}$), should theoretically behave similarly to completely dissociated ones. The main difference is that the ionic strength I in the above equations refers to the actual concentrations of ions and so is proportional to the degree of dissociation α, but otherwise α does not affect the transference number for it cancels out between numerator and denominator in an equation such as (4.14) where $M_{H^+} = \alpha M$ and $M_{HOx^-} = \alpha M$. This is in marked contrast to the equivalent conductance which is directly proportional to the degree of dissociation [10] [cf. (4.5)]. Transference numbers of weak electrolyte solutions should therefore vary far less with concentration than do their conductances, and also change less than transference numbers of strong electrolyte solutions.

Tests of the applicability of (4.22) and (4.24) to weak electrolyte solutions are sparse and largely restricted to a few slightly associated 1:1 salts in non-aqueous media [14, 16] and higher valence ones in water [20]. The transference numbers of certain acids such as iodic and phosphoric in aqueous solution, which might be obvious examples, deviate considerably from the concentration dependence expected as a result of the presence of hydrogen-bonded triple ions [21].

Electrolytes, symmetrical or not, producing complex ions exhibit diverse behavior. A change in the proportion of the different ionic species often has a greater influence than ionic atmosphere effects and may lead to a marked change in the transference number. The example of triple ions in aqueous acid solutions has just been mentioned, and similar effects are expected in solvents of low dielectric constant [22]. A classic instance is provided by concentrated aqueous zinc or cadmium halide solutions in which the cation-constituent transference number is negative, indicating net transfer of metal from cathode to anode, because a large fraction of the metal ion-constituent is present in the form of electrically mobile and negatively charged complexes such as $ZnCl_3^-$ and $ZnCl_4^{2-}$. In polyelectrolyte solutions the formation of micelles surrounded by counter ions leads to a sudden drop in the transference number of the counter ion-constituent and a corresponding rise in that of the polymeric ion-constituent (Fig. 4.3).

For *mixtures of electrolytes* the transference numbers are still given by (4.10) and may be calculated from ionic conductances with the aid of the very approximate rule that the latter are functions of the ionic strength in dilute

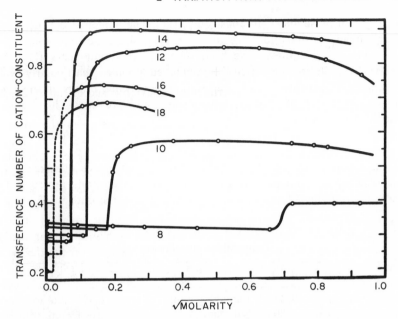

Fig. 4.3 Variation with concentration of the transference numbers of some primary amine hydrochlorides in water at 60°C [23]. The number on each curve indicates the number of carbon atoms in the electrolyte.

solution. Theory [24] shows, however, that while the electrophoretic effect, at low concentrations, depends only upon the ionic strength, the relaxation effect of each ion is a function of the conductances of all the ions present in the ionic atmosphere and on the ratio of their concentrations. Quantitatively, even existing theory is in poor agreement with experiment [25], and therefore it is not surprising that at higher concentrations ionic conductances are no longer additive [26].

Variation with Solvent, Temperature, and Pressure

According to Stokes' "law," derived from hydrodynamics in 1845, a large spherical ion of radius r and charge $z_i e$ moving through a homogeneous continuum of viscosity η under an applied electric field X attains a steady velocity

$$v_i = |z_i| \, eX/6\pi r_i \eta,$$

provided there is no slip at the interface. It thus possesses a limiting ionic conductance [see (4.2) and (4.3)]

$$\lambda_i^0 = \frac{Fv_i}{X} = \frac{|z_i| \, eF}{6\pi\eta r_i} = \frac{0.820 \, |z_i|}{\eta r_i}, \tag{4.26}$$

where η is in poise and r_i in ångströms [12c]. The model used is of course rather unrealistic—most ions are about the same size as solvent molecules— and several modifications have been suggested in recent years [12c, 27]. Nevertheless, Stokes' law itself has remained popular, particularly when combined with the further assumption that the solvated radius of any given ion is independent of solvent, temperature, and pressure. This leads to Walden's rule, formulated in 1906,

$$\lambda_i{}^0\eta = \text{constant} \tag{4.27}$$

and the corollary [cf. (4.19)] that the limiting transference number of any given electrolyte is independent of solvent, temperature, and pressure. The approximate nature of this rule is demonstrated by the data in Table 4.1,

Table 4.1   Variation of Transference Number with Solvent, Temperature, and Pressure

P(atm)	t(°C)	Solvent	$T_{K^+}{}^0$ in KCl
1	−33	Liquid ammonia	0.493
1	30	Sulfolane	0.303
1	25	Formamide	0.427
1	25	N-Methylformamide	0.529
1	25	N,N-Dimethylformamide	0.359
1	25	Methanol	0.501
1	25	Ethanol	0.519
1	25	Water	0.491
1	25	Water	0.490 (0.1 N)
1000	25	Water	0.480 (0.1 N)
25	100	Water	0.481 (0.1 N)

obtained directly or by calculation from the references cited in the Appendix. It should be pointed out that the transference numbers of KCl vary less with changes in physical conditions than do those of many other electrolytes; HCl, for example, exhibits far greater variations because the proton-jump mechanism of the H^+ ion is strongly structure dependent. It is true, all the same, that transference numbers, being functions of ratios of ionic conductances, are relatively insensitive to changes in solvent, temperature, and pressure when compared with the ionic conductances themselves. For instance, in water at 25°C most ionic conductances increase by ca. 2% per 1°C, while transference numbers change by amounts of the order of 0.1% per 1°C.

A rather different theoretical approach has been the application of the theory of transition states which postulates that the ion migrates through the liquid by a series of jumps of length L from one equilibrium position to another and that each jump requires a characteristic standard partial molar free energy of activation, ΔG_0. In the presence of an electric field, the energy of activation is lowered in the direction of the field and raised in the opposite direction, so that the ions move preferentially one way. The resulting equation of Stearn and Eyring [28], modified by Brummer and Hills [29], is

$$\lambda_i^0 = \frac{z_i e F L^2}{6h} \exp\left(\frac{-\Delta G_0}{RT}\right), \tag{4.28}$$

where h is Planck's constant, R the gas constant, and T the absolute temperature. It follows that [29]

$$\left(\frac{\partial \ln \lambda_i^0}{\partial T}\right)_P = \frac{E_P}{RT^2} = \frac{\Delta H_0^{\ddagger}}{RT^2} + \frac{2\alpha}{3} \tag{4.29}$$

$$\left(\frac{\partial \ln \lambda_i^0}{\partial P}\right)_T = -\frac{\Delta V_0^{\ddagger}}{RT} + \frac{2\beta}{3}, \tag{4.30}$$

in which α and β are the coefficients of cubical expansion and of compressibility, respectively, of the system (i.e., of the solvent), and arise out of the variation of L with temperature and pressure. $\Delta H_0^{\ddagger}$ and $\Delta V_0^{\ddagger}$ are the partial molar enthalpies and volumes of activation. It is a consequence of the above equations that

$$\left[\frac{\partial \ln\left(\frac{T_+^0}{T_-^0}\right)}{\partial T}\right]_P = \frac{(\Delta H_0^{\ddagger})_+ - (\Delta H_0^{\ddagger})_-}{RT^2} \tag{4.31}$$

$$\left[\frac{\partial \ln\left(\frac{T_+^0}{T_-^0}\right)}{\partial P}\right]_T = -\frac{(\Delta V_0^{\ddagger})_+ - (\Delta V_0^{\ddagger})_-}{RT}. \tag{4.32}$$

From (4.31) a plot of $\log (T_+^0/T_-^0)$ should be linear in $1/T$ provided the activation energies of transport of cation and anion differ by an amount that is independent of temperature. This conclusion was first pointed out by Allgood et al. [30]. A slightly curved plot has now been found [31] to fit KCl and NaCl data in water over a temperature range exceeding 100°C. The corresponding test of (4.32), the linearity of $\log (T_+^0/T_-^0)$ against pressure, awaits the determination of more data; earlier, less accurate figures [32] are consistent with it but more recent ones [33] are not. Equations (4.31) and (4.32) are useful for interpolation and extrapolation, and it is to be hoped that eventually tables of ionic activation enthalpies and volumes will become available so that the signs and magnitudes of the temperature and pressure coefficients can be predicted. We may note, however, as regards sign, that the faster the ion the smaller usually is its activation energy of transport and thus the temperature coefficient of its ionic conductance, a generalization reached empirically by Kohlrausch as early as 1902. As a result, the transference numbers of many strong symmetrical electrolytes (except potassium halides) approach 0.5 as the temperature rises.

3 APPLICATIONS

Fundamental Information

Solvation and other aspects of ion-solvent interaction are understood most easily if we can determine some property of individual ions rather than of the complete electrolyte. Such a property is the ionic conductance, arrived at by combining limiting equivalent conductances and limiting transference numbers. In every solvent and at every temperature and pressure, the transference numbers of at least one electrolyte must be known before limiting conductances can be split up into the contributing ionic conductances. Unfortunately this cannot often be done, for the literature, rich as it is in conductance data, is poor as regards good transference numbers. This is rather surprising when we bear in mind that in the case of associated electrolytes the extrapolation to zero concentration is far shorter for transference numbers than for conductances. Once a table of limiting ionic conductances has been drawn up for a given solvent, conductances and transference numbers of many other electrolytes can be predicted on the basis of Kohlrausch's law of independent ionic mobilities.

Ion-ion interaction can be studied by the variation of many electrolyte properties with concentration. Transference numbers are particularly useful here because their concentration dependence, unlike that of conductances, yields the electrophoretic effect without the relaxation effect as well [13, 14]. The electrophoretic term of the D-H-O theory and its modifications (see Section 2) can therefore be independently tested.

Another method of studying ion-ion interaction is through activity coefficients. These are easily determined from the e.m.f.s of galvanic cells provided electrodes exist that are reversible to both the cation and the anion of the electrolyte in question. If, however, accurate transference numbers are available, cells with transference requiring only *one* type of reversible electrode can be employed [12d]. The activity coefficients of inorganic electrolytes [12d, 34] and of polyelectrolytes [35] have been obtained in this way. This method is the reverse of that described in Section 4 for the determination of transference numbers from the e.m.f.s of cells with transference.

Of cardinal importance in the field of irreversible thermodynamics is the validity of the Onsager reciprocal relations. A useful test of these, for the case of isothermal electrical and matter transport, is the numerical equality at high concentrations of the transference numbers determined by the Hittorf method and by the e.m.f.s of cells with transference. The reciprocal relation in question has recently been tested and confirmed by such experiments [36].

Structural Investigations

The transference number has served as an indicator of structure for over a century and, more recently, so has its concentration variation when this departed markedly from that predicted by the D-H-O theory. A few examples indicate the range of applications. In 1859, Hittorf electrolyzed potassium silver cyanide solutions and found that the transference number of the silver ion-constituent was negative and that its absolute magnitude was always one-half that of the transference number of the cyanide ion-constituent. In other words, silver and cyanide both migrated from cathode to anode in the gram-equivalent ratio of $1:2$ [cf. (4.10)]. This clearly pointed to the existence of the complex ion $Ag(CN)_2^-$. The same method was later employed to discover the existence and properties of complexes of the transuranium metals [37]. A moving boundary version of this type of "migration" experiment is also available and has been applied, for instance, to the study of the hydrolysis of rare earth chelates [38]. As mentioned in Section 2, the presence of complex anions in concentrated aqueous zinc and cadmium halide solutions was revealed by the negative transference numbers of the metal ion-constituent, and the theory of micelle formation in polyelectrolyte solutions was tested and confirmed by transference measurements (Fig. 4.3). Similarly, it has been shown that complex ions exist in solutions of aluminium halides in nonaqueous solvents [39] and of Grignard reagents [40], and these observations help to elucidate the mechanisms of certain reactions important in synthetic work.

Although much information can be derived from transference measurements alone, both transference and conductance determinations have been used together in the "mobility" approach to structural questions. Equations (4.4) and (4.10) can then be applied to give quantitative information on the ionic conductances (or mobilities), concentrations, and charges of the various species, since the mobilities of some of the ions present are already known from work with simpler electrolytes. This method of attack has been successfully employed in the study of the constitution of ion-exchange resins [41] and of solutions of soaps and detergents, dyes, proteins, and other polyelectrolytes [5]. An interesting recent example is an investigation [42] of solutions of the sodium salt of deoxyribonucleic acid (DNA) from which properties such as the polymer charge fraction and the fraction of free gegenions were derived. The "mobility" approach to structural problems is complementary to the "thermodynamic" approach; in the former mobilities and in the latter activity coefficients must be determined, and both methods are often used together.

One special type of transference experiment (page 264) has been adapted to the determination of the rate of exchange of a common ion-constituent between two species in a solution. In this way information has been obtained on the binding between counter ions and micelles in polyelectrolyte solutions [43].

Separations and Other Applications

The direct moving boundary method (Section 4) is the logical forerunner of the moving boundary method for investigating electrophoresis. Studies of mixtures of strong electrolytes and of buffer mixtures [44] have provided much insight into the significance of the schlieren patterns recorded in the electrophoresis of proteins and other substances [45]. Just as its electrophoretic counterpart, the moving boundary method has been used to separate mixtures of ions (e.g., caproate and valerate ions) both in solutions [46] and in ion-exchange resins [41]. Since isotopic ions display small conductance differences [47], these too can be separated by appropriate ion migration techniques in either solutions or melts [48].

4 MEASUREMENT OF TRANSFERENCE NUMBERS

Comparison of Methods

The decision as to which experimental method should be used depends upon a number of factors. Chief among these are the accuracy desired, the concentration range of interest, and the properties of the solution in question. Comparative information regarding the first two points is given in Table 4.2, in which m.b. stands for moving boundary method. The principal requirements for any given method are outlined in Table 4.3, and they must be compatible with the properties of the solution under investigation. For aqueous solutions, results accurate to about 2 to 5% can usually be readily produced with almost any method, and it must be pointed out here that many precautions and corrections enjoined in the text below are meant only for cases in which the precision warrants their adoption. The simplest method is probably the Hittorf or the analytical boundary one. The m.b. methods, or the Hittorf method for concentrated solutions, are the best choices if high accuracy is desired. The e.m.f. method is most useful for a survey over a wide concentration and temperature range. The centrifugal e.m.f. method requires complex equipment and has so far been applied only to iodide solutions.

Transference measurements in molten salts have been reviewed elsewhere [8] and are not discussed in this chapter.

Table 4.2 The Concentration Ranges of Aqueous Solutions for Which the Main Transference Methods are Best Suited, and the Usual Maximum Accuracy Obtainable*

			Transference Method			
Approximate Optimum	*Hittorf*	*Direct m.b.*	*Indirect m.b.*	*Analytical Boundary*	*E.m.f.*	*Centrifugal e.m.f.*
Upper concentration	No limit	0.2 *N*†	No limit‡	0.2 *N*	No limit§	No limit§
Lower concentration ‖	0.01 *N*	0.001 *N*†	0.001 *N*	0.01 *N* #	0.001 *M*§	0.001 *M*§
Error limits	±0.1%	±0.03%	±0.03%	±0.2%	±0.2%	±0.2%

* In solvents of lower dielectric constant, difficulties are introduced by the higher electrical resistance of the solutions and the resulting Joule heating and convection, the relative scarcity of suitable electrodes, the decreased solubility of many electrolytes, and certain additional specific points mentioned in other footnotes. The accuracy is often lower than with aqueous solutions.

† In most nonaqueous solvents lower concentrations must be used. The volume correction restricts the accuracy at high concentrations, the solvent correction at low concentrations.

‡ The method works satisfactorily at very high concentrations, and is restricted mainly by the availability of data for the leading solution.

§ Provided the electrodes remain reversible, reproducible, and insoluble. The stability and/or reproducibility of electrodes tend to decrease in nonaqueous solutions. Some electrodes become quite soluble at high concentrations.

‖ Lower concentrations can often be investigated but only at the expense of decreased accuracy.

Surface conductance on the disk introduces errors at low concentrations.

Hittorf Method

Introduction and Theory

In the method named after Hittorf, who introduced it in 1853, a known quantity of electricity is passed through a cell filled with the solution whose transference numbers are to be determined. The solutions in various sections of the cell are then directly or indirectly separated, weighed, and quantitatively analyzed. The method has therefore sometimes been called the analytical or gravimetric method.

Figure 4.4 illustrates schematically a general Hittorf cell, with E_A and E_C representing the anodic and cathodic electrode sections or compartments and M_A, M_M, and M_C three middle sections (anode middle, middle middle, and

Table 4.3   Basic Requirements for the Application of the Main
Transference Methods

	Transference Method					
Necessity for Suitable:	*Hittorf*	*Direct m.b.*	*Indirect m.b.*	*Analytical Boundary*	*E.m.f.*	*Centrifugal e.m.f.*
Analytical method*	Yes	—	Yes	Yes	—	—
Leading solution†	—	—	Yes	Yes	—	—
Following solution‡	—	Yes	—	—	—	—
Method of following m.b.	—	Yes	—	—	—	—
Electrode(s)§	Yes	Yes	Yes	Yes	Yes	Yes
Electrical measurements ‖	Yes	Yes	—	Yes	Yes	Yes
Activity coefficients	—	—	—	—	Yes	—
Partial molar volumes #	—	Yes	—	Yes	—	Yes

* In the indirect m.b. and the analytical boundary methods, the transference number is directly proportional to the one analysis made in each run. In the Hittorf method more than one analysis is performed per run, and the transference number is proportional to the relatively small difference between two such analyses.

† Containing a faster non-common ion-constituent. In the indirect m.b. method, the transference number in the leading solution must be known. In the analytical boundary method, an appropriate tracer can be used instead.

‡ Containing a slower non-common ion-constituent. For autogenic boundary formation this requirement merges with the conditions for a suitable electrode.

§ In the e.m.f. methods the electrode reaction must be reversible on the passage of very small and momentary currents; in all other methods the electrodes need not act reversibly but they must stand up to the passage of appreciable quantities of electricity.

‖ Of potential in the e.m.f. methods, of the quantity of electricity (current and time) in the other methods.

Enter directly into the centrifugal e.m.f. equation, but affect the m.b. methods only through the volume correction.

cathode middle). The dotted lines stand for the section boundaries which, for the reason given in Section 1, are considered fixed with respect to the solvent throughout the experiment. The basic principle follows directly from the definition of transference number: for every faraday of electricity passed through the cell, T_R equivalent of cation-constituent R migrate across every section boundary in the direction of the cathode, that is, out of E_A and into M_A, out of M_A and into M_M, out of M_M and into M_C, out of M_C and into E_C.

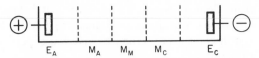

E_A M_A M_M M_C E_C

Fig. 4.4   Schematic diagram of a Hittorf cell.

Clearly the compositions of the middle sections remain the same and only those of the electrode sections change. Analysis of either of these for the constituent R, therefore, yields T_R. A similar description can be given for an anion-constituent migrating in the opposite direction. In practice we must also take into account the electrode reactions since these may involve the constituent we are interested in; an example demonstrates the reasoning employed.

Suppose the anode is a thick silver wire, the cathode a silver wire thickly coated with AgCl, and that the cell is filled with a solution of $BaCl_2$. The following chemical changes then take place when 1 faraday is allowed to flow through the cell. In the anode compartment, T_{Ba} equivalents (or $\frac{1}{2}T_{Ba}$ moles) of barium ion-constituent (which carries T_{Ba} faradays of charge) migrate out and T_{Cl} equivalents (or T_{Cl} moles) of chloride ion-constituent migrate in. Moreover, 1 equivalent of chloride ion-constituent disappears from the solution through the electrode reaction

$$Cl^- + Ag \rightarrow AgCl + e^-.$$

This produces a net loss in the anode solution of $1 - T_{Cl}$, or T_{Ba} [see (4.8)], equivalents of chloride and, because an equal number of equivalents of barium ion-constituent have migrated away, the solution remains electrically neutral with a net loss of $\frac{1}{2}T_{Ba}$ moles of $BaCl_2$. The converse occurs in the cathode section: T_{Ba} equivalents of barium migrate in, T_{Cl} equivalents of chloride migrate out, and 1 equivalent of chloride enters the solution by virtue of the electrode process

$$e^- + AgCl \rightarrow Ag + Cl^-.$$

There is thus a net gain of $\frac{1}{2}T_{Ba}$ moles of $BaCl_2$. It must be emphasized, however, that the chemical changes in the anode and cathode compartment solutions are not equal and opposite in every Hittorf experiment and quite a different result would have emerged if, for example, a cadmium anode had been chosen for the present example instead of a silver one. In the middle sections as much chloride or barium ion-constituent enters by migration on one side as leaves by migration on the other, and there is therefore no net change in solution composition. The purpose of these middle sections is to ensure that all concentration changes have taken place entirely within the two electrode compartments and that no intermixing has occurred. One middle compartment alone is not sufficient since equal diffusion from the more concentrated cathodic side and toward the more dilute anodic one would leave no net change in concentration and so give the experimenter the false impression that all is well.

An illustration taken from the literature will clarify the procedure used. Jones and Dole [49], in their sixteenth run, electrolyzed an aqueous 0.24745 M (4.949 wt %) solution of $BaCl_2$ (molecular weight 208.27) with a silver anode and a AgCl cathode in a cell similar to that in Fig. 4.5. They stopped the experiment when 0.024644 faradays of electricity had passed and then isolated the electrode compartments by closing the stopcocks S and S'. The stoppers were removed and the solutions in the sections M_C, M_M, and M_A carefully pipetted out and gravimetrically analyzed. They were found to contain 4.946, 4.947, and 4.946 wt % $BaCl_2$, respectively, so that no significant mixing had occurred and the initial concentration can be taken as 4.947 wt %. The solutions in the electrode compartments were then weighed and similarly analyzed. The anode solution weighed 121.58 g and contained 4.148 wt % $BaCl_2$ or, to put it another way, it was made up of 116.54 g water and 5.043 g (or $5.043/208.27 = 0.02421_3$ mole) $BaCl_2$. In order to find the decrease in the number of moles in the anode compartment, we must now calculate the number of moles of $BaCl_2$ that were present initially in the *same mass of water*—transference numbers are defined in terms of migration across a plane fixed with respect to the solvent. Since 4.947 g $BaCl_2$ were present originally in $(100 - 4.947)$ or 95.053 g water, it is clear that 116.54 g water must at the start have contained

$$(116.54 \times 4.947)/(95.053 \times 208.27)$$

or 0.02912_1 mole $BaCl_2$. Thus on the passage of 0.024644 faradays, the anode compartment lost $(0.02912_1 - 0.02421_3)$ or 0.00490_8 mole $BaCl_2$. The theoretical discussion above showed that 1 faraday would have led to the loss of $\frac{1}{2}T_{Ba}$ mole and so the cation-constituent transference number is

$$2 \times 0.00490_8/0.024644 = 0.398.$$

An analogous treatment of the cathodic department figures gives $T_{Ba} = 0.399$. The agreement is very good and illustrates how analysis of both electrode sections provides an excellent check on the internal consistency of a run. The calculation also highlights the principal weakness of the Hittorf method, the fact that the final result depends upon the relatively small difference between two quantities.

It was tacitly assumed in the above treatment that all the current is carried by the solute ions. In fact the solvent itself contains some ions, partly from self-dissociation and partly from dissolved impurities, and if their concentrations and conductances are not altered by introducing the solute, the corrected transference number is given by [50]

$$T_R(\text{corr.}) = T_R(\text{obs.})[1 + (\kappa_{\text{solvent}}/\kappa_{\text{solute}})] \qquad (4.33)$$

where κ is the specific conductance in $\text{ohm}^{-1}\,\text{cm}^{-1}$. Such a *solvent correction* is appreciable only in fairly dilute solutions. It must be modified if there is any chemical interaction between the solute and the solvent ions, as is the case with aqueous solutions of acids, bases, and carbonates because water contains the ions H^+, OH^-, and HCO_3^-. An acid solute, for example, suppresses the dissociation of dissolved carbon dioxide, and the specific conductance of the solvent must be changed accordingly [51].

More recently it has been pointed out [52] that the solvent correction may not be simply that given by (4.33) because fast impurity ions such as H^+ and OH^- migrate in different proportions to those of the solute, and the safest procedure for dilute solutions is to reduce the solvent conductance to as low a level as possible.

Experimental

ANALYSIS

The transference number depends upon a relatively small difference between two concentrations, and thus the accuracy of the final result depends mainly upon the accuracy of the analyses. Four- to five-figure analyses yield three-figure transference numbers. Any suitable analytical procedure may be used—gravimetric, volumetric, electrometric, spectrophotometric, counting methods for radioactive isotopes, and so on. If the solute consists of a mixture of substances an analysis should be performed for each one.

After a run, samples from the various middle sections are analyzed first, and if the concentrations in these differ appreciably from each other or from the initial concentration the run should be rejected. If, however, the concentrations are the same, then, depending upon the system, the number of gram-equivalents of one or more ion-constituents or of the electrolyte as a whole, in one or both electrode compartments, is determined by one of three main methods:

1. If the cell is made in several pieces attached during the run, then the solutions in the middle sections can be drained off and the electrode section separated from the remainder of the cell. The section is dried on the outside and weighed. Either the whole solution is washed out for analysis, or the solution in the section is carefully homogenized by tilting and shaking without spilling any solution and a sample is taken, weighed, and analyzed. In either case the electrode section is then washed out completely, dried, and weighed, unless the weight of the empty section has been determined before the run. Thus the total weight of the electrode compartment solution is known, as well as the concentration of a weighed aliquot, so that both the mass of

solvent and the number of gram-equivalents of ion-constituent can be calculated [53].

2. The entire solution in the electrode section is transferred to a weighed flask. The section is carefully rinsed out, not with solvent, but with some of the original electrolyte solution with which the cell had been filled, and the wash liquid is also added to the flask [54]. This is equivalent to including some of the unchanged solution in the middle compartments with the electrode solution, so that in effect the boundary line of the electrode compartment is taken somewhat further away from the electrode; this has no effect at all on the results. The flask is reweighed, and either the whole or a weighed aliquot of the solution is analyzed. Alternatively, if the density of the solution is known, volume rather than weight can be measured by transferring the solution and the wash liquid to a volumetric flask and filling to the mark with the original solution. In accurate work or in work with volatile solvents, it is necessary to minimize solvent evaporation during these operations.

3. The compartments are isolated from each other after the run by closing stopcocks or some other procedure. The solutions are carefully homogenized in each section and are then analyzed *within* the cell. This can be done, for example, by measuring the conductivity of the solutions with platinum electrodes which are already sealed into each compartment, provided the conductivity-concentration relationship is known [52]. The total mass of solution in at least one electrode compartment must be found also, either by weighing or, if the density of the solution is known, from a previous calibration of the volume of the electrode section itself.

If possible, both anode and cathode compartment solutions should be weighed and analyzed, as this provides an excellent check on the internal consistency of the run.

ELECTRODES

The choice of suitable electrodes is very important, and the following criteria should be applied in making a selection:

1. The electrode reaction should be known, so that the transference number can be derived by an examination of the cell processes (see Theory section). Secondary reactions can often be eliminated by using electrodes of large surface area and sometimes by excluding oxygen from the solution.

2. An ideal electrode reaction introduces no foreign ions into the solution [53]. If any foreign ions are introduced, they must not interfere with the analysis of the solution nor must they reach the middle compartments. This restricts the quantity of electricity that can be passed, and therefore the concentration changes that may occur in the electrode section are smaller. Electrodes producing highly mobile ions, such as H^+ and OH^- in aqueous solution, should therefore be avoided.

3. If possible, the electrode reaction should not be associated with any physical action tending to stir and mix the solution, such as gas evolution or the formation of any precipitate that does not adhere to the electrode [53]. If these disturbances are unavoidable, their effect can be reduced by suitable cell design.

Anodes in common use are ones composed of the metal whose cation-constituent is present in the solute (e.g., Cd in $CdCl_2$ solution), or of a metal whose cation reacts with an anion in solution to form an insoluble salt that adheres to the anode (e.g., Ag in NaCl solution), or of a metal whose cation-constituent, although foreign to the solution, moves slowly and does not complicate the analysis (e.g., Zn or Cd). In difficult cases (e.g., solutions of soaps or detergents), electrodes with suitable guard solutions around them can be employed, but the electrode compartment should be designed to minimize mixing with the main solution. If a gassing electrode must be used, an appropriate guard substance can prevent the formation of very mobile ions (e.g., in aqueous solution, Li_2CO_3 surrounding a Pt anode substitutes slow Li^+ ions for mobile H^+ ions). Ion-exchange beads around the electrode, with perhaps a membrane as well, prevent the introduction of undesirable ions into the solution.

Cathodes in common use are those that liberate an anion-constituent already present in the solution (e.g., silver halide electrodes in halide solutions), or ones composed of a metal upon which a metallic ion in the solution can plate out (e.g., Pt in $AgNO_3$ solution). A guard solution in a well-designed electrode compartment may be necessary (e.g., Pt surrounded by concentrated ferric acetate solution, or Hg in strong $Zn(NO_3)_2$ solution). If a gassing cathode must be employed in aqueous solution, a suitable substance (e.g., benzoic acid crystals, or anion-exchange resin beads in the nitrate form) around the electrode can prevent the escape of the fast OH^- ions.

At least one electrode reaction in the cell should be free from objections so that the analysis of one electrode compartment is reliable. However, from the point of view of handling the cell after the run, as described under Analysis (p. 227), the simpler the electrode compartments the better, so that guard tubes should not be used unless necessary. Aqueous chloride solutions have often been electrolyzed with Ag anodes and AgCl cathodes, and there is much useful advice in the literature [49, 52, 53, 55, 56] on the construction of such electrodes.

DISTURBING EFFECTS

The various effects that cause mixing between compartments can be decreased as follows.

Diffusion can be reduced by increasing the length of the cell and decreasing the time of electrolysis.

Electrical Joule heating sets up a radial temperature distribution with a maximum temperature at the axis of the tube [57], and the resulting density differences cause convective stirring. The effect is decreased if the current is small and if tubing of uniform and moderately large bore is used throughout the whole cell [53]. A series of right-angled bends in the tubing has been recommended for breaking up convective flow [55]. The thickness of the glass wall does not alter the radial temperature gradient but does influence the mean temperature inside the tube [57]. Thermostatting reduces temperature fluctuations. If the experimental temperature is not important, advantage can be taken of the almost complete absence of convective mixing in aqueous solutions near 4°C, where the density of water is a maximum [45]. Convective stirring becomes pronounced in nonaqueous solutions of low conductance, and a preliminary investigation by means of colored indicators may indicate the extent of the effect [58].

Electrical migration of ions introduced by the electrode reactions into the middle compartments is reduced by increasing the volumes of the electrode compartments and by decreasing the quantity of electricity passed through the cell. Other undesirable electrode effects are discussed above.

Vibrational disturbances are considerably reduced by hanging either the cell [59] or the thermostat [55] from a special suspension or by mounting the thermostat motor on a stand separate from the thermostat. Vibrations from the motor are not then transmitted directly to the apparatus.

The changed solubility of air in the solution at the experimental temperature may lead to the formation of air bubbles. This is overcome by partially degassing the solution, just prior to filling, by shaking it for a few moments at a reduced pressure. Sometimes dissolved oxygen must be removed to avoid side reactions at the electrodes.

CELL DESIGN

In a large cell many of the above disturbing effects are small. However, the changes in molarity in the electrode compartments per faraday are also small, especially in concentrated solutions [60], and the analyses need to be very precise. A compromise cell size is therefore selected in practice. The cell should be made of Pyrex rather than soda glass so as not to increase the solvent correction.

Good cell design is very important, and it is unfortunate that much of the early Hittorf work at the end of the nineteenth century was carried out with cells so poorly designed that the resulting data are now of little value. It is even more unfortunate that diagrams of such early apparatus are still faithfully reproduced in general textbooks and even in laboratory manuals. Some examples of well-designed cells are given in Figs. 4.5–4.9, in which most of the lettering follows that in Fig. 4.4. The cells in Figs. 4.7 and 4.8 are easily

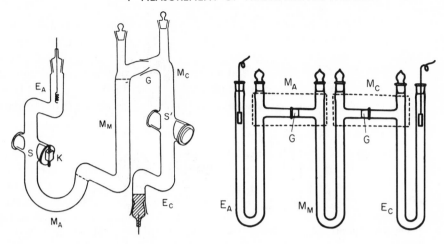

Fig. 4.5 Hittorf cell used by MacInnes and Dole [55] (reprinted with permission from Ref. 61).

Fig. 4.6 Hittorf U-tube cell used by Laing [62].

modified to include two middle compartments. As far as possible, the bore of the tubing is fairly large and uniform throughout each cell and includes several bends so as to reduce the disturbances caused by convection and Joule heating.

Cells differ mainly in the methods devised for separating the compartments and removing the solutions. If a stopcock is used, the bore should be the same

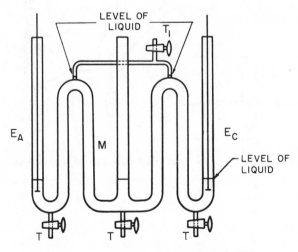

Fig. 4.7 Hittorf cell used by Wall et al. [63].

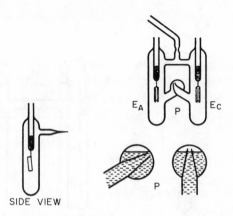

SIDE VIEW

Fig. 4.8 Hittorf cell used for nonaqueous solutions by Fairbrother and Scott [64].

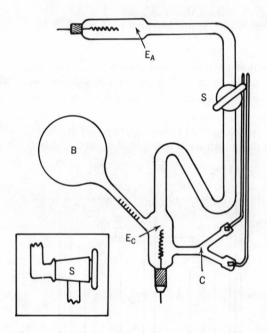

Fig. 4.9 Hittorf cell designed by Steel and Stokes [52, 59] for internal conductometric analysis.

as in the remainder of the cell and the barrel should be hollow and open to allow the thermostat liquid to circulate freely through it and prevent local heating. The handle should also be hollow with openings in it, or else both ends of the barrel can be open and a removable metal key (K in Fig. 4.5), used for turning. The choice of a suitable lubricant may prove difficult with nonaqueous solvents or emulsifying solutes [65]. Diaphragms are not as good as their popularity indicates: they all cause electroosmosis unless one electrode compartment is closed, relatively low currents must be used to avoid local heating and convection, and most membranes other than sintered glass frits lead to selective transmission of ions [66]. Even with glass diaphragms surface conductance effects may appear at concentrations below about 0.01 N. With the pressure-head device in Fig. 4.7, the solutions are joined initially by suction at T_1 and are separated again after the run by opening T_1. The tilting device P in Fig. 4.8 is useful for solutions that should not be exposed to the atmosphere, although some local heating may develop at the junction. The solutions in the middle compartments can be removed in Figs. 4.5 and 4.6 by pipetting off the requisite amount of solution, and in the simple cell in Fig. 4.6 this automatically isolates the electrode solutions. Alternatively, the solutions may be forced out of a side arm by gas pressure as in Fig. 4.8, or drawn off through taps at the bottom as in Fig. 4.7. If lubricant problems arise, the taps can be replaced by Pyrex and plastic taps or needle valves now available commercially, or by simple Bunsen valves. These consist of a piece of rubber or plastic tubing with a glass or metal ball or short rod in the middle to act as a seal. Squeezing the tubing around the solid obstruction with thumb and forefinger allows liquid to pass through. Ground-glass joints (G) allow the electrode sections to be separated for weighing or rinsing as described under Analysis. If the solution reacts with the lubricant, the latter can be replaced by a thin Teflon sleeve made to fit over the standard taper joint, although a thin smear of silicone grease is often preferable [59]. If necessary, the joints can be pressed together during the run with the usual glass hooks and steel springs arrangement or by means of special clamps. Electrode guard or baffle systems for special purposes often consist of upright and inverted tubes inside each other, but regions of varying electrical resistance should be avoided. A given electrode must be introduced at the top if the products of electrolysis are lighter than the main solution, and vice versa, so as to eliminate mixing effects caused by gravitational instability (see also Figs. 4.21 and 4.22).

Figure 4.9 shows a design allowing internal analysis by means of the conductivity cell C. The bulb B is a mixing chamber sufficient to hold the entire contents of the rest of the apparatus, and its interior and that of its calibrated stem are coated with a silicone layer to ensure complete drainage. With B full of air, the volume of E_C and C and the N-shaped tube up to the large stopcock S is first measured by weighing the amount of water needed to fill these parts with the meniscus standing at a known position on the calibrated stem. For a run the cell is filled with solution so that the level of liquid is within the range of the calibrated stem (with B containing air). The entire solution is then tipped several times into the mixing bulb and returned to its original position until adsorption changes are complete and the conductance is constant. S is then opened, a known current passed for a known time, and S is closed again. The contents of the lower compartment are once more thoroughly mixed via the bulb B, and the conductance remeasured. The volume should be the same as before. As a check, S can be opened and the conductance of the entire cell solution determined; if the chemical changes in the anode and cathode compartments have been equal and opposite, the reading should be the same as the original one. From the known, previously determined, conductance-concentration relationship, the change in concentration in E_C can now be calculated and, bringing in the measured volume, so can the change in the number of equivalents and hence the transference number. The main defect of this otherwise elegant apparatus is that it allows one to determine the concentration change in only one electrode compartment (in Fig. 4.9, of the one whose contents increase in density during electrolysis, and this is either E_A or E_C depending on the circumstances) and there are no middle compartments. Introducing middle compartments, and the appropriate conductance attachments, might complicate the cell unduly and there is therefore no direct proof that convection and diffusion do not contribute to ionic transport in the region of the stopcock. One must therefore rely on the indirect evidence provided by constancy of the transference number when the current and time of electrolysis are varied.

QUANTITY OF ELECTRICITY

A larger current and a longer time of electrolysis produce, on the one hand, larger changes in concentration in the electrode compartments and therefore a more precise transference number and, on the other hand, increases in many of the disturbing effects. A compromise must be reached on the basis of the magnitudes of the factors involved; normally currents ranging from about 1 to 100 mA are passed for several hours to give concentration changes in the electrode solutions of around 5 to 50%. A 50% variation in current or time should not affect the transference number.

Current leakages in the cell obviously vitiate the measurements, and by far the safest procedure is to employ an oil-filled thermostat. In water thermostats leakage is much reduced by coating all joints with wax [53] or collodion, and the quantity of electricity is often measured at both ends of the cell as a safeguard.

Three different methods of measurement are available:

1. *Chemical coulometers.* These are based on Faraday's laws of electrolysis, and the quantity of electricity passed is calculated from the extent of the chemical reaction at one or both electrodes in the coulometer. The silver and the water coulometers have become classic, and these and more recent ones have been described in detail previously [67]. Since then, descriptions of several more chemical coulometers, largely for the micro-range, have been published [68]. Nowadays, however, most workers choose instead one of the following electronic devices.

2. *Electronic and mechanical coulometers or current-time integrators.* Milner and Phillips [69] have given the circuits and designs of a variety of these instruments, and a few are commercially available. The latest developments have been, on the one hand, the description [70] of a fairly inexpensive mechanical integrator with a compensating circuit capable of measuring to better than $\pm 0.1\%$ the quantity of electricity over a wide current range and, on the other, the publication [71] of two electronic constant-current coulometers accurate to ± 30 p.p.m.

3. *Separate current and time measurements.* The concentration, hence the resistance, changes in a Hittorf experiment are relatively small, and any simple electronic device can maintain a constant current. Numerous circuits have been reported in the literature [69, 71], and several such supplies can be purchased commercially. The current is determined by reading a potentiometer connected across a calibrated resistor in series with the cell, remembering that one side of normal potentiometers must be at ground (earth) potential. The time of electrolysis may be measured by electric or electronic means or, more simply, with a watch or stopwatch.

Direct Moving Boundary Method

Introduction and Theory

In the direct m.b. method, an electric current maintains and moves a fairly sharp boundary between the solution under investigation (the leading solution) and a solution of a suitable indicator electrolyte (the following solution). The volume traversed by the boundary is measured when a known quantity of electricity has passed through both solutions.

The theory is explained with reference to Fig. 4.10, in which AZ is the leading and BZ the following electrolyte, so that B is the non-common indicator constituent. The m.b. moves in the direction of the arrow, and a rising (Fig. 4.10a) or a falling (Fig. 4.10b) boundary is employed depending upon whether the indicator solution behind the boundary is heavier or lighter than the leading solution. The two-salt boundary AZ/BZ is termed a cation boundary if A and B are cation-constituents (E_1 in Fig. 4.10 is then the cathode and E_2 the anode), and an anion boundary if A and B are anion-constituents (E_1 is then the anode and E_2 the cathode). The following examples from the literature refer to aqueous solutions and should clarify these classifications of the AZ/BZ boundary:

Rising cation boundaries	$KCl/CdCl_2$; $K_3Fe(CN)_6$/methylene blue ferricyanide
Falling cation boundaries	$KCl/LiCl$; $K_2C_2O_4$/cetylpyridinium oxalate
Rising anion boundaries	KCl/KIO_3; H_3PO_4/citric acid
Falling anion boundaries	$KCl/KOOCCH_3$; sodium salt of Benzopurpurine 4B dye/sodium benzoate.

It is a rather remarkable phenomenon that these boundaries exist at all, for one might expect diffusion to work toward their destruction. The explanation lies in the so-called electrical restoring effect [72]. This comes into operation automatically whenever, in both solutions, the leading non-common ion-constituent A has a higher mobility than the following non-common ion-constituent B, that is,

$$\bar{u}_A > \bar{u}_B. \tag{4.34}$$

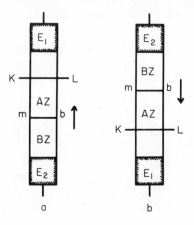

Fig. 4.10 Schematic diagram of a general m.b. cell.

For suppose that some A ion-constituent drifts into the BZ solution: at once it finds itself in an environment (usually of higher potential gradient) in which it moves more swiftly than its fellow B ion-constituent and so it shoots ahead and overtakes the boundary again. Again, a trace of B ion-constituent diffusing into the leading AZ solution is slowed down by the conditions there (usually a lower potential gradient, or possibly a change in pH) and is eventually overtaken by the boundary [3]. In practice a steady state is soon set up by the tug-of-war between the electrical restoring effect and diffusion, and boundaries therefore possess a small but finite "thickness" δ of the order of 0.01 cm. MacInnes and Longsworth [72] have derived an equation for this thickness:

$$\delta = \frac{4RT}{Fv} \cdot \frac{\bar{u}_A \bar{u}_B}{(\bar{u}_A - \bar{u}_B)}, \tag{4.35}$$

where v is the velocity of the boundary. It is clear that the boundary is the "sharper" the more A and B differ in their mobilities, and also the faster the boundary. High currents therefore sharpen the boundary [see (4.36)], a well-known experimental fact.

The relation between the boundary velocity and the transference number of the leading non-common ion-constituent A can be simply derived with reference to Fig. 4.10. Let 1 faraday of electricity pass. The boundary then moves along the tube through a region of volume V_F and in so doing "sweeps forward" $C_A^{AZ} V_F$ equivalents of ion-constituent A. All this must in turn have migrated across the imaginary horizontal plane KL ahead of the boundary and is therefore, by definition, equal to T_A^{AZ}, that is,

$$T_A^{AZ} = C_A^{AZ} V_F = \frac{C_A^{AZ} V F}{It}.$$ (4.36)

As regards the last term, in most experiments less than 1 faraday passes and the volume V traversed by the boundary is then proportionately smaller than V_F by the ratio of the number of coulombs passed (current $I \times$ time t) to the number in 1 faraday (96487). No assumptions about the degree of dissociation in either solution were made in deriving (4.36) because both the transference number and the concentration are properties, not of the ion, but of the ion-constituent. Another important point is that the velocity of the boundary depends upon the properties of the leading solution only and is usually independent of the type and initial concentration of the following electrolyte (the indicator) within fairly wide limits if the cell is properly designed. In fact, the concentration of the indicator solution near the boundary adjusts itself automatically to a new value, and if the properties of this adjusted solution are characterized by an asterisk, it can be shown [72] by a similar argument to the above that

$$\frac{*C_B^{BZ}}{C_A^{AZ}} = \frac{*T_B^{BZ}}{T_A^{AZ}} = s.$$ (4.37)

This is the Kohlrausch equation, s the Kohlrausch ratio, and $*C_B^{BZ}$ the Kohlrausch concentration which usually differs from the initial concentration C_B^{BZ}. The diffuse "concentration boundary" that exists between the original and the adjusted indicator solutions forms the basis of the "differential moving boundary method" discussed later.

Moving boundary theory has now been generalized and extended to systems of mixed strong electrolytes [73] and of buffer [74] mixtures. The basic equation common to all m.b. systems is, for a given ion-constituent R,

$$T_R^\alpha - T_R^\beta = V_F(C_R^\alpha - C_R^\beta)$$ (4.38)

where α and β designate the leading and following solutions, respectively. Equation (4.36) is clearly a special case for an ion-constituent present on one side of the boundary only.

These equations are subject to two corrections. The *solvent correction* is applied as in (4.33), bearing in mind that the conductance of the solvent is appreciably greater inside the cell [75]. This is due in part to contamination during filling and to leaching of electrolytes from the cell walls and partly to preferential migration of impurities. The solvent correction increases with decrease in concentration, and the uncertainty inherent in it sets a lower limit on the applicability of the m.b. method. The *volume correction* is necessary because, from the definition of transference number, the boundary movement should have been determined relative to a plane in the AZ solution fixed with respect to the solvent instead of relative to plane KL in Fig. 4.10 which is fixed with respect to the cell. Since solvent is displaced relative to the cell by the volume changes at the electrodes, it follows that

$$T_A^{AZ}(\text{corr.}) = T_A^{AZ}(\text{obs.}) \pm C_{AZ}\,\Delta V, \tag{4.39}$$

where ΔV is the volume increase, in liters per faraday of electricity, between whichever electrode is closed and a point in the AZ solution which the boundary does not pass [76]. The plus or minus sign refers to the boundary moving toward or away from the closed electrode, respectively. The case of two open electrodes is discussed by Lewis [76]. The correction clearly becomes more important as the concentration increases, and the uncertainty in it limits the accuracy of the m.b. method at higher concentrations.

As an example, consider the effect of 1 faraday of electricity passing through the cell in Fig. 4.11, in which XY is a hypothetical plane fixed relative to the solvent. Between XY and the Ag-AgCl cathode, there is a gain (in gram-equivalents) of $+1$ of Ag(s), -1 of AgCl(s), and $+1$ of Cl$^-$ attributable to the electrode reaction, and of $+T_K^{KCl}$ of K$^+$ and $-T_{Cl}^{KCl}$ of Cl$^-$ as a result of migration across XY. Thus if the cathode is closed and the anode open to the air:

$$\Delta V = +V_{\text{Ag(s)}} - V_{\text{AgCl(s)}} + T_K^{KCl}\bar{V}_{KCl},$$

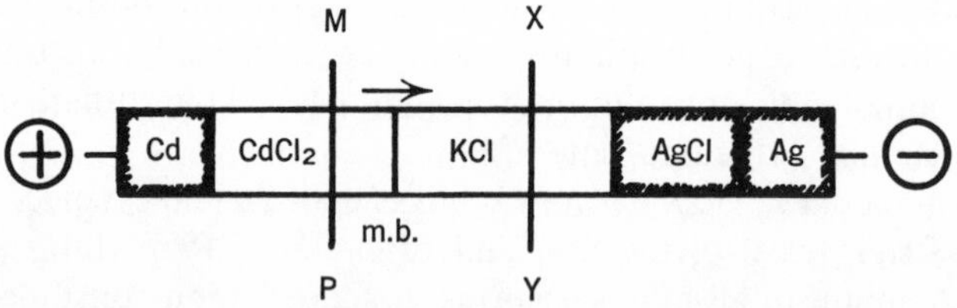

Fig. 4.11 Schematic sketch of a specific m.b. cell to illustrate the calculation of the volume correction.

since

$$T_K^{KCl} + T_{Cl}^{KCl} = 1.$$

V and $\bar{V}$ are the molar and partial molar volumes, respectively. Between XY and the Cd anode, there is a gain (in gram-equivalents) of -1 of Cd(s) and $+1$ of Cd^{2+} because of the electrode reaction, and of $-T_K^{KCl}$ of K^+ and $+T_{Cl}^{KCl}$ of Cl^- because of migration across XY. There is also a transfer, relative to the m.b., of 1 of Cl^- from KCl to $CdCl_2$. Hence if the anode is closed and the cathode open:

$$\Delta V = -\tfrac{1}{2}V_{Cd(s)} + \tfrac{1}{2}\bar{V}_{Cd}^{CdCl_2} - T_K^{KCl}\bar{V}_K^{KCl} + T_{Cl}^{KCl}\bar{V}_{Cl}^{KCl} + \bar{V}_{Cl}^{CdCl_2} - \bar{V}_{Cl}^{KCl}$$

$$= -\tfrac{1}{2}V_{Cd(s)} + \tfrac{1}{2}\bar{V}_{CdCl_2}^{CdCl_2} - T_K^{KCl}\bar{V}_{KCl}^{KCl}.$$

Various texts describe analytical methods of calculating $\bar{V}$ from solution density data [77]. Values of $\bar{V}$ for common electrolytes in aqueous solution may be found in the literature, not infrequently in m.b. papers.

A concealed assumption in these calculations is that the partial molar volumes of the solvent [78] and solute [79] are constant between a point in the leading solution which the boundary does not pass and the closed electrode. A somewhat modified volume correction [78] is based on the slightly less restrictive assumption that the partial molar volumes of the components are constant on only that side of the m.b. that is nearer the closed electrode. It follows that large localized concentration changes around the closed electrode should be avoided [78] if the volume correction is to be meaningful.

Experimental

OBSERVATION OF THE BOUNDARY

Three main techniques are known and a fourth has been suggested.

Optical Methods. The boundary readily stands out if one of the solutions is brightly colored (e.g., potassium permanganate or tetraiodofluorescinate). If the solutions differ in pH and if high accuracy is not required, the same effect may be achieved by adding a little acid-base indicator to both solutions although this does decrease the transference number slightly [80]. However, a simple, more general, and better procedure is to utilize the difference in refractive index between the solutions, and the essential equipment is shown in Fig. 4.12. The telescope T, which can be slid up and down on the stand D, is focused on the calibrated m.b. tube C in the thermostat B, which is fitted with parallel glass windows W and W'. At temperatures below that of the room, double windows are advisable [81]. The person peering through the telescope can raise or lower the light source A either by means of

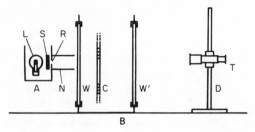

Fig. 4.12 Simple optical assembly for m.b. work (not to scale).

a counterweight and string passing over pulleys on the ceiling or by a more complicated thread-and-screw arrangement. L is a lamp bulb (25 W is usually ample), and N is a cylinder (an empty can, blackened inside) which prevents interference from surrounding lights. S is a ground-glass screen which can be made by grinding the faces of a microscope slide and which diffuses the light so that the rectangular opening R is uniformly illuminated. R should be about 4 mm high and at least 6 cm long. A should be set up so that R and C are exactly at right angles to each other as in Fig. 4.13 where R appears as a rectangular bar of light. G and G' are the glass sides of tube C. The inside walls of the tube often appear thick and black, and this can be overcome either by employing a bath liquid of similar refractive index to the solution [82] or, more easily, by a suitable lens system. A vertical 12 cm wide cylindrical glass tube filled with water, between the light and the thermostat, is most effective [82]. The boundary becomes visible only when it is close to one of the edges of bar R; if it is near one edge, it appears as a dark line (X), and if it is near the other as a bright line (X'). The former is the easier to use experimentally and it can often be seen directly with the naked eye. These images are caused by the refraction of the light beam in the optically inhomo-geneous boundary region, and for a given electrolyte system they are sharper the greater the current. Although not all boundaries give visible images by this method (e.g., in water KCl/LiCl and NaCl/LiCl give visible images,

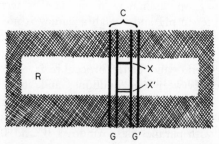

Fig. 4.13 View of the boundary through the telescope T of Fig. 4.12.

KCl/NaCl does not), it is generally more economical of time and money to try out several different indicators before turning to one of the more complex optical methods [83], such as the Longsworth schlieren scanning procedure, which are described elsewhere [45] and by means of which practically all types of boundary can be followed (cf. Fig. 4.24). Tracking with polarized light is an obvious modification of the optical technique should one of the solutions be optically active [84].

The optical method of following boundaries, despite its inherent simplicity, suffers from some serious limitations [85]. Thus it requires optically transparent windows in the thermostat bath, and this makes it difficult to apply at very low or very high temperatures and impossible to use at high pressures, where the cell must be kept inside a sealed container. Moreover, in solutions more dilute than about 0.005 N, the difference in refractive index becomes very small and the boundary is then too faint for visual observation. Yet for work in nonaqueous solvents of lower dielectric constant, it is only the very dilute solutions that can be investigated at all, boundaries at higher concentrations being disrupted by Joule heating [86]. Another, though relatively minor, restriction is that some solutions that form easily visible boundaries are somewhat light-sensitive; certain picrate solutions are an example [82]. These difficulties are overcome by the next method which, however, requires rather more elaborate instrumentation.

Electrical Methods. The indicator solution always possesses a higher electrical resistance than the leading one (cf. (4.34)). This fact enables one to follow the passage of a boundary along the tube by electrical signals from tiny platinum probe electrodes imbedded in the cell wall. Four main types of electrical detection have been reported, of which the first two depend on resistance measurements and the others on changes in electrical potential. Each will be described in turn. Number 4 has led to results of the highest accuracy and is the one most recommended.

1. Each section of the m.b. channel is constructed by stretching a thin platinum ribbon 1 mm wide across the upper end of a 2.5 mm i.d. glass tube, pressing another glass tube down on it, and fusing the pieces together. The piece of ribbon inside the tube is removed by cutting with a sharp steel rod and rubbing with carborundum powder. These sections are then sealed together to form a m.b. tube containing about five pairs of horizontally opposite probes. As the boundary passes each probe pair, a 20,000 Hz signal across it, after amplification and rectification, is traced on a recording instrument (Fig. 4.14a). The time of passage of the boundary is taken as that corresponding to the average of the signals when only leading solution or indicator is between the probes. Capacitors of 0.02 μF on each probe lead

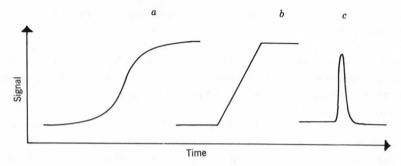

Fig. 4.14 Types of electrical recording obtained on passage of the boundary (schematic).

protect the a.c. bridge circuit from the d.c. flowing through the transference cell [87].

2. At least two platinum probe electrodes are mounted, several centimeters apart, into a precision bore glass tube. (The best method for doing so is described under method 4.) The resistance between these probes is then measured periodically during a run with an a.c. conductance bridge. The original workers [32] switched off the d.c. just before each measurement to avoid interaction between the two circuits, but it seems preferable to use large protective capacitances as in method 1. The resulting plot is shown in Fig. 4.14b. The initial horizontal portion represents the resistance of the leading solution, the linearly rising portion the movement of the boundary between the probes, and the final horizontal line is characteristic of the following solution. The time needed to traverse the space between the probes can be obtained by projecting each linear portion and reading off the times corresponding to the points of intersection.

3. In a rather different method [88], two pairs of thin platinum foil are sealed into the cell wall, the members of each pair being 0.3 mm apart. They are connected through a 1 μF condenser (to eliminate electrolysis at the electrodes) to a galvanometer or other current-sensitive device. As long as the probes face only one kind of solution, they are subject to a constant potential difference. As the boundary passes by, this potential difference changes from the value characteristic of the leading solution to that in the following solution. The resulting readjustment of electrical charge on the condenser plates gives rise to a current peak (Fig. 4.14c) and the velocity of the boundary can thus be measured as it passes the sets of probes.

4. In the last method [85, 89], the potential between probes spaced vertically along the m.b. tube—or even between one probe and the cathode or anode if either is reversible [90]—is monitored with a high-impedance electrometer connected to a recorder. The resulting voltage-time plot

resembles the lower half of Fig. 4.14b, the time of passage of the boundary being obtained from the intersection of the base line and the line drawn through the first straight section of the rising portion. High accuracy is attainable with a carefully designated electrical circuit [85]. The most satisfactory probes are produced [85] by 0.2 mm diameter platinum wire sealed into uranium (not Pyrex) glass tubes. The degassed but unbeaded piece of wire is inserted into a tiny hole, and heat applied to one edge until the glass melts on to the wire. It is advantageous if the wire hangs downward. That part of the wire inside the tube is sheared off with a sharp metal rod, and the probe reheated so that the glass melts and fills any crevices. The whole tube is finally annealed.

In all these electrical methods, great care must be taken to avoid breaking off any probes and to insulate the probe leads extremely well. Calibration of the tube may be carried out with a leading solution of known transference number.

As already mentioned, an electrical method of boundary detection is likely to be the only suitable one for very dilute solutions (as in many non-aqueous solvents) and for work at high temperatures [89, 90] and high pressures [32, 33] where the cell must be inside another container.

Radioactive Tracer Method. This procedure has been used for ion-exchange resins, where m.b. systems such as $AZ/^{\ddagger}BZ$ or $^{\ddagger}AZ/BZ$ (‡ indicates radioactivity) were followed by Geiger-Müller counters [91]. Marx and his co-workers have now succeeded in applying this technique to solutions, and have followed the movement of boundaries such as:

$HCl \leftarrow {}^{42}KCl$	(following ion labeled [92])
$KH_2{}^{32}PO_4 \leftarrow KOOC \cdot C_6H_4 \cdot NH_2$ (p)	(leading ion labeled [93])
$KNO_3 \leftarrow {}^{42}KNO_3$	(isotopic boundary [94])

It is the transference number of the labeled ion that is measured in the last three examples, and in the case of ${}^{42}K^+$ there is a significant difference (ca. 0.2%) with normal K^+. In $H_2{}^{32}PO_4^-$ the central atom is too well shielded by surrounding oxygens for the mobility to be altered appreciably. The isotopic boundary possesses the advantage that no new indicator substance need be chosen, and this is particularly valuable for those solutions in which hydrolysis rules out the use of conventional boundaries. The corresponding disadvantage is the absence of any electrical restoring effect so that the boundary becomes more diffuse as it travels along [94].

Radioactive tracers recommended [95, 96] include fairly long-lived and hard β-emitters (maximum energy > 1.5 MeV) such as ${}^{32}P$, for which blackened Perspex (Plexiglas) screening is needed, or β-emitters with a weak γ-background such as ${}^{42}K$. These require Al/Pb/Al sandwich shielding although Perspex is adequate if the amount of γ-radiation is very small. Geiger-Müller end-window counters are suitable detectors in both cases. Boundaries involving γ-emitting isotopes can be followed by means of a γ-scintillation counter with a rectangular NaI crystal (ca. 25 mm wide, 4 mm high, and 2 mm thick [96, 97]). In all cases, sections of the m.b. cell (other than the m.b. tube itself) that contain radioactive solution and that are near the counter should be screened to reduce the background count. The concentration of radioactivity required lies in the range 0.1–5 μCi cm^{-3} [98].

The geometry for β-emitting tracers is schematically illustrated in Fig. 4.15. The thickness of the wide screen B should be 10% less than that needed

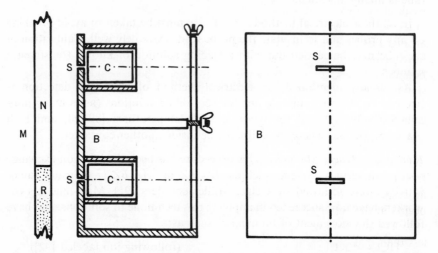

Fig. 4.15 Schematic diagram [95] of the arrangement for following boundaries between a radioactive solution and a non-radioactive one.

to absorb all the radiation [95]. The counters C, which may be either movable or fixed, are screened and placed as close as possible to the m.b. tube M. The horizontal slits S are several centimeters long with a slit height of about 2.8 mm [95]. As the boundary (rising or falling) between the radioactive solution R and the non-radioactive solution N passes each slit S, the counting rate changes. The same is true when R contains a γ-emitter: the counting rate then changes as the boundary passes the rectangular NaI crystal mounted in front of the scintillation counter [96]. The shapes of the counting plots obtained, either on a recorder or digitally, resemble those in Fig. 4.14a. The boundary is deemed to have "passed" at the point of inflection of the curve [92, 96]. Its determination is more difficult with isotopic boundaries which broaden as they move [94]. Isotopic boundary systems, moreover, should be arranged with the radioactive solution as the following one; if the tracer leads, its adsorption on the glass walls produces asymmetric counts-versus-time curves whose midpoints are difficult to locate [97].

Calibration of the tube can be carried out either with cathetometer readings if the inner diameter of the tube is known accurately [92] or by runs with solutions of known transference number. So far, air thermostats have been employed for temperature regulation.

Thermal method. Joule heating developed by the electric current causes the temperature of both solutions to rise, especially that of the more poorly conducting following solution. The effect is particularly pronounced with weak electrolyte solutions in which it can amount to several degrees centigrade with currents of the order of 1 mA [84]. Normally conditions are so arranged as to minimize this temperature rise, but two Russian workers [99] have pointed to the possibility of utilizing it for detecting the passage of the boundary. The response of a microthermistor in the tube should give a plot similar to that in Fig. 4.14*a*.

INDICATOR AND INITIAL INDICATOR CONCENTRATION

A suitable indicator must meet the following requirements:

1. It must not react chemically with the leading solution. However, the term chemical reaction is not intended to exclude a rapid reversible shift in an association-dissociation equilibrium such as $H^+ + OAc \rightleftharpoons HOAc$ which occurs in the NaOAc/HOAc system.

2. The mobility of the following non-common ion-constituent B must be less than that of the leading non-common ion-constituent A in both solutions and in any possible mixture of the solutions that can exist in the boundary (see (4.34)). Take the case of a completely dissociated electrolyte, such as KCl in water, for which a cation indicator is required. LiCl is very suitable since the mobility of Li^+ is much less than that of K^+, but HCl is not since the H^+ ion is faster than K^+. As a second example, consider an incompletely dissociated electrolyte such as aqueous H_3PO_4 for which an anion indicator is needed. Picric acid is not a suitable indicator even though the mobility of the picrate *ion* is less than that of the $H_2PO_4^-$ *ion* because picric is a stronger acid than phosphoric and the mobility of the picrate *ion-constituent* is greater than that of the phosphate *ion-constituent* (see (4.17)). However, formic acid could be selected as a following electrolyte: it is a weaker acid than phosphoric and, despite the fact that the formate *ion* is faster than the $H_2PO_4^-$ *ion*, the mobilities of the respective *ion-constituents* obey condition (4.34). A third and quite famous example [99] is the stable aqueous boundary system NaOAc ← HOAc in which the intrinsically slow but free Na^+ ion leads hydrogen, a very fast ion considerably hampered by strong association [3].

3. The leading and indicator solutions should differ appreciably in some property (conductance, refractive index increment, radioactivity) that enables the boundary to be followed. An indicator giving an easily visible boundary with the assembly in Fig. 4.12 requires the simplest apparatus.

4. The indicator solution (or, for that matter, the leading solution) should not be appreciably hydrolyzed [100]. The hydrolysis can sometimes be repressed, for example, in KOAc solutions by adding a little HOAc [101]. Alternatively, isotopic boundaries can be resorted to.

Certain other points should also be considered in choosing an indicator substance, such as the interaction between it and the selected electrode in the indicator solution (see below). It should be possible to purify the indicator so as to remove from it any fast ion-constituents which might overtake the boundary and so affect its velocity. Boundaries are sharper if there is a greater mobility difference between the leading and following ion-constituents (see (4.35)), and they tend to be more stable if there is a greater density difference between the leading solution and the indicator solution at the Kohlrausch concentration. If the latter solution is lighter than the leading solution, a falling boundary must be used, and vice versa.

Detailed work on certain systems has shown [102] that, provided the cell is carefully designed, indicator solutions of widely differing concentrations adjust to the Kohlrausch concentration behind the m.b. However, every system must be tested on its own merits [103], and much trouble may be saved if the following rules are adopted:

1. The initial concentrations should not differ from the Kohlrausch values by more than about 25% in aqueous solutions and by less in other solvents. A rough estimate of the Kohlrausch concentration can normally be derived from (4.37), (4.10), and (4.5) and from whatever transference and conductance data are available.

2. For falling boundaries it is safer to use initial concentrations equal to or less than the Kohlrausch concentration so that the system in Fig. 4.10b—BZ(init.)/BZ(Kohlr.)/AZ—is gravitationally stable. This system is usually thermally stable also, for the solution of highest resistance, which is warmed most by Joule heating, will lie at the top of the tube.

3. For rising boundaries it is wise to use initial concentrations equal to or greater than the Kohlrausch concentration, so that the system in Fig. 4.10a—AZ/BZ(Kohlr.)/BZ(init.)—is gravitationally stable. This system is generally not thermally stable since cooler solutions lie above warmer ones, and convective disturbances are somewhat more likely.

4. At least two indicator concentrations, differing by a few percent, should be used; if the transference numbers do not agree, further work is necessary

in order to find a range of indicator concentrations over which the transference number is constant and independent of current. Should no such range be discovered then another indicator substance is called for.

It is not essential that the indicator and leading electrolytes have an ion-constituent in common, for the ion-constituent in the leading solution moving in the opposite direction to the boundary becomes the common ion-constituent. Thus AZ/BY becomes $AZ/BZ/BY$ when the boundary moves. The only restriction is the avoidance of an unstable density difference [104]. The idea can be used to advantage if discharge of the Z ion at the electrode would cause some disturbance there or generate an ion that overtakes the boundary [105].

ELECTRODES

One of the electrode compartments should be open to allow for the expansion or contraction of the whole system, and the other should be closed to avoid net electroosmosis and to permit calculation of the volume correction. Good temperature control must be maintained in the thermostat as volume fluctuations in the closed side of the cell seriously affect the boundary velocity.

The electrode in the closed compartment must meet certain requirements:

1. No gas evolution must occur.

2. No foreign ions generated by the electrode reaction must reach the boundary, as this progressively decreases the boundary velocity. Since the time taken by such an ion to migrate to the boundary is proportional to the volume of solution it has to pass through, it is sometimes helpful to use a large or extended electrode compartment. Improved temperature regulation is then advisable.

3. The electrode reaction and the physical state of the products should be known so that the volume correction can be calculated. The extent of any appreciable side reaction can often be found by weighing the electrode before and after a run.

For the open electrode, only condition 2 is applicable. The mixing effect of gas evolution can be reduced by appropriate cell design.

Silver halide electrodes, made electrolytically or by dipping a platinum wire at the end of a glass tube repeatedly into the molten salt, are often used as cathodes. Platinum electrodes immersed in an aged slurry of silver succinate, or some other insoluble silver or lead salt, are convenient cathodes for work with aqueous acid solutions [106, 82]. Pure silver or cadmium are normally excellent anodes. Palladium membranes have been suggested [107] as non-gassing hydrogen electrodes but have not yet been found entirely satisfactory [82, 108]. Various other electrodes described in the Hittorf

Section can be employed equally well here and since the currents and times of electrolysis are usually smaller than in Hittorf work the electrode areas need not be as large. If a simple non-gassing electrode cannot be found for either the leading or the following solution, an intermediate guard solution may have to be introduced, for example, a KCl solution can serve around a cadmium anode or an AgCl cathode. However, care must be taken during filling to prevent mixing between the guard solution and the main solution, density instabilities must be avoided, foreign ions from the guard solution must not catch up with the boundary, and the guard solution must be taken into account when the volume correction is calculated.

Methods of introducing the electrodes into the cell are discussed under Cell Design.

CELL DESIGNS AND FILLING TECHNIQUES

Tube Marking and Calibration. The present paragraph refers to the optical method of boundary detection. For work of relatively low precision, the m.b. tube can be a graduated commercial pipet of uniform cross-section and of 1 or 2 ml volume [109] (the etched marks are generally too broad for very accurate definition of the boundary). Alternatively, a glass tube of constant and known internal area (0.05 to 0.1 cm^2) can serve if the telescope in Fig. 4.12 is mounted on a cathetometer stand. For accurate work, fine parallel lines are etched completely around a well-aged uniform piece of Pyrex tubing 15 to 25 cm long and 2 to 5 mm i.d. (a volume of about 1 ml is convenient). In Le Roy's [110] procedure, the tube is drawn to a fine point at one end, and the part to be graduated is covered with a thin film of wax (1 part hard paraffin and 1 part beeswax). The tube is then placed horizontally on two V-shaped metal cradles and the pointed end is held tightly against a flat piece of metal. While the tube is rotated slowly, the end of a small needle held in a metal arm is brought down on the tube to trace a fine line through the wax. Finally, the tube is rotated for 1 min in contact with a few drops of a 1:1 HF-HCl solution. Some plain tubing must be left at each end for sealing to the rest of the cell. Enough lines should be etched on to permit several independent measurements of the boundary velocity per run. For example, with 12 lines (called 1, 2, 3, . . . , 12) spaced at roughly equal intervals, six independent determinations are possible (the differences 7 − 1, 8 − 2, . . . , 12 − 6), and any change in the velocity of the boundary is easily detected. If desired, a larger interval can be left in the middle of the tube (with 12 lines, between lines 6 and 7); this increases the relative precision of each determination but makes it harder to detect any trend in the boundary velocity. The thin, etched lines appear horizontally in G and G' in Fig. 4.13, and their position is more readily spotted during the run if after the cell is filled a black grease pencil mark is made on the tube a few millimeters above or below each line. Calibration is achieved either by mercury weighing [111] or by doing some runs with a substance whose transference numbers are known accurately, for example, KCl in water. In the latter case, the literature values should be "uncorrected" again for the solvent and volume corrections to give the directly observed value that is appropriate in (4.36).

Calibration for the radiochemical method of following the boundary can be carried out either by using precision bore tubing and a cathetometer, or preferably by employing a solution of known transference number. The latter procedure is the only one possible with the electrical method of boundary detection. Techniques of sealing in the platinum probes have been described in an earlier section.

Formation of the Boundary. Most boundaries of the so-called "sheared" type are formed initially by suddenly joining the leading and indicator solutions by some mechanical device. Four of these are described below. The advantage of sheared boundaries is that the initial indicator concentration can be varied at will—the exact Kohlrausch concentration not usually being known in advance—until a range of concentrations is found over which the velocity of the boundary is constant. For "autogenic" boundaries, however, the initial junction is between the leading solution and an electrode, and the indicator is formed by the electrode itself. Thus KCl over a Cd anode forms an indicator solution of $CdCl_2$, and the boundary KCl $\leftarrow CdCl_2$, on passage of current. Autogenic boundary formation is particularly favored for experiments at high temperatures [89, 90] and high pressures [32, 33] in which a minimum of outside manipulation is desirable. The autogenic technique has so far been applied only to rising cation boundaries. Experiments in a number of instances have shown [112] that the solution following behind an autogenic boundary automatically adjusts itself to the Kohlrausch concentration, and this is now taken for granted in every case. Independence of current is one indirect way of checking that convective mixing has not occurred behind the boundary with the more concentrated solution in the vicinity of the electrode. The concentration stratification that normally exists above an autogenic electrode introduces some uncertainty into the value of the volume correction [50, 78].

Several devices have been invented for the creation of sheared boundaries. The first, described in detail by MacInnes and Longsworth [111], consists of plane (glass) disks or plates to which tubes filled with the two solutions are attached; to start a run the plates are slid over each other until the openings coincide (Fig. 4.16). This equipment is only occasionally employed [113] nowadays, but a similar technique is extensively used in Tiselius electrophoresis cells [45]. The second shearing method employs a stopcock. To

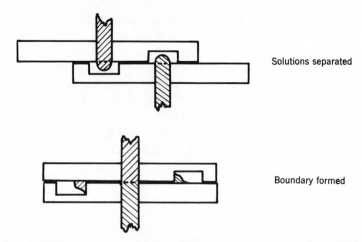

Solutions separated

Boundary formed

Fig. 4.16   Schematic diagram showing the formation of a sheared boundary by the sliding of plates over each other.

reduce local Joule heating, the barrel should be hollow and open to permit circulation of the thermostat liquid, and the inner bore should have the same cross-sectional area as the m.b. tube. The shape of the stopcock depends on the type of cell (Figs. 4.17 and 4.18). The choice and purity of the lubricant in both these shearing methods is important. Many simple lubricants are essentially hydrocarbons: where these fail, recourse may be had to silicones (which can be removed with triethylamine) [82], fluorocarbons [114], special formulations such as lithium stearate—oil for methanolic solutions [115], or special techniques to deal with solutions of detergents [116]. The suitability of any given grease should be tested, either by comparing it with another and/or by seeing if there is any effect on the boundary velocity when more grease has been deliberately smeared inside the uncalibrated portion of the m.b. tube. The last two methods do not require a lubricant. One is the "air bubble" method whose name describes the device by which the solutions are initially separated; to start a run the air bubble is forced up a side tube and the boundary is produced without appreciable mixing [104, 117] (Fig. 4.19). The fourth and final method is the "flowing junction." One solution streams up the tube, the other down it, and at the point where the boundary is to form the mixed solution leaves through an auxiliary tube. At the appropriate moment the latter is removed—by simply withdrawing it [85] or by sliding a plate [118]—and the boundary remains.

Specific Cells.   The cells in the diagrams can be used for any method of boundary detection. The tiny marks on the m.b. tubes M can be imagined as etch marks for optical observation or as platinum probes for electrical

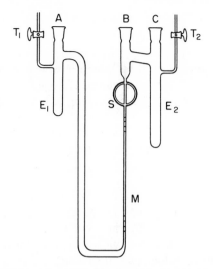

Fig. 4.17 Sheared falling boundary cell. The handle of stopcock S has been omitted from the drawing, and should be on the side facing the reader.

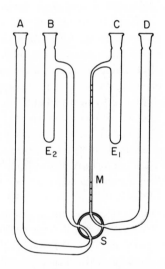
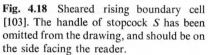

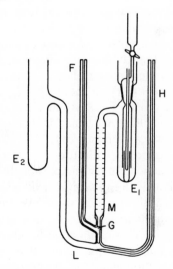

Fig. 4.18 Sheared rising boundary cell [103]. The handle of stopcock S has been omitted from the drawing, and should be on the side facing the reader.

Fig. 4.19 Sheared rising boundary cell used by Hartley and Donaldson [104] and employing the air bubble method of boundary formation.

detection. The designs in Figs. 4.17 and 4.18 are based on a careful study [119, 120] of boundary stability.

In Fig. 4.17, tap T_1, tap T_2, and stopcock S are opened, and leading solution is poured into A until the liquid level is just above S. Air bubbles clinging to the lubricant around S can be removed by successively squeezing and releasing a piece of clean flexible tubing inserted just inside A. Jiggling a platinum wire through B is also effective. S is then closed, and the E_1 compartment filled up with leading solution. Solution above S is removed by suction with a drawn-out piece of glass tubing, and the indicatorcompartment above and to the right of S is repeatedly rinsed, first with solvent, then with indicator solution, the liquid being sucked out each time. The compartment is finally filled with indicator solution, the electrodes are introduced through A and C, B is closed with a ground-glass plug, and the cell is allowed to come to temperature equilibrium in the thermostat. The side-arm tap of the closed electrode section is then shut, S is opened so that bore and tube are well lined up, and the current is switched on immediately.

In Fig. 4.18, with stopcock S aligned as shown, indicator solution is added through A and leading solution through D until the cell is full. Air bubbles may be removed from inside S as described above. The electrodes are inserted into B and C, and the cell is allowed to reach thermal equilibrium. Then, to connect E_1 with E_2, S is turned through 90°, either clockwise for "top shearing" or counterclockwise for "side shearing." In the latter case the boundary has to round a corner, but stable indicator adjustment appears to occur more easily [120]. The current is turned on immediately. The cells in Figs. 4.17 and 4.18 can be combined to form a cell suitable for both falling and rising boundaries, but it is often found more satisfactory to use two separate cells.

Figure 4.19 illustrates the air-bubble or air-lock method of boundary formation. To the end of capillary tube F is attached a short piece of pressure tubing with two screw clips (screw clamps), the top one closed and the lower one half open. The whole apparatus is first filled with leading solution, the electrode is inserted into E_1 and that section is closed, and the lower screw clip on F is compressed so as to force out of F the air bubble G which separates the solution into two parts. The sections E_2, L, and H are emptied by suction at E_2 and H and are washed through several times with solvent from E_2 with controlled suction at H, care being taken to keep liquid in contact with G so that the air bubble remains intact. E_2, L, and H are then filled with indicator solution, the second electrode is put into E_2, and the voltage applied. The lower screw clip on F is unscrewed until the air bubble has receded sufficiently far into F to allow current to flow, and the bubble is completely withdrawn into F only when the boundary has passed G. This ensures that no leading electrolyte remains behind the m.b. in a region through which no current flows.

In the corresponding falling boundary cell, L and M are connected by a plain bend, the upper part of M is narrow and has the tube F sealed into it, and tube H becomes an extension of the tube joining M to E_1. The air column forced into the top of M through F in this case expands upward [104].

Figure 4.20 depicts a "flowing junction" cell, which can be used for either rising or falling boundaries. With the joints at F and tap T_2 closed, solution

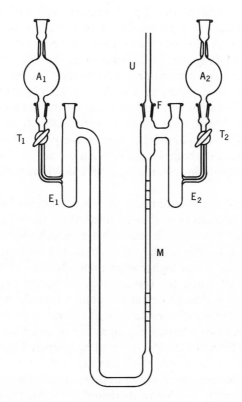

Fig. 4.20 Rising or falling boundary cell designed by Kay et al. [85]. The boundary is formed in the tube M by means of a flowing junction.

I (following solution for a rising boundary, leading solution for a falling one) is added to E_1 and stops after filling about 1 cm of the narrow tube M. The openings on the left-hand side are then closed and the other solution, II, is poured into E_2 until it just overflows into M. With aqueous solutions a bubble usually forms in M, for other systems the solution normally flows slowly down the tube and mixes with the first solution. In either case the open tube U is inserted, and a thin-walled capillary pipet, rigidly held on a traveling cathetometer and connected to a siphoning system, is inserted through U into M.

If a bubble has formed, the pipet is slowly lowered to allow solution II to flow down its sides until a point about 1 cm above solution I is reached. At this stage the siphoning system is turned on and the gas bubble completely removed, bringing the two solutions into contact. The right-hand side of the cell is then completely filled, and the 75 cm^3 bulb A_2 is added with solution II filled well up into the narrow capillary at the top. The level of solution in U is brought to this same height, tap T_2 opened, and the solution is slowly siphoned out through the pipet which is now placed at the point where the boundary is to be formed—near the bottom of tube M for a rising boundary, near the top for a falling one. In the meantime, the 75 cm^3 bulb A_1 has been filled with solution I and placed in position. When the solution level in U and A_2 reaches that in the capillary neck in A_1, tap T_1 is opened and both solutions are permitted to flow into the cell and, at the boundary, out into the siphoning system. When, after about 15 min, bulbs A_1 and A_2 are almost empty, the taps T_1 and T_2 and the siphon on the pipet are closed. The pipet is very slowly retreated about 5 mm, T_2 again opened, and solution II in A_2 permitted to flow again through the pipet very slowly as the latter is removed from the cell.

If a bubble is not initially formed in M when solutions I and II have been poured in, some solution II is allowed to overflow from E_2 into M and the pipet is lowered to a point well below that at which the boundary is to be formed. Most of this added solution is removed and the process repeated but with the pipet about 5 mm higher after each addition. If this procedure is repeated at least 10 times, the top solution is entirely free of the bottom solution. The pipet is then placed at the point where the boundary is to be formed and the solutions allowed to flow as before.

Figure 4.21 gives two alternative designs for an *autogenic* cell. The closed electrode is often formed [32, 34, 101, 109] by a tapered plug of metal (e.g., Cd, Ag) ground to fit tightly into the lower end of the m.b. tube M and sealed in with a suitable cement which must not interact with the solution. Araldite, for instance, is slightly soluble in methanol [97]. Alternatively, the metal anode can be pressed firmly on to a glass flange at the end of the calibrated tube by a Lucite clamp, a leak-proof seal being effected with a small rubber O-ring [33]. In the type [115, 119] shown in Fig. 4.21 a long slightly tapered male joint is sealed on to the bottom of M. A tungsten wire is sealed through a small well at the base of the corresponding female joint E_2, and the well is filled with metal shavings which are fused in a hydrogen atmosphere. The metal surface is scraped clean with a steel wire between runs. Since cavities can form between metal and glass, any air should be removed before runs by filling E_2 with solution and evacuating. To fill the cell, E_2 is slightly greased

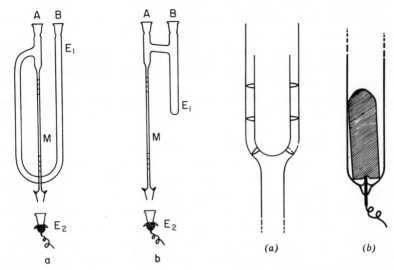

Fig. 4.21 Autogenic rising boundary cells.

Fig. 4.22 Electrode section designs. (*a*) Device for accommodating a dense electrode solution near the top of the cell. (*b*) A simple device [115] for introducing a metal anode.

and pressed home, secured with steel springs on the glass hooks shown, and leading solution is poured in. Any air that is trapped in the narrow tube M can be allowed to escape through a fine capillary inserted through A. The other electrode is placed in position, A plugged up, and the current started.

General Points.   Cells should be made of Pyrex so that the solvent correction is not appreciably increased in the cell. For the sake of the calibrated tube, cells should be dried by a stream of nitrogen and not by heating. The cells can be strengthened by a bar of glass tubing sealed horizontally across the back, and a stout vertical glass tube can be sealed on to this to act as a handle. Vibrational disturbances can be counteracted as described on page 230.

Gravitational stability must be maintained in the electrode compartments; an electrode should be at the bottom if the products of electrolysis are heavier than the main solution, and vice versa. Figures 4.21*a* and 4.21*b* illustrate one way of catering, respectively, for lighter and heavier electrode solutions in E_1. A dense electrode solution can be accommodated near the top of a cell by placing the electrode at the bottom of an inner tube as in Fig. 4.22*a*. A small electrode section similar to this has been used [121] for a tiny autogenic m.b. cell needing only 10 to 15 cm³ of solution. This type of design also reduces mixing effects around gassing electrodes. A satisfactory method [115, 119] of introducing a metal anode into sheared cells is to slide a rod of the metal gently down on top of a thick tungsten wire sealed through the bottom of the compartment, as in Fig. 4.22b. The electrical contact is sufficiently good to prevent gassing as electrolysis occurs mainly at the top of the rod. Bare electrode leads can trail through an oil-filled thermostat with little risk of current leakage.

Adsorption of electrolyte on the cell walls may lead to small errors in work with dilute solutions. It is then advisable to rinse out the m.b. tube with leading solution before finally filling it with a fresh sample. This is automatically done with the flowing junction described earlier.

Solutions should be partially degassed (see p. 230) to avoid air bubbles appearing in the closed compartment or the m.b. tube. The cell should always be carefully examined for tiny gas bubbles before and after each run.

Leakage of solution in the closed part of the cell must be prevented. The cell openings should be closed by well-ground standard taper joints, either in the form of plugs or electrodes, lightly greased and held down with springs and glass hooks or with special clamps. Leakage can be tested for by filling the cell with solvent and adding mercury so as to maintain a positive or negative pressure head of a few centimeters. At constant temperature, a variation in the position of the mercury surface in the m.b. tube indicates a leak. Irreproducibility is sometimes traceable to cracks in the glass or insecurely stoppered openings.

The cell should be placed in a liquid, not in an air, thermostat, so that the Joule heat is dissipated sufficiently rapidly to prevent destruction of the m.b. by convection [109]. The novel recommendation [122] that convection in the tube be cured by packing it with quartz sand cannot be endorsed for accurate transference measurements despite its usefulness in isotope separation experiments [123]. Although the temperature coefficients of transference numbers are relatively small, very good constancy of the thermostat temperature is essential in m.b. work because a small expansion or contraction of the solution in the relatively large closed electrode compartment causes an appreciable shift in the position of the boundary in its narrow tube. The analogy with an ordinary mercury thermometer is obvious.

For work at higher temperatures, a layer of mineral oil on top of the electrolyte solution in the open compartment reduces evaporation losses [89]. In high pressure experiments, a flexible Teflon extension of a sidearm allows pressure to be transmitted to the system without contaminating the solution with the pressurizing oil [33].

CURRENT AND TIME

A by-product of the passage of electricity is Joule heating. A theoretical analysis of the resulting temperature rise in a m.b. tube has been given in several papers [57, 93, 111], and experiments show [84, 98] that in solutions of low conductivity (low concentration, ion association) temperatures may rise by several degrees. In these cases the transference number must be measured at several currents and extrapolated to zero current. A twofold current variation should be investigated for every system; in most cases the transference number remains constant. Too high a current produces a curved boundary or even more drastic changes. Too low a current gives a very diffuse boundary (see (4.35) and (4.36)). Currents used seldom exceed 5 mA and need be only a few microamperes. As a rough guide for aqueous solutions, if the transference number is around 0.5, the current should be such that the boundary takes 1 to 3 hr to traverse 1 cm^3 in a tube of 2 to 3 mm i.d.

One of the classic experiments [111] was to interrupt an m.b. run and then to demonstrate, on restarting the current, that the boundary formed again and moved forward at its old rate. Closer investigation indicates [124], however, that current interruption does disturb the boundary afterward and should therefore be avoided.

Current leakage in the cell is most simply and effectively avoided by employing an oil thermostat.

Coulometers are not favored in m.b. work, for one must be thrown into and another out of circuit every time the boundary passes a calibration mark [101]. Separate current and time measurements are used instead. Since the replacement of one electrolyte in the m.b. tube by another leads to large increases in the cell resistance, an electronic current regulator is essential. There are as many circuit designs as there are groups of workers in the field, in part because different researches have different requirements of current range, cell resistance, and desired current stability. Many excellent constant current devices have been described in m.b. papers (see recent references listed in the Appendix) and elsewhere [69], and several are available commercially.

A simple timing method for optical observation requires only an accurate watch and a metronome clicking at a convenient rate, such as 2 beats/sec; when the experimenter notices that the boundary is just coincident with a calibration mark, he starts counting aloud and continues to do so until he can look at the watch and note some easily remembered time. For example, if he has counted 23 half-second beats until the watch registers 11/14/30, then the boundary passed the mark at 11/14/18$\frac{1}{2}$. No appreciable error is introduced, since coincidence of boundary and tube mark can seldom be judged to better than $\pm\frac{1}{2}$ sec. Another simple device [34] is a "split-second" stopwatch with two second hands, one of which can be stopped and, after the time has been read, resynchronized with the other hand. In electrical methods of following the boundary, a time mark, as determined to 0.2 sec by an electric clock, can be placed on the recorder chart by an event marker. The exact time of boundary passage is then calculated from this mark and the recorder chart speed [85]. Many other timing techniques are described in the literature (see papers listed in the Appendix). For electric timers the fluctuations in a.c. supply frequency may have to be watched.

Transference numbers obtained by the direct m.b. method can be very precise and are correspondingly sensitive to faulty technique. If accurate data are desired, it is essential to have evidence of reproducibility, absence of progression [125], independence of current and, for sheared boundaries, independence of initial indicator concentration (p. 246). The transference number should also be independent of the indicator if different indicators are tried, and if the transference numbers both of the cation- and of the anion-constituent are measured they should add up to unity [see (4.8)] within experimental error. Lest it should seem that too much caution is being urged, it may be pointed out that in a number of cases apparently reliable work was shown to be in error because independence of current and initial indicator concentration had been taken for granted [103]. Reproducibility and absence of progression do not by themselves guarantee the reliability of the measurements.

The Differential Moving Boundary Method

A slow diffuse "concentration boundary" moves between two solutions of the same salt but of different concentrations. Theory shows [111] that for the boundary $AZ(\text{soln.}')/AZ(\text{soln.}'')$:

$$T'_A - T''_A = V_F(C'_A - C''_A) = \frac{VF(C'_A - C''_A)}{It} \qquad (4.40)$$

where the symbols are the same as for (4.36) and where V is positive if the boundary moves in the same direction as the ion-constituent A. The heavier solution, which is usually the more concentrated, must be the lower one to avoid gravitational instability.

Such a system is useful if only the concentration variation of the transference number is wanted or if the transference number is already known at one concentration. The main advantage is that no indicator is necessary, and the method has therefore been used for polyelectrolyte solutions for which indicators are hard to find [126] (see, however, the indirect m.b. and analytical boundary methods, and the use of isotopic boundaries on p. 243). There are two disadvantages. First, the boundary is diffuse because the restoring effect is weak, and the boundary cannot be followed by the simple optical method [127]. Second, the method appears to be unsatisfactory below 0.02 N [127], and the volume corrections (equation (4.39)) are of the same order of magnitude as the boundary displacements. Nevertheless, accurate results can be obtained, particularly in aqueous solutions at lower temperatures [127].

Indirect Moving Boundary Method

Introduction and Theory

This method is complementary to the direct m.b. method in that the solution following the boundary is the solution under investigation. The transference number in this solution is determined by measuring the adjusted or Kohlrausch concentration behind the boundary. Indeed, the method has sometimes been called the adjusted indicator technique.

The theory is based on the Kohlrausch equation (4.37) and Fig. 4.10. If a leading solution AZ of known normality C_A^{AZ} and known transference number of the A ion-constituent, T_A^{AZ}, forms a stable m.b. with the following solution BZ, then analysis of the adjusted solution of BZ behind the boundary gives $*C_B^{BZ}$. Rearranging (4.37) gives:

$$*T_B^{BZ} = T_A^{AZ} \cdot *C_B^{BZ}/C_A^{AZ} = T_A^{AZ} \cdot s, \qquad (4.41)$$

where $*T_B^{BZ}$ is the transference number of the B ion-constituent in a BZ solution of normality $*C_{BZ}$, and where s is the Kohlrausch ratio. Thus once a suitable leading solution of known properties is available (see Appendix for examples), the experimental problem is merely one of analyzing the BZ solution behind the boundary. It must be pointed out, however, that (4.41) is not strictly correct, for it ignores the *volume correction* that must be applied to both transference numbers. Although this can be done [128] as in (4.39), the neglect of changes in the partial molar volume of the solvent itself introduces appreciable errors at higher concentrations [129]. Luckily, a rigorous expression that completely eliminates the effect of volume changes is applicable in this case:

$$*T_B^{BZ} = T_A^{AZ} \cdot *c_B^{BZ}/c_A^{AZ}, \qquad (4.42)$$

where c is the concentration in equivalents per kilogram of solvent (C, it will be remembered, represents equivalents per liter of solution). Originally arrived at by Hartley [130], this equation—forgotten for many years—was recently resurrected and more elegantly derived [129, 131]. Its re-introduction heralds a new era for the indirect m.b. method which can now be applied with confidence to studies at high concentrations at which the results of the direct method, because of uncertainties in the volume correction, become less reliable.

As with the Hittorf and direct m.b. methods, there is need for a small *solvent correction*. Here it takes the form [131]

$$T_R(\text{corr.}) = T_R(\text{obs.})\left[1 + \kappa_{\text{solvent}}\left(\frac{1}{\kappa_{BZ}} - \frac{1}{\kappa_{AZ}}\right)\right], \qquad (4.43)$$

where κ is specific conductivity in ohm^{-1} cm^{-1}. The derivation assumes that $\kappa_{solvent}$ is the same in both solutions and that there is no preferential migration of dissolved impurities; for these reasons one might well heed an earlier warning [52] that the only safe solvent correction is one very close to zero. Attention should also be drawn to the likelihood of contamination in the cell itself and in the sampling pipet employed in methods of external analysis [132, 133].

A phenomenon that is relatively rare, but which has been observed [134] in some polyelectrolyte solutions, is that the ratio $*T_B^{BZ}/*c_B^{BZ}$ may have the same value at two different concentrations. Under these circumstances two different values of $*c_B^{BZ}$, hence of $*T_B^{BZ}$, may be obtained with the same leading AZ solution but with different initial concentrations of BZ.

In the direct m.b. method one of the main requirements for the following electrolyte is that it contain a slower non-common ion-constituent. In the indirect method the corresponding requirement for the leading solution is that it contain a faster non-common ion-constituent. The two methods therefore complement each other, so that m.b. work in general is virtually free from a restriction of this sort. The indirect method has been applied frequently to compounds such as polyelectrolytes in which the organic ion-constituents are large and slow. Moreover, as already mentioned, the indirect m.b. method is more suitable than the direct method for the accurate determination of transference numbers at high concentrations. The removal of doubts about the volume correction is one reason, another is the greater tolerance by the indirect method of diffuse boundaries. Smaller currents may therefore be employed, and convective disturbances are reduced [135].

Experimental

GENERAL CONDITIONS

A suitable leading electrolyte (AZ) must meet the following conditions:

1. It must have an ion-constituent (Z) in common with the electrolyte under investigation.

2. It must not react chemically with the solution under investigation, although rapid reversible changes in association-dissociation equilibria such as $H^+ + OAc^- \rightleftharpoons HOAc$ may take place in the boundary.

3. The mobility of the leading non-common ion-constituent (A) must be greater than that of the following non-common ion-constituent (B) in both solutions and in any possible mixture of the solutions that can exist in the boundary.

4. The transference number of the leading non-common ion-constituent (T_A^{AZ}) should be known over an appropriate concentration range (see Appendix for examples).

5. The leading solution (and the following solution) should not be appreciably hydrolyzed, although hydrolysis can sometimes be repressed (p. 246).

6. The choice of the leading electrolyte depends also on certain other considerations, such as purification, interaction with electrodes, and gravitational stability; these were discussed earlier (p. 246).

The initial concentration of the following solution should be slightly lower than the Kohlrausch concentration given by (4.42) for falling boundaries, and slightly greater for rising ones, in order to reduce mixing effects between the initial and the Kohlrausch solutions. In either case it is essential to check that the adjusted BZ concentration obtained is independent of a few percent variation in the initial BZ concentration. The Kohlrausch concentration is, of course, not known to begin with, but a good approximation to it can usually be obtained from a preliminary run [132] using a relatively dilute solution of BZ for a falling boundary or a relatively concentrated solution for a rising one.

The main restriction governing the choice of electrodes is that no foreign ions formed by the electrode reactions must reach the boundary or the BZ solution near the boundary. Various suggestions regarding electrodes were made on pages 229 and 247.

In contradistinction to the direct m.b. method, the position of the boundary need be determined only approximately, either by following the boundary visually or electrically or by calculating its position from (4.36) if the cross-sectional area of the (unmarked) m.b. tube is known and if current and time are measured. However, it is quite feasible [85] to build cells that permit application of the direct as well as the indirect m.b. method and that enable the transference numbers of both the leading and the following solutions to be measured; in the same run, if desired.

The main experimental techniques used for finding $*c_B^{BZ}$ are summarized below. It is necessary in all cases to test whether or not undisturbed adjustment to the Kohlrausch concentration has occurred by repeating the experiment with another current and again with a somewhat different initial concentration of BZ.

EXTERNAL ANALYSIS

The removal of a small sample of solution for external analysis must be accomplished without disturbing the m.b. system. This difficulty can be minimized in work with aqueous solutions near $0°C$ [136] where substantially convection-free electrolysis can be carried out in a wide Tiselius cell [45]. If certain precautions are used, a thin-walled capillary attached to a syringe can be lowered into the cell while the current is flowing, and a sample of solution can be removed without appreciably disturbing the boundary or mixing the solutions [136]. A more general method of surmounting the problem was given by Muir, Graham, and Gordon [132].

These workers used a falling boundary cell incorporating the long narrow tube depicted in Fig. 4.23. The boundary is formed above S_1, as described for Fig. 4.17, the current is started, and the boundary moves down the tube.

Fig. 4.23 Moving boundary tube used for the external analysis of the Kohlrausch solution.

When it is certain that the boundary has passed through S_2, the current is shut off and both stopcocks are closed, thus isolating a column of the following solution between S_1 and S_2. The solution above S_1 is then removed under suction with a fine pipet, S_1 is opened, and the solution in the tube down to a level 2 cm below S_1 is also sucked out. A second long capillary pipet (or a needle at the end of a syringe [135]) is next inserted well into the tube, and the solution is removed and analyzed. Conductometric analysis using tiny conductance cells has usually been favored, but any accurate analytical procedure suitable for small samples is satisfactory. An analogous experimental procedure can be devised for rising boundaries.

Results agreeing with direct m.b. data to better than $\pm 0.03\%$ have been obtained for the KCl/NaCl system when the operations were carefully standardized and when certain small handling errors were corrected for [132]. The method has since been successfully applied to dilute aqueous solutions of sodium lauryl sulfate [133] and to concentrated solutions of several alkali halides [137, 135]. In one paper [135] the method was reversed, and the transference numbers of the leading solution were determined by the Kohlrausch equation from a knowledge of the transference numbers of the following solution.

In another version giving results accurate to about $\pm 0.2\%$ [120], a removable section is mounted as part of the m.b. tube (in Fig. 4.17 or 4.18) by means of standard taper ground-glass joints. When the boundary has passed completely through this part of the tube, the cell is taken apart at the ground-glass joints and a platinum electrode on a complementary joint is inserted into either end of the removable section, thus turning it into a small conductance cell filled with the Kohlrausch solution.

INTERNAL CONDUCTOMETRIC ANALYSIS

This idea was developed by Hartley and his co-workers [138] with a "balanced boundary cell" in which a glass piston, moved by a synchronous motor, produces a slow counterflow of solution in the moving boundary tube. This causes the boundary to remain almost stationary while the following Kohlrausch solution is gradually pushed into a side arm containing platinum wire electrodes; its a.c. conductance is then measured. The resulting transference numbers are accurate to about $\pm 1\%$.

Hartley's cell, though slightly cumbersome, is advantageously designed to enable the conductance measurements to be carried out in a side arm through which the main cell current does not flow. Similar experiments 20 years later [87], with platinum probes sealed into the moving boundary tube itself, at first gave results several percent in error, even though the a.c. bridge was protected by large capacitors from the direct current passing through the cell. Only recently has this problem been overcome [139], and it is now possible to determine the concentration of the following solution to 0.1% with platinum microprobes in the m.b. tube. The careful sealing of the probes into the glass is most important and has been briefly described on page 243. During a run, the resistances between pairs of microprobes are measured with a d.c. conductance circuit. In essence, the potential between any pair (such as 1–5, 2–6, 3–7, 4–8, if the probes are numbered 1, 2, . . . , 8 in sequence) is balanced out by dry batteries placed across variable resistors and, by means of a simple voltage divider network, an accurately known small fraction of this balancing potential is measured with a potentiometer. Dividing the balancing potential by the current flowing through the cell gives the resistance across the probe pair in question. To convert the resistance to specific conductivity and hence to concentration, one can either determine a "cell constant" for every probe pair or else, more simply, measure the ratio of the balancing potential of the leading solution to that of the following one. Current and cell constant then cancel out, and the ratio of the specific conductivities of the two solutions remains and can easily be converted into the ratio of concentrations demanded by (4.42). Kay et al. [139] emphasize that the potential drop across the probe combinations must be kept large (of the order of 100 V) so that the contribution of electrode polarization becomes negligible.

INTERNAL REFRACTOMETRIC ANALYSIS

Figure 4.24 shows the type of photograph obtained [140, 141] when the m.b. system between AZ and BZ is observed by the schlieren scanning optical method [45]. Peak L records the sharp $AZ/*BZ$ two-salt m.b. The broad peak N is produced by the diffuse concentration boundary (p. 258) between the Kohlrausch solution of BZ behind the boundary ($*BZ$) and the initial solution of BZ. The area under peak N is proportional to the difference in the refractive index between the two BZ solutions; if the refractive index increment of BZ has been determined previously, then the Kohlrausch concentration can be calculated from a knowledge of the initial concentration of BZ and the area of peak N. The error involved is probably less than $\pm 1\%$ [140].

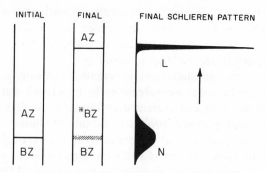

Fig. 4.24   Typical optical schlieren patterns for moving boundary system $AZ \leftarrow {}^{*}BZ \leftarrow BZ$ [141].

Analytical Boundary Method

Introduction and Theory

In this method the solution under investigation is initially separated from an indicator solution by a mechanical plane such as a porous glass disk. A known quantity of electricity is passed through the solutions, and the amount of electrolyte under investigation which has migrated past the plane is determined by analysis of the indicator solution.

In Fig. 4.25 MP represents the mechanical plane. The leading electrolyte AZ acts as indicator as in the indirect m.b. method, and the following electrolyte BZ is the one being investigated. According to the definition of transference number in Section 1, T_B^{BZ} equivalents of B ion-constituent migrate across plane MP on the passage of 1 faraday of electricity. It follows that if n_B equivalents of B ion-constituent are found in the AZ solution after

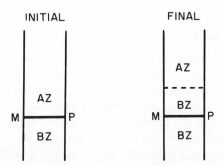

Fig 4.25   Schematic diagram illustrating the analytical boundary method.

the passage of q faradays of electricity, then:

$$T_B^{BZ} = \frac{n_B}{q} = \frac{Fn_B}{It},\tag{4.44}$$

where I is the current in amperes and t the time in seconds. The analytical boundary method resembles both the Hittorf and the m.b. methods, and it is really a quantitative development [142] of the so-called migration method which has often been used to detect the presence of complex ions in solutions [143].

The transference number in (4.44) is subject [144] to both a solvent correction as in (4.33) and a volume correction analogous to that in (4.39), namely,

$$T_B^{BZ}(\text{corr.}) = T_B^{BZ}(\text{obs.}) \pm C_{BZ} \Delta V,\tag{4.45}$$

where C_{BZ} is the normality of the BZ solution and ΔV the volume increase, in liters per faraday of electricity, between plane MP and whichever electrode is closed. The plus or minus sign refers, respectively, to the boundary moving toward or away from the closed electrode. As an example, if BZ, AZ, and the closed electrode are, respectively, $CdCl_2$, KCl, and a Cd anode as in Fig. 4.11, then it can be shown by a procedure similar to that following (4.39) that:

$$\Delta V = -\tfrac{1}{2}V_{Cd(s)} + \tfrac{1}{2}T_{Cl}^{CdCl_2}\bar{V}_{CdCl_2}^{CdCl_2},$$

where V and $\bar{V}$ stand for the molar and the partial molar volumes.

All analytical boundary work falls into two categories, in which the application of the above general theory is somewhat modified.

1. A and B are chemically different ion-constituents.

a. If the ion-constituent A is more mobile than the ion-constituent B in both solutions, then the two-salt system AZ/BZ is the same as that discussed on pages 235 and 259. In this case, the dotted line in Fig. 4.25 represents a moving boundary between the AZ solution and an adjusted or Kohlrausch solution of BZ, and in addition a concentration boundary between the adjusted and the original BZ solutions will move slowly away from MP. With this more detailed model, Brady [142] showed that (4.44) gives T_B^{BZ} at the initial BZ concentration if the concentration boundary in Fig. 4.25 moves upward (according to (4.40), this occurs when T_B^{BZ} increases with increasing BZ concentration), but that (4.44) gives T_B^{BZ} at the Kohlrausch concentration of BZ in all other cases. This might introduce some ambiguity into the interpretation of the results were it not for the experimental finding [144] that a transference number agreeing with the accepted literature value is obtained only when the initial BZ concentration is close to the Kohlrausch concentration. Thus T_B^{BZ} from (4.44) can always be taken as a property of the initial BZ solution, and the difficulty becomes the experimental one of deciding what concentration of AZ to employ so that the initial and Kohlrausch concentrations of BZ are almost the same. Two steps are necessary. First, an approximate figure for the indicator concentration c_{AZ} can be calculated by inserting into the Kohlrausch equation (see (4.42)) the initial BZ concentration instead of $*c_{BZ}$, a rough value of T_B^{BZ} furnished by a preliminary run, and an estimate of T_A^{AZ} either derived from ionic mobilities as in (4.10) or taken from the literature (cf. Appendix). Second, several runs should be made with indicator concentrations varying from about 10% above to about 10% below this calculated value until a short range of AZ concentrations is discovered over which a constant value of T_B^{BZ}, independent of current, is obtained. This, then, is the sought-after transference number.

b. If the ion-constituent A is less mobile than the ion-constituent B in both solutions, then no stable moving boundary can form. Although this provides no a priori reason for discrediting an equation such as (4.44), which was not derived from moving boundary theory, experiments [144] nevertheless suggest that in this case the correct transference number is obtained only with an initial concentration of AZ such that the specific conductances of the initial AZ and BZ solutions are equal.

2. A and B are chemically identical constituents, but the latter is "tagged." Figure 4.25 then represents a "tagged boundary" system ($AZ/^{\ddagger}AZ$ or $BZ/^{\ddagger}BZ$) in which the compositions of both solutions are identical except that one ion-constituent in the solution under investigation is tagged by a tracer, such as a radioactive isotope or a dye as described in the Experimental section. The resemblance with the isotopic boundaries mentioned on page 243 may be noted. Experimentally, the amount of tagged material transported into the untagged solution is measured. This ingenious modification of the general method, which was suggested by Brady [142], not only eliminates the search for a method of analysis and simplifies the choice for a suitable indicator substance, but also circumvents the difficulty in case 1a because the initial and Kohlrausch concentrations are now the same. There is of course no electrical restoring effect as in case 1a so that no real boundary exists, and greater care must be taken to avoid mixing by diffusion and convection. It should also be pointed out that the use of tagged boundaries necessarily involves two assumptions: (i) The mobilities of tagged and untagged ion-constituents are identical. (ii) The rate of exchange of the tagged ion-constituent between any two species incorporating this ion-constituent is infinitely large. This is not always true; it has been found [145], for example, that in sodium polyacrylate solutions there is a finite rate of exchange of sodium between the free sodium cations and the sodium cations moving with the micelles, and in fact the rate of exchange can be calculated from the results of a planned series of analytical boundary experiments [146].

To summarize, it seems advisable, in the absence of further information, to apply cases 1a and 2 only. Case 2 is probably the more useful of these.

Experimental

QUANTITY OF ELECTRICITY

It is desirable to obtain a transference number that is both precise and free of errors. The former end is best served by passing a large current for a long time to secure a large transport of electrolyte; the latter end is promoted by the use of a small current for a short time to reduce convectional and diffusive mixing between the solutions and to prevent loss of transported electrolyte in the electrode section of the indicator solution. Moderation should thus be the keynote in practice. The current and time of electrolysis should not exceed a few milliamperes and a few hours, respectively, and the results should be independent of at least a 50% change in either variable.

Experimental details were discussed on pages 234 and 257.

THE MECHANICAL PLANE

Either a porous glass frit (disk, diaphragm) or a stopcock can be used. A frit reduces mixing between the solutions, but it should be used neither with high currents, which give rise to local heating and subsequent convection, nor with solutions more dilute than about 0.01 N since surface conductance effects become prominent at low concentrations. Net electroosmosis is eliminated by closing one side of the cell, and differential electroosmosis within the pores is small because the ions diffuse across each pore much more rapidly than they move along it [147]. The chief disadvantage of a large, well-designed stopcock (Figs. 4.5 and 4.17) is possible interaction between the lubricant and the solution, although this can be reduced by suitable measures (p. 250).

ELECTRODES

The comments made under this heading on page 247 are applicable here also, with some slight qualifications. Ions formed at the electrodes must neither reach the mechanical plane nor interfere with the analysis, and it may be preferable to have the closed electrode compartment in the solution under investigation. Condition 3 on page 247 is of minor importance as the analytical boundary method is less precise than the direct moving boundary method.

Suggestions for possible electrodes were made on pages 229 and 247.

ANALYSIS, INDICATOR, AND CELL DESIGN FOR TWO-SALT SYSTEMS

Any analytical procedure may be used that can determine quantitatively a small amount of BZ in an excess of AZ.

The specifications for a suitable indicator electrolyte AZ are very similar to those listed on page 260. If A is faster than B, the conditions are the same except for condition 4: this is not necessary but is an advantage for estimating the right AZ concentration to use. The effect of AZ on the analytical procedure should also be considered. Finally, since rising boundary cells for the present method are a little easier to handle than falling boundary ones, an indicator forming a light rather than a heavy solution may be preferred.

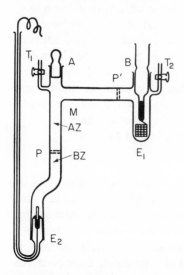

The necessity for regulating the AZ concentration was explained in the Theory section.

A knowledge of Hittorf and direct m.b. cell design should prove helpful in the construction of a cell, and the different disturbing effects (p. 229) should be taken into account.

Figure 4.26 shows a cell for rising systems, equipped with a frit P and a closed electrode E_2 in the solution under investigation. The cell is held upside down and the E_2 compartment and the frit are filled with BZ solution. The electrode E_2 is inserted, and the cell is turned up as in Fig. 4.26 and allowed to come to thermal equilibrium. The middle section M is rinsed several times with solvent and then with AZ solution and is finally filled with AZ solution. The stopper is put into A, T_1 is closed,

Fig. 4.26 Analytical rising boundary cell used by Spiro and Parton [144].

either AZ solution or guard solution for the open electrode E_1 is poured in through B, E_1 is inserted, and the current is switched on. The coarse guard frit P' reduces interdiffusion between E_1 and M. At the end of a run, T_2 is closed, T_1 opened, the stopper removed, and the whole solution in M is quantitatively washed out or drawn out for analysis. A tube diameter of about 12 mm ought to be convenient.

TRACER, ANALYSIS, AND CELL DESIGN FOR TAGGED SYSTEMS

The most general tracer is a radioactive isotope of the ion-constituent whose migration is measured [142]. Only if the ion is small is there any appreciable difference in the mobilities of the isotopes [47]. Alternatively, a solvent-insoluble tracer can be used if the ion-constituent concerned exists in the form of solubilizing particles; for example, a small quantity of an insoluble dye is solubilized by, and therefore tags, the micelles in a polyelectrolyte solution

[148]. There is evidence [149] that in such a case the dye tracer does not modify the micelle properties and that the tracer migrates only by riding in the micelles. The tag distribution, however, is skewed toward the larger micelles [150].

The method of analysis is either a radiochemical or a spectrophotometric one, depending upon the tracer used.

Because of the very small density differences between tagged and untagged solutions, a cell as in Fig. 4.26 can, if desired, be replaced by one incorporating a longer, narrower, and horizontal tube. Figure 4.27 illustrates two basic designs.

Figure 4.27a is similar to Fig. 4.25 and the technique resembles that described above; after the run the right-hand compartment is quantitatively

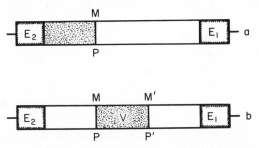

Fig. 4.27. Schematic diagrams of analytical boundary cells for tagged systems. MP and $M'P'$ are mechanical planes, E_1 and E_2 are electrodes. The initial positions of the tagged solutions are indicated by the shaded regions. The right-hand compartment of b may, however, contain either tagged or untagged electrolyte initially.

rinsed out and the solution is analyzed to determine how much tagged material the compartment has *gained*.

In Fig. 4.27b, however, the central solution is examined after the run to determine how much tagged material it has *lost*. This can be done [151] either by collecting all the central solution with washings of untagged solution and analyzing the whole or an aliquot or, if the volume V of the middle compartment is known, by carefully homogenizing the central solution within the cell and determining the tracer concentration in a sample. Hoyer et al. [151] have considered in detail the influence of convection, diffusion, gravitation, and differential electroosmosis in the tube. They concluded that with the use of stopcocks as MP and $M'P'$, and a tube of 3 to 4 mm i.d., no perceptible systematic error affects the results provided less than about 70% of the tagged electrolyte is allowed to leave the central compartment. However, when porous glass frits were used as MP and $M'P'$, it was found advisable to modify the procedure and to stir the central solution magnetically all through the run [147]. The frits must be sufficiently fine to prevent solution being pumped through them [152]. Equation (4.44) must then be replaced by:

$$T_B^{BZ} = \frac{*n_B{}^0}{q} \ln \left(\frac{*n_B{}^0}{*n_B}\right) = \frac{*C_B{}^0 V}{q} \ln \left(\frac{*C_B{}^0}{*C_B}\right), \tag{4.46}$$

where $*n_B$ and $*n_B{}^0$ are the final and initial numbers of equivalents of tagged B ion-constituent in the central compartment, $*C_B$ and $*C_B{}^0$ the corresponding normalities, and V the volume of the compartment in liters.

E.m.f. Method Using Cells with Transference

Introduction and Theory

In the methods already described, electricity actually flows and the subsequent transport of an ion-constituent is measured. Here, however, the tendency for an ion-constituent to migrate where an electric field applied affects the Gibbs free energy change of certain processes and is thus detected by an electromotive force. We need a galvanic (or voltaic) concentration cell with transference and activity data.

A concentration cell with transference is a cell that includes a liquid junction between two solutions which contain the same electrolyte but at different concentrations. In each solution there is immersed a chemically identical electrode reversible to one of the ions in solution. If, as is usually the case, this ion is the same as one of the ion-constituents of the solute, we may write the cell:

electrode reversible to B $\left|A_{\nu_A} B_{\nu_B}\right|$ $\left|A_{\nu_A} B_{\nu_B}\right|$ electrode reversible to B (4.47)

$\qquad\qquad\qquad\quad m_1 \qquad\quad m_2$

where m is the molality of the solute—it is a tradition in e.m.f. work to express concentrations in moles per kilogram of solvent. The formula of the binary electrolyte indicates that 1 mole consists of ν_A gram-formula-weights of A ion-constituent and ν_B of B ion-constituent. If the algebraic charge numbers of A and B are z_A and z_B, respectively, then by the condition of electroneutrality:

$$\nu_A z_A + \nu_B z_B = 0. \tag{4.48}$$

The equation for the e.m.f. is derived by the standard thermodynamic procedure of imagining the cell to be discharged infinitely slowly. Since the cell reaction includes the transfer of ion-constituents across the liquid junction, the transference number T_A of the ion-constituent A is included in the resulting equation:

$$E_t = \left(\frac{z_B - z_A}{z_B z_A}\right) \frac{RT}{F} \int_2^1 T_A \, d \ln (m\gamma), \tag{4.49}$$

where E_t is the e.m.f. in volts of the cell as written, R the gas constant, T the absolute temperature, F Faraday's constant, and γ the mean stoichiometric activity coefficient [153]. Numerical values of $2.3026RT/F$ at various temperatures have been tabulated [154a, 155b]; at 25°C this factor is 0.059158 absolute volt. As an example, for the cell

$$\text{Ag} \mid \text{AgCl} \mid \underset{m_1}{\text{BaCl}_2} \mid \underset{m_2}{\text{BaCl}_2} \mid \text{AgCl} \mid \text{Ag},$$

$$E_t = \frac{3RT}{2F} \int_2^1 T_{\text{Ba}} \, d \ln (m\gamma)$$

because here the electrodes are reversible to the Cl ion-constituent so that $A = \text{Ba}$, $B = \text{Cl}$, $z_A = +2$, $z_B = -1$. It should be added that diffusion at the liquid junction in cell (4.47) is not taken into account in the above classic thermodynamic treatment, but the latter is justified by the fact that the thermodynamics of irreversible processes lead to the same equation provided the Onsager reciprocal relations hold [156]. Since the experimental results are overwhelmingly in favor of the validity of these relations [156, 157], we can accept the classic results with some confidence. Experimentally, too, it is well known that the e.m.f.s of cells with transference do not change with time nor do they depend upon the manner in which the junction is made provided the composition of the transition solution varies continuously from one side to the other [158].

Equation (4.49) can be employed as it stands if some functional relationship between T_A and m is assumed and if the experimental data are needed solely to fix the numerical values of the parameters involved. It is much more useful, however, to obtain an explicit value for the transference number without any a priori postulate as to the form of its concentration variation, and this can be done through the differential form

$$T_A = \left(\frac{z_B z_A}{z_B - z_A} \right) \frac{F}{RT} \cdot \frac{dE_t}{d \ln (m\gamma)}. \tag{4.50}$$

The transference number can therefore be obtained from the e.m.f.s, provided activity coefficients are procured from another source. A wide choice of methods is available for their determination [154b] and there already exist tables, largely of isopiestic origin, for some 200 electrolytes in aqueous solution at 25°C [154c]. Whenever electrodes exist that are reversible to both ion-constituents of the electrolyte, it is possible to obtain the required activity coefficients by means of a back-to-back combination of cells without liquid junctions:

$$\text{electrode reversible to } B \mid \underset{m_1}{A_{v_A} B_{v_B}} \mid \text{electrode reversible to } A$$

$$\text{electrode reversible to } A \mid \underset{m_2}{A_{v_A} B_{v_B}} \mid \text{electrode reversible to } B \tag{4.51}$$

whose e.m.f. E, in volts, is given by

$$E = \left(\frac{z_B - z_A}{z_B z_A}\right)\frac{RT}{F}\int_2^1 d\ln(m\gamma) = \left(\frac{z_B - z_A}{z_B z_A}\right)\frac{RT}{F}\ln\left(\frac{m_1\gamma_1}{m_2\gamma_2}\right). \quad (4.52)$$

It is usual to measure the e.m.f.s of the individual cells and to produce the effect of the back-to-back combination by subtracting the e.m.f.s on paper.

The differential form of (4.52), together with (4.50), gives:

$$T_A = \frac{dE_t}{dE}. \quad (4.53)$$

This simple form is a popular alternative to (4.50) because E can always be calculated from (4.52) when the activity coefficients are derived from some other physicochemical measurement.

For those relatively few cases in which the electrodes in cell (4.47) are reversible, not to the simple ion B, but to some complex ion such as $A_x B_y$, the above equations require modification [159].

The problem involving the concentration to which T_A in (4.50) or (4.53) refers is solved by selecting one molality (m_2, for example) as a reference concentration and by keeping it constant in all runs. The transference number then becomes a property of the solution of variable concentration (m_1). An equivalent plan is to use a small number of intermediate reference concentrations and to add up the appropriate e.m.f.s so that they all refer to only one reference concentration. Such a procedure is in fact essential whenever a large heat of mixing at the liquid junction makes it necessary to place an upper limit on the difference $|m_1 - m_2|$ [160].

There are two main computational methods for obtaining transference numbers from (4.50) or (4.53):

1. E_t and either $\log(m\gamma)$ or E are related to each other by a simple, empirical, polynomial expression (such as $E_t = a + bE + cE^2$) in which the coefficients are calculated by means of least squares or some other curve-fitting procedure, and the equation is differentiated to give the transference number. However, a relation of this sort does not always fit the experimental figures accurately, and the method may therefore fail [161].

2. A more general method, devised by Rutledge [162], makes it possible to find, with a known limit of error, the derivative function dy/dx of a differentiable function y from values of the latter observed for equally spaced values of the independent variable x. A fourth-degree polynomial is employed as a differentiating tool and is applied successively to sets of five consecutive points of the data. It is not implied, however, that the data as a whole can be adequately represented by a polynomial or any other elementary type of function. Suppose that five given values of x are x_{-2}, x_{-1}, x_0, x_1, and x_2,

spaced apart by equal intervals h so that $h = x_2 - x_1 = x_1 - x_0 = x_0 - x_{-1} = x_{-1} - x_{-2}$, and suppose the five corresponding values of y are y_{-2}, y_{-1}, y_0, y_1, and y_2. The derivatives at -1, 0, and 1 are then given by [162]:

$$\left(\frac{dy}{dx}\right) = \frac{1}{12h}(C_{-2}y_{-2} + C_{-1}y_{-1} + C_0 y_0 + C_1 y_1 + C_2 y_2), \qquad (4.54)$$

where the C values are integers listed in Table 4.4. For example,

$$\left(\frac{dy}{dx}\right)_{x=x_{-1}} = \frac{1}{12h}(-3y_{-2} - 10y_{-1} + 18y_0 - 6y_1 + y_2).$$

This relation can be applied to (4.50) and (4.53) as they stand, or else to the inverse forms of these equations. In other words, E_t can be taken either as y

Table 4.4 Coefficients of Rutledge Derivative Function [162]

Derivative at	C_{-2}	C_{-1}	C_0	C_1	C_2
-1	-3	-10	$+18$	-6	$+1$
0	$+1$	-8	0	$+8$	-1
$+1$	-1	$+6$	-18	$+10$	$+3$

or as x, whichever is more convenient. Usually more than five results are available, and the procedure is then generalized by shifting the five points successively along the series of experimental data. In more detail, suppose the series of consecutive paired results, E_t and either log (my) or E, is represented by the Roman numerals I, II, III, IV, and so on. Then the transference numbers at II, III, and IV are calculated by applying the Rutledge formula (see (4.54)) from I to V, those at III, IV, and V by applying it from II to VI, those at IV, V, and VI by applying it from III to VII, and so on. The derivative is therefore calculated three times for each point (except at the extreme ends of the series where certain other equations [163] are more appropriate), but these values are not identical because of experimental errors in the data and small residual errors attributable to the method itself. The agreement between the three values is poor if the experimental errors are large or extremely erratic, and it has been/emphasized [164] that the Rutledge method cannot properly be applied to scattered data. Various schemes for interpolation and for smoothing have been described in the literature [160, 165, 166], as well as one for checking the calculations [161].

The differentiation can be appreciably simplified [157] by introducing a dummy transference number $\bar{T}$ defined by $\bar{T} = E_t/E$, and then transforming (4.53) to

$$T_A = \bar{T} + E\frac{d\bar{T}}{dE}. \qquad (4.55)$$

The advantage in this equation is that $\bar{T}$ is a much more slowly varying function of E than is T_A. An equivalent relation can be written for (4.50).

The process of differentiation inevitably magnifies any slight errors in the observed quantities, and it has been reported that changes of 0.01 mV in E_t or E affect the transference number to the extent of 0.0004 in the case of HNO_3 [166] and 0.001 in the case of HCl [165] solutions. E.m.f. transference numbers are therefore rarely more accurate than $\pm 1\%$ and probably the most useful application of the e.m.f. method is for a relatively rapid survey over a wide concentration and temperature range. A note of warning: a few cases are now known [167] in which e.m.f. and m.b. transference numbers agree well at low concentrations but diverge widely at higher ones, a result that raises the question of the reliability of certain electrodes in concentrated solutions [131].

The solvent correction has been discussed in the literature [168].

Experimental

ELECTRODES

Electrodes are the prima donnas of e.m.f. opera; their performance makes or mars the results and they are apt to be temperamental unless they are carefully handled. Suitable electrodes should be not only reversible to one of the ion-constituents in solution, but also insoluble, stable, and reproducible. Much advice on the preparation and care of a great variety of electrodes can be found in the book by Ives and Janz [155a]. Membrane electrodes giving a Nernstian response, not only to H^+ in the traditional glass electrode but also—fairly selectively—to cations such as Ag^+, Na^+, and Cu^{2+} and even to anions such as F^-, S^{2-}, and ClO_4^-, are now known. Their operation depends largely on ion-exchange processes, and they contain solid membranes, solid-state devices, or internal liquid ion-exchangers [169]. Many are commercially produced [170] (but are expensive). The precision of these electrodes, however, is as yet not nearly as high as those of the traditional homemade variety (± 0.1 mV to several millivolts, compared with 0.01 to 0.1 mV), and the accuracy of the resulting transference numbers is correspondingly reduced.

Difficulties are sometimes encountered until the technique of making and using electrodes has been perfected, and the following hints may help to track down a source of trouble. Electrodes prepared by plating tend to behave erratically if the whole surface is not covered and respond sluggishly if the deposit is too thick. Many electrodes are adversely affected by the presence of oxygen, by traces of grease, by shock, by stress, and by certain impurities (e.g., very small amounts of Br^-, I^-, or S^{2-} change the potential of a silver–silver chloride electrode because the more insoluble salt is preferentially formed on the surface). Too large an electrode can adsorb a significant percentage of solute from very dilute solutions, and the possibility of chemical interaction between electrode and solute must not be overlooked [171]. It is rare, but not unknown, for electrodes to be attacked by parasitic molds [172]. All-Pyrex equipment is advisable, and care should be taken in the sealing of the electrode into the glass [155c].

For several types of electrodes (e.g., silver–silver halide), a batch of individual electrodes must be made from time to time. Suitable electrodes can be selected from among these by placing them all in one solution and comparing their bias potentials. Two electrodes are chosen whose potential difference is small, and this small difference is then eliminated either by correcting for it directly [173] or by interchanging the electrodes (or the solutions) in the two compartments of the cell with transference and recording the mean e.m.f. [168, 173, 174]. In another method, which is perhaps not quite as reliable as the above, several selected electrodes are inserted into each compartment and the average cross e.m.f. is taken as the best value. The e.m.f.s of all cells, no matter what the electrodes, should be constant with time once thermal and chemical equilibria have been established, and they should be reproducible and additive [168].

One of the main features of the e.m.f. method is the ease and rapidity with which results can often be obtained over a wide temperature range. The cell is either left in the thermostat while the thermoregulator setting is changed or transferred to another thermostat, and when the temperature and the cell potential are again constant their values are once more recorded. The readings should be checked by returning to the original temperature at the end of the run. If the starting temperature is an intermediate one on the range to be covered then two checks are possible, one after the temperature has been decreased and one after it has been increased. Most electrodes behave well from about 0 to about 50°C.

CELL DESIGN

The e.m.f. of a cell with transference is independent of the way in which the liquid junction is formed, provided that no large heat effects occur when the two solutions meet and provided further that the mixed solution does not reach an electrode [175]. The first restriction is overcome, where necessary, by using two solutions whose concentrations do not differ too much (see Theory section). And as a consequence of the second restriction, the solutions should be brought into contact in such a way that the rate of intermixing is reduced as much as possible. This can be done with a flowing junction [176], or with a gravitational free-diffusion junction in which the lighter solution rests on top of the heavier one. The latter is simple and is frequently employed; two examples of cells based on this principle are shown in Figs. 4.28 and 4.29.

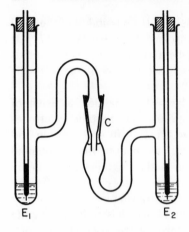

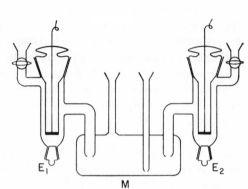

Fig. 4.28 Cell with transference used by Stokes and Levien [177]. In a later version [164] the stoppers were replaced by ground-glass joints. An almost identical cell was employed by Stonehill [166].

Fig. 4.29 Cell with transference used by Horni-brook et al. [174].

The cell in Fig. 4.28 is set up by pouring less dense solution into the E_1 electrode compartment and then fitting the electrode E_1 while the bent side tube is temporarily closed with a cap. The right-hand compartment is similarly filled with the denser solution to the level shown, and the two parts are joined at the ground-glass joint C. The air space in the bulb below C prevents liquid from touching the lubricant on C.

The electrode compartments in Fig. 4.29 can be filled with the appropriate solutions when the cell is held upside down. The cell is then inverted to its normal position as in Fig. 4.29, and the middle compartment M is filled with the heavier solution. During this operation care should be taken that gas bubbles are not formed in the entrances of the electrode compartment side tubes [173].

A visual check on the stability of the liquid junction in a given cell is obtained by putting a colored test solution into one side of the cell and a colorless one into the other [177]. In long runs over a large temperature range, it is an advantage to minimize mixing by connecting the two compartments with a three-way stopcock inside which the liquid junction is formed [165]. The stopcock is opened only when e.m.f. readings are being made.

ELECTRICAL MEASUREMENTS

The e.m.f.s are measured to ± 0.01 mV or better (see Theory section) on a digital potentiometer or voltmeter which is regularly standardized. Much useful practical advice is given by Ives and Janz [155c], and the following points may be particularly emphasized. Dirty contacts must be avoided as they give rise to spurious potentials, and all leads should be well insulated as a precaution against electrical leakage. In humid weather there is need for protection against moisture. The thermostat must be grounded (earthed), and as an additional safeguard all the measuring instruments can be placed on separate wax blocks resting on a common metal sheet which acts as an equipotential shield [178]. Interference by motors or other apparatus can be overcome by enveloping the offending piece of equipment with a wire mesh screen or, more generally, by employing electrically shielded wire for all leads, with the metal shielding connected to ground or to the equipotential sheet. A very sensitive method of testing the adequacy of the latter arrangement is to observe the e.m.f. balance point when a glass rod previously rubbed with silk is brought near one of the wires. There is no risk of picking up stray potentials around the cell when oil is used as the thermostat liquid.

E.m.f. Method using Cells in Force Fields

Introduction and Theory

The e.m.f. of a cell containing two chemically identical electrodes bathed by the same solution is zero. It is not zero, however, if the cell is placed in a gravitational or centrifugal force field; a potential difference of the order of microvolts develops when one electrode is several meters above the other in the earth's gravitational field and e.m.f.s of a millivolt or so are found when the cell is placed in a centrifuge and spun at a few thousand revolutions per minute. These e.m.f.s are functions of the transference numbers of the constituents in the solution and so permit the latter to be determined [154d, 179].

The only type of cell employed so far this century has consisted of platinum electrodes immersed in an iodide–iodine solution, and we can write as a typical example:

$$\text{Pt} \mid \quad \underset{\substack{\text{height above ground} \\ \text{or} \\ \text{distance from center of centrifuge}}}{I_2, \quad KI} \quad \mid \text{Pt} \qquad (4.56)$$

The right-hand end is at the "top" of the force field, that is, above the other end in a gravitational field or nearer the center of rotation of a centrifuge. The source of the cell's e.m.f., that is, of its electrical energy, lies in the spontaneous tendency of the chemicals involved in the cell reaction to fall downward in the gravitational field or to be thrown outward in the centrifugal field. Let us imagine the cell to be discharged infinitely slowly, and equate energies:

$$EF_{\text{grav}} = Wgh, \qquad (4.57)$$

where g is the acceleration due to gravity and h the height between the electrodes, and

$$EF_{centrif} = \int_{r_1}^{r_2} \frac{Wv^2}{r} \, dr = \int_{r_1}^{r_2} W\omega^2 r \, dr = 2\pi^2 f^2 W(r_2^2 - r_1^2), \quad (4.58)$$

in which r_1 and r_2 are the distances of the electrodes from the center of the centrifuge, v is the velocity, ω the angular velocity, and f the frequency of rotation of the centrifuge. In both cases W is the net effective mass of the cell reaction per faraday (of thermodynamically reversible discharge), and this consists of the transfer of $\frac{1}{2}$ mole I_2 "down" the force field and of T_{K+} moles of the salt KI in the opposite direction. Thus

$$W = \frac{1}{2}(M_{I_2} - \bar{V}_{I_2}\rho) - T_{K+}(M_{KI} - \bar{V}_{KI}\rho), \quad (4.59)$$

where M is molecular weight and where the buoyancy of the medium has been allowed for by subtracting the products of partial molar volume and solution density ρ. Combination of (4.59) with (4.57) or (4.58) yields the desired relation between e.m.f. and transference number.

No account has been taken in this derivation of the species I_3^- in which form most of the iodine is likely to be found. The appropriate modifications to the above equations have been given in the literature [180, 181] and it has been shown there that the effect of any such complex between I^- and I_2 can be eliminated by extrapolating the e.m.f. to zero concentration of dissolved iodine.

Although cell (4.56) is not, of course, discharged during an actual experiment, the process of sedimentation in the force field gradually does take place until concentration gradients are set up that reduce the e.m.f. to zero [180]. Luckily this process is a slow one and there is adequate time for the initial e.m.f.s to be read accurately [181, 182]. The presence of irreversible sedimentation has another and theoretical effect and that is to cast doubt on the validity of a derivation based on classic equilibrium thermodynamics. Fortunately, however, a treatment using the thermodynamics of irreversible processes leads to the same final equations provided the Onsager reciprocal relations hold [180], and there is substantial evidence that they do [156, 157].

Experimental

Gravity cells, because of the tiny e.m.f.s developed, have received relatively little attention [182], and for transference number determinations the centrifugal method has been preferred despite its greater experimental complexity. The many practical difficulties have now been overcome by MacInnes and his co-workers [181, 183–186] and a schematic diagram of their equipment is shown in Fig. 4.30. The rotor R-R is turned in the horizontal plane by the pressure of the disk D on the plate P which is rotated by the synchronous

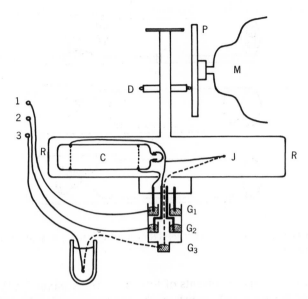

Fig. 4.30 Schematic diagram of apparatus for centrifugal e.m.f. work, after MacInnes and Dayhoff [184].

motor M. The rotor speed is measured stroboscopically and it can be varied at will by altering the position of D with respect to the plate P. A vacuum chamber surrounds the rotor so that no heat is developed by gas friction and the constancy of the rotor temperature is checked by means of thermocouples similar to J. The glass cell C is surrounded by a semifluid such as Vaseline or silicone oil to prevent it being shattered by the centrifugal force, and it is counterpoised by a similar cell to avoid precession. The e.m.f. is measured via the mercury wells G_1 and G_2. One of the major practical problems has been irreproducibility in the e.m.f. caused by minute suspended particles of dust being centrifuged onto the outer electrode (Fig. 4.31a), and this can be overcome [184] by making the platinum electrodes into rings (Fig. 4.31b). The particles then sweep harmlessly past and collect at the end of the cell. Because of mechanical difficulties of construction, another design was later developed [185] in which the electrodes are small platinum spheres sealed into the side walls (Fig. 4.31c). In the last two types of cell the factor $(r_2^2 - r_1^2)$ in (4.58) cannot be employed as it stands and needs empirical calibration with a solution of known transference number. The other auxiliary data that the equation requires are accurate densities from which the partial molar volumes can be derived.

The centrifugal e.m.f. method involves no flow of current and thus no Joule heating, and is therefore a promising tool for the determination of

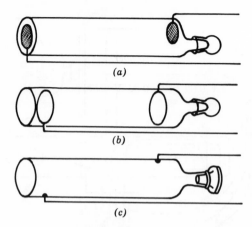

Fig. 4.31   Schematic diagrams of cells used for centrifugal e.m.f. measurements, after Kay and MacInnes [185].

transference numbers in solvents of low dielectric constant [187]. The same can be said for cells with transference, but the centrifugal e.m.f. method, although it requires more complex equipment, is superior as regards accuracy. All recent work has been confined to the use of iodide solutions and iodine-iodide electrodes, a restriction that is to some extent self-imposed. Workers in the nineteenth century tried different systems [188, 189] and, in view of the complications caused by the presence of the triiodide ion, other reversible electrodes such as quinhydrone or silver might well be equally successful.

5 APPENDIX

Table 4A.1   Accurate Transference Numbers of Some
Electrolytes in Water at 25°C

Solute*	Normalities†	Cation-Constituent Transference Number				Method‡	Ref.§
		0.00 N**	0.01 N	0.10 N	1.00 N		
HCl	0.01–0.1	0.8209	0.8251	0.8314	—	m.b.	a, b, c
	0.002–14.7	—	0.825	0.832	0.841	e.m.f.	d, e
HClO₄	0.003–5.0	0.8386	0.8438	0.8452	0.778	m.b.	b, f, g
HNO₃	0.004–9.4	0.8304	—	0.8408	0.833	m.b.	b, g, h
HIO₃	0.01–0.08	0.8961	0.9017	—	—	m.b.	i
HSCN	0.005–0.09	0.8409	0.846	0.8497	—	m.b.	j
HPic	0.005–0.02	0.9201	0.9242	—	—	m.b.	k

Table 4A.1   (contd.)

Cation-Constituent Transference Number

Solute*	Normalities†	0.00 N**	0.01 N	0.10 N	1.00 N	Method‡	Ref.§
H_2SO_4	0.1–34.0	0.8139	—	0.819	0.815	e.m.f.	*l*
H_2T	0.01–0.1	0.9269	0.926_7	0.926_9	—	m.b.	*k*
H_3PO_4	0.02–0.09	0.9155	—	—	—	m.b.	*m*
LiCl	0.003–0.1	0.3361	0.3289	0.3168	—	m.b.	*a, n*
	0.023–2.95	—	—	0.318_5	0.286_4	Hittorf	*o*
NaCl	0.01–2.3	0.3963	0.3918	0.3853	—	m.b.	*a, p, q*
NaOAc	0.02–0.2	0.5507	—	0.5596	—	m.b.	*r, s*
Na_2SO_4	0.01–0.2	0.3852	0.3848	0.3828	—	m.b.	*r*
KCl	0.001–1.0	0.4905	0.4902	0.4899	0.488_2	m.b.	*a, n, t-v*
	0.02–3.0	—	—	0.4898	0.487_1	Hittorf	*w, x*
KBr	0.015–0.1	0.4846	—	0.4852	—	m.b.	*y*
KI	0.01–0.2	0.4889	0.4884	0.4883	—	m.b.	*r, z*
KNO_3	0.01–0.2	0.5070	0.5084	0.5103	—	m.b.	*r, aa*
KH_2PO_4	0.005–0.1	0.695	0.710	0.730	—	m.b.	*bb*
KOAc	0.01–0.25	0.6425	0.6498	0.6609	—	m.b.	*cc*
K_2SO_4	0.003–0.5	0.4788	0.4829	0.4890	—	m.b.	*u*
$K_3Fe(CN)_6$	0.01–0.1	0.4215	0.4315	0.4410	—	m.b.	*u*
RbCl	0.25–4.7	0.5027	—	—	0.498	m.b.	*dd*
CsCl	0.25–3.5	0.5013	—	—	0.493	m.b.	*dd*
NH_4Cl	0.01–0.2	0.4907	0.4907	0.4907	—	m.b.	*r*
$AgNO_3$	0.01–2.0	0.4642	0.4648	0.4682	0.495_4	m.b.	*g, ee*
	0.01–10.9	—	0.464	0.468	0.490	Hittorf	*ff, gg, hh*
	0.05–14.0	—	—	0.468	0.486	e.m.f.	*hh*
$TlClO_4$	0.05–0.26	0.525_8	—	0.533	—	m.b.	*ii*
$CaCl_2$	0.01–0.15	0.4380	0.4277	0.4070	—	m.b.	*jj*
$BaCl_2$	0.02–2.0	0.4546	—	0.425	0.379	Hittorf	*kk*
$CoCl_2$	0.0001–0.1	0.4121	0.396	0.376	—	m.b.	*ll*
$CuSO_4$	0.25–1.0	0.414_4	—	—	0.304	m.b.	*mm*
$ZnSO_4$	0.01–0.09	—	0.3830	—	—	m.b.	*nn*
$LaCl_3$	0.006–0.5	0.477_2	0.4624	0.438	—	m.b.	*oo, pp*
Rare earth salts	0.01–0.1	—	—	—	—	m.b.	*qq*
$Co(en)_3Cl_3$	0.006–0.07	0.4945	—	—	—	m.b.	*rr*

* The values for HCl, NaCl, NaOAc, and KCl obtained by different schools are in good agreement and are therefore particularly reliable. HPic is picric acid; H_2T, tartaric acid; NaOAc, sodium acetate.

† The concentrations for most e.m.f. results are in gram-equivalents per kilogram of solvent.

** Calculated from modern values of limiting ionic conductances.

‡ The relatively high precision of moving boundary determinations accounts for the preponderance of m.b. references.

§ Key to references:

[a] L. G. Longsworth, *J. Am. Chem. Soc.*, **54**, 2741 (1932).

[b] A. K. Covington and J. E. Prue, *J. Chem. Soc.*, 1567 (1957).

[c] G. Marx, L. Fischer, and W. Schulze, *Radiochim. Acta*, **2**, 9 (1963).

[d] H. S. Harned and E. C. Dreby, *J. Am. Chem. Soc.*, **61**, 3113 (1939).

[e] S. Lengyel, J. Giber, and J. Tamás, *Acta Chim. Acad. Sci. Hung.*, **32**, 429 (1962).

[f] K. Banerjee and R. D. Srivastava, *Z. Physik. Chem. N. F.*, **38**, 234 (1963).

[g] R. Haase, G. Lehnert, and H.-J. Jansen, *Z. Physik. Chem. N. F.*, **42**, 32 (1964).

[h] K. Banerji, R. D. Srivastava, and R. Gopal, *J. Indian Chem. Soc.*, **40**, 651 (1963).

[i] M. Spiro, *J. Phys. Chem.*, **60**, 976, 1673 (1956).

[j] H. E. Bartlett, A. Jurriaanse, and K. de Haas, *Canad. J. Chem.*, **47**, 2981 (1969).

[k] M. Shamim and M. Spiro, *Trans. Faraday Soc.*, **66**, 2863 (1970).

[l] W. J. Hamer, *J. Am. Chem. Soc.*, **57**, 662 (1935).

[m] M. Selvaratnam and M. Spiro, *Trans. Faraday Soc.*, **61**, 360 (1965).

[n] R. L. Kay, G. A. Vidulich, and A. Fratiello, *Chem. Instrum.*, **1**, 361 (1969).

[o] G. Jones and B. C. Bradshaw, *J. Am. Chem. Soc.*, **54**, 138 (1932).

[p] R. W. Allgood and A. R. Gordon, *J. Chem. Phys.*, **10**, 124 (1942).

[q] D. J. Currie and A. R. Gordon, *J. Phys. Chem.*, **64**, 1751 (1960); recalculated by L. J. M. Smits and E. M. Duyvis, *J. Phys. Chem.*, **70**, 2747 (1966).

[r] L. G. Longsworth, *J. Am. Chem. Soc.*, **57**, 1185 (1935).

[s] D. J. Le Roy and A. R. Gordon, *J. Chem. Phys.*, **7**, 314 (1939).

[t] D. A. MacInnes and L. G. Longsworth, *Chem. Revs.*, **11**, 171 (1932).

[u] G. S. Hartley and G. W. Donaldson, *Trans. Faraday Soc.*, **33**, 457 (1937).

[v] R. W. Allgood, D. J. Le Roy, and A. R. Gordon, *J. Chem. Phys.*, **8**, 418 (1940).

[w] B. J. Steel, *J. Phys. Chem.*, **69**, 3208 (1965).

[x] D. A. MacInnes and M. Dole, *J. Am. Chem. Soc.*, **53**, 1357 (1931).

[y] A. G. Keenan and A. R. Gordon, *J. Chem. Phys.*, **11**, 172 (1943).

[z] J. N. Sahay, *J. Sci. Ind. Research*, **18B**, 235 (1959).

[aa] L. Fischer and K. Hessler, *Ber. Bunsenges.*, **68**, 184 (1964) used $^{42}KNO_3$, and their cation transference numbers are 0.2% lower.

[bb] G. Marx, L. Fischer, and W. Schulze, *Z. physik. Chem. N. F.*, **41**, 315 (1964).

[cc] D. J. Le Roy and A. R. Gordon, *J. Chem. Phys.*, **6**, 398 (1938).

[dd] J. Tamás, O. Kaposi, and P. Scheiber, *Acta Chim. Acad. Sci. Hung.*, **48**, 309 (1966).

[ee] D. A. MacInnes and I. A. Cowperthwaite, quoted by Ref. *t*.

[ff] H. Strehlow and H.-M. Koepp, *Z. Elektrochem.*, **62**, 373 (1958).

[gg] A. N. Campbell and K. P. Singh, *Canad. J. Chem.*, **37**, 1959 (1959).

[hh] M. J. Pikal and D. G. Miller, *J. Phys. Chem.*, **74**, 1337 (1970).

[ii] W. G. Breck, *Trans. Faraday Soc.*, **52**, 247 (1956).

[jj] A. G. Keenan, H. G. McLeod, and A. R. Gordon, *J. Chem. Phys.*, **13**, 466 (1945).

[kk] G. Jones and M. Dole, *J. Am. Chem. Soc.*, **51**, 1073 (1929).

[ll] G. Marx, W. Riedel, and J. Vehlow, *Ber. Bunsenges.*, **73**, 74 (1969).

[mm] J. J. Fritz and C. R. Fuget, *J. Phys. Chem.*, **62**, 303 (1958).

[nn] J. L. Dye, M. P. Faber, and D. J. Karl, *J. Am. Chem. Soc.*, **82**, 314 (1960); **83**, 5047 (1961).

[oo] L. G. Longsworth and D. A. MacInnes, *J. Am. Chem. Soc.*, **60**, 3070 (1938).

[pp] F. H. Spedding, P. E. Porter, and J. M. Wright, *J. Am. Chem. Soc.*, **74**, 2778 (1952).

[qq] F. H. Spedding and G. Atkinson, "Properties of Rare Earth Salts in Electrolytic Solutions", in *The Structure of Electrolytic Solutions*, W. J. Hamer, Ed., Wiley, New York, 1959, Chapter 22, and references cited therein.

[rr] D. J. Karl and J. L. Dye, *J. Phys. Chem.*, **66**, 550 (1962).

Table 4A.2 Selected Transference Data in Several Solvents and at Various Temperatures

Solute	Solvent	Temperature (°C)	Concentration Range*	Method	Ref.†
HCl	H_2O	5–50	0.002–3.0 m	e.m.f.	a
HCl	H_2O	0	0.022–0.1 N	m.b.	b
H_2SO_4	H_2O	0–60	0.05–17.0 m	e.m.f.	c
NaCl	H_2O	15–45	0.015–0.1 N	m.b.	d
KCl	H_2O	15–45	0.01–0.1 N	m.b.	e
KCl	H_2O	0	0.01–0.05 N	m.b.	f
$CaCl_2$	H_2O	15–35	0.01–0.15 N	m.b.	g
NaCl, KCl	99.36 mol % D_2O–H_2O	25	0.02–0.1 N	m.b.	h
NaCl, KCl	50 mol % MeOH–H_2O	25	0.005–0.08 N	m.b.	i
NaCl	MeOH	25	0.003–0.01 N	m.b.	j
KCl	MeOH	25	0.005–0.02 N	m.b.	j, k
KBr	MeOH	25	0.004–0.01 N	m.b.	l
KSCN	MeOH	25	0.002–0.01 N	m.b.	l
HCl	50 mol % EtOH–H_2O	25	0.001–2.0 m	e.m.f.	m
LiCl, NaCl	EtOH	25	0.001–0.0025 N	m.b.	n

Electrolyte	Solvent	Temp.	Concentration	Method	Ref.
KSCN	EtOH	25	0.002–0.008 N	m.b.	l
HCl	20 wt. % dioxane–H_2O	0–50	0.002–2.8 m	e.m.f.	a
HCl	45 wt. % dioxane–H_2O	0–50	0.004–3.0 m	e.m.f.	a
HCl	70 wt. % dioxane–H_2O	5–50	0.004–1.3 m	e.m.f.	a
HCl	82 wt. % dioxane–H_2O	5–45	0.005–0.46 m	e.m.f.	a
NaBr	10 wt. % mannitol–H_2O	25	0.02–0.05 N	Hittorf	o
KBr	10 wt. % mannitol–H_2O	25	0.01–0.05 N	Hittorf	o
KBr	20 wt. % sucrose–H_2O	25	0.015–0.06 N	Hittorf	p
KBr	20 wt. % glycerol–H_2O	25	0.02–0.05 N	Hittorf	p
KCl	Ethylene glycol	25	0.005–0.05 N	m.b.	q
KCl, KF, KOH, HCl	Glycerol	5, 25	0.01 N	m.b.	r
Na (Na$^+$–)	Liquid ammonia	−37	0.02–0.14 N	m.b.	s
KNO$_3$, NH$_4$NO$_3$	Liquid ammonia	−65 to −45	0.006–0.15 N	m.b.	t
AgNO$_3$	Acetonitrile	25	0.01–1 N	Hittorf	u
Me$_4$NClO$_4$	Acetonitrile	25	0.0006–0.013 N	m.b.	v
R$_4$NCl, R$_4$NBr (R = Me, Et, Pr, Bu)	Nitromethane	25	0.0002–0.01 N	m.b.	w
KCl	Formamide	25	0.01–0.1 N	m.b.	x
KSCN	Dimethylformamide	25	0.006–0.013 N	m.b.	y
LiCl	Dimethylformamide	25	0.086–0.55 N	Hittorf	z
KBr	N-Methylacetamide	35–50	0.1–0.3 N	Hittorf	aa
KCl	N-Methylacetamide	40	0.02–0.063 N	Hittorf	bb
Sr(HSO$_4$)$_2$, Ba(HSO$_4$)$_2$	Sulfuric acid	25	0.2–0.8 m	Hittorf	cc
AgClO$_4$	Dimethylsulfoxide	25	0.02–0.1 N	Hittorf	dd
AgClO$_4$	Sulfolane (tetramethylene sulfone)	30	0.02–0.1 N	Hittorf	ee

* N and m mean, respectively, gram-equivalents of solute per liter of solution and moles of solute per kilogram of solvent. References are not normally given to papers in which one concentration only was used.

† Key to references:

^{a} H. S. Harned and E. C. Dreby, *J. Am. Chem. Soc.*, **61**, 3113 (1939).

^{b} A. K. Covington and J. E. Prue, *J. Chem. Soc.*, 1930 (1957).

^{c} W. J. Hamer, *J. Am. Chem. Soc.*, **57**, 662 (1935). However, doubt has been cast on the accuracy of the results by W. H. Beck, K. P. Singh, and W. F. K. Wynne-Jones, *Trans. Faraday Soc.*, **55**, 331 (1959). It should also be pointed out that the transference numbers of H_2SO_4 in MeOH at 25°C reported by E. W. Kanning and J. E. Waltz, *J. Am. Chem. Soc.*, **63**, 2676 (1941) cannot now be accepted in view of the results obtained later by E. W. Kanning and M. G. Bowman, *ibid.*, **68**, 2042 (1946).

^{d} R. W. Allgood and A. R. Gordon, *J. Chem. Phys.*, **10**, 124 (1942). M.b. values for 0.1 N NaCl from 54° to 125°C have been given by J. E. Smith, Jr. and E. B. Dismukes, *J. Phys. Chem.*, **68**, 1603 (1964).

^{e} R. W. Allgood, D. J. Le Roy, and A. R. Gordon, *J. Chem. Phys.*, **8**, 418 (1940). M.b. values for 0.1 N KCl from 70° to 115°C have been given by J. E. Smith, Jr. and E. B. Dismukes, *J. Phys. Chem.*, **67**, 1160 (1963).

^{f} B. J. Steel, *J. Phys. Chem.*, **69**, 3208 (1965). M.b. values at 1° and 10° for 0.01 N KCl have been obtained by R. L. Kay and G. A. Vidulich, *J. Phys. Chem.*, **74**, 2718 (1970).

^{g} A. G. Keenan, H. G. McLeod, and A. R. Gordon, *J. Chem. Phys.*, **13**, 466 (1945).

^{h} L. G. Longsworth and D. A. MacInnes, *J. Am. Chem. Soc.*, **59**, 1666 (1937).

^{i} L. W. Shemilt, J. A. Davies, and A. R. Gordon, *J. Chem. Phys.*, **16**, 340 (1948). M.b. transference numbers of 0.05 N LiCl and NaCl at 25°C were measured by L. G. Longsworth and D. A. MacInnes, *J. Phys. Chem.*, **43**, 239 (1939), in 10, 20, 40, 60, and 80 mol. % MeOH–H_2O. The Hittorf transference numbers of 0.53 m $CaCl_2$ in 30 and 60 mol % MeOH–H_2O at 25°C were determined by H. Schneider and H. Strehlow, *Z. Elektrochem.*, **66**, 309 (1962).

^{j} J. A. Davies, R. L. Kay, and A. R. Gordon, *J. Chem. Phys.*, **19**, 749 (1951). The value for 0.01 N NaCl in MeOH has been confirmed by G. Marx and D. Hentschel, *Talanta*, **16**, 1159 (1969).

^{k} T. Erdey-Grúz and L. Majthényi, *Acta Chim. Acad. Sci. Hung.*, **16**, 417 (1958).

^{l} J. Smisko and L. R. Dawson, *J. Phys. Chem.*, **59**, 84 (1955).

^{m} H. S. Harned and M. H. Fleysher, *J. Am. Chem. Soc.*, **47**, 92 (1925). These authors also reported transference numbers in pure EtOH, but J. W. Woolcock and H. Hartley, *Phil. Mag.*, [7], **5**, 1133 (1928), later showed that their solvent had probably contained some water. J. Lin and J. J. De Haven, *J. Electrochem. Soc.*, **116**, 805 (1969), have measured the transference numbers of 0.01 m HCl from 0 to 92 wt. % EtOH–H_2O by the m.b. method at 25°C, and G. Kortüm and A. Weller, *Z. Naturforsch.*, **5a**, 590 (1950), determined the transference numbers of dilute lithium picrate in various EtOH–H_2O mixtures at 25°C by the Hittorf method.

^{n} J. R. Graham and A. R. Gordon, *J. Am. Chem. Soc.*, **79**, 2350 (1957).

^{o} B. J. Steel, J. M. Stokes, and R. H. Stokes, *J. Phys. Chem.*, **62**, 1514 (1958).

p B. J. Steel and R. H. Stokes, *J. Phys. Chem.*, **62**, 450 (1958). M.b. transference numbers, without volume correction, have been reported by S. Fredriksson, *Acta Chem. Scand.*, **23**, 1993 (1969), for 0.1 *N* KCl and three other salts at 0.5°C in water containing up to 500 g l⁻¹ sucrose.

q M. Carmo Santos and M. Spiro, in course of publication. The m.b. transference numbers of T. Erdey-Grúz and L. Majthényi, *Acta Chim. Acad. Sci. Hung.*, **20**, 175 (1959), differ from these by 0.02 but their solvent contained a little water.

r T. Erdey-Grúz, L. Majthényi, and I. Nagy-Czako, *Acta Chim. Acad. Sci. Hung.*, **53**, 29 (1967). Their solvent was, however, not entirely freed from water.

s J. L. Dye, R. F. Sankuer, and G. E. Smith, *J. Am. Chem. Soc.*, **82**, 4797 (1960); **83**, 5047 (1961). Much earlier m.b. data for several salts were published by E. C. Franklin and H. P. Cady, *ibid.*, **26**, 499 (1904).

t J. B. Gill, *Chem. Commun.*, 7 (1965).

u H. Strehlow and H.-M. Koepp, *Z. Elektrochem.*, **62**, 373 (1958). Experiments were also carried out with various acetonitrile-water mixtures.

v C. H. Springer, J. F. Coetzee, and R. L. Kay, *J. Phys. Chem.*, **73**, 471 (1969).

w S. Blum and H. I. Schiff, *J. Phys. Chem.*, **67**, 1220 (1963).

x J. M. Notley and M. Spiro, *J. Phys. Chem.*, **70**, 1502 (1966).

y J. E. Prue and P. J. Sherrington, *Trans. Faraday Soc.*, **57**, 1795 (1961).

z R. C. Paul, J. P. Singla, and S. P. Narula, *J. Phys. Chem.*, **73**, 741 (1969). The difference between the limiting ionic conductances derived in refs. *y* and *z* is due to the different extrapolation procedures employed.

aa R. Gopal and O. N. Bhatnagar, *J. Phys. Chem.*, **69**, 2382 (1965). These workers have also reported Hittorf transference numbers for KBr in *N*-methylformamide [*ibid.*, **70**, 3007 (1966)] and *N*-methylpropionamide [*ibid.*, **70**, 4070 (1966)]; their values for KCl in formamide [*ibid.*, **68**, 3892 (1964)] differ by ca. 0.01 from those in Ref. *x*.

bb G. P. Johari and P. H. Tewari, *J. Phys. Chem.*, **70**, 197 (1966). The results are, however, not consistent with those of Ref. *aa* by ca. 0.01, and their Hittorf data for KCl in formamide differ by 0.02 from those of Ref. *x*.

cc R. J. Gillespie and S. Wasif, *J. Chem. Soc.*, 209 (1953). The transference numbers of AgHSO₄, LiHSO₄, NaHSO₄, and KHSO₄, at one or two concentrations only, were also measured, in the case of KHSO₄ from 25° to 61°C. No solvent corrections were applied.

dd M. Della Monica, D. Masciopinto, and G. Tessari, *Trans. Faraday Soc.*, **66**, 2872 (1970).

ee M. Della Monica, U. Lamanna, and L. Senatore, *J. Phys. Chem.*, **72**, 2124 (1968). However, the resulting ionic conductances are completely at variance with a m.b. value for NaClO₄ reported by J. Lawrence and R. Parsons, *Trans. Faraday Soc.*, **64**, 751 (see p. 761) (1968).

Transference Number Bibliographies

McBain* has given a complete list, together with critical comments, of all quantitative transference determinations that appeared prior to 1906. Noyes

* J. W. McBain, *Proc. Wash. Acad. Sci.*, **9**, 1 (1907).

and Falk† presented tables of "best" values selected from literature results between 1888 and 1910, and Le Roy‡ continued McBain's valuable survey work by summarizing most of the relevant papers published between 1906 and 1939. A recent Russian review§ contains a not quite complete collection of transference numbers in aqueous solutions; the references, partly as a result of translation, require checking. The most useful list of data is that given in Landolt-Börnstein,‖ although their accuracy is variable and results obtained in the last decade are not included.

† A. A. Noyes and K. G. Falk, *J. Am. Chem. Soc.*, **33**, 1436 (1911).
‡ D. J. Le Roy "Transference Numbers of Electrolytes. A Bibliography, 1906–1939", *Univ. Toronto Studies, Papers Chem. Lab.*, No. 156 (1939).
§ E. A. Kaimakov and N. L. Varshavskaya, *Russian Chem. Rev.*, **35**, 89 (1966).
‖ *Landolt-Börnstein Zahlenwerte und Funktionen*, J. Bartels et al., Eds., 6th ed., Vol. II, Part 7, Springer-Verlag, Berlin, 1960.

References

1. M. Spiro, *J. Chem. Educ.*, **33**, 464 (1956), and references quoted therein.
2. H. Svensson, *Sci. Tools*, **3**, 30 (1956).
3. M. Spiro, *Trans. Faraday Soc.*, **61**, 350 (1965).
4. See, however, R. Haase, *Z. physik. Chem. N.F.*, **39**, 37 (1963); *Angew. Chem. Intern. Ed.*, **4**, 485 (1965).
5. T. Kurucsev and B. J. Steel, *Rev. Pure Appl. Chem.*, **17**, 149 (1967).
6. R. Haase, *Z. physik. Chem. N.F.*, **14**, 292 (1958); *Z. Elektrochem.*, **62**, 279 (1958).
7. A brief review of the evidence is given by M. Spiro, *J. Inorg. Nucl. Chem.*, **27**, 902 (1963).
8. H. Bloom, *Rev. Pure Appl. Chem.*, **9**, 139 (1959); A. Klemm, "Transport Properties of Molten Salts," in *Molten Salt Chemistry*, M. Blander, Ed., Interscience, New York, 1964; B. R. Sundheim, "Transport Properties of Liquid Electrolytes," in *Fused Salts*, B. R. Sundheim, Ed., McGraw-Hill, New York, 1964.
9. M. Spiro, *Educ. Chem.*, **3**, 139 (1966).
10. T. Shedlovsky and L. Shedlovsky, "Conductometry," in *Physical Methods of Chemistry*, A. Weissberger and B. W. Rossiter, Eds., Part II, Interscience, New York, 1971.
11. D. A. MacInnes, *The Principles of Electrochemistry*, Reinhold, New York, 1939, p. 333.
12a. R. A. Robinson and R. H. Stokes, *Electrolyte Solutions*, 2nd ed., Butterworths, London, 1959, Chapter 7.
12b. Ref. 12a, Appendix 7.1.
12c. Ref. 12a, pp. 43–44 and Chapter 6.
12d. Ref. 12a, Chapter 8.

13. R. H. Stokes, *J. Am. Chem. Soc.*, **76**, 1988 (1954).

14. R. L. Kay and J. L. Dye, *Proc. Natl. Acad. Sci. U.S.*, **49**, 5 (1963).

15. E. Pitts, B. E. Tabor, and J. Daly, *Trans. Faraday Soc.*, **65**, 849 (1969); **66**, 693 (1970).

16. M. Spiro, in *Physical Chemistry of Organic Solvent Systems*, A. K. Covington and T. Dickinson, Eds., Plenum Press, London, 1971.

17. L. G. Longsworth, *J. Am. Chem. Soc.*, **54**, 2741 (1932).

18. A. G. Keenan, H. G. McLeod, and A. R. Gordon, *J. Chem. Phys.*, **13**, 466 (1945).

19. J. L. Dye and F. H. Spedding, *J. Am. Chem. Soc.*, **76**, 888 (1954); D. J. Karl and J. L. Dye, *J. Phys. Chem.*, **66**, 550 (1962).

20. J. L. Dye, M. P. Faber, and D. J. Karl, *J. Am. Chem. Soc.*, **82**, 314 (1960).

21. M. Selvaratnam and M. Spiro, *Trans. Faraday Soc.*, **61**, 360 (1965); A. D. Pethybridge and J. E. Prue, *ibid.*, **63**, 2019 (1967); M. Shamim and M. Spiro, *ibid.*, **66**, 2863 (1970).

22. M. Dole, *Trans. Electrochem. Soc.*, **77**, 385 (1940).

23. A. W. Ralston, *Fatty Acids and Their Derivatives*, Wiley, New York, 1948, p. 663.

24. H. S. Harned and B. B. Owen, *The Physical Chemistry of Electrolytic Solutions*, 3rd ed., Reinhold, New York, 1958, pp. 200–203; H. L. Friedman, *J. Chem. Phys.*, **42**, 462 (1965).

25. G. S. Kell and A. R. Gordon, *J. Am. Chem. Soc.*, **81**, 3207 (1959).

26. W. H. Martin and Q. Van Winkle, *J. Phys. Chem.*, **63**, 1539 (1959); R. A. Horne and R. A. Courant, *J. Chem. Soc.*, 3548 (1964).

27. R. M. Fuoss, *Proc. Natl. Acad. Sci. U. S.*, **45**, 807 (1959); R. H. Boyd, *J. Chem. Phys.*, **35**, 1281 (1961); R. Zwanzig, *ibid.*, **52**, 3625 (1970).

28. A. E. Stearn and H. Eyring, *J. Phys. Chem.*, **44**, 955 (see esp. p. 976) (1940).

29. S. B. Brummer and G. J. Hills, *Trans. Faraday Soc.*, **57**, 1816 (1961) and later publications, for example, F. Barreira and G. J. Hills, *ibid.*, **64**, 1359 (1968).

30. R. W. Allgood, D. J. Le Roy, and A. R. Gordon, *J. Chem. Phys.*, **8**, 418 (1940).

31. J. E. Smith, Jr., and E. B. Dismukes, *J. Phys. Chem.*, **67**, 1160 (1963); **68**, 1603 (1964); R. L. Kay and G. A. Vidulich, *ibid.*, **74**, 2718 (1970).

32. F. T. Wall and S. J. Gill, *J. Phys. Chem.*, **59**, 278 (1955); F. T. Wall and J. Berkowitz, *ibid.*, **62**, 87 (1958).

33. R. L. Kay, K. S. Pribadi, and B. Watson, *J. Phys. Chem.*, **74**, 2724 (1970).

34. A. K. Covington and J. E. Prue, *J. Chem. Soc.*, 1567, 1930 (1957).

35. N. Ise and T. Okubo, *J. Phys. Chem.*, **69**, 4102 (1965); **70**, 1930, 2400 (1966); *Macromolecules*, **2**, 401 (1969); T. Okubo and N. Ise, *ibid.*, **2**, 407 (1969); D. Dolar and H. Leskovsek, *Makromol. Chem.*, **118**, 60 (1968).

36. M. J. Pikal and D. G. Miller, *J. Phys. Chem.*, **74**, 1337 (1970).

37. G. K. McLane, J. S. Dixon, and J. C. Hindman, "Complex Ions of Plutonium. Transference Measurements," in *The Transuranium Elements*, G. T. Seaborg, J. J. Katz, and W. M. Manning, Eds., Part 1, McGraw-Hill, New York, 1949.

38. L. C. Thompson and J. C. Nichol, *Inorg. Chem.*, **2**, 222 (1963).
39. R. E. Van Dyke, *J. Am. Chem. Soc.*, **72**, 3619 (1950); Z. A. Sheka, I. A. Sheka, and E. I. Pechenaya, *Ukr. Khim. Zh.*, **17**, 911 (1951); *Chem. Abstr.*, **48**, 6790 (1954); F. Fairbrother and N. Scott, *J. Chem. Soc.*, 452 (1955).
40. R. E. Dessy and G. S. Handler, *J. Am. Chem. Soc.*, **80**, 5824 (1958).
41. K. S. Spiegler, *J. Electrochem. Soc.*, **100**, 303c (1953).
42. T. Okubo and N. Ise, *Macromolecules*, **2**, 407 (1969).
43. F. T. Wall and P. F. Grieger, *J. Chem. Phys.*, **20**, 1200 (1952); F. T. Wall, P. F. Grieger, J. R. Huizenga, and R. H. Doremus, *ibid.*, p. 1206; F. T. Wall and R. H. Doremus, *J. Am. Chem. Soc.*, **76**, 868, 1557 (1954); S. J. Gill and G. V. Ferry, *J. Phys. Chem.*, **66**, 995 (1962); G. V. Ferry and S. J. Gill, *ibid.*, p. 999.
44. J. C. Nichol, E. B. Dismukes, and R. A. Alberty, *J. Am. Chem. Soc.*, **80**, 2610 (1958), and earlier papers.
45. D. H. Moore, "Electrophoresis," in *Physical Methods of Chemistry*, A. Weissberger and B. W. Rossiter, Eds., Part II, Interscience, New York, 1971.
46. L. G. Longsworth, *Natl. Bur. Standards Circ.* No. 524, 59 (1953).
47. R. W. Kunze and R. M. Fuoss, *J. Phys. Chem.*, **66**, 930 (1962); L. Fischer and K. Hessler, *Ber. Bunsenges.*, **68**, 184 (1964).
48. K. Wagener, *Ber. Bunsenges.*, **71**, 627 (1967); H. D. Freyer and K. Wagener, *Angew. Chem. Intern. Ed.*, **6**, 757 (1967).
49. G. Jones and M. Dole, *J. Am. Chem. Soc.*, **51**, 1073 (1929).
50. L. G. Longsworth, *J. Am. Chem. Soc.*, **54**, 2741 (1932).
51. C. W. Davies, *The Conductivity of Solutions*, Chapman and Hall, London, 1930, Chapter IV; M. Spiro, *J. Phys. Chem.*, **60**, 976, 1673 (1956).
52. B. J. Steel and R. H. Stokes, *J. Phys. Chem.*, **62**, 450 (1958).
53. E. W. Washburn, *J. Am. Chem. Soc.*, **31**, 322 (1909).
54. A. L. Levy, *J. Chem. Educ.*, **29**, 384 (1952).
55. D. A. MacInnes and M. Dole, *J. Am. Chem. Soc.*, **53**, 1357 (1931).
56. L. G. Longsworth and D. A. MacInnes, *Chem. Rev.*, **24**, 271 (1939).
57. M. Mooney, "Minimizing Convection Currents in Electrophoresis Measurements," in *Temperature, Its Measurement and Control in Science and Industry*, Reinhold, New York, 1941, p. 428.
58. L. P. Hammett and F. A. Lowenheim, *J. Am. Chem. Soc.*, **56**, 2620 (1934).
59. B. J. Steel, Ph.D. Thesis, Univ. of New England, 1960.
60. A. N. Campbell and K. P. Singh, *Can. J. Chem.*, **37**, 1959 (1959).
61. M. Dole, *Principles of Experimental and Theoretical Electrochemistry*, McGraw-Hill, New York, 1935, p. 134.
62. M. E. Laing, *J. Phys. Chem.*, **28**, 673 (1924).
63. F. T. Wall, G. S. Stent, and J. J. Ondrejcin, *J. Phys. Colloid Chem.*, **54**, 979 (1950).
64. F. Fairbrother and N. Scott, *J. Chem. Soc.*, 452 (1955).
65. H. W. Hoyer, K. J. Mysels, and D. Stigter, *J. Phys. Chem.*, **58**, 385 (1954).
66. J. W. McBain, *Proc. Wash. Acad. Sci.*, **9**, 1 (1907).

67. M. Spiro, "Determination of Transference Numbers," in *Physical Methods of Organic Chemistry*, A. Weissberger, Ed., 3rd ed., Interscience, New York, 1959, Chapter 46.

68. J. Proszt and J. Kis, *Acta Chim. Acad. Sci. Hung.*, **9**, 191 (1956); J. Proszt and L. Poos, *Periodica Polytech.*, **1**, 25 (1957); P. L. Bourgault and B. E. Conway, *J. Electroanal. Chem.*, **1**, 8 (1959); W. Helbig, *ibid.*, **3**, 146 (1962); S. Barnartt and R. G. Charles, *J. Electrochem. Soc.*, **109**, 333 (1962).

69. G. W. C. Milner and G. Phillips, *Coulometry in Analytical Chemistry*, Pergamon Press, Oxford, 1967, Chapters 3 and 4.

70. J. J. Lingane, *Anal. Chim. Acta*, **44**, 199 (1969).

71. J. C. Quayle and F. A. Cooper, *Analyst*, **91**, 355 (1966); J. A. Pike and G. C. Goode, *Anal. Chim. Acta*, **39**, 1 (1967).

72. D. A. MacInnes and L. G. Longsworth, *Chem. Rev.*, **11**, 171 (1932).

73. V. P. Dole, *J. Am. Chem. Soc.*, **67**, 1119 (1945); H. Svensson, *Arkiv Kemi Mineral. Geol.*, **22A**(10) (1946); E. B. Dismukes and E. L. King, *J. Am. Chem. Soc.*, **74**, 4798 (1952); J. C. Nichol and L. J. Gosting, *ibid.*, **80**, 2601 (1958).

74. J. C. Nichol, E. B. Dismukes, and R. A. Alberty, *J. Am. Chem. Soc.*, **80**, 2610 (1958), and earlier papers.

75. D. R. Muir, J. R. Graham, and A. R. Gordon, *J. Am. Chem. Soc.*, **76**, 2157 (1954).

76. G. N. Lewis, *J. Am. Chem. Soc.*, **32**, 862 (1910).

77. G. N. Lewis and M. Randall, *Thermodynamics*, McGraw-Hill, New York, 1923, p. 36; F. H. MacDougall, *Thermodynamics and Chemistry*, 3rd ed., Wiley, New York, 1939, pp. 30–32; D. A. MacInnes, *The Principles of Electrochemistry*, Reinhold, New York, 1939, pp. 82–83; H. S. Harned and B. B. Owen, *The Physical Chemistry of Electrolytic Solutions*, 3rd ed., Reinhold, New York, 1958, pp. 358–360.

78. P. Milios and J. Newman, *J. Phys. Chem.*, **73**, 298 (1969).

79. R. J. Bearman and L. A. Woolf, *J. Phys. Chem.*, **73**, 4403 (1969).

80. G. Baca and R. D. Hill, *J. Chem. Educ.*, **47**, 235 (1970).

81. J. L. Dye, R. F. Sankuer, and G. E. Smith, *J. Am. Chem. Soc.*, **82**, 4797 (1960).

82. M. Shamin and M. Spiro, *Trans. Faraday Soc.*, **66**, 2863 (1970).

83. L. G. Longsworth, *Ind. Eng. Chem. Anal. Ed.*, **18**, 219 (1946).

84. K. N. Marsh, M. Spiro, and M. Selvaratnam, *J. Phys. Chem.*, **67**, 699 (1963).

85. R. L. Kay, G. A. Vidulich, and A. Fratiello, *Chem. Instrum.*, **1**, 361 (1969).

86. J. A. Davies, R. L. Kay, and A. R. Gordon, *J. Chem. Phys.*, **19**, 749 (1951).

87. J. W. Lorimer, J. R. Graham, and A. R. Gordon, *J. Am. Chem. Soc.*, **79**, 2347 (1957).

88. E. Passeron and E. Gonzalez, *J. Electroanal. Chem.*, **14**, 393 (1967).

89. J. E. Smith, Jr., and E. B. Dismukes, *J. Phys. Chem.*, **67**, 1160 (1963).

90. J. E. Smith, Jr., and E. B. Dismukes, *J. Phys. Chem.*, **68**, 1603 (1964).

91. K. S. Spiegler, *J. Electrochem. Soc.*, **100**, 303c (1953).

92. G. Marx, L. Fischer, and W. Schulze, *Radiochim. Acta*, **2**, 9 (1963).

93. G. Marx, L. Fischer, and W. Schulze, *Z. physik. Chem. N. F.*, **41**, 315 (1964).

94. L. Fischer and K. Hessler, *Ber. Bunsenges.*, **68**, 184 (1964).

95. G. Marx and W. Schulze, *Kerntechnik*, **7**, 13 (1965).
96. W. Schulze, M. Hornig, and G. Marx, *Z. physik. Chem. N. F.*, **53**, 106 (1967).
97. G. Marx, W. Riedel, and J. Vehlow, *Ber. Bunsenges.*, **73**, 74 (1969).
98. G. Marx and D. Hentschel, *Talanta*, **16**, 1159 (1969).
99. E. A. Kaimakov and V. I. Sharkov, *Russian J. Phys. Chem.*, **38**, 893 (1964).
100. G. S. Hartley and J. L. Moilliet, *Proc. Roy. Soc. (London)*, **A140**, 141 (1933).
101. D. J. Le Roy and A. R. Gordon, *J. Chem. Phys.*, **6**, 398 (1938).
102. A. R. Gordon and R. L. Kay, *J. Chem. Phys.*, **21**, 131 (1953).
103. For example, R. W. Allgood, D. J. Le Roy, and A. R. Gordon, *J. Chem. Phys.*, **8**, 418 (1940); A. G. Keenan and A. R. Gordon, *ibid.*, **11**, 172 (1943).
104. G. S. Hartley and G. W. Donaldson, *Trans. Faraday Soc.*, **33**, 457 (1937).
105. For example, J. N. Sahay, *J. Sci. Ind. Res.*, **18B**, 235 (1959).
106. M. Selvaratnam and M. Spiro, *Trans. Faraday Soc.*, **61**, 360 (1965); M. Selvaratnam, private communication.
107. R. Neihof and S. Schuldiner, *Nature*, **185**, 526 (1960); A. Küssner and E. Wicke, *Z. physik. Chem. N. F.*, **24**, 152 (1960); H. J. Cleary and N. D. Greene, *Electrochim. Acta*, **10**, 1107 (1965).
108. M. Shamim, Ph.D. Thesis, University of London (1967).
109. L. G. Longsworth, *J. Chem. Educ.*, **11**, 420 (1934).
110. D. J. Le Roy, Ph.D. Thesis, University of Toronto, 1939.
111. D. A. MacInnes and L. G. Longsworth, *Chem. Rev.*, **11**, 171 (1932).
112. H. P. Cady and L. G. Longsworth, *J. Am. Chem. Soc.*, **51**, 1656 (1929); L. W. Shemilt, J. A. Davies, and A. R. Gordon, *J. Chem. Phys.*, **16**, 340 (1948).
113. For example, R. Haase, G. Lehnert, and H.-J. Jansen, *Z. physik. Chem. N.F.*, **42**, 32 (1964).
114. A. M. Sukhotin, *Russian J. Phys. Chem.*, **34**, 29 (1960).
115. J. A. Davies, R. L. Kay, and A. R. Gordon, *J. Chem. Phys.*, **19**, 749 (1951).
116. H. W. Hoyer, K. J. Mysels, and D. Stigter, *J. Phys. Chem.*, **58**, 385 (1954).
117. R. H. Davies, *Nature*, **161**, 1021 (1948).
118. C. A. Coulson, J. T. Cox, A. G. Ogston, and J. St. L. Philpot, *Proc. Roy. Soc. (London)*, **A192**, 382 (1948); M. B. Smith and S. J. Roe, *Anal. Biochem.*, **17**, 236 (1966).
119. R. L. Kay, Ph.D. Thesis, Univ. of Toronto (1952).
120. A. R. Gordon and R. L. Kay, *J. Chem. Phys.*, **21**, 131 (1953).
121. L. G. Longsworth and D. A. MacInnes, *J. Am. Chem. Soc.*, **59**, 1666 (1937).
122. E. A. Kaimakov and V. B. Fiks, *Russian J. Phys. Chem.*, **35**, 873 (1961).
123. B. P. Konstantinov and E. A. Balukin, *Russian J. Phys. Chem.*, **39**, 1, 315 (1965).
124. J. C. Nichol, *J. Phys. Chem.*, **66**, 830 (1962).
125. A boundary is said to show progression up (or down) when the velocity of the boundary, in cm^3 coulomb^{-1}, increases (or decreases) as the boundary moves along the tube.
126. H. Lal, *Nature*, **171**, 175 (1953).
127. L. G. Longsworth, *J. Am. Chem. Soc.*, **65**, 1755 (1943).
128. M. Spiro, *J. Chem. Phys.*, **42**, 4060 (1965).

129. L. J. M. Smits and E. M. Duyvis, *J. Phys. Chem.*, **70**, 2747 (1966); **71**, 1168 (1967).

130. G. S. Hartley, *Trans. Faraday Soc.*, **30**, 648 (1934).

131. R. L. Kay, private communication.

132. D. R. Muir, J. R. Graham, and A. R. Gordon, *J. Am. Chem. Soc.*, **76**, 2157 (1954).

133. P. Mukerjee, *J. Phys. Chem.*, **62**, 1397 (1958).

134. G. S. Hartley, B. Collie, and C. S. Samis, *Trans. Faraday Soc.*, **32**, 795 (1936).

135. J. Tamás, O. Kaposi, and P. Scheiber, *Acta Chim. Acad. Sci. Hung.*, **48**, 309 (1966).

136. L. G. Longsworth, *J. Am. Chem. Soc.*, **67**, 1109 (1945).

137. D. J. Currie and A. R. Gordon, *J. Phys. Chem.*, **64**, 1751 (1960). The results have been recalculated in Ref. 129.

138. G. S. Hartley, E. Drew, and B. Collie, *Trans. Faraday Soc.*, **30**, 648 (1934); G. S. Hartley, B. Collie, and C. S. Samis, *ibid.*, **32**, 795 (1936); C. S. Samis and G. S. Hartley, *ibid.*, **34**, 1288 (1938).

139. R. L. Kay, G. A. Vidulich, and A. Fratiello, *Chem. Instrum.*, **1**, 361 (1969).

140. L. G. Longsworth, *J. Am. Chem. Soc.*, **66**, 449 (1944).

141. R. A. Alberty, *J. Chem. Educ.*, **25**, 619 (1948).

142. A. P. Brady, *J. Am. Chem. Soc.*, **70**, 911 (1948).

143. For example, E. G. Jones and F. G. Soper, *J. Chem. Soc.*, 802 (1935); E. R. Nightingale, Jr., *Anal. Chem.*, **28**, 281 (1956).

144. M. Spiro and H. N. Parton, *Trans. Faraday Soc.*, **48**, 263 (1952).

145. J. R. Huizenga, P. F. Grieger, and F. T. Wall, *J. Am. Chem. Soc.*, **72**, 2636 (1950).

146. F. T. Wall and P. F. Grieger, *J. Chem. Phys.*, **20**, 1200 (1952); F. T. Wall, P. F. Grieger, J. R. Huizenga, and R. H. Doremus, *ibid.*, **20**, 1206 (1952); F. T. Wall and R. H. Doremus, *J. Am. Chem. Soc.*, **76**, 868, 1557 (1954); S. J. Gill and G. V. Ferry, *J. Phys. Chem.*, **66**, 995 (1962); G. V. Ferry and S. J. Gill, *ibid.*, p. 999.

147. K. J. Mysels and H. W. Hoyer, *J. Phys. Chem.*, **59**, 1119 (1955).

148. H. W. Hoyer and K. J. Mysels, *J. Phys. Colloid Chem.*, **54**, 966 (1950).

149. D. Stigter and K. J. Mysels, *J. Phys. Chem.*, **59**, 45 (1955).

150. K. J. Mysels, *J. Colloid Interface Sci.*, **23**, 474 (1967).

151. H. W. Hoyer, K. J. Mysels, and D. Stigter, *J. Phys. Chem.*, **58**, 385 (1954).

152. G. V. Ferry and S. J. Gill, *J. Phys. Chem.*, **66**, 999 (1962).

153. C. W. Davies, *J. Chem. Soc.*, 349 (1939); *Electrochemistry*, Newnes, London, 1967, p. 72.

154a. R. A. Robinson and R. H. Stokes, *Electrolyte Solutions*, 2nd ed., Butterworths, London, 1959, p. 469.

154b. Ref. 154a, Chapter 8.

154c. Ref. 154a, Appendix 8.

154d. Ref. 154a, pp. 114–116.

155a. D. J. G. Ives and G. J. Janz, *Reference Electrodes: Theory and Practice*, Academic Press, New York, 1961.

155b. Ref. 155a, p. 67.

155c. Ref. 155a, pp. 56–63.
156. D. G. Miller, *Chem. Rev.*, **60**, 15 (1960).
157. M. J. Pikal and D. G. Miller, *J. Phys. Chem.*, **74**, 1337 (1970).
158. E. A. Guggenheim, *Thermodynamics*, 4th ed., North-Holland, Amsterdam, 1959, Chapter 9.
159. M. Spiro, *Trans. Faraday Soc.*, **55**, 1207 (1959).
160. W. J. Hamer, *J. Am. Chem. Soc.*, **57**, 662 (1935).
161. R. H. Stokes and B. J. Levien, *J. Am. Chem. Soc.*, **68**, 1852 (1946).
162. G. Rutledge, *Phys. Rev.*, **40**, 262 (1932). A slightly different presentation of the method is given by H. Margenau and G. M. Murphy, *The Mathematics of Physics and Chemistry*, 2nd ed., Van Nostrand, Princeton, N.J., 1956, p. 473.
163. Methods of obtaining the slopes at the extreme ends of a curve are given by C. Lanczos, *Applied Analysis*, Pitman, London, 1957, p. 323; J. C. Amphlett and E. Whittle, *Trans. Faraday Soc.*, **64**, 2130 (1968).
164. E. P. Purser and R. H. Stokes, *J. Am. Chem. Soc.*, **73**, 5650 (1951).
165. H. S. Harned and E. C. Dreby, *J. Am. Chem. Soc.*, **61**, 3113 (1939); H. S. Harned and B. B. Owen, *The Physical Chemistry of Electrolytic Solutions*, 3rd ed., Reinhold, New York, 1958, pp. 478–482.
166. H. I. Stonehill, *J. Chem. Soc.*, 647 (1943).
167. J. L. Dye, M. P. Farber, and D. J. Karl, *J. Am. Chem. Soc.*, **82**, 314 (1960); **83**, 5047 (1961); L. J. M. Smits and E. M. Duyvis, *J. Phys. Chem.*, **70**, 2747 (1966).
168. A. S. Brown and D. A. MacInnes, *J. Am. Chem. Soc.*, **57**, 1356 (1935).
169. G. Eisenman, ed., *Glass Electrodes for Hydrogen and Other Cations*, Dekker, New York, 1967; A. K. Covington, *Chem. in Britain*, **5**, 388 (1969); R. A. Durst, ed., *Ion-Selective Electrodes*, N.B.S., Washington, D.C., 1969, Special Publication 314.
170. From Beckman Instruments Inc., Fullerton, Calif., Corning Glass Works, Corning, New York, and Orion Research Inc., Cambridge, Massachusetts.
171. M. Spiro and A. B. Ravnö, *J. Chem. Soc.*, 78 (1965); M. D. Archer and M. Spiro, *J. Chem. Soc. A*, 68 (1970).
172. D. A. MacInnes and K. Parker, *J. Am. Chem. Soc.*, **37**, 1445 (1915).
173. T. Shedlovsky and D. A. MacInnes, *J. Am. Chem. Soc.*, **58**, 1970 (1936).
174. W. J. Hornibrook, G. J. Janz, and A. R. Gordon, *J. Am. Chem. Soc.*, **64**, 513 (1942).
175. D. A. MacInnes, *Chem. Rev.*, **18**, 347 (1936).
176. A. B. Lamb and A. T. Larson, *J. Am. Chem. Soc.*, **42**, 229 (1920); E. J. Roberts and F. Fenwick, *ibid.*, **49**, 2787 (1927); J. V. Lakhani, *J. Chem. Soc.*, 179 (1932); G. N. Ghosh, *J. Indian Chem. Soc.*, **12**, 15 (1935).
177. R. H. Stokes and B. J. Levien, *J. Am. Chem. Soc.*, **68**, 333 (1946).
178. W. P. White, *J. Am. Chem. Soc.*, **36**, 2011 (1914).
179. D. A. MacInnes, *The Principles of Electrochemistry*, Reinhold, New York, 1939, Chapter 9.
180. D. G. Miller, *Am. J. Phys.*, **24**, 595 (1956).
181. D. A. MacInnes and B. R. Ray, *J. Am. Chem. Soc.*, **71**, 2987 (1949).

182. S. W. Grinnell and F. O. Koenig, *J. Am. Chem. Soc.*, **64**, 682 (1942).
183. B. R. Ray and D. A. MacInnes, *Rev. Sci. Instr.*, **20**, 52 (1949).
184. D. A. MacInnes and M. O. Dayhoff, *J. Chem. Phys.*, **20**, 1034 (1952).
185. R. L. Kay and D. A. MacInnes, *J. Phys. Chem.*, **61**, 657 (1957).
186. B. R. Ray, D. M. Beeson, and H. F. Crandall, *J. Am. Chem. Soc.*, **80**, 1029 (1958).
187. R. M. Stanton, *Dissertation Abstr.*, **28**, 637B (1967).
188. R. Colley, *Pogg. Ann.*, **157**, 370, 624 (1876); *Wied. Ann.*, **17**, 55 (1882).
189. Th. Des Coudres, *Ann. Phys. u. Chem.*, **49**, 284 (1893); **57**, 232 (1896).

General

Kaimakov, E. A., and N. L. Varshavskaya, "Measurement of Transport Numbers in Aqueous Solutions of Electrolytes," *Russian Chem. Revs.*, **35**, 89 (1966).

MacInnes, D. A., *The Principles of Electrochemistry*, Reinhold, New York, 1939.

MacInnes, D. A., and L. G. Longsworth, "Transference Numbers by the Method of Moving Boundaries," *Chem. Rev.*, **11**, 171 (1932).

Robinson, R. A., and R. H. Stokes, *Electrolyte Solutions*, 2nd ed., Butterworths, London, 1959.

Chapter V

POLAROGRAPHY

Otto H. Müller

I INTRODUCTION

Polarography is an electrochemical method of analysis incorporating features of electrolysis and potentiometry but distinct from both. In *electrolysis* the aim is to remove completely a chosen constituent from the solution by passing an electric current through it for a sufficient length of time; the electrodes have relatively large surfaces, and the solution is stirred to facilitate transport of the electroactive material to the electrode. In contrast to this, the electrolysis in polarography is of short duration and the electrode on which the constituents of the solution are plated out is a dropping mercury electrode or other microelectrode so that the currents are very small. Hence the changes produced by polarographic electrolysis are normally (with 5 to 20 ml of test solution) not measurable, and the polarographed solution can be recovered virtually unchanged. In *potentiometry* the unknown potential of an electrode is measured against a reference electrode with known potential by balancing the potential difference between the two against a known but opposite voltage. During this balancing process every effort is made to avoid any flow of current since this might polarize an electrode, that is, change its potential from the value it has naturally. In contrast to this, in polarography the microelectrode is purposely polarized to different potentials by a voltage applied from an outside source between it and a reference electrode. Naturally, this results in the flow of current if an electroactive depolarizer is present which, by definition, tends to prevent such polarization. If the currents and the circuit resistance are small and if the potential of the reference electrode is constant, the potential of the micro-electrode will be equal to the applied voltage. By plotting the currents obtained when the applied voltage is gradually altered, one obtains a current-voltage curve which under suitable conditions not only indicates the nature but also the concentration of the electroactive material or materials present. Thus polarography makes possible the simultaneous qualitative and quantitative analysis of one or more compounds in solution.

The polarographic method of analysis was invented in 1922 by Jaroslav Heyrovský of Prague [1] who, with his co-workers from the Physico-chemical Institute of Charles University, also established most of its fundamentals during the following 15 years. After that, considerable interest in the new method was also aroused outside of Czechoslovakia, but major contributions continue to come from Prague. The nucleus of this activity is now in the J. Heyrovský Polarographic Institute of the Czechoslovak Academy of Sciences, so named after Heyrovský received the Nobel prize in 1959.

Literature on polarography is presently growing at a rate of more than 1500 papers per year. Because of the magnitude of the task of collecting titles, most of the previously available annual bibliographies have been discontinued. Even the bibliography published annually by Heyrovský in *Collection of Czechoslovak Chemical Communications* since 1938 has ceased with its last volume covering papers of 1964 [2]. At present, *Biblio-grafia Polarografica*, started by Semerano in 1949 [3], is the only up-to-date bibliography, which by the end of 1967 had listed 21,798 papers on polarography [4]. Titles of these papers are printed in their original language if they are in English, Italian, German, or French; if not, an English trans-lation of the title is given. In the latter case, the corresponding reference in *Chemical Abstracts* is often added. This bibliography is particularly valuable because it also contains an amazingly comprehensive subject index, in English, which refers not only to the titles but also to the contents of the papers.

A modern development in the bibliographic field is the preparation of punch cards, in English, by the Polarographic Society of Japan [5]. These "Polarographic Literature—Data Cards" should permit the easy punching of authors' initials and abbreviated journal references, as well as field, methodology, and classification of samples. The direct coding for organic chemistry is of special interest because it serves to list the various kinds of organic compounds that may be subjects of polarographic studies. They are: aliphatic (alicyclic), aromatic, and heterocyclic compounds; alcohols (phenols), aldehydes, ketones (quinones), acids (acid derivatives), O-hetero, nitro (nitroso), and amino compounds ($C\!=\!N$ compounds), azo, and N-hetero compounds, S-compounds (SH-compounds), halogen compounds, unsaturated hydrocarbons, oxides (peroxides), sugars, proteins (enzymes), alkaloids, antibiotics, vitamins (fat-soluble vitamins), and hormones (steroids). So far only 800 cards have become available and these still reflect some of the uncertainties and inaccuracies of such a new venture. Unfortunately, the annual output of these cards is only about 400 at present, so that but a fraction of the current papers on polarography can be covered and very few of the older papers can be included.

Polarography developed from studies of electrolyses with a slowly dropping mercury electrode. Heyrovský [1] made this electrode part of a cell to which he applied different voltages. Plotting the resulting current against the applied voltage, he obtained current-voltage curves which could be used for analytical purposes. These curves show great reproducibility because of the regularity with which each newly forming drop exposes to the solution a fresh surface of mercury with identical properties. Under such conditions electrolyses are independent of time and are well suited for automatic recording. Realizing this, Heyrovský and Shikata [6] constructed a machine that increased the applied voltage at a steady rate and simultaneously recorded the corresponding current-voltage curve. Since the curves obtained by the instrument are a graphical representation of the polarization of the electrode, the machine was called a *polarograph* and the records obtained with it *polarograms*.

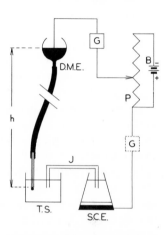

Fig. 5.1 Diagram of essentials for polarographic analysis. *DME*, dropping mercury electrode; *TS*, test solution; *J*, agar bridge; *SCE*, saturated calomel half-cell; *P*, potentiometer or polarograph; *B*, battery; *G*, galvanometer, either in cathodic or anodic circuit; *h*, height of mercury column.

The principle of such an arrangement is shown schematically in Fig. 5.1. The essential part is the dropping mercury electrode (*DME*) which in its simplest form is a thick glass capillary with a very small internal diameter (about 0.05 mm) connected to a mercury reservoir by a 40- to 80-cm-long rubber or plastic tubing. This permits ready alteration of the total pressure *h*, forcing mercury through the capillary. Individual drops of mercury issuing from this capillary into the test solution (*TS*) at a slow rate (1 drop in several seconds) serve as electrodes during their lifetime, that is, while they are still connected with the mercury reservoir. The other electrode may be a layer of mercury at the bottom of the vessel or, better, any one of many nonpolarizable reference electrodes. As an example, a saturated calomel half-cell (*SCE*) is shown in Fig. 5.1. This is connected to the test solution by a potassium chloride–agar bridge or other liquid junction J. Different known voltages can then be applied between these two electrodes by means of a battery B and voltage divider (potentiometer) P. To avoid contamination of the mercury, platinum contacts are used to connect the circuit to the cell. The current, corresponding to each applied voltage is read from a suitable device, such as a galvanometer G, which is customarily inserted in the cathodic leads but may just as well be in the anodic leads (as indicated by the dotted lines). When this current is plotted on graph paper against the applied voltage, a manually obtained polarogram results. With a recording polarograph the change in applied voltage is brought about by a motor that is also geared to a current-recording device so that the "polarogram" is obtained automatically.

In the original polarograph a beam of light reflected from a mirror galvanometer was made to fall on photographic paper which was pulled past a slit in a light-tight box by a mechanism geared to the motor which caused the change in applied voltage. Although this photographic method of recording is still used (most of the polarograms reproduced in this chapter were recorded in this way), it has nowadays been largely replaced by ink writers and similar recording instruments. In these the galvanometer G of Fig. 5.1 is replaced by a suitable known resistance; the voltage drop across it is proportional to the current and can be amplified and recorded against the applied voltage. It should be emphasized, however, that for preliminary tests of the applicability of polarography to a given problem or for learning this technique no automatically recording apparatus is needed. A cell with a dropping mercury electrode and a reference electrode plus a galvanometer and simple potentiometer is quite adequate.

The success of polarography was largely attributable to the fact that the dropping mercury electrode combines a number of advantageous features:

1. It provides a continually renewed electrode of a very pure metal not affected by any previous electrolysis or aging process and with a perfectly reproducible, fresh surface.

2. During the reduction of many metallic ions, the deposited metal does not accumulate on its surface but diffuses into the body of the mercury drop as an amalgam.

3. The dropping mercury electrode is a microelectrode. Hence the electrolytic currents remain small, leaving the composition of the test solution unaltered. This permits repeated polarographic analysis of the same solution with identical results.

4. By changing the pressure head on a given capillary, the rate of growth of the mercury drop and its surface can be altered without changing the final

size of the drop. The resulting changes in the observed current give clues to the mechanism of the electrode reaction. Similarly, valuable information is obtained from changes in the drop size produced by artificially knocking off the mercury drops at chosen intervals of time. Other extreme conditions occur (a) when the mercury flow is arrested after a drop has been formed, as in the so-called hanging drop electrode or (b) when a constantly fresh surface of fixed area is exposed to the solution, as in the streaming mercury electrode.

5. The largest hydrogen overvoltage is found on mercury. This means that the potential at which hydrogen ions are reduced at a mercury electrode is far more negative than at any other electrode. Hence the dropping mercury electrode is suitable for analyses in acid solutions and in regions of more negative potentials than electrodes of other metals. It is thus the electrode par excellence for the study of many irreversible reactions, particularly of organic compounds.

The actual recording of a polarogram requires no special training, and quite often the interpretation of results is extremely simple. For instance, the organic chemist may wish to evaluate a distillation procedure by locating the fraction in which a polarographically active impurity is most concentrated or where it is absent. As long as he obtains polarographic waves of different magnitude which correlate with expectations based on experience, he does not even have to know the nature of the impurity to obtain valuable information from the polarograms. There have been a few instances in which the analysis of a specific compound has been described in sufficient detail in the literature even for a neophyte in polarography, but such cases are rare. Therefore, as in all physical methods described in these volumes, an understanding of fundamentals is essential if one wishes to avoid numerous pitfalls and especially if one hopes to develop a new procedure.

For an understanding of polarographic principles, one needs to know, on the one hand, all the factors that govern the formation and size of the drop of mercury serving as microelectrode. On the other hand, one must be cognizant of the numerous forces and mechanisms that bring material to this electrode that can either accept electrons or part with them, or which may be adsorbed at the electrode-solution interface without an exchange of electrons. In addition, one needs to know the conditions that regulate the electron transfer between electrode and electroactive solute, or the factors that determine electrode potentials. The speed with which these factors operate is so great in the so-called reversible systems that equilibrium conditions are maintained even though the electrode is a drop of mercury steadily growing into a medium that is constantly changing in composition in the immediate vicinity of the electrode. Whenever the rate of electron transfer or the potential-determining chemical reaction in the solution is less than this, one finds potentials that become a function of kinetics also and one deals with irreversible systems. In connection with the latter, the effect of certain catalysts that can accelerate specific reactions at the electrode surface deserves special attention.

Since the earlier editions of this book, polarography has developed in many directions. Although electrochemical analysis is still its major field of application, recent developments suggest that its scope is much more extensive; in fact, polarography has started a renaissance of interest in electrochemistry. Similar to microscopy in biology, polarography with its emphasis on microelectrodes has provided a generally applicable tool in electrochemistry for the investigation of reaction mechanisms and kinetics at an electrode. There have been many modifications in the design of mercury microelectrodes. These range from a stationary hanging mercury drop, through the simple dropping mercury electrode and a variety of rapidly dropping electrodes with artificially controlled drop times, to a streaming mercury electrode in which a fresh surface of known area is constantly exposed to the solution. Furthermore, microelectrodes of metals other than mercury and even of carbon have been extensively used. The polarograph has grown from a simple instrument with all wires showing, to a typical black-box-type apparatus which may contain the latest in electronic equipment and design. Unfortunately, the extensive branching of polarography into such fields as mathematics, electronics, and physics has made it practically impossible today for a single individual to understand, let alone treat with some authority, all aspects of polarography. As a consequence, the subject matter has been broken up in this new edition into five chapters. In each, the selection of topics introduces a considerable amount of personal bias, which hopefully should enhance the value of this volume. Thus the special fields of voltammetry, oscillographic and ac polarography, pulse polarography, tensammetry, controlled-potential and controlled-current electrolysis are treated in later chapters. This chapter deals mainly with what has lately been called "classic polarography," restricted to mercury microelectrodes, in which mathematical derivations are reduced to a minimum.

2 BASIC PRINCIPLES

Graphical Representation of Data

Any chemist contemplating research involving polarography will find it essential to be familiar with the graphical representation of data. Of particular importance are graphs depicting conditions of a perfectly reversible equilibrium, as expressed by the reaction:

$$B + nA \underset{k_2}{\overset{k_1}{\rightleftharpoons}} A_nB \tag{5.1}$$

in which a mole of the reactant B associates with n moles of reagent A to form a mole of product A_nB. According to the law of mass action, the rate of the forward reaction (the association) is proportional to the product of the concentrations (strictly speaking, the activities) of the reactant and of the reagent, the latter raised to the nth power. This is written as $k_1[B][A]^n$, where k_1 is the proportionality constant. The rate of the backward reaction (the dissociation) is proportional to the concentration (activity) of the product, or equal to $k_2[A_nB]$. At equilibrium the rates of the forward and backward reactions are equal so that

$$\frac{[B][A]^n}{[A_nB]} = \frac{k_2}{k_1} = K, \qquad (5.2)$$

where K is the equilibrium constant. It can also be called the dissociation constant since it becomes the greater the more the tendency of A_nB to dissociate into A and B.

This relationship can be further simplified by letting [B] and $[A_nB]$ stand for the components of a *system* with concentration [S]. If the concentration of A_nB is given as the fraction x of [S], then the concentration of [B] must be $(1 - x)[S]$ and (5.1) becomes

$$\frac{(1 - x)[S] \cdot [A]^n}{x[S]} = K \qquad \text{or} \qquad K\frac{x}{(1 - x)} = [A]^n. \qquad (5.3)$$

It is then easy to plot the relationship between the two unknowns, x and [A], for different values of K and n. In Figs. 5.2 to 5.4 this has been done in three different ways, each of which has found extensive application, although usually restricted to special fields. In each figure the effect of varying K is shown at the bottom, and the effect of varying n is shown at the top. The basic curve with K and n equal to unity is drawn with a heavy line.

When x *is plotted against* [A] as in Fig. 5.2, there is at first a rapid increase in x with relatively small increases in [A] until the value of $x = 0.5$ is reached, that is, until about one-half of the system has been changed from pure B into A_nB. This is then followed by a slower increase in the fraction x which asymptotically approaches unity at very high values of [A]. Whenever $x = 0.5$, the ratio of $x/(1 - x)$ becomes unity and $K = [A]^n$. It may be seen that the smaller K, the steeper the curve and the faster the asymptotic approach to $x = 1$. A decrease in n below unity produces a more rapid rise in the curve up to $x = 0.5$, followed by a markedly diminished tendency to reach $x = 1.0$, while n values greater than unity produce a slower rise in the formation of x at low concentrations of [A], followed by a very rapid rise near $x = 0.5$ and a fast approach to the asymptotically reached value of $x = 1.0$.

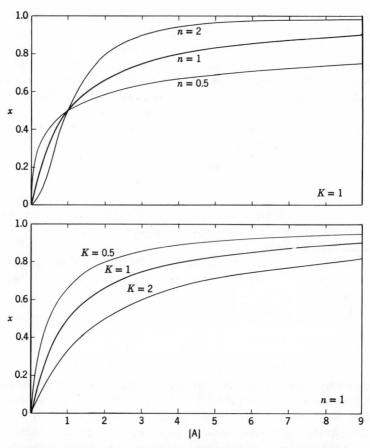

Fig. 5.2   Graph of $[A]^n = K \cdot x/(1 - x)$ for different values of K and n.

Curves of this type are used to demonstrate the equilibrium expressed by the well-known *Langmuir adsorption isotherm* [7], adapted to solutions, which is usually written as $x/m = abc/(1 + ac)$. In this equation a and b are constants, x is the number of grams of substance adsorbed by m grams of solid, while c is the concentration of the solution. Using the terms of (5.3), one can substitute $x[S]$ for x/m, $[S]$ for b, K^{-1} for a, and $[A]^n$ for c to obtain

$$x[S] = K^{-1}[S][A]^n/(1 + K^{-1}[A]^n). \tag{5.4}$$

This can also be written as

$$x[S] + x[S]K^{-1}[A]^n = [S]K^{-1}[A]^n \tag{5.4a}$$

or

$$x[S] = (1 - x)[S]K^{-1}[A]^n, \tag{5.4b}$$

which is the same as (5.3). The term [S] here represents a two-dimensional surface concentration at full saturation, or it can be interpreted as total surface area of which fraction x is covered, while $(1 - x)$ is still available for further adsorption. The equilibrium constant K in this case is equal to the concentration $[A]^n$ at which one-half of the surface is covered and one-half is free. An increase in K thus signifies a diminution in the adsorbability of a given system.

Another equilibrium that is depicted by curves of the type shown in Fig. 5.2 is that involving enzymes, as given by the *Michaelis-Menten equation* [8]. This is usually written as $v = V_{max}[\text{Sub}]/(K_m + [\text{Sub}])$ where v is the observed velocity of a reaction (assumed to be proportional to the concentration of the product or enzyme-substrate complex), [Sub] is the concentration of substrate on which the enzyme acts, V_{max} is the maximum velocity of the reaction (assumed to occur when all of the enzyme is bound as the complex), and K_m is the so-called Michaelis constant. (Although K_m is not actually the dissociation constant of the enzyme–substrate complex, it is formally treated as such by combining several constants into k_2.) When v is taken as a fraction x of V_{max}, this equation becomes $xK_m + x[\text{Sub}] = [\text{Sub}]$ or $K_m x/(1 - x) = [\text{Sub}]$ which is again identical to (5.3) with [A] representing the substrate concentration. K_m is that substrate concentration at which the velocity of the enzymic reaction is one-half maximal. An increase in K_m signifies a diminished tendency of the enzyme to form a complex with the substrate. One may consequently anticipate that of several enzymes acting on a common substrate the one with the smallest K_m may be the "natural" one.

One difficulty often encountered in these graphs is that V_{max} (or $x = 1.0$) is not approached closely enough at practically feasible concentrations of the substrate. This disadvantage is partly overcome by a graph of the inverse functions $1/x$ versus $1/[A]$, the so-called *Lineweaver-Burk plots* [9], shown in Fig. 5.3, in which the emphasis is on data obtained with more dilute solutions. Inversion of (5.3) leads to the following equation:

$$\frac{1}{x[S]} = \frac{K}{[S]}\frac{1}{[A]^n} + \frac{1}{[S]} \qquad \text{or} \qquad \frac{1}{x} = \frac{K}{[A]^n} + 1 \qquad (5.5)$$

where [S] again stands for V_{max} and $x[S]$ for v of the Michaelis-Menten equation. It may be seen that as long as only one substrate molecule combines with one enzyme molecule ($n = 1$) straight lines are obtained which cross the vertical axis (when $1/[A] = 0$) at $1/x = 1$ (or at $1/V_{max}$) and which extrapolate to a point $-1/[A]$ on the horizontal axis (when $1/x = 0$). The latter indicates the value of $1/K_m$. These lines become curved and K_m can no longer be obtained by extrapolation when the ratio of enzyme to substrate in the complex differs from unity ($n \neq 1$).

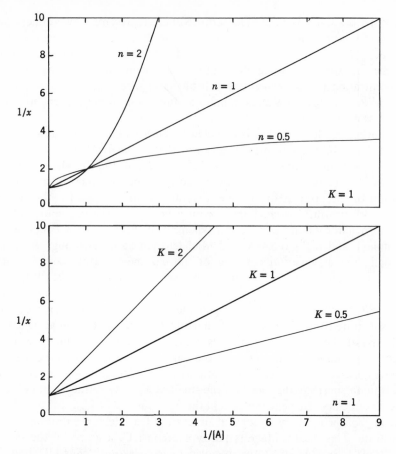

Fig. 5.3   Graph of $1/[A]^n = (1/K)((1 - x)/x)$ for different values of K and n.

Other equilibria, such as that between the acid and salt forms of a weak acid and that between the oxidant and reductant of an oxidation-reduction system can be similarly plotted. The quantity x represents the fraction of the weak acid in its associated (protonated) form or the fraction of the oxidation-reduction system present in its reduced form, while [A] stands for the concentration of free hydrogen ions or for the "concentration" or "pressure" of electrons. However, because of the tremendous ranges practically available for hydrogen ion concentrations or for electron "pressures," it is customary to plot x as a function of the *logarithm* of [A], as in Fig. 5.4. Equation (5.3) here takes the form:

$$\log [A] = \log K' + 1/n \log x/(1 - x), \tag{5.6}$$

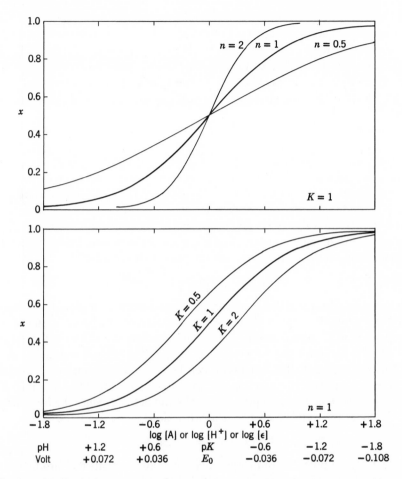

Fig. 5.4 Graph of $\log [A] = \log K' + 1/n \log x/(1 - x)$ for different values of K and n. The corresponding pH and voltage scales are also indicated.

where $K' = K^n$. Note that a symmetrical S-shaped curve results with a point of inflection when $\log [A] = \log K'$. Altering K' merely shifts the curve on the horizontal axis, while a change in n results in alteration of the slope of the S-curve. Another familiar way of labeling these curves is to use the negative logarithms and substitute for them a new term with prefix "p," such as pA for $-\log [A]$ or for $\log (1/[A])$. Thus with proper corrections of signs before all log terms and conversion to new terms, (5.6) is changed into

$$pA = pK + 1/n \log (1 - x)/x. \qquad (5.7)$$

Equation (5.7) becomes directly applicable to the equilibrium existing between a weak acid and its salt. For the case of $n = 1$, it is identical with the *Henderson-Hasselbalch equation* [10]: $pH = pK_a + \log ([A^-]/[HA])$, if one substitutes pH for pA, pK_a for pK, and [HA] (the concentration of the associated form of the weak acid) for x, pK_a is that pH at which the weak acid is one-half dissociated. pK_a is the negative logarithm of the dissociation constant K_a, hence an increase in K_a, which signifies an increase in the tendency of the acid to dissociate (i.e., a stronger acid), results in a decrease in pK_a.

Not so obvious is the application of (5.7) to the equilibrium of oxidation-reduction systems which are usually characterized by the following form of the *Nernst equation* (for 30°C): $E = E_0 + (0.060/n) \log ([ox]/[red])$. Here E is the potential, in volts, observed when an oxidation-reduction system with standard potential E_0 has its components in the concentration ratio [ox]/[red]. An equation analogous to the Henderson-Hasselbalch equation, but for the electron dissociation rather than for the proton dissociation, is

$$p\epsilon = pK_0 + 1/n \log [ox]/[red]. \tag{5.8}$$

This suggests that $E/0.060 = p\epsilon$ and $E_0/0.060 = pK_0$, or that the voltage scale is essentially a log scale of electron "pressure" in which a change of 0.060 V (at 30°C) produces a 10-fold increase or decrease in electron pressure ϵ. Here K_0 represents the "dissociation constant" of the reductant.

The correctness of this conclusion can be confirmed by means of the law of mass action. According to this, the *free energy of a reaction* is equal to $RT \ln K_0$ when all reactants and products are at unit activity. Since the free energy is also equal to $-nFE_0$ (the electrical work done), one can write

$$RT \ln K_0 = -nFE_0, \tag{5.9}$$

where E_0 is the potential of an electrode when all reactants and products are at unit activity. R is the gas constant (8.314 joules per degree), T the absolute temperature (°C + 273.1), F is the faraday (96,496 coulombs per equivalent, or joules per volt per equivalent), and n the number of equivalents. Upon conversion from natural (ln) to common (log) logarithms and substitution of proper values, one then finds

$$\frac{1}{n} pK_0 = -\frac{1}{n} \log K_0 = \frac{96496}{(2.303)(8.314)(303.1)} E_0 = 16.67E_0. \tag{5.10}$$

The maximum voltage span available in the electrochemistry of aqueous solutions is about 3 V, which on the basis of the above indicates a range of electron pressure of 50 log units, or from 10^0 to 10^{50} pressure units.

By analogy with the pH scale, one can see that many oxidation-reduction systems can be characterized by their E_0 values, just as acids are characterized by their pK_a values. This principle serves as the basis of qualitative analysis in polarography.

For quantitative purposes it becomes necessary to know the quantity [S] of which x is but a fraction. No matter how large or how small this [S], the curves relating x to log [A] should all be the same. For instance, in the quantitative determination of the concentration of the salt of a weak acid by means of titration, it is necessary to know the quantity of hydrogen ions added (obtained from the number of moles of a strong acid added) in order to produce an end point with a desired pH (indicated by the color change of a suitable dye or by a pH meter). No matter what the concentration, one can always predict that at a pH that is 2 units higher than the known pK_a, the ratio of associated acid to free anion $x/(1 - x)$ will be 99/1, while at a pH 3 units higher than the pK_a this ratio will be 999/1. It becomes clear that for practical purposes the titration was finished at a pH 2 units higher than pK_a and that theoretically it will never be completely finished since the pH would have to be infinity (see Fig. 5.4).

The quantitative determination of the oxidant of an oxidation-reduction system can similarly be carried out by titration with a known quantity of the reductant of another suitable oxidation-reduction system (see Chapter IA1) until an end point is reached with an E value that is 120 mV more negative than its E_0 value (assuming $n = 1$). Another way is to *titrate this oxidant directly with electrons* coming from an electrode surface, polarized to the same potential which is the end point of the titration. If the quantity of oxidant reaching this electrode is a known function of its concentration in the solution, one can correlate the current flowing at the end point to the concentration of the oxidant. This principle serves as the basis of quantitative analysis in polarography.

Since only reversible reactions have so far been considered, it is obvious that the reverse of the above, the dissociation of hydrogen ions from an associated weak acid or the removal of electrons from a reductant, could be treated in a similar way. In general, polarography is concerned with electrode reactions, that is, reactions involving a transfer of electrons between the electrode and components of the solution. These components are called *oxidants* when they accept electrons, or *reductants* when they lose electrons. The electrode is a *cathode* when a reduction occurs and an *anode* when an oxidation takes place on its surface. During the reduction of an oxidant at the cathode, electrons leave the electrode and enter the solution with the formation of an equivalent amount of reductant. Conversely, during the oxidation of a reductant at the anode, electrons pass from the solution to the electrode and leave behind an equivalent amount of oxidant. Since free electrons cannot exist in solution, any process of reduction at the cathode requires a simultaneous oxidation at the anode as well as an external electrical circuit by which electrons are led from anode to cathode. The dropping mercury electrode can serve either as cathode or anode, hence the large reference or working electrode must be able to receive electrons as well as supply them, but it must do so without alteration of its own potential. The various types of reactions may be summarized by a single equation:

$$\text{oxidant} + n\epsilon \rightleftharpoons \text{reductant} \tag{5.11}$$

The reductant obviously differs from the oxidant merely by n electrons (ϵ); together they form an *oxidation-reduction system*. It should be pointed out that not only cations, but also anions, as well as uncharged molecules, can be reduced at the cathode and that oxidation at the anode is similarly possible for all three species. This reaction, similar to all reactions so far considered, is perfectly reversible, and the starting and end products are stable. Cases in which this is not true are considered later.

Potential Measurements

In organic polarography, all components of an oxidation-reduction system are dissolved in solution. The dropping mercury electrode is indifferent and participates in the reactions only to the extent of furnishing or removing electrons. Since it acquires a charge indicative of the potential of the oxidation-reduction system in the surrounding solution, it is called an *indicator* electrode. When mercury ions are involved or when metals are plated out or amalgamated with the dropping mercury electrode, it may also become an *active* electrode whose potential is a function of only one ionic species in solution. Only a small number of electrons must be added or taken away in order to charge a microindicator electrode such as the dropping mercury electrode to the potential of the surrounding solution. Any change in potential is resisted by the components of the solution that can accept or furnish electrons and hence act as *depolarizers* or as stabilizers of a given potential. The power of an oxidation-reduction system to stabilize the electrode potential is called its *poising action*. Obviously, it increases with increasing concentration of the system; at any given concentration it is greatest when the system contains equal parts of oxidant and reductant because then changes in the concentration of each have the smallest effect on their ratio. It has already been shown that this is also the condition at which the electrode potential is E_0 which is characteristic for the given oxidation-reduction system.

A direct determination of E_0 or of $-\log \epsilon$ is not feasible; only the difference between two such systems or half-cells, suitably separated, can be measured.

If one of these has a known potential, that of the other can then be determined. In order to have a point of reference, *the potential of the normal hydrogen electrode (NHE) has been arbitrarily assigned a value of 0 V at all temperatures.* It is the potential of a platinized platinum electrode in a solution of unit hydrogen ion activity, bathed by hydrogen gas at 1 atm of pressure. Potentials of other systems referred to it are designated E_h. Calomel half-cells or silver–silver chloride electrodes are more convenient to use than the hydrogen electrode, and their E_h potentials are well known. In polarography the saturated calomel electrode (SCE) has been particularly popular; its potential [11] is $E_h = +0.241$ V at 30°C ($+0.244$ V at 25°C). Potentials of the dropping mercury electrode referred to it are designated E_{SCE} and can be readily converted to E_h values by subtracting the potential of the saturated calomel electrode. [Other often used reference electrodes are: the normal calomel electrode (NCE) with $E_h = +0.280$ V and the 1 M silver–silver chloride electrode with $E_h = +0.222$ V, both at 25°C [11].]

If one considers the effects of an electric current on the potential of the dropping mercury electrode when it is poised by reversible systems one may distinguish three different situations:

1. Its *potential remains unchanged* while the currents are of short duration and at a small current density (current per unit area of electrode surface). Here the composition of the electrode-solution *interface* (the very thin layer of solution in direct contact with the electrode which determines the potential) is not materially altered and is in equilibrium with the body of the solution.

2. Its *potential is altered* while the currents are of long duration although at a small current density, (especially if the volume of solution is small). Here there is not only an alteration of the composition of the electrode-solution interface by the current, but also of the composition of the body of the solution with which this interface is still in equilibrium. Because of this latter condition, the potential indicates at all times the changing composition of the solution. To aid in maintaining equilibrium between interface and body of the solution, the solutions are generally stirred and the potential indicates the progress of the electrolysis.

3. Its *potential is altered* if the currents are of short duration but at a relatively large current density. This is the situation usually found in polarography. Here the composition of the interface is markedly altered while the body of the solution is essentially unchanged. Hence there is no longer any equilibrium between the latter and the interface so that the potential no longer reflects the composition of the solution: the electrode is *polarized*. Agents that tend to prevent this polarization are called *depolarizers*. Their action depends on their concentration and the manner in which they reach

the electrode and results in limiting currents which become especially signifi-cant in polarography where relatively dilute solutions (low depolarizer concentrations) are used. Depolarizer concentrations of 10^{-6} M are generally still sufficient to produce measurable effects.

In order to ascertain which one of the electrodes in a cell is polarized, it is best to determine the potential of each electrode potentiometrically against a third electrode of known and constant potential. This method is particularly advisable in cells with high resistance, as found in organic solvents. However, a more common technique is the use of a cell in which one electrode is of known potential, well poised, and with a large surface, while the other electrode is the dropping mercury electrode. During the brief currents of small magnitude commonly found in polarography, the large electrode has a constant potential and can serve as reference electrode. Such an arrangement has the advantage that the composition of the body of the solution remains essentially unaltered and that the variations in current with changes in the applied voltage apparent in a current-voltage curve can be directly correlated with processes going on at the microelectrode.

Given the potential of the nonpolarizable reference electrode E_r, one can calculate the potential of the polarized dropping mercury electrode E_{DME} from the applied voltage V and the observed current I by means of Ohm's law,

$$E_{DME} = E_r + V + IR, \tag{5.12}$$

where R is the resistance of the electric circuit. In polarography it is customary to consider the current positive when the microelectrode is cathode and negative when it is anode. [If the opposite terminology is used, a minus sign must be placed before IR in (5.12).]

The correction for IR becomes necessary whenever the product of current and resistance exceeds 1 mV (at the usual resistance of 2000 to 3000 Ω, this means where currents are greater than 0.2 to 0.3 μA). Hence many polaro-graphic *current-voltage* curves must be corrected for IR before they represent *current-potential* curves [12]. This correction may be very large in the presence of organic solvents which increase the resistance of the electrolyte, wherefore apparatus has been designed which, by means of the three-electrode system mentioned above, can plot current-potential curves directly regardless of the resistance. Fortunately, it is seldom necessary to recalculate every point on a current-voltage curve; usually only those points on the curve that are of special interest are corrected for IR.

Ideal Current-Voltage Curves (Polarograms)

The standard method of drawing current-voltage curves is illustrated by Fig. 5.5. Here current is plotted on the vertical and the applied voltage on

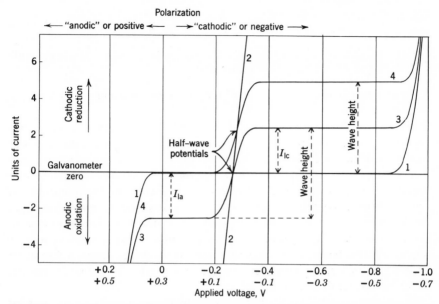

Fig. 5.5   Types of current-voltage curves and standard coordinate system used in polarography.

the horizontal axis. The line marked, "galvanometer zero" shows the position of the galvanometer at rest, that is, when no current is flowing. Current above this line is positive and indicates a reduction at the microelectrode which is cathode. Negative current below this line represents an oxidation at the microelectrode which is then anode. The abscissa corresponding to zero applied voltage is of special importance because it represents the known potential of the nonpolarizable electrode to which the polarization of the microelectrode is referred. To the right of this abscissa, the microelectrode is polarized to more negative potentials, and to the left of it to more positive potentials than the reference electrode. These polarizations have sometimes been called *cathodic and anodic polarizations*, respectively, a terminology which has led to some confusion because it implies a function of the electrode which may be incorrect. For instance, a cathodically polarized electrode can still function as anode as is illustrated by curve 3 of Fig. 5.5. For our purpose it is important to remember that the electron pressure of the microelectrode is logarithmically increased from left to right on the polarogram, and at one point along this curve it is equal to the electron pressure of the reference electrode.

If a different reference electrode is used, identical current-voltage curves will result with a shift of the abscissa. For example, if the E_h of the first used reference electrode were +0.300 V, and the experiment had been repeated with a normal hydrogen electrode as reference electrode, the voltage axis would have become as indicated in italics in Fig. 5.5. In either case (5.12) gives identical values for the potential of the microelectrode at any point on the curve. This example illustrates clearly that a *statement of applied voltage is meaningless unless it is accompanied by information about the potential of the reference electrode used.*

Figure 5.5 shows four very simple current-voltage curves under ideal conditions which illustrate the principles of polarographic analysis. They have been drawn with the assumption that $IR = 6.66$ mV per unit of current. Curve 1 represents a pure *indifferent electrolyte*, so-called because it serves largely to conduct current and preserve electroneutrality in solution while other substances contained therein react at the electrode. As may be seen, no measurable current flows through such a solution over a wide range of applied voltage, proving the absence of any depolarizer which could act in that region. At the two extreme ends of the curve, however, even this solution becomes electroactive and is decomposed as indicated by the large currents which do not reach any limits.

If a reversible oxidation-reduction system with equal proportions of oxidant and reductant is added in sufficient amount to this solution, one obtains curve 2. In this unusual case even the microelectrode does not become polarized at the current strengths shown. A straight line is obtained with a slope determined by the resistance of the circuit. This line passes through "galvanometer zero" at a potential characteristic of the oxidation-reduction system; we have at that point a typical example of a potentiometric potential determination (see Chapter IA1 in original plan, by Wawzonek). This same potential is obtained from every point along this linear current-voltage curve when it is corrected for IR according to (5.12), indicating that the electrode potential has remained constant in spite of the flow of current. However, this ceases to be the case if the concentration and thus the poising action of the oxidation-reduction system is greatly reduced, and curve 3 is obtained during the electrolysis.

As could be expected, curve 3 passes through the same point on the galvanometer zero line as curve 2 because the potentiometrically determined potential depends only on the ratio [ox]/[red]. However, the system is so poorly poised that even a small current alters the composition of the interface significantly and produces a polarization of the microelectrode, as indicated by the departure of curve 3 from the straight line, curve 2. This polarization increases with an increase in current and finally is complete, as the current reaches a limit. Here the depolarizer is used up by the electric current as fast as it reaches the interface. From this point on, polarization of the microelectrode is a linear function of the applied voltage as in curve 1, until the indifferent electrolyte is decomposed. The magnitude of the observed anodic and cathodic *limiting currents* I_{la} and I_{lc} depends on the effectiveness of the different depolarizers and on the rate at which they reach the microelectrode in a given time. Under certain conditions the limiting current is directly proportional to the concentration of the reactive material in solution and thus serves for *quantitative analysis*. In curve 3, for instance, the anodic limiting current I_{la} indicates how much reductant, and the cathodic limiting current I_{lc} how much oxidant, is in solution, while the sum of these currents is a function of the total concentration of the oxidation-reduction system. Obviously, in this instance the oxidant and reductant concentrations must be about equal.

The symmetrical S-shaped curve connecting the limiting currents is called a *wave*, and the sum of the two limiting currents the *wave height*. The midpoint of the wave represents its steepest part, or the point of inflection, where the electrode potential is maximally poised because oxidant and reductant are present in equal concentration at the interface. This midpoint, or *half-wave potential* characterizes the reacting material and serves for *qualitative analysis*.

Solutions in which both the oxidant and reductant of the same system are present in measurable amounts are rare in analytical polarography. Usually, at the beginning of a measurement the substances present are either wholly oxidized or wholly reduced, and the corresponding polarographic waves are either all cathodic or all anodic. For instance, curve 4 in Fig. 5.5 would be obtained instead of curve 3 if the oxidation-reduction system were all in the oxidized form in the bulk of the solution. Wave 4 is completely cathodic but has the same half-wave potential as wave 3.

Ideal curves of this type were not known until 1937 when Müller and Baumberger [13] discovered them during the investigation of well-buffered solutions of quinhydrone or its components quinone and hydroquinone. Since that time they have served as a basis of comparison for other polarographic curves, the majority of which deviate in a variety of ways from these ideal types.

3 FORCES GOVERNING FORMATION AND SIZE OF THE DROPPING MERCURY ELECTRODE

In contrast to all other electrodes, the dropping mercury electrode has a continually changing surface area whose ultimate size and rate of growth must be known in order to interpret the polarographic data. In addition to frictional forces which limit the flow of mercury through the narrow capillary, there are surface forces which tend to reduce this flow and which control the size of the "ripe" drops. These surface forces vary with the potential of the electrode and depend on the composition of the solution. Only a brief discussion of these forces can be given here.

Poiseuille's Law and Back Pressure

In the most common form of the dropping mercury electrode, shown in Fig. 5.1, mercury is forced through a vertically mounted glass capillary of uniform bore cut off square at its lower end. This capillary has such a narrow bore that, even with considerable pressure, drops form at its tip only at intervals of several seconds. The pressure is measured as the height h of a column of mercury (see Fig. 5.1) between the tip of the capillary and the mercury level in the reservoir. The latter is large enough to prevent measurable changes in the mercury level during an experiment in which only 0.01 to 0.02 ml of mercury is forced through the capillary per minute.

Because the capillary has a uniform bore, it is possible to apply Poiseuille's law: $m = P(\pi d\rho^4/8l\eta)$, where m is the rate of flow, P the pressure difference between beginning and end of the capillary, d the density, and η the viscosity of mercury, while ρ is the radius and l the length of the capillary. This equation becomes

$$m = P \times 4.64 \times 10^9 \times \rho^4/l = P/\kappa \qquad (5.13)$$

when m is expressed in milligrams per second, P in centimeters of mercury (where 1 cm mercury $= 13.3 \times 10^3$ dynes/cm^2), and ρ and l in centimeters. The last two factors do not change for any given capillary and can be combined with the other constants as shown. The capillary constant κ characterizes a given capillary, as it represents the resistance of the capillary to the flow of mercury; it indicates the pressure necessary to produce a flow of 1 mg of mercury per second. κ is readily determined for any given capillary by measuring the ratio P/m when the mercury flows into a pool of mercury [14]. In this case $P = h - h'$ where h' is the depth to which the capillary dips into the mercury. (When the capillary dips into a solution with a specific gravity near unity, the quantity h' must be divided by 13.53 and usually becomes negligible.)

Whenever there is drop formation, as in a solution, surface forces involved in their formation exert a *back pressure*, P_{back}, on the column of mercury that can become appreciable. Here the effective pressure P is given by

$$P = h - P_{back}. \qquad (5.14)$$

According to Kučera [15],

$$P_{back} = 2\sigma/gdr_d, \qquad (5.15)$$

where σ is the surface tension (in dynes per centimeter), g the gravitational constant, (980.6 cm/sec^2), d the density of mercury (13.53 gm/cm^3), and

r_d the radius of the drop (in centimeters). It may be seen that P_{back} diminishes as the drop grows and has its maximum value when the drop first issues from the capillary orifice as a hemisphere with radius ρ.

The *maximum back pressure* is readily determined as the maximum value of h at which there is still no flow of mercury through the capillary while its tip is dipping into a solution. When this solution is 0.1 N potassium chloride, open to air, and the electrode is not connected to an electric circuit, the maximum back pressure is called the *critical pressure* P_c of the dropping mercury electrode. Under these conditions σ has a value of 380 dynes/cm and, according to (5.15)

$$P_c = 1.57 \times 10^{-4} \, \sigma \cdot \rho^{-1} = 0.0573/\rho \tag{5.16}$$

where P_c is expressed in centimeters of mercury and ρ in centimeters [14]. This serves as an *indirect means for determining the value of ρ*. Sometimes, when h is only slightly greater than P_c, a capillary may stop dropping during the recording of part of a polarogram when an increase of surface tension over a certain range of the applied voltage causes a maximum P_{back} that exceeds the value of h. Obviously, diminishing h below P_c by lowering the mercury reservoir serves as a convenient means to arrest the flow of mercury after an experiment; in practice this is only feasible when $P_c > 10$ cm of mercury or when $\rho < 5 \times 10^{-3}$ cm.

A *minimum back pressure* exists when r_d has reached its maximum size, that is, just before the drop falls. The weight of such a ripe drop W (in milligrams) is then suspended from the thread of mercury in the capillary and pulls on it with a force of $Wg/1000$ dynes. However, the thread of mercury at the capillary orifice has a circumference of $2\pi\rho$ cm over which the surface tension acts to produce a total force of $2\pi\rho\sigma$ dynes, which must balance the gravitational force of the drop. During the measurement of the critical pressure, this same force balances the gravitational force of the cylinder of mercury *inside* the capillary, with its base of $\pi\rho^2$ and a height equal to the maximum back pressure. Thus it is clear that for a given cross-sectional area the surface tension either supports the weight of a critical volume of mercury above this area by preventing its flow through the orifice, or below this area by keeping it attached in the form of a sphere. Any slight excess above this particular volume causes either flow through the capillary or separation of the drop.

The foregoing considerations bring out the interesting relationship,

$$W = 2000\pi\rho\sigma/g. \tag{5.17}$$

Thus the drop weight W may serve as a *second indirect means for the determination of ρ*. The drops must be collected in a 0.1 N solution of potassium chloride, open to the air, so that the value of 380 dynes/cm for σ is applicable. One then finds [14]:

$$\rho = 0.411 \times 10^{-3} \, W, \qquad (5.18)$$

where ρ is in centimeters and W in milligrams.

Since W is equal to $4/3\pi r_d^3 d$, the radius of the drop r_d can be expressed in terms of ρ [14], so that (5.15) becomes

$$\text{Minimum } P_{\text{back}} = 3.12 \times 10^{-3} \, \sigma^{2/3} \rho^{-1/3}, \qquad (5.19)$$

where P is expressed in centimeters of mercury, σ in dynes per centimeter, and ρ in centimeters.

It becomes clear that variations in the back pressure, occurring during the growth of each drop, may become very large (e.g., between 29.0 and 1.7 cm of mercury for a capillary with $\rho = 0.002$ cm [16]) and have a considerable effect on P, especially when h is relatively small [see (5.14)]. There are, furthermore, changes in surface tension accompanying changes in applied voltage which affect the back pressure. For these reasons it is advisable to work with an arrangement in which h is considerably larger than the critical pressure.

For general polarographic work, the *most suitable capillaries* have orifices with diameters varying between 0.003 and 0.010 cm. (Capillaries with larger orifices tend to form drops that hang from a neck of mercury and are no longer spherical as Smith [17] has clearly shown in a series of photographs.) Such capillaries have critical pressures varying from 40 to 12 cm, respectively, so that the maximum ratio of P_c/h varies between 0.40 and 0.12 if one allows a maximum practical value of 100 cm for h.

Variations in P naturally affect m [see (5.13)], especially during the early stages of drop growth. This requires special consideration when the drops are knocked off mechanically before they are ripe. The temporal interrelations between mercury flow and back pressure during the growth of a single drop are rather complicated; they have been described and their effect on polarographic currents has been evaluated by Barker and Gardner [18].

Normally, one calculates a mean value $\bar{P}_{\text{back}}$ and subtracts this from h to obtain an equivalent constant pressure $\bar{P}$ that causes the same mean flow $\bar{m}$ of mercury through the capillary as does the oscillating pressure. In the past [14] this mean value has erroneously been obtained by integration of r_d instead of the P_{back} values. The correct *mean back pressure* should be 3/2 of the minimum back pressure, or:

$$\bar{P}_{\text{back}} = 4.68 \times 10^{-3} \, \sigma^{2/3} \rho^{-1/3}, \qquad (5.20)$$

where P is expressed in centimeters of mercury, σ in dynes per centimeter, and ρ in centimeters. Combining this with (5.13) and (5.14) one obtains $\bar{m}$

defined by the constants ρ and κ and the variables h and σ:

$$\bar{m} = (h - 4.68 \times 10^{-3}\,\sigma^{2/3}\rho^{-1/3})\kappa^{-1}, \qquad (5.21a)$$

which with a value of 380 dynes/cm for σ becomes

$$\bar{m} = (h - 0.246\rho^{-1/3})\kappa^{-1}. \qquad (5.21b)$$

Surface Tension and the Electrocapillary Curve

The surface tension σ is an important variable controlling the size of the dropping mercury electrode. It varies with the applied voltage and with the solution surrounding the electrode. Lippmann [19] and especially Gouy [20] made extensive studies of it with a capillary electrometer which measures the maximum back pressure from which σ is calculated. They showed that as the mercury is polarized negatively σ increases to a maximum value near -0.5 V. The curve of σ against the potential has the form of a parabola and is called an electrocapillary curve. Similar results were obtained by Kučera [15], who developed the drop-weight method for measuring σ. According to (5.17), the weight of the ripe drop W is dependent entirely on the radius of the capillary orifice ρ and the surface tension σ, while it is completely independent of the flow of mercury m. Consequently, W is a most reliable criterion for measuring σ when ρ is known. However, the collecting and weighing of drops after they have been washed and dried is a rather tedious process. A much simpler method is the measurement of the drop time t_d that is, the time necessary for the formation of a ripe drop. Since $t_d = W/\bar{m}$, this gives only approximate values of σ because $\bar{m}$ is not independent of σ. However, as is indicated by (5.21a), the errors inherent in this technique can be minimized if h and ρ are chosen as large as possible.

Examples of electrocapillary curves obtained this way are shown in the upper portion of Fig. 5.6. Note that the drop time reaches a maximum value near -0.5 V (versus SCE) in a solution of 0.1 N potassium nitrate. The generally accepted reason is that in such a "capillary-inactive" solution the mercury surface is uncharged at the electrocapillary maximum or "electrocapillary zero" potential. Polarized to more negative potentials the drops become negatively charged, while they acquire a positive charge on the positive side of electrocapillary zero. Whenever the dropping mercury electrode has a charge, ions with opposite charges are attracted from the solution and an electrical double layer is formed resulting in a diminution of the surface tension and consequently smaller drops. Normally large currents, caused either by dissolution of mercury at positive potentials, or by amalgam formation at negative potentials, limit the extent to which the potentials can be altered and to which the drop time can be reduced. In solutions of pure tetraalkylammonium salts, however, it is possible to reach potentials as

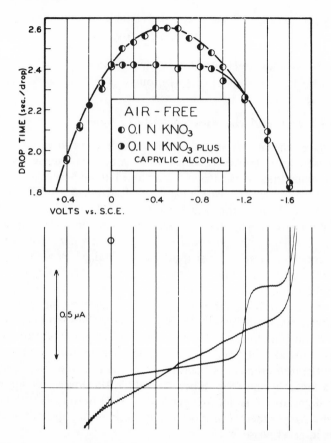

Fig. 5.6 Upper portion: Electrocapillary curve of mercury in 0.1 N potassium nitrate solution with and without caprylic alcohol. The solutions were freed from air by purging with nitrogen gas. Lower portion: Polarogram of the same solutions. The abrupt changes in current are caused by the adsorption and desorption of caprylic alcohol.

negative as -3.0 V without reducing the cations present. At such negative potentials the surface tension is reduced almost to zero, with the result that mercury no longer forms drops but issues from the capillary in a fine vapor-like stream [21].

Any substance present in solution that can alter the electrocapillarity is called capillary-active. In addition to ions that form complexes with mercury (e.g., I^-, CN^-, S^{2-}), many organic substances—amines, alcohols, ethers—belong to this group. Their effect may differ depending on their nature. The electrocapillary maximum may be shifted to more positive potentials (as with camphor or caffeine), or to more negative potentials (as with thiourea), or it may not be shifted at all but reached at a lower surface tension (as with sucrose). Adsorption of capillary-active substances usually takes place over a certain range of potential, after which desorption takes place. This results in a flattening of the electrocapillary curve, as seen in the case of caprylic alcohol in 0.1 N potassium nitrate shown also in Fig. 5.6. As could be expected, adsorbed substances often seriously affect the normal reduction processes at the electrode; this is demonstrated later.

Sometimes it is the *product* of an electrode reaction, rather than the starting material, which is adsorbed and then causes a change in the electrocapillary curve. Benesch and Benesch [22] found that in the reduction of phenyl-mercuric hydroxide, which is known to proceed in two steps, only the product obtained in the first step is capillary-active.

Surface Area of the Dropping Mercury Electrode and Its Rate of Formation

Naturally Formed Drops

Polarographic currents depend not only on the available electrode surface but also on the rate of formation of fresh surface. In the case of the normal dropping mercury electrode, the determination of these parameters is simple because a ripe drop is always allowed to form. The investigator must ascertain the total pressure h (see Fig. 5.1) and measure either the *characteristics of the capillary*, ρ and κ, or the *characteristics of the drop*, $\bar{m}$ and t_d, where t_d is the time necessary for the formation of a ripe drop. These quantities are best determined while the electrode is disconnected from the electrical circuit and dropping in an 0.1 N potassium chloride solution, open to air, because then a value of 380 dynes/cm can be used for the surface tension σ. Equations (5.18) and (5.21a) show the interrelation between the capillary and drop characteristics. Hence (5.20) can also be written as

$$\bar{P}_{\text{back}} = 4.68 \times 10^{-3} \times 380^{2/3} \times (0.411 \times 10^{-3})^{-1/3} \times W^{-1/3}$$
$$= 3.3m^{-1/3}t_d^{-1/3}, \quad (5.20a)$$

where $\bar{P}_{\text{back}}$ is in centimeters of mercury, $\bar{m}$ is in milligrams per second, and t_d is in seconds. Since W/d gives the volume of a drop and this is equal to $\frac{4}{3}\pi r_d^3$, the area of a ripe drop is

$$A = 0.008517W^{2/3} \text{ cm}^2, \quad (5.22)$$

where W is in milligrams. At a given surface tension, the surface area of the drop, similar to its weight or volume, depends entirely on the diameter of the capillary orifice. Upon substitution of mt for W, this area can be differentiated with respect to time, so that

$$dA/dt = 0.00565\bar{m}^{2/3}t^{-1/3} \text{ cm}^2/\text{sec} \quad (5.23)$$

if the mean rate of flow is applicable. However, this condition is approximated only after about 1 sec of flow; it changes from

$$dA/dt = 0.00565(h - 0.0573\rho^{-1})^{2/3}\kappa^{-2/3}t^{-1/3}$$

at the beginning of the drop to

$$dA/dt = 0.00565(h - 0.164\rho^{-1/3})^{2/3}\kappa^{-2/3}t_d^{-1/3}$$

at the end of the drop.

These relationships, together with corresponding data for A, m, and P_{back}, are represented graphically in Fig. 5.7 for an electrode with a capillary 5 cm long and with an internal diameter of 0.0050 cm. Mean values of each parameter have been indicated by a vertical line. The data have been calculated for total pressures of 60.0 and 40.0 cm of mercury to show the effect of h. Since ρ is the same for both capillaries, the drop size is the same, but the drop time t_d is 2.90 sec at the higher and 4.43 sec at the lower pressure. For easy comparison the parameters have been plotted as a function of fractions of each "natural" drop time. The quantity m was calculated for sufficiently small increments of these fractions so as to obtain enough points for a

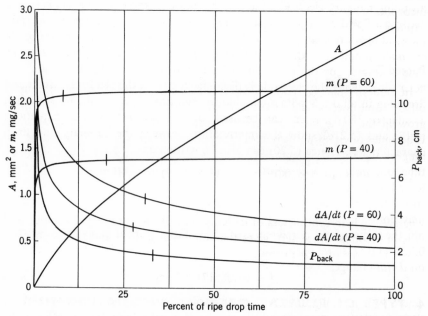

Fig. 5.7 Effect of time on A, m, dA/dt, and P_{back} of a capillary 5 cm long and with an internal diameter of 0.0050 cm. Data are calculated (1) for a total pressure, h, of 60 cm with a drop time t_d of 2.90 sec, and (2) for $h = 40$ cm with $t_d = 4.43$ sec. P_{back} and A values are the same in both cases since time is plotted as percent of drop time (see text).

reliable curve. For this, (5.13) was used after P_{back} had been determined by means of (5.15) with a value of 380 dynes/cm for σ. The surface area was found at each of these points from the size of the corresponding drop, while the dA/dt values were derived by means of (5.23). In this way of plotting, the A and P_{back} data for the two pressures are identical. It is quite apparent that there is a rapid change in the m values over the first tenth of t_d, followed by relative constancy. Because of this change in m, the curve for A deviates considerably from a 2/3-order parabola during the first tenth of t_d. In contrast to this, the marked changes in the dA/dt values persist in a diminished form throughout the growth of the drop. If the same capillary had been 7.5 cm long and h had been kept at 60 cm of mercury, the resulting curves would have been identical with the 5.0-cm long capillary at a pressure of 40 cm of mercury. However, increasing the internal diameter to 0.005534 cm of the 5-cm capillary at 40 cm pressure would have given curves that closely parallel the 0.0050-cm diameter capillary of the same length at 60 cm pressure. The A and m values would be somewhat higher while the dA/dt value would be somewhat lower. The relationship between the length and diameter of the capillaries and their drop time and mercury flow has been incorporated into a nomogram [23] on the basis of which one may select capillaries of desired characteristics, since capillary tubing of a variety of uniform bores is commercially available today.

Forced Drops

In modern polarography electrodes are often used whose drops are forced off by some mechanical or other means before they are ripe. These drops grow in the same fashion as normal drops so that Fig. 5.7 is applicable. It becomes obvious that one may expect little difficulty from forcing drops as long as the forced drop time is in excess of one-tenth of the natural drop time, since m, P_{back}, and dA/dt values are still close to their mean values. As the forced drop time is shortened to $<0.1t_d$ for a particular capillary, deviations from the mean become so great that general polarographic relationships may no longer hold. However, in view of the fact that these deviations are not the same for all parameters, one may expect forced drop times between 0.01 and $0.1t_d$ to be particularly useful for studies of phenomena that depend on dA/dt changes.

4 TYPES OF CURRENT OBTAINED WITH A DROPPING MERCURY ELECTRODE

Drop formation and with it the continued presence of some *fresh* surface are characteristics of the dropping mercury electrode that result in a variety of currents not found with any other electrode. In the following sections seven different types are discussed in some detail. They may be distinguished by the forces that bring electroactive material to the electrode such as migration, diffusion, convection, and adsorption, or they may be governed by the speed of chemical reactions at the electrode surface. Although one of them does not even involve any reduction or oxidation of material, all can be used for analytical purposes. However, to be meaningful, these different currents must be recognized by the polarographer and isolated where possible from other currents. To do this, the effects of alterations in the performance of the dropping mercury electrode are evaluated and materials are either added or removed from the test solution. For instance, since dissolved oxygen is reduced at the dropping mercury electrode, all solutions must be deaerated before polarography unless the effect of oxygen is under investigation.

Charging Current (Condenser Current)

In order to polarize or charge any electrode to a desired potential in the absence of a depolarizer, a number of electrical charges (electrons) must be added or removed from its surface. This quantity is very small and, in the case of a stationary electrode with fixed surface, is transported to it through the connecting wires at the instant when the electrical contact is first made. It is thus just a current pulse that rapidly returns to zero value at a fixed surface and hence contributes essentially nothing to the current-voltage relationships. This is not true in the case of the dropping mercury electrode. Here the surface is continually increasing and requiring additional electrons to maintain a desired charge. Furthermore, the drops fall off, taking any charges with them so that in effect an electron transport results that is large enough to be recorded as a current. This current is truly a charging or capacity current (often also called "condenser" or "nonfaradaic" current) that does not involve reductions or oxidations of material in solution.

Such a polarographic charging current is readily demonstrated with air-free *indifferent* or "supporting" electrolytes, so-called because they contain no oxidizable or reducible material over a large voltage span. A solution of $0.1 N$ potassium chloride from which the oxygen has been removed by bubbling hydrogen gas through it may serve as example. As may be seen in Fig. 5.8, there are small changes in current with applied voltage over the range of $+0.1$ to $-1.0 V$ versus *SCE* in spite of the fact that there is no oxidant or reductant present that might react in the voltage range.

In Fig. 5.8 and all subsequent figures, S indicates the fraction of the maximum galvanometer sensitivity used, the horizontal line represents "galvanometer zero," and the lowercase "o" on the abscissa indicates zero applied voltage for the curve shown. The reference electrode in all instances is a saturated calomel electrode. The oscillations in current correspond to the formation and growth of each drop of mercury. Actually, the current starts

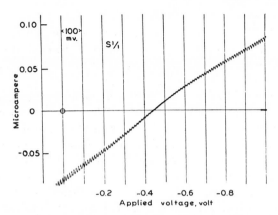

Fig. 5.8 Polarogram of 0.1 N potassium chloride solution freed from air by purging with hydrogen gas, illustrating the charging current.

from zero with each new drop, but the galvanometer response is too slow to record the whole current sweep and merely oscillates around a mean value. The charging current is anodic (negative) at zero applied voltage, passes through zero at about -0.45 V, and then becomes cathodic as the applied voltage is made more negative. Also the small current oscillations vanish near -0.45 V. In agreement with the earlier discussion of electrocapillary curves (see Section 3, p. 320), at the electrocapillary zero potential the mercury is uncharged, while at other potentials it has to acquire a charge (in the case of zero applied voltage, the mercury has to become positively charged by the removal of electrons as indicated by the anodic current). The anodic portion has a somewhat greater slope than the cathodic charging current. This was first observed by Ilkovič [24], who determined values of 22.3 and 42.2 $\times$ 10^{-6} F/cm^2, respectively, for the cathodic and anodic polarization capacities of the dropping mercury electrode.

Since the electrocapillary curve is affected by capillary-active substances, one can expect changes in the charging current as well. This is demonstrated in the lower portion of Fig. 5.6 where caprylic alcohol, because of its adsorption at 0 V and desorption at -1.2 V, causes marked changes in the charging current. (The return to the original charging current value at -1.2 V is not perfect because of the presence of some oxidation products of caprylic alcohol that are hard to remove.) If instead of this charging current (or specific static capacity [25]) its derivative with respect to the potential (differential capacity [25]) is plotted, a curve with pronounced peaks at the points of maximum capacity changes results. Oscilloscopic ("tensammetric" [26]) techniques, based on this phenomenon have been developed which make it possible to analyze substances that are neither reducible nor oxidizable at the dropping mercury electrode.

While the charging current is very small (less than 0.1 μA), it is nevertheless significant and must be subtracted from or added to the total current observed during the reduction of some reducible material in order that the actual current used in the reduction may be known. In very dilute solutions, neglect of this point may cause a large error.

Residual Current. Often the indifferent electrolyte contains traces of several impurities, so that small, almost imperceptible, waves are superimposed on the charging current (compare the polarogram of the nitrogen-washed 0.1 N potassium nitrate solution in Fig. 5.6 with that of the relatively pure hydrogen-washed 0.1 N potassium chloride solution in Fig. 5.8). It is customary to include these in the *residual current.* Since organic test substances may often be capillary-active, it is not always correct to subtract the "blank" curve of the indifferent electrolyte, the residual current, from the final test polarogram. Errors from this source may be suspected whenever a graph of diffusion currents plotted versus concentration of reacting material does not extrapolate to zero. As is shown later, in practical polarographic work the residual current is best subtracted automatically from the total observed current by the proper placement of tangents to the wave.

Migration Current

In an electrolyte solution all currents are carried by the ions present, the anions migrating to the anode and the cations migrating to the cathode. However, while there is always an equal number of cations and anions to preserve electrical neutrality, these ions do not necessarily carry equal fractions of the current. In the case of a single dissolved salt, the respective fractions of current that each type of ion carries are given by the so-called transport numbers (see Chapter IC in original plan, by Spiro). If the ions of this single salt also determine the reaction at the anode and cathode, there exists a pure migration current. However, if these reacting ions are part of a solution containing another salt with nonreacting ions (indifferent electrolyte) in much higher concentration, the latter will carry practically all the current in solution but will not participate in the electrode reaction. Under these conditions the reacting ions must reach the electrode by means other than migration, usually by diffusion in a concentration gradient generated by their removal at the electrode surface by the transfer of electrons. In practice the concentration of the added indifferent electrolyte must be at least 25 times as large as the concentration of the reacting material to produce a pure diffusion current.

In the reduction of cations, the migration current resulting from the attraction of positively charged ions to the negatively charged cathode produces an increase in the number of reducible ions at the electrode and therefore a current greater than the diffusion current. However, during the *reduction* of anions, these negatively charged ions are repelled by the negatively charged cathode, which results in a diminution of the reducible ions at the electrode, hence in a current smaller than the diffusion current. The reduction of uncharged substances does not involve a migration current and is not affected by the presence of an indifferent electrolyte.

A more complicated situation arises if two different materials are reduced in the absence of added electrolyte since the migration current is a function of the *total* limiting current and not of the individual diffusion current. The migration current observed in the reduction of the cation would therefore be increased if it is preceded by another reduction current on the polarogram, while the migration current found during the reduction of an anion would be diminished under the same circumstances. These effects have been called *exaltation* and have found some analytical applications [27].

The main value of the migration current studies has been to bring out the significance of indifferent or supporting electrolytes for producing diffusion currents. Buffer mixtures themselves, if concentrated enough, conduct the current well and thus act as supporting electrolytes and eliminate the migration current.

Convection Current

Maxima in General

Polarographic analyses are usually carried out with an experimental setup that ensures drop formation undisturbed by vibration or stirring. Thus convection in the solution is kept at a minimum. However, under a variety of conditions, a convection or streaming of solution results from the electrode reaction itself and is therefore very reproducible. This phenomenon exists in a thin layer around the mercury drop and can be seen under the microscope (and sometimes by the naked eye) if traces of solid material or dyestuffs are added to the solution. Since much more material can be brought to the electrode by convection than by diffusion, the currents become very much larger than pure diffusion currents. As a rule, they are easily recognizable since they are not maintained over a large voltage span but diminish more or less abruptly to the smaller value of the diffusion currents, giving rise to sharp or rounded *maxima* on the polarographic curves. Because addition of capillary-active substances (*maximum suppressors*) abolishes them, these maxima can be distinguished from other types of maxima, such as those resulting from anodic film formation or from catalytic reactions. (See Section 4, p. 347.)

Maxima are observed most frequently at the beginning of a wave, but on occasion may develop in the middle of the straight portion of a limiting current. A maximum may be found during a reduction or oxidation of ionized or nonionized, organic or inorganic substances. More than one maximum may be observed on a single current-voltage curve if the solution contains two or more electroactive substances. The maxima may be acute or rounded, or may be a combination of an acute maximum followed by a rounded one, depending not only on the nature and concentration of the material reduced but also on the concentration of the indifferent electrolyte and the resistance of the circuit. The heights of the maxima are variable, but it is usually possible to find optimum conditions for the production of a maximum for each electroactive substance. In general, the maxima become smaller the longer the drop time of the electrode; they usually increase with an increasing concentration of the electroactive material. Maxima obtained during organic reductions and oxidations may depend on the pH of the solution. Two different maxima may be observed if a single substance is reduced in steps at two different potentials, or the maximum may be present only on the second wave. It is likely that not all such maxima are identical in nature because they are affected to a different extent by the addition of maximum suppressors. At present, distinction is made between two types, the maxima of the first and of the second kind, which are described in more detail. Perhaps a maximum of the third kind should be added to distinguish those maxima where electrolyte streaming is produced by the buoyancy of the product, as in the reduction of hydrogen ions to hydrogen gas.

Maxima were originally attributed to the accumulation of excess material by adsorption, and the resulting current was called "adsorption current." However, this interpretation of the currents was abandoned after Antweiler first demonstrated streaming of electrolyte close to the electrode surface during maxima [28]. By an interferometric method or by microscopic observation of dust particles or of added talcum or carbon black, one may note that the streaming is the more marked the higher the maximum and that it stops abruptly with the fall of the maximum to the limiting current. The streaming is downward when the maximum is on the positive side of the electrocapillary maximum ("positive maximum"); otherwise it is upward, that is, toward the capillary orifice ("negative maximum"). Such a streaming effect easily accounts for the excess material available for reduction and for the absence of accumulated reaction products which would have an effect on the potential.

Maxima of the First Kind

Maxima of the first kind are usually acute and appear at the beginning of a wave in relatively dilute solutions. Figure 5.9 shows the maximum obtained during the reduction of oxygen dissolved in a 0.001 N solution of potassium chloride when it is in equilibrium with room air. This maximum is very prominent; it has a sharp peak and, as shown by curve b, is completely

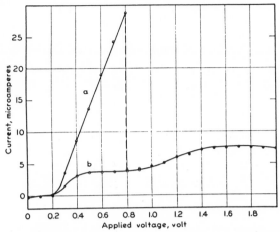

Fig. 5.9 Reduction of dissolved oxygen in 0.001 N potassium chloride solution (*a*) in absence and (*b*) in presence of a trace of acid fuchsin as maximum suppressor [29].

suppressed by the addition of a small amount of suppressor. Such an acute maximum is of special interest because the current increases linearly with the applied voltage until the peak is reached, when it drops discontinuously to the level of the limiting current. The slope of this straight line is equal to the reciprocal of the resistance of the cell in accordance with Ohm's law. This means that the potential of the dropping mercury electrode remains constant throughout this interval of applied voltage and that no further polarization occurs from the start of the straight line until the peak of the maximum is reached (in Fig. 5.9 from about −0.3 to −0.8 V). However, at that critical point, a further increase in the applied voltage of a few millivolts produces a sudden marked polarization of the electrode which may amount to 1 V (in Fig. 5.9 it is 0.5 V). Maxima of the first kind can also be obtained with stationary but not with streaming mercury electrodes.

The velocity of electrolyte streaming around the mercury drop can be estimated from three different measurements: (1) direct observation (photography), (2) calculation of volume elements of solution that must have reached the dropping mercury electrode during its lifetime in order to produce a current of the observed magnitude, and (3) calculation of the increase in drop size as the result of buoyancy produced by the streaming of the solution. According to v. Stackelberg and Doppelfeld [30], this buoyancy is effective only in the case of positive maxima where the solution streams downward from the drop; it results in a small increase in drop weight since it assists the surface tension. In the case of negative maxima, the opposite effect should be found; the upward streaming of solution should exert a pull on the drop, compete with the surface tension, and result in a smaller drop, but so far no one has investigated this possibility. The highest velocities found in the case of a maximum of mercurous ions by methods (1) and (2) have been about 20 to 25 cm/sec [30].

Streaming is believed to result from potential differences existing between the point of attachment of the drop and its opposite pole because of differences in current density caused by the shielding effect of the capillary tip. The sudden change from the maximum to the limiting current results from accumulation in the interface of indifferent ions which block further access of electroactive substances. A diffusion layer once formed, or an adsorbed capillary-active substance, is presumed to have a damping effect on this streaming. It is beyond the scope of this chapter to discuss the progress that has been made in the quantitative formulation of the effects [30].

Maxima of the Second Kind

Maxima of the second kind are obtained with concentrated solutions and at high flow rates of the mercury. They are usually rounded and return gradually to the value of the limiting current at more negative potentials. The example shown in Fig. 5.10 demonstrates that such maxima can also be suppressed by surface-active material. These maxima are particularly dependent on the rate of mercury flow. They disappear when the linear velocity of the mercury inflow into the drop is less than 2 cm/sec, or when

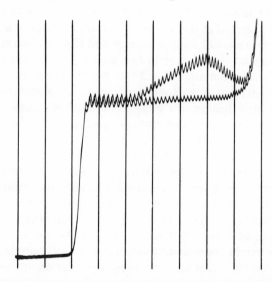

Fig. 5.10   Maximum of the second kind in 1 mM cadmium chloride solution in 1 N potassium chloride, suppressed by a trace of 0.1 % fresh gelatin solution. Solution was deaerated by nitrogen gas. Curves start at −0.2 V versus SCE; interval between abscissas is 0.2 V; the sensitivity used is $\frac{1}{50}$.

2.35 $\times$ $10^{-5}m\rho^{-2}$ < 2 cm/sec. By substituting for m from (5.13), one finds that capillaries with this property must meet the condition: 1.09 $\times$ $10^{5}P\rho^{2}l^{-1}$ < 2 cm/sec, which in the case of $P = 40$ cm of mercury and $\rho = 2 \times 10^{-3}$ cm means that its length l must be 10 cm (assuming, of course, uniform bore). Maxima of the second kind are obviously not found with stationary or streaming mercury electrodes.

Since they are so different from other maxima, they were ascribed to a variety of special reactions until Kryukova found that electrolyte streaming took place whenever they occurred [31]. She therefore called them "maxima of the second kind." They are believed to be caused by a stirring effect produced inside the drop by the inflowing mercury, which communicates itself to the surrounding solution layer and thus causes an increased transport of electroactive material to the electrode.

The possibility of maxima of the second kind should always be considered whenever work is done at electrolyte concentrations greater than 0.1 N since they may easily be overlooked. They have been used for the quantitative analysis of as little as 10^{-9} M surface-active material [32].

Maximum Suppressors

As a rule, maxima are an obstacle to quantitative polarographic analyses, and it is fortunate that they can be eliminated by the addition of a small quantity of a maximum suppressor to the solution, as shown in Figs. 5.9 and 5.10. Most popular have been organic indicators and dyes, but alkaloids, purine derivatives, organic acids, and gelatin, agar, and other colloids, as well as some detergents, have also been used with success. While some of these may be reduced themselves at the dropping mercury electrode, they do not interfere with the analyses as long as they are present in very small concentration. At higher concentrations, however, they render interpretation of polarograms difficult. First, the current attributable to the reduction of the maximum suppressor itself becomes appreciable and, second, the limiting current of the substance under investigation may actually be diminished by the suppressor for reasons unknown. Correction for the first effect is relatively simple, but the second is neither easily recognized nor corrected for unless a series of calibration curves has been prepared. Careless and excessive addition of maximum suppressors can therefore lead to serious errors.

The effectiveness of a suppressor can be expressed as the maximum dilution or the minimum concentration necessary for complete suppression of a given maximum. A better way, because the concentration of the suppressor remains small, is to note the concentration necessary for a 50% reduction of the maximum [33]. More often, the reciprocal of concentration, that is, the dilution of a compound at which the maximum is half suppressed is given in publications. This is then called the *adsorption coefficient*.

Diffusion Current

General

The establishment of a limiting current at the dropping mercury electrode signifies that all of a given reactive substance that reaches the electrode surface during its life time is altered by the electric current. In the presence of sufficient supporting electrolyte and maximum suppressor, currents resulting from migration and convection (maxima) are eliminated and the electroactive material can reach the electrode only by diffusion in a concentration gradient caused by the removal of the material at the electrode. The current is then called a diffusion current. The gradient extends through a diffusion layer which varies in concentration from practically zero at the electrode surface to that existing in the body of the solution.

In the case of the dropping mercury electrode, the diffusion layer is a spherical shell, the thickness of which determines the concentration gradient for any given concentration in the solution; the thinner the layer, the steeper the concentration gradient. As the current removes more and more material, the diffusion layer should become thicker, and thus the concentration gradient should decrease with time were it not for the fact that the diffusion shell expands in size and becomes thinner as the mercury drop grows. The observed current is therefore a complex function of the growth of the mercury drops and of the diffusion coefficient of the reactive material. Ilkovič [34], who first defined these relationships mathematically, derived an approximate equation in which a spherical diffusion layer of negligible thickness was assumed. This Ilkovič equation is in excellent agreement with experimental data and has become the fundamental equation for most quantitative polarography. Although more rigid mathematical treatments have shown that a correction factor should properly be added to the Ilkovič equation, specific tests have indicated that this correction factor is largely canceled out by other errors (e.g., depletion of material from the solution by preceding drops, shielding of the mercury by the tip of the glass capillary, and so on).

Quite generally, at any electrode with area A, the intensity of the diffusion current at time t is given by

$$I_t = nFD \cdot A\delta, \tag{5.24}$$

where n is the number of faradays required per mole of substance reacting at the electrode, F is the faraday, and D the diffusion coefficient of the reacting substance (in square centimeters per second), while δ is the concentration gradient $(\partial C/\partial x)$ near the electrode, that is, when the distance from the electrode x approaches zero.

There is also a volume of solution near the electrode which, as the result of the electrode reaction, has undergone a change in the concentration of electroactive material. Instead of this volume with nonuniform concentration, one may calculate an equivalent volume V which has been "completely exhausted" of current-producing material, while the remaining solution is considered to be at the unaltered concentration C. This exhausted volume must have originally contained all of the material that has reacted in time t; it can therefore be found by integration of the current between starting time 0 and final time t:

$$V n F C = \int_0^t I \, dt \qquad \text{or} \qquad V = D/C \int_0^t \delta A \, dt. \qquad (5.25)$$

This volume is also equal to $\bar{I}t$ if $\bar{I}$ is the mean current and t is the duration of electrolysis.

By means of (5.24) and (5.25) and the respective values of A and δ it has been possible to evaluate the diffusion currents for the following four different situations [35]:

1. *Linear diffusion to an electrode with one plane surface of area A*

$$A\delta_1 = AC/\sqrt{\pi D t}. \qquad (5.26)$$

2. *Symmetrical diffusion to a spherical electrode with radius r and fixed surface area A.* This may be considered to have a linear component (as in case 1) and a spherical component:

$$A\delta_2 = AC(1/\sqrt{\pi D t} + 1/r) = AC/\sqrt{\pi D t}\left(1 + \frac{\sqrt{\pi D t}}{r}\right). \qquad (5.27)$$

The quantity $\sqrt{\pi D t}$ for a typical diffusion coefficient of 8.6×10^{-6} cm^2/sec and for 3 sec (a typical drop time) becomes equal to 9×10^{-3} cm. Hence the factor $\sqrt{\pi D t}/r$ can be expected to cause an addition of more than 10% to the linear diffusion when the radius of the spherical electrode is smaller than 0.9 mm. Very early in the electrolysis, when t is very small, the spherical component is negligible.

3. *Linear diffusion to an electrode with a plane surface of area A that is growing at the same rate as the surface of the dropping mercury electrode.* The area A at any given time t is given by $4\pi\gamma^{2/3}t^{2/3}$, where γ represents the rate of growth of the drop (in cubic centimeters per second). At a constant flow of mercury m (in milligrams per second) $\gamma = 3m/4\pi d$. The density of mercury d at 25°C is 13534 mg/cm^3 so that $\gamma = 17.64 \times 10^{-6}m$. Because the expanding surface area continuously exposes fresh solution to the electrode, the diffusion gradient is about 50% greater than that at an electrode of constant size as in case 1; the mathematically derived ratio is actually $\sqrt{\frac{7}{3}}:1$. Hence

$$A\delta_3 = 4\pi\gamma^{2/3}t^{2/3} \cdot C \frac{\sqrt{\frac{7}{3}}}{\sqrt{\pi D t}} = 7.34 \times 10^{-3} C D^{-1/2} m^{2/3} t^{1/6}. \qquad (5.28)$$

This is the relationship first developed by Ilkovič [34]; when incorporated into (5.24), it leads to the *Ilkovič equation.*

4. *Symmetrical diffusion to a spherical electrode growing at the same rate as the dropping mercury electrode.* Here the surface area A increases exactly as in case 3, but the diffusion gradient is greater because in spherical diffusion the area of the diffusion boundary increases even more than the surface area of the drop. One can write in analogy with case 2:

$$A\delta_4 = 4\pi\gamma^{2/3}t^{2/3}C\left(\frac{\sqrt{\frac{7}{3}}}{\sqrt{\pi Dt}} + \frac{K}{r}\right)$$

$$= 7.34 \times 10^{-3}CD^{1/2}m^{2/3}t^{1/6}\left(1 + \frac{K\sqrt{\pi Dt}\sqrt{\frac{3}{7}}}{r}\right).$$

The factor K has been introduced in this equation to account for any deviation from the $1/r$ relationship of case 2. Such a deviation could result from the fact that the continuous growth of the electrode area tends to counteract the development of the spherical component. K therefore should be less than unity.

Substitution of $\gamma^{1/3}t^{1/3}$ or $0.026m^{1/3}t^{1/3}$ for r leads to:

$$A\delta_4 = 7.34 \times 10^{-3}CD^{1/2}m^{2/3}t^{1/6}(1 + 44.6KD^{1/2}m^{-1/3}t^{1/6}). \tag{5.29}$$

This relationship incorporated into (5.24) leads to the *modified Ilkovič equation.*

Ilkovič Equation

In his original derivation, Ilkovič expressed the concentration C in moles per cubic centimeter and the flow of mercury m in grams per second; he measured the diffusion current I_d in amperes:

$$I_d = 0.734nFCD^{1/2}m^{2/3}t^{1/6}. \tag{5.30}$$

By combining (5.24) and (5.28) and substituting 96,496 coulombs for F, one obtains the more common version of the Ilkovič equation:

$$I_d = 708nCD^{1/2}m^{2/3}t^{1/6}, \tag{5.31}$$

in which I_d is expressed in microamperes ($1\ \mu A = 10^{-6}$ A), C in millimoles per liter and m in milligrams per second. The number of faradays involved in the reduction is given by n, while D is the diffusion coefficient (in square centimeters per second) and t is the time (in seconds) during the life of the drop. (If $t = t_d$, i.e., the drop time of the electrode, then I_d is the maximal diffusion current obtained during the growth of the drop.)

The direct proportionality between the concentration of the reacting material and the diffusion current is a fact that had been known since the earliest days of polarography. It is the basis for quantitative polarography. The effect of n had been equally well known, but the relationship between the diffusion current and diffusion coefficient, and the dependence of the diffusion current on the characteristics of the dropping mercury electrode, were brought out for the first time by this equation.

The number of electrons, n, involved in the reduction of an electroactive compound is usually two in organic chemistry, especially in reversible reactions. However, oxidations and reductions are also known in which each electron is removed or added separately with the formation of an intermediate radical, called semiquinone. (See Section 4, p. 360.) In irreversible reductions sometimes as many as six or eight electrons may be involved in the complete reduction of the molecule, and two or more steps are observed on the polarogram. The value of n is determined by comparison with a suitable similar reaction for which the same diffusion coefficient may be assumed, or by detailed analysis of the current-potential relationship and comparison with a theoretical curve of the type shown in Fig. 5.4. Sometimes an estimate of n may be made from diffusion current measurements on the basis of the Ilkovič equation when reasonable values for the diffusion coefficient are available. The problem is simplified when several reduction steps are found with fixed ratios of their wave heights, but one must make sure that the separate waves are indeed diffusion controlled (i.e., that each is proportional to $P^{1/2}$). The most reliable values for n are probably obtained by coulometric analysis and controlled-potential electrolysis which permits the identification of the product (see Chapter IIE in original plan, byMeites).

The direct proportionality between diffusion current and concentration holds for inorganic and organic ions and molecules as long as there is no polymerization or tautomerism of the reagents. Furthermore, there should be no effect on the drop time as the result of unusual surface activity or adsorption of the materials, nor should there be any interaction of products of the reaction with electroactive material. The relationship between diffusion current and concentration is usually tested by making analyses over as large a concentration range as possible and checking the constancy of the I_d/C values obtained. Another way is to plot the data on a graph with double logarithmic coordinates. The points must then all lie on a straight line with a slope of 45°, passing through $\log C = 0$. This linear relationship is so well established that one should look for errors if any deviations are found. At the concentrated end this might be caused by insufficient elimination of the migration current, while at the dilute end some interfering impurity may be suspected.

The diffusion coefficient, D, is another factor of the Ilkovič equation, but its value has remained uncertain in spite of considerable efforts. In general, D for metal and organic ions and molecules is of the order of 6 to 10 $\times$ 10^{-6} cm²/sec, although there are many exceptions, notably those of OH⁻ and H⁺ ions, which are about 5 and 10 times larger, respectively.

In the case of ions, diffusion coefficients can be calculated from λ^0, their equivalent conductances at infinite dilution, by means of the Nernst relationship

$$D^0 = 2.67 \times 10^{-7} \lambda^0/z \text{ cm}^2/\text{sec} \qquad \text{(at 25°C)}, \qquad (5.32)$$

where z is the charge of the ion. In the case of uncharged molecules, the Stokes-Einstein law may be applied on the assumption that the molecules are spherical. The following equation then holds, again at infinite dilution:

$$D^0 = 2.96 \times 10^{-7} \eta^{-1}d^{1/3}M^{-1/3} \text{ cm}^2/\text{sec} \qquad \text{(at 25°C)}, \qquad (5.33)$$

where η is the viscosity of the solvent (0.89 and 1.10 cP for water and ethanol, respectively), d is the density of the molecules, and M their molecular weight.

These values at infinite dilution can only serve as first approximations for polarographic diffusion coefficients since the latter are obtained in media containing indifferent electrolyte or buffers. Attempts to bring about more valid correlation by extrapolation of polarographic data to infinite dilution have not been very successful. Even corrections applied to the Ilkovič equation in order to account for spherical diffusion have only pointed out additional discrepancies. One must therefore regard polarographically determined diffusion coefficients as approximate.

The factor $m^{2/3}t^{1/6}$ is characteristic for diffusion currents obtained with the dropping mercury electrode. Typical of case 3, discussed earlier, it is the product of the electrode surface $(k'm^{2/3}t^{2/3})$ and the diffusion gradient $(k''t^{-1/2})$. To study the effect of this factor, one uses different electrodes, or one varies the height h of the mercury level above the tip of a given capillary and notes the corresponding diffusion current at a constant applied potential. At the same time a suitable number of drops of mercury are collected beneath the solution and weighed after being dried, and the time necessary for their formation is measured with a stop watch. Thus one obtains m and t values from which $m^{2/3}t^{1/6}$ can be calculated.

There is also a simpler test which is often used to ascertain the existence of a diffusion current. As was shown in Section 3, p. 317, m is proportional to the effective pressure P, and the weight of each drop $(W = mt_d)$ at a given potential is constant. Thus t_d is inversely proportional to P and

$$I_d = km^{2/3}t_d^{1/6} = k'P^{1/2}, \qquad (5.34)$$

which shows that the diffusion current is proportional to the square root of the effective pressure. The latter is readily obtained by subtracting the mean

$\bar{P}_{back}$ from h. If not already known, the mean back pressure of a given capillary is calculated by means of (5.20), while the radius of the capillary orifice ρ is obtained from the drop weight according to (5.18).

It is well to remember that at constant voltage a change in pressure alters only the rate of dropping, not the size of the individual mercury drops. However, the surface tension of mercury is a function of the electrode potential and the surrounding solution. Hence W and t_d must vary with the applied voltage (Section 3, p. 320). However, since t_d appears only as the sixth root in the Ilkovič equation, changing t_d has a relatively small effect on the diffusion current. In fact, over the potential range from 0 to about -1.0 V (versus SCE), the resulting change in $m^{2/3}t_d^{1/6}$ and therefore in I_d is usually so small ($\pm 1\%$) that it can be neglected. At more negative potentials these changes become appreciable and must be taken into account. Also, in reactions of organic compounds, surface activity must be expected which can have an influence on t_d that significantly alters I_d. The quantity $m^{2/3}t_d^{1/6}$ should therefore always be determined at the potential at which the diffusion current is measured.

The factor $m^{2/3}t^{1/6}$ further indicates that at constant rates of mercury flow the diffusion current should increase with the sixth root of time during the life of a drop. Only a very fast string galvanometer or cathode ray oscilloscope can follow such current changes; most galvanometers and recording instruments used are much too slow. As a consequence, the usual polarograms represent recordings of current oscillations around some mean current which is approached more and more closely the longer the period of the galvanometer. Unfortunately, when the latter is so slow as to record only a mean current and no oscillations, it is also too slow to record current changes resulting from changes in the applied voltage. Hence some optimum rate of oscillation of the galvanometer is desirable.

It can be proved mathematically that the mean diffusion current is equal to $\frac{6}{7}$ of the maximal current obtained just before the drop falls off. Thus the Ilkovič equation for the mean diffusion current becomes:

$$I_d = 607nCD^{1/2}m^{2/3}t_d^{1/6}. \tag{5.35}$$

On the recorded polarograms, any part of the oscillations can be measured with equal accuracy. For practical analyses it makes little difference which part of the oscillations is used for comparisons as long as the investigator is consistent and works with the same setting of the capillary and with the same galvanometer. However, if the factors of the the Ilkovič equation are to be evaluated or the results with different capillaries are to be compared, every attempt must be made to obtain current values as near as possible to the mean current. As long as the oscillations are very small, the minimum and maximum deflections as well as their average are very close to the true mean currents, and any of them can be used for measurement without involving appreciable errors. When the oscillations increase in size, however, the errors of all three measurements also increase. Since with a very fast galvanometer the current at the beginning of a drop is practically zero while the maximum deflection reaches $\frac{7}{6}$ of the mean current, the average deflection should diverge faster from the mean than the maximum deflection. Hence one might expect that the true mean current would be nearer to $\frac{6}{7}$ than to $\frac{1}{2}$ the observed current oscillations. Yet this condition is apparently not approached with the slow galvanometers usually employed in polarography. For instance, the curves of α-oxyphenazine (Fig. 5.11) obtained with three different capillaries, one of them at two different pressures of mercury, have been evaluated by measuring average (as shown in curve B) and maximum galvanometer deflections (as shown in curve D), as well as the current corresponding to $\frac{6}{7}$ of the oscillations [36]. Table 5.1 indicates that

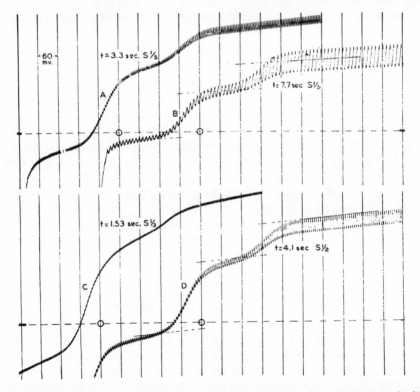

Fig. 5.11 Polarograms of 0.0001 M α-oxyphenazine in 0.1 N sulfuric acid obtained with capillaries of different drop times [36]. Curve B illustrates measurement of *average* galvanometer deflections, curve D, measurement of *maximum* galvanometer deflections. (See Table 5.1.)

Table 5.1 Application of the Ilkovič Equation to the Diffusion Current of α-Oxyphenazine Measured with Different Dropping Mercury Electrodes and Using Different Methods of Measurement

Curve in Fig. 5.11	Capillary No.	κ	P (cm)	t (sec)	$m^{2/3}t^{1/6}$ (mg$^{2/3}$ × sec$^{-1/2}$)	Recorded Galvanometer Deflection					
						Average		Maximum		$\frac{6}{7}$ of Total	
						I_d (×10^{-8} A)	$I_d/m^{2/3}t^{1/6}$	I_d (×10^{-8} A)	$I_d/m^{2/3}t^{1/6}$	I_d (×10^{-8} A)	$I_d/m^{2/3}t^{1/6}$
A	1	29.7	64	3.3	2.042	55.6	27.2	56.6	27.7	56.3	27.6
B	1	29.7	29	7.7	1.380	36.4	26.4	40.9	29.6	39.6	28.7
C	2	44.6	74	1.53	1.506	39.1	26.0	39.2	26.1	39.2	26.1
D	3	182.0	83	4.1	0.748	20.1	26.9	20.7	27.7	20.5	27.4
					Average		26.6		27.8		27.4

the measurement of the average galvanometer deflection is best since the four results agree with an accuracy of $\pm 3\%$. Note that the difference between the three methods of measurement becomes negligible when the fast capillary, no. 2, is used. Available data for the diffusion current of lead, measured with galvanometers of different half-periods, similarly point to the average as the best measurement [37]. One reason for this is that the current-time relationship during the formation of an individual drop differs somewhat from the one-sixth-order parabola predicted by the Ilkovič equation [38].

The study of current-time curves of individual drops during the application of a constant voltage has yielded some interesting results. By means of a device that permits the application of a desired polarizing voltage at the moment when a new drop starts to form, Hans et al. [39] demonstrated a difference in current between this "first drop" and subsequent drops. The current for the first drop was always larger and did indeed follow the one-sixth-order parabola while subsequent drops did not. This verified a suggestion made by Airey and Smales [40] that each drop of mercury during the recording of an ordinary polarogram starts in a solution that has been partially depleted of depolarizer by the preceding drop.

Smoler [41], in a different approach to the problem, constructed *horizontal* capillaries and capillaries with a tip slanting at $45°$ from which the drops rolled off along the capillary tip and thus carried along some solution across the capillary orifice. This prevented the depletion of depolarizer in the surrounding solution by preceding drops and resulted in current-time curves with a parabolic form very close to that obtained with the first drop on an ordinary "vertical" capillary. These and similar studies have also given new insight into the effects of maximum suppressors, polarographic maxima, and other types of current in addition to the diffusion current. An excellent discussion of the potentialities of this type of investigation is given by Kůta and Smoler [42]. Since ordinary vertical thick-walled capillaries have a lot of glass that effectively shields the drop from the solution at its neck, it seemed promising to study dropping mercury electrodes in which this would not be the case. Flemming and Berg [43] succeeded in drawing out micro-capillaries in which the wall thickness was less than the internal diameter; however, in contrast to the slanted Smoler capillaries, they still showed differences between first and subsequent drops.

Modified Ilkovič Equation

According to (5.35), the ratio of the mean diffusion current to the quantity $Cm^{2/3}t_d^{1/6}$ should be a constant. Unfortunately, this is not quite true. First, below the so-called critical drop time, which varies with capillaries and experimental conditions, the Ilkovič equation is not applicable. Second, even with long drop times and slow flow of mercury the constancy of the above relationship is only approximated. This has led to a modification of the Ilkovič equation in which account is taken of the fact that symmetrical spherical rather than linear diffusion is involved. As discussed earlier under case 4, the diffusion gradient in spherical diffusion is greater than that of linear diffusion. A combination of (5.24) and (5.29) leads to the modified Ilkovič equation:

$$I_d = 708nCD^{1/2}m^{2/3}t^{1/6}(1 + 44.6KD^{1/2}m^{-1/3}t^{1/6}), \tag{5.36}$$

in which the same units are used as in (5.31). At a given drop time t_d, the mean current $\bar{I}_d$ then becomes:

$$\bar{I}_d = 607nCD^{1/2}m^{2/3}t_d^{1/6}(1 + 39.0KD^{1/2}m^{-1/3}t_d^{1/6}). \tag{5.37}$$

This equation contains the uncertain term K or the quantity $39.0K$ which shall be designated by A. Several approximate and two rigorous mathematical treatments have so far failed to yield a uniform value for A. Instead of 39.1 used by Lingane and Loveridge [37] and implying $K = 1.0$, Strehlow and v. Stackelberg [44] found $A = 17 \pm 3$ as the best fit to a whole series of data [45]. This A value corresponds to a K value of $\frac{3}{7}$ [35]. There have been other proposals for A values between these extremes, but all have been approximations. Two rigorous mathematical derivations have added a third term to the parentheses. The Koutecky [46] equation,

$$\bar{I}_d = 607nCD^{1/2}m^{2/3}t^{1/6}[1 + 34D^{1/2}m^{-1/3}t_d^{1/6} + 100(D^{1/2}m^{-1/3}t_d^{1/6})^2], \tag{5.38}$$

contains an A value of 34. The Kimla and Štráfelda [47] equation, when restated in terms commonly used in polarography, becomes:

$$\bar{I}_d = 1.040[607nCD^{1/2}m^{2/3}t_d^{1/6}(1 + 50D^{1/2}m^{-1/3}t_d^{1/6} + 0.00193D^{-1/2}m^{1/3}t_d^{-1/6})]. \tag{5.39}$$

The A value of 50 in this equation is exceptionally high, which is probably the reason why diffusion currents with regular thick-walled polarographic capillaries were consistently lower than calculated. However, data obtained with specially drawn fine-tip capillaries producing drop times greater than 20 sec, were in good agreement with theory [47].

Until these variants of the Ilkovič equation have a common solution, there will be uncertainty concerning the polarographical determination of diffusion coefficients. This will be true even if every care is taken to measure only first drops and to use capillaries with large κ values so that changes in m during drop formation are minimized. In general, the depletion effect and other deviations from ideal conditions (constant m, perfect sphericity of the drop, its shielding from the solution at its neck, and so on) seem to make an almost perfect compensation for the error resulting from the assumption of linear diffusion to the dropping mercury electrode. As a result, *the original Ilkovič equation* (5.35) *is still the best for describing diffusion currents in practical polarography.*

Kinetic Current (Current Limited by Reaction Rates)

A basic requirement for the diffusion current, which so far has not yet been discussed, is that reactions other than diffusion which are involved in the electrode process must take place with speeds far different from that of diffusion. If this is not the case, mixed currents may result partly controlled by diffusion and partly by the rate of a chemical reaction. With the exception of catalytic reactions at the electrode surface, which will be treated in the next section, the reaction-controlled "kinetic" currents look very much like ordinary diffusion currents on the polarogram, but they do not obey the Ilkovič equation.

Pure kinetic currents are proportional to the surface area of the electrode and hence to $m^{2/3}t^{2/3}$. Since W for a given capillary is independent of pressure, kinetic currents do not vary with changes of pressure on a capillary. Catalytic currents are likewise independent of the pressure. Hence if such currents are reported as current densities, that is, in microamperes per square millimeter of electrode surface, results obtained with different capillaries agree very well [48]. The surface area is calculated by means of (5.22) from the drop weight W, which should be determined at the applied voltage at which the current is measured.

Kinetic Current Caused by Reestablishment of Disturbed Equilibrium

This type of kinetic current is caused by the reestablishment of a disturbed equilibrium between two substances R_p and R_n of which R_p is reduced more easily (at a more positive potential) than R_n. Three different situations may arise, depending on the rate with which this equilibrium is reestablished.

1. *If the equilibrium is established very slowly*, the two compounds will be reduced independently of each other. Each will give rise to its own diffusion current if both are reactive, or only one will give rise to its own diffusion current if the other does not react within the available voltage range. In the former case one may see the effects of the slow equilibrium that exists in the bulk of the solution by repeating polarograms at suitable time intervals: one diffusion current may grow at the expense of the other and each polarogram will show the progress made in the establishment of the new equilibrium.

2. *If the equilibrium is established very rapidly*, only a single wave of R_p will be obtained because as soon as any R_p is removed by the current, an equivalent amount of R_n is used to replace it until all available R_n is also used up, no matter whether the latter is reducible within the available voltage range

or not. The current of the single wave is then composed of the sum of the diffusion currents of the two compounds and will be in accordance with the Ilkovič equation.

3. *If the equilibrium is established at an intermediate rate,* the wave of R_p will no longer be diffusion controlled. To the diffusion current, determined by the concentration of R_p, will be added a kinetic current produced by the amount of R_n that changes to R_p at the interface during the life of the drop. If R_n is also reducible, its diffusion current will be diminished by a corresponding amount. Neither current will then be in accordance with the Ilkovič equation, but their sum will still be diffusion controlled.

The first demonstration of such a system was given in 1939 by Müller and Baumberger [49], who studied pyruvic acid in buffers of different pH. Their polarogram is reproduced in Fig. 5.12. Note that the single wave, obtained at pH 4.35, gradually diminishes as the solution is made more alkaline, to be replaced by a second wave with a more negative half-wave potential. At pH 5.8 the two waves are about equal. The sum of the waves of course is constant. The effect of mercury pressure on these currents indicates their nature; it is illustrated in Fig. 5.13 [50]. Obviously, the first wave must be kinetic since it is independent of pressure. In contrast, the sum of the first and second waves is a linear function of the square root of the pressure and thus proves to be diffusion controlled. The significance of the middle portion of the second wave, which has an almost constant height but tends to form a maximum at slower dropping rates, has not yet been ascertained.

The original interpretation that the slow equilibrium in this case concerned the keto and enol forms of pyruvic acid [49] was found to be in error 8 years later when Brdička discovered that phenylglyoxylic acid, which cannot enolize, gave strictly analogous results when it was similarly reduced [51]. He concluded that the slow equilibrium pertained to the association of pyruvate or phenylglyoxylate ions with hydrogen ions and that the first wave in either case was produced by the molecular form of the acid while the second wave was produced by the anion of the acid. The fact that the pH at

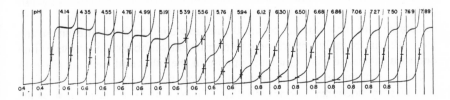

Fig. 5.12 Current-voltage curves of 0.001 M pyruvic acid, buffered at the pH indicated [49]. The initial applied voltage is shown for each curve, the voltage increment is 0.2 V for each abscissa; the sensitivity used is $\frac{1}{50}$.

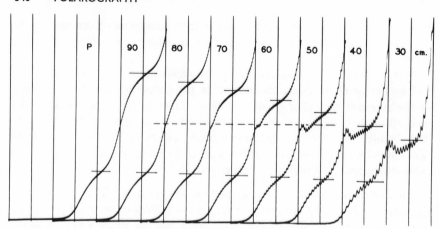

Fig. 5.13 Effect of changing the mercury pressure P on the current-voltage curves of a 0.0025 M solution of pyruvic acid buffered at pH 6.21. All curves start at -0.6 V versus SCE, the voltage increment is 0.2 V for each abscissa, the sensitivity used is $\frac{1}{150}$ [50].

which both waves are equal was several pH units higher than the pK of the respective acids was considered to be a result of the slowness of recombination of anion with hydrogen ions. This idea has been widely accepted and has formed the basis for most of the additional studies in this particular field. Elaborate mathematical equations have been derived by several groups of investigators [52]. With these equations and tabulated values of functions [53], it has been possible to correlate experimental data with theory and to calculate rate constants [54].

For instance, k, the rate constant of recombination of similarly behaving acids has been calculated on the basis of the equation

$$\log k = 2\mathrm{p}K' + \log 1.31 - \mathrm{p}K - \log t_d. \qquad (5.40)$$

Here pK' (the "apparent pK") corresponds to the pH at which the first and second waves are equal, pK is the negative logarithm of the known dissociation constant of the acid, 1.31 is a constant derived on theoretical grounds by Koutecky [53], and t_d is the drop time of the electrode. As a rule, the calculated rate constants fall within the range of 10^7 to 10^{13} mole^{-1} liter sec^{-1}. Since the latter may differ even in the order of magnitude from values obtained otherwise [55], one must apply this method cautiously.

Other reactions giving rise to kinetic waves are the reductions of sugars and of formaldehyde. Certain reducing sugars give a small polarographic wave, much smaller than could be expected on the basis of their concentration. This has been ascribed to the reduction of the aldehydo form with which these sugars are in equilibrium [56]. Wiesner [57] showed that this wave is independent of the pressure on the capillary and thus represents a true kinetic current. The small quantity of electroactive sugar obviously must have been regenerated from the inactive fraction as soon as it was reduced.

Formaldehyde in aqueous solution is present almost exclusively in the hydrated form, methylene glycol, which is nonreducible. Yet a small current is found with solutions of formaldehyde that varies markedly with pH and temperature. It is independent of capillary pressure, hence a kinetic current. The reducible formaldehyde must therefore be formed from the hydrate as soon as it is removed by the electric current [58].

Kinetic Current Caused by Products of the Electrode Reaction

REACTIONS OF OXIDATION PRODUCTS

During the analysis of organic oxidation-reduction systems with the dropping mercury electrode, it is of value to test for reversibility by observing half-wave potentials before and after chemical reduction of the oxidant with hydrogen gas in the presence of platinized asbestos or palladium black (see Section 6, p. 356). If the hydrogen gas is not washed out of the solution with pure nitrogen gas, the anodic waves of the system will be abnormally large [59]. This is caused by a kinetic current which results from the chemical reduction at the electrode surface of part of the oxidant that has just been formed by electrolytic oxidation. An interesting extension of this reaction is the kinetic current of ferrous ion produced through the intermediary of an organic oxidation-reduction system by means of palladium black and hydrogen gas [60].

REACTIONS OF REDUCTION PRODUCTS

In the writer's opinion, the product of the reduction of hydrogen ions, namely, atomic or nascent hydrogen, may well be involved in many of the irreversible reactions of organic compounds and thus give rise to kinetic currents.

Another reduction product, hydrogen peroxide, which is generated during the reduction of oxygen, has received more attention in this respect. Normally, this hydrogen peroxide is further reduced to hydroxyl ions at a considerably more negative potential than that at which it was formed. However, certain iron-containing biological substances, such as hemoglobin and heme, even in very dilute solution catalyze this reaction so that it takes place at a more positive potential [61]. One may therefore see on the polarograms three waves during the reduction of oxygen. The first one, diffusion controlled and of the same height as the sum of the second and third waves, is caused by the reduction of oxygen to hydrogen peroxide. The second wave, a kinetic current, grows at the expense of the third and depends on the amount of catalyst present. The third wave represents the remaining uncatalyzed reduction of hydrogen peroxide to hydroxyl ions.

If catalase is used, the kinetic wave is superimposed on the first wave which increases at the expense of the second [62]. Brdička et al. [63] think that here the kinetic reaction in some way involves regeneration of oxygen which is once more reduced to give the increased current. This is not likely, however, because the second wave would then also be higher.

Reactions of this sort may also be studied in oxygen-free solutions to which hydrogen peroxide has been added. In this way it was shown that ferric iron [55] and other inorganic systems [64] also catalyze the reduction of hydrogen peroxide. An interesting additional finding [65] was that in the presence of acrylonitrile the kinetic current was reduced by one-half because this compound interacts with the hydroxyl free radicals and polymerizes. The formation of such radicals at the dropping mercury electrode was thus indicated.

If the hydrogen peroxide concentration in the solution is increased sufficiently, any kinetic currents observed will no longer be limited by its diffusion. They will then fall into the category of catalytic currents discussed in the next section.

Catalytic Current (Peak Current)

General

The kinetic currents previously discussed were all limited in some way by diffusion of substances present in relatively small concentration. The currents considered in this section differ in that the reacting material is present in such high concentration that it is not a limiting factor. The reaction is brought about by the presence of relatively small quantities of catalysts which are probably adsorbed at the electrode surface. A distinctive feature of these catalysts is that they react only over a limited range of potential so that the current-voltage curves show peak currents resembling maxima which, however, are not suppressed by ordinary maximum suppressors. The peaks are more-or-less rounded, depending on how close several catalytic currents come together or how far the catalytic current is from a diffusion current.

This often makes measurements of such peak currents difficult. One should strive to measure the current between peak and preceding minimum values, but in successive peaks or other overlaps this may be impossible. In such a case, one measures the second peak value from the minimum preceding the first peak or on the basis of a blank curve, but neither method is entirely reliable.

The height of the catalytic peaks is not a linear function of the concentration of the catalyst, but seems to depend on its adsorption. Hence concentration-peak-height relationships resemble Langmuir's adsorption isotherms (see Fig. 5.2) [66]. The peak heights are proportional to the surface area of the mercury drops and therefore independent of the drop time of a given capillary. When quantitative relationships are compared, the results are preferentially plotted as inverse functions, resulting in straight lines as shown in Fig. 5.3.

The reason for the peaks is not yet known with certainty. In general, it is agreed that they involve a lowering of the hydrogen overvoltage on mercury. This means that they are produced by the reduction of hydrogen ions from relatively concentrated buffers and that this is catalyzed, or shifted to a more positive potential, by the presence of organic or inorganic substances or their complexes. If simple surface catalysis is occurring, the peak currents may terminate abruptly at a potential at which the catalyst is desorbed. Such catalytic maxima are often accompanied by streaming around the electrode so that they resemble convection currents. However, many of these catalytic currents are less abrupt, suggesting a more gradual disappearance of the catalyst either by desorption or by electrochemical removal (reduction) starting at the potential at which the maximum catalytic current is reached.

Cobalt-Catalyzed Current

Certain catalytic currents produced by various proteins and sulfhydryl compounds in buffered cobalt (or nickel) solutions have become of special interest because of their importance in biological systems. An example is given in Fig. 5.14 [67]. This illustrates a typical protein double-wave (actually

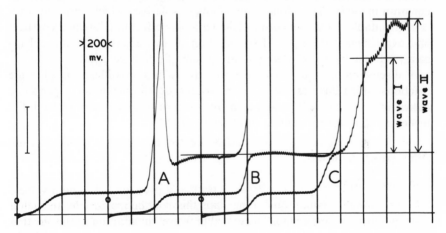

Fig. 5.14 Polarogram demonstrating Brdička reaction [67]. (A) 0.001 M Co (NH$_3$)$_6$Cl$_3$, 0.1 N NH$_4$Cl, and 0.1 N NH$_4$OH; (B) solution A, containing 0.02% caffeine; (C) solution A, containing 0.1 ml dog plasma. All solutions had a pH of 9.3 and were deaerated with nitrogen gas for 5 min. Short vertical line at left indicates 10 μA. Method of measurement of proteins waves I and II is shown on curve C.

two peaks) in curve C and indicates how they are measured. In the absence of protein, the same solution produces a marked cobalt maximum of the first kind (curve A) which can be suppressed by caffeine (curve B). Suppression of this maximum by means of added serum led Brdička to the discovery of the catalytic reaction which bears his name. Cysteine and cystine react similarly but produce only a single wave (peak) on the polarogram. Brdička found that the catalyzed wave in the case of cystine is from 100 to 500 times as large as that obtained in the normal reduction of cystine in the absence of metals [68]. He thinks of it as a catalytic reduction of hydrogen, loosened from the sulfur atom by a coordination bond between the sulfhydryl group and the cobalt. This idea is substantiated by the existence of stable complexes between cobalt and sulfhydryl-containing compounds. Furthermore, cystine and cysteine produce a catalytic wave only in the presence of divalent cobalt, while sulfhydryl-containing compounds of higher molecular weight produce similar catalytic waves with di- or trivalent cobalt. However, an increasing number of sulfur-free compounds has been found to produce similar cobalt-catalyzed currents which promise to help elucidate this interesting polarographic current [67]. One of these, sodium nitrohydroxylaminate, has been particularly suitable since the polarographic waves can be analyzed in terms of their individual components: (1) the current caused by the ammoniacal complex of cobalt, (2) the current caused by the catalytically active complex, (3) the catalytic current (H^+ reduction) itself, and (4) the current resulting from the reduction of the basic electrolyte (uncatalyzed H^+ reduction). An equation involving these four components was derived which upon substitution of suitable parameters produced an almost perfect fit between theoretical and experimental data [69]. Extrapolation of this equation to the cobalt-catalyzed cysteine wave produced an excellent fit even here [70].

A discussion of the extensive applications and modifications of these reactions which have led to some interesting developments in the fields of biochemistry, enzymology, and medicine, cannot be given here. Instead, the reader is referred to a number of reviews [67, 71].

Adsorption Current

Anomalous Waves

Under suitable conditions some reversible organic oxidation-reduction systems show an "anomalous" wave or "prewave" in addition to the one that might be expected on the basis of potentiometric data. This was first demonstrated by Müller and Baumberger in 1937 for rosinduline GG [13] and was subsequently investigated in more detail with other systems [36, 72]. The anomalous wave is the only wave existing at the lowest concentrations of rosinduline GG and it grows to a limiting height at about 10^{-4} M (Fig. 5.15).

As this concentration is exceeded, another more negative wave appears which grows with further increases in concentration. This second wave has a half-wave potential that agrees with the potentiometrically determined oxidation-reduction potential. The sum of the two waves fulfills the requirements of the Ilkovič equation but the anomalous wave does so only when the normal wave is absent. Otherwise, its height becomes a function of the rate of growth of the surface of the dropping mercury electrode, or of $m^{2/3}t_d^{-1/3}$. This can be written as $(m^{2/3}t_d^{2/3})t_d^{-1}$, indicating that the anomalous wave is an inverse function of t_d (since mt_d is constant for a given capillary) or a linear function of the effective mercury pressure on the capillary P. It can thus be readily distinguished from a diffusion current.

The anomalous wave is not the result of an impurity and appears on the polarograms whether one starts with the oxidized or reduced form of the oxidation-reduction system (see bottom of Fig. 5.15). It is known that definite changes in the absorption spectra of these dyes occur at the same concentration range at which the limiting current of the anomalous wave is found. Hence Müller [72] suggested as a common cause an unusually high "activity" of the dye in dilute solutions which is demonstrated by the available "freshest" surface of the dropping mercury electrode. Brdička, investigating similar prewaves of methylene blue [73], concluded that they are caused by adsorption of the reduction product at the electrode surface and that the energy set free by this adsorption is responsible for the greater ease of reduction (more positive potential) [74]. Accordingly, he developed an equation, analogous to the Langmuir adsorption isotherm (Fig. 5.2), in which the prewave was proportional to $dA/dt \cdot z \cdot \omega_{red}[red]_i/(1 + \omega_{red}[red]_i)$. Here dA/dt is the rate of growth of the electrode surface, z is the maximum number of moles adsorbed per unit of electrode surface, and ω_{red} is the adsorption coefficient of the reductant which has the average concentration $[red]_i$ at the electrode during the life of a drop. There was excellent agreement between the experimental data and this equation when proper values were substituted. The anomalous wave has therefore also been called *adsorption current*, a term that should not be confused with the convection current maxima (see Section 4, p. 328) to which it was formerly applied. In agreement with Brdička's theory is the fact that the adsorption is limited by the available space for which several systems may compete. An interesting application of this idea to the calculation of the number of molecules adsorbed per unit area during reduction of phenylmercuric hydroxide was made by Benesch and Benesch [22].

Although general maximum suppressors often have no effect on anomalous waves, marked blocking may occur over specific voltage ranges under certain conditions, which are by no means clearly understood. One example is the abolition of the anomalous wave of indigo trisulfonate by eosine, which

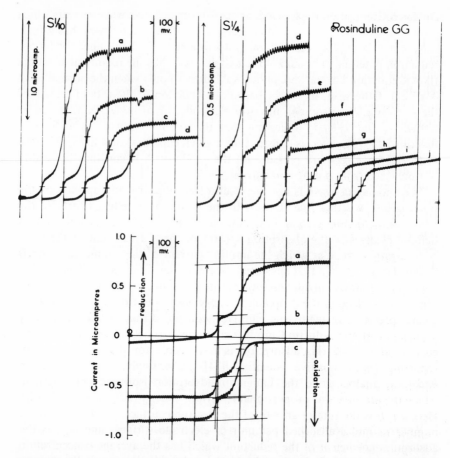

Fig. 5.15 Curves of rosinduline GG in a phosphate buffer of pH 6.8, showing a typical anomalous wave. Upper polarogram: concentrations $a - j$: 7.65, 5.10, 3.83, 3.06, 2.19, 1.70, 1.13, 0.85, 0.68, 0.57 × 10^{-4} M. Lower polarogram: concentration 3.63 × 10^{-4} M; (a) fully oxidized, (b) partly reduced, and (c) fully reduced with hydrogen gas in the presence of platinized asbestos [72].

apparently has only a very small depressing effect on the main reduction wave of the indigo trisulfonate [75]. Another example is the blockage of the reduction of methylene blue near the limit of its anomalous wave by dissolved oxygen for a range of about 300 mV when nitrate ions are also present [76]. This blockage was effective against certain organic and inorganic reductions as well, but it was ineffective against all reversible organic oxidation-reduction systems tested. If film formation is involved, such a film must have very peculiar properties.

Diffusion Current Blocked by Adsorption

It has been stated earlier (Section 4, p. 332) that the addition of surface-active materials for the suppression of maxima has to be made cautiously since too much maximum suppressor also suppresses the diffusion current. The first specific study of this effect was made by Wiesner with eosine additions to several reversible oxidation-reduction systems [75]. Its quantitative analysis was carried out by Koryta [77], who made an important basic assumption: The adsorption is instantaneous so that the adsorbed material generates a similar diffusion gradient as if it had been removed by the electric current. Hence the Ilkovič equation is applicable also for the supply of adsorbate to the dropping mercury electrode. Koryta postulated that during the growth of a drop there is room on its surface for $0.85m^{2/3}t^{2/3}\Gamma_m$ moles of the adsorbate, where Γ_m is its maximum surface concentration in moles per square centimeter. If the concentration of the adsorbate in solution is sufficient to create a high enough diffusion rate, the surface will be covered, before the drop falls, at the instant in time δ defined by the following equation

$$0.85m^{2/3}\delta^{2/3}\Gamma_m = 0.627 \times 10^{-3}C_aD_a^{1/2}m^{2/3}\delta^{7/6}, \tag{5.41}$$

where C_a and D_a are the concentration and diffusion coefficient of the adsorbate. Hence

$$\Gamma_m = 7.38 \times 10^{-4}C_aD_a^{1/2}\delta^{1/2} \tag{5.42}$$

and

$$\delta = 1.84 \times 10^6\Gamma_m^2D_a^{-1}C_a^{-2}. \tag{5.43}$$

When the surface is completely covered, any further reduction of reducible material is blocked so that the recorded current is diminished to a new limit, I_l. The ratio of this to the true diffusion current I_d is then given by

$$I_l/I_d = 2 \times 10^7D_a^{-7/6}\Gamma_m^{7/3}t_d^{-7/6}C_a^{-7/3}. \tag{5.44}$$

If $\delta = > t_d$, there should be no interference with I_d and the relative coverage: Γ/Γ_m is equal to $\sqrt{t_d/\delta}$. Hence fast dropping mercury electrodes have less interference from this adsorption than electrodes with long drop times. When the current is limited by the blockage of an adsorbate, it behaves as an "adsorption" current and is therefore directly proportional to P. These theoretical deductions have shown very good agreement with experimental data [77]. Various mechanisms may operate to produce these phenomena, which can be further complicated by irreversibility of the reaction and slow adsorption [78]. Details are best studied by means of current-time curves on single drops (see Chapter IIC in original plan by Adams and Piekarski).

5 EFFECT OF TEMPERATURE ON POLAROGRAPHIC CURRENTS

It has been found experimentally that the diffusion currents of most ions and molecules increase by about 1 to 2% per degree between 20 and 50°C. Thus one can conclude that a temperature control to at least ±0.5°C is necessary to keep variations in the diffusion current below 1%.

In the case of kinetic currents, the temperature coefficients may become much greater and they depend on pH. One may even approach the well-known rule that a reaction rate doubles with a temperature increase of 10°C. However, this is not always the case; the kinetic current of pyruvic acid has a temperature coefficient similar to that of a diffusion current. The catalytic currents likewise may have a very high or a very low temperature coefficient.

The only waves relatively unaffected by temperature are the anomalous waves (the adsorption current).

6 POTENTIAL OF THE DROPPING MERCURY ELECTRODE IN REVERSIBLE OXIDATIONS AND REDUCTIONS

General

As stated in the introduction, the potential to which the dropping mercury electrode must be polarized in order to produce an electrolysis serves for qualitative analysis. As long as the solution contains only one electroactive component, this analysis is simple; it is still possible when the solution contains several components if each is reducible at a potential sufficiently different from the other. However, as soon as the reduction potentials of two or more components come so close together that their waves on the current-voltage curves overlap, the analysis becomes difficult. Sometimes the reduction potentials can be sufficiently separated if they are altered to different degrees by changing the pH of the medium or by forming complexes in solution, but more often a preliminary chemical separation of the components into suitable fractions is necessary before an analysis is possible. The latter is especially true in the case of organic compounds.

As explained in connection with Fig. 5.5, at any point on the S-shaped portion (the wave) of the curve, the potential of the electrode indicates the composition of the electrode-solution interface. Before the significance of the polarographic potentials was fully understood, arbitrary points on the current-voltage curve (e.g., a 45° tangent) were chosen to characterize reducible substances and thus make the polarographic method suitable for qualitative analysis. These have become obsolete since 1935 when Heyrovský and Ilkovič [79] clarified the significance of the polarographic wave and suggested the *half-wave potential* for the characterization of the reacting material. The half-wave potential is independent of the concentration of the reacting material, the drop time of the electrode, and the galvanometer sensitivity. It is, furthermore, closely related to the potentials obtained in potentiometric studies, as was first demonstrated by Müller and Baumberger [13]. Although its major value is for reversible reactions at the dropping mercury electrode, it is widely used even for irreversible reactions.

Oxidations and Reductions in Well-Buffered Solutions

The polarographic wave in a thermodynamically reversible reaction is a perfect equivalent of a potentiometric oxidation-reduction titration curve, as may be seen by inspection of Fig. 5.16. Yet both types of curves signify different conditions [80]. The potentiometric titration curve, *b*, indicates the alteration of a solution brought about by the addition of a suitable oxidant or reductant. The polarographic curve, *a*, on the other hand, indicates a change in the composition of the interface brought about by the addition or removal of electrons in an essentially unaltered solution. The polarographic wave may therefore be considered as representing an *electron titration of the*

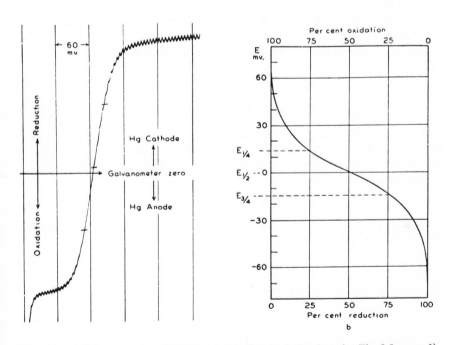

Fig. 5.16 (*a*) Polarogram of quinhydrone in a buffered solution (see also Fig. 5.5, curve 3) representing a direct electron titration of the electrode-solution interface, while body of solution remains practically unchanged. (*b*) Typical potentiometric oxidative or reductive titration curve representing changes throughout body of solution [80].

interface, where the oxidizing or reducing agent is the electrode itself, removing electrons from or adding them to the electroactive material in the interface [80]. Just as potentiometric potentials are defined by the relative concentrations of oxidant and reductant *in solution*, so the polarographic potentials are defined by the relative concentration of oxidant and reductant *at the interface*. Consequently, the previously developed equations (5.8) and (5.10) may be applied to polarographic curves by substituting for the concentrations existing in the body of the solution those existing at the interface, indicated by the subscript i as follows:

$$E_{DME} = E_0 + RT/nF \ln [\text{ox}]_i/[\text{red}]_i. \tag{5.45}$$

When current is governed only by the rate of diffusion of a reactant, it is a function of the difference in concentration between the interface and the body of the solution. In the case of a reduction, the current, by definition, is labeled positive and can be expressed as

$$I = kD_{\text{ox}}^{1/2}([\text{ox}] - [\text{ox}]_i),$$

where I is the current, D_{ox} the diffusion coefficient of the oxidant, [ox] the concentration in the body of the solution, and $[\text{ox}]_i$ the concentration at the interface. The constant k is given by the Ilkovič equation. When the cathodic diffusion current reaches its limit, $[\text{ox}]_i$ becomes negligible and

$$I_{dc} = kD_{\text{ox}}^{1/2}[\text{ox}].$$

Therefore

$$[\text{ox}]_i = (I_{dc} - I)/kD_{\text{ox}}^{1/2}. \tag{5.46}$$

Conversely, the concentration of reductant at the interface is equal to that existing in the body of the solution plus that fraction of the additional reductant formed by the current that has not diffused away. This can be expressed as

$$[\text{red}]_i = [\text{red}] + I/kD_{\text{red}}^{1/2}$$

The negative anodic diffusion current $-I_{da}$ is reached when the concentration of reductant at the interface is zero, or:

$$-I_{da} = kD_{\text{red}}^{1/2}[\text{red}].$$

A combination of these last two equations yields:

$$[\text{red}]_i = (I - I_{da})/kD_{\text{red}}^{1/2}. \tag{5.47}$$

Substitution of (5.46) and (5.47) into (5.45) leads to

$$E = E_0 + \frac{RT}{nF} \ln \frac{(I_{dc} - I)}{(I - I_{da})} + \frac{RT}{nF} \ln \frac{D_{\text{red}}^{1/2}}{D_{\text{ox}}^{1/2}}, \tag{5.48}$$

which is the most general equation for the polarographic wave involving reversible reactions. It holds for anodic waves, where $I_{dc} = 0$ and I is negative, and for cathodic waves, where $I_{da} = 0$ and I is positive. In addition, it is applicable to any single wave in which the current is partly anodic and partly cathodic, that is, where I_{dc} and I_{da} have values other than zero.

By definition, the half-wave potential corresponds to the midpoint of the wave, that is, where $I = (I_{dc} + I_{da})/2$. When this value is substituted for I in (5.48), the middle term drops out:

$$E_{1/2} = E_0 + RT/nF \ln D_{red}^{1/2}/D_{ox}^{1/2}. \tag{5.49}$$

The last factor in this equation is a constant for any given case and is usually negligibly small. Equation (5.49) then means that the half-wave potential, whether measured on a cathodic wave or an anodic wave, or on a wave that is partly cathodic and partly anodic, should be equal to the normal electrode potential of the reacting system.

This conclusion is of great importance because it serves as a *criterion of reversibility in polarography:* A substance is polarographed once when it is present in the oxidized form and once when it is present in the reduced form. If the system is reversible, the half-wave potentials of the cathodic and anodic waves obtained are identical and equal to the potentiometrically determined normal potential of the same system [13].

This is illustrated by the hand-drawn polarogram shown in Fig. 5.17, obtained with the simplest of equipment [12]. It was prepared in three stages: (1) Curve *a* was obtained first with a solution of quinone in a McIlvaine buffer of pH 7 purged of air by means of nitrogen gas; this wave is all cathodic. (2) The quinone was then reduced by bubbling hydrogen gas through the solution in which platinized asbestos had been present throughout; after the chemical reduction had proceeded for a short time, curve *b* was prepared, showing that about one-half of the quinone has been reduced to hydroquinone. This wave is now partly anodic and partly cathodic. (3) The reduction by means of hydrogen gas was then continued until all of the remaining quinone was changed into hydroquinone; at this stage the hydrogen was replaced by nitrogen gas and curve c was obtained which is entirely anodic. All three curves have been corrected for IR; the half-wave potentials, indicated on the polarogram, are identical.

The wave heights of the quinone and hydroquinone curves of Fig. 5.17 are in a ratio of 15.3 to 14.1. If none of the quinone was destroyed during the reducing process and if adsorption of oxidant and reductant on the platinized asbestos was equal, this ratio must express the ratio of the square roots of the diffusion coefficients of quinone and hydroquinone. Hence according to (5.49) the polarographic half-wave potential and the potentiometrically determined potential of quinhydrone in the same buffer should differ by about

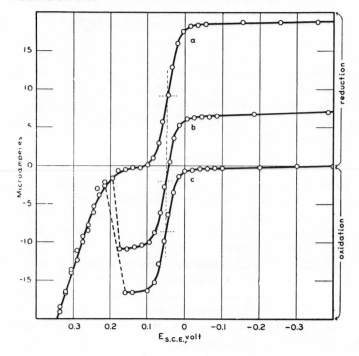

Fig. 5.17 Current-potential curves obtained at pH 7 [12]. (*a*) quinone; (*b*) quinone partly reduced, and (*c*) quinone fully reduced by hydrogen gas in the presence of platinized asbestos.

1 mV. This difference is so small that it falls within the experimental error of the measurements.

For a complete analysis of current-potential curves, it is customary to plot the logarithmic factor of the middle term of (5.48) (converted to common logarithms) against the electrode potential. A straight line should result, with a slope of 2.3 RT/nF, from which can be ascertained the number of electrons involved in the reaction. At 30°C the slope is equal to 0.0601 V if n is unity; it is twice as steep, or 0.030 V, if $n = 2$. The curves in Fig. 5.17 were obtained at 30°C. The corresponding graph of $\log (I_{dc} - I)/(I - I_{da})$ versus the potential of the dropping mercury electrode, E, shown in Fig. 5.18 gives the expected straight line with the theoretically correct slope of 0.030 V.

The evaluation of the continuous anodic-cathodic curve b of Fig. 5.17 is tedious because the current changes sign and two different diffusion currents have to be considered. Since it is much simpler and gives identical results for purposes of this graph, the writer has in the past treated such a polarographic wave as completely cathodic; in other words, the current I was considered

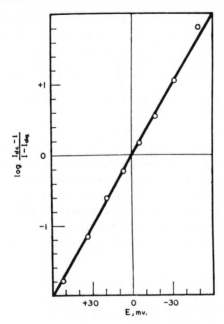

Fig. 5.18 Graph of log $(I_{dc} - I)/(I - I_{da})$ versus the electrode potential [12]. Data from Fig. 5.17.

equal to zero at the start or bottom of the wave and equal to I_d at the end or top of the wave. Equation (5.48) then takes on the simpler form:

$$E = E_0 + RT/nF \ln (I_d - I)/I, \tag{5.50}$$

where I and I_d are always positive. In this case log $(I_d - I)/I$, plotted against the potentials E, gives the same straight line as above.

The results obtained with the quinone-hydroquinone system were duplicated with other reversible organic oxidation-reduction systems and eventually also with reversible inorganic systems and certain dropping amalgam electrodes. The identity of polarographic and potentiometric potentials demonstrates that all rates involved in the actual electrode reaction are so fast that equilibrium is established in spite of the fact that considerable current is flowing as the mercury drops grow to their full size. Thus the above examples serve as *ideal* patterns which will be useful for achieving an understanding of any deviations from this ideal behavior.

Generalizing the quinone-hydroquinone system, one has to consider that the reaction actually involves the addition of two electrons to the oxidant R to change it into the corresponding reductant R^{2-}. Here R^{2-} represents a divalent ion, or the completely ionized form of a dibasic acid. Its concentration is therefore subject to the hydrogen ion concentration in the solution, which governs the degree of ionization of the acid according to the equations:

$$Ka_1 = [H^+](RH^-]/[RH_2] \quad \text{and} \quad Ka_2 = [H^+](R^{2-}]/[RH^-], \quad (5.51)$$

in which Ka_1 and Ka_2 are the two dissociation constants of the acid. Equation (5.45) could thus be expressed in terms of R, RH_2, H^+, Ka_1, and Ka_2, but the new equation would still be impractical because the concentrations of R^{2-} or RH_2 cannot be directly determined. Clark and Cohen [81] have shown that these difficulties can be overcome if the equation is expressed in terms that represent the *sum* of all the forms in which an oxidant or a reductant can appear. These summation terms are designated by capital letters, Ox, and Red; in the above case this means that $RH_2 + RH^- + R^{2-} = $ Red, and R = Ox. It is thus possible to substitute for $[ox]_i$ and $[red]_i$ in (5.45) the corresponding values of $[R]_i$ and $[R^{2-}]_i$ as defined by the summation terms and (5.51):

$$E = E_0 + RT/2F \ln [Ox]_i/[Red]_i + RT/2F \ln (Ka_1Ka_2 + Ka_1[H^+]_i + [H^+]_i^2)$$

$$(5.52)$$

Assuming as a first approximation that the diffusion coefficients of Ox and Red are those of a single species and that their ratio is near unity at any pH, one can derive the equation:

$$E = E_0 + \frac{RT}{2F} \ln \frac{(I_{dc} - I)}{(I - I_{da})} + \frac{RT}{2F} \ln (Ka_1Ka_2 + Ka_1[H^+]_i + [H^+]_i^2) \quad (5.53)$$

If the pH of the interface is kept constant by a suitable buffer, the last term in this equation is also constant and can be incorporated in the E_0 term:

$$E = E_0' + RT/2F \ln (I_{dc} - I)/(I - I_{da}). \quad (5.54)$$

At the half-wave potential, the logarithmic term is zero and $E_{1/2} = E_0'$. That this is true has already been demonstrated in Fig. 5.17.

Clark and Cohen [81], in their potentiometric studies, have shown that it is possible to determine the dissociation constants of the reductant or oxidant from a graph in which E_0' is plotted against pH. In potentiometry this is somewhat tedious because the concentrations of Ox and Red have to be made equal to obtain $E = E_0'$ [see (5.52)]. In polarography such graphs are much simpler to prepare, since $E_{1/2} = E_0'$, regardless of the relative concentrations of Ox and Red in the body of the solution. It must be borne in mind, however, that the potential measurements in polarography are as a rule much less accurate than in potentiometry. Since they give much needed information and are relatively simple to obtain polarographically, the preparation of pH-potential curves over as large a pH range as possible is recommended strongly.

Stepwise Oxidations and Reductions in Well-Buffered Solutions (Semiquinone Formation)

In the organic reactions so far considered, the transfer of the two electrons was assumed to be simultaneous. This assumption was correct since the experimental data fitted the theoretical titration curves constructed on the basis that $n = 2$. However, a number of compounds have been found which in acid solution add the electrons in two separate steps, forming first an intermediate radical with an odd number of electrons, called a semiquinone [82]. It is possible that all organic reductions and oxidations of reversible systems take place by way of semiquinones, but in most of them the semiquinones are so unstable that their existence cannot be demonstrated.

A good example of the effect of semiquinones on polarographic curves is given by α-oxyphenazine [36, 83]. Figure 5.19 clearly illustrates that a single

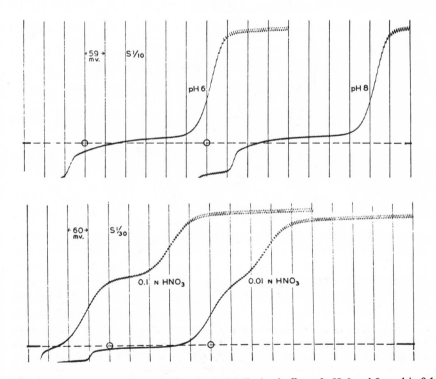

Fig. 5.19 Polarogram of α-oxyphenazine in McIlvaine buffers of pH 6 and 8, and in 0.1 and 0.01 N nitric acid solutions, showing two-step reduction in the acid solutions [83].

wave is obtained at pH 6 or 8, while in acid solution two distinct waves of equal height appear which are more widely separated the stronger the acid. The half-wave potentials (or the $E_{1/4}$ and $E_{3/4}$ potentials during semiquinone formation) obtained when the concentration of α-oxyphenazine was 5 × 10^{-4} M or greater are plotted against pH in Fig. 5.20. As may be seen, they show excellent agreement with data obtained potentiometrically. In more dilute solutions of α-oxyphenazine, however, an anomalous wave appears at a more positive potential.

The following equation for the polarographic potential takes into account the formation of semiquinone:

$$E = E_0' + \frac{RT}{2F} \ln \frac{(I_d - I)}{I}$$

$$+ \frac{RT}{2F} \ln \frac{\sqrt{4K_d I_d^2 - (4K_d - I)(I_d - 2I)^2} + (I_d - 2I)}{\sqrt{4K_d I_d^2 - (4K_d - I)(I_d - 2I)^2} - (I_d - 2I)}, \quad (5.55)$$

where K_d is the dismutation constant of the process:

$$2\,\text{Sem} \rightleftharpoons \text{Ox} + \text{Red} \qquad (5.56)$$

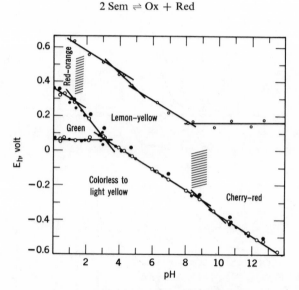

Fig. 5.20 Comparison of potentiometric and polarographic data for α-oxyphenazine. ●, Potentials determined potentiometrically. ○, Polarographic half-wave potentials obtained with 5 × 10^{-4} and 10^{-3} M solutions (main wave). ◐, Polarographic half-wave potentials of anomalous wave, obtained with 10^{-4} M and less concentrated solutions. ○, Polarographic half-wave potentials of small anodic waves, the significance of which is still unknown [36].

Anomalous Waves of Oxidation-Reduction Systems

Anomalous waves (see Section 4, p. 349) have been observed mainly with reversible systems such as phenazines, thiazines, oxazines, riboflavin and indigo disulfonate [72]. The difference between their half-wave potentials and that of the normal wave varies from system to system and depends on the pH in any given system. For instance, the pH-half-wave potential relationship of α-oxyphenazine [83] from pH 3 to 13 is linear for the anomalous wave, whereas several points of inflection and different slopes are found for the normal wave (see Fig. 5.20). So far, these differences have remained unexplained.

It should be mentioned that anomalous waves are not found exclusively in reversible systems. For instance, anomalous waves have been found in the undoubtedly irreversible reductions of colchicine [84] and phenylmercuric hydroxide [22].

Reversible Oxidations and Reductions in Unbuffered or Poorly Buffered Solutions

A consideration of the equation (5.53) for organic substances makes it obvious that the half-wave potential is only significant if the pH at the interface is known and constant, which means that adequate buffering is essential. It is therefore of practical importance, as well as of theoretical interest, to be able to recognize the effect of inadequate buffering on the shape of the current-voltage curve. This problem has been investigated in detail by Müller who used solutions of quinhydrone to which increasing amounts of suitable buffers were added [85].

Figure 5.21 may serve as an example. Here, to a solution of 0.1 N potassium nitrate (curve a) was added quinhydrone. This resulted in the unbuffered curve of quinhydrone consisting of an anodic wave b widely separated from the cathodic wave c. This separation is brought about by the fact that hydrogen ions are produced during oxidation of hydroquinone and are consumed during the reduction of quinone, so that the pH at the electrode interface is very much lower at b than at c in this unbuffered solution. If sufficient buffer had been present, a single continuous anodic-cathodic wave (as a in Fig. 5.16 and b in Fig. 5.17) would have been obtained.

Next, a phosphate buffer, composed of equal amounts of its acid and basic forms (NaH_2PO_4 and Na_2HPO_4) was added, but in a quantity insufficient to buffer all the reacting quinhydrone. This resulted in a curve with three waves (d-e-f). The unbuffered waves d and f are no longer as far apart as a and b because the unbuffered portion of the quinhydrone is smaller than before. The middle wave c illustrates the action of the buffer; it is divided equally between the reduction of quinone and the oxidation of hydroquinone. If either the acid or the basic component of the buffer is added, either the

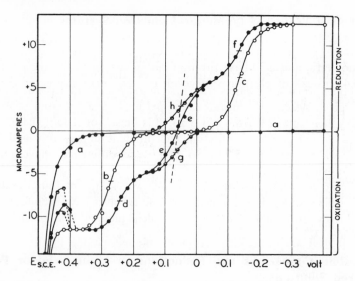

Fig. 5.21   Unbuffered and poorly buffered curves of quinhydrone showing the quantitative analysis of a nonelectroactive substance by means of its buffer action [86] (see text).

cathodic and anodic portions of the reaction are buffered. For instance, when only the disodium phosphate half of the buffer was added, a curve with three waves (*d-g-c*) resulted; obviously, the reduction of quinone was completely unbuffered. However, when only the dihydrogen phosphate half of the buffer was added, the curve with three waves (*b-h-f*) resulted and the oxidation of hydroquinone was completely unbuffered.

One can conclude that the middle wave *e* of Fig. 5.21, which can only be caused by the action of the buffer must represent an *acid-base titration of the interface* [80], the magnitude of which is controlled by the amount of buffer diffusing to the interface. Thus results a diffusion current of a substance which is itself not reducible or oxidizable. Wave *e* is almost identical with a potentiometric acid-base titration curve, except that a small change in the ratio of reductant/oxidant, that is $(I - I_{da})/(I_{dc} - I)$, has been superimposed on it. (This can be minimized by making the quinhydrone concentration as large as possible.) The titrating agents are either free hydrogen ions formed during the anodic oxidation of hydroquinone or hydroquinone anions, acting as a strong base, formed during the cathodic reduction of quinone. The disodium phosphate anion (HPO_4^{2-}) is titrated by the former, while the acid phosphate anion ($H_2PO_4^{-}$) is titrated by the latter. The whole of the wave *e* is therefore proportional to the total concentration of buffer, and it represents an *indirect method of quantitative analysis of a nonelectroactive substance by means of its buffer action* [86, 87]. This method should be useful for the determination of pK values of acids (equal to the midpoint of the buffer waves).

Waves e, g, and h of Fig. 5.21 also demonstrate the similarity between potentiometric and acid-base titration curves discussed in connection with Fig. 5.4. Instead of plotting voltage on the abscissa, one could substitute a pH scale with each 59 mV (at 25°C) representing 1 pH unit for these three waves. However, this pH scale would have to be reversed in sign since here one does not plot the protonated product x, as in Fig. 5.4. Instead, the cathodic part of the wave represents the dissociation of the acid component ($H_2PO_4^-$), and the anodic portion represents the protonation of the basic component (HPO_4^{2-}) of the buffer (see also Fig. 5.24).

In calculations of the pH existing at the mercury-solution interface during the reaction of unbuffered solutions of quinhydrone of different concentrations, constant discrepancies were found between expected and observed values [85]. In the case of anodic waves, the observed pH was 0.5 pH unit larger, and in cathodic waves the observed pH was 1.0 unit smaller than the expected values. According to Müller [85], the most likely cause for the apparent loss in acidity and basicity is the diffusion of hydrogen and hydroxyl ions away from the electrode, since both have mobilities exceeding those of all other ions. Thus the discrepancy found in the anodic waves should be about twice as large as that of the cathodic waves, while actually the opposite is the case. The cause of these discrepancies was in doubt until Kolthoff and Orlemann [88] proved conclusively that at least in the oxidation of hydroquinone the above hypothesis was correct. By including the diffusion coefficient of hydrogen ions in their considerations, these authors were able to account quantitatively for the discrepancy of 0.5 pH unit found at the half-wave potential of the hydroquinone oxidation in unbuffered solution. Still unexplained is the abnormally high discrepancy in the case of the cathodic waves.

Equation (5.53) can be modified as follows, to hold at least for the *anodic* portion of unbuffered waves:

$$E = E_0 + \frac{RT}{2F} \ln \frac{(I_{dc} - I)}{(I - I_{da})} + \frac{RT}{F} \ln \frac{2I}{I_{da}} + \frac{RT}{F} \ln [\text{Red}] + \frac{RT}{F} \ln \frac{D_{\text{Red}}^{1/2}}{D_{\text{H}^+}^{1/2}}.$$

$$(5.57)$$

From this equation are omitted: (1) the hydrogen ion concentration of the body of the solution, (2) the dissociation constants Ka_1 and Ka_2, and (3) the factor involving the ratio of the diffusion coefficients of oxidant and reductant, because in unbuffered anodic waves all these terms are negligibly small.

It is obvious from the foregoing that adequate buffering is essential if a constant and significant half-wave potential is to be obtained in a reversible reaction at the dropping mercury electrode. This should be kept in mind whenever the constancy of the half-wave potential is doubted during changing concentration of the reacting material. In general, a 100-fold concentration of buffer over that of the reacting material assures adequate buffering [89].

Deviations from Ideal Curves of Reversible Systems Caused by Insolubility or Instability of Products

It has been assumed in the foregoing discussion that the two components of an oxidation-reduction system are both sufficiently soluble and stable to produce well-defined anodic and cathodic waves. This assumption was justified in dilute solutions of most systems studied. In more concentrated solutions, however, after chemical reduction of the oxidized form of a system, part of the reduced compound, for example, the leuko form of a dye, may precipitate out because it is less soluble than the oxidized form, or may form a nonreactive complex with the medium in which it is dissolved. The anodic wave will therefore be considerably smaller than the corresponding cathodic wave or may be completely absent. Methylene blue in a phosphate buffer [72] may serve as an example of this.

If a compound is stable in the oxidized form, but very unstable in the reduced form, it may give an ideal cathodic wave, every point of which fits the theoretical equation for a reversible reaction, but it will not give the corresponding anodic wave. If the reduction product decomposes slowly, its presence may still be revealed by a quick polarographic analysis after chemical reduction. The presence of an anodic wave, although smaller than the cathodic wave, but with the same half-wave potential, may serve as assurance that the primary electrode reaction is reversible. Adrenochrome can serve as an example of this [90]. When this fails, one can study the reversibility of the electrochemically formed product by means of a fast reversal of the electrode process. For this purpose a switching circuit has been designed by Kalousek [91] which is discussed in more detail in Section 10. Many reactions whose reversibility was questionable have been found to be reversible by this technique.

Effects of Ionic Strength and Temperature on the Half-Wave Potential

Ionic Strength

The half-wave potential depends on the activity of the components of the oxidation-reduction system in solution, rather than on their concentration. So far, the assumption has been made that the two are practically identical in the dilute solutions used so that simplified equations could be derived.

As different buffers were used and refinements in the potential determinations were worked out, it became apparent that the effect of ionic strength on the activities of the reacting material could no longer be neglected.

Early studies by Lingane [92] on several metallic ions showed that the potential became more negative by as much as 43 mV as the concentration of indifferent electrolyte was increased 10-fold. Elving et al. [93] were the first to carry out a systematic study of α-bromo-n-butyric acid over the pH range of 1 to 12 and a range of 0.1 to 3 M of ionic strength, using many of the commonly encountered buffer systems. The ionic strength μ was calculated on the basis of the concentration M (in moles per liter) of the buffer components at a given pH (obtained by reference to its pK):

$$\mu = \tfrac{1}{2} \sum M z^2, \tag{5.58}$$

where z is the number of unit charges on the ionized component. The effects of ionic strength varied widely, but there was a region for each buffer in which changes in ionic strength had only a small effect on the half-wave potential. Such regions should prove best for the comparison of results by different investigators.

These effects have demonstrated the need for adequate description in published articles, not only of the pH of the buffer, but also of its composition and concentration.

Tables for acetate buffers of different pH but constant ionic strength have been available for some time [94]. To these has been added a table for McIlvaine buffers of constant ionic strength [95].

Temperature

The effects of temperature on the half-wave potential can be expected to be similar to those known from potentiometry for oxidation-reduction systems. For instance, the potential of quinhydrone is generally given as

$$E_h \text{ (or } E_{1/2}) = 0.7177 - 0.00074T, \tag{5.59}$$

where T is measured in degrees Centigrade. Irreversible organic reductions are affected similarly by an increase in temperature, that is, they become from 0.7 to 1.3 mV more negative per degree rise in temperature.

For the reduction of free hydrogen ions, in which the temperature coefficient is higher (3 mV per degree) and is of opposite sign, the half-wave potential becomes more positive with an increase in temperature [96].

Reversible Reactions Made Apparently Irreversible by Slow Secondary Reactions

The results of dilute buffers (Fig. 5.21) indicate that the diffusion coefficients of the components of the buffer are about the same as those of the

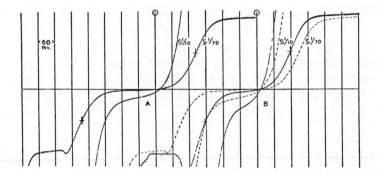

Fig. 5.22   Polarogram of 0.001 M solutions of quinhydrone: (A) 0.1 N potassium nitrate solution; (B) 0.01 M carbonate buffer of pH 7.5. Broken line is curve A superimposed on curve B [85].

components of the oxidation-reduction system. Since equilibrium is established at the interface, the speed with which buffers react to furnish or remove hydrogen ions must have been as fast as the other electrode reactions. Most common buffers fulfill this condition with the notable exception of bicarbonate buffer [85]. As shown in Fig. 5.22, quinhydrone in a bicarbonate buffer of pH 7.5 gives separate anodic and cathodic waves (curve B), similar but not identical with those found in an unbuffered solution (curve A), which means that its buffering action is not fast enough to maintain a constant pH at the electrode interface. Hence this represents an ideally reversible system which, by the effect of secondary reactions, has been made apparently irreversible, that is, oxidation and reduction are not taking place at the same potential.

The buffering action of the bicarbonate buffer is more complex than most, because it involves two equilibria

$$CO_2 + H_2O \rightleftharpoons H_2CO_3 \rightleftharpoons HCO_3^- + H^+ \tag{5.60}$$

which together give the buffer a pK of 6.4. The second of these equilibria is relatively fast with a pK of 3.5 [97], showing that carbonic acid is a fairly strong acid, but the first equilibrium is attained so slowly that in biological systems an enzyme, carbonic anhydrase, is found wherever acceleration of this reaction is of importance.

As can be seen from Fig. 5.23, the separation of anodic and cathodic waves in a quinhydrone solution in bicarbonate buffer (curve A) is completely abolished if sufficient carbonic anhydrase is added (curve E), indicating that a constant pH is now maintained at the electrode interface. This effect can be reversed by the addition of certain enzyme inhibitors [98].

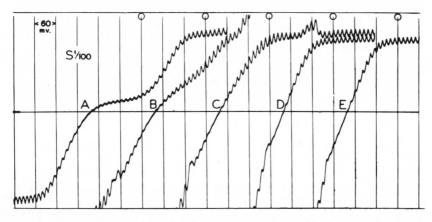

Fig. 5.23 A, 10 ml 0.001 M quinhydrone in 0.01 M sodium bicarbonate solution plus 1 drop caprylic alcohol, equilibrated with carbon dioxide, kept near $0°C$. To this was added, in B, 0.1 ml (115 units), in C, 0.2 ml (230 units), and in D and E, 0.5 ml (575 units) of carbonic anhydrase prepared from dog erythrocytes [98].

Without question, the enzyme has accelerated the first of the equilibria of (5.60), but it appears that it also may have accelerated the second equilibrium.

7 APPLICATION OF THE THEORY OF ABSOLUTE REACTION RATES TO THE EQUATION FOR THE POLAROGRAPHIC WAVE

Electron Transfer

As polarography advanced to the state in which rate-controlled reactions were given major attention, it became necessary to evaluate wave equations in terms of kinetics. It was possible to show that the Nernst equation, which had heretofore been the basis of most wave equations, is obtained as one of the limiting values from the newer kinetic equations.

Eyring et al. [99] were the first to apply the theory of absolute reaction rates to the equation for the polarographic wave. Since that time, other investigators have made similar applications with such success that kinetic studies have become a major field in polarography. The rather involved and lengthy mathematical derivations that have been described [100] are beyond the scope of this chapter. However, the following relatively simple analysis for irreversible and partially reversible cathodic currents should suffice to state the problem and its implications. For simplicity, concentrations (indicated by brackets) of the reagents are used in the equations instead of activities.

In the theory of absolute reaction rates, the transfer of electrons across the electrode-solution boundary is considered to be a reversible process in which n electrons are added in a single step: ox $+ n\epsilon =$ red. Here the reduction of ox is the forward reaction and the oxidation of red is the backward reaction. Both reactions go on simultaneously at the following velocities (in centimeters per second).

velocity of forward reaction: $k_f = [\text{ox}]k_{f_0} \exp\,[-\alpha EnF/RT]$ (5.61)

velocity of backward reaction: $k_b = [\text{red}]k_{b_0} \exp\,[(1 - \alpha)EnF/RT]$ (5.62)

where [ox] and [red] stand for the concentrations of ox and red, respectively, k_{f_0} and k_{b_0} are the forward and backward velocities across the electrode-solution boundary when $E = 0$ and the concentrations are unity, while n, F, R, and T are the number of electrons involved in the reduction, the faraday, the gas constant, and the absolute temperature, respectively. $-\alpha E$ represents the portion of the potential across the boundary that facilitates the forward reaction, while $(1 - \alpha)E$ is the remainder which facilitates the backward reaction. The parameter α is called the *transfer coefficient* of the electrode reaction. Equations (5.61) and (5.62) demonstrate that the reaction rate is determined by three factors. The first of these is the activity of the reactant, the second is an expression for the velocity in the absence of external factors, while the third represents the effect of an external factor (a potential) on the reaction rate.

When the velocities of the forward and backward reactions are equal, the net transport of electrons is zero and the system is in equilibrium:

$$[\text{ox}]k_{f_0} \exp\,[-\alpha E_e nF/RT] = [\text{red}]k_{b_0} \exp\,[(1 - \alpha)E_e nF/RT]. \quad (5.63)$$

The *equilibrium potential* E_e is therefore given by

$$E_e = RT/nF \ln k_{f_0}/k_{b_0} + RT/nF \ln [\text{ox}]/[\text{red}], \quad (5.64)$$

from which one obtains E_0, the *standard potential* when [ox] and [red] are equal. Substituting E_0 for $RT/nF \ln k_{f_0}/k_{b_0}$ then leads to

$$E_e = E_0 + RT/nF \ln [\text{ox}]/[\text{red}], \quad (5.65)$$

which is essentially the same as (5.45) but which can now be extended to systems that are not in equilibrium. As long as equilibrium exists, the concentrations at the electrode interface $[\text{ox}]_i$ and $[\text{red}]_i$ are identical with the bulk concentrations [ox] and [red].

When the rate of the forward reaction is greater than that of the backward reaction, more electrons leave the electrode than enter it, the electrode is cathode, and a current flows. The magnitude of this current I, is found by multiplying the difference in rates by nFA (where A is the area of the electrode):

$$I = nFA([ox]_i k_{f_0} \exp[-\alpha EnF/RT] - [red]_i k_{b_0} \exp[(1-\alpha)EnF/RT]),$$

$$(5.66a)$$

where $[ox]_i$ and $[red]_i$ stand for concentrations at the cathode-solution interface which may be different from the bulk concentrations.

This current is the sum of the positive cathodic and negative anodic currents:

$$I = I_c + I_a. \qquad (5.66b)$$

It becomes equal to zero at E_e where $I_c = -I_a$.

E in (5.66a) differs from E_e by an amount determined by the current, even when all other factors are kept constant. This difference $(E_e - E)$ is called *overvoltage;* it should exist whenever there is a flow of current. However, in many systems this overvoltage becomes so small at the dropping mercury electrode that it can be neglected. Furthermore, as Delahay has pointed out [101], the backward reaction can be neglected during cathodic current whenever the overvoltage exceeds $0.12/n$ V. When this is the case, the current becomes an exponential function of the potential, that is, the logarithm of the current varies linearly with the potential. Since this potential can be considered a logarithmic function of the electron pressure (see Section 2, p. 309), the current under these conditions of overvoltage must be directly proportional to the electron pressure. Graphs of log I versus E indeed show linear portions which intersect, when extrapolated, at the equilibrium potential of the irreversible system and have slopes determined by the transfer coefficient α [101] according to the equation for the cathodic current at 30°C:

$$E = E_e - 0.12/n + 0.060/\alpha \log nFAk_{f_0}[ox]_i - 0.060/\alpha \log I_c. \quad (5.67)$$

Since $[ox]_i$ is assumed to be constant, (5.67) represents a more modern version of the well-known Tafel equation: overvoltage $= a + b \log I$ [102].

It should be pointed out that so far the discussion has covered only the factors affecting *electron* transport across the electrode-solution boundary. The concentration of the oxidant or reductant at the interface has been taken as constant, which implies that the mass transfer of electroactive material must have been infinite so that concentration polarization is impossible. In polarography this situation is found only during the reaction of the main components of a supporting electrolyte since these are usually present in high concentration. During the reduction of sodium ions from 0.1 N sodium chloride solutions or during the reduction of hydrogen ions from strong acid solutions, one should expect a 10-fold increase in current for each 120-mV change in potential if the transfer coefficient is 0.5. This should also be true during the reduction of hydrogen ions from certain buffers present in high concentration where, in spite of the flow of large currents, the concentration of hydrogen ions at the interface, although very small, is kept constant by the dissociation of the acid form of the buffer. Equation (5.67) further predicts that 10-fold concentration changes of the reducible substance would shift at least the beginning of such polarographic currents by 120 mV on the voltage axis. This is shown in Fig. 5.25 where the potential of $H_{\lim}$ is plotted against pH. Here $H_{\lim}$ is the experimentally determined potential at which the currents attributable to the reduction of hydrogen ions are so large ($\approx 10\ \mu A$) that they interfere with the study of other reducible reagents [50].

For ordinary polarographic waves, which are characterized by concentration polarization, it is necessary to relate the rates of electron transport across the electrode-solution boundary to the rate of mass transport from the bulk of the solution to the electrode. The latter is reasonably well defined by the original Ilkovič equation (5.28) or (5.30), from which the following relationships have been taken for a cathodic current:

$$I_{dc} = nFAk'D_{ox}^{1/2}[ox] \qquad \text{where } k' = (\tfrac{3}{7}\pi t)^{-1/2} = 0.865t^{-1/2}$$

and

$$I_c = nFAk'D_{ox}^{1/2}([ox] - [ox]_i) = nFAk'D_{red}^{1/2}[red]_i \qquad \text{when [red]} = 0.$$

These equations lead to:

$$[ox]_i = (I_{dc} - I_c)/nFAk'D_{ox}^{1/2} \qquad \text{and} \qquad [red]_i = I_c/nFAk'D_{red}^{1/2}$$

which can be substituted into (5.66a). If, furthermore, the exponential terms, $\exp[-\alpha EnF/RT]$ and $\exp[(1 - \alpha)EnF/RT]$ are replaced by E_f and E_b, respectively, (5.66a) changes into

$$I_c = (I_{dc} - I_c)k_{f_0}E_f/k'D_{ox}^{1/2} - I_c k_{b_0}E_b/k'D_{red}^{1/2} \qquad (5.68)$$

or

$$I_c = I_{dc}\frac{k_{f_0}E_f/k_{b_0}E_b}{k'D_{ox}^{1/2}/k_{b_0}E_b + k_{f_0}E_f/k_{b_0}E_b + D_{ox}^{1/2}/D_{red}^{1/2}}, \qquad (5.69)$$

which is the basic equation for rate-controlled reductions.

Since $k_{f_0}E_f/k_{b_0}E_b = \exp[(E_0 - E)nF/RT]$ and $D_{ox}^{1/2}/D_{red}^{1/2}$ is usually close to unity in polarography,

$$I_c = I_{dc}\frac{\exp[(E_0 - E)nF/RT]}{k'D_{ox}^{1/2}/k_{b_0}E_b + \exp[(E_0 - E)nF/RT] + 1} \qquad (5.70)$$

or

$$E = E_0 - \frac{RT}{nF}\ln\frac{I_c}{I_{dc} - I_c} - \frac{RT}{nF}\ln\frac{k_{b_0}E_b + k'D_{ox}^{1/2}}{k_{b_0}E_b} \qquad (5.71)$$

In order to make this equation more useful, it is best to introduce a *constant*, E_{b_0} ($= \exp [(1 - \alpha)E_0 nF/RT]$) into the fraction of the last logarithmic term by first multiplying it by E_b/E_{b_0} and then subtracting $RT/nF \ln E_b/E_{b_0}$. Equation (5.71) then takes on the form

$$E = E_0 - RT/nF \ln I_c/(I_{dc} - I_c) - RT/nF \ln [(E_b/E_{b_0}) + Q]$$
$$+ RT/nF \ln E_b/E_{b_0} \quad (5.72a)$$

or

$$E = E_0 - RT/\alpha nF \ln I_c/(I_{dc} - I_c) - RT/\alpha nF \ln [(E_b/E_{b_0}) + Q], \quad (5.72b)$$

since $E_b/E_{b_0} = \exp [(1 - \alpha)(E - E_0)nF/RT]$. Q stands for $0.865 D_{ox}^{1/2} t^{-1/2}/k_{b_0} E_{b_0}$. The quantity $k_{b_0} E_{b_0}$ is also equal to $k_{f_0} E_{f_0}$ and represents the reaction rate at the standard potential (where [ox] = [red]), at which forward and backward reaction rates are equal; it can therefore be combined into a single velocity constant k_s (in centimeters per second), which characterizes the irreversibility of the system. Q then stands for $0.865 D_{ox}^{1/2} t^{-1/2}/k_s$.

Equation (5.72) allows a distinction between three types of waves, depending on the relative magnitudes of E_b/E_{b_0} and Q.

1. If $E_b/E_{b_0} \gg Q$, or as Q approaches zero, the last two terms of (5.72a) cancel out and the wave is typically *polarographically reversible*. This condition is met when $k_s \gg 0.865 D_{ox}^{1/2} t^{-1/2}$ or, in practice, when $Q < 0.02$, which means that the velocity k_s is about 50 times faster than the "diffusion velocity." The wave is then called "diffusion controlled" because it is always the slowest velocity that controls the reaction. The "diffusion velocity" can be estimated for $t = 4$ sec and $D = 10^{-5}$ cm²/sec to be about 1.3×10^{-3} cm/sec; hence a reaction velocity of $>6 \times 10^{-2}$ cm/sec should produce an apparent reversible wave. Changing t should have no effect on this type of wave.

2. If $E_b/E_{b_0} \ll Q$, one can neglect E_b/E_{b_0} in (5.72b) so that its last term becomes a constant. This results in a *totally irreversible* wave, which is still symmetrical about its midpoint but has a slope $1/\alpha$ times that of the reversible wave. Its half-wave potential ($E_{1/2}$) is more negative than that of the reversible wave by $RT/\alpha nF \ln Q$, which (at 30°C) is 0.12 V for each 10-fold increase in Q when $\alpha = 0.5$ and $n = 1$. Also, since Q contains the factor $t^{-1/2}$, $E_{1/2} = k + RT/2\alpha nF \ln t$. This means that a doubling of t causes a shift in the half-wave potential of 18 mV toward more positive values (at 30°C, when $\alpha = 0.5$). This shift with change in drop time distinguishes irreversible from reversible waves. The values of α (if n is known) can be determined from the slope of the wave (by plotting $\log I_c/(I_{dc} - I_c)$ versus E), or from the shift in the half-wave potential resulting from a change in t.

Whenever $Q > 50$, such totally irreversible waves can be expected. This means that the reaction velocity k_s is less than 3×10^{-5} cm/sec and therefore the waves are "reaction controlled."

3. If E_b/E_{b_0} and Q are of the same order of magnitude, a wave results that may be called *partially reversible*. Its major characteristic is that it is asymmetrical because the parameter E_b/E_{b_0} varies with the applied voltage. Since $(E - E_0)$ cannot be very large over the significant portions of the wave, the relationship $E = E_0 + RT/(1 - \alpha)nF \ln E_b/E_{b_0}$ cannot be much different from the "reversible" $E = E_0 + RT/nF \ln (I_{dc} - I_c)/I_c$ so that one can, as a first approximation, substitute $[(I_{dc} - I_c)/I_c]^{0.5}$ for E_b/E_{b_0}, taking $\alpha = 0.5$. Upon conversion to common logarithms, substitution of numerical values (for 30°C), and assuming $n = 1$, (5.72b) then changes to

$$E = E_0 - 0.12 \log I_c/(I_{dc} - I_c) - 0.12 \log ([(I_{dc} - I_c)/I_c]^{0.5} + Q). \quad (5.73)$$

A number of cathodic curves have been calculated on the basis of (5.73) for different values of Q and are shown at the top of Fig. 5.24. They are in fair agreement with curves constructed on the basis of much more elaborate mathematics and rigorous derivations [53, 54, 100, 101]. These curves clearly demonstrate the gradual shift from a symmetrical reversible wave ($Q = 0$), first to a partially reversible asymmetrical wave ($0.02 < Q < 50$), and finally to a totally irreversible, again symmetrical, wave ($Q > 50$). The corresponding anodic waves are shown in the lower half of Fig. 5.24. Complete anodic-cathodic curves found as the sum of the separate anodic and cathodic

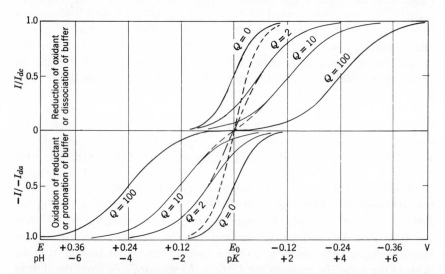

Fig. 5.24 Calculated curves showing spreading of anodic and cathodic portions of polarographic waves, hence apparent polarographic irreversibility, because of rate-controlled reactions, characterized by different Q values (see text).

waves have also been plotted. Note that while the separate anodic and cathodic waves for $Q = 2$ are asymmetrical, the corresponding combined anodic-cathodic wave is symmetrical and shows no sign of any separation into two waves. At values of $Q > 2$, this separation becomes apparent and is complete when $Q > 50$.

Proton Transfer

It should be emphasized that this discussion has so far been concerned only with reactions that either do not involve hydrogen ions or are so well buffered that the hydrogen ion concentration at the interface is maintained constant. However, the reverse situation can also be created by having in solution large concentrations of oxidant and reductant so that the ratio of [ox]/[red] is not significantly altered even during the flow of appreciable current. One can then study the effect of small concentrations of buffers on the wave, a condition that has been called the "acid-base titration of the interface" [80].

The quinhydrone system may again serve as a typical example. For each molecule of quinone reduced, two hydrogen ions are consumed in the formation of hydroquinone. Thus for n electrons added, n protons are added as well. These have to come from a buffer acid which must diffuse to the electrode and dissociate there. Hence cathodic current in this case of acid-base titration of the interface represents dissociation rather than association of buffer acid so that the pH scale of Fig. 5.4 has to be inverted in order to be applicable here. Otherwise the analogy is so close that (5.73) can be rewritten:

$$\text{pH} = \text{p}K + \log I_c/(I_{dc} - I_c) + \log \left([(I_{dc} - I_c)/I_c]^{0.5} + Q\right), \quad (5.74)$$

where each 0.060 V corresponds to 1 pH unit and where Q now stands for $0.865 D_{\text{HA}}^{1/2} t^{-1/2}/k_{d_0}(1 - \alpha)\text{p}K$. The velocity of dissociation k_{d_0} is equivalent to the velocity of the backward reaction at the standard potential; its subscript indicates that it holds for pH = 0 and for the condition that the concentrations of the protonated acid and of its anion are equal ([HA] = [A$^-$]). Jointly with $(1 - \alpha)\text{p}K$ it forms a characteristic rate constant for proton transfer (similar to k_s) which, divided into the "diffusion constant" $(0.865 D_{\text{HA}}^{1/2} t^{-1/2})$ determines the quantity Q. The curves of Fig. 5.24 are therefore equally applicable to cases of reversible and irreversible proton transfer if one plots pH on the abscissa instead of voltage, each increase in pH corresponding to a change of 0.060 V toward more negative potentials (as indicated in Fig. 5.24).

It is therefore obvious that either a slow electron transfer or a slow proton transfer can give rise to a separation of the cathodic and anodic portions of a wave. Distinction between the two possibilities is easy by substitution of buffers that are known to react fast, or by the addition of suitable catalysts.

As has been shown in Fig. 5.23, acceleration of the rate of proton transfer can bring about a complete unification of such separate waves into a single anodic-cathodic wave.

8 REDUCTION OF HYDROGEN IONS

Many polarographic studies would be impossible were it not for the fact that the reduction of hydrogen ions at the dropping mercury electrode is irreversible and takes place at a potential far more negative than that of the reversible hydrogen electrode. This difference in potential is called the overvoltage of hydrogen; it depends on the metal used as electrode and seems to be largest at a fresh surface of mercury. Figure 5.25 may serve to indicate the potential ranges involved [50]. All experimental points shown were obtained with a dropping mercury electrode. The heavy line marked H shows the pH dependence of the reversible hydrogen electrode (where the electrode is platinized platinum bathed with hydrogen gas); this line has a slope of 0.060 V per pH unit (at 30°C). The heavy broken line marked H_{lim} indicates the limitation at the negative end of the potential range of the dropping mercury electrode in buffered solutions; its slope is 0.120 V per pH unit. Here the current caused by reduction of hydrogen ions reaches values of 10 μA which would interfere with the study of more negative reactions. Between these two lines is the range of overvoltage in which apparently all irreversible organic reductions take place. The slope of the pH-potential curves in this region is predominantly 0.090 or 0. All known polarographic reversible reactions occur at potentials more positive than the reversible H line. Their pH-potential slopes are determined by the dissociation constant of the components of the reaction in accordance with (5.53). The short heavy line marked L-P shows the pH-potential relationship of the lactate-pyruvate system measured potentiometrically in the presence of the proper enzyme and oxidation-reduction indicator which makes this system reversible [103]. Its potential is far different from the half-wave potentials, labeled PA, which were obtained during the irreversible reduction of pyruvate at the dropping mercury electrode.

Heyrovský [104] recognized the importance of hydrogen overvoltage on mercury even before the construction of the first polarograph. He suggested that hydrogen atoms amalgamate with mercury and together with the hydrogen ions in solution form an oxidation-reduction system with a constant half-wave potential [79]. However, later work from the Prague school was at variance with this conclusion. Tomeš [105] found that the half-wave potential became more positive with an increase in [H+], while Kůta [106] found the opposite to be true (as did Tamamushi [107]). Tomeš (and Tamamushi) also found the waves to be asymmetrical, while Kůta found them symmetrical.

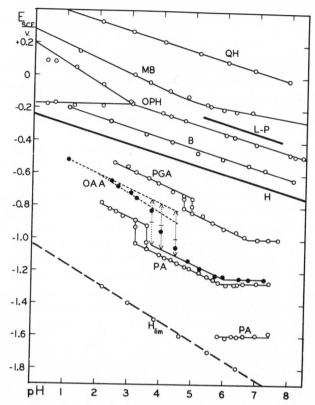

Fig. 5.25 Potential-pH curves of reversible systems: QH, quinhydrone; MB, methylene blue; OPH, α-oxyphenazine; L-P, enzyme-catalyzed lactate-pyruvate system; and B, benzil. The heavy line marked H represents the potential of the reversible hydrogen electrode. Half-wave potential-pH curves of the irreversible reductions of: PGA, phenyl-glyoxylic acid; OAA, *trans*-oxaloacetic acid; PA, pyruvic acid. The interrupted line marked H_{lim} indicates the limits of polarographic measurements because of hydrogen evolution, which then produces a current of 10 μA.

In an attempt to correlate these contradictory results, Müller [108] reinvestigated the reduction of hydrogen ions from hydrochloric acid solutions in different concentrations of supporting electrolytes. By going to hydrochloric acid concentrations greater than 1 mN he obtained hydrogen waves with two unequal steps which turned out to be two different kinds of waves which could be obtained in pure form under different conditions. One of these, usually found in dilute solutions, was a symmetrical wave, typical of (5.50) when $n = 0.5$; this wave was diffusion controlled because its height was proportional to the square root of the effective pressure on the mercury. The other was an asymmetrical wave which fitted the following equation, first developed for the unbuffered oxidation of hydroquinone [80]:

$$E = E_0 - RT/F \ln I/(I_d - I) - RT/F \ln I/I_d. \tag{5.75}$$

This wave was obtained in 1 mN solutions of hydrochloric acid in 1 N potassium chloride. It behaved as a typical anomalous or prewave since its height was directly proportional to the effective pressure. At concentrations higher than 1 mN, the currents showed an abrupt change from the prewave to the main wave, the sum of the two again being diffusion controlled. It thus appears as if in the reduction of hydrogen ions similar anomalous waves appear as in the case of dyestuffs (see Section 4, p. 349), but they exist up to concentrations that are greater by one order of magnitude.

In connection with these studies, Müller [108] investigated the quantitative relationship between current and anticipated product (hydrogen gas) by volumetric analysis of the latter. The very simple arrangement used is shown in Fig. 5.26. As long as the cathode was a platinum microelectrode, the volume of hydrogen gas evolved during the passage of current through a solution of 0.1 N hydrochloric acid was about 97% of theory; none evolved following the electrolysis. The situation was quite different when either a dropping or a stationary mercury microelectrode served as cathode. Here only about 30 to 40% of the theoretical volume of hydrogen gas was evolved

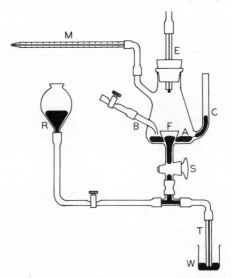

Fig. 5.26 Arrangement for volumetric analysis of hydrogen gas produced by electrolysis. E, Dropping mercury electrode; A, layer anode; C, contact to cathodic lead; B, side tube for deaeration before experiment; F, funnel for catching drops; S, stopcock; R, mercury reservoir; T, narrow outflow tube for mercury; W, mercury outflow weighing bottle; M, horizontally placed 0.2-ml pipet graduated in 0.001 ml [108].

during the electrolysis which was maintained for about 30 to 40 min at an intensity of 1 mA. The remainder of the gas (up to 87% of theory) formed over a period of 24 hr or more *after* the electrolysis, as shown in Fig. 5.27. Curve *E* of this figure demonstrates the interference of potassium ions. If these are reduced along with the hydrogen ions, the volumetric experiments

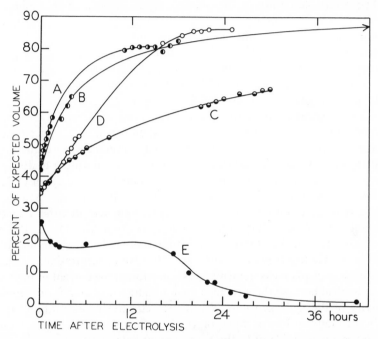

Fig. 5.27 Volume changes in percent of expected volume observed *after* electrolysis of 0.1 *N* hydrochloric acid solutions. (*A* and *B*) After electrolysis with a dropping mercury electrode of 49 and 30 min, respectively, at a current intensity of 1 mA. The drops were eventually dropping on the anode. (*C*) After electrolysis with a dropping mercury electrode of 33 min at a current intensity of 1 mA. No drops ever fell on the anode. (*D* and *E*) After electrolysis with a stationary mercury electrode in the funnel for 40 and 35 min, respectively, at a current intensity of 0.2 mA, but in the presence of 0.1 *N* potassium chloride in the case of *E* [108].

give erroneous results since the potassium amalgam formed causes a shrinkage of the mercury with time, which counteracts the volume increase produced by the evolution of hydrogen gas.

These results were interpreted as follows [108]: The hydrated hydrogen ion, H_3O^+, is reduced reversibly to a radical, H_3O, which, similar to the reduction product of ammonium ions, forms an amalgam with the electrode mercury. The amalgam may diffuse into the mercury where it slowly decomposes with the formation of hydrogen gas, or it may decompose more rapidly at the electrode surface to form hydrogen atoms. These atoms are either adsorbed at the electrode surface or combine with H_3O^+ ions of the solution to form H_4O^+ or H_2^+ ions. Depending on the pH of the solution, the latter could then exist in a series of concentrations maintained by the buffer present and could react with other substances in solution.

The reversible nature of the primary reduction step in the reduction of hydrogen ions was confirmed by Smith and Wells [109], who made triangular voltage measurements with a mercury electrode after 40-hr cathodization in a phosphate buffer of pH 7.5. They found evidence not only for cathodic desorption at the mercury electrode containing hydrogen atoms (H-Hg), but also for anodic ionization of these hydrogen atoms to form H_3O^+ ions.

In general, the limiting current of hydrogen ions is diffusion controlled. This diffusion may involve free H^+ (or H_3O^+) ions or it may involve the acid component of a buffer which dissociates when it reaches the electrode. The former is dominant at pH < 4 and at equal concentrations results in much higher currents than the latter, since hydrogen ions diffuse 5 to 10 times as fast as most other ions and molecules. Calculated on the basis of (5.32), the diffusion coefficient of hydrogen ions is $9.34 \times 10^{-5} \, cm^2/sec$. Müller [110] found that the diffusion currents of hydrochloric acid in either aqueous or alcoholic 0.1 N potassium chloride solutions are a function of the added hydrochloric acid, but those in the comparable (47.5%) ethanolic solution are much smaller. However, when an "effective concentration," calculated on the basis of pH measurements, was substituted, the current in both media was identical because this provided an automatic correction for differences in activity of the hydrogen ions. When the "effective concentration" is used, only the modified Ilkovič equation (5.37) with a K value of $\frac{3}{7}$ (an A value of 17) gives a reasonable fit to the experimental data [110].

Addition of hydrochloric acid to solutions containing anions of weak acids causes currents that are determined by the diffusion of the associated weak acid, hence can be used to determine the diffusion coefficients of the latter [111]. This could have been expected on the basis of the buffer action demonstrated for quinhydrone waves (see Fig. 5.21). In both cases the diffusion of the weak acid determines the magnitude of the current, although the current is caused by the reduction of hydrogen ions in one case and by the reduction of quinone in the other.

When hydrochloric acid is added to solutions containing reducible organic nonacids, (e.g., acetophenone or azobenzene), a positive wave results whose magnitude is a function of the diffusion of free hydrogen ions [111, 112]. This seems to indicate that the reduction at the positive potential must take place via atomic hydrogen (or H_2^+ ions) generated at the dropping mercury electrode during the reduction of hydrogen ions. The situation becomes less clear when hydrochloric acid is added to anions of weak acids that are themselves reducible. The diffusion coefficient is that of the weak acid, rather than that of the free hydrogen ions, and the current may become rate-controlled [113].

As could be expected, the hydrogen wave is distorted in the case of bicarbonate buffer unless sufficient enzyme (carbonic anhydrase) is added [98].

Since the final product of the reduction of oxygen at the dropping mercury electrode is hydroxyl ion and this reduction takes place before the reduction of hydrogen ions, a wave attributable to the latter is diminished in the presence of oxygen [114]. This diminution is quantitative and may be used as a confirmatory test if the existence of a hydrogen wave is questioned.

9 IRREVERSIBLE REACTIONS AT THE DROPPING MERCURY ELECTRODE

Introduction

Whenever it is impossible to oxidize (or reduce) the end product of a reduction (or oxidation) at the dropping mercury electrode at the same potential at which it was formed, the reaction must be considered polarographically irreversible. On the basis of this criterion the majority of reactions that have been studied polarographically belong to this group. Even a perfectly reversible reaction may appear irreversible if any component of the total system (such as a buffer) does not react with sufficient speed to maintain equilibrium conditions at the electrode interface (see Fig. 5.22).

In the presence of buffers that have proved themselves fast enough with reversible reactions, the slow steps may involve:

1. Direct transfer of electrons from electrode to reagent as in reversible reactions but complicated by: (a) slow establishment of equilibrium, and (b) irreversible further reaction (instability) of product.

2. Modification of the reagent to make it more reducible, followed by direct transfer of electrons to it.

3. Indirect reduction of reagent by means of products formed during the reduction of hydrogen ions. These products, atomic hydrogen, or H_2^+ ions, may act (a) directly, and (b) after being activated by adsorption on the electrode material, or by catalysts from the solution.

Definite evidence for the irreversibility of a reaction is always the separation of anodic and cathodic polarographic waves, as shown in Fig. 5.24, but in most instances only the cathodic wave can be obtained since the anodic wave falls into a positive potential region where mercury goes into solution.

A major difficulty concerning irreversible reactions is that in most of them the end products are not known with certainty, since not enough material is formed at the dropping mercury electrode to permit an analysis. Sometimes a reasonable guess can be made on the basis of results obtained when the electrolysis is carried out over a long period of time and with a large stationary electrode. However, the products under these conditions are not necessarily identical with those formed at the dropping mercury electrode, and any generalization must be made with reservation. In some instances one can predict the product from the number of electrons involved in the reaction of a compound. To estimate this number, use is made of the Ilkovič equation in which an appropriate diffusion coefficient can be approximated. This is not applicable when the relation between concentration and diffusion current fails to be linear or when the half-wave potential changes with changing concentration of the reactant. Here, and in cases in which the wave loses its symmetry, the half-wave potential becomes meaningless and serves only as an arbitrary characteristic.

While the irreversibly formed product is usually electroinactive, some exceptions have been reported. For instance, the oxidation of hydroxy-chromans and hydroxycoumarans yields products that are irreversibly changed to quinones which can be reversibly reduced to the corresponding hydroquinones but not to the original hydroxychromans or hydroxycou-marans [115]. The counterpart to these oxidations is the reduction of the dye T-1824, also called Evans blue, which involves a reversible step followed by an irreversible transformation to a different but reversible system [116].

In many cases, the half-wave potentials of irreversible reactions depend to a considerable degree on the concentration and kind of other electrolytes present in the solution. Higher valent cations especially have often shown a marked effect on the electrode reaction and can certainly not be considered "indifferent," although they are discharged at more negative potentials than the organic compound. Such phenomena are especially striking in the reduction of inorganic anions [117], where they were first observed.

Effect of pH on Irreversible Reductions

Many irreversible reductions at the dropping mercury electrode give rise to symmetrical S-shaped waves with half-wave potentials that are a function of pH. To account for this type of current-voltage curve, Müller and Baumberger [49] suggested that the polarographic wave represents a reversible step in the otherwise irreversible reductions. This made possible an analogy with some irreversible reactions, investigated by Conant [118] with an entirely different technique, in which a reversible step was similarly postulated. It seems that Conant's "apparent reduction potential" (ARP) coincides approximately with the beginning of the rise in current on the polarogram and not with the half-wave potential.

The effect of pH on the irreversible reduction of pyruvic acid is particularly interesting because here it is possible to compare the polarographic half-wave potentials with the potentiometrically determined reversible potentials of the lactate-pyruvate system. At a platinum electrode in the presence of the proper enzyme and coupling reagent, the lactate-pyruvate system gives potentials that are a function of pH and of the ratio of [lactate]/[pyruvate]. While the potentials thus measured are not in perfect agreement [103, 119], one can conclude that at pH 6 the oxidation-reduction potential of this system is near -0.3 V versus *SCE* (see Fig. 5.28). The oxidation of lactic acid has not been possible at the dropping mercury electrode, but lactic acid has been chemically oxidized by potassium permanganate in acid solution

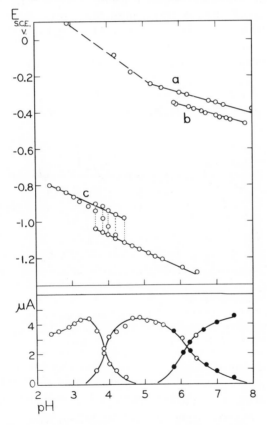

Fig. 5.28 Upper portion: Potential-pH curves of the lactic acid–pyruvic acid system in the presence of enzyme and mediator. (*a*) Data of Baumberger et al. [103]. (*b*) Data of Barron and Hastings [119]. (*c*) Half-wave potentials of pyruvic acid reduced from phthalate buffers. Lower portion: Polarographic current changes observed with changes in pH. Open circles, phthalate buffer; filled circles, McIlvaine citric acid–phosphate buffers [50].

[49] so that a rough estimate can be made by Conant's method of the half-wave potential of a hypothetical anodic wave at pH 6 with a value of +0.8 V versus *SCE*. When this is compared with the real cathodic wave of the reduction of pyruvate at the same pH, with a half-wave potential near −1.3 V (versus *SCE*), the reversible potential falls approximately half-way between the two irreversible values.

Müller and Baumberger's study [49] of the reduction of pyruvic acid and pyruvate as a function of pH revealed the following phenomena that even today are not fully explained:

1. Graphs of half-wave potentials versus pH show inflections near pK values of pyruvic acid (2.45) and lactic acid (3.85) (see Fig. 5.28).

2. The pH-dependent wave (I) is changed into a pH-independent wave (II) with more negative half-wave potentials at higher pH (Fig. 5.12). Near pH 6 both waves are of equal height.

3. The pH-independent wave (II) is complex and may exhibit maxima (Fig. 5.13).

4. Near pH 4 the pH-dependent wave (I) loses its symmetry and is apparently composed of two overlapping waves which are of equal height at pH 3.8 (Fig. 5.29).

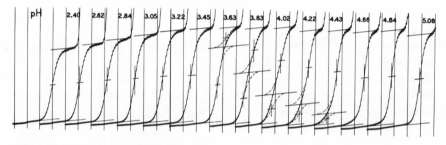

Fig. 5.29   Polarogram of 0.001 M solutions of pyruvic acid in Clark and Lubs phthalate buffers. The final pH of each solution is indicated. Note the asymmetry of the waves around pH 4, attributed to overlap of two waves, drawn in as interrupted lines. Also note the decrease in wave height below pH 4. At lower pH each curve starts at −0.2 V, above pH 3.63, each curve starts at −0.4 V. The voltage increment is 0.2 V for each abscissa and the sensitivity used is $\frac{1}{30}$ [50].

5. Although the solutions had been prepared with identical concentrations of pyruvic acid, the wave height diminished considerably as the pH decreased below pH 4 (Fig. 5.29).

Of these findings, 3 has not yet found a suitable explanation, while 5 has been ascribed originally to polymerization [49] later to hydration [120] and to hydration and enolization [121]. Findings 1 and 4 have been considered evidence for the effect of ionization of pyruvic and lactic acids on the potential [49] and have so far not received additional scrutiny.

Most attention has been given to finding 2 which had originally been attributed to keto-enol tautomerism of pyruvate. When Brdička [51] found that phenylglyoxylic acid, which cannot enolize, behaved similarly, this explanation had to be abandoned. In a new theory, developed by Brdička and Wiesner [52], wave (I) is ascribed to the reduction of the associated or nonionized acid, while wave (II) is thought to be caused by the reduction of the anion. Brdička considered that the difference between the "apparent pK," that is, the pH at which wave (I) is diminished by one-half (near 6 in Fig. 5.28) and the true pK of pyruvic acid (2.45) is caused by regeneration of the associated form from anions and hydrogen ions as the former is removed from the electrode interface by electrolysis. The rate at which this association can occur is limited and determines the pH at which the associated acid is finally exhausted and dissociated anions are available for reduction. The latter are thought to be less easily reduced than the associated acid because the conjugation of the carbonyl group with the C=O group of the carboxyl is no longer maintained. As mentioned in Section 4, p. 345 this explanation has found wide acceptance and has led to the derivation of equations for kinetic currents in general, of which (5.40) can be used to determine the rate constant for the recombination of acids behaving in this fashion.

Müller [50] has objected to Brdička's explanation largely because it does not encompass the various other findings listed above and because the rate constants calculated on the basis of it may be in error by several orders of magnitude. However, recent experiments of the writer have demonstrated that wave (I) indeed involves the reduction of the associated pyruvic acid molecules as postulated by Brdička [51]. In these experiments various organic compounds were reduced in the presence of a fixed concentration of nitric acid. It was found that the reduction of quinone, acetophenone, benzil, or azobenzene required hydrogen ions and thus diminished the height of the more negative hydrogen wave which followed the reduction wave of the organic compound. This diminution was a function of the concentration of the organic compound, as expected, and implied the following reactions at the electrode surface:

$$O:C_6H_4:O \quad + 2\epsilon + 2H^+ \rightleftharpoons HO \cdot C_6H_4 \cdot OH$$
$$CH_3COC_6H_5 \quad + 2\epsilon + 2H^+ \rightarrow C_3CHOHC_6H_5$$
$$C_6H_5COCOC_6H_5 + 2\epsilon + 2H^+ \rightarrow C_6H_5CHOHCOC_6H_5$$
$$C_6H_5N:NC_6H_5 \quad + 2\epsilon + 2H^+ \rightarrow C_6H_5NH \cdot NHC_6H_5$$

However, the reduction of pyruvic acid and other keto acids actually increased the hydrogen wave that followed, apparently by contributing hydrogen ions from dissociation of the reduction product. This indicates that the product must have been an acid in which the keto group has received electrons but no protons, so that a divalent radical is formed:

$$R—CO—COOH + 2\epsilon \rightarrow [R—CO—COOH]^{2-}$$

Thus the product at the electrode surface is not lactic acid as previously assumed.

Brdička's only concern has been with the ratio of the wave heights (I)/(II), with pK', and with its shift as a consequence of slow dissociation and recombination rates as indicated in Fig. 5.24. The electrode mechanism and the effect of pH on the half-wave potential is thereby not clarified. Tamamuchi and Tanaka [122] derived equations from which the effect on pH could be calculated. These equations were based on the following three equilibria which have been rewritten for the case of pyruvate reduction:

$$CH_3COCOO^- + H^+ \rightleftharpoons CH_3COCOOH$$
$$CH_3COCOOH + 2\epsilon \rightleftharpoons [CH_3COCOOH]^{2-}$$
$$[CH_3COCOOH]^{2-} + 2H^+ \rightleftharpoons CH_3CHOHCOOH$$

After substitution of suitable values for the various rate constants, Tamamushi and Tanaka were able to calculate from their equation half-wave potentials for the pyruvate wave which were in good agreement with experimental data from pH 4.5 to 6.5. However, their equation failed to account for the deviations below pH 4. This is not surprising since it now appears that the third of the above equilibria is established so slowly that it does not have any effect on the potentials.

Effect of Substitution in the Organic Molecule on the Half-Wave Potential

The effect of substitution in the organic molecule has interested polarographers for many years. Even before the half-wave potential was postulated as a characteristic of the reacting substances, the reduction potentials of organic compounds were compared and correlations were tried between different substituents in a given series of compounds. Thus Shikata and Tachi formulated their "electronegativity rule" of reduction potentials [123].

In recent years it has been possible to show that a correlation existed between half-wave potentials and *Hammett's equation*. This lends additional support to the measurement of half-wave potentials in irreversible reductions.

Hammett [124] established certain quantitative relationships between systems of reactions. For instance, the hydrolysis rate of esters and the ionization constants of benzoic acid substituted in the *meta* and *para* (but not *ortho*) positions changes as a function of the substituent. This was expressed mathematically as:

$$\log k - \log k^0 = \rho\sigma, \tag{5.76}$$

where k is any rate or equilibrium constant, and k^0 is the rate of the unsubstituted reaction. σ is a *substituent constant* which is independent of the reaction and may be positive or negative, while ρ is the *reaction constant* which is the same for all substituents and depends only on the reaction series.

Brockmann and Pearson [125] found that the Hammett equation could be extended to polarographic data and that for *meta-* and *para-*substituted benzophenone, its oxime, and semicarbazone the following relationship held:

$$E_{1/2} - E^0_{1/2} = \rho\sigma, \qquad (5.77)$$

where $E^0_{1/2}$ is the half-wave potential of the unsubstituted organic substance. Since that time, much has been accomplished in further correlations of polarographic half-wave potentials and substituent effects. Not only polar effects, covered by the Hammett equation, but also steric effects, treated by the similar *Taft equation* [126], have been considered. Details need not be included here, especially since Zuman devoted a large monograph entirely to this subject [127].

However, mention should be made of Hückel's molecular orbit method which has been extended by Maccoll [128] to polarography. In the case of conjugated hydrocarbons, he found a linear relationship between the half-wave potential and a, where a is the coefficient of molecular orbital resonance integral in the expression for the energy of the lowest unoccupied orbital. This correlation has been further established for many alternate and non-alternate hydrocarbons containing both aromatic nuclei and olefinic double bonds. See also the quantum mechanical relationship between structure and half-wave potential of carbonyl compounds found by Coulson and Crowell [129].

Anyone working in this field will welcome the very extensive list of half-wave potentials that has been prepared by Zuman [130].

10 TESTS FOR REVERSIBILITY AND FOR REACTION PRODUCTS

The quinhydrone system in well-buffered solution serves as an ideal example of polarographic reversibility (see Fig. 5.17). Here the anodic wave of hydroquinone has the same half-wave potential as the cathodic wave of quinone, leaving no doubt about the starting material and end products of the reaction. Such systems are relatively scarce, however, and in most organic reactions one can only speculate about the reversibility of a reaction and its products. Coulometry (see Chapter IID in original plan, by Murray) has now progressed to a state where one can be reasonably sure about the number of electrons involved, but this information alone cannot be used to ascertain the product and it certainly gives us no indication about reversibility. Speculations have sometimes been made that at least a part of the reaction is reversible on the basis of: (1) the shape of the polarographic wave, especially the so-called log-plots, as shown in Fig. 5.18; (2) the fact that the half-wave potential is independent of the concentration of the depolarizer; and (3) the dependence of the wave height on drop time and drop size in agreement with the Ilkovič equation. Any more definite, direct method is of course to be preferred.

Heyrovský and Forejt [131] have developed an oscillographic method which clearly demonstrates reversibility by identical potentials for anodic and cathodic reactions occurring during any single alternating current cycle of suitable frequency. However, these reactions would have to come to equilibrium within 0.01 sec. Reactions that take longer, but less than the drop time of the dropping mercury electrode, to come to equilibrium appear reversible on the polarogram but irreversible on the oscillograph. However, reactions that are oscillographically reversible may undergo secondary reactions during the drop time and thus be irreversible polarographically.

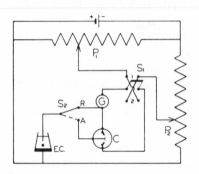

Fig. 5.30 Kalousek switch circuit for testing the reversibility of a reaction at the dropping mercury electrode (see text).

A very valuable additional tool was therefore introduced by Kalousek [91], who used a commutator method which made the dropping mercury electrode alternately cathode and anode. The principle of this method is illustrated in Fig. 5.30. The slide-wire of the polarograph P_1 and a variable resistance of similar magnitude P_2 are connected in parallel to a battery. The reference electrode of the electrolytic cell $E.C.$ is connected to the positive terminal. The dropping mercury electrode is connected to switch S_2. G is the galvanometer with shunt and C is a commutator or electronic switch. With switch S_2 in position R, regular polarograms are obtained when the double-pole, double-throw switch S_1 is in position 1 (S_1 in position 2 serves to find the proper setting for P_2). With S_2 in position A, the commutator is included in the circuit. Depending on the setting of S_1, the galvanometer records processes at the dropping mercury electrode relating to P_1 as compared to P_2, or vice versa. (An electronically operating *drop life timer*, commercially available, has been used by the writer as a satisfactory substitute for Kalousek's commutator. See also Ref. 132.)

By switching the applied voltage 5 to 8 times per second, Kalousek [91] obtained anodic waves of the freshly reduced products with half-wave potentials identical to those of the material that was reduced when the reaction was reversible. When the reaction was irreversible, either no corresponding anodic wave, or an anodic wave with a different half-wave potential, was obtained. This method showed clearly that a step in the reduction of oxygen to hydrogen peroxide is reversible in acid, neutral, and alkaline media, while the reduction of hydrogen peroxide to water is irreversible. The reduction of fumaric and maleic acids and the oxidation of ascorbic acid are also irreversible. Tamamushi [107] employed this method to show that the reduction of hydrogen ions from strong acids is not reversible, and Kalousek and Rálek [133] demonstrated its possible application to the identification of reaction products of irreversible processes. For instance, in the irreversible reduction of hexachlorocyclohexane, one of the products of the reaction was definitely established as chloride ion. Another modification of this technique was used by Suzuki [134], who investigated the various steps in complicated reactions such as the reduction of nitro compounds, ketones, and aldehydes.

The principle of the Kalousek switch has also found applications in the field of analytical polarography. For instance, in reversible reactions the switching technique can double the sensitivity of the polarographic method because of the continued reoxidation and re-reduction of reactant at the same drop. If an arrangement for differential polarography is used, peak currents may thus be obtained with reversible systems that are about four times as large as ordinary diffusion currents [135]. Furthermore, the anodic wave produced by switching permits the determination of a reversible species in the presence of an irreversible species reducible at the same potential, or when its normal wave would be masked by the reduction of hydrogen ions [136].

It has already been pointed out that in most instances one can only speculate concerning the primary products formed during irreversible reactions at the dropping mercury electrode. End products actually obtained during prolonged electrolyses under supposedly similar conditions may help in an interpretation but even under relatively ideal conditions different conclusions may be reached. For instance, Semerano and Bettinelli [137] electrolyzed a few milligrams of aconitic acid at a dropping mercury electrode for about 2 weeks and then determined succinic acid fluorimetrically as one of the end products. However, Siebert [138], using multiple mercury drops (formed by letting mercury flow through a glass filter plate), electrolyzed 175 mg of aconitic acid for 40 hr under otherwise supposedly identical conditions and obtained tricarballylic acid as the end product, with a yield of 97%.

Similarly, conclusions reached on the basis of coulometric analyses are not necessarily final. This may be illustrated by the reduction of picric acid in 0.1 N hydrochloric acid, which produces a complicated wave. Reducing such a solution at -0.65 V (versus SCE), Lingane [139] found by means of coulometry that 17 electrons per mole are involved, so that the most probable reduction product was di-(4-hydroxy-3,5-diaminophenyl)-hydrazine. However, Meites and Meites [140] found that when the concentration of picric acid was kept low, coulometric analysis under similar conditions gave a value of 18 electrons per mole for the reduction of picric acid. The end product then should be 2,4,6-triaminophenol.

II   REDUCTION OF METALLIC IONS

The reduction of metallic ions has probably contributed more than any other reaction to the development of polarography; it certainly has found the widest practical application. Metallic ions can be reduced to a lower valence ion or to the metallic state, which usually forms an amalgam with the mercury, and they may be present as free ions or as a complex in the solution. When Heyrovský and Ilkovič [79] developed the half-wave potential for the characterization of reactions at the dropping mercury electrode, Majer in an addendum to their paper published the first table of half-wave potentials for the reduction of many cations from neutral, acid, and alkaline solutions, as well as from complexing media such as 10% potassium tartrate and 1 N potassium cyanide. Although Heyrovský and Ilkovič [79] speculated that the half-wave potential should represent the "normal oxidation-reduction potential" in electroreduction, there was no evidence available at that time for the corresponding oxidation of the reduction product. It was not until 1938, a year after the discovery of the reversible quinone-hydroquinone system [13], that Strubl [141] found an anodic wave for Ti^{3+} ion quite different from the cathodic wave for Ti^{4+} ion, which could be made to form one continuous anodic-cathodic wave when sufficient potassium tartrate was added to the solution. This is clearly demonstrated by Fig. 5.31, taken from Spalenka [142]. The first papers on the polarographic oxidation of dilute amalgams appeared in 1939 [143]. Lingane found that cadmium amalgam gave an anodic wave, while Heyrovský and Kalousek proved that the reduction of copper, lead, cadmium, and zinc is reversible, and that the anodic-cathodic half-wave potential is shifted to more negative values if the solution contains complex formers.

The equations for the polarographic reduction of metallic cations are analogous to those for the reduction of organic compounds, with complexing agents taking the place of the hydrogen ions. Again, for a model system, a perfectly reversible reaction must be assumed between free metallic ions and complex-forming anions:

$$MX_p^{(pm-n)^-} \rightleftharpoons M^{n+} + pX^{m-},$$ (5.78)

which can be expressed by a dissociation constant K_c as:

$$\frac{[M^{n+}][X^{m-}]^p}{[MX_p^{(pm-n)^-}]} = K_c.$$ (5.79)

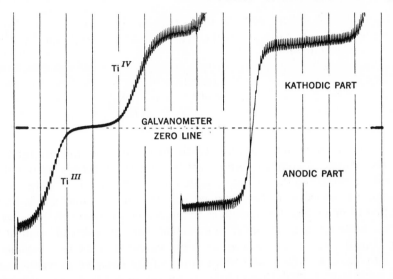

Fig. 5.31 At left: 12 ml of a mixture of $TiCl_3$ and $TiCl_4$ in 0.07 N hydrochloric acid, deaerated. At right: same solution after addition of 0.2 gm tartaric acid. Curves start at 0.4 V, voltage increment is 0.2 V for each abscissa and the sensitivity used is $\frac{1}{5}$ [142].

The concentration of free metallic ions, the oxidant in the reaction, is then given by

$$[M^{n+}] = \frac{K_c[MX_p^{(pm-n)^-}]}{[X^{m-}]^p} = [\text{ox}]. \tag{5.80}$$

Since the quantities $[M^{n+}]$ and $[MX_p^{(pm-n)^-}]$ are not directly measurable, they are summated into the form [Ox] which includes all the metallic components of the solution and is therefore measurable. Equation (5.80) thus becomes:

$$[\text{ox}] = [M^{n+}] = \frac{K_c[\text{Ox}]}{K_c + [X^{m-}]^p}, \tag{5.81}$$

which when incorporated into (5.45) leads to

$$E = E_0 + \frac{RT}{nF}\frac{[\text{Ox}]_i}{[\text{red}]_i} + \frac{RT}{nF}\ln\frac{K_c}{K_c + [X^{m-}]_i^p}. \tag{5.82}$$

Analogous to (5.46):

$$[\text{Ox}]_i = (I_{dc} - I)/kD_{\text{Ox}}^{1/2}, \tag{5.83}$$

where D_{Ox} represents an average diffusion coefficient of the free metallic ions and of the complex, which depends on their relative concentration and diffusion coefficients. Putting this value into (5.82), plus $[\text{red}]_i = I/kD_{\text{red}}^{1/2}$

since [red] = 0 (where [red]$_i$ is the concentration of the amalgam at the electrode surface), one obtains the following general equation for the reduction of metallic ions:

$$E = E_0 + \frac{RT}{nF} \ln \frac{I_{dc} - I}{I} + \frac{RT}{nF} \ln \frac{D_{red}^{1/2}}{D_{Ox}^{1/2}} - \frac{RT}{nF} \ln \frac{1}{K_c}$$

$$- \frac{RT}{nF} \ln (K_c + [X^{m-}]_i^p). \quad (5.84)$$

A number of important conclusions can be drawn from this equation: (1) In the absence of complex former, the last two terms become zero and the equation reduces to that of the reduction of free metal ions. (2) The wave is identical in shape with that obtained with free metallic ions and is merely shifted to more negative potentials by the complex former. (3) The dissociation constant of the complex K_c or its stability constant $1/K_c$ can be calculated from the difference between the half-wave potentials of the free metallic ion and the complex when the concentration of the complex former is unity, since the last term will then become essentially zero. (4) The middle term is usually so small as to fall within the experimental error, and can therefore be neglected. (5) The concentration of $[X^{m-}]_i$ changes as more is liberated from the complex at the interface during reduction, hence it is best kept constant by a large excess or by a bufferlike action. (6) The coordination number p can be determined by varying the concentration of X^{m-}; since K_c is very small compared to $[X^{m-}]$, it can be neglected for this purpose. The half-wave potentials then vary linearly with log $[X^{m-}]$, the straight line obtained having a slope of $-p(2.303RT/nF)$.

Lingane [144] tested this relationship by measuring the half-wave potentials of lead in alkaline media as a function of the concentration of hydroxyl ions. He found a p value of 3 and concluded that the ionic state of lead at hydroxyl ion concentrations from 0.01 to 1 N corresponds to the biplumbite ion $HPbO_2^-$. Just as the potential-pH curve for organic substances has inflection points when different dissociation constants are involved, so the potential–log $[X^{m-}]$ curve has inflection points if more than one stability constant exists for a given metallic ion and its complex former [145]. However, all these relationships are only valid if the rates of all the reactions involved (electron transfer, complex formation and dissociation, and complex former availability which is akin to buffer action) are fast enough to represent an essentially reversible process at the dropping mercury electrode. The same criteria hold for this reversibility as in the case of organic reductions, and spreading of anodic and cathodic portions of a wave occur when the reaction rate k_s, as defined in Section 7, p. 372, is of the same order of magnitude or smaller than the "diffusion velocity" (see Fig. 5.24). For details and examples the reader is referred to special reviews [144, 146].

12 APPARATUS

General Introduction

For polarographic analysis, any apparatus is suitable that permits polarizing a microelectrode to a desired potential and measuring the current flowing at that potential. A sensitive and properly shunted galvanometer and a potentiometer, together with a dropping mercury electrode and a large calomel half-cell, are perfectly adequate. However, much simpler as well as more elaborate equipment is available. The choice of apparatus depends on the work for which it is intended.

In this article no attempt is made to describe the many manual and automatically recording polarographs which are now available commercially. Their value varies largely with the quality of the components (resistors, slide-wire, and so on) built into the apparatus. In general, the writer feels that those instruments built primarily as polarographs are to be preferred to those that are built primarily for some other purpose but have been converted into polarographs by means of some attachment.

Recording Apparatus

Two basically different methods are used for the measurement of current: the direct, galvanometric, and the indirect, potentiometric, methods. In the former it is most desirable to have a very sensitive galvanometer with a low internal resistance, a relatively high critical damping resistance, and a fairly long period of swing. If the galvanometer is too fast, the oscillations in current with each drop become too large; if it is too slow, its reading becomes too time-consuming. The *Nejedly* moving coil galvanometer which has been used for the automatically recorded polarograms reproduced in this article, has a sensitivity of 2.1×10^{-9} Ampere per millimeter per meter, an internal resistance of $470 \ \Omega$, a critical damping resistance of $6000 \ \Omega$, and a half-period of swing of 8 sec.

The sensitivity of the galvanometer is regulated conveniently by means of an Ayrton shunt, which should have a total resistance equal to the critical damping resistance of the galvanometer. A galvanometer is critically damped when it indicates the flow of current in as short a time as possible for the type of galvanometer used without overshooting its mark. Obviously, in polarography, where the galvanometer is used to measure currents and not as a null instrument, it should be critically damped at all times. This is especially important in automatically recorded polarograms. The effect of the damping resistance may be visualized by remembering that a galvanometer, deflected from its zero position, generates electricity in its coil while it is moving in the stationary magnetic field on its return to the zero position. A critical resistance or Ayrton shunt can then be found which, when placed across the terminals of the galvanometer, slows down the movement of the coil just enough so that it reaches its zero position without overshooting. Regardless of the setting of the shunt, the galvanometer is then always critically damped.

The shunt can be a variable resistance or a series of fixed resistances. With the variable resistance, a galvanometer can be calibrated to read directly in percent or in milligrams of a particular compound that is analyzed, but each new setting of the resistance requires a new calibration. The fixed resistances, however, if precision-made and properly chosen, make possible the selection of a suitable known fraction of the maximal sensitivity by the mere turning of a switch.

While galvanometer deflections are easily observed and measured visually, they require photographic paper and development if they are to be recorded automatically. The early polarographs all used a photographic recording geared to the continuously changing applied voltage [147]. This allowed extreme simplicity of design, ready accessibility of wiring and connections, and easy adaptability to other purposes [148]. The writer has used one such polarograph continuously since 1935 and finds it still satisfactory. A number of attempts have been made to convert galvanometer deflections into ink recordings, but they have not become popular because of the development of recorders that depend on an entirely different method of current measurement.

In this second method, current is measured indirectly, on the basis of Ohm's law, by observing the small potential drop across a fixed resistance as current flows. The manual application of this method is somewhat tedious because a potentiometer has to be manipulated until the galvanometer, which here serves only as null-point indicator, reads zero near the end of the current surge during the growth of each drop. However, this method is particularly well suited for the adaptation of modern recording equipment, specially built to indicate changes in potential. For satisfactory performance, of course, even these recorders must have the special property of proper response time, corresponding to the critical damping of the galvanometer.

One might list the following requirements for a research model of a modern polarograph:

1. The applied voltage can be calibrated against a standard cell.
2. The applied voltage or the potential of the microelectrode is actually recorded on the polarogram.
3. The voltage range covered is flexible (e.g., from $+1.0$ to -3.0 V; from 0 to -2.0 V; and from -1.0 to -2.0 V).
4. The applied voltage can be increased or decreased continuously (forward and backward polarization).

5. At constant potential, current-*time* curves can be obtained.
6. The current sensitivity of the instrument can be altered in small fixed steps.

In addition there should be the possibility for incorporation of:

a. *Residual current compensation*, which may be necessary in very dilute solutions. The circuit of Ilkovič and Semerano [149], shown in Fig. 5.32, has three resistances that can be so chosen that the charging current (as long as it is linear) is automatically compensated at any applied voltage by a current of equal magnitude but opposite direction. To prevent any appreciably effect

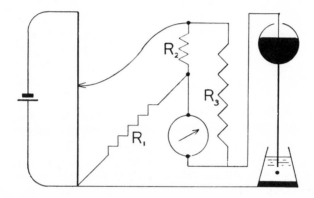

Fig. 5.32   Circuit for compensation of the charging current [149].

of these additional resistances on the applied voltage and the total circuit resistance, R_1 must be large and R_2 small. The value of resistance R_3 is best determined empirically. Optimum values given by Ilkovič and Semerano are $R_1 = 1000 \, \Omega$, $R_2 = 10 \, \Omega$, and $R_3 = 75,000 \, \Omega$.

b. *Compensation of a large current preceding a small one.* Whenever a polarographic wave is preceded by a much larger diffusion current of a more easily reducible substance, the galvanometer sensitivity must be reduced to such an extent that it becomes too small for accurate measurement. By means of an additional electric circuit with battery, it is possible to send through the galvanometer current equal in magnitude to the first wave but opposite in direction (see Fig. 5.33). One can then use higher galvanometer sensitivities for measurement of the more negative wave.

c. *Damping of current oscillations.* When currents are compensated as under (b) the magnification of the more negative currents also magnifies the current oscillations with each drop. These can be damped by connecting an electrolytic condenser of high capacitance (2000 to 5000 μF) across the galvanometer shunt as indicated in Fig. 5.33. This reduces the oscillations tremendously without influencing the sensitivity of the galvanometer. The effectiveness of the capacitor is maximal when the net parallel resistance of the galvanometer and shunt is greatest, that is, when the shunt reduces the galvanometer sensitivity to one-half [150]. Introduction of a variable resistor (1 to 300 Ω) in series with the center tap of the shunt and substitution of two 1000-μF electrolytic condensers back to back for the single one of Fig. 5.33

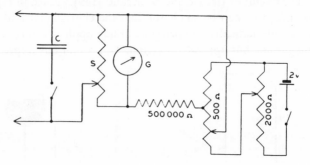

Fig. 5.33 Circuit for compensating an interfering diffusion current and for minimizing the galvanometer oscillations. *C*, Capacitor of 2000 to 5000 μF; *G*, galvanometer; and *S*, Ayrton shunt.

results in a more effective damping over a wider range of sensitivity and drop time values [151].

  d. Kalousek switch (see Section 10, p. 386).
  e. Differential polarography (see Section 13, p. 411).
  f. Derivative polarography (see Section 13, p. 412).

Most polarographs record currents against the applied voltage; the experimenter must then convert the latter into‘ potentials by means of corrections for *IR* at different parts of the current-voltage curve. With small currents and good electrolytes, this correction is often negligible, but it becomes of great significance in organic solvents or when the currents become large. With some modern polarographs this problem has been overcome by the use of a three-electrode system: the dropping mercury electrode, a large working anode, and a reference electrode. The polarograph applies an increasing voltage between the dropping mercury electrode and the working anode but records the potentials between the dropping mercury electrode and the reference electrode against the existing current.

Any other requirements depend on particular needs and personal preferences. The size of the records should be considered from the standpoint of ease and accuracy of reading of data as well as ease of handling and storing.

Automatic recording apparatus has the advantage of being a great time saver; furthermore, the resulting polarograms are continuous curves on which small abnormalities become readily apparent. With the manual apparatus usually relatively few points are plotted unless one expects some abnormalities. The ease with which reproducible curves are obtained with automatic recorders may give false confidence to data which may include artifacts of the apparatus. It is therefore a good policy to make occasional checks of automatically recorded data with a manual procedure.

All recording instruments have in common one disadvantage, namely, a certain lag between the applied voltage and the recorded current. The magnitude of this lag depends on the rate with which the applied voltage is increased and on the speed of the current-indicating device. In the photographically recording polarograph, this lag is usually small and can be easily determined by first polarizing forward, that is, making the dropping mercury electrode more and more negative, and then reversing the process, that is, polarizing backward. The difference between the two half-wave potentials thus obtained is equal to twice the lag. Whenever this technique is not possible, one must prepare current-voltage curves of a solution containing two simple ions with known, well-established, half-wave potentials. After correction for IR, the difference between the two half-wave potentials, compared with the difference theoretically expected, gives an indication of the lag. To minimize this lag, the rate at which the applied voltage is increased should be slow. In general, the time required for the preparation of a polarogram varies from 2 to 20 min, depending on the voltage range covered and on the speed of the recording device.

Dropping Mercury Electrode

The type of dropping mercury electrode used in polarography was originally introduced by Kučera [15] for surface tension measurements of polarized mercury where the weight of the drops was used as criterion. In its simplest form it consists of a very narrow capillary connected by means of thick-walled rubber tubing to a mercury reservoir (leveling bulb), the height of which can be adjusted to produce a desired drop time. Contact between the mercury and the electrical circuit is made by means of platinum wire to prevent contamination of the mercury. If the mercury is pure enough it will exhibit no "tailing" when dry and it will form a fairly stable foam when shaken with pure water [152]. The rubber tubing is about 60 to 70 cm long and must be boiled first in a strong sodium hydroxide solution and subsequently several times in distilled water to remove traces of sulfur and other impurities. After that it should be thoroughly dried overnight. For general work a capillary is recommended that delivers in 3 to 6 sec a drop of mercury weighing between 3 and 6 mg, but considerably faster electrodes have also given satisfactory results. It should be emphasized that *the electrode must be mounted on a vibrationless stand so that the drops fall under their own weight.*

Where this is not possible, one of the now commercially available arrangements in which the drops are mechanically forced off from the tip of the capillary should be used.

The simplest construction of the dropping mercury electrode is also the most flexible and has been used for practically all the developmental work. Certain special capillaries have been described in Section 4, p. 341. Space does not permit a discussion of the many additional modifications that have been proposed, largely with the purpose of substituting glass, neoprene, or plastic for the rubber tubing, or of shielding the electrode against light, convection, or electrostatic disturbances. Many varieties are now commercially available.

While a desirable capillary can be drawn by anyone with some experience in glassblowing, it is probably best to purchase commercial "marine barometer" tubing of very fine, uniform bore (50μ or less). This can be cut to a desirable size and fastened into the rubber tubing if it is large enough, or it can be first sealed to ordinary glass tubing if it is so narrow that only a very short piece of capillary tubing is needed. It can then be calibrated according to the critical pressure and drop weight methods described in Section 3, p. 318, and characterized by means of its capillary constant; these tests can again serve later on to check its behavior after actual use.

Replacement of a defective capillary with a new one not only involves the trouble and time necessary for its installation and calibration but often necessitates an involved and not too accurate recalculation of the results obtained with the discarded capillary. It is therefore of paramount importance that a capillary be kept in good working order as long as possible. To do this, Heyrovský recommends on the basis of many years of experience, that the capillary remain dropping at all times unless it is immersed in distilled water. The beginner usually does not appreciate this important direction until, through neglect of it, he has had to discard a few days' or weeks' work.

After an experiment, the capillary *while dropping* should be washed thoroughly with distilled water, cleaned off with filter paper, and then immersed in distilled water. Only then should the flow of mercury be stopped by lowering the mercury reservoir. (It can be calculated that as long as the capillary orifice is not larger than 0.05 mm in diameter the flow of mercury will stop in water when the mercury pressure is reduced to 22 cm or less.) Since water or an aqueous solution wets glass, while mercury does not, some solution always enters the tip of the capillary and should be washed out after conclusion of the experiment. Keeping the tip of the electrode in distilled water when stopping the flow of mercury keeps it washed out, and the capillary will remain in working order for many years, as has been found by actual experience. Sometimes, especially in studies of biological fluids, water is not enough to wash out the tip. According to the type of solution studied, the tip can then be cleaned, while the mercury is still flowing, by immersion in the proper organic solvent or in concentrated acid or alkali, followed by a rinse with water. It should then have the same capillary constants as when first assembled. If this is not the case, the dropping mercury electrode should be taken apart and the capillary cleaned more thoroughly. First, aqua regia, then water, and finally filtered air is passed through the capillary; a blowing and sucking procedure is usually more effective than either one alone. Unless this cleaning process restores the original characteristics to the capillary, it must be discarded.

To guard against *mercury poisoning*, the greatest care should be taken not to spill any mercury on the desks or floor of the laboratory. A simple precaution is to have a large tray of water beneath the electrode stand so that any spilled mercury is immediately covered with water. A short vigorous airing of the room at the start of work has been found very effective in reducing mercury vapor in a room that has been closed overnight and has had floor cracks full of mercury.

Nonpolarizable Reference Electrode

In early polarographic work carried out with inorganic solutions, a large pool of mercury at the bottom of the electrolysis vessel served as nonpolarizable anode, the potential of which was measured against a standard calomel electrode. The advantages were that the internal resistance of the cell remained small and that reduction products dissolved in the mercury drops were returned to the solution as soon as these drops merged with the anode. The effect of size and other factors on the constancy of such a "pool" electrode has been studied by Majer [153]; as could be expected, the potential of the anode was most stable when the solution contained halide ions.

Instead of the anodic mercury, Schwarz [154] used a silver wire coated with silver chloride which was wound around the tip of a dropping mercury electrode. Since this reference electrode neither takes up any appreciable space nor requires special vessels, it is particularly well suited for microanalyses and routine work.

Since halide ions are usually absent when organic reactions are studied in a buffer solution, Müller and Baumberger [13] introduced the use of an external reference electrode of known and constant potential which is connected to the test solution in the electrolytic cell by means of an agar bridge containing a suitable electrolyte. This technique, which has been customary in potentiometric studies of organic oxidation-reduction systems, offers the advantage that all polarographic curves are referred to the same reference electrode and thus can be compared directly. A further practical advantage is that much less mercury is needed for the measurements.

The most widely used reference electrode is the saturated calomel half-cell, but other standard half-cells are equally satisfactory. Of these, the hydrogen electrode is especially convenient to use because its potential can be varied at will by the experimenter simply by using different buffers [155]. If these electrodes are also used as working electrodes, they must have large surfaces to prevent polarization, that is, any change in their potential, while current is flowing.

The *agar bridges* are usually saturated with potassium chloride to minimize the liquid junction potential. However, such bridges are not suitable when potentials more positive than $E_h + 0.250$ V are studied, since chloride ions interfere [13]. In this case it is best to work with two bridges, first a potassium chloride bridge which preserves the calomel half-cell from contamination and leads to a junction vessel containing saturated potassium chloride that can be replaced frequently. The second bridge then leads to the test solution; it can be fitted with ground-glass plugs at either end and filled with the test solution, or may simply be a potassium nitrate–agar bridge.

Agar bridges may be prepared by warming in a water bath a suspension of 3 gm of powdered agar-agar in 100 ml of either saturated potassium chloride or 13 % potassium nitrate solution. When the solution is homogeneous, it is sucked into glass U-tubes of suitable dimensions, where it gels after cooling.

It must be remembered that the use of these bridges introduces an additional resistance into the circuit which changes when the bridges become old or when part of the salt diffuses out from the bridges. The resistance should therefore be checked at frequent intervals. The error attributable to liquid junction potentials is usually so small that it can be neglected. If necessary, it can be determined by adding to the solution an internal indicator, that is, a substance that produces a good polarographic curve with a known half-wave potential.

Vessels Used with the Dropping Mercury Electrode

The most important considerations in designing polarographic vessels are: (1) the volume of the solution analyzed; (2) whether or not the analyses are to be carried out in the absence of air; (3) whether an external or internal reference electrode is used; (4) whether or not the work involves routine analyses; (5) how easily the vessel can be cleaned; and (6) what kind of temperature control is available. Of the many designs already in use and commercially available, only a few typical ones can be mentioned. Modifications of these, meeting individual requirements, will suggest themselves. It should be remembered, however, that when a "pool" reference electrode is used it must be connected to the circuit by means of a platinum wire.

The simplest type of vessel is an open beaker or test tube with a pool of mercury on the bottom as reference electrode. For best results the tip of the capillary should dip at least 2 mm below the surface of the solution and be at least that far removed from the walls of the vessel and from the bottom layer. For work in air-free solutions, the vessel must be stoppered, and an inlet and exit tube for the passage of an indifferent gas, such as nitrogen or hydrogen, must be provided. The most common vessel for such work is the conical vessel A in Fig. 5.34, in which the platinum contact and the inlet and exit tubes for the indifferent gas are built in. When external reference electrodes are used, the platinum contact becomes unnecessary and an agar bridge is used in connection with vessels of type B (Fig. 5.34). They may be supplied either with or without the outlet at the bottom which is closed with a standard vial stopper. Through the latter a fine hypodermic needle can be

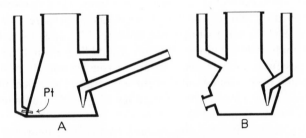

Fig. 5.34 Types of vessels used for polarographic analysis.

inserted which permits removal or addition of air-free solutions in exact quantities by means of commercially available microsyringes.

For routine analyses, an external reference electrode is unquestionably preferable to internal electrodes, not only for convenience in operation but also because all measurements are thus automatically referred to the same potential. The writer has found very convenient an arrangement in which the dropping mercury electrode is clamped firmly and permanently in a chuck on a heavy stand. A calomel half-cell with a large surface (at the bottom of a 50-ml Erlenmeyer flask), two agar bridges, and a junction vessel containing saturated potassium chloride solution are also mounted on this stand. The first bridge, in the form of a U-tube, is fastened to the capillary by means of a two-hole rubber stopper in such a way that its one tip is level with the tip of the dropping mercury electrode. A series of solutions, contained in small beakers, can then be analyzed in rapid succession simply by moving each beaker onto a little platform beneath the capillary. Between different samples the capillary and bridge are washed with distilled water and dried with filter paper, thus preventing any measurable contamination of the solutions. A large evaporating dish filled with water is placed at the bottom of the stand to catch any spillage. When the analyses have to be carried out in the absence of air, vessels of type B (Fig. 5.34) are used and shoved onto the rubber stopper. A number of solutions can be freed simultaneously from atmospheric oxygen before the analysis. This prevents any delay resulting from the purging process.

In other designs the calomel half-cell and test solution are separated by a porous diaphragm. Lingane and Laitinen [156] recommend an H-shaped vessel in which the calomel half-cell on one side is separated from the test solution on the other side by a sintered glass disk and an agar plug. In the similar "Kalousek" vessel [141], the reference electrode is separated from the test solution by an ungreased stopcock. In a modification of the H-type vessel, the calomel half-cell and sintered glass part are connected to the test solution side by means of a standard taper joint; thus the vessel can be cleaned more easily and it is no longer necessary to prepare a new calomel electrode for each vessel. Laitinen and Burdett [157] designed a vessel that permits the analysis of solutions while gas is bubbling through them. This is particularly useful in amperometric titrations (see Section 13, p. 410).

It has already been pointed out that the temperature at which analyses are carried out should be known with an accuracy of $\pm 0.5°C$. Room temperature seldom varies more than that during the time necessary for an experiment, but from day to day it may vary enough to make comparison of results laborious. An automatically controlled thermostat is therefore of great convenience for routine work, and most of the designs mentioned above are such that the cell and reference electrode can be immersed in a water bath. The stirrer of the water bath should be so mounted that its vibrations do not jar the dropping mercury electrode, or it should be turned off during the short time necessary for an actual recording of the current-voltage curve. When a range of temperatures is investigated, it is simpler to use as a thermostat a large beaker filled with water which can be kept at sufficiently constant temperature by the addition of either hot or cold water as required.

13   EXPERIMENTAL TECHNIQUE

General Technique and Applications

Introduction

The number of organic and inorganic compounds reducible or oxidizable at a dropping mercury electrode is extremely large, and the potential range in which these electrochemical reactions take place is rather limited. Hence it is almost always necessary that a preliminary chemical treatment and separation into suitable fractions be carried out before the polarographic analysis is performed. Often this may become so complicated that other available methods are preferable. In some instances, however, polarography has proved the only workable method of analysis; in others it has replaced older methods because of its excellent reproducibility, high sensitivity, and other advantageous features. Among these may be mentioned: (1) the rapidity of operation; (2) simultaneous quantitative and qualitative analysis of several components of a solution; (3) the fact that in most cases the solution remains unaltered during the analysis and can be used for other purposes; (4) the possibility of analyzing quantities the size of a drop; and (5) the automatic graph of a recording apparatus which provides a permanent record of the analysis. Simple manual equipment is recommended for preliminary studies and for the novice at polarography, who can gain experience by carrying out a variety of typical practice experiments [158]. Once a particular procedure has been fully developed, the polarographic analyses can be carried out by anyone with some chemical training. It is well, however, to have expert advice available, since experience has shown that even in the simplest analyses irregularities of unforeseen origin are bound to occur.

To be suitable for direct polarographic analysis, a substance must be electroreducible or -oxidizable within the range of the electrode. It must be in true solution and it must be stable for the duration of the measurement so that a pure diffusion current can be obtained. In addition, substances affecting the surface tension of mercury and substances that have buffer or catalytic activity may be analyzed indirectly.

While nearly all inorganic ions are reducible, in general, only those organic compounds are reducible at the dropping mercury electrode that are reducible by chemicals (including dissolving metals), as contrasted with catalytic reductions [159]. Thus, for instance, simple purely aliphatic olefins are not reducible at the dropping mercury electrode, while the presence of an aryl, carbonyl, or ethylene group, conjugated with the unsaturation, makes this reduction possible. With the extension of the potential range of the dropping mercury electrode to potentials as negative as -2.9 V (versus SCE) measurable polarographic waves have been obtained even with a number of phenyl-substituted olefins and acetylenes [160] and aromatic polynuclear hydrocarbons [161], although their half-wave potentials are all more negative than -2.0 V.

Aside from the reversible oxidation-reduction systems, the following groups of substances have been studied polarographically with good success: unsaturated hydrocarbons, aldehydes, ketones, acids and their derivatives; organic nitro, nitroso, and azoxy compounds; sulfur compounds; organic peroxides and hydroperoxides; and oxygen-, sulfur-, and nitrogen-containing heterocyclic compounds.

In general, organically bound halogen is reducible. This reduction is independent of the pH of the solution, a fact that can be used to recognize the type of reaction in complicated cases [162]. The ease of reduction is usually in the order iodide > bromide > chloride [163], but in the case of p-fluoroiodobenzene, the fluorine is reduced before the iodine [164]. If there is more than one halogen on a carbon atom, there are more waves and at more positive potentials [165]. Allylic halides are more easily reduced than vinyl halides. The reduction of a halogen becomes more difficult the longer the aliphatic chain to which it is attached; an aromatic ring behaves similarly to a long chain.

Sometimes it is possible that an organic substance that is not reducible at a dropping mercury electrode can be altered in such a way that it becomes electroactive and, consequently, determinable. The reactions include oxidations [166], nitrations [167], formation of complexes with Girard reagents [168] and others.

Outside of these organic reactions, there are of course many metalloorganic reactions which have served to identify and analyze organic compounds. One should also mention reactions involving the anodic oxidation of the mercury of a dropping electrode which has analytical value when it is limited by complex formation or precipitation with some organic compound. For additional detail the reader is referred to a special monograph by Zuman [169].

Range of Potentials

The potential range over which the dropping mercury electrode can be used depends on the composition of the supporting electrolyte. In the absence of ions with which mercury forms complexes or precipitates, potentials as positive as $+0.4$ V (versus SCE) can be attained before appreciable oxidation of mercury to mercurous ions takes place. The presence of some anions of organic acids, of halogen, hydroxyl, cyanide, and especially sulfide ions reduces this limit considerably (to as much as -0.6 V, versus SCE). The study of reactions with fairly positive potentials should therefore always be carried out in the absence of these ions and an external reference electrode should be used, connected to the cell by a potassium nitrate bridge. For studies of still more positive potentials, other microelectrodes, preferably platinum, must be used.

As far as the negative potential limits are concerned, the dropping mercury electrode is unexcelled because of the large hydrogen overvoltage. The most negative potentials attainable with it in the presence of buffers have been indicated in Fig. 5.25. At pH > 8, the alkali and alkaline earth metals limit the potential range (at about -1.8 to -2.2 V), but Pech [170] found that quaternary amines permitted an extension of this limit to -2.6 V (versus SCE). (Laitinen and Wawzonek [160], using tetrabutylammonium iodide, were able to study reductions as negative as -2.8 V versus SCE.) In such media the interfacial tension of mercury approaches zero as the potential becomes very negative, so that the mercury is dropping very rapidly from the capillary in an almost continuous stream [21]. Naturally, these solutions cannot be buffered, and one must consider possible changes in the pH at the interface brought about by the reaction. Some information about the effect of pH may be obtained, however, by using different concentrations of the hydroxides of the tetraalkylammonium salts. These hydroxides can be prepared by treating solutions of the purified tetraalkylammonium halides with an excess of moist silver oxide [170].

Range of Concentrations

The polarographic method is applicable over a concentration range of 10^{-6} to 10^{-2} M; for best results, however, a 10^{-4} to 10^{-2} M solution is desirable. In the latter case, an accuracy of $\pm 1 \%$ can be reached by repeating polarograms of the same solution several times and averaging the individual measurements of the diffusion current. This can easily be done if the volume of the solution is 1 ml or more because the quantity of material electrolyzed is so small that the body of the solution remains unaltered and subsequent polarograms are identical. When smaller volumes of solution are analyzed, special precautions are necessary since the quantity of material removed by electrolysis and also the end products formed become significant under these conditions. In general, the accuracy and sensitivity of the method are as follows: In practice, 1 ml of a 10^{-5} M solution, or 1 μg of material with a molecular weight of 100, can be analyzed with an accuracy of ± 1 to 3%. Under extreme conditions, 0.01 ml of a 10^{-6} M solution, or 0.001 μg of the same material, can be analyzed but the error is about $\pm 100 \%$.

Current Measurements

The measurement of the diffusion current as a rule presents no difficulties. A preliminary "blank" curve obtained with the supporting electrolyte indicates the residual current that must be subtracted from the test curve to obtain the net diffusion current. This technique is of little help if the unknown produces several waves on the polarogram or changes the electrocapillary curve of mercury. In these instances it is best to draw parallel lines to the beginning and end of the waves, as shown in Fig. 5.11. The slope of these lines automatically subtracts the charging or residual current.

For quantitative work, it is convenient to prepare calibration curves in which diffusion current is plotted versus concentration. The concentration of an unknown can then be read directly from the graphs, which must be straight lines passing through zero if the Ilkovič equation holds. Some authors [171] measure wave heights by various empirical graphical methods, with results differing from the theoretically correct diffusion currents. Hence calibration curves prepared from such data are often not straight lines and do not pass through zero. The so-called "increment-method" [172], in which the wave height is measured as the difference in current at two applied voltages equidistant from the half-wave potential, is equally arbitrary. Nevertheless, these empirical calibration curves are valuable for practical work, provided that they are prepared by analyzing standard solutions with a composition approximating that of the unknown.

If the unknown is more concentrated than 10^{-3} M, it is usually possible to work with the solutions open to air, but the current of the oxygen curve must be subtracted from the "test" curve. In more dilute solutions, the atmospheric oxygen must be removed by chemical means (with sodium sulfite in neutral or alkaline solution) or by bubbling an inert gas (hydrogen, nitrogen) through the solution. The latter method is preferable because it does not alter the composition of the solution except in the presence of volatile constituents. In such cases complications are avoided by passing the gas through a wash bottle containing some of the solution to be analyzed before bubbling it through the test solution. In all other instances the gas should be saturated with water vapor before passage through the electrolysis vessel to prevent evaporation of water from the test solution. The gases used should be tested polarographically for their oxygen content; traces of oxygen which may be present can be removed by passing the gas through heated copper turnings or through oxygen-absorbing solutions. Of course, the flow of gas through the test solution must be stopped during the recording of a polarogram because the drops of mercury should never be forced off by motion in the solution but should fall under their own weight.

In addition to the comparison with calibration curves, two other methods are available for quantitative analysis. The *method of standard addition* [173] is particularly advantageous in isolated determinations. As the name implies, a known standard solution is added to the test solution and polarograms before and after are compared. The accuracy of this method depends on the precision with which the two volumes of solution and the corresponding diffusion currents are measured. The *method of step quotients* [174] simplifies polarographic work with different capillaries and in solutions of different viscosity because the *ratio* of the wave heights of two ions in any given solution is independent of the capillary used and of the viscosity of the medium. Once this ratio, the "step quotient," of two substances of equivalent concentrations has been determined, any other concentration of either substance can be found. A standard solution of the mate, the "pilot substance," is prepared and a known volume of it is added to a known volume of the test solution. The mixture is then polarographed and the concentration of the unknown calculated from the ratio of the observed diffusion currents, the known concentration of the pilot substance, and the step quotient. For greatest accuracy, the wave height of the two reactants should be about equal.

It is essential of course that the unknown solution originally contain none of the material used as the pilot substance.

In addition to this quantitative application, the pilot substance technique has been used for some time for potential measurements in which the half-wave potential of the pilot substance, or indicator, was taken as the standard of reference [175]. As mentioned previously, this is permissible only when the potential of the large electrode remains constant.

Effect of Overlapping Waves

Qualitative as well as quantitative analysis in general becomes increasingly difficult as the number of reducible or oxidizable substances in a given solution increases because the waves often overlap or come so close together that their resolution is impossible. Figures 5.35 and 5.36 have been prepared to demonstrate this need for separation and for a favorable relation of the heights of individual waves [176]. The curves shown are theoretical current-potential curves of two reversible reductions at 30°C, calculated for a two-electron change per molecule. Since these are current-*potential* curves, they are independent of the actual scale of the current axis, that is, the curves should look the same whether the sum of the two waves is 10 or 1000 μA.

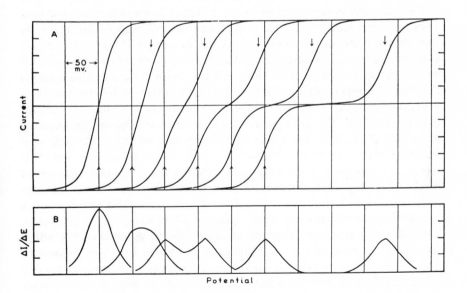

Fig. 5.35 (*A*) Theoretical current-potential curves of two waves of equal height with half-wave potentials (indicated by arrows) separated by 0, 30, 60, 90, 120, and 180 mV. (*B*) $\Delta I/\Delta E$ of the above curves plotted against the potential [176].

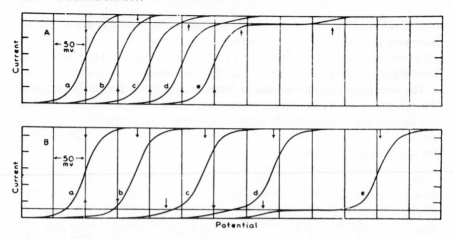

Fig. 5.36 Theoretical current-potential curves of two waves with half-wave potentials (indicated by arrows) separated by 0, 30, 60, 90, and 180 mV [176]. The heights of these waves are in ratios of (*A*) 10:1 and (*B*) 1:10.

In Fig. 5.35 the ratio of the two waves is unity; in Fig. 5.36, the more positive wave is 10 times as high as the more negative wave in the upper series *A*, while the opposite is true in the lower series *B*.

The half-wave potentials of the waves of Fig. 5.35 are indicated by arrows. They are separated by 0, 30, 60, 90, 120, and 180 mV, respectively. It may be seen that a difference of at least 90 mV between the half-wave potentials is necessary to produce measurable waves. If the difference is only 30 mV, the waves overlap to such an extent that a determination of wave heights becomes impossible. A separation of 60 mV in this case is just enough to see that there are two inflections, but no accurate measurements can be made. If the change of current per unit change in potential, $\Delta I/\Delta E$, is plotted against the potential as in part *B* of Fig. 5.35, a sharp peak will be obtained at the half-wave potential of each wave. In a double wave two such peaks are observed if the waves are separated sufficiently. These peaks are proportional to wave height when only single waves are measured. Such increment ratios are often used in potentiometry; they have been adapted to polarography in the so-called "derivative polarography" (Section 13, p. 412).

Curves with similar separations of the half-wave potentials, but with a less favorable ratio of the two waves, are shown in Fig. 5.36. It may be observed that even at a separation of 90 mV the measurement of the waves is somewhat difficult. The small waves can be measured with greater accuracy if they are magnified by increasing the galvanometer sensitivity; in the upper series *A*, the first diffusion current must be compensated to make this possible.

These same figures can be used for a consideration of reactions involving one electron per molecule, by doubling the scale of the abscissa. A minimum of 180 mV is then found necessary for a satisfactory separation of the half-wave potentials, leading to the general conclusion that for measurements of simultaneous reductions the midpoints of waves must be separated by at least $RT/nF \ln 1000$.

The above considerations hold as long as there is no interaction between the end products of the more positive reaction and the starting material of the more negative reaction. Otherwise, there may be some effect on either curve. A clear-cut example of such interactions is the effect of hydroxyl ions, formed during the reduction of oxygen, on the hydrogen wave of a dilute solution of hydrochloric acid, first observed by Kemula and Michalski [114].

Measurement of Circuit Resistance

The resistance of the circuit, which must be known for all IR corrections, can be determined in several ways. Undoubtedly, the conventional measurement by means of a Wheatstone bridge is the most accurate [177]. With this method one finds a minimum resistance just before the drop falls off, and Ilkovič [21] has shown that this value must be multiplied by $\frac{4}{3}$ to represent the mean resistance necessary in our calculations. The latter can also be obtained directly from the polarographic curves. For instance, (1) the slope of a steep maximum, as in the case of the oxygen reduction (Fig. 5.9), (2) the slope of a line combining the anodic and cathodic half-wave potentials of a reversible system, polarographed both in the oxidized and the reduced forms, and (3) the slope of a line passing through the half-wave potentials of a given substance reduced at different concentrations under otherwise identical conditions, are all equal to the reciprocal of the resistance of the circuit, in accordance with Ohm's law. In still another method [178] different voltages are applied to the cell from a B battery and the corresponding current is noted on a milliammeter. By means of Ohm's law, resistances are calculated that approach a limit as the applied voltage increases past 9 or 10 V. Even better results can be obtained if the voltage is plotted against the observed current; it will be found that after a certain minimum voltage is exceeded the current increases almost linearly with a slope equal to the reciprocal of the cell resistance, as above.

Since it is not necessary to know the circuit resistance more accurately than to a few hundred ohms, the resistance of the capillary itself (about $50\ \Omega$) and that of the potentiometer and lead wires can be neglected. However, the resistance of the galvanometer and shunt may become appreciable, depending on the type used and on the setting of the shunt.

Polarometric Determinations

Polarometric Studies of Reaction Kinetics

Studies with the manual apparatus suggest that whenever a plateau is reached at the end of a polarographic wave a constant applied voltage can be used for the determination of the substance. At this fixed voltage the diffusion current should be a linear function of the concentration. In making this simplified assumption, however, one must be sure that (1) the diffusion current is corrected for the residual current and any current that might be caused by the preceding discharge of another ion, and (2) no significant error is produced by small variations in the applied voltage or by changes in the electrode potential attributable to different *IR* corrections with changes in current. One should therefore select for such measurements an applied voltage at which a level plateau in the polarographic curve indicates that small changes in the applied voltage produce no measurable change in the diffusion current.

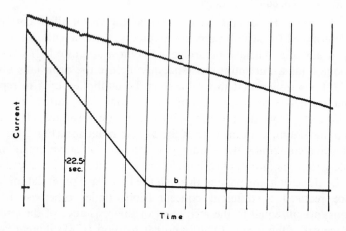

Fig. 5.37 Polarographic record of current-time curves representing rates of oxygen consumption of yeast cells: (*a*) in the absence of substrate; (*b*) in the presence of glucose [176].

In polarometric studies of reaction kinetics, a suitable fixed voltage is applied to the cell and the change in current is observed over a period of time. By means of the polarograph, a record can be prepared automatically. This method has been most widely used in the study of oxygen consumption by biological materials. An example is shown in Fig. 5.37, where the rate of oxygen consumption of yeast cells is given by the slope of the current-time curves [176]. (The applied voltage was constant at −0.5 V versus *SCE*.) Note that the rate increases in the presence of substrate but is independent of the oxygen concentration which is constantly falling. This technique has also been applied successfully in organic chemistry to such diversified studies as: the coupling of diazotized *p*-anisidine and 3-methyl-1-phenyl-5-pyrazolone [179]; the diazotization of aniline hydrochloride [179]; the formation of complexes between benzil and boric acid [180]; and the cyclization (to p-cymol) of citral [181].

Instead of continuous analysis at constant potential, it is often more advantageous, especially for slow reactions, to make complete polarograms of a reaction mixture at definite intervals of time during the reaction. This procedure has been used for years. The following studies may serve as examples: the hydrolysis of acyl derivatives of nitrophenols and nitranilines [182]; the alkaline splitting (to acetaldehyde and methyl heptenone) and the alkaline condensation (with acetone to pseudoionone) of citral [181]; the catalytic hydratation of acetylene in sulfuric acid solutions [183] and the effect of the half-wave potential of dyestuffs on their rate of reduction by phenyl-hydrazine [184].

Amperometric Titrations

In amperometric titrations, originally called "polarometric titrations" by Majer [185], the current at a constant applied voltage is observed while known quantities of a titrating agent are added to the solution. From a graph of the data, the end point of the titration can be determined.

This type of analysis is suitable for precipitations, neutralizations, or oxidation-reduction titrations. To account for the change in volume of the solution during the titrations, the observed currents should be multiplied by the factor, $(V + X)/V$, where V is the starting volume of the solution and X is the volume of titrating agent added [185]. Thus straight lines are usually obtained which permit extrapolations to the end points.

Figure 5.38 shows four of the most common types of curves obtained in amperometric titrations. In A, the electroactive substance is removed from the solution by precipitation with an inactive substance (e.g., lead titrated with sulfate ion [185]). In B, to an inactive substance an electroactive precipitating agent is added (e.g., sulfate titrated with barium ion [185]). In C both the component of the solution and the precipitant are reducible at the given applied voltage (e.g., lead titrated with dichromate ion [186]). Finally, in D, the titrating agent produces a linear decrease in current until negative values are reached. This type of curve can be obtained by an oxidative-reductive titration (e.g., titanium titrated with dichromate ion [142]) or by precipitation of a polarographically determinable anion with a suitable cation (e.g., chloride titrated with silver ion [176]). The end point in this case is reached when the current becomes zero (corrected for any residual current), that is, where the straight titration curve intersects the galvanometer zero line. These polarometric titrations are fast and accurate but have limitations similar to conductivity measurements.

For further examples and details, the reader is referred to the special reviews on the subject by Laitinen [187] and the book by Stock [188].

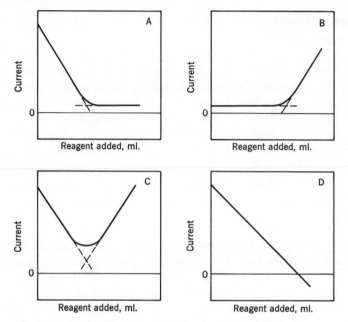

Fig. 5.38   Types of amperometric titration curves (see text).

Differential Polarography

A most promising technique of differential polarography has been developed by Semerano and Riccoboni [189]. In this procedure two matching capillaries connected to the same mercury reservoir dip into two solutions, one of which contains the unknown, the other one only the supporting electrolyte. The two anodes, which are necessary, are connected through suitable resistances (1) with each other, (2) with the galvanometer, and (3) with the proper end of the slide-wire of the polarograph. With this arrangement all currents resulting from the supporting electrolytes balance out; it is therefore not necessary to free the solutions from atmospheric oxygen unless its reduction products, hydrogen peroxide and hydroxyl ions, cause trouble. Even at voltages where the alkali metals are deposited, the current flow through the galvanometer is zero if the composition of the two solutions is identical. A practical disadvantage of this method is that because of small differences in surface tension of the two solutions the original synchronism of the two dropping mercury electrodes is thrown out of phase so that the current, instead of being steady, shows large undulations around some mean value. This should be readily overcome by use of a streaming mercury electrode or other dropping mercury electrodes where synchronization can be enforced mechanically (see Section 3, p. 324).

Derivative Polarography

As shown earlier (Fig. 5.35), if the rate of change of current with respect to potential is recorded rather than the current itself, the curves show marked peaks at the half-wave potentials and the height of these peaks is proportional to the concentration of the reactive material if there is no overlapping. Such curves can be automatically recorded with the polarograph by using either one or two dropping mercury electrodes [190].

Derivative polarography with two dropping mercury electrodes. The electrodes in this method must have identical characteristics and must be perfectly synchronized with regard to their drop time. Voltage.applied to them is made to differ always by a small amount; hence the current flowing between them, which is recorded, reaches a maximum value at the half-wave potential.

Derivative polarography with one dropping mercury electrode. Heyrovský [190] has described several ways of recording derivative curves by this method. In one of these, the current from a dropping mercury electrode is sent through a resistance (e.g., 300 Ω) which is in parallel with the galvanometer and a large condenser (e.g., 2000 μF). A constant rate of change of the applied voltage is essential. Lingane and Williams [191] have analyzed in detail the necessary requirements for making such circuits most reliable from a qualitative and quantitative standpoint. Instead of this circuit a double-coil galvanometer or an induction coil can be used to produce the derivative curves [190]. As Ishibashi and Fujinaga [135] have shown, the Kalousek switch (see Section 10) can also be used to prepare derivative curves. For analytical purposes these have the advantage of enhanced sensitivity.

Potentiometric (Controlled Current-Scanning) Polarography

Adams et al. [192] have developed a form of polarography in which the current is kept constant and the varying electrode potentials are observed. This has been called potentiometric polarography or controlled current-scanning polarography. The circuit is essentially identical with that used by Ilkovič [24] in the determination of the polarization capacity of the dropping mercury electrode. Voltage of 50 V or more is applied to the cell through a resistance of several megohms; thus the current is not affected by small losses in voltage resulting from polarization of the electrode. The potentials at the electrode are measured against a known reference electrode by means of a vacuum tube voltmeter in a separate circuit. Ilkovič [24] automatically recorded the voltage fluctuations during the growth of each drop by means of a string galvanometer. Adams et al. [192] showed that current-voltage curves determined by the current-scanning method were essentially identical with conventional polarographic curves. Fujinaga [193] has constructed a special apparatus by means of which such applied voltage-potential curves can be recorded automatically. The recorded polarograms have contours similar to those of ordinary polarographic waves but without maxima or minima. Also, certain abnormalities occur when two reducible substances are analyzed simultaneously [193].

Other Polarographic Developments

A number of developments that properly could be included here have become so extensive that they are treated in separate chapters in this book. They include oscillographic and alternating current polarography, polarography with microelectrodes other than dropping mercury, and polarography in nonaqueous media.

References

1. J. Heyrovský. *Chem. Listy*, **16**, 256 (1922); *Phil. Mag.*, **45**, 303 (1923).
2. J. Heyrovský, *Bibliography of Publications Dealing with the Polarographic Method in 1964*, Academia, Prague, 1966.
3. G. Semerano, *Bibliografia Polarografica (1922–1949)*, *La Ricerca Scientifica*, **19**, Suppl. A (1949).
4. L. Jellici and L. Griggio, *Bibliografia Polarografica (1922–1967)*, Consiglio Nazionale delle Ricerche, Roma, Suppl. 20 (1968).
5. *Polarographic Literature—Data Cards*, Vols. I and II, Polarographic Society of Japan, Ed., Gikensha Co., Ltd. Tokyo.
6. J. Heyrovský and M. Shikata, *Rec. Trav. Chim.*, **44**, 496 (1925).
7. I. Langmuir, *J. Am. Chem. Soc.*, **40**, 1384 (1918).
8. L. Michaelis and M. L. Menten, *Biochem. Z.*, **49**, 333 (1913).
9. H. Lineweaver and D. Burk, *J. Am. Chem. Soc.*, **56**, 658 (1934).
10. K. A. Hasselbalch, *Biochem. Z.*, **78**, 112 (1916).
11. D. J. G. Ives and H. J. Janz, *Reference Electrodes, Theory and Practice*, Academic Press, New York, 1961.
12. O. H. Müller, *J. Chem. Educ.*, **18**, 227 (1941).
13. O. H. Müller and J. P. Baumberger, *Trans. Electrochem. Soc.*, **71**, 181 (1937).
14. O. H. Müller, *J. Am. Chem. Soc.*, **66**, 1019 (1944).
15. G. Kučera, *Ann. Physik*, **11**, 529 (1903).
16. O. H. Müller, *The Polarographic Method of Analysis*, 2nd ed., Chem. Education Publishing Co., Easton, Pennsylvania, 1951, p. 191.
17. G. S. Smith, *Trans. Faraday Soc.*, **47**, 63 (1951).
18. C. G. Barker and A. W. Gardner, "The Effect of the 'Back Pressure' on the Diffusion Current," in *Advances in Polarography*, I. S. Longmuir, Ed., Vol. I, Pergamon Press, Oxford, 1960, p. 330.
19. G. Lippmann, *Ann. Physik. Chem.*, **149**, 547 (1873).
20. G. Gouy, *Ann. Chim. Phys.*, [7], **29**, 145 (1903); [8], **8**, 291, and **9**, 75(1906).

21. D. Ilkovič, *Collection Czech. Chem. Commun.*, **4**, 480 (1932).
22. R. E. Benesch and R. Benesch, *J. Phys. Chem.*, **56**, 648 (1952).
23. O. H. Müller, *Anal. Chem.*, **23**, 1175 (1951).
24. D. Ilkovič, *Collection Czech. Chem. Commun.*, **8**, 170 (1936).
25. D. C. Grahame, *J. Am. Chem. Soc.*, **63**, 1207 (1941).
26. B. Breyer and H. H. Bauer, *Alternating Current Polarography and Tensammetry*, Interscience, New York, 1963.
27. J. Heyrovský and M. Bureš, *Collection Czech. Chem. Commun.*, **8**, 446 (1936).
28. H. J. Antweiler, *Z. Elektrochem.*, **43**, 596 (1937); **44**, 719, 831, and 888 (1938).
29. O. H. Müller, *J. Chem. Educ.*, **18**, 172 (1941).
30. M. v. Stackelberg and R. Doppelfeld, "Die Polarographischen Maxima," in *Advances in Polarography*, I. S. Longmuir, Ed., Vol. I, Pergamon Press, Oxford, 1960, p. 68.
31. T. A. Kryukova, *Zavodskaya Lab.*, **9**, 699 (1940); *Chem. Abstr.*, **37**, 2679 (1943); *J. Phys. Chem. (USSR)*, **21**, 365 (1947); *Chem. Abstr.*, **41**, 6160 (1947).
32. T. A. Kryukova, *Zavodskaya Lab.*, **14**, 767 (1948); *Chem. Abstr.*, **44**, 5757 (1950).
33. J. Rasch, *Collection Czech. Chem. Commun.*, **1**, 560 (1929).
34. D. Ilkovič, *Collection Czech. Chem. Commun.*, **6**, 498 (1934).
35. O. H. Müller, "Development of Constants in Polarography: A Correction Factor for the Ilkovič Equation," Chapter 27, in *Electrochemical Constants*, *Natl. Bur. Standards (U.S.) Circ.*, No. 524 (1953).
36. O. H. Müller, *J. Biol. Chem.*, **145**, 425 (1942).
37. J. J. Lingane and B. A. Loveridge, *J. Am. Chem. Soc.*, **66**, 1425 (1944).
38. J. J. MacDonald and F. E. W. Wetmore, *Trans. Faraday Soc.*, **47**, 533 (1951).
39. W. Hans, W. Henne, and E. Meurer, *Z. Elektrochem.*, **58**, 836 (1954).
40. L. Airey and A. A. Smales, *Analyst*, **75**, 287 (1950).
41. I. Smoler, *Collection Czech. Chem. Commun.*, **19**, 238 (1954).
42. J. Kůta and I. Smoler, "The Instantaneous Currents (i-t Curves) on Single Drops," in *Progress in Polarography*, P. Zuman and I. M. Kolthoff, Eds., Vol. I, Interscience, New York, 1962, p. 43.
43. J. Flemming and H. Berg, *J. Electroanal. Chem.*, **8**, 291 (1964).
44. H. Strehlow and M. v. Stackelberg, *Z. Elektrochem.*, **54**, 51 (1950).
45. H. Strehlow, O. Mädrich, and M. v. Stackelberg, *Z. Elektrochem.*, **55**, 244 (1951).
46. J. Koutecky, *Czech. J. Phys.*, **2**, 50 (1953).
47. A. Kimla and F. Stráfelda, *Collection Czech. Chem. Commun.*, **28**, 3206 (1963).
48. O. H. Müller and J. S. Davis, Jr., *J. Biol. Chem.*, **159**, 667 (1945); *Arch. Biochem.*, **15**, 39 (1947).
49. O. H. Müller and J. P. Baumberger, *J. Am. Chem. Soc.*, **61**, 590 (1939).
50. O. H. Müller, "Rate-controlled Reactions as Illustrated by the Reduction of Pyruvic Acid," in *Advances in Polarography*, I. S. Longmuir, Ed., Vol. I, Pergamon Press, Oxford, 1960, p. 251.
51. R. Brdička, *Collection Czech. Chem. Commun.*, **12**, 212 (1947).

52. R. Brdička and K. Wiesner, *Collection Czech. Chem. Commun.*, **12**, 138 (1947); J. Koutecky and R. Brdička, *ibid.*, **12**, 337 (1947); J. Koutecky, *ibid.*, **18**, 11, 183, 597 (1953); *ibid.*, **19**, 857, 1045, 1093 (1954); P. Delahay, *J. Am. Chem. Soc.*, **73**, 4944 (1951); *ibid.*, **74**, 3506 (1952); T. Berzins and P. Delahay, *ibid.*, **77**, 6448 (1953); P. Delahay and G. L. Stiehl, *ibid.*, **74**, 3500 (1952); H. Strehlow, "Electrochemical Methods," in *Investigation of Rates and Mechanisms of Reactions*, S. L. Friess, E. S. Lewis, and A. Weissberger, Eds., 2nd ed., Interscience, New York, 1963, p. 799.

53. J. Koutecky, *Collection Czech. Chem. Commun.*, **18**, 311 (1953); J. Weber and J. Koutecky, *Chem. Listy*, **49**, 562 (1955).

54. R. Brdička, *Collection Czech. Chem. Commun.*, **19**, Suppl. II, S41 (1954).

55. I. M. Kolthoff and E. P. Parry, *J. Am. Chem. Soc.*, **73**, 3718 (1951).

56. S. M. Cantor and Q. P. Peniston, *J. Am. Chem. Soc.*, **62**, 2113 (1940).

57. K. Wiesner, *Collection Czech. Chem. Commun.*, **12**, 64 (1947).

58. R. Bieber and G. Trümpler, *Helv. Chim. Acta*, **30**, 706 (1947); K. Vesely and R. Brdička, *Collection Czech. Chem. Commun.*, **12**, 313 (1947).

59. K. Wiesner, *Z. Elektrochem.*, **49**, 164 (1943).

60. E. Svátek, *Proc. Intern. Polarog. Congr. Prague*, **1**, 789 (1951); *Chem. Abstr.* **46**, 10956 (1952).

61. R. Brdička and C. Tropp, *Biochem. Z.*, **289**, 301 (1937); R. Brdička and K. Wiesner, *Naturwissenschaften*, **31**, 247 (1943); *Collection Czech. Chem. Commun.*, **12**, 39 (1947).

62. R. Brdička, K. Wiesner, and K. Schäferna, *Naturwissenschaften*, **31**, 390 (1943).

63. R. Brdička, K. Wiesner, and K. Schäferna, *Věstnik Králov. České Společnosti Nauk, Třida Mat. Přirod.*, **1944**, No. 4.

64. I. M. Kolthoff and E. P. Parry, *J. Am. Chem. Soc.*, **73**, 5315 (1951).

65. I. M. Kolthoff and E. P. Parry, *Proc. Intern. Polarogr. Congr. Prague*, **1**, 145 (1951); *Chem. Abstr.*, **46**, 6967 (1952).

66. E. Jurka, *Collection Czech. Chem. Commun.*, **11**, 243 (1939); R. Brdička, *ibid.*, **11**, 614 (1939).

67. O. H. Müller, "Polarographic Analysis of Proteins, Amino Acids, and Other Compounds by Means of the Brdička Reaction," in *Methods of Biochemical Analysis*, D. Glick, Ed., Vol. 11, Interscience, New York, 1963, p. 329.

68. R. Brdička, *Collection Czech. Chem. Commun.*, **5**, 112, 118 (1933); *Biochem. Z.*, **272**, 104 (1934).

69. A. Căluşaru, *Compt. Rend.*, **262C**, 4 (1966).

70. A. Căluşaru, *Compt. Rend.*, **262C**, 676 (1966).

71. O. H. Müller, "Polarographic Behavior of Various Plasma Protein Fractions," in *Electrochemistry in Biology and Medicine*, T. Shedlovsky, Ed., Wiley, New York, 1955, p. 301; M. Březina and P. Zuman, *Polarography in Medicine, Biochemistry, and Pharmacy*, revised Engl. ed., (S. Wawzonek, transl.), Interscience, New York, 1958; S. G. Mairanovskii, *J. Electroanal. Chem.*, **6**, 77 (1963).

72. O. H. Müller, *Trans. Electrochem. Soc.*, **87**, 441 (1945).

73. R. Brdička, Z. Elektrochem., **48**, 278 (1942).

74. R. Brdička, Collection Czech. Chem. Commun., **12**, 522 (1947).

75. K. Wiesner, Collection Czech. Chem. Commun., **12**, 594 (1947).

76. O. H. Müller and B. Ogden, Abh. Deutsch. Akad. Wiss. Berlin, Kl. Med., **1966**, No. 4, 617.

77. J. Koryta, Collection Czech. Chem. Commun., **18**, 206 (1953).

78. P. Delahay and I. Trachtenberg, J. Am. Chem. Soc., **79**, 2355 (1957); J. Weber, J. Koutecky, and J. Koryta, Z. Elektrochem., **63**, 583 (1959); J. Weber and J. Koutecky, Collection Czech. Chem. Commun., **25**, 2993 (1960); C. N. Reilley and W. Stumm, "Adsorption in Polarography," in Progress in Polarography, P. Zuman and I. M. Kolthoff, Eds., Vol. I, Interscience, New York, 1962, p. 81; B. Kastening, "Beiträge zur Adsorptionsverdrängung als Ursache der Inhibitionswirkung Grenzflächen-aktiver Stoffe bei Polarographischen Vorgängen," in Polarography 1964, G. J. Hills, Ed., Vol. I, Macmillan, London, 1966, p. 359.

79. J. Heyrovský and D. Ilkovič, Collection Czech. Chem. Commun., **7**, 198 (1935).

80. O. H. Müller, Cold Spring Harbor Symp. Quant. Biol., **7**, 59 (1939).

81. W. M. Clark and B. Cohen, U.S. Public Health Rept., **38**, 666 (1923).

82. E. Friedheim and L. Michaelis, J. Biol. Chem., **91**, 355 (1931); L. Michaelis, ibid., **92**, 211 (1931); R. Elema, Rec. Trav. Chim., **50**, 807 (1931).

83. O. H. Müller, Ann. N.Y. Acad. Sci., **40**, 91 (1940).

84. R. Brdička, Collection Czech. Chem. Commun., **12**, 522 (1947).

85. O. H. Müller, J. Am. Chem. Soc., **62**, 2434 (1940).

86. O. H. Müller, The Polarographic Method of Analysis, 2nd ed., Chem. Education Publishing Co., Easton, Pennsylvania, (1951), p. 108.

87. J. C. Abbott and J. W. Collat, Anal. Chem. **35**, 859 (1963).

88. I. M. Kolthoff and E. F. Orlemann, J. Am. Chem. Soc., **63**, 664 (1941).

89. O. H. Müller, Chem. Rev., **24**, 95 (1939).

90. K. Wiesner, Biochem. Z., **313**, 48 (1942).

91. M. Kalousek, Collection Czech. Chem. Commun., **13**, 105 (1948).

92. J. J. Lingane, J. Am. Chem. Soc., **61**, 2099 (1939).

93. P. J. Elving, J. C. Komyathy, R. E. van Atta, C. S. Tang, and I. Rosenthal, Anal. Chem., **23**, 1218 (1951).

94. A. A. Green, J. Am. Chem. Soc., **55**, 2336 (1933).

95. P. J. Elving, J. M. Markowitz, and I. Rosenthal, Anal. Chem., **28**, 1179 (1956).

96. J. Novak, Collection Czech. Chem. Commun., **9**, 207 (1937).

97. F. J. W. Roughton, Harvey Lectures, **39**, 96 (1944).

98. O. H. Müller, Proc. Intern. Polarog. Congr. Prague, **1**, 159 (1951); Chem. Abstr. **46**, 6968 (1952).

99. H. Eyring, L. Marker, and T. C. Kwoh, J. Phys. Colloid Chem., **53**, 1453 (1949).

100. N. Tanaka and R. Tamamushi, Proc. Intern. Polarog. Congr. Prague, **1**, 486 (1951); Chem. Abstr. **46**, 10008 (1952); H. Gerischer, Z. Elektrochem., **59**, 604 (1955); R. Brdička, V. Hanuš, and J. Koutecky, "General Theoretical

Treatment of the Polarographic Kinetic Currents," in *Progress in Polarography*, P. Zuman and I. M. Kolthoff, Eds., Vol. I, Interscience, New York, 1962, p. 145; P. Delahay, *Double Layer and Electrode Kinetics*, Interscience, New York, 1965.

101. P. Delahay, *New Instrumental Methods in Electrochemistry*, Interscience, New York, 1954, p. 32.

102. J. Tafel, *Z. Physik. Chem.*, **50**, 641 (1905).

103. J. P. Baumberger, J. J. Jürgensen, and K. Bardwell, *J. Gen. Physiol.*, **16**, 961 (1933).

104. J. Heyrovský, *Trans. Faraday Soc.*, **19**, 785 (1924).

105. J. Tomeš, *Collection Czech. Chem. Commun.*, **9**, 150 (1937).

106. J. Kůta, *Collection Czech. Chem. Commun.*, **16**, 1 (1951).

107. R. Tamamushi, *Bull. Chem. Soc. Japan*, **25**, 293 (1952); **26**, 56 (1953).

108. O. H. Müller, "The Reduction of Hydrogen Ions at the Dropping Mercury Electrode," in *Polarography 1964*, G. J. Hills, Ed., Vol. I. Macmillan, London, 1966, p. 319.

109. F. R. Smith and A. E. Wells, *Nature*, **215**, 1165 (1967).

110. O. H. Müller, *J. Polarogr. Soc.*, **14**, 45 (1968).

111. O. H. Müller, *J. Polarogr. Soc.*, **14**, 50 (1968).

112. P. Rüetschi and G. Trümpler, *Helv. Chim. Acta*, **35**, 1021 (1952).

113. P. Rüetschi and G. Trümpler, *Helv. Chim. Acta*, **35**, 1957 (1952).

114. W. Kemula and M. Michalski, *Roczniki Chem.*, **16**, 535 (1936).

115. L. I. Smith, I. M. Kolthoff, S. Wawzonek, and P. M. Ruoff, *J. Am. Chem. Soc.*, **63**, 1018 (1941).

116. H. Gilder, O. H. Müller, and R. A. Phillips, *Am. J. Physiol.*, **129**, 362 (1940).

117. M. Tokuoka, *Collection Czech. Chem. Commun.*, **4**, 444 (1932); M. Tokuoka and J. Růžička, *Collection Czech. Chem. Commun.*, **6**, 339 (1934).

118. J. B. Conant, *Chem. Rev.*, **3**, 1 (1926).

119. E. S. G. Barron and A. B. Hastings, *J. Biol. Chem.*, **107**, 567 (1934).

120. M. Becker and H. Strehlow, *Z. Elektrochem. Ber. Bunsenges. Phys. Chem.*, **64**, 813, 818 (1960).

121. S. Ono, M. Takagi, and T. Wasa, *Collection Czech. Chem. Commun.*, **26**, 141 (1961).

122. R. Tamamushi and N. Tanaka, *Bull. Chem. Soc. Japan*, **24**, 127 (1951).

123. M. Shikata and I. Tachi, *J. Chem. Soc. Japan*, **53**, 834 (1932); *Collection Czech. Chem. Commun.*, **10**, 368 (1938).

124. L. P. Hammett, *Physical Organic Chemistry*, McGraw-Hill, New York, 1940.

125. R. W. Brockman and D. E. Pearson, *J. Am. Chem. Soc.*, **74**, 4128 (1952).

126. R. W. Taft, Jr., "Separation of Polar, Steric, and Resonance Effects in Reactivity," in *Steric Effects in Organic Chemistry*, M. S. Newman, Ed., Wiley, New York, 1956, p. 556.

127. P. Zuman, *Substituent Effects in Organic Polarography*, Plenum Press, New York, 1967.

128. A. Macoll, *Nature*, **163**, 178 (1949). See also J. Koutecky, *Electrochim. Acta*, **13**, 1079 (1968).

129. D. M. Coulson and W. R. Crowell, *J. Am. Chem. Soc.*, **74**, 1294 (1952).

130. P. Zuman, *Collection Czech. Chem. Commun.*, **15**, 1107 (1950).

131. J. Heyrovský and J. Forejt, *Z. Physik. Chem.*, **193**, 77 (1943).

132. M. Rálek and L. Novák, *Chem. Listy*, **49**, 557 (1955); *Collection Czech. Chem. Commun.*, **21**, 248 (1956).

133. M. Kalousek and M. Rálek, *Collection Czech. Chem. Commun.*, **19**, 1099 (1954).

134. M. Suzuki, *J. Electrochem. Soc.*, *Japan*, **22**, 63, 112, 162, 220 (1954); *Chem. Abstr.* **48**, 13472-3 (1954).

135. M. Ishibashi and T. Fujinaga, *Bull. Chem. Soc. Japan*, **25**, 238 (1952).

136. W. F. Kinard, R. H. Philp, and R. C. Propst, *Anal. Chem.*, **39**, 1556 (1967).

137. G. Semerano and G. Bettinelli, *Mem. Reale Accad. Italia, Classe Sci. Fis. Mat. Nat.*, **8**, 243 (1937). See also A. Miolati and G. Semerano, *Z. Elektrochem.*, **45**, 226 (1939).

138. H. Siebert, *Z. Elektrochem.*, **44**, 768 (1938); **45**, 228 (1939).

139. J. J. Lingane, *J. Am. Chem. Soc.*, **67**, 1916 (1945).

140. L. Meites and T. Meites, *Anal. Chem.*, **28**, 103 (1956).

141. R. Strubl, *Collection Czech. Chem. Commun.*, **10**, 475 (1938).

142. M. Spalenka, *Collection Czech. Chem. Commun.*, **11**, 146 (1939).

143. J. J. Lingane, *J. Am. Chem. Soc.*, **61**, 976 (1939); J. Heyrovský and M. Kalousek, *Collection Czech. Chem. Commun.*, **11**, 464 (1939).

144. J. J. Lingane, *Chem. Rev.*, **29**, 1 (1941).

145. M. v. Stackelberg and H. v. Freyhold, *Z. Elektrochem.*, **46**, 120 (1940); D. D. DeFord and D. H. Hume, *J. Am. Chem. Soc.*, **73**, 5321 (1951).

146. H. Irving, "The Stability of Metal Complexes and Their Measurement Polarographically," in *Advances in Polarography*, I. S. Longmuir, Ed., Vol. I, Pergamon, Oxford 1960, p. 42; A. A. Vlček, "Mechanism of the Electrode Processes and Structure of Inorganic Complexes," in *Progress in Polarography*, P. Zuman and I. M. Kolthoff, Eds., Vol. I, Interscience, New York, 1962, p. 269; D. R. Crow and J. V. Westwood, *Quart. Rev.*, **19**, 57 (1965); M. Kopanica, J. Doležal, and J. Zýka, "Chelates in Inorganic Polarographic Analysis: Fundamentals," in *Chelates in Analytical Chemistry*, H. A. Flaschka and A. J. Barnard, Jr., Eds., Vol. I, Marcel Dekker, New York, 1967, p. 145.

147. O. H. Müller, *J. Chem. Educ.*, **41**, 320 (1964).

148. O. H. Müller, *Ind. Eng. Chem.*, *Anal. Ed.*, **14**, 99 (1942).

149. D. Ilkovič and G. Semerano, *Collection Czech. Chem. Commun.*, **4**, 176 (1932).

150. J. J. Lingane and H. Kerlinger, *Ind. Eng. Chem.*, *Anal. Ed.*, **12**, 750 (1940).

151. G. E. Philbrook and H. M. Grubb, *Anal. Chem.*, **19**, 7 (1947).

152. O. H. Müller, *Chem. Eng. News*, **20**, 1528 (1942).

153. V. Majer, *Collection Czech. Chem. Commun.*, **7**, 146 (1935).

154. K. Schwarz, *Z. Anal. Chem.*, **115**, 161 (1939).

155. J. P. Baumberger and K. Bardwell, *Ind. Eng. Chem.*, *Anal. Ed.*, **15**, 639 (1943).

156. J. J. Lingane and H. A. Laitinen, *Ind. Eng. Chem.*, *Anal. Ed.*, **11**, 504 (1939).

157 H A. Laitinen and L. W. Burdett, *Anal. Chem.*, **22**, 833 (1950).

158. O. H. Müller, *The Polarographic Method of Analysis*, 2nd ed., Chem. Education Publishing Co., Easton, Pennsylvania, 1951.

159. K. N. Campbell and B. K. Campbell, *Chem. Rev.*, **31**, 77 (1942).
160. H. A. Laitinen and S. Wawzonek, *J. Am. Chem. Soc.*, **64**, 1765 (1942).
161. S. Wawzonek and H. A. Laitinen, *J. Am. Chem. Soc.*, **64**, 2365 (1942).
162. R. Pasternak and H. v. Halban, *Helv. Chim. Acta*, **29**, 190 (1946).
163. M. v. Stackelberg and W. Stracke, *Z. Elektrochem.*, **53**, 118 (1949).
164. E. L. Colichman and S. K. Liu, *J. Am. Chem. Soc.*, **76**, 913 (1954).
165. P. J. Elving, *Record Chem. Progr.* (*Kresge-Hooker Sci. Lib.*), **14**, 99 (1953).
166. M. J. Boyd and K. Bambach, *Anal. Chem.*, **15**, 314 (1943); B. Warshowsky and P. J. Elving, *ibid.*, **18**, 253 (1946).
167. A. S. Landry, *Anal. Chem.*, **21**, 674 (1949); D. Monnier and Y. Rusconi, *Helv. Chim. Acta*, **34**, 1297 (1951); P. E. Wenger, D. Monnier, and H. Schmidgall, *ibid.*, **35**, 1192, 2242 (1952).
168. E. B. Hershberg, J. K. Wolfe, and L. F. Fieser, *J. Biol. Chem.*, **140**, 215 (1941); J. K. Wolfe, L. F. Fieser, and H. B. Friedgood, *J. Am. Chem. Soc.*, **63**, 582 (1941); V. Prelog and O. Häfliger, *Helv. Chim. Acta*, **32**, 2088 (1949); J. R. Young, *J. Chem. Soc.*, **1955**, 1516.
169. P. Zuman, *Organic Polarographic Analysis*, Macmillan, New York, 1964.
170. J. Pech, *Collection Czech. Chem. Commun.*, **6**, 126 (1934).
171. H. Hohn, *Z. Elektrochem.*, **43**, 127 (1937); G. T. Borcherdt, V. W. Meloche, and H. Adkins, *J. Am. Chem. Soc.*, **59**, 2171 (1937); J. K. Wolfe, E. B. Hershberg, and L. F. Fieser, *J. Biol. Chem.*, **136**, 653 (1940).
172. H. G. Petering and F. Daniels, *J. Am. Chem. Soc.*, **60**, 2796 (1938); R. H. Müller and J. F. Petras, *ibid.*, **60**, 2990 (1938).
173. H. Hohn, *Chemische Analysen mit dem Polarographen*, Springer, Berlin, 1937.
174. H. E. Forche, *Mikrochemie*, **25**, 217 (1938).
175. J. Heyrovský and O. H. Müller, *Collection Czech. Chem. Commun.*, **7**, 281 (1935).
176. O. H. Müller, *J. Chem. Educ.*, **18**, 320 (1941).
177. M. R. Pesce, S. L. Knesbach, and R. K. Ladisch, *Anal. Chem.*, **25**, 979 (1953).
178. O. H. Müller, *J. Chem. Educ.*, **18**, 111 (1941).
179. R. M. Elofson, R. L. Edsberg, and P. A. Mecherly, *J. Electrochem. Soc.*, **97**, 166 (1950).
180. R. Pasternak, *Helv. Chim. Acta*, **30**, 1984 (1947).
181. K. Schwabe and H. Berg, *Z. Physik. Chem.*, **203**, 318 (1954).
182. L. Holleck and G. A. Melkonian, *Z. Elektrochem.*, **58**, 867 (1954).
183. K. Schwabe and J. Voigt, *Z. Physik. Chem.*, **203**, 383 (1954).
184. K. Schwabe and H. Berg, *Z. Physik. Chem.*, **204**, 78 (1955).
185. V. Majer, *Z. Elektrochem.*, **42**, 120, 123 (1936).
186. A. Neuberger, *Z. Anal. Chem.*, **116**, 1 (1939).
187. H. A. Laitinen, *Anal. Chem.*, **21**, 66 (1949); **24**, 46 (1952); **28**, 666 (1956); **30**, 657 (1958); **32**, 180R (1960); **34**, 307R (1962); J. T. Stock, *ibid.*, **36**, 355R (1964); **38**, 452R (1966); **40**, 392R (1968).
188. J. T. Stock, *Amperometric Titrations*, Interscience, New York, 1965.
189. G. Semerano and L. Riccoboni, *Gazz. Chim. Ital.*, **72**, 297 (1942).
190. J. Heyrovský, *Analyst*, **72**, 229 (1947); *Chem. Listy*, **43**, 149 (1949).
191. J. J. Lingane and R. Williams, *J. Am. Chem. Soc.*, **74**, 790 (1952).

192. R. N. Adams, C. N. Reilley, and N. H. Furman, *Anal. Chem.*, **25**, 1160 (1953).

193. T. Fujinaga, "Constant-Current Polarography and Chronopotentiometry at the Dropping Mercury Electrode," in *Progress in Polarography*, P. Zuman and I. M. Kolthoff, Eds., Vol. I, Interscience, New York, 1962, p. 201.

General

Březina, M., and P. Zuman, *Polarography in Medicine, Biochemistry, and Pharmacy*, revised Engl. ed., (S. Wawzonek, transl.), Interscience, New York, 1958.

Delahay, P., *New Instrumental Methods in Electrochemistry*, Interscience, New York, 1954.

Delahay, P., *Double Layer and Electrode Kinetics*, Interscience, New York, 1965.

Harley, J. H., and S. E. Wiberley, *Instrumental Analysis*, Wiley, New York, 1954.

Heyrovský, J., *Polarographie*, Springer, Vienna, 1941 (Reprinted by J. W. Edwards, Inc., Ann Arbor, Michigan).

Heyrovský, J., *Polarographisches Praktikum*, 2nd ed., Springer, Berlin, 1960.

Heyrovský, J., and J. Kůta, *Principles of Polarography*, Publishing House of the Czechoslovak Academy of Sciences, Prague, 1965.

Kolthoff, I. M., and J. J. Lingane, *Polarography*, Vols. I and II, 2nd ed., Interscience, New York, 1952.

Lingane, J. J., *Electroanalytical Chemistry*, Interscience, New York, 1953.

Meites, L., *Polarographic Techniques*, 2nd ed., Interscience, New York, 1965.

Meites, L., "Voltammetry at the Dropping Mercury Electrode (Polarography)," in *Treatise on Analytical Chemistry*, Part I, Vol. 4, I. M. Kolthoff and P. J. Elving, Eds., Interscience, New York, 1963.

Milner, G. W. C., *The Principles and Applications of Polarography and Other Electroanalytical Processes*, Longmans, Green, London, 1957.

Müller, O. H., *The Polarographic Method of Analysis*, 2nd ed., Chem. Education Publishing Co., Easton, Pennsylvania, 1951.

Rulfs, C. L., "Polarographic Analysis," in *Selected Topics in Modern Instrumental Analysis*, D. F. Boltz, Ed., Prentice-Hall, New York, 1952.

Schmidt, H., and M. v. Stackelberg (R. E. W. Maddison, transl.), *Modern Polarographic Methods*, Academic Press, New York, 1963.

Stackelberg, M. v., *Polarographische Arbeitsmethoden*, DeGruyter, Berlin, 1950.

Strobel, H. A., *Chemical Instrumentation. A Systematic Approach to Instrumental Analysis*. Addison-Wesley, Reading, Massachusetts, 1960.

Tachi, I., Ed., *Polarography*, (in Japanese), Iwanami Book Co., Tokyo, 1954.

Zuman, P., *Organic Polarographic Analysis*, Macmillan, New York, 1964.

Zuman, P., *Substituent Effects in Organic Polarography*, Plenum Press, New York, 1967.

Chapter **VI**

CYCLIC VOLTAMMETRY, AC POLAROGRAPHY AND RELATED TECHNIQUES

Eric R. Brown and Robert F. Large

I INTRODUCTION

Cyclic voltammetry and alternating current (ac) polarography are modern electrochemical techniques which provide the means to examine the nature, or "pathway," of an electrochemical reaction in detail. From a detailed investigation of an electrochemical reaction, the chemist obtains information which not only provides a firmer basis for control of the reaction but which also opens the door to studies of reactive intermediates and the fundamental chemistry that underlies the reaction system of interest.

It should be emphasized that the two techniques receiving prime attention in this chapter are not the only ones at our disposal for mechanistic studies. The other techniques described in this volume have their own unique utility and in particular cases may indeed provide more pertinent information than that obtainable by cyclic voltammetry or ac polarography. Thus the prudent chemist should be familiar with all the techniques available and should employ those yielding the most fruitful measurements for a specific study.

The selection of the best technique to apply to a stated problem is unfortunately not always as obvious as might be hoped. Resorting to those methods that provide information that is readily interpreted in breadth, though not in depth, seems the proper practical alternative as a point of departure. Thus dc polarography is still the first tool to be employed in the majority of mechanistic investigations and most likely will remain so for some time. Unfortunately, the information obtainable from dc polarography (without resorting to sophisticated experimentation) provides limited basis for a characterization of the nature of a charge transfer process. Conversely, cyclic voltammetry and ac polarography are of great value for this purpose. Therefore the objective of this chapter is to describe the theory and experimentation of these two techniques in sufficient detail that those currently unfamiliar with their utility will seek to apply them.

Because of its relative experimental simplicity, cyclic voltammetry is perhaps the more readily applied of these two techniques. Thus the approach to a mechanistic study still employed by many chemists includes a simple overview by dc polarography or voltammetry, a qualitative examination by cyclic voltammetry, and quantitative studies by that technique affording the best chance of obtaining reliable and unambiguous data pertinent to the problem at hand. Those individuals who are especially versed in a particular technique, ac polarography, for example, might well employ that technique at an earlier stage in the investigation. However, the inexperienced worker should still proceed with some defined plan of attack ranging from readily interpretable experiments to those that require a greater degree of familiarization with the theory to be of value.

For the most part, cyclic voltammetry and ac polarography provide access to the same information. There are a few specific instances in which one of the techniques clearly is more useful than the other, and comments to that effect are found at the appropriate point in the text. In general, both techniques are relatively new and have not yet received the benefit of extensive application to a diversity of practical systems.

2 A MODEL OF ELECTROCHEMICAL REACTIONS

To attempt to understand a process as complex as a series of reactions at an electrode surface requires some measure of a point of view. Two alternatives are apparent in selecting a useful perspective from which experimental data can be rationalized. The first of these is a mathematical model complete insofar as experimental capabilities allow its authenticity to be verified. To those well schooled in mathematical operations and their significance, this is perhaps the only model of real meaning. However, to the majority of chemists, such an approach leaves room for vast misunderstanding. Thus an intuitive model which affords ready incorporation of new concepts seems a better perspective from which to view electrochemical reactions. Such a model should never be assumed to be an exact representation of the physical or chemical processes that are occurring but rather should serve as a basis on which charge transfer processes can be rationalized within existing knowledge. It is within this framework that the conceptual model below is presented and is employed throughout this chapter.

The model to be used was first presented in detail by Vlcek [1, 2] in describing the nature of charge transfer reactions of coordination compounds. However, this model is general enough to be valid for all types of charge transfer reactions. It consists of seven sequential steps:

1. The reactant is brought to the vicinity of the electrode by mass transfer.
2. A chemical reaction occurs which yields the species actually entering the inner part of the electrical double layer.
3. A structural rearrangement occurs yielding the species that takes part in the actual electron transfer reaction.
4. The electron transfer *per se* takes place.
5. A structural reorganization occurs leading to the immediate product of charge transfer.

6. A chemical reaction of the immediate product of charge transfer occurs which yields the species stable in solution.

7. The final product of charge transfer is removed to the bulk of the solution by mass transfer.

The following discussion of each of the separate processes in this model should afford the reader the opportunity to note the approximations that are being made and thereby develop a better understanding of the relationships between experimental facts and the present capability to explain them.

Mass Transport Processes

In the theoretical development of polarography and voltammetry, diffusion is considered as the sole means of mass transport. During the reaction at an electrode surface, material is depleted and a concentration gradient is set up. Reactant from the bulk of solution then diffuses toward the electrode surface in response to this gradient. In a similar manner products of the electrode reaction diffuse away from the electrode.

Two other common modes of mass transfer to the electrode surface are convection, or stirring, and migration. Migration, the movement of charged species in the presence of an electric field, is eliminated by "swamping" the solution with some chemically inert and electroinactive ionic species termed the supporting electrolyte. The inert ions present in large excess are the major charge carriers in the cell when current flows.

In some types of experiments, forced convection is used as a means of mass transport; rotating disk, rotating wire, and stirred solution voltammetry are examples. In voltammetric studies in which diffusion is considered the only means of mass transfer, convective effects must be minimized. The two most common causes of convection are mechanical vibrations of the cell and movement of solution caused by temperature or density gradients within the cell. To eliminate them is impossible, but these effects can be minimized. The cell can be shock mounted, much as an accurate balance must be, so that building vibrations are minimized. Temperature gradients can be offset by jacketing the cell either with air or water from a constant-temperature bath.

The diffusion equations used to describe the concentration gradients of the reactant species as a function of time and distance from the electrode surface are the same as those used in heat transfer [3]. The well-known Fick's laws of diffusion give this relation. The most commonly used law is that for linear diffusion to a stationary plane from an essentially infinitely large bulk:

$$\frac{\delta C(x, t)}{\delta t} = D \frac{\delta^2 C(x, t)}{\delta x^2}. \tag{6.1}$$

The concentration of the species C is expressed as a function of both time t and distance x. The diffusion coefficient D is constant with respect to both time and distance and is a function of other parameters which are generally constant in the cell. These include solution viscosity and temperature. Equation (6.1) is then solved using the appropriate boundary and initial conditions for the surface concentration as a function of t, that is, $C(0, t)$. This is done in Section 4 for both forms of the redox couple.

Equation (6.1) is generally not an accurate description for the types of electrodes used in many cases, particularly polarography in which the dropping mercury electrode (DME) is used. It is neither planar nor stationary. To account for electrode growth with the DME, the diffusion equation to an expanding plane is often used [4]:

$$\frac{\delta C(x, t)}{\delta t} = D \frac{\delta^2 C(x, t)}{\delta x^2} + \frac{2x}{3t} \frac{\delta C(x, t)}{\delta x}. \tag{6.2}$$

In the theory presented here for ac polarography, this diffusion model is used, whereas the cyclic voltammetric theory uses (6.1) as its basis. Suitable corrections can be made to the final current expressions when the electrode is not planar but spherical, such as a hanging mercury drop.

The effects attributable to sphericity of a mercury drop can be accounted for by using the equation for diffusion to a sphere, but the mathematics becomes more cumbersome:

$$\frac{\delta C(r, t)}{\delta t} = D \frac{\delta^2 C(r, t)}{\delta r^2} + \frac{2}{r} \frac{\delta C(r, t)}{\delta r}. \tag{6.3}$$

The distance r is now measured from the center of the sphere in a radial direction.

In general, the effects of both drop growth and sphericity are small in the ac polarographic technique, amounting to at most a 3 to 4% correction in current magnitudes over the case of simple planar diffusion [5]. A major exception to this statement occurs when one of the forms of the redox couple is soluble in the electrode itself. The correction to current magnitudes may then exceed 10%.

Finally, yet another phenomenon, adsorption, may contribute to control of concentrations at the electrode surface. When adsorption is present, material also exists on the electrode in a quantity governed by some adsorption isotherm, the term Γ denoting the amount of material adsorbed in moles per square centimeter. Normally, the flux of reactant to the electrode surface is controlled only by diffusion, but if adsorption occurs the flux is also controlled by the rates of adsorption and desorption of the electroactive species, with a corresponding effect on the current. The appropriate diffusion equation is used to solve for the surface concentrations but (6.5) is used as a boundary value instead of (6.4). This type of problem is discussed in Section 12 for cyclic voltammetry.

Diffusion flux

$$\frac{i(t)}{nFA} = D \frac{\delta C(x, t)}{\delta x} \qquad (6.4)$$

Adsorption and diffusion flux

$$\frac{i(t)}{nFA} = D \frac{\delta C(x, t)}{\delta x} - \frac{\delta \Gamma}{\delta t} \qquad (6.5)$$

Coupled Chemical Reactions

It has been held by some chemists that reactions that occur at an electrode surface are "unusual" and should not be considered in the same manner as homogeneous chemical reactions. This may well be true of reactions that occur during the time interval in which a molecular species is actually very close to the electrode. However, many chemical reactions occur as a result of the presence of a charge transfer process, that is, they are "coupled" to the charge transfer process, but are purely homogeneous reactions having no reaction parameters specific to the charge transfer process. Such are the processes considered in steps 2 and 6 of the above model.

Rather than the charge transfer process altering the course of the chemical reaction, the result of coupled chemical reactions is to alter the flux of species involved in the charge transfer process. Perhaps the example of the reduction of a proton from a weak acid is illustrative. The equilibrium

$$HA \rightleftharpoons H^+ + A^-$$

exists in the solution and the predominant form is HA. If the free proton concentration of the solution is altered by any means, the chemical system will respond to maintain equilibrium. Such is the case when, at an electrode surface, the free proton concentration is markedly diminished by reduction of the proton to hydrogen. A molecule of HA approaching the electrode encounters solution conditions other than those existing in the bulk of the solution and reacts according to the laws of mass equilibria. If the rate of the chemical reaction is very rapid in comparison with the rate of diffusion of electroactive species to the electrode surface, no effect on the flux of reactant for the charge transfer process will be observed. Conversely, if the rate of the chemical reaction is not sufficiently rapid with respect to diffusional processes, the flux of electroactive species will be some function of the rate of the chemical reaction. The nature of the chemical reaction and its kinetics may be examined by studying the effect it has on the flux of reactant for the charge transfer process.

The Charge Transfer Process

Having moved from the bulk of the solution, a reactant species now enters the portion of solution under direct influence of the electrode. This solution element, termed the electrical double layer, has unique properties which are impressed upon a reacting species. The nature of the electrical double layer is a subject of its own and beyond the scope of this chapter. Let it suffice to note that reactions that occur in the double layer are not exactly equivalent to homogeneous reactions. The reader is advised, perhaps after studying the material in this and similar volumes, to pursue the subject of the double layer from detailed writing [6]. The recent chapter by Mohilner [7] provides a summary of important aspects with references to the full literature.

Direct knowledge of the processes reacting species undergo at an electrode surface is limited, and the theoretical treatment of these processes is beyond the present interests. In a purely qualitative manner at least, a portion of what these reactions involve can be presumed. One process is the act of electron transfer between the reactant and the electrode, which implies at least some contact with the electrode. This contact may be visualized as an overlap of a molecular orbital of the reactant with an orbital of the electrode. For such overlap to occur, some structural reorganization of the reactant species is assumed to be required. This process can be viewed as the formation of a transition state for electron transfer in direct analogy to transition states for homogeneous reactions. The immediate product of electron transfer most likely does not have precisely the structure of the species that moves away from the electrode. So the processes of structural reorganization to satisfy mutually the requirements of the species existing immediately prior to and following electron transfer can be viewed as the formation of the transition state.

Little if anything is known of the actual act of electron transfer. By analogy to the Franck-Condon effect, it is assumed that the transfer is an extremely rapid adiabatic process occurring in a time interval significantly less than that of the formation of the transition state. Use of the term "electron transfer" to describe the overall charge transfer process is probably inaccurate and should be discouraged.

The model employed for the charge transfer process consists of at least three distinct steps; a structural reorganization, electron transfer, and another structural reorganization. In a qualitative sense this model explains why various charge transfer processes differ significantly in rate. If only minor reorientations were required to reach the transition state, only a limited time interval would probably be required. Alternatively, if a molecule must undergo severe structural reorientation to reach the configuration required for electron transfer, it seems reasonable that the process may occur at a slower rate.

Although the subject is treated quantitatively later, a discussion of the concept of electrochemical reversibility is in order at this time. If a charge transfer process is said to be "reversible," one is in reality only stating that the process occurs at a significantly more rapid rate than the rate of diffusion. A charge transfer process governed by both diffusion and charge transfer kinetics is termed "quasi-reversible." A reaction in which the charge transfer process is much slower than the diffusion rate is termed "irreversible."

The best usage of the irreversible terminology is also worthy of mention. Overall electrochemical reactions may not be reversible in a gross sense for a variety of reasons. For example, if a reversible charge transfer process is followed by a rapid irreversible chemical reaction, a preliminary examination would suggest that the charge transfer process is irreversible. A characterization of this electrochemical reaction as irreversible, implying a slow charge transfer process, is not correct. The rate of the charge transfer process was sufficiently rapid that the reaction was totally reversible; however, with limited experimentation, information for definition of this fact was not available. Thus in this chapter an irreversible electrochemical reaction is confined to the description of a process in which the charge transfer act is rate determining.

3 THE EXPERIMENTAL FOUNDATIONS OF THE TECHNIQUES

For the same reasons that have required the presentation of a working model of electrochemical reactions, a qualitative understanding of the manner in which electrochemical experiments, cyclic voltammetric and ac polarographic experiments in this case, are conducted is in order.

All electrochemical experimentation involves the application of some perturbation to the system, monitoring the resulting response, and interpreting the data so obtained. There are four fundamental parameters of electrochemical experimentation: current, potential, time, and concentration of reactant. All techniques, either directly or indirectly, are governed by these four parameters. In cyclic voltammetry and ac polarography, we are primarily concerned with three of these parameters, current, potential, and time, since they are of most value in elucidating reaction mechanisms. Concentration of reactant is considered only for those reactions in which it assumes added importance, that is, second-order processes.

The Cyclic Voltammetric Experiment

Consider a system comprised of a reversible redox couple, the oxidized form of which is present in a solution containing an excess of a supporting

$$O + ne^- \rightleftharpoons R$$

electrolyte. A stationary working electrode is employed, the potential of which is monitored versus some reference electrode and, through the use of appropriate instrumentation, there is no IR drop in the solution. If the electrode has imposed upon it a potential less cathodic than required to reduce the oxidized form of the redox couple, only a small current will flow because of faradaic processes resulting from trace components and the nonfaradaic process of charging of the double layer. If the potential of the electrode is made increasingly more cathodic (more negative) at an infinitesimally slow rate and the current is monitored, the familiar S-shaped current-potential curve shown in Fig. 6.1 will be obtained. If, again at an

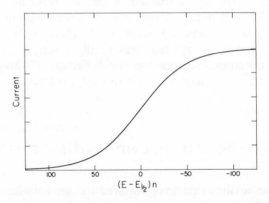

Fig. 6.1 Generalized current-potential curve for a diffusion controlled reaction.

infinitesimally slow rate, the potential is brought back to the initial value, the S-shaped curve is retraced. From this current-potential curve, the maximum current flow, which is related to concentration, and the standard potential for the reaction can be observed. By a close analysis of the shape of the current-potential curve, some information about the nature of the electrochemical reaction taking place could be deduced. However, the information so obtained is quite tentative by itself since a number of different mechanistic pathways could provide very similar results. Repeating the experiment but varying the potential at a finite rate as a linear function of time results in a significantly different response, illustrated in Fig. 6.2. Two differences between the response in Fig. 6.1 and that in Fig. 6.2 are apparent. First, the current-potential curve in Fig. 6.2 no longer has the familiar S-shape but exhibits peaks. Second, the current-potential curves observed for the forward and reverse processes are no longer identical. These examples illustrate the effect of the variable of time on the nature of the electrochemical experiment and its manifestations in the responses obtained.

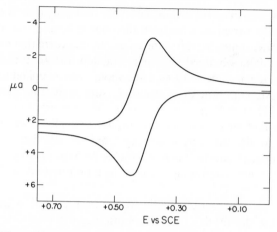

Fig. 6.2 Reversible cyclic voltammetric curve. Experimental conditions: 1 mM ferrocene in acetonitrile, 0.1M TBAP, Pt electrode, scan rate, 10 V/min.

The current-potential curve in Fig. 6.2 provides four measurable parameters; the net current at the peak of the cathodic response, the potential at the peak of the cathodic response, and the corresponding parameters for the anodic or reverse response. It is obvious that some relationship should exist between the magnitudes of the cathodic and anodic currents and between the potential values at the respective anodic and cathodic peak currents. It is with these parameters and the parameter of time that we deduce mechanistic information from cyclic voltammetric data.

In the experiment shown in Fig. 6.1, all changes in potential were made at a rate such that the rates of diffusion of oxidized and reduced forms to and from the electrode surface kept the system in equilibrium with the bulk of the solution at all times. However, in the experiment illustrated in Fig. 6.2, the rate of variation of potential was too rapid for diffusional processes to maintain equilibrium with the bulk of the solution. In the course of the cathodic variation in potential, the reduced form of the reactant was produced in the vicinity of the electrode while a depletion of the oxidized form occurred. Given sufficient time, the reduced form would have diffused into the bulk of the solution; but the potential was taken back to the initial value at a rate such that some of the reduced form was still present at the electrode surface and underwent a process of oxidation back to the form of the couple initially present in the solution.

The Ac Polarographic Experiment

The ac polarographic experiment, although slightly more complex, also represents a similar process of perturbation of electrochemical equilibrium as a function of time. In this experiment a dc potential is applied to the electrode and is varied as a function of time, it is assumed, at an infinitesimally slow rate. Thus by this process alone the current-potential curve in Fig. 6.1 would be obtained. However, in addition to the dc potential, a potential periodic in time, such as a sine wave, is also applied to the electrode. This sinusoidal variation in potential is similar to that employed in the experiment illustrated in Fig. 6.2 with the exception that the potential excursions are of much smaller magnitude, typically 10-mV peak-to-peak. In addition to the dc current flowing from the applied dc potential, an alternating or ac current flows corresponding to the applied sinusoidal potential. The net alternating current is monitored as a function of dc potential.

A typical ac polarogram is illustrated in Fig. 6.3, where the current oscillations are the result of the dropping mercury electrode used. At a dc potential more anodic than required to effect reduction, no ac current is expected to flow. At a dc potential corresponding to the limiting current

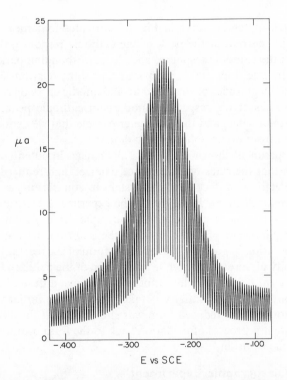

Fig. 6.3 Reversible ac polarogram. Experimental conditions: 1 mM ferric oxalate in 0.5 M K$_2$C$_2$O$_4$, DME electrode, applied potential is 10 mV p-p at 150 Hz, dc potential scan, 60 mV/min.

plateau in Fig. 6.1, the application of a small-amplitude sinusoidal potential causes no net change in concentrations at the electrode surface and no ac current flows. At the extremes in potential of the dc wave where only one form of the redox system is present at the electrode surface, there is no ac response. By reference to the Nernst equation, it should be noted that at the point in potential corresponding to the standard potential, equal concentrations of oxidized and reduced forms of the redox system are present at the electrode surface. Under such circumstances, the application of a superimposed sinusoidal variation in potential produces the greatest change in relative concentrations of the two forms of the redox couple and a maximum ac current is observed. At other dc potentials along the wave, the sinusoidal potential variation causes less of a perturbation of the dc surface concentrations and the ac current is correspondingly less. It can be seen that because the ac current occurs when the surface concentrations of the redox couple change as a result of the applied ac potential, the ac current as a function of dc potential has the form of the derivative of the dc polarographic wave.

Two parameters for use are immediately obvious; the peak magnitude of the ac current and the dc potential at this peak, which is related to the standard potential for the redox couple. These two parameters coupled again with variations in time (frequency of our potential excursions) provide the basis for the utility of this technique for charactering the nature of charge transfer processes. The exact shape of the curve is also examined carefully to note deviations from reversibility. However, one other fundamental relationship, the phase angle between the applied ac potential and the resulting ac current, must also be included. This parameter is discussed in more detail later.

4 REVERSIBLE CHARGE TRANSFER

Complete mathematical treatments of cyclic voltammetry and ac polarography are available. In particular, the series of papers by Shain [8–13] and co-workers present a detailed development of the theory of cyclic voltammetry. The excellent review by Smith [5] of the theory and practice of ac polarography presents a complete development of the fundamental relationships of the ac polarographic process. No attempt is made in this chapter to present completely the theoretical basis of these techniques.

However, since the reversible charge transfer case provides the foundations upon which much of the subsequent material is based, the means by which quantitative expressions for the cyclic voltammetric and ac polarographic responses are obtained are reviewed in sufficient detail so that the reader can note where differences appear in the other cases and more fully recognize the effects of other processes, a chemical reaction, for example, on the current-potential profile.

Derivation of Surface Concentration Expressions

If the rate of a charge transfer process is very fast and there are no coupled chemical reactions involved in the total electrochemical reaction, mass transfer of the electroactive species is the only process that determines the flux at the electrode surface. The limiting current that flows in the cell is related to the flux of the electroactive species at the electrode surface. The current-potential profile is determined only by the thermodynamic considerations of the Nernst equation and such a reaction is termed reversible.

The surface concentrations at the electrode are obtained by solving the appropriate Fick's second law expression (see Section 2) with the proper boundary conditions. The expressions for the surface concentrations of both forms of the redox couple are then substituted in the Nernst equation and the appropriate expression for the potential function is used to obtain the current as a function of potential.

Of particular interest are the current expressions resulting when the potential function is a dc potential with a superimposed ac ripple (ac polarography) or a linearly varying dc potential (linear scan or cyclic voltammetry). Both of these problems have been treated extensively, and only the outlines of the derivations will be given.

The diffusion equation is given by

$$\frac{\delta C(x, t)}{\delta t} = D \frac{\delta^2 C(x, t)}{\delta x^2}. \tag{6.6}$$

The boundary conditions that at $t = 0$, any x,

$$C_O(x, 0) = C_O^* \tag{6.7a}$$

$$C_R(x, 0) = C_R^*, \tag{6.7b}$$

are applied to (6.6). Subscript O and R denote the oxidized and reduced forms, respectively, and the asterisks denote the value of a parameter in the bulk of the solution. This means that initially the concentrations are bulk concentrations and uniformly distributed in the solution.

Also, at $t > 0$, $x \to \infty$

$$C_O(x, t) \to C_O^* \tag{6.8a}$$

$$C_R(x, t) \to C_R^* \tag{6.8b}$$

This is the approximation of semi-infinite linear diffusion.

Finally, for the condition $t > 0$, $x = 0$, (6.9) applies:

$$D_O \frac{\delta C_O(0, t)}{\delta x} = -D_R \frac{\delta C_R(0, t)}{\delta x} = \frac{i(t)}{nFA}. \tag{6.9}$$

Equation (6.9) states that the flux of the oxidized species at the electrode surface is equal but opposite to the flux of the reduced species and also equates the current to the flux. The electrode area is given by A, n is the number of electrons transferred per molecule, and F is the Faraday constant.

Laplace transforms can be used to solve (6.9) for the surface concentrations $C_O(0, t)$ and $C_R(0, t)$:

$$C_O(0, t) = C_O{}^* - \int_0^t \frac{i(t)\, d\tau}{nFA(\pi D_O\{t - \tau\})^{1/2}} \tag{6.10a}$$

$$C_R(0, t) = C_R{}^* + \int_0^t \frac{i(t)\, d\tau}{nFA(\pi D_R\{t - \tau\})^{1/2}}. \tag{6.10b}$$

Using the same method of Laplace transforms, the corresponding surface concentrations for diffusion to an expanding plane and diffusion to a sphere can be obtained. These expressions are summarized in Table 6.1.

Table 6.1 Surface Concentration Expressions for Various Diffusion Models

---

Planar Diffusion

$$C_O(0, t) = C_O{}^* - \int_0^t \frac{i(t)\, d\tau}{nFA(\pi D_O\{t - \tau\})^{1/2}}$$

$$C_R(0, t) = C_R{}^* + \int_0^t \frac{i(t)\, d\tau}{nFA(\pi D_R\{t - \tau\})^{1/2}}$$

Expanding Plane Diffusion

$$C_O(0, t) = C_O{}^* - \sqrt{\frac{7}{3}} \int_0^t \frac{\tau^{2/3} i(t)\, d\tau}{nFA(\pi D_O\{t^{7/3} - \tau^{7/3}\})^{1/2}}$$

$$C_R(0, t) = C_R{}^* + \sqrt{\frac{7}{3}} \int_0^t \frac{\tau^{2/3} i(t)\, d\tau}{nFA(\pi D_R\{t^{7/3} - \tau^{7/3}\})^{1/2}}$$

Spherical Diffusion

$$C_O(r_0, t) = C_O{}^* - \int_0^t \frac{\left[\dfrac{1}{(\pi t)^{1/2}} - \dfrac{D_O^{1/2}}{r_0} \exp\left(\dfrac{D_O \tau}{r_0^2}\right) \operatorname{erfc}\left(\dfrac{\sqrt{D_O \tau}}{r_0}\right)\right] i(t - \tau)\, d\tau}{nFAD_O^{1/2}}$$

$$C_R(r_0, t) = C_R{}^* + \int_0^t \frac{\left[\dfrac{1}{(\pi t)^{1/2}} - \dfrac{D_R^{1/2}}{r_0} \exp\left(\dfrac{D_R \tau}{r_0^2}\right) \operatorname{erfc}\left(\dfrac{\sqrt{D_R \tau}}{r_0}\right)\right] i(t - \tau)\, d\tau}{nFAD_R^{1/2}}$$

The Reversible Cyclic Voltammetric Response

Derivation of the Current Expression

The Nernst equation can be written in the form

$$C_O(0, t) = \left(\frac{D_R}{D_O}\right)^{1/2} C_R(0, t) \exp [j(t)], \tag{6.11}$$

where

$$j(t) = \frac{nF}{RT}(E(t) - E_{1/2}), \tag{6.12}$$

and $E_{1/2}$ is the polarographic reversible half-wave potential of the couple.

Substituting (6.10) into (6.11) and rearranging slightly results in (6.13), relating current to potential. The assumption is also made that $C_R^* = 0$, a common situation in electrochemical experiments.

$$\int_0^t \frac{i(t)\, d\tau}{nFAC_O^*(\pi Do\{t - \tau\})^{1/2}} = \frac{1}{1 + e^{j(t)}} \tag{6.13}$$

Equation (6.13) can now be solved for various experimental techniques which determine the form of $E(t)$ used in (6.12).

In the linear sweep or cyclic voltammetric experiment, the potential-time function is

$$E(t) = \begin{cases} E_i - vt & \text{for } 0 \leq t \leq t_s \\ E_i - 2vt_s + vt & \text{for } t \geq t_s, \end{cases} \tag{6.14}$$

where E_i is the initial potential, v is the sweep rate in volts per second and t_s is the time when the potential scan is reversed. The expression for $j(t)$ becomes

$$j(t) = \begin{cases} \dfrac{nF}{RT}(E_i - E_{1/2}) - at & 0 \leq t \leq t_s \\ \dfrac{nF}{RT}(E_i - E_{1/2}) + at - 2at_s & t \geq t_s, \end{cases} \tag{6.15}$$

where

$$a = \frac{nFv}{RT}. \tag{6.16}$$

Since the goal in solving (6.13) is to obtain an expression for current as a function of potential, not time, an appropriate transformation of variables is made. By reference to (6.15) and (6.16), it can be seen that the term at is dimensionless although it is proportional to potential:

$$at = \frac{nFvt}{RT} = \frac{nF}{RT}[E_i - E(t)]. \tag{6.17}$$

This suggests the following variable changes in (6.13) to make it dimensionless:

$$\tau = z/a \qquad (6.18)$$

$$i(t) = g(at), \qquad (6.19)$$

where

$$\chi(at) = \frac{g(at)}{nFAC_O^*(\pi D_O a)^{1/2}}. \qquad (6.20)$$

Equation (6.13) becomes

$$\int_0^{at} \frac{\chi(z)\,dz}{(at - z)^{1/2}} = \frac{1}{1 + e^{j(at)}}, \qquad (6.21)$$

where $j(at) = j(t)$.

The solution of (6.21) gives values of $\chi(at)$ as a function of at and ultimately furnishes current values as a function of potential, according to (6.22):

$$i = nFAC_O^*(\pi D_O a)^{1/2}\chi(at). \qquad (6.22)$$

Equation (6.21) has been solved analytically [14–17], but a numerical solution is more generally applicable to cyclic voltammetric experimentation. The technique involves dividing the integration range into many small equal intervals to obtain n algebraic equations. The nth equation involves all the solutions to the $n - 1$ previous equations, so this is an iterative technique. Nicholson and Shain [8] have used this technique to calculate $\chi(at)$ to an accuracy of ± 0.001 for a single cyclic scan as a function of potential.

The solution of (6.22) for the peak current of the forward potential scan is given by (6.23):

$$i_p = 2.69 \times 10^5 n^{3/2} A D_O^{1/2} C_O^* v^{1/2}. \qquad (6.23)$$

This is often called the Randles-Sevcik equation after the two investigators who first independently derived the equation [17, 18]. A good discussion of the historical development of this equation has recently been given by Adams [19].

Diagnostic Criteria for the Reversible Cyclic Voltammetric Response

The general shape of the linear scan voltammetric response is embodied in the term $\chi(at)$. Nicholson and Shain [8] have given a table of $(\pi)^{1/2}\chi(at)$ with values every 5 mV. A maximum occurs at $(E_{dc} - E_{1/2})n = -28.5$ mV. Thus the peak potential is

$$E_p = E_{1/2} - \frac{28.5}{n} \text{ mV}. \qquad (6.24)$$

Because the peak is quite broad, it is often easier to measure the half-peak potential, which is related to the half-wave potential by (6.25):

$$E_{p/2} = E_{1/2} + \frac{28.0}{n}\,\mathrm{mV}. \qquad (6.25)$$

These equations are for a reduction reaction. For oxidation the sign of the numerical term is reversed.

On the reverse scan or reoxidation response, the position of the peak depends on vt_s, the switching potential. As this potential moves more negative, the position of the anodic (reoxidation) peak becomes constant at $29.5/n$ mV anodic of the half-wave potential. With the switching potential more than $100/n$ mV cathodic of the reduction peak, the separation of the two peaks will be $59/n$ mV and independent of the rate of potential scan. This is a commonly used criterion of reversibility.

It is also noted from (6.22) that the current at any potential increases linearly with the square root of the potential scan rate. Normally, the peak current is chosen for experimental use because it is easier to measure. For convenience, the peak current divided by the square root of scan rate, to be termed the "current function," is examined as a function of scan rate. If (6.22) is exactly obeyed, there should be no variation in the current function with changes in the rate of potential scan. The value of the anodic and cathodic peak currents is the same if the decaying cathodic current is used as the base line for the subsequent reoxidation cycle. To measure the anodic peak, the decay of the cathodic response must be measured past the switching time. Polcyn and Shain [11] have used an estimation technique in which the decay is equated with the simple potentiostatic decay curve. Table 6.2 gives a summary of the parameters used to identify a reversible cyclic voltammetric response.

Numerous examples are currently known of electrochemical reactions that behave reversibly. The ferrocene-ferrocinium ion couple illustrated in Fig. 6.2 is a typical example since many organic compounds undergo reversible electrochemical reactions in inert solvents [19].

It should be emphasized that the reaction medium frequently determines the nature of the overall electrochemical reaction. Thus reactions that are complex under one set of conditions may be simplified by the proper choice of solvent system. It is often beneficial to a study of a complex reaction to search for conditions under which the reaction approaches the behavior of the reversible charge transfer case and to subsequently note which addenda result in the overall reaction approaching that of prime interest.

The theoretical cyclic voltammetric response curves for reversible electrochemical reactions of nonunity reaction order have been studied by Shuman [20]. Qualitatively, as the reaction order increases, the response becomes less

Table 6.2 Diagnostic Criteria for Reversible Charge Transfer
$$O + ne^- \rightleftharpoons R$$

Cyclic Voltammetry
Properties of the potential of the response: $v = $ sweep rate
   E_p is independent of v
   $E_p^c - E_p^a = 59/n$ mV at 25°C and is independent of v
Properties of the current function:
   $i_p/v^{1/2}$ is independent of v
Properties of the anodic to cathodic current ratio:
   i_p^a/i_p^c is unity and independent of v

Ac Polarography
Properties of the potential of the response:
   E_p is independent of ω
   Half-peak width is $90/n$ mV at 25°C and independent of ω
Properties of the ac current:
   $I(\omega t)$ is proportional to $\omega^{1/2}$ with a zero intercept
Properties of the phase angle:
   ϕ is 45° and constant with respect to both ω and E_{dc}
Others:

   Linear plot of E_{dc} versus $\log \left[\left(\dfrac{I_P}{I}\right)^{1/2} - \left(\dfrac{I_P - I}{I}\right)^{1/2} \right]$
   with a slope of $118/n$ mV at 25°C

---

peak-shaped in nature and lower in peak current. The maximum value of the calculated peak cathodic current parameter, $\pi^{1/2}\chi(at)$, for the reaction

$$mO + ne^- \rightleftharpoons R$$

decreases from 0.446 for m equal to 1 to 0.353 for m equal to 2 and to 0.303 for m equal to 3. The peak potential also shifts cathodically of $E_{1/2}$ as m increases ($E_p - E_{1/2} = 36.0/n$ mV for $m = 2$ and $49.8/n$ mV for $m = 3$).

Experimental data for a typical system of nonunity reaction order, the dissolution of mercury in cyanide media, were in good agreement with the theoretical treatment [20]. Thus extension of the theory to all reaction orders of common interest appears possible.

The Reversible Ac Polarographic Response

Derivation of the Current Expression

In ac polarography the expression for the time-dependent potential is

$$E(t) = E_{dc} - \Delta E \sin \omega t, \tag{6.26}$$

where ΔE is the amplitude and ω is the frequency in radians per second of the ac ripple applied to the cell. The expression for $j(t)$ becomes

$$j(t) = j - \frac{nF}{RT} \Delta E \sin \omega t, \tag{6.27}$$

where

$$j = \frac{nF}{RT} (E_{\text{dc}} - E_{1/2}). \tag{6.28}$$

Equation (6.13) is rearranged in the following manner:

$$e^{-j(t)} - e^{-j(t)} \int_0^t \frac{\psi(t)\, d\tau}{(\pi\{t - \tau\})^{1/2}} = \int_0^t \frac{\psi(t)\, d\tau}{(\pi\{t - \tau\})^{1/2}}, \tag{6.29}$$

where

$$e^{-j(t)} = e^{-j} \exp\left(\frac{nF\,\Delta E}{RT} \sin \omega t\right) \tag{6.30}$$

and

$$\psi(t) = \frac{i(t)}{nFAC_O^* D_O^{1/2}}. \tag{6.31}$$

The power series expansion for $\exp[(nF\,\Delta E/RT)\sin(\omega t)]$ is made with the equivalent power series representation for $\psi(t)$, as first proposed by Matsuda [21].

$$\exp\left(\frac{nF\,\Delta E}{RT} \sin \omega t\right) = \sum_{p=0}^{\infty} \left(\frac{nF\,\Delta E}{RT}\right)^p \frac{(\sin \omega t)^p}{p!} \tag{6.32}$$

and

$$\psi(t) = \sum_{p=0}^{\infty} \psi_p(t)\left(\frac{nF\,\Delta E}{RT}\right)^p. \tag{6.33}$$

Substituting (6.30) through (6.33) in (6.29) and equating coefficients of equal powers of $(nF\,\Delta E/RT)^p$, one obtains a system of integral equations:

$$\frac{e^{-j}(\sin \omega t)^p}{p!} - \sum_{r=0}^{p} \frac{e^{-j}(\sin \omega t)^r}{r!} \int_0^t \frac{\psi_{p-r}(t)\, d\tau}{(\pi\{t - \tau\})^{1/2}} = \int_0^t \frac{\psi_p(t)\, d\tau}{(\pi\{t - \tau\})^{1/2}}. \tag{6.34}$$

Equation (6.34) can now be solved for various values of p. For $p = 0$, one obtains

$$\int_0^t \frac{\psi_0(t)\, d\tau}{(\pi\{t - \tau\})^{1/2}} = \frac{1}{1 + e^j}, \tag{6.35}$$

which could have been obtained by setting $E(t) = E_{\text{dc}}$ in (6.13) and substituting in (6.31). The solution of this equation by Laplace transforms yields, with the use of (6.31),

$$i(t)_{\text{dc}} = \frac{nFAC_O^* D_O^{1/2}}{(1 + e^j)(\pi t)^{1/2}}. \tag{6.36}$$

This is the well-known Cottrell equation for any value of dc potential (embodied in e^j).

For $p = 1$, one obtains the equation for the fundamental harmonic ac current:

$$e^{-j}\left[1 - \int_0^t \frac{\psi_0(t)\,d\tau}{(\pi\{t - \tau\})^{1/2}}\right] \sin \omega t = (1 + e^{-j})\int_0^t \frac{\psi_1(t)\,d\tau}{(\pi\{t - \tau\})^{1/2}}. \quad (6.37)$$

Notice that this expression contains an integral with the expression $\psi_0(t)$. This is the expression for the dc current solution when $p = 0$, given by (6.36). This is a general phenomenon of this method of solution. All the lower-order terms are needed in order to solve for the higher-order current expressions [22]. Similarly, the expressions for $\psi_0(t)$ and $\psi_1(t)$ are needed to obtain the expression for the second harmonic current $\psi_2(t)$.

Substituting (6.35) in (6.37) yields the expression

$$\int_0^t \frac{\psi_1(t)\,d\tau}{(\pi\{t - \tau\})^{1/2}} = \frac{\sin \omega t}{4 \cosh^2 (j/2)}, \quad (6.38)$$

given by Smith [5], which can be solved to yield the fundamental harmonic ac current expression:

$$I(\omega t) = \frac{n^2 F^2 A C_O^*(\omega D_O)^{1/2} \Delta E}{4RT \cosh^2 (j/2)} \sin (\omega t + \pi/4). \quad (6.39)$$

Equation (6.39) has been obtained assuming only planar diffusion. It is interesting that this same expression is obtained assuming diffusion to an expanding plane. This fact applies to the reversible case only and is a manifestation of the time dependence of the dc process. There is no change in the type of time dependence of the dc process when the expanding plane model is used, so the time dependence of the ac process remains the same, as does the expression for the current magnitude. For other reaction mechanisms the current expression is different for the planar and the expanding plane model, but again it is attributable to changes in the dc process. This is discussed in more detail in later sections.

Diagnostic Criteria for the Reversible Ac Polarogram

Referring to (6.39), one will note that when the ac current is plotted against dc potential the shape of the wave is given by $1/\cosh^2 (j/2)$, since from (6.28) j is the expression of dc potential:

$$j = \frac{nF}{RT} (E_{dc} - E_{1/2}). \quad (6.40)$$

When j is equal to zero, $1/\cosh^2 (j/2)$ has a maximum value of 1; therefore there is a peak in $I(\omega t)$ at $E_{dc} = E_{1/2}$. It can be shown that the shape of the reversible ac polarographic wave is actually the derivative of the dc polarographic wave. From (6.36) we can see $i_{dc} \propto 1/(1 + e^j)$ and

$$\frac{d}{dj}\left(\frac{1}{1 + e^j}\right) = \frac{1}{4\cosh^2(j/2)}. \qquad (6.41)$$

It can also be shown from (6.39) that the width of the ac wave at the half-peak height is $90/n$ mV at 25°C. Substituting for the definition of j, one can further show that

$$E_{dc} = E_{1/2} + \frac{2RT}{nF} \ln\left\{ \left(\frac{I_P}{I}\right)^{1/2} - \left(\frac{I_P - I}{I}\right)^{1/2} \right\}. \qquad (6.42)$$

The term I_p is the peak current of the ac wave and I is the value at any potential E_{dc}. Thus one obtains a linear plot of E_{dc} versus $\log\{[I_P/I]^{1/2} - [(I_P - I)/I]^{1/2}\}$ with a slope of $118/n$ mV at 25°C [5].

Examination of (6.39) shows that there is a phase shift between the applied potential and the resulting current. For the reversible case the phase angle is constant over the whole wave at $\pi/4$ radians or 45°. The phase angle is a very important parameter in ac polarography for mechanistic studies. See Section 14 for a more detailed discussion of the origin of the phase angle.

Historically, the experimental parameters of interest with respect to phase angle measurement are the resistive and capacitive components of the faradaic impedance, generally measured using a Wien bridge. The technique of ac polarography, however, yields the same experimental information in a slightly different way. Any combination of two of the four possible parameters—total current, resistive current, capacitive current, and phase angle—yields the other two. The theory of ac polarography uses the total current and phase-angle information for diagnostic study, whereas faradaic impedance theory makes use of the capacitive and resistive components of the total current. A further discussion of faradaic impedance is given in Section 14. The significance of the resistive and capacitive impedance components discussed there applies equally to the phase angle parameter in ac polarography.

It can also be seen from (6.39) that $I(\omega t)$ is proportional to $\omega^{1/2}$ and has a zero intercept at $\omega = 0$. Table 6.2 summarizes the properties of the reversible ac polarogram useful for diagnostic purposes.

It should be noted that all of the above criteria regarding the shape of the wave are valid only for small amplitudes of applied potential. The value of the applied potential should be less than $16/n$ mV peak to peak. If this criterion is not followed there will be a significant contribution to the fundamental current by the higher harmonic components. As Smith [5] has shown, all odd values of p used in solving (6.34) yield a fundamental current component. These higher-order current components are negligible only if ΔE is small.

As for the cyclic voltammetric response, many systems are known that behave reversibly. The trisoxalatoiron(III) reduction is perhaps the best example of an aqueous system [23]. The reduction of nitrobenzene in dimethylformamide or acetonitrile to form the anion radical also shows ac polarographic reversibility [24].

5 QUASI-REVERSIBLE CHARGE TRANSFER

Derivation of Current Expressions

In the previous section the case of a purely diffusion controlled reaction (reversible case) was considered. The quasi-reversible case is the mechanism in which the current is controlled by a mixture of diffusion and charge transfer kinetics. Because diffusion plays a part in controlling the surface concentrations of the redox couple, these expressions are the same as those outlined for the reversible case in Section 4.

The surface concentration expressions are now substituted into the absolute rate expression instead of the Nernst equation [25] to yield an equation that relates the current to the charge transfer rate of the reaction. This rate is potential dependent, so the total current can be expressed by (6.43) [26]:

$$\frac{i(t)}{nFA} = C_O(0, t)k_s \exp\left(\frac{-\alpha nF}{RT}\{E(t) - E^0\}\right)$$

$$- C_R(0, t)k_s \exp\left(\frac{\beta nF}{RT}\{E(t) - E^0\}\right), \quad (6.43)$$

where k_s is the charge transfer rate constant at the standard potential, E^0, of the redox couple, α is the charge transfer coefficient of the reduction step and,

$$\beta = 1 - \alpha. \tag{6.44}$$

A review of the significance of the charge transfer coefficient has been given recently by Bauer [27].

Equation (6.43) can be rewritten in terms of the potential dependent parameter $j(t)$, defined by (6.12), which yields

$$\frac{i}{nFAk_s} = C_O(0, t)e^{-\alpha j(t)} - \left(\frac{D_R}{D_O}\right)^{1/2} C_R(0, t)e^{(1-\alpha)j(t)}. \tag{6.45}$$

The appropriate potential-time relationship and the concentration expressions are substituted into (6.45), and the current-potential relationships are determined as before for ac polarography and cyclic voltammetry.

The Cyclic Voltammetric Response

The surface concentration expressions given by (6.10) are substituted in (6.45) with the potential expression $j(at)$ now given by (6.15) for this technique. The resultant expression, in dimensionless parameters, is given by

$$\frac{\chi(at)}{\psi} e^{\alpha j(at)} = 1 - (1 + e^{j(at)}) \int_0^{at} \frac{\chi(z)\, dz}{(at - z)^{1/2}}. \tag{6.46}$$

Equation (6.46) has been made dimensionless by the substitutions [28]:

$$\psi = \frac{k_s}{(\pi a D)^{1/2}}, \tag{6.47}$$

where

$$D = D_O^\beta D_R^\alpha, \tag{6.48}$$

$$a = \frac{nFv}{RT}, \tag{6.49}$$

and

$$\chi(at) = \frac{i(at)}{nFAC_O^*(\pi a D_O)^{1/2}}. \tag{6.50}$$

If (6.46) can be solved for $\chi(at)$, the cyclic voltammetric current can be obtained from (6.50) as a function of at, which is related to potential [see (6.17)].

Nicholson [28] has presented (6.46) in a slightly different form and has obtained a numerical solution for different values of k_s and α. The dimensionless parameter ψ contains both k_s and α, and Nicholson has shown that when $\psi \geq 7$ the solution of $\chi(at)$ becomes equal to the Nernstian case discussed in Section 4. When $\psi < 0.001$, the solution of (6.46) is the same as for the irreversible case discussed in Section 6. The region in between $\psi = 0.001$ and $\psi = 7$ is that of the so-called quasi-reversible mechanism of interest here.

Through several calculations of theoretical curves, Nicholson [28] has found that in the region of $0.3 \leq \alpha \leq 0.7$ the separation of anodic and cathodic peaks is almost independent of α, being mainly a function of k_s. This approximation becomes less accurate, however, as ψ becomes small. At $\psi = 0.5$ there is only a 5% variation in peak separation as α is varied from 0.3 to 0.7. The peak separation at this point is $105/n$ mV, which is significantly different from the reversible peak separation of about $59/n$ mV.

The exact peak separation as a function of ψ is given in the working curve in Fig. 6.4. The data used to construct this curve are given by Nicholson [28]. The useful working range of Fig. 6.4 for determination of k_s is $0.5 \leq \psi \leq 5.0$. Since ψ is also a function of scan rate, this experimental parameter can be varied to make ψ fall within the required range. If the scan rate is slow

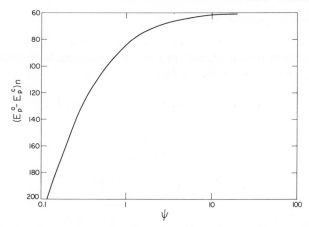

Fig. 6.4 Cyclic voltammetric working curve for the quasi-reversible reaction. Abscissa: kinetic function, ψ. Ordinate: potential separation of cathodic and anodic peak, mV.

enough, the peak separation approaches reversible behavior, and if it is very fast, irreversible behavior is approached.

Experimentally, a scan rate is chosen to obtain a peak separation somewhere between $120/n$ and $60/n$ mV. From Fig. 6.4 the corresponding value of ψ is determined. Equation (6.47) is then used to determine k_s. To use this method rigorously D_O, D_R, and α must be known, but except for α values very different from 0.5 and D_R very different from D_O, the approximation $D_O^{1/2} = D_R^{1/2} = D^{1/2}$ can be used.

The value of α can only be estimated by matching experimental curves with theoretical curves for various values of α. For large values of α, the reduction peak is much sharper and the current larger than that of the corresponding anodic peak. The opposite is true for small values of α.

Nicholson [28] has used this technique to study the reduction of Cd(II) at a mercury electrode in 1 M Na$_2$SO$_4$. He used published values of D_O and D_R for this system and an estimated value of $\alpha = 0.25$ obtained by comparing various theoretically calculated curves for shape and symmetry with experimental curves. The value of $k_s = 0.24$ cm/sec remains fairly constant with scan rate variations from 48 to 120 V/sec.

Recently, Perone [29] has also used this technique of peak separation measurements to study reactions at very high scan rates. Because some of his experimental data yielded $\psi < 0.5$, it became necessary for him to have good α values to calculate k_s. For the same Cd(II)/Cd(Hg) system in 1 M Na$_2$SO$_4$, Perone reports $k_s = 0.30$ cm/sec and $\alpha = 0.58$, which is in reasonable agreement for this system, considering the multitude of other values of k_s and α that have been reported [30]. Perone also obtained α by comparison of the shapes and positions of experimental and theoretical waves.

The Ac Polarographic Response

By the same method as outlined for the reversible case, a system of integral equations is formed, and for $p = 1$ the fundamental ac polarographic current expression can be obtained:

$$I(\omega t) = I_{rev} F(\lambda t^{1/2}) G(\omega^{1/2} \lambda^{-1}) \sin (\omega t + \phi), \qquad (6.51)$$

where

$$I_{rev} = \frac{n^2 F^2 A C_O^* (\omega D_O)^{1/2} \Delta E}{4RT \cosh^2 (j/2)}, \qquad (6.52)$$

$$F(\lambda t^{1/2}) = 1 + \frac{(\alpha e^{-j} - \beta) \psi(\lambda t^{1/2}) D^{1/2}}{k_s e^{-\alpha j}}, \qquad (6.53)$$

$$G(\omega^{1/2} \lambda^{-1}) = \left[\frac{2}{1 + \frac{(1 + \{2\omega\}^{1/2})^2}{\lambda}} \right]^{1/2}, \qquad (6.54)$$

$$\phi = \cot^{-1} \left(1 + \frac{(2\omega)^{1/2}}{\lambda} \right), \qquad (6.55)$$

$$\lambda = \frac{k_s}{D^{1/2}} (e^{-\alpha j} + e^{\beta j}), \qquad (6.56)$$

$$D = D_O^\beta D_R^\alpha, \qquad (6.57)$$

$$\psi(\lambda t^{1/2}) = \frac{k_s e^{-\alpha j}}{D^{1/2}} \exp (\lambda^2 t) \operatorname{erfc} (\lambda t^{1/2}). \qquad (6.58)$$

The expression for $\psi(\lambda t^{1/2})$ will be recognized as that for the dc current expression at constant potential, that is, chronoamperometry [26]. The term I_{rev} is the reversible current magnitude obtained in Section 4.

Some general comments can be made about (6.51) that give a better insight into exactly what it means. The expression for $F(\lambda t^{1/2})$ gives the deviation from reversibility attributable to effects of the charge transfer process on the dc response. When $\lambda t^{1/2} > 50$ the dc process is essentially reversible, $F(\lambda t^{1/2})$ is unity [5, 31], and the time dependence of (6.51) is eliminated. For the normal polarographic drop life of 5 to 10 sec, this means that the dc process is reversible for $k_s > 10^{-2}$ cm/sec.

The expression $G(\omega^{1/2} \lambda^{-1})$ is a measure of the deviation from reversibility of the ac process. This term is unity when $(2\omega)^{1/2} \lambda^{-1} \ll 1$ [5]. For any given value of rate constant, the deviations from reversibility become greater as the frequency is increased. Thus as faster charge transfer rates are studied, higher frequencies must be used to obtain reliable results.

The deviations discussed above are actually surface concentration deviations. If the charge transfer is very fast, essentially thermodynamic (reversible) surface concentrations are observed. As the charge transfer rate becomes slower, it takes time to reach these equilibrium values, hence the time dependence of the $F(\lambda t^{1/2})$ term for the dc process.

The time dependence of the ac process is embodied in the frequency ω. As the frequency increases there is less time for the concentration to relax to the equilibrium value before a new cycle begins. Thus there is a lag which decreases the current magnitude as expressed by $G(\omega^{1/2}\lambda^{-1})$. This lag also causes a phase shift between the applied ac potential and the resulting current as seen by (6.55). The ratio $(2\omega)^{1/2}\lambda^{-1}$ again expresses this deviation from reversible behavior. The reversible case has a constant phase shift of $\pi/4$ and $\cot \pi/4 = 1$. In the quasi-reversible case there is an additional shift represented by the ratio $(2\omega)^{1/2}\lambda^{-1}$ attributable to the finite rate of the charge transfer process (see also the discussion in Section 14 on faradaic impedance). As the charge transfer rate constant decreases, this deviation in phase angle from $\pi/4$ manifests itself at lower and lower frequencies.

The expression for $\psi(\lambda t^{1/2})$ has been derived assuming planar diffusion. The corresponding expression assuming diffusion to an expanding plane also yields a time-dependent term [32, 33]:

$$\psi(\lambda t^{1/2})_{\text{ep}} = \left(\frac{7}{3\pi}\right)^{1/2} \frac{(1.61 + \lambda t^{1/2})}{(1.13 + \lambda t^{1/2})^2}. \tag{6.59}$$

Thus although the drop growth influences the magnitude of the time dependence it is not the sole cause. It should be mentioned again that no time dependence is observed for the quasi-reversible case when the dc process is reversible, that is, $k_s > 10^{-2}$ cm/sec. One concludes that the time dependence of the ac polarographic wave is only caused by nonequilibrium of the dc process [5].

With these thoughts in mind, several criteria can be developed which describe the quasi-reversible system and differentiate it from other mechanisms. For any value of k_s, $F(\lambda t^{1/2})$ becomes unity when $\alpha e^{-j} - \beta = 0$, as can be seen from (6.53), even though the dc process may be non-Nernstian. This occurs at only one potential E_{dc}^{*}

$$E_{\text{dc}}^{*} = E_{1/2} + (RT/nF) \ln \left(\frac{\alpha}{1 - \alpha}\right). \tag{6.60}$$

This "crossover" potential has been proposed as a means of measuring the charge transfer coefficient because the current is the same at this potential regardless of drop life [31]. Current anodic of this potential increases as the drop life increases, and current cathodic of this potential decreases as the drop life increases.

Because the phase angle is independent of drop life, yet depends on the charge transfer rate, it offers an excellent means of measuring k_s. The cot ϕ at any point along the wave is linear with $\omega^{1/2}$ with an intercept of unity and a slope of $(2/\lambda)^{1/2}$. The charge transfer rate constant can be determined from λ if the diffusion coefficients and α are known. Examples of a plot of the phase angle at $E_{1/2}$ versus $\omega^{1/2}$ for two different values of k_s are given in Fig. 6.5.

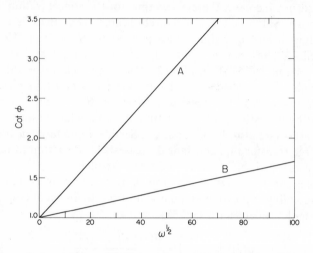

Fig. 6.5 Theoretical cot ϕ vs. $\omega^{1/2}$. Value of cot ϕ is measured at $E_{dc} = E_{1/2}$. Curve A, $k_s = 0.02$ cm/sec. Curve B, $k_s = 0.10$ cm/sec.

At any frequency cot ϕ varies with the dc potential with a maximum value reached at $(E_{dc})_{max}$:

$$(E_{dc})_{max} = E_{1/2} + \frac{RT}{nF} \ln \left(\frac{\alpha}{1 - \alpha} \right). \tag{6.61}$$

Note that this is the same potential as the crossover potential and is frequency independent. The value of cot ϕ at $(E_{dc})_{max}$ is given by (6.62):

$$\cot \phi_{max} = 1 + \frac{(2\omega D)^{1/2}}{k_s[(\alpha/\beta)^{-\alpha} + (\alpha/\beta)^{\beta}]}. \tag{6.62}$$

By using these two equations, both k_s and α can be obtained [5, 24], but the value of k_s is dependent on the value of α obtained from either (6.60) or (6.61). In (6.62) the value of D can be obtained by determining the diffusion coefficients D_O and D_R from some other technique, perhaps dc polarography, if k_s is not too small. Figure 6.6 shows the plot of cot ϕ versus $(E_{dc} - E_{1/2})$ for the same values of k_s given in Fig. 6.5 at $\omega = 200$ radian/sec.

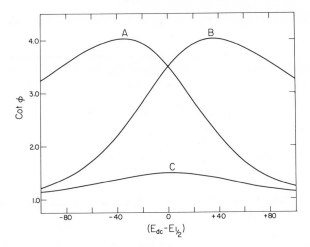

Fig. 6.6   Theoretical cot ϕ vs. dc potential. Curve A; $\omega = 5000$ sec^{-1}, $k_s = 0.02$ cm/sec, $\alpha = 0.2$. Curve B; $\omega = 5000$ sec^{-1}, $k_s = 0.02$ cm/sec, $\alpha = 0.8$. Curve C; $\omega = 5000$ sec^{-1}, $k_s = 0.10$ cm/sec, $\alpha = 0.5$.

Expressions for the peak current magnitude and the position of the peak cannot be obtained except for the high-frequency limit, and these are given elsewhere [35]. At low frequencies the position of the peak approaches $E_{1/2}$ [as $G(\omega^{1/2\lambda})$ approaches unity]. The peak current approaches linearity with $\omega^{1/2}$.

As the frequency decreases, the experimental parameters all approach the behavior of the reversible case since there is more time for equilibrium to occur. Of course, for very slow charge transfer reactions, this behavior is not reached until the frequency becomes so low that the steady-state approximation used in deriving (6.51) is no longer valid [36, 37]. Table 6.3 summarizes the observable criteria for the quasi-reversible reaction using ac polarography.

Studies of the quasi-reversible mechanism have been the most extensive use of ac polarography to date. Smith [38] has studied the reduction of Ti(IV) to Ti(III) in 0.2 M H$_2$C$_2$O$_4$ under polarographic conditions (net dc current flow) and reported $k_s = 4.6 \times 10^{-2}$ cm/sec and $\alpha = 0.35$. Using a value of $D_O = 6.6 \times 10^{-6}$ cm^2/sec and these values of k_s and α, he plotted curves of cot ϕ versus $\psi^{1/2}$ and cot ϕ versus E_{dc} at several frequencies. The results of these theoretical curves agreed well with the experimental points [5, 38].

Timnick and co-workers [39] have studied the reduction of Cd(II) in various supporting electrolytes by ac polarography and have shown how the rate parameters depend on the medium. Smith and Hung [40] have measured k_s and α for the formation of the free radical in nonaqueous media for several group-IV organometallic species.

Table 6.3 Diagnostic Criteria for Quasi-Reversible Charge Transfer

$$O + ne^- \rightleftharpoons R$$

Cyclic Voltammetry
Properties of the potential of the response:
  E_p shifts with v
  $E_p^c - E_p^a$ may approach $60/n$ mV at low v but increases as v increases
Properties of the current function:
  $i_p/v^{1/2}$ is virtually independent of v
Properties of the anodic-to-cathodic current ratio:
  i_p^a/i_p^c equal to unity only for $\alpha = 0.5$
Others:
  The response visually approaches that of the irreversible charge transfer
  as v is increased

Ac Polarography
Properties of the potential of the response:
  E_p varies with ω if $\alpha \neq 0.5$
Properties of the ac current:
  $I(\omega t)$ is nonlinear with $\omega^{1/2}$. Linearity with $\omega^{1/2}$ may be noted at low
  frequencies and a limiting amplitude may be noted at high frequencies
Properties of the phase angle:
  Cot ϕ varies linearly with $\omega^{1/2}$, approaching unity at low ω
  Cot ϕ versus E_{dc} exhibits a maximum
Others:
  The wave is dependent on drop time. The ac current increases with drop
  time at potential anodic of a "crossover" potential and decreases
  with drop time cathodic of this potential

The reduction of $Cr(CN)_6^{-3}$ in 1 M KCN has been studied by Smith and co-workers and values of k_s have been reported as 0.22 cm/sec [41] and 0.24 cm/sec [42]. Values for α were 0.45 and 0.59, respectively. These agree with Randles and Somerton's [43] study using faradaic impedance where they report $k_s = 0.25$ cm/sec. More recently, McCord and Smith [44] have used second harmonic ac polarography to study this system; they report $k_s = 0.25 \pm 0.02$ cm/sec and $\alpha = 0.55 \pm 0.03$, in excellent agreement with previous results.

Hung and Smith [31] have made use of the time dependence of the ac wave to calculate α based on the "crossover" potential for the reduction of V(III) to V(II) in 1 M H_2SO_4. Their reported α value of 0.46 ± 0.03 is in excellent agreement with the value reported by Randles [45] for this system. Based on their experiments with other systems, Hung and Smith state that this method is probably useful only for systems in which

k_s is between 10^{-4} cm/sec and 5×10^{-3} cm/sec [31]. Larger values do not show sufficient time dependence, and smaller values yield currents too small for accurate measurements.

6 IRREVERSIBLE CHARGE TRANSFER

An irreversible electrochemical reaction is one for which the charge transfer rate constant is small. Thus the potential at which a significant extent of reaction occurs may be quite cathodic of the standard potential. At such potentials the reverse charge transfer process does not occur. If the absolute rate expression is then employed to describe the current, the second term on the right side of (6.43) becomes zero. Thus for an irreversible reaction, (6.43) becomes

$$\frac{i(t)}{nFA} = C_O(0, t)k_s \exp\left[\frac{-\alpha nF}{RT}\{E(t) - E^0\}\right]. \tag{6.63}$$

If the reaction is irreversible, the peak separation is so large that there is often no current observed on the return potential sweep in cyclic voltammetry and in ac polarography the observed total current is very small. The potential of the response in both techniques is shifted very cathodically for reductions and very anodically for oxidations with respect to the standard potential.

By using the same boundary conditions as employed for the reversible case, with the exception that the concentration of the product species is assumed to be zero, an expression for the surface concentration of the reactant species is obtained. The current is then related to the charge transfer rate by (6.63), using the value for $C_O(0, t)$ from (6.10a). The appropriate value of $E(t)$ is substituted into (6.63) as well and the resultant integral equation is solved.

If the product of a reversible charge transfer reaction is destroyed by a rapid chemical reaction such that no reverse reaction can proceed, the overall reaction sequence will reduce to a response having the same qualitative form as that of the irreversible charge transfer reaction and will yield the same response if α is equal to unity. Such reactions can be treated by the equations derived for the irreversible system. However, it should be emphasized that slow charge transfer and rapid charge transfer coupled to a very rapid and irreversible chemical reaction are not conceptually equivalent processes. Thus care should be exercised in terming a reaction irreversible in the absence of knowledge of the overall reaction. The use of supposed heterogeneous rate constants and α values to describe mechanistic subtleties for a reaction which in reality does not involve a slow charge transfer process should be strongly discouraged.

The Cyclic Voltammetric Response

The appropriate integral equation describing the response of the irreversible reaction for the cyclic voltammetric technique is given by (6.64):

$$\frac{\chi(at)}{\psi} e^{\alpha j(at)} = 1 - \int_0^{at} \frac{(z)\, dz}{(at - z)^{1/2}}, \tag{6.64}$$

where now $j(at)$ is given by (6.65) for the single potential scan:

$$j(at) = \frac{nF}{RT}(E_i - vt - E^0). \tag{6.65}$$

The initial potential is E_i and scan rate is given by v in volts per second. Equation (6.64) has been made dimensionless by making the following substitutions:

$$\chi(at) = \frac{i(at)}{nFAC_O{}^*(\pi D_O a)^{1/2}} \tag{6.66}$$

$$\psi = \frac{k_s}{(\pi D_O a)^{1/2}}, \tag{6.67}$$

where a is given by (6.16).

These equations differ slightly from those given by Nicholson and Shain [8] because no distinction is made between the total number of electrons consumed per mole and the number consumed in the rate-determining charge transfer step, these two quantities being assumed to be the same.

Delahay [46] and Matsuda and Ayabe [14] have given a numerical solution for (6.64), but the data of Nicholson and Shain [8] are more accurate, being given to within ± 0.001.

From the potential scale given by Nicholson and Shain [8] in their tabulation, the peak potential can be calculated for an irreversible response:

$$E_p = E^0 - \frac{RT}{\alpha n F}[0.780 + \ln(D_O \alpha a)^{1/2} - \ln k_s]. \tag{6.68}$$

There is about a $30/\alpha n$ mV cathodic shift in peak potential for every 10-fold increase in scan rate, and this criterion can be used to characterize an irreversible response. An example of this shift is shown in Fig. 6.7 in which the current response $\pi^{1/2}\chi(at)$ is plotted for various scan rates. The same amount of shift per 10-fold scan rate increase obtains for the half-peak potential also [8]. Since this parameter is frequently measured more accurately, it can be used as well to characterize the irreversible response. Thus the

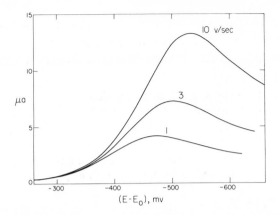

Fig. 6.7 Theoretical irreversible cyclic voltammetric curves. $k_s = 10^{-6}$ cm/sec, $\alpha = 0.5$, $n = 0.1$, $D_O = D_R = 10^{-6}$ cm²/sec, area = 0.02 cm², $C_O{}^* = 10^{-6}$ M, $E^0 = 0.0$ V, scan rate as shown.

shape of the response is independent of scan rate and the separation of E_p and $E_{p/2}$ is given by (6.69):

$$E_p - E_{p/2} = \frac{-1.857RT}{\alpha nF} .$$ (6.69)

An alternative expression to (6.68) can be obtained relating peak current to peak potential, from which the parameters k_s and α can be found [8, 15]:

$$i_p = 0.227nFAC_O{}^*k_s \exp\left(\frac{-\alpha nF}{RT}\{E_p - E^0\}\right)$$ (6.70)

A plot of log i_p versus $E_p - E^0$ (or $E_{p/2} - E^0$) for different scan rates yields α from the slope and k_s from the intercept provided E^0 is known.

Reinmuth [16] has shown that at the foot of the response the current is independent of scan rate and can be related to the potential and the initial potential, E_i. The initial potential is chosen at the foot of the response where no current flows.

$$i = nFAC_O{}^*k_s \exp\left[\frac{-\alpha nF}{RT}(E - E_i)\right]$$ (6.71)

Nicholson and Shain [8] have shown that this equation is valid only for current values less than $0.1i_p$, but (6.71) does offer a convenient way to obtain k_s and α even when the standard potential E^0 for the reaction is unknown.

Table 6.4 Diagnostic Criteria for Irreversible Charge Transfer

$$O + ne^- \xrightarrow{\;k_s\;} R$$

Cyclic Voltammetry

Properties of the potential of the response:

E_p shifts cathodically by $30/\alpha n$ mV for a 10-fold increase in v

Properties of the current function:

$i_p/v^{1/2}$ is constant with scan rate

Properties of the anodic to cathodic current ratio:

There is no current on the reverse scan

Ac Polarography

Properties of the potential of the response:

E_p shifts cathodically by $15/\alpha n$ mV for a 10-fold increase in ω

Properties of the ac current:

I_p is independent of ω at high frequencies and proportional to $t^{-1/2}$ (drop time)

Properties of the phase angle:

ϕ is much less than $45°$

Others:

The ac response is very small, and in some cases, may not be observable.

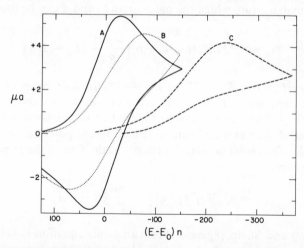

Fig. 6.8 Theoretical cyclic voltammetric curves. $A = 0.02\,\text{cm}^2$, $C_O = 10^{-6}\,M$, $D_O = D_R = 10^{-6}\,\text{cm}^2/\text{sec}$, scan rate $= 1$ volt/sec, $E^0 = 0$. Curve A, reversible curve. Curve B, quasi-reversible, $k_s = 0.03$, $\alpha = 0.5$. Curve C, irreversible, $k_s = 10^{-4}\,\text{cm/sec}$, $\alpha = 0.5$.

Table 6.4 summarizes the cyclic voltammetric diagnostic criteria for the irreversible reaction. Also, by way of summary, Fig. 6.8 shows the effect of the charge transfer rate constant on the cyclic voltammetric response as the rate is varied from essentially reversible (diffusion-controlled) behavior to irreversible behavior.

Experimental study of an irreversible system using cyclic voltammetry has been recently reported. Sundholm [47] has studied the reduction of $PtCl_4^{2-}$ on mercury in $1\ M$ $NaClO_4$ and has shown that $\dfrac{i_p}{v^{1/2}}$ is independent of scan rate while the peak shifts cathodically as the scan rate increases. Knowing the value of E for this reduction, Sundholm has reported a value of $\alpha = 0.23$ and $k_s = 9 \times 10^{-6}$ cm/sec, an essentially irreversible reaction.

The Ac Polarographic Response

Until quite recently, it was believed that there was no measurable ac current for an irreversible reaction [31, 48], since no reverse reaction can occur as the sine wave applied potential oscillates. Recently, however, Timmer et al. have shown, both theoretically [49] and experimentally [50], that such a signal does exist for an irreversible reaction caused by slow charge transfer. Smith and McCord [51] independently showed theoretically that a signal could exist for this reaction and extended their work to fast following chemical reactions including the catalytic regeneration of starting material.

Smith and McCord [51] used the expanding plane model and the absolute rate expression in their calculation. Making appropriate approximations to account for irreversibility, they obtained (6.72) for the ac current. This same result could also have been obtained by using the modified absolute rate expression given by (6.63).

$$I(\omega t) = \left(\frac{7}{3\pi}\right)^{1/2} \frac{1.349\alpha n^2 F^2 A C_O{}^*(2\omega D_O)^{1/2}\,\Delta E}{RTe^{-J}(1 + e^{1.091J})[1 + (1 + Qe^J)^2]^{1/2}}$$
$$\times \sin\left[\omega t + \cot^{-1}(1 + Qe^J)\right], \quad (6.72)$$

where

$$Q = 1.349(2\omega t)^{1/2}, \qquad (6.73)$$

$$J = \alpha j - \ln(1.349 k_s t^{1/2}/D^{1/2}), \qquad (6.74)$$

$$D = D_O{}^\beta D_R{}^\alpha. \qquad (6.75)$$

All other terms have their usual significance. Equation (6.72) involves an approximation to the dc solution [33]. Remember that to solve the fundamental harmonic expression requires a knowledge of the appropriate dc current expression. This approximation is very good except near the anodic foot of the ac wave where the error is only a few percent [33, 51].

With the aid of some simplifying assumptions, it is possible to obtain an expression for the peak current of the ac wave [51]:

$$I_p(\omega t) = \left(\frac{7}{3\pi}\right)^{1/2} \frac{1.349\alpha n^2 F^2 A C_O{}^*(2\omega D_O)^{1/2}\,\Delta E}{RTQ^{1/2}(1 + Q^{0.545})[1 + (1 + Q^{1/2})^2]^{1/2}} . \qquad (6.76)$$

When $Q^{1/2} \gg 1$ or at fairly high frequencies, above 1 KHz, (6.76) reduces to the expression given by Timmer et al. [49, 51]:

$$I_p(\omega t) = \left(\frac{7}{3\pi t}\right)^{1/2} \frac{\alpha n^2 F^2 A C_O{}^* D_O^{1/2}\,\Delta E}{RT} . \qquad (6.77)$$

Equation (6.77) predicts that the peak current is independent of frequency and proportional to $t^{-1/2}$, where t is the drop life. Deviations from this behavior will probably be experimentally measurable at low frequencies where the more general expression, (6.76), should be used. For example, there is about a 15% increase in $I_p(\omega t)$ going from $\omega = 10^2$ to $\omega = 10^4$ radians per second based on (6.76) [51]. An expression for the dc potential of the peak can also be obtained [51]:

$$(E_{dc})_{peak} = E_{1/2} + \frac{RT}{\alpha nF} \ln \left[\frac{1.349 k_s t^{1/2}}{D^{1/2}}\right] - \frac{RT}{2\alpha nF} \ln Q. \qquad (6.78)$$

The same equation was obtained by Timmer et al., with the numerical coefficient of k_s equal to 1.251 [49, 51]. If the reversible half-wave potential can be determined, both k_s and αn can be found by plotting $[(E_{dc})_{peak} - E_{1/2}]$ against log ω. The straight line has a slope of $14.7/\alpha n$ mV at 25°C and an intercept from which k_s can be determined.

From the definition of $(E_{1/2})_{irr}$, the half-wave potential of the irreversible dc polarographic wave, (6.78) can be rewritten [51].

$$(E_{dc})_{peak} = (E_{1/2})_{irr} - \frac{RT}{2\alpha nF} \ln Q \qquad (6.79)$$

Thus it appears that no information regarding the value of k_s can be obtained from ac polarography that cannot be obtained from dc polarography for the irreversible case, although it may be possible to determine αn more accurately.

Referring again to (6.72), one can see that while the current is directly proportional to α it is independent of the charge transfer rate constant k_s. The only effect k_s has is on the dc potential of the ac wave; see (6.63). These diagnostic criteria are summarized in Table 6.4.

Sluyters and co-workers have studied the reduction of Eu(III) to Eu(II) in 1 M NaClO$_4$ solution using faradaic impedance measurements and ac polarography [50]. Using a value of $\alpha = 0.41$ and $k_s = 2.9 \pm 0.2 \times 10^{-4}$ cm/sec, they were able to match their experimental ac polarograms to the theoretically calculated ones between 320 and 2000 Hz. The rate parameters were calculated from the faradaic impedance measurements (see Section 14). These rate parameters agree well with results of other workers [50, 52].

7 STUDIES OF COUPLED CHEMICAL REACTIONS

If all electrochemical reactions were merely the reversible, quasi-reversible, or irreversible cases described in the preceding sections, the characterization of electrochemical reaction mechanisms would not be difficult. However, there are many electrochemical reactions in which chemical complications are known to exist and many others are likely to become evident as the capability for detecting the presence of chemical reactions increases.

It is emphasized again that the chemical reaction manifests itself in the charge transfer process, not the reverse. Thus there are two objectives of studies of electrochemical reactions which include coupled chemical reactions: (1) a characterization of the nature of the overall electrochemical reaction and the nature of the charge transfer process, and (2) use of the overall electrochemical reaction to evaluate the kinetic parameters of chemical reactions. Cyclic voltammetry and ac polarography can in many cases provide information useful for reaching both these objectives.

Consider a reversible redox reaction in which the product of charge transfer undergoes a reversible chemical reaction:

$$O + ne^- \rightleftharpoons R$$

$$R \underset{k_b}{\overset{k_f}{\rightleftharpoons}} Z.$$

If the product of reaction is continually removed, the energy necessary to carry out the reaction is altered and a deviation of the polarographic $E_{1/2}$ from the reversible value is noted. If the product of the chemical reaction is nonelectroactive, the net current flow required to reduce the concentration of reactant to zero at an electrode surface will not be affected. Thus a dc polarographic experiment provides little information from the current magnitude. However, in the cyclic voltammetric and ac polarographic experiments, the nature of the current response is not at all the same as that of the uncomplicated reversible case. In the cyclic experiment, for example, the magnitude of the current flow detected on the reverse potential scan is less than that observed in the absence of the chemical reaction. The degree to which the current is decreased depends on the rate of the chemical reaction relative to the time scale of the experiment. As the rate of a following chemical reaction decreases with respect to the time scale of the experiment, the less the effect of the chemical reaction is noted. The same reasoning applies to the reverse small amplitude cycle in the ac polarographic experiment in which the time scale is represented by the frequency of the alternating potential applied.

The detailed nature of the reaction exerts an influence on the extent the response differs from that of the uncomplicated system. For example, the change in product concentration for a first-order reaction of a given rate is different from that noted for a second-order reaction of an equivalent rate. Likewise, a second-order reaction involving only the product of the charge transfer reaction (a dimerization, for example) is expected to exert an influence differing from that of a reaction that is second-order in the product of reaction but involves other reactants as well. Differences of this type lead to rather severe complications in the theoretical description of the processes and have indeed been obstacles to a complete description of the effects of all conceivable mechanistic schemes. However, again to our eventual benefit, these differences afford the means by which we obtain mechanistic characterizations and hence study kinetic processes in detail.

This chapter does not attempt to cover all cases of coupled chemical reactions that have received sufficient attention to be at least partially understood but focuses on those cases that are sufficiently well described that practical applications are possible.

Derivation of Surface Concentration Expressions

The reversible charge transfer process followed by a reversible chemical reaction described earlier is employed to illustrate how expressions describing the nature of the current-potential relationships for a reaction sequence complicated by a chemical reaction are developed. For convenience in the treatment of ac polarography, the equilibrium constant of a following chemical reaction is defined as

$$K = \frac{k_b}{k_f},$$

its value being the reciprocal of the value of the equilibrium constant as conventionally defined.

The development of mathematical expressions for the current-potential curve parallels the methods described in the previous sections for charge transfer reactions without coupled chemistry. The manifestation of the chemical reaction is embodied in the description of the transport of the electroactive species to and from the electrode surface. The terms describing the changes in surface concentrations of the electroactive species resulting from the effects of the chemical reaction are added to Fick's law of diffusion. Thus for the following chemical reaction mechanism, the appropriate mass transfer expressions become

$$\frac{\delta C_O(x, t)}{\delta t} = D_O \frac{\delta^2 C_O(x, t)}{\delta x^2}, \tag{6.80}$$

$$\frac{\delta C_R(x, t)}{\delta t} = D_R \frac{\delta^2 C_R(x, t)}{\delta x^2} - k_f C_R(x, t) + k_b C_Z(x, t), \tag{6.81}$$

$$\frac{\delta C_Z(x, t)}{\delta t} = D_Z \frac{\delta^2 C_Z(x, t)}{\delta x^2} + k_f C_R(x, t) - k_b C_Z(x, t). \tag{6.82}$$

The boundary conditions are those for the reversible case with three additions. The condition is applied that at $t = 0$, any x,

$$\frac{C_R(x, t)}{C_Z(x, t)} = \frac{C_R^*}{C_Z^*} = \frac{k_b}{k_f} = K, \tag{6.83}$$

where the asterisk represents bulk concentrations. Equation (6.83) is the definition of the equilibrium constant K and a statement that chemical equilibrium is obeyed. Further, the condition is applied that at

$$t > 0, \qquad x \to \infty, \qquad C_Z(x, t) \to 0, \tag{6.84}$$

which states that the equilibrium concentration of Z in the bulk of the solution, which is zero, is approached with increasing distance from the electrode surface. Finally, the condition that at $t > 0$, $x = 0$,

$$D_Z \frac{\delta C_Z(x, t)}{\delta x} = 0 \tag{6.85}$$

is applied, which states that the flux of the product of the chemical reaction at the electrode surface is zero.

The modified diffusion equations can now be solved using the modified boundary conditions to obtain surface concentrations of the electroactive species. The only stipulations required are that $D_Z = D_R$ and that $C_R^* = 0$, the latter being the normal experimental situation. This also means of course that $C_Z^* = 0$ for this particular mechanism. Employing the Laplace transform and making suitable variable changes eventually results in surface concentration expressions of the form

$$C_O(0, t) = C_O^* - W \int_0^t \frac{i(t)\, d\tau}{nFA(D_O\pi\{t - \tau\})^{1/2}} - X \int_0^t \frac{e^{-\ell\tau} i(t)\, d\tau}{nFA(D_O\pi\{t - \tau\})^{1/2}}, \tag{6.86}$$

$$C_R(0, t) = C_R^* + Y \int_0^t \frac{i(t)\, d\tau}{nFA(D_R\pi\{t - \tau\})^{1/2}} + Z \int_0^t \frac{e^{-\ell\tau} i(t)\, d\tau}{nFA(D_R\pi\{t - \tau\})^{1/2}}. \tag{6.87}$$

The constants W, X, Y, Z, and ℓ are determined by the type of coupled chemical reaction under consideration. For mathematical simplicity, the term ℓ has been defined as the summation of the rate constants k_f and k_b and embodies the effect of the kinetics of the chemical process. For the following chemical reaction under discussion here, $W = 1$, $X = 0$, $Y = K/(1 + K)$, $Z = 1/(1 + K)$, and $\ell = (k_f + k_b)$.

These concentration expressions can then be substituted into either the Nernst equation or the absolute rate expression depending on the type of charge transfer reaction that is being considered. The values of the constants W, X, Y, Z and ℓ are given in Table 6.5 for several of the mechanisms that are treated below.

If the expanding plane diffusion model is used, the integrals in (6.86) and (6.87) have a slightly different form (refer to Table 6.1). The integral

$$\int_0^t \frac{i(t)\,d\tau}{nFA(\pi D\{t - \tau\})^{1/2}}$$

is replaced by

$$\left(\tfrac{7}{3}\right)^{1/2} \int_0^t \frac{\tau^{2/3} i(t)\,d\tau}{nFA(\pi D\{t^{7/3} - \tau^{7/3}\})^{1/2}},$$

and the integral

$$\int_0^t \frac{e^{-\ell\tau} i(t)\,d\tau}{nFA(\pi D\{t - \tau\})^{1/2}}$$

is replaced by

$$\left(\tfrac{7}{3}\right)^{1/2} \int_0^t \frac{e^{-\ell(t-\tau)}\tau^{2/3} i(t)\,d\tau}{nFA(\pi D\{t^{7/3} - \tau^{7/3}\})^{1/2}}.$$

Table 6.5   Values of Constants in Generalized Theoretical Expressions for Coupled Chemical Reactions

Mechanism	W	X	Y	Z	Kinetic Parameter
Preceding reaction	$K/(1 + K)$	$1/(1 + K)$	1	—	$\ell = k_f + k_b$
$Y \underset{k_b}{\overset{k_f}{\rightleftharpoons}} O + ne^- \rightleftharpoons R$					
$K = k_f/k_b$					
Following reaction	1	—	$K/(1 + K)$	$1/(1 + K)$	ℓ
$O + ne^- \rightleftharpoons R \underset{k_b}{\overset{k_f}{\rightleftharpoons}} Z$					
$K = k_b/k_f$					
Catalytic regeneration					
$O + ne^- \rightleftharpoons R + Z$	0	1	0	1	k_c

Derivation of Current Expressions

Cyclic Voltammetry

The surface concentration expressions derived above, the Nernst equation or the absolute rate equation, and the expression for the applied potential as a function of time are then combined to yield a single integral equation. These integral equations, to be solved for the current function $\chi(at)$ are given in Table 6.6 for preceding, following, and catalytic chemical reactions coupled to the appropriate charge transfer reaction. The current can then be obtained from the calculated $\chi(at)$ from (6.88) or (6.89), depending on whether the charge transfer process is reversible or irreversible:

$$\text{reversible } i = nFAC_O{}^*(\pi D_O a)^{1/2}\chi_R(at), \tag{6.88}$$

$$\text{irreversible } i = nFAC_O{}^*(\pi D_O \alpha a)^{1/2}\chi_I(at). \tag{6.89}$$

This treatment [8] of coupled chemical reactions assumes that the charge transfer step is either reversible or irreversible, the somewhat more complicated case of an intermediate charge transfer rate having been considered only for the catalytic mechanism [53]. The case of an irreversible charge transfer followed by a chemical reaction yields the same result as a simple irreversible reaction and is thus trivial.

The diagnostic criteria and working curves for quantitative kinetic measurements derived from solutions of the integral equation for a given mechanism are discussed in the appropriate section.

Ac Polarography

In the ac polarographic experiment, a closed form solution for the current expressions can be obtained, the general form of which is the same as that given for the quasi-reversible case, (6.51):

$$I(\omega t) = I_{rev}F_{dc}(t)G_{ac}(\omega) \sin(\omega t + \phi), \tag{6.90}$$

where

$$I_{rev} = \frac{n^2 F^2 A C_O{}^* (D_O\omega)^{1/2} \Delta E}{4RT \cosh^2(j/2)}. \tag{6.91}$$

The $F_{dc}(t)$ term represents the deviations from reversibility of the dc process and $G_{ac}(\omega)$ represents the effect of the deviations from reversibility of the ac process on the current magnitude.

The $G_{ac}(\omega)$ and ϕ terms can be expressed in terms of two variables V and S such that

$$\phi = \frac{V}{S} \tag{6.92}$$

and

$$G_{ac}(\omega) = \left[\frac{2}{V^2 + S^2}\right]^{1/2}. \tag{6.93}$$

Chemical Reaction	Charge Transfer	Integral Equation
Preceding	Reversible	$$1 - \int_0^{at} \frac{\chi(z)\,dz}{(at-z)^{1/2}} = \frac{e^{J(at)}}{W}\int_0^{at}\frac{\chi(z)\,dz}{(at-z)^{1/2}} + \frac{1}{K}\int_0^{at}\frac{e^{-\ell/a(at-z)}\chi(z)\,dz}{(at-z)^{1/2}}$$
Preceding	Irreversible	$$1 - \int_0^{at}\frac{\chi(z)\,dz}{(at-z)^{1/2}} = \frac{e^{\alpha J(t)}\chi(at)}{W\psi} + \frac{1}{K}\int_0^{at}\frac{e^{-\ell/\alpha a(at-z)}\chi(z)\,dz}{(at-z)^{1/2}}$$
Following (reversible)	Reversible	$$1 - \int_0^{at}\frac{\chi(z)\,dz}{(at-z)^{1/2}} = Ye^{J(at)}\int_0^{at}\frac{\chi(z)\,dz}{(at-z)^{1/2}} + Ze^{J(at)}\int_0^{at}\frac{e^{-\ell/a(at-z)}\chi(z)\,dz}{(at-z)^{1/2}}$$
Following (irreversible)	Reversible	$$1 - \int_0^{at}\frac{\chi(z)\,dz}{(at-z)^{1/2}} = e^{J(at)}\int_0^{at}\frac{e^{-k_f/a(at-z)}\chi(z)\,dz}{(at-z)^{1/2}}$$
Catalytic	Reversible	$$1 - \int_0^{at}\frac{e^{-k_c/a(at-z)}\chi(z)\,dz}{(at-z)^{1/2}} = e^{J(at)}\int_0^{at}\frac{e^{-k_c/a(at-z)}\chi(z)\,dz}{(at-z)^{1/2}}$$
Catalytic	Irreversible	$$1 - \int_0^{at}\frac{e^{-k_c/a(at-z)}\chi(z)\,dz}{(at-z)^{1/2}} = \frac{e^{\alpha J(at)}\chi(at)}{\psi}$$
Catalytic	Quasi-reversible	$$1 - \int_0^{at}\frac{e^{-k_c/a(at-z)}\chi(z)\,dz}{(at-z)^{1/2}} = \frac{e^{\alpha J(at)}\chi(at)}{\psi} + e^{J(at)}\int_0^{at}\frac{e^{-k_c/a(at-z)}\chi(z)\,dz}{(at-z)^{1/2}}$$

$$\text{where} \qquad J(at) = \begin{cases} \dfrac{nF}{RT}(E_i - E^0) - at & 0 \le t \le t_s \\[2mm] \dfrac{nF}{RT}(E_i - E^0) + at - 2at_s & t_s \le t \end{cases}$$

* The constants W, Y, Z, ℓ, and K are given in Table 6.5 for the appropriate reaction scheme and the other terms are defined in the discussion of the appropriate charge transfer reaction.

The exact expressions for $F_{dc}(t)$, V, and S depend on the type of reaction scheme being considered. McCord and Smith [54] have recently given a generalized solution for the fundamental dc polarographic current in terms of the constants of Table 6.5. They have also given the complex expression for $F_{dc}(t)$.

The general expressions for V and S for cases involving a single coupled chemical reaction are

$$
V = \frac{(2\omega)^{1/2}}{\lambda} + \frac{1}{(1 + e^j)}\left[W + Ye^j + X\left(\frac{\{1 + g_P{}^2\}^{1/2} + g_P}{1 + g_P{}^2}\right)^{1/2}\right.
$$
$$
\left. + e^j Z\left(\frac{\{1 + g_F{}^2\}^{1/2} + g_F}{1 + g_F{}^2}\right)^{1/2}\right] \quad \text{and} \tag{6.94}
$$

$$
S = \frac{1}{(1 + e^j)}\left[W + Ye^j + X\left(\frac{\{1 + g_P{}^2\}^{1/2} - g_P}{1 + g_P{}^2}\right)^{1/2}\right.
$$
$$
\left. + e^j Z\left(\frac{\{1 + g_F{}^2\}^{1/2} - g_F}{1 + g_F{}^2}\right)^{1/2}\right], \tag{6.95}
$$

where

$$
g_P = \ell/\omega \text{ for a preceding reaction} \tag{6.96}
$$

and

$$
g_F = \ell/\omega \text{ for a following reaction.} \tag{6.97}
$$

The constants W, X, Y, Z, and ℓ are as given in Table 6.5 for the appropriate reaction. These equations assume one first-order reaction coupled to the charge transfer step, but the equations can be extended to include any number of chemical steps coupled to either the reactant and/or the product of the charge transfer step [5, 54, 55].

In (6.94) the first term on the right accounts for the rate of the slow charge transfer step and becomes small as k_s increases (reversibility). For an irreversible charge transfer reaction, the $(2\omega)^{1/2}/\lambda$ contribution to the phase angle becomes very large and completely obscures the effects of the chemical reaction. Thus chemical reactions coupled to an irreversible charge transfer reaction cannot be studied by ac polarography.

More detail is given in the following sections for each coupled chemical reaction scheme. Special emphasis is placed on using cot ϕ as an experimental parameter in determining rate constants. Diagnostic criteria, such as peak current shifts with changes in frequency and drop life, are also described.

8 CHEMICAL REACTIONS FOLLOWING CHARGE TRANSFER

The sequence including a chemical reaction following the charge transfer constitutes a most important segment of electrochemical reactions, especially for organic compounds.

The manner in which a following chemical reaction manifests itself in the overall electrochemical reaction is a function of the nature of the charge transfer process and the nature of the chemical reaction. If the charge transfer process is irreversible, a following chemical reaction will have no effect on it. The quasi-reversible charge transfer will be influenced by a chemical reaction, but the theoretical treatment is of such complexity that it has received limited attention to date. With a quasi-reversible charge transfer, qualitative information may be obtained denoting the persence of a coupled chemical reaction, but the diagnostic process is not simple and quantitative studies are unquestionably difficult. Thus the discussion that follows is centered on studies of following chemical reactions coupled to reversible charge transfer reactions.

The effects of reversible and irreversible chemical reactions on a charge transfer process are not quantitatively the same and are thus treated separately.

Reversible Chemical Reaction Following Charge Transfer

The reaction mechanism employed above for a general development of the theoretical foundations of the cyclic voltammetric and ac polarographic responses is discussed in more detail in this section. The chemical reaction is assumed to be a first-order process. Both the position of equilibrium and the rate at which equilibrium is reached are important parameters with respect to the influence of the chemical reaction on the overall electrochemical reaction. Since the definition is somewhat unusual, the reader is reminded that the equilibrium constant has been defined as

$$K = k_b/k_f$$

so that the ac polarographic working curves are the same for the following and preceding chemical reactions.

The Cyclic Voltammetric Response

With a following chemical reaction, the magnitude of the cathodic current response is not expected to vary significantly and is thus of little value for either qualitative or quantitative use. Conversely, relatively large changes in the magnitude of the anodic current are expected at an appropriate balance of the chemical kinetic and experimental parameters. Therefore the variations in anodic current, or the ratio of anodic to cathodic current as a function of the time scale of the experiment, serve as a prime variable for both qualitative and quantitative measurements.

The value of the equilibrium constant and the values of the rate constants determine the applicability of the treatment that follows. If the rate constant for the forward chemical reaction is small and the equilibrium constant is large, that is, equilibrium shifted toward the reactant, the reaction system reduces to the reversible case. As the rate constant for the forward reaction increases, treatment of the system as charge transfer with a reversible chemical reaction becomes applicable. If the equilibrium constant is small, the system reduces to the irreversible following reaction with the exception of the position of the response in potential. Thus instances in which the rate constant of the chemical reaction is significant and the equilibrium constant is not excessively small are of the most importance.

Illustrative cyclic voltammetric responses for cases of interest are shown in Fig. 6.9. At high values of the rate of potential scan relative to the chemical

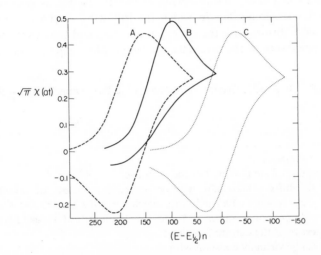

Fig. 6.9 Theoretical cyclic voltammetric curves for following chemical reaction. Curve A—Chemical reaction is fast; wave has reversible shape with potential shifted by RT/nF ln $(1 + K)/K$, where $K = 10^{-3}$. Curve B—Intermediate chemical reaction rate, $K = 10^{-3}$, $l/a = 10^4$. Curve C—Chemical reaction is slow; essentially reversible behavior.

kinetics, the reversible response is obtained at the expected potential. With sufficiently rapid kinetics, a response that appears reversible in nature is observed, but shifted in potential by an amount reflecting the magnitude of the equilibrium constant. The intermediate cases of balance of the kinetic parameter and time scale of the experiment yield a response that is qualitatively different from the reversible case and appears at a potential other than the reversible potential.

For these cases the ratio of anodic to cathodic current can serve as a useful parameter for quantitative measurements. However, the magnitude of the anodic current is a function of the switching potential. A working curve of the ratio of anodic to cathodic current as a function of $(a/\ell)^{1/2}/K$ is employed, but such curves must be calculated with an arbitrarily selected switching potential [8]. The ratio $i_p{}^a/i_p{}^c$ approaches unity as K increases and/or $(a/\ell)^{1/2}$ decreases. The ratio decreases at higher values of $(a/\ell)^{1/2}$ and/or as K decreases. If the equilibrium constant is known, kinetic data can thus be calculated from a single current potential curve.

The ratio of anodic to cathodic peak current is useful for diagnostic purposes. At low values of the scan rate, the ratio approaches unity. A decrease in the ratio $i_p{}^a/i_p{}^c$ is noted as the scan rate is increased. This behavior alone almost serves to differentiate this case from all others.

The behavior of the cathodic peak potential over the range of kinetic parameters for a reversible following chemical reaction is somewhat more complex. Under conditions in which the following chemical reaction is in complete equilibrium with the system at all times and the observed response has the appearance of the reversible system in other respects, the response is

Table 6.7 Diagnostic Criteria for a Reversible Following Chemical Reaction

$$O + ne^- \rightleftharpoons R \underset{k_b}{\overset{k_f}{\rightleftharpoons}} Z; \qquad K = k_b/k_f$$

Cyclic Voltammetry
   Properties of the potential of the response:
      E_p shifts cathodically with an increase in v by an amount that approaches $60/n$ per 10-fold increase in v if ℓ is large and K is small. Intermediate values of K and ℓ lead to a shift of a lesser magnitude
   Properties of the current function:
      $i_p/v^{1/2}$ virtually constant with v
   Properties of the anodic to cathodic current ratio:
      $i_p{}^a/i_p{}^c$ decreases from unity as v increases
   Others:
      If K is small and the chemical kinetics rapid, a response typical of reversible charge transfer, except shifted in potential, will be noted

Ac Polarography
   Properties of the potential of the response:
      E_p shifts cathodically as ω is increased
      E_p shifts anodically as the drop time is increased
   Properties of the ac current:
      $I(\omega t)$ increases by less than $\omega^{1/2}$ as ω is increased if ℓ is not large and K is not quite small
      $I(\omega t)$ decreases as the drop time is increased
   Properties of the phase angle:
      Cot ϕ has sigmoidal shape with a limiting value at anodic potentials and a value of unity at cathodic potentials

---

shifted anodically on the potential axis by the quantity $(RT/nF) \ln [(1 + K)/K]$. When the rate of the chemical reaction is slow, the response occurs at the reversible potential.

In the cases in which ℓ/a is large and the kinetic parameter $(a/\ell)^{1/2}/K$ small, the equation,

$$E_p = E_{1/2} - (RT/nF)[0.780 + \ln (a/\ell)^{1/2} - \ln (1 + K)], \qquad (6.98)$$

has been shown to be obeyed [8]. Thus under this relationship the peak potential shifts more cathodically by $60/n$ mV for a 10-fold increase in $(a/\ell)^{1/2}/K$.

Intermediate values of $(a/\ell)^{1/2}/K$ lead to an intermediate variation in the position of the response on the potential axis. At large values of $(a/\ell)^{1/2}/K$, the peak potential is independent of $(a/\ell)^{1/2}/K$; that is, reversible plus $\ln [(1 + K)/K]$ term. A working curve can be constructed from published data [8] relating the potential to the kinetic parameter. Both $E_{1/2}$ and K must be known for use of this relationship.

The diagnostic criteria for the reversible charge transfer reaction followed by a reversible chemical reaction are summarized in Table 6.7.

The Ac Polarographic Response

A theoretical discussion of the current response [56] and the phase angle behavior [57] has been given for the general case of a following chemical reaction coupled to a quasi-reversible charge transfer reaction using the expanding plane diffusion model. In this section, only the case of reversible charge transfer with a coupled reversible chemical reaction is considered.

Where the chemical rate is sufficient to maintain equilibrium, the following expression for the peak potential of the ac wave is obtained [58]:

$$E_p = E_{1/2} + \frac{RT}{nF} \ln \frac{1 + K}{K} . \qquad (6.99)$$

As the equilibrium shifts toward formation of product from R, K becomes small and the potential of the wave becomes more anodic. When R predominates, K is large, and there is little shift in the peak potential.

At the opposite extreme in chemical kinetics, where the chemical rate is very slow, the peak potential is the reversible half-wave potential $E_{1/2}$. Thus as the kinetic parameter ℓ/ω is varied from large values (large rate constant or low frequency) to smaller values, the peak potential shifts cathodically, in a manner similar to the shift in peak potential of the cyclic voltammetric response as a function of the balance of the chemical kinetics and the time scale of the experiment. This cathodic shift in the peak potential as the frequency is increased can be used to indicate the possible presence of a following chemical reaction.

Another means to test for a chemical reaction involves monitoring the peak current as a function of frequency. The term $i_p/i_{p,d}$, where i_p is the observed peak current and $i_{p,d}$ is the calculated diffusion-controlled peak current, is plotted as a function of $\omega^{1/2}$. A reversible charge transfer with no coupled chemical reactions has a value of $i_p/i_{p,d}$ of unity over the whole frequency range. In the presence of either a following or a preceding reaction, the term $i_p/i_{p,d}$ decreases as the frequency increases [56, 59]. If the chemical rate is very rapid and the equilibrium favors the electroactive species, there may be little change in the ratio over a moderate frequency range, but the value of $i_p/i_{p,d}$ will be less than unity. As the rate of the chemical reaction goes to zero, the system approaches diffusion control and the ratio again approaches unity. Note that this test indicates the presence of a coupled chemical reaction but says nothing about the type of reaction. Further information from either shifts in the peak potential with frequency or phase angle data must be obtained to characterize the nature of the chemical reaction.

The effects of a following chemical reaction on phase angle data show up most strongly on the anodic side of the wave. The variation in $\cot \phi$ as a function of E_{dc} is shown in Fig. 6.10 for typical values of K and ℓ. The sigmoidal curve reaches a limiting value of $\cot \phi$ at anodic potentials, denoted as $\cot \phi^{lim}$. Unfortunately, this behavior may be experimentally difficult to detect because of the low current values at these potentials. This $\cot \phi$ versus E_{dc} behavior is the mirror image of the behavior of a system with a preceding chemical reaction (discussed below) about the value $E_{dc} = E_{1/2}$.

The value of $\cot \phi^{lim}$ at positive potentials is also a function of the kinetic

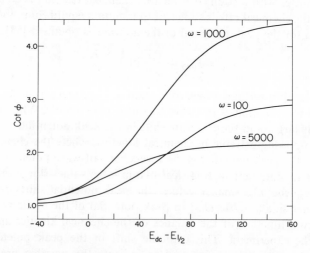

Fig. 6.10 Theoretical $\cot \phi$ vs. E_{dc} for a following chemical reaction. $K = 0.10$, $\ell = 5000$ sec^{-1}, reversible charge transfer.

parameter ℓ/ω, going through a maximum, cot ϕ_{max}^{lim}, as the value of ℓ/ω increases. The exact value of cot ϕ_{max}^{lim} depends on both ℓ and K. An example of this behavior is shown in Fig. 6.11 for two values of K.

Since the value of cot ϕ_{max}^{lim} depends on K and ℓ/ω, working curves of cot ϕ_{max}^{lim} versus K and versus ℓ/ω can be used to determine both the equilibrium

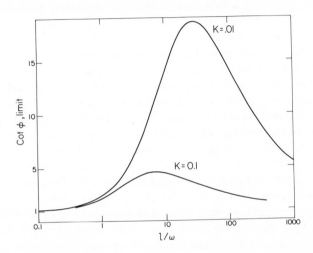

Fig. 6.11 Theoretical limiting cot ϕ as a function of the kinetic parameter ℓ/ω. The equilibrium values are shown.

constant and rate parameters for either a following or a preceding reaction. These two curves are shown in Figs. 6.12 and 6.13. The experimentally determined cot ϕ_{max}^{lim} is used to determine K from Fig. 6.12 and, based on the frequency of cot ϕ_{max}^{lim}, ℓ can be determined from Fig. 6.13. Additional discussion of obtaining rate parameters by this method is given in Section 9, where the preceding reaction is discussed.

Since it may be difficult to measure cot ϕ^{lim}, McCord and Smith [59] have suggested using the value of cot ϕ at the peak potential. Working curves may actually be calculated at any potential by using the expression

$$\cot \phi = \frac{1 + K + Ke^j + e^j\left[\dfrac{(1 + g^2)^{1/2} + g}{1 + g^2}\right]^{1/2}}{1 + K + Ke^j + e^j\left[\dfrac{(1 + g^2)^{1/2} - g}{1 + g^2}\right]^{1/2}}, \tag{6.100}$$

where

$$g = \frac{\ell}{\omega} \tag{6.101}$$

and

$$j = \frac{nF}{RT}(E_{dc} - E_{1/2}).$$ (6.102)

In addition to the shift in peak potential with changes in frequency and the behavior of cot ϕ, a following chemical reaction may be characterized by the negative time dependence of the ac current [56]. This means that as the drop time increases the ac current decreases. The peak potential also shifts to more anodic potentials as the drop time increases.

There can be several causes of a time-dependent ac polarographic wave which are similar to the time dependence of a following chemical reaction, so this behavior is not by itself a sufficient diagnostic criterion. However, coupled with the effects previously mentioned it affords a fairly reliable indication of this mechanism. The diagnostic criteria for reversible charge transfer followed by a reversible chemical reaction are summarized in Table 6.7.

McCord et al. [56] have recently described a general method for obtaining kinetic parameters from ac polarographic data. At a selected frequency and drop time, the peak current dependence on the chemical rate ℓ can be described by a plot of the theoretical values of $i_p/i_{p,d}$ versus ℓ for various values of K, yielding a series of curves. In addition, the behavior of cot ϕ at any potential, most conveniently the peak potential, can be calculated as a function of ℓ for the same frequency and drop time, for various values of K, yielding a second series of curves.

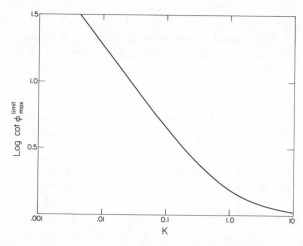

Fig. 6.12 Maximum limiting cot ϕ. Working curve as a function of the equilibrium constant (Courtesy of Anal. Chem., Ref. 59).

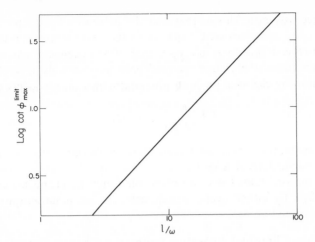

Fig. 6.13 Maximum limiting cot ϕ working curve as a function of the kinetic parameter ℓ/ω. (Courtesy of Anal. Chem., Ref. 59.)

Experimentally, cot ϕ at the peak potential and i_p are measured at this selected frequency and drop time with $i_{p,d}$ estimated from a knowledge of the diffusion coefficients from some independent measurement. This experimental value of cot ϕ now intersects the curves of cot ϕ versus ℓ and the experimental value of $i_p/i_{p,d}$ intersects the series of curves of $i_p/i_{p,d}$ versus ℓ for each value of K. These intersections yield two sets of points of ℓ versus K for the cot ϕ and $i_p/i_{p,d}$ values. If the two sets of ℓ versus K values are plotted, two curves will be obtained and their intersection gives an experimental value of ℓ and K for the system. This procedure is now repeated for different values of frequency and/or drop time to determine an average value.

When the equilibrium constant is known from some other technique, its value may be inserted in the calculation to obtain theoretical curves of cot ϕ versus ℓ and $i_p/i_{p,d}$ versus ℓ. The experimentally observed values of cot ϕ or $i_p/i_{p,d}$ then provide ℓ directly. Again, several values of frequency and/or drop time should be used. This technique works equally well for preceding chemical reactions (see Section 9).

The above discussion has assumed that the charge transfer step is reversible. Complications arise when this is not true, and an indication of what to look for if slow charge transfer is suspected seems in order.

In general, the behavior of cot ϕ versus E_{dc} gives the best clue. Instead of the sigmoidal curve, a curve will be observed with a maximum in cot ϕ somewhere near the potential of the peak current. If the maximum is not pronounced, the effects of slow charge transfer may not be important at the frequency employed.

It has recently been shown [56] that the presence of a coupled chemical reaction places more stringent demands on the charge transfer rate sufficient to assume reversibility than in the absence of any chemical reaction. Under normal conditions the charge transfer behaves reversibly if $k_s \gg (\omega D)^{1/2}$. With a following chemical reaction, the system behaves reversibly only when

$$k_s \left(\frac{K}{1+K} \right)^\alpha \gg (\omega D)^{1/2} \qquad (6.103)$$

This means that k_s must be larger than would normally be considered sufficient, particularly if $K \ll 1$.

To the authors' knowledge, no reversible following chemical reaction has been studied by either cyclic voltammetry or ac polarography and the methods outlined above.

Irreversible Chemical Reaction Following Charge Transfer

This section treats reactions described by the scheme

$$O + ne^- \rightleftharpoons R$$

$$R \xrightarrow{k_f} Z$$

and is merely an extension of the previous case to conditions wherein the equilibrium between R and Z lies far to the right. In fact, experimental conditions can often be adjusted so that a reaction that is in equilibrium can be effectively reduced to this somewhat less complicated case. This reaction sequence is among the most useful for quantitative work of all those amenable to study.

As discussed earlier, many reactions that appear to be irreversible charge transfer processes by a cursory examination may actually be described by this scheme. Full use of the capabilities of cyclic voltammetry and ac polarography can ascertain the applicability of this treatment.

The Cyclic Voltammetric Response

Irreversible following chemical reactions produce only minor variations in the magnitude of the cathodic peak current compared with the reversible charge transfer mechanism in the absence of a chemical reaction. Thus this parameter has no real use for either qualitative characterization or quantitative measurement. The magnitude of the anodic response, however, is markedly influenced by the coupled chemical reaction except at very low values of the rate constant relative to the scan rate (low values of k_f/a), where the response is essentially that of the reversible case. At large values of the rate constant, the anodic response is essentially that of the irreversible charge transfer and little information concerning the chemical reaction is obtained. At intermediate values of k_f/a, however, the magnitude of the anodic current response provides the basis for excellent measurements of the kinetics of the chemical reaction.

Through the calculation of a large number of theoretical response curves varying both k_f/a and the switching potential E_s, Nicholson and Shain [8] have shown that the problem of variation of the anodic current with switching potential can be circumvented by use of the parameter $k_f\tau$, where τ is the time in seconds required to scan from the $E_{1/2}$ to the switching potential. A working curve of $i_p{}^a/i_p{}^c$ as a function of $\log k_f\tau$ (Fig. 6.14) can

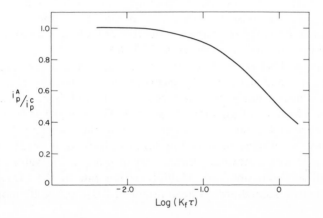

Fig. 6.14 Anodic to cathodic peak current ratio as a function $k_f\tau$ for irreversible following chemical reaction. (Courtesy of Anal. Chem., Ref. 8.)

be constructed which is useful throughout the range in which the response can be distinguished from the reversible case ($k_f \approx 0.02$) or for which an anodic response can be measured ($k_f \approx 1.6$). Of course, $E_{1/2}$ for the reaction must be known accurately for use of this relationship. If $E_{1/2}$ is not available from an independent measurement, obtaining a value from the limiting behavior at rapid scan rates is possible, assuming that behavior very nearly that of the reversible process is experimentally obtainable.

The variation in anodic response as a function of the rate of potential scan is also quite useful for qualitative purposes. As the scan rate is increased, $i_p{}^a/i_p{}^c$ increases toward unity (reversible case) as a limit. Since other mechanisms in which the anodic current can be measured have a behavior markedly different from that of this case, the variation of $i_p{}^a/i_p{}^c$ with scan rate is a particularly valuable diagnostic tool. However, the magnitude of the cathodic current as a function of scan rate must be monitored to avoid confusion with the ECE cases (Section 11).

The variation in peak potential with the kinetic parameters for the irreversible following reaction is also of value. At higher values of k_f/a, the relationship

$$E_p = E_{1/2} - (RT/nF)[0.780 - \ln (k_f/a)^{1/2}] \qquad (6.104)$$

has been obtained [8], which describes an anodic shift of $30/n$ mV for a 10-fold increase in k_f/a.

For qualitative characterization the variation in E_p is also useful if a sufficiently wide range of scan rates can be employed. At low values of the scan rate, E_p shifts cathodically by $30/n$ with a 10-fold increase in the scan rate. As the scan rate is increased, lesser magnitudes of this shift are observed until E_p becomes constant with the scan rate, as in the reversible case. Diagnostic criteria for reversible charge transfer followed by an irreversible chemical reaction are summarized in Table 6.8.

Kuempel and Schaap [60] have studied the rate of ligand exchange between cadmium ion and calcium ethylenediaminetetraacetate by cyclic voltammetry. Cadmium amalgam is oxidized to form $Cd(II)$ in calcium–EDTA solutions. The rate of formation of the Cd–EDTA complex is very rapid and the equilibrium lies far to the right (favoring the complex). Under these conditions the working curve of Fig. 6.14 can be used. A pseudo-first-order rate constant with a maximum uncertainty of $\pm 8\%$ at a k_f value of approximately 10 sec^{-1} was obtained. Multiplying this rate by the EDTA concentration yields a second-order rate constant of $2.6 \pm 0.7 \times 10^8$ mole^{-1} sec^{-1},

Table 6.8   Diagnostic Criteria for an Irreversible Chemical
Reaction Following Charge Transfer

Cyclic Voltammetry
  Properties of the potential of the response:
    E_p shifts cathodically, by $30/n$ mV at low v with a lesser shift noted at
      higher values of v
  Properties of the current function:
    $i_p/v^{1/2}$ is independent of v
  Properties of the anodic-to-cathodic current ratio:
    $i_p/{}^a i_p{}^c$ increases toward unity as v is increased

Ac Polarography
  Properties of the potential of the response:
    E_p shifts cathodically as ω is increased and anodically as the drop time is
      increased
  Properties of the ac current:
    $I(\omega t)$ decreases at low values of ω or as the drop time is increased
  Properties of the phase angle:
    Cot ϕ increases without limit as ω is decreased
    Cot ϕ has sigmoidal shape with E_{dc} as for a reversible chemical reaction

---

in agreement with the value of $6.1 \pm 0.7 \times 10^8$ reported by Aylward and Hayes [61] using ac polarography and faradaic impedance. A rate of 2.3×10^9 liter mole^{-1} sec^{-1} was obtained by Matsuda and Tamamushi [62] using ac polarography by reducing the complex at the electrode surface. This work is discussed in the next section on preceding chemical reactions.

The rate constant for the ligand exchange reaction between pyridine and cyanopyridine hemochrome yielding pyridine hemochrome has also been evaluated by Davis and Orleron [63], using cyclic voltammetry. The more rapid reaction of pyridine with the cyanide hemochrome ($k_f \approx 20$ sec^{-1} compared with 0.5 sec^{-1} for the cyanopyridine hemochrome reaction) could also be examined by this technique. It thus appears that the cyclic voltammetric examination of ligand exchange kinetics is broadly applicable.

Stapelfeldt and Perone [64] have applied cyclic voltammetry and the working curve in Fig. 6.14 to a study of the reduction of benzil. At pH values greater than 11, benzil is reduced to the stilbene diolate which undergoes an

$$\phi—C—C—\phi + 2e^- + H^+ \rightleftharpoons \phi—C{=}C—\phi$$
$$\quad\ \ \|\ \ \|\qquad\qquad\qquad\qquad\quad\ \ |\ \ \ |$$
$$\quad\ \ O\ \ O\qquad\qquad\qquad\qquad\quad HO\ \ O^-$$

irreversible rearrangement to the benzoin anion

$$\phi—C{=}C—\phi \xrightarrow{\ k_f\ } \phi—CH—C—\phi$$
$$\quad\ \ |\ \ \ |\qquad\qquad\qquad\quad\ |\ \ \ \ \|$$
$$\quad HO\ \ O-\qquad\qquad\qquad O-\ \ O$$

with a rate constant of approximately 2 sec^{-1}.

Perone and Kretlow [65] have also evaluated the rate constant for the hydration of the oxidation product of ascorbic acid by means of cyclic voltammetric data. The rate constant of 1.4×10^3 sec^{-1} they obtained was in excellent agreement with the value from potential step measurements.

Although few second- or higher-order following reactions have been treated, the second-order mechanism

$$O + ne^- \rightleftharpoons R$$
$$2R \xrightarrow{\ k_f\ } Z,$$

a dimerization of the product of charge transfer, has been treated by Olmstead et al. [66]. A combination of the integral equation procedures and a finite difference method for the solution of the nonlinear differential equations, similar to a treatment employed by Saveant et al. [67], was utilized.

In analogy with the first-order case, a working curve of the ratio of anodic to cathodic peak currents as a function of $\log k_f C\tau$ can be employed to evaluate the rate constant. The dependence of $i_p{}^a/i_p{}^c$ on concentration provides a convenient means to distinguish between the first- and higher-order following reactions. In general, the diagnostic criteria for the higher-order reactions parallel those of the first-order case with the addition of the concentration dependencies.

The successful treatment of this case forecasts the extension of the theory of cyclic voltammetry to other second-order kinetic schemes.

The Ac Polarographic Response

The general case of a reversible chemical reaction was treated earlier, but several simplifications arise when the chemical reaction can be considered to proceed in only one direction. McCord et al. [56] have shown that this situation arises when $K(\ell t)^{1/2} \leqslant 10^{-3}$. For example, if $K = 10^{-4}$ and $t = 5$ sec, the chemical reaction is essentially irreversible if $k_f > 20$ sec^{-1}.

The previous discussion of drop time dependence and potential shifts is generally applicable to the irreversible case also. The behavior of cot ϕ with potential is also similar although the value of cot ϕ^{lim} may be quite large. The main difference lies in the variation of cot ϕ^{lim} with the chemical rate constant k_f. In the reversible chemical reaction, cot ϕ reaches a maximum value as ℓ is varied, the value of ℓ depending on the equilibrium constant K. With an irreversible chemical reaction, cot ϕ increases without limit as k_f increases and thus techniques other than the working curves from which both ℓ and K are obtained for the reversible case must be used.

Aylward et al. [68] have obtained a working curve for the case in which the chemical reaction rate is much less than the applied frequency. This curve relates the ratio $i_p/i_{p,d}$ to the dimensionless parameter $k_f t$, where t is the drop time. McCord and co-workers [56] obtained similar curves for the planar diffusion model and the expanding plane model. These working curves are illustrated in Fig. 6.15, taken from Ref. 56, and show the importance of using the more exact expanding plane model for this type of analysis.

Aylward and Hayes [61] have used this technique to measure the rate of formation of Cd–EDTA complex by the reaction scheme,

$$Cd(Hg) \rightleftharpoons Cd^{2+} + 2e^-$$

$$Cd^{2+} + HY^{3-} \xrightarrow{k_f} Cd\,HY^-,$$

where Y^{4-} is the EDTA tetravalent anion. The equilibrium constant for the second reaction is small, $K = 10^{-3.67}$ [62], and the pseudo-first-order rate constant is kept small by having a low concentration of HY^{3-}. Under these experimental conditions, the constant $K(k_f t)^{1/2} \approx 10^{-3}$ and the reaction is essentially irreversible. Aylward and Hayes measured $i_p/i_{p,d}$ by faradaic impedance techniques (discussed in Section 14) at varied frequencies, drop lives, and concentrations of HY^{3-} and their value of $6.1 \pm 0.7 \times 10^8$ mole^{-1} sec^{-1} agreed well with those obtained by other workers [62].

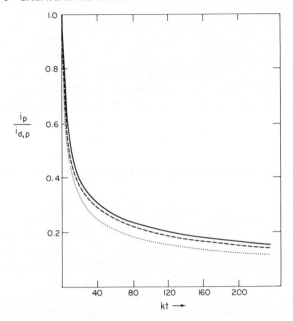

Fig. 6.15 Ac polarographic working curve for irreversible following chemical reaction with reversible charge transfer. Expanding plane theory. Aylward, Hayes, and Tamamushi theory [68]. Stationary plane theory (Courtesy of Elsevier Publishing Co., Ref. 56).

With the unique behavior of $\cot \phi$ for the irreversible reaction, this parameter can be used to determine the rate of the following reaction if the reversible half-wave potential is known. The appropriate expression for the cotangent of the phase angle under these conditions is given by (6.105). A working curve of $\cot \phi$ versus k_f/ω can be generated from (6.105) for some convenient potential, and the experimental $\cot \phi$ at this potential yields k_f since the frequency is known. By varying the applied frequency, a series of values of k_f is obtained to determine the uncertainty of the measurements:

$$\cot \phi = \frac{1 + e^j \left[\dfrac{(1 + g^2)^{1/2} + g}{1 + g^2} \right]^{1/2}}{1 + e^j \left[\dfrac{(1 + g^2)^{1/2} - g}{1 + g^2} \right]^{1/2}}. \tag{6.105}$$

A more general approach to data analysis in which $E_{1/2}$ need not be known nor the special condition applied that $k_f \ll \omega$ arises through relating the dc polarographic current to the ac polarographic current [56]. A series of working curves is constructed by calculating the ratio $i_{ac}/(nF \, \Delta E/RT)i_{dc}$ versus the potential $(E_{dc} - E_{1/2})$ for several values of the kinetic parameter k_f/ω.

Experimental curves of the current ratio as a function of E_{dc} are compared with the theoretical curves. The shape of the theoretical curve that best matches that of the experimental curve yields k_f, and the position of the experimental curve on the potential axis yields $E_{1/2}$.

The cot ϕ parameter can also be used in this manner to obtain k_f and $E_{1/2}$. Theoretical curves using (6.105) can be constructed of cot ϕ versus ($E_{dc} - E_{1/2}$) for various values of k_f/ω. The shape of the experimental curve of cot ϕ versus E_{dc} yields k_f, and the position yields $E_{1/2}$.

9 CHEMICAL REACTIONS PRECEDING CHARGE TRANSFER

Sequences in which the reactant is produced by a homogeneous chemical reaction constitute an important class of electrochemical reactions. Many reactions are known to proceed through such a sequence and others will undoubtedly be characterized as the recently defined theoretical relationships are applied to practical systems.

Only reversible chemical reactions are applicable for this general mechanism, but the effect of such a reaction can be observed for both reversible and irreversible charge transfer processes. Application to quasi-reversible charge transfer is also possible if the complexity of the theory can be surmounted.

Chemical Reaction Preceding Reversible Charge Transfer

This reaction sequence is discussed on the basis of the equations

$$Z \underset{k_b}{\overset{k_f}{\rightleftharpoons}} O$$
$$O + ne^- \rightleftharpoons R,$$

with the chemical reaction assumed to be first order. The forward chemical reaction is that which produces the electrochemically reactive species, and the equilibrium constant is defined as

$$K = k_f/k_b.$$

The relationship between this sequence and that for the following chemical reaction as defined in Section 8 should be noted, particularly with respect to the treatment for ac polarography.

The Cyclic Voltammetric Response

With a preceding chemical reaction, the magnitude of the cathodic current response is expected to be sensitive to the rate of the chemical reaction. If the rate of the chemical reaction is slow compared with the time scale of the experiment, no significant conversion of the electroinactive form into the active form can take place. Under such conditions the cathodic response has the characteristics of the reversible case but the current magnitude is determined by the equilibrium concentration of the electroactive species. If a firm assessment of n has been reached, the lower current is of significant diagnostic value. At the opposite extreme in rate of the chemical reaction, the current has the characteristics of the reversible case with variations in position on the potential axis.

Under conditions intermediate to the above cases, Nicholson and Shain [8] have shown through the calculation of a large number of theoretical response curves that a working curve very nearly fits the equation

$$i_k/i_d = \frac{1}{1.02 + 0.471(a/\ell)^{1/2}/K}, \tag{6.106}$$

where i_k is the observed peak current and i_d is the diffusion-controlled peak current expected in the absence of the chemical complications. The i_d value is calculated or determined from an experiment on a time scale slow with respect to the rate of the chemical reaction.

The anodic current response is not quite as sensitive to variations in the kinetic parameters as is the cathodic response and therefore offers little additional data. However, for cases in which the i_d value cannot be obtained, the ratio of i_a to i_c can be used for kinetic measurements through a working curve of i_a/i_c as a function of $(a/\ell)^{1/2}/K$. The anodic response is a function of the switching potential and such working curves must be calculated for a particular value of the switching potential.

The variation of i_a/i_c as a function of scan rate is quite useful for diagnostic purposes, approaching a value of unity at slow scan rates and increasing at higher scan rates.

The position of the response on the potential axis is also determined by the kinetic parameters. If the chemical reaction is slow, the response occurs at the potential of the uncomplicated reversible charge transfer, even though the current magnitude is lower. At low values of $(a/\ell)^{1/2}/K$ the response is shifted from the reversible potential by an amount equal to $(RT/nF) \ln [K/(1 + K)]$ and is independent of scan rate. At intermediate values of $(a/\ell)^{1/2}/K$, the anodic shift approaches $60/n$ mV for a 10-fold increase in $(a/\ell)^{1/2}/K$. Working curves have been presented [8] for this behavior that afford kinetic measurements, but the potential shift does not seem to offer the advantages of the current measurements. The potential behavior offers diagnostic data for differentiation from the similar case with irreversible charge transfer only at low values of the kinetic parameter, that is, when the response approaches reversible behavior. Thus the utility of the potential variations appears trivial in light of the other quite clear diagnostic information, summarized in Table 6.9.

Table 6.9   Diagnostic Criteria for a Preceding Chemical Reaction

$$Z \underset{k_b}{\overset{k_f}{\rightleftharpoons}} O + ne- \rightleftharpoons R; \qquad K = k_f/k_b$$

Cyclic Voltammetry
Properties of the potential of the response:
  E_p shifts anodically with an increase in v
Properties of the current function:
  $i_p/v^{1/2}$ decreases as v increases
Properties of the anodic to cathodic current ratio:
  $i_p{}^a/i_p{}^c$ is generally greater than unity and increases as v increases with a
  value of unity approached at lower values of v
Others:
  A response similar to the reversible case, but lower in magnitude, is
  obtained when the chemical kinetics are slow and K is moderate in
  value

Ac Polarography
Properties of the potential of the response:
  E_p shifts anodically as ω is increased and cathodically as the drop time
  is increased
Properties of the ac current:
  $I(\omega t)$ increases by less than $\omega^{1/2}$ as ω increases
  $I(\omega t)$ increases with drop time
Properties of the phase angle:
  Cot ϕ as E_{dc} has sigmoidal shape with a limiting value at cathodic
  potentials and value of unity at anodic potentials

Figure 6.16 shows theoretically calculated voltammograms for the cathodic
scan only to illustrate the general shape of the response. The curve becomes
very flat as the scan rate increases and the peak current is much less than that
obtained for the corresponding reversible case [cf. (6.106)].

Bailey et al. [69] have applied this treatment in a qualitative manner to an
estimation of the equilibrium constant for the second protonation of phenazine (P),

$$PH^+ + H^+ \rightleftharpoons PH_2{}^+,$$

through the electrochemical reaction,

$$PH_2{}^+ + e^- \rightleftharpoons PH_2.$$

Values of $K(\ell)^{1/2}$ were obtained from experimental data for i_k/i_d, the value of
i_d being determined from the limiting value of the peak current at lower scan
rates. By estimation of the upper limit of the forward reaction, an estimate of
the equilibrium constant was obtained.

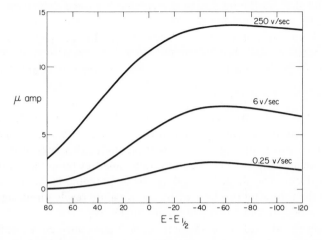

Fig. 6.16 Theoretical cathodic linear scan voltammograms for chemical reaction preceding are versible charge transfer; $n = 1$, $A = 0.02 \text{ cm}^2$, $C_O{}^* = 10^{-6}\, M$, $D_O = D_R = 10^{-6}$ cm^2/sec, $K = 1$, $\ell = 100 \text{ sec}^{-1}$, scan rate shown on curve.

The Ac Polarographic Response

The general case of a chemical reaction preceding a quasi-reversible charge transfer has been treated theoretically at some length by McCord and Smith [70] using the expanding plane model. A discussion of the phase angle expressions has also been given earlier for this mechanism [57].

When the preceding chemical reaction is very fast, a diffusion-controlled wave results because the chemical reaction has no influence on the flux of the electroactive species. In this case, however, the potential of the peak is shifted cathodically, depending on the value of the equilibrium constant K:

$$E_p = E_{1/2} + \frac{RT}{nF} \ln \frac{K}{1 + K}. \qquad (6.107)$$

If the equilibrium constant is very large, there is no thermodynamic potential shift, as can be seen from (6.107). This means that O is the predominant species in solution. The current magnitude is also controlled by the sum of the bulk concentration of C_Z, the electroinactive species, and C_O, the electroactive species.

The ac polarographic kinetic parameter of interest is ℓ/ω where, as before, ℓ is the sum of the rate constants of the forward and reverse chemical reactions. As ℓ/ω becomes smaller (decreasing rate constant or increasing frequency), two effects are observed. The ac polarographic current begins to decrease and the wave shifts to more anodic potentials. This shift is similar to that observed in the cyclic experiment as the scan rate increases. The anodic shift of E_p as the frequency is increased is a good diagnostic criterion for the preceding reaction mechanism.

The peak current values are very sensitive to the rate parameters of the chemical reaction and become more so as the frequency is increased. This behavior is discussed in Section 8 for any type of coupled chemical reaction.

The phase angle also has properties unique to the preceding chemical reaction. At a given frequency cot ϕ plotted against dc potential has a sigmoidal shape with a limiting value at cathodic potentials. Examples of this behavior are shown in Fig. 6.17. Experimentally, this may not be easily observed because the current becomes quite small at potentials cathodic of the peak current. The shape of cot ϕ versus dc potential is the mirror image of that of the following reaction where cot ϕ values reach a limiting value at potentials anodic to the half-wave potentials [5, 57].

This limiting value of the cotangent cot ϕ^{lim} goes through a maximum as the kinetic parameter ℓ/ω is varied. The limit occurs at different values of

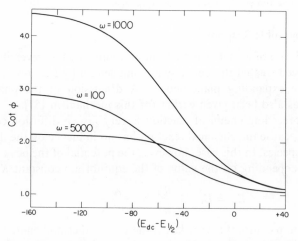

Fig. 6.17 Theoretical cot ϕ vs. E_{dc} for a preceding chemical reaction. $K = 0.10$, $\ell = 5000 \text{ sec}^{-1}$, reversible charge transfer.

ℓ/ω, depending on the equilibrium constant, so it may be necessary to cover a frequency range of two orders of magnitude or more to observe this phenomenon. This maximum value of cot ϕ^{lim} can be used to determine the kinetic parameters of the chemical system, as described in Section 8 for following chemical reactions. Using the two working curves in Fig. 6.12 and 6.13, one can determine both K and ℓ when the experimentally determined value of cot ϕ_{max}^{lim} is used, since the frequency of cot ϕ_{max}^{lim} is also known. The individual values of k_f and k_b can be determined from (6.108):

$$\ell = k_f \frac{(1 + K)}{K} = k_b(1 + K). \tag{6.108}$$

An assumption implicit in the above analysis is that the charge transfer reaction remains essentially reversible at the frequency necessary to determine the cot ϕ_{max}^{lim}. If this is not true, phase shifts associated with the charge transfer rate complicate the analysis in an undetermined way.

McCord and Smith [70] have suggested that the cot ϕ at the peak potential can be used in this analysis instead of the limiting cot ϕ. Cot ϕ is easier to measure because the current magnitudes are larger at this potential. Unfortunately, the influence of the chemical reaction is also less at these potentials.

The time dependence of the ac polarographic wave for the preceding reaction mechanism offers an excellent means of determining whether or not such a mechanism is operative. The current increases as the drop time increases and the peak potential becomes more cathodic [70]. This positive time dependence is exactly opposite that observed for a following chemical reaction. Diagnostic criteria for a preceding chemical reaction are summarized in Table 6.9.

The variations in current magnitude as a function of frequency can be used to determine rate parameters if some estimate of the diffusion current can be obtained. A plot of $i_p/i_{p,d}$ against log $\omega^{-1/2}$ yields a curve the shape of which can be compared to that of the working curve shown in Fig. 5 of Ref. 70 to obtain an estimate of K. The separation of the experimental curve from the working curve along the x-axis can be used to obtain a measure of ℓ. If K is known from some other source, this method can yield fairly good results. Again a large range of frequencies must be used for best results.

McCord et al. [56] have given a general method of obtaining kinetic parameters from ac polarographic data. This method is described in Section 8 for the case of following chemical reactions but can be applied to any kinetic system. It is particularly useful when the charge transfer rate may not be reversible or when the reaction mechanism is quite complex. Thus a system that has both a preceding and a following chemical reaction [5, 56] or multiple coupled chemical reactions [71] might be analyzed by this method.

Recently, Matsuda and Tamamushi [62] have used ac polarography to study the reduction of Cd(II) in EDTA which is preceded by the dissociation of the Cd–EDTA complex. This same reaction has been studied by Aylward and Hayes [61] by forming Cd(II) from Cd(Hg), as discussed in Section 8:

$$Cd\,Y^{2-} + H^+ \underset{k_b}{\overset{k_f}{\rightleftharpoons}} Cd^{2+} + HY^{3-}$$

$$Cd^{2+} + 2e^- \overset{DME}{\rightleftharpoons} Cd(Hg).$$

Matsuda and Tamamushi used phase angle data to calculate the value of k_b (formation of Cd Y^{2-}) in the above reaction scheme. The equilibrium constant was known for this system, and theoretical curves of $\cot \phi$ versus frequency were calculated for various values of $g = \ell/\omega$ using (6.109):

$$\cot \phi = \frac{K + (1 + K)e^j + \left[\dfrac{(1 + g^2)^{1/2} + g}{1 + g^2}\right]^{1/2}}{K + (1 + K)e^j + \left[\dfrac{(1 + g^2)^{1/2} - g}{1 + g^2}\right]^{1/2}}. \tag{6.109}$$

The $\cot \phi$ values were measured at the peak potential, which was cathodic of the reversible half-wave potential but relatively constant over the frequency range from 70 to 2000 Hz. The theoretically calculated $\cot \phi$ versus frequency curve at this potential that most closely matched the experimental curve was for $g = 1000$. By using a value of $K = (Cd^{2+})/(Cd\ Y^{2-}) = 10^{-3.67}$ [62], the pseudo-first-order formation rate k_b can be calculated from (6.108):

$$\ell = 1000 = k_b(1 + 10^{-3.67}). \tag{6.110}$$

Dividing k_b by the concentration of HY^{3-} yields the appropriate second-order formation rate constant. The value of 2.3×10^9 liter/mole-sec obtained by this technique agrees with values reported by other workers (see Table I of Ref. 62).

Although only the case of a preceding reaction coupled to a reversible charge transfer reaction has been considered in detail in this discussion, it is interesting to note the trends in experimental parameters as the charge transfer rate decreases. That the charge transfer is not fast can be determined most easily from the time dependence of the ac polarographic wave. With reversible charge transfer the chemical reaction shows a positive time dependence (increasing current with increasing drop life) over the whole wave. The magnitude of the time dependence is greatest on the cathodic side of the wave.

In the case of slow charge transfer a "crossover" point may be observed in the time dependence [70]. The time dependence is positive on the anodic side of the crossover point and negative on the cathodic side. This behavior is similar to that of the quasi-reversible charge transfer case discussed in Section 5.

The effect of slow charge transfer is also observed in the phase angle behavior as a function of E_{dc}. Instead of the sigmoidal curve there is a peak in $\cot \phi$ somewhere near the potential of the peak current. This behavior is characteristic of a quasi-reversible charge transfer step (see Section 5).

McCord and Smith [70] have pointed out that the preceding chemical reaction increases the influence of the charge transfer rate on the ac polarographic wave. The kinetic parameter is no longer just the charge transfer rate constant k_s, but the term $k_s[(K)^\beta/(1 + K)]$, where K is the equilibrium constant of the preceding reaction. Thus for the system to behave reversibly, the charge transfer rate has to be larger than would be expected in the absence of the preceding reaction.

In spite of the restriction of fast charge transfer, there are many systems that can be conveniently studied by this technique [72].

Chemical Reaction Preceding Irreversible Charge Transfer

The mechanistic scheme for this reaction is that of the previous section with the charge transfer now irreversible:

$$Z \underset{k_b}{\overset{k_f}{\rightleftharpoons}} O$$

$$O + ne^- \xrightarrow{k_s} R$$

This case describes a number of electrochemical reactions occurring in aqueous solution and has received considerable study with dc polarography. For ac polarography the case of a chemical reaction preceding an irreversible charge transfer reaction yields the same equations as for the irreversible charge transfer reaction alone. Therefore ac polarography offers no information concerning the coupled chemical reaction.

The Cyclic Voltammetric Response

With an irreversible charge transfer reaction, only variations in the cathodic current and in the position of the response on the potential axis are available for use. Nicholson and Shain [8] have treated this case, and the correlations below are from their work.

If the time scale of the experiment is great relative to the rate of the chemical reaction, that is, if ℓ/b is small ($b = \alpha a$, as in Section 6), the response is that of the uncomplicated irreversible charge transfer with the exception of a displacement on the potential axis.

If ℓ/b is large and $(b/\ell)^{1/2}/K$ is also large, the response obtained is no longer peak-shaped and both the potential of the response and the magnitude of the current are independent of b. Under these conditions the current is directly proportional to $K(\ell)^{1/2}$.

At intermediate values of $(b/\ell)^{1/2}/K$ with ℓ/b large, numerical solution of the integral equation for values of the kinetic parameters leads to the empirical equation

$$i_k/i_d = \frac{1}{1.02 + 0.531b^{1/2}/K(\ell)^{1/2}},$$

where again i_k is the observed current and i_d is the current expected in the absence of the chemical reaction.

Since other correlations are not available, the variation of the potential of the response with variation in the kinetic parameter assumes a greater importance. This shift decreases to a relative independence of $(b/\ell)^{1/2}/K$ for small values of the kinetic parameter.

Qualitatively, this case is rather easy to note with the lack of anodic response, the decrease in current function $(i_p/v^{1/2})$ with an increase in scan rate, and the general nature of the irreversible charge transfer case.

10 CATALYTIC REGENERATION OF REACTANT

The catalytic mechanism refers to the case in which the product of an electrode reaction undergoes a homogeneous irreversible first-order or pseudo-first-order chemical reaction to regenerate the starting material. This scheme is represented by

$$O + ne^- \rightleftharpoons R + Z$$

$$\uparrow \boxed{ k_c }$$

Generally, R must react with some reagent Z in solution to regenerate O. By keeping this reagent in large excess, a pseudo-first-order reaction rate constant k_c is determined from which the second-order rate constant k_2 can be determined using the known concentration of the reagent C_Z,

$$k_2 = k_c(C_Z)^{-1}$$

The second-order regeneration of O by the dismutation of R has not been treated theoretically for ac polarography but has been for cyclic voltammetry and is briefly mentioned.

The Cyclic Voltammetric Response

The catalytic mechanism has been treated theoretically by Saveant and Vianello [73] for the single-sweep technique and by Nicholson and Shain [5] for the cyclic experiment, assuming the charge transfer is reversible. Saveant and Vianello [53] also considered the case of quasi-reversible charge transfer and presented the integral equation for this case which is given in Table 6.6. Their subsequent discussion, however, dealt only with the case of reversible charge transfer.

Numerical values of $\chi(at)$ have been given [8, 73] for various values of the kinetic parameter k_c/a. These calculations show that as the term k_c/a becomes small (rapid scan rates or slow kinetics), the current approaches that for the simple reversible case. As k_c/a becomes larger the current decreases but is always larger than for the reversible case. Ultimately, the current becomes constant for increasing values of k_c/a, and the peak of the response disappears. The response assumes a sigmoidal shape, and the limiting current at cathodic potentials is given by (6.111) [73]. This is the case of pure catalytic control of the current [53]:

$$i_{\text{lim}} = nFAC_O^*(D_O k_c)^{1/2}. \tag{6.111}$$

The half-wave potential now is also the reversible half-wave potential for the redox couple being studied. Under these conditions of current independent of scan rate, an expression for the current at any point along the wave is given by

$$i = \frac{nFAC_O^*(D_O k_c)^{1/2}}{1 + e^j}, \tag{6.112}$$

where j is given by

$$j = \frac{nF}{RT}(E - E_{1/2}). \tag{6.113}$$

Saveant and Vianello [53] have given the corresponding pure catalytic current expression for a quasi-reversible charge transfer:

$$i = \frac{nFAC_O^*(D_O k_c)^{1/2}}{1 + e^j + \dfrac{(D_O k_c)^{1/2}}{k_s} e^{\alpha j}}. \tag{6.114}$$

The current is still independent of scan rate and yields the same limiting current at cathodic potentials given by (6.111), but the half-wave potential is cathodic of the reversible $E_{1/2}$ and dependent on the kinetic parameters of the system.

The ratio of the limiting catalytic current to the diffusion current in the absence of the catalytic reaction can be used to determine the catalytic rate constant k_c. Combining (6.111) and (6.22) yields (6.115) since $\pi^{1/2}\chi(at) = 0.446$ at the peak of the diffusion-controlled response [53]:

$$\frac{(i_\infty)_c}{(i_p)_d} = \frac{k_c^{1/2}}{0.446a^{1/2}}. \tag{6.115}$$

Nicholson and Shain [8] have examined $(i_\infty)_c/(i_p)_d$ as a function of the kinetic parameter for the general case of mixed diffusion and catalytic current. The linear portion of the curve $(k_c/a > 1)$ is where (6.111), (6.112), and (6.115) are applicable.

The case of mixed catalytic and diffusion control is similar to the reversible case in that the anodic peak current is equal to the cathodic peak current when measured using the cathodic response as a base line. However, in contrast to the reversible response, the potential of the cathodic peak is not constant with scan rate but shifts anodically by about $60/n$ mV for a 10-fold increase in scan rate. Moreover, the value of the peak current, or plateau current if there is no peak, for the catalytic case is always larger than that of the corresponding reversible current, as mentioned above. Because the peak potential does shift anodically as the scan rate increases, a measure of the catalytic rate can be obtained by plotting the peak potential against the kinetic parameter k_c/a. Because the half-peak potential is easier to measure, it is the recommended parameter to use. $E_{p/2}$ varies between the value of the reversible case, $E_{p/2} - E_{1/2} = 28/n$ mV, for small k_c/a and that of the pure catalytic case, $E_{p/2} - E_{1/2} = 0$, when k_c/a becomes large. Experimentally, the scan rate is adjusted so that there is still a peak in the response, $E_{p/2}$ is measured for various values of $\log k_c/a$, and results are compared with a working curve to obtain k_c [8].

The wave shape also begins to change markedly going from essentially diffusion control to catalytic control of the current. Examples of this are shown in Fig. 6.18. Diagnostic criteria for the catalytic mechanism are summarized in Table 6.10.

The catalytic reaction following an irreversible charge transfer reaction has also been treated by Nicholson and Shain [8]. The behavior is similar to

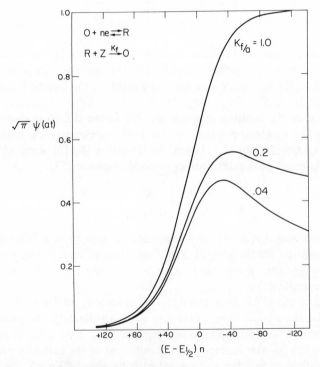

Fig. 6.18 Theoretical cyclic voltammetric curves for various values of the catalytic rate parameter. The charge transfer step is reversible.

Table 6.10   Diagnostic Criteria for a Catalytic Reaction

$$O + ne^- \rightleftharpoons R + Z$$
$$\underset{k_c}{\underline{\qquad\qquad\qquad}}$$

Cyclic Voltammetry
   Properties of the potential of the response:
      E_p shifts anodically with an increase in v by an extent which goes
      through a maximum of $60/n$ mV per 10-fold increase in v. No depend-
      ence of E_p on v is noted at either extreme in k_c/a
   Properties of the current function:
      $i_p/v^{1/2}$ increases at lower values of v and becomes independent of v
   Properties of the anodic to cathodic current ratio:
      $i_p{}^a/i_p{}^c$ is unity
   Others:
      As k_c/a becomes large, the response approaches a sigmoidal shape
Ac Polarography
   Properties of the ac current:
      $I(\omega t)$ is larger than $\omega^{1/2}$ dependency at lower ω and approaches
      proportionality to $\omega^{1/2}$ at higher ω
   Properties of the phase angle:
      Cot ϕ is independent of E_{dc}
      Cot ϕ increases greatly as $\omega \to 0$

---

that of the reversible charge transfer case, but the kinetic parameter is now
$k_c/\alpha a$. Many of the same criteria can be used to diagnose the catalytic
reaction.

In the case of complete catalytic control, a closed-form solution for the
wave is obtained similar to (6.112):

$$i = \frac{nFAC_O{}^*(D_Ok_c)^{1/2}}{1 + \dfrac{(\alpha D_Ok_c)^{1/2}}{k_s}e^{\alpha j}}.$$
(6.116)

At very cathodic potentials the limiting current expressed by (6.111) is
obtained and a plot of $(i_\infty)_c/(i_p)$ as a function of the kinetic parameters can
be used to obtain k_c. The linear portion of the plot for the irreversible case is
now given by

$$\frac{(i_\infty)_c}{(i_p)} = \frac{k_c^{1/2}}{0.496(\alpha a)^{1/2}}.$$
(6.117)

In the region of partial charge transfer control and catalytic control, the
potential of the wave shifts cathodically as the kinetic parameter $k_c/\alpha a$ is

increased. A measure of the catalytic rate could be obtained from this shift in the same manner as for the reversible case.

Saveant and Vianello [53] have reported an experimental study of three catalytic reactions coupled to a reversible charge transfer reaction. They used the method of calculating the pseudo-first-order rate constant according to (6.115) for various scan rates and various concentrations of the catalytic regenerating agent. To calculate the second-order rate constant k_2, (6.115) can be rewritten to include the concentration of the catalytic species Z:

$$\frac{(i_\infty)_c}{(i_p)_d} = \frac{(C_Z k_2)^{1/2}}{0.446 a^{1/2}}. \tag{6.118}$$

The scan rates used for these studies ranged from 0.1 to 1.0 V/sec, and the regenerating agent was present in 30-fold excess or more.

The reversibility of the systems was checked at scan rates up to 10 V/sec, and peak separations (anodic and cathodic) in the absence of any catalytic agent were about $60/n$ mV at all scan rates used in the subsequent catalytic studies.

The three systems studied were the reaction of H_2O_2 with Fe(II)–EDTA in acetate buffer at pH 4.7, the reaction of NH_2OH with Ti(III) in 0.2 M oxalate, and the reaction of NH_2OH with Ti(III) in 2 M H_3PO_3. The rate constant for the reaction of the Ti(III) oxalate complex was found to be 42 liters/mole-sec, in good agreement with other studies. Blazek and Koryta [74] reported a value of 42 liters/mole-sec obtained polarographically, while Delahay et al. [75] and Fisher et al. [76] obtained results of 30 liters/mole-sec and 46 liters/mole-sec, respectively, using chronopotentiometry.

Ac Polarography

Smith [5, 51] has treated the catalytic case theoretically for ac polarography, using the planar diffusion model to obtain a general current expression and the expression for the phase angle. He has assumed that the charge transfer process may not be reversible, so the expression for cot ϕ contains terms for both k_s and k_c. Only the expression for cot ϕ,

$$\cot \phi = \frac{\dfrac{(2\omega)^{1/2}}{\lambda} + \left[\dfrac{(1 + g^2{}^{1/2} + g}{1 + g^2}\right]^{1/2}}{\left[\dfrac{(1 + g^2)^{1/2} - g}{1 + g^2}\right]^{1/2}}, \tag{6.119}$$

where $g = k_c/\omega$ and λ is given by (6.56), is discussed here as it is sufficient to obtain kinetic parameters for the mechanism. The term $(2\omega)^{1/2}/\lambda$ expresses the cot ϕ dependence on the charge transfer rate and the terms involving g describe the effects of the catalytic reaction on cot ϕ.

It can be seen that as the frequency becomes small cot ϕ becomes very large, going to infinity as $\omega \to 0$. This is an excellent criterion to use to indicate the presence of such a reaction, as noted in Table 6.10.

Observe that the potential dependence of cot ϕ is embodied only in the term containing λ (6.56) because the effect of the catalytic reaction is constant over the potential range. Thus if the charge transfer rate is fast enough $[(2\omega)^{1/2}/\lambda \to 0]$, cot ϕ is a constant over the potential range of the wave. Examples of the dependence of cot ϕ on frequency are shown in Fig. 6.19 for several values of the catalytic rate constant.

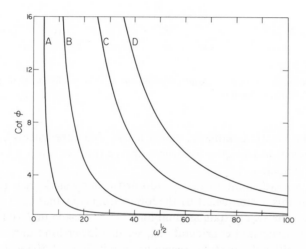

Fig. 6.19 Theoretical cot ϕ as a function of frequency for catalytic regeneration of starting material, for various values of the catalytic rate. Curve A, $k_c = 100$ sec^{-1}. Curve B, $k_c = 1000$ sec^{-1}. Curve C, $k_c = 5000$ sec^{-1}. Curve D, $k_c = 10,000$ sec^{-1}.

The large value of cot ϕ at low frequencies in the catalytic case means that the equivalent faradaic impedance becomes purely resistive. The total amount of R formed at the electrode surface is immediately reoxidized to O, causing a very steep concentration gradient. This means that diffusion now plays a minimal role in mass transport, from which the capacitive component of the impedance is obtained (see Section 14).

When the charge transfer rate is fast enough to be considered reversible, the cot ϕ depends only on the catalytic reaction, that is, $(2\omega)^{1/2}/\lambda \ll 1$ in (6.119). Under these conditions a working curve of cot ϕ plotted against the kinetic parameter k_c/ω can be used to obtain the rate constant k_c. This curve is shown in Fig. 6.20. The linear portion of the curve represents essentially complete kinetic control of the wave. Note that this region occurs for large

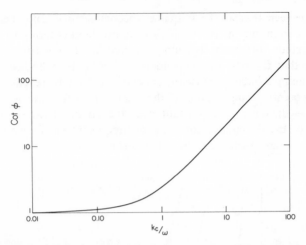

Fig. 6.20 Ac polarographic working curve of cot ϕ vs. k_c/ω for catalytic regeneration with reversible charge transfer.

values of the kinetic parameter, which means that experiments can be conducted at fairly low frequencies where charge transfer has a minimal effect on the phase angle.

There is a lower limit to the pseudo-first-order kinetic rate that can be measured by this technique set by the lowest frequency that can conveniently be obtained. The steady-state assumptions [37, 38] made in deriving (6.119), as well as experimental considerations, set this lower limit at about 10 Hz; that is, it is difficult to study reactions in which $k_c < 30$ sec^{-1}. For slow catalytic rates it may be necessary to use fairly high concentrations of the catalytic reagent to make $k_c \geq 30$ sec^{-1}.

Smith [38] has tested (6.119) experimentally by determining k_c for a system using dc polarographic data and obtaining k_s and α from ac polarographic data. Phase angle data were then obtained using the catalytic system and the values were compared with theory and with the values of k_c, k_s, and α previously obtained. The experimental plots of cot ϕ against frequency and dc potential at several frequencies were compared with the theoretical calculations, with excellent agreement.

The system Smith used was the Ti(IV)/Ti(III) couple in 0.2 M K$_2$C$_2$O$_4$, with KClO$_3$ as the catalytic regenerating agent. Charge transfer data were obtained in the absence of ClO$_3^-$ ion and then a 50-fold excess of ClO$_3^-$ ion was added to obtain phase angle data on the pseudo-first-order catalytic system. A value of $k_s = 0.046$ cm/sec with $\alpha = 0.35$ was obtained by ac polarography and the second-order catalytic rate of 4.25×10^3 liters/mole-sec was obtained from dc polarography at three chlorate concentrations. Data for cot ϕ using ac polarography and these constants gave excellent agreement with theoretical curves. Because of the somewhat low value of k_s, it might be difficult to study this system by using Fig. 6.20 and the assumption of charge transfer reversibility.

II THE ECE MECHANISM

Electrochemical reaction sequences in which the product of charge transfer undergoes a chemical reaction yielding an electrochemically active species were excluded from the consideration of following chemical reactions given earlier. However, this reaction sequence, the ECE mechanism, is a most important one. Historically, many reactions of organic compounds have been shown to involve two electrons and one or two protons. Many of these reactions proceed by the ECE mechanism in one of its various forms and a characterization of this mechanism is essential to the full use of modern electroanalytical techniques. Unfortunately, the ECE mechanism has not yet been treated for ac polarography and the treatment for cyclic voltammetry does not include the full complement of reaction schemes. Thus the reaction sequence

$$O + ne^- \rightleftharpoons R$$
$$R \xrightarrow{k_f} X$$
$$X + ne^- \rightleftharpoons Y$$

under the conditions that the first charge transfer reaction is reversible, both charge transfers are one-electron reactions, and the chemical reaction is a first-order irreversible process, receives prime consideration. Deviations to be expected as the nature of the chemical reaction is changed are described.

Complex reaction sequences such as the ECE mechanism provide emphasis concerning the utility of cyclic voltammetry in qualitative characterization of electrochemical reactions. It is appropriate at this point to review the extent to which qualitative data should in practice lead an experimental study. Qualitative information can include the simple observation of the presence of a new electroactive species as a result of the reaction of prime interest, as well as define a behavior of the response as a function of scan rate which definitely indicates a particular reaction pathway. Perhaps the point is obvious, but the reader is reminded that an initial overview of the problem should proceed any attempt to test a reaction for a particular diagnostic criterion. A few cyclic voltammetric scans covering a large potential range at a few widely differing scan rates may provide valuable information. Although the quantitative treatment is based on the initial scan, a complex system may show useful qualitative information from multiple scans.

A comparison of the current magnitude at a selected scan rate with that of known reactions is often instructive although care must be exercised if different solvents are employed since diffusion coefficients are a function of the medium. Purposeful addition of suspected chemical reactants to the medium can frequently provide a strong indication of the nature of the overall reaction.

Thus cyclic voltammetry provides experiments that are inexpensive in time and effort and often can reduce the total number of reactions that must be considered in detail.

ECE Mechanism with Both Charge Transfers Reversible

Even with the charge transfer reactions defined as reversible, there are two additional considerations that have a significant effect on the quantitative nature of the cyclic voltammetric response for an ECE mechanism. The more obvious of these is the relative position on the potential scale of the two redox couples. If the species formed by the chemical reaction is reduced as readily as the original compound or more readily, the current-potential curve observed at potentials required for the reaction of the original compound will include a component for the product species. This response will differ markedly from that obtained for the case in which the potential required for the second charge transfer is quite cathodic of that of the original compound. Thus the treatment presented here is on the three intuitively important and theoretically treated cases in which the E^0 of the product species (E_2^0) is very cathodic of the E^0 of the original species (E_1^0), the two E^0 values are equal, and E_2^0 is very anodic of E_1^0.

In addition, the number of electrons involved in the two redox reactions is important [9]. This discussion is restricted to cases in which the number of electrons transferred is the same for both charge transfer reactions.

E_2^0 Cathodic of E_1^0

Under the conditions defined above, the first charge transfer process has the properties of the reversible charge transfer, irreversible following chemical reaction case discussed in Section 8. Thus the quantitative relationships as well as the diagnostic criteria for this case can be used. At small values of the chemical rate constant with respect to the time scale of the experiment (k_f/a is small), the response approaches reversible behavior in all respects.

As the value of k_f/a increases, the second process begins to appear as a broad, relatively ill-defined response. This response increases to the limiting case of two fully developed cathodic responses at large values of k_f/a [9]. The relative magnitude of the two current responses can be employed for quantitative measurements through data presented by Nicholson and Shain [9] for a working curve of i_p^c (O to R)/i_p^c (X to Y) as a function of k_f/a.

The nature of the total anodic response is also a function of the rate of the chemical reaction. With the second charge transfer reversible and uncomplicated by chemical kinetics, the anodic response is simply that of a reversible charge transfer. However, at intermediate values of k_f/a, the magnitude of the anodic response for the second charge transfer is a function of both the separation in the two processes and the switching potential. With the quantitative relationships noted above, this anodic process offers little of value for other than qualitative purposes. At smaller values of the kinetic parameter, the anodic response of the first process can be employed for quantitative purposes. Simulated cyclic current-potential curves are shown in Fig. 6.21 to illustrate the qualitative nature of the response as a function of k_f/a.

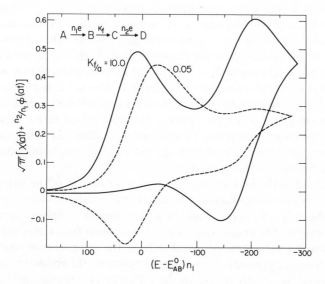

Fig. 6.21 Theoretical cyclic voltammetric curves for ECE mechanism. E_2^0 is cathodic of E_1^0 and the chemical reaction is irreversible. Values of K_f/a are shown (Courtesy of Anal. Chem., Ref. 9).

The variations in potential for these coupled processes are of no great value for quantitative use in view of the excellent properties of the current measurements. As the rate constant for the chemical reaction increases, the variation in potential of the first process can be employed for qualitative purposes to provide assurance that the mechanism is that treated.

Differentiation between an ECE mechanism and an uncomplicated multistep charge transfer reaction is also of interest. Polcyn and Shain [11] have shown that if the potentials of the two processes are sufficiently separated

($> 100/n$ mV), the two processes behave independently as uncomplicated charge transfer systems in all respects. Excellent agreement with the theory was obtained by these authors for the two-step reduction of copper(II) in an ammonium chloride–ammonium hydroxide medium. Thus examination of the composite response as a function of scan rate affords ready differentiation of the simple charge transfer system from the ECE mechanism wherein the response is dependent on the time scale of the experiment.

E^0 for the Two Couples Equal

As the separation in E_1^0 and E_2^0 decreases, the nature of the response becomes more complicated. Two responses may be observed if the kinetic parameter is large, but the nature of the current-potential curve becomes less recognizable as the kinetic parameter decreases.

Calculations [9] for the case in which the separation in E^0 is zero provide insight into the quantitative aspects of the response. The response for the product couple is likely to occur at potentials cathodic of the reversible value since the concentration of X is determined by the production of R (and the chemical reaction) and the concentration of R does not reach a limiting value at the E_2^0. The magnitude of this effect also depends on the rate constant of the chemical reaction. At the appropriate value of the rate constant of the chemical reaction, two peaks may actually be observed since the product couple may be shifted cathodically and the original couple shifted anodically by the following chemical reaction.

The nature of the anodic response is also a function of the two processes. At lower values of the kinetic parameter, the anodic response is primarily that of the original couple, while at high values the response is that of the product couple. The anodic response is a compound function of the switching potential as a result of the potential shifts in the two processes, and thus it is difficult to employ quantitatively. More important, the anodic current cannot be measured accurately owing to the difficulty in establishing the cathodic base line.

Quantitative measurements for an ECE mechanism that does not involve couples separated markedly from each other in potential appears to be impractical by any means other than computer-aided comparison of experimental and theoretically calculated responses.

E_2^0 Anodic of E_1^0

With the product couple more readily reduced than the parent couple, only a single cathodic response is observed. If the rate constant for the chemical reaction is rapid, essentially all the initial species that reacts will be converted to the final product of the mechanism. As the potential is returned to anodic values, an anodic response is observed at more anodic potentials.

A subsequent cathodic scan from the anodic limit reveals the reduction portion of the product couple.

This reaction sequence is very common for organic compounds. Adams and co-workers have noted the presence of this sequence for a number of compounds, and the recent book by Adams [19] summarizes this body of information. The oxidation of triphenylamine studied by these workers [78] may be considered a classic example of this reaction. A typical cyclic voltammetric response for the oxidation of triphenylamine is illustrated in Fig. 6.22

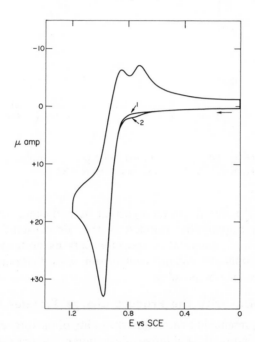

Fig. 6.22 Cyclic voltammetric curve of triphenylamine in acetonitrile. Supporting electrolyte—0.1 M TBAP, $5 \times 10^{-4}\ M$ triphenylamine. Scan rate = 18 mV/sec. Numbers refer to first and second scan. Large pyrolytic graphite electrode.

As discussed above, essentially no anodic response for the initial oxidation is observed, but a redox system at more cathodic potentials is apparent. The product redox couples result from reactions of N,N,N',N'-tetraphenylbenzidine produced in a bimolecular coupling of the triphenylamine cation radicals.

The peak current function $i_p/v^{1/2}$ for this reaction obeys the prediction for this type of mechanism. It is large at low scan rates and decreases to a constant value at higher scan rates as shown in Fig. 6.23.

The appearance of the reaction for the product couple as well as the

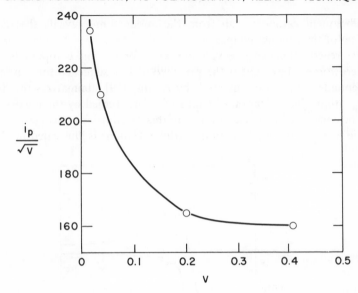

Fig. 6.23 Current function behavior of oxidation peak for triphenylamine as a function of scan rate. Experimental conditions: same as Figure 6.22.

magnitude of the initial cathodic current as a function of scan rate can readily provide a qualitative estimate of the rate constant of the chemical reaction. However, quantitative measurements by cyclic voltammetry for this case are difficult. Again, comparison with theoretically calculated response curves may be possible.

ECE Mechanism with the Product Charge Transfer Irreversible

This case is presented to emphasize the utility of the total cyclic experiment in obtaining a thorough qualitative description of the reaction. The cathodic portion of the response for this case is not markedly different from the previous case. With the less distinct nature of the irreversible response, the cathodic response for the case in which E_2^0 is very cathodic of E_1^0 is even less distinct at low values of the kinetic parameter. As k_f/a increases, the nature of the response becomes more well defined, but characterization as an irreversible process is likely to be difficult. With no separation in E^0 for the two processes, the response is difficult to differentiate from the previous case [9]. When E_2^0 is anodic of E_1^0, the irreversibility of the product couple cannot be denoted by the cathodic scans.

However, the anodic scan provides direct observation of the irreversibility of the product charge transfer, except at low values of k_f/a. At high values of k_f/a with E_2^0 cathodic of E_1^0, two processes with no anodic current are observed. With E_1^0 equal to E_2^0, a single process with no anodic current is observed. When E_2^0 becomes anodic of E_1^0, no anodic response is observed but the cathodic portion of the $X \to Y$ process is seen. The response is quite small, however, since X can only accumulate when the potential is anodic of E_2^0 [9].

Further Properties of the ECE Mechanism

In the above discussion it is assumed that the intermediate chemical reaction is irreversible, but in fact many reactions that include a reversible chemical reaction must exist. The general sequence

$$O + ne^- \rightleftharpoons R$$

$$R + Z \rightleftharpoons X$$

$$X + ne^- \rightleftharpoons Y$$

has been considered by Mastragostino et al. [67] in the context of acid-base reactions coupled to charge transfer processes. Working curves for the variation in the current function and the peak potential as a function of the kinetic parameters may be found in the original literature. If the equilibrium constant for the chemical reaction is known from independent measurements, the kinetics of the chemical reaction can be examined. As the equilibrium of the chemical reaction shifts more toward X, the reaction sequence reduces to that discussed above with an irreversible chemical reaction.

In the reaction sequence described above, it has been assumed that no interactions between species O, R, X, and Y are operative. However, for the case in which the E^0 of the $X \to Y$ couple is anodic of that for the $O \to R$ couple, the reaction

$$R + X \overset{K}{\rightleftharpoons} O + Y$$

becomes important. Feldberg et al. [79] have shown that the behavior of such systems is a function of the equilibrium constant for the cross-redox reaction. Thus this reaction must be considered as a possible reaction sequence.

If X is more readily reduced than O, the disproportionation

$$2X \rightleftharpoons O + Y$$

must also be considered. Mastragostino and Saveant [80] have presented means to differentiate between the ECE reaction proceeding through charge transfer and the reaction proceeding through the disproportionation. Concentration of the reactant becomes a prime variable in this differentiation as in all second-order processes. The variation of the current function $i_p/v^{1/2}$ with scan rate shows a concentration dependency if the disproportionation is involved, the current function increasing for a given scan rate with an increase in concentration. The ECE process shows no concentration dependency. In the general, and important, case of charge transfer reactions with intermediate protonation reactions, the ECE mechanism should be favored at lower concentrations, at higher acidities, and with rapid charge transfer rates. The disproportionation path becomes more important with lower acidity, high concentration, and less rapid charge transfer [80].

If the number of electrons transferred in the separate charge transfer processes of an ECE mechanism differ, the nature of the overall response will vary significantly with the appropriate combination of reaction parameters. The reader is referred to a more complete discussion of this matter by Nicholson and Shain [9].

For an ECE process, the current magnitude is a function of $n_1 + n_2$ rather than the $n^{3/2}$ function of the Randles-Sevcik equation. Nelson [81] has suggested that some ECE processes may be differentiated from a direct multielectron transfer by the current magnitude since the current for a direct two-electron transfer is 2.83 times that for a one-electron transfer, while the current for the ECE reaction with n_1 and $n_2 = 1$ is only twice that for a one-electron transfer.

I2 ADSORPTION

As discussed in Section 2, supply of the reactant to the electrode surface may be governed by an adsorption process as well as by diffusion. All too frequently in practice adsorptive phenomena complicate the overall electrochemical reaction to be examined. Although adjustment of experimental conditions can many times reduce the adsorptive complications, an appreciation of the effects of adsorption of an electroactive species is mandatory to an adequate use of cyclic voltammetry and ac polarography. Most importantly, a recognition of the presence of an adsorption process is absolutely necessary. The treatment of adsorption presented here is primarily for this latter purpose.

A complete discussion of adsorption phenomena in electrochemical reactions is beyond the scope of this chapter and is not attempted. Even a qualitative discussion, however, requires that the parameters necessary to describe an adsorption be identified. The primary information one must have is knowledge about which of the electroactive species is adsorbed. Diagnostic criteria are presented which afford this information for most situations.

The exact nature of the adsorption isotherm must be known to describe the process accurately. For the relative simplicity of the treatment, the Langmuir isotherm given by

$$\Gamma = \Gamma^*C/(K + C) \qquad (6.121)$$

(where Γ = surface concentration; Γ^* = saturation value of the surface concentration; C = solution concentration; and K = proportionality constant) is frequently employed. It should be noted that this isotherm assumes that a limiting surface coverage will be reached. The free energy of adsorption is then given by

$$\Delta G = -RT \ln K, \qquad (6.122)$$

with the value of K dependent on a number of experimental conditions including the particular electrode material, the solvent system, and the presence of other adsorbable materials.

The value of K may be a function of the electrode potential, and some modification of the free energy expression to yield an expression this is an appropriate function of potential is required [82].

The "strength" of the adsorption is embodied in the value of the free energy and, through K, the extent to which a material is adsorbed. Thus "strong" adsorption denotes both a greater free energy of adsorption and a greater extent of adsorption.

The kinetics of the adsorption reaction are also important and must be included in a precise characterization of the process. For the present discussion, it is assumed that the adsorption-desorption reactions are rapid.

Again, the present treatment is intended primarily for qualitative use. Neither cyclic voltammetry nor ac polarography provide a particularly convenient means to study adsorption phenomena in detail. Several of the other techniques discussed in this volume afford more direct routes to an understanding of adsorption processes.

The Cyclic Voltammetric Behavior

The nature of the effects of adsorption processes on the cyclic voltammetric response has been treated recently by Wopschall and Shain [19]. For convenience, the separation of possible cases into weak or strong adsorption of either reactant or product employed by these authors is utilized. No consideration is given to adsorption involving both reactant and product.

Strong Adsorption of the Product of Charge Transfer

Processes involving strong adsorption of the product of charge transfer are frequently encountered in electrochemical reactions of organic compounds. Wopschall and Shain have treated this case theoretically [13] and experimentally [83] for the reduction of methylene blue. A typical cyclic voltammetric response involving strong adsorption of the product of charge transfer is illustrated in Fig. 6.24. Under conditions in which both adsorption and diffusion control are significant, a response prior to the diffusion-controlled response is obtained for the reduction to the adsorbed state. The reverse

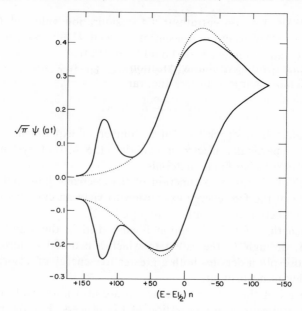

Fig. 6.24 Theoretical cyclic voltammetric curve with strong adsorption of product. Dotted curve shows normal reversible curve.

potential scan also contains a response related to the adsorption process, in this case following the diffusion-controlled response. Observation of both the cathodic and anodic responses is mandatory for characterization of the process as the sequence

$$O + ne^- \rightleftharpoons R$$

$$R \rightleftharpoons R_{(ads)}$$

The separation in the peak potentials of the adsorption and diffusion-controlled responses is a function of the free energy of adsorption. As the energy of adsorption increases, the separation in the peak potential of the two responses ΔE_p increases. However, this separation is also a function of the bulk concentration of reactant and cannot be used directly to calculate the free energy of adsorption. The reader is referred to the original literature for a discussion of the measurements required to relate ΔE_p to ΔG. The shape of the adsorption-controlled response is a function of the potential dependency of the isotherm. Qualitatively, a dependency of the adsorption on potential such that there is greater adsorption with more cathodic potential would result in a sharper adsorption response than the response of potential-independent adsorption. The sharpness of the response is also concentration dependent with the width of the response increasing at low concentrations, that is, conditions under which diffusion cannot maintain the surface concentrations in complete equilibrium with the solution concentration.

For further qualitative characterization of the case of strong adsorption of the product of charge transfer, the variation of the response with bulk concentration and scan rate are important. At very low concentrations the reduction to the adsorbed state is the primary process. As the concentration is increased, both processes are observed, and the relative height of the adsorption response to the diffusion-controlled response decreases with an increase in concentration. The potential of the adsorption process becomes more anodic as the concentration increases by approximately $60/n$ mV per 10-fold increase in concentration in the absence of a potential dependency of the isotherm.

At very low values of the scan rate, the adsorption response may not appear. As the scan rate is increased, the adsorption response increases relative to the diffusion-controlled response. At very rapid scan rates, only the adsorption response is noted. This behavior illustrates the point that although the peak current is an increasing function of the scan rate the total coulombs involved in the experiment decrease with an increase in scan rate. Thus at very rapid scan rates the total coulombs transferred becomes equal to (or less than) the amount that can be transferred to adsorbed product.

Strong Adsorption of the Reactant

Processes involving a strong adsorption of reactant are also commonly encountered in electrochemical reactions of organic compounds. In this case, a response cathodic of the diffusion-controlled response is obtained as is illustrated in Fig. 6.25. Again, appearance of both the cathodic and anodic adsorption-controlled response is required for the sequence

$$O \rightleftharpoons O_{(ads)}$$

$$O_{(ads)} + ne^- \rightleftharpoons R$$

to be that which describes the overall reaction.

The variation of the response with experimental parameters parallels that of the previous case. Complete details of the treatment have not been presented, but extension of the detailed treatment of the above case to reactant adsorption should lead to a determination of the energy of adsorption from ΔE_p values.

The relative magnitudes of the adsorption and diffusion responses are a function of concentration, with the adsorption response predominant at lower concentrations. Integration of the total area under the adsorption response (coulombs) under these conditions provides convenient access to surface coverage values. The limiting value of the integrated area as the

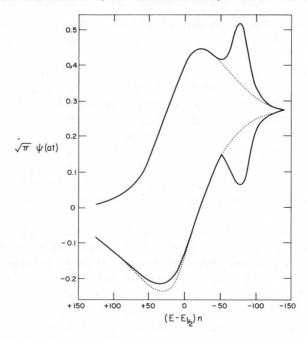

Fig. 6.25 Theoretical cyclic voltammetric curve with strong adsorption of reactant. Dotted curve shows normal reversible curve.

concentration is increased is a convenient means to estimate the maximum or saturation surface coverage. As the bulk concentration is increased, the relative magnitude of the adsorption response decreases.

The variation of the relative magnitudes of the two responses with the rate of potential scan is again such that the diffusion process predominates at low scan rates and the adsorption process predominates at high scan rates.

With adsorption of the reactant, it should be convenient to study the kinetics of slow adsorption phenomena through carefully timed delays between introduction of the material and the potential scan.

Weak Adsorption of the Reactant

When the electroactive species is only weakly adsorbed, the nature of the cyclic voltammetric response is not as markedly different from that of the uncomplicated case as it is with strong adsorption. With the free energy of adsorption low, the difference in potential between that required for reduction of the adsorbed reactant and that required for the soluble species is too small to show a separate response for the adsorbed material. However, the nature of the single response obtained reflects the presence of adsorption, as illustrated in Fig. 6.26. The response exhibits a more pronounced peak shape

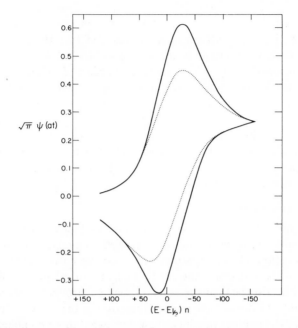

$\sqrt{\pi}\ \psi\ (at)$

$(E - E_{l_2})\ n$

Fig. 6.26 Theoretical cyclic voltammetric curve with weak adsorption of reactant. Dotted curve shows normal reversible curve.

for both the cathodic and anodic scans. Differentiation of this response from that of a multi-electron charge transfer reaction is difficult, and a single experiment provides little basis for detecting the presence of weak adsorption. However, variation in the response as a function of scan rate and concentration of reactant may be employed to afford a characterization of the process.

As discussed for the case of strong adsorption of the reactant, as the time scale of the experiment is decreased, a greater portion of the total charge is passed through the adsorbed reactant. Since adsorption implies a surface excess of reactant, the response with increasing scan rate should increase in the presence of adsorption relative to that of the purely diffusion-controlled case. Wopschall and Shain [13] have presented theoretical calculations of this behavior. Since this behavior is unique, a significant increase in the current function at faster scan rates is a strong indication of the presence of weak adsorption.

The concentration of reactant may also provide diagnostic information for the detection of the presence of weak adsorption. As the concentration of reactant is increased, a greater portion of the total response occurs through diffusion control. A limiting behavior at high concentrations, essentially that of the uncomplicated charge transfer, should be observed. The theoretical response as a function of concentration has been presented by Wopschall and Shain [13]. These authors have emphasized that the full variation in behavior from a predominantly adsorption-controlled to a predominantly diffusion-controlled reaction involves several orders of magnitude in concentration, which is frequently beyond experimental practicality. Thus the variation of the response with concentration must be used with discretion. Perhaps the best experimental compromise for diagnostic purposes is to use the variation in response with scan rate and confirm the behavior at several concentrations. Use of cyclic voltammetry for the study of the adsorption per se for weak adsorption appears not to be fruitful, particularly in comparison with other techniques.

Other Adsorption Cases

The preceding discussion has been limited to those cases that are both commonly encountered in practice and, at least on a relative basis, amenable to characterization. Other situations are of course encountered in practical studies, that is, both reactant and product may be adsorbed, only the product may be weakly adsorbed, or the reactant may be strongly adsorbed and the product weakly adsorbed. The reader is referred to the original literature for discussions of the manner in which these somewhat more complex situations are analyzed.

The above treatment has also assumed the absence of any complicating chemical reactions. Unfortunately, many electrochemical reactions of organic compounds involve both coupled chemical reactions and adsorption. The manner in which these two processes reflect their presence on each other has been presented by Wopschall and Shain [84] for the case of an irreversible following chemical reaction with weak adsorption of the reactant. A semiempirical method was developed by these authors and applied successfully to the reduction of azobenzene. More complicated kinetic systems and/or more complex adsorptive behavior obviously present more difficult problems, but the promise for resolution of such cases is apparent. In summary, the effects of an adsorption process on cyclic voltammetric data can be observed and applied, even in the presence of coupled chemical reactions. However, the nature of the effects of adsorption must be carefully examined with respect to the effects of any chemical reactions and their interactions included in arriving at diagnostic conclusions.

The Ac Polarographic Behavior

Adsorption of the Electroactive Species

Adsorption behavior of electroactive species has not received an extensive theoretical treatment for ac polarography. Most work to date has involved a qualitative explanation of experimental results. For example, Breyer and Bauer [85] have observed nonlinear current versus concentration profiles for several organic compounds. In addition, the temperature coefficient of the current changed as the concentration was increased indicating that two different processes were occurring. These observations suggested that current is obtained from both the adsorbed and the unadsorbed electroactive species.

For systems involving strong adsorption of the electroactive species where there is a definite poststep or prestep in the dc polarogram, an ac wave is observed at both the main wave (diffusion-controlled) and the smaller wave (reaction of the adsorbed species). This result is similar to that observed in the cyclic voltammetric process described above, however, no theoretical study has been made of the ac polarographic response under the influence of strong adsorptive phenomena.

Several studies of methylene blue in aqueous buffer solutions have been made using ac polarography [86–88], all showing similar adsorptive behavior. The concentration dependence of the current is nonlinear; in fact, an easily measured wave is observed at concentrations of dye as low as 10^{-5} M. In addition, there was an ac polarographic wave at both the prestep and the normal wave observed in dc polarography. From current data, Bauer [86] reported adsorption coefficients based on Langmuir isotherms. Base current depression in the presence of methylene blue was further indication of adsorption of the dye and the leuco dye [88].

A similar study of chloranilic acid [89] using ac polarography showed that both forms of the redox couple were adsorbed. Addition of cyclohexanol destroyed the response, indicating that chloranilic acid has to be adsorbed to be reduced at the mercury surface or that cyclohexanol is quite efficient in blocking the electrode surface to the chloranilic acid.

Tensammetry

In addition to ac polarographic waves that have been observed for electroactive species, another type of wave can be observed when a nonelectroactive species is adsorbed or desorbed at the electrode surface. This wave has been called a tensammetric wave [90] and occurs when the potential-dependent adsorption-desorption process results in a change in the double-layer capacitance. This process gives rise to a peak-shaped ac polarographic wave over the dc potential range where it occurs. Many organic surfactants, both charged and uncharged, exhibit this behavior even though they do not react in any faradaic reactions. An example of a tensammetric wave is shown in Fig. 6.27 for the adsorption and desorption of t-butanol on mercury in 0.5 M KNO$_3$.

A detailed discussion of experimental observations of this phenomenon is not in keeping with the purpose of this chapter and the interested reader

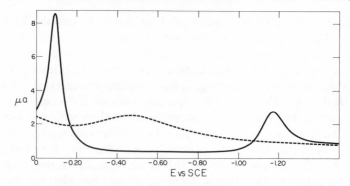

Fig. 6.27 Experimental tensammetric wave of iso-butyl alcohol. Solid curve is 0.42 M iso-butyl alcohol in 0.5 M KNO_3 supporting electrolyte. Dotted curve is 0.5 M KNO_3 only. $\Delta E = 5$ mV at 80 Hz. The drop oscillations are not shown.

is referred to the monograph of Breyer and Bauer [91] in which this process is discussed more thoroughly. A qualitative examination by cyclic voltammetry provides ready differentiation of tensammetric waves from waves for adsorption-controlled faradaic processes.

13 EXPERIMENTAL TECHNIQUES

This section is restricted to a discussion of some of the special problems that must be considered in performing electrochemical experiments with faster relaxation techniques such as cyclic voltammetry and ac polarography. These techniques generally have more stringent experimental requirements than slower techniques such as dc polarography and coulometry. The reader is referred to other chapters of this volume for general experimental discussion.

Instrumentation

While this chapter is concerned mainly with cyclic voltammetry and ac polarography, the need to use other techniques in mechanistic studies has been noted. Generally, no one technique used exclusively is as powerful in the study of an unknown system as is the judicious use of several techniques. Thus in recent years an effort by instrument makers to satisfy the demand for reasonably priced, accurate and, above all, versatile electrochemical instrumentation has been evident.

For the chemist setting out to assemble an electrochemical instrument, there are three paths to follow, the choice between them being based primarily on his knowledge of and interest in instrumentation.

Those with no background in instrumentation can purchase a completely assembled instrument [92–94]. The manufacturer's literature generally provides some background information on design considerations for acceptable performance of the product. For the worker with some background in instrumentation, a laboratory-assembled, modular type of instrument may be satisfactory. With such instruments the operator simply plugs in various amplifier "black boxes" to construct a potentiostat, function generator, or signal-conditioning device [95–98]. Hayes and Bauer [99] have tested one of these commercial instruments for phase-sensitive ac polarography.

As a last resort, the experienced electrochemist must build his own instrument, as many have done [5, 100–108]. Many detailed circuit diagrams have been published [100–103, 106, 109], together with operational explanations, and it is not too difficult to adapt these to the instrumentation requirements of a particular application.

It is convenient to divide the electrochemical instrument into three units for discussion purposes. The first unit is the function generator which creates the input potential signals. The second unit, the potentiostat, applies these signals to the cell and monitors the cell response. This response is amplified, filtered, or processed in some other way by the third unit, the signal-conditioning circuit, before output to some recording device. These sections of the instrument and their relation to the chemical cell are shown in Fig. 6.28. The

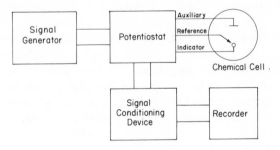

Fig. 6.28 Generalized electrochemical instrument.

desired potential function is applied to the cell through the driving electrode. The cell current resulting from this applied potential is monitored by the potentiostat at the indicator electrode.

Some essential components of cyclic voltammetric and ac polarographic instrumentation are described for each of the three basic units mentioned above.

Signal Generation

It is convenient to have one instrument that can generate both triangular and sine waves, and several instrument companies manufacture them. A triangular wave is formed by integrating a square wave, and the sine wave is formed by "shaping" the triangular wave with the diode circuit. This shaping causes a 2 to 3% distortion in the wave which appears mainly at the second and third harmonic frequencies but does not interfere with fundamental ac polarographic measurements. However, the harmonic components of the input signal generate signals in the cell that are larger than the second harmonic current signal generated from the fundamental input signal. This makes accurate measurements of higher harmonic signals impossible. Generally, a second harmonic content of 0.2% or greater cannot be tolerated in an oscillator. Such an input signal can be used only if it is filtered by an amplifier tuned to the fundamental frequency. The amplifier passes the fundamental frequency signal with gain and attenuates the higher harmonic components of the signal. Sine wave oscillators are available currently from Krohn-Hite and Hewlett-Packard, for example, with sine wave outputs that have harmonic distortions of less than 0.1% occurring mainly at the second harmonic.

A triangular wave ideally should allow the scan rate to be adjusted independently of the amplitude setting. In most instruments, however, it is the frequency that is controlled, not the scan rate. This means that as the amplitude is increased at a fixed frequency, the scan rate or slope of the wave also increases. Of course, if the amplitude is held constant, the scan rate will be directly proportional to the frequency of the triangular wave. One would also like to be able to generate only one cycle or triangular potential function at a time for some experiments, that is, single-cycle voltammetry. In the instruments made by Exact, and Hewlett-Packard, such a single cycle can be initiated by an external pulse, or by a manual push button for slow sweeps. The initial dc potential and the initial direction of the sweep can be controlled, making these instruments ideally suited to the single-cycle voltammetric experiment.

Several circuit diagrams have been published in the chemical literature and elsewhere describing triangular wave generators [101, 102, 109] and sine wave oscillators [101, 110–112]. Sine wave oscillators are generally designed using a tuned passive element such as twin-T network or a Wien Bridge element and an amplifier employing positive feedback. Large amounts of negative feedback are also used to help stabilize the output amplitude. Discussions of these circuits are given in the manufacturer's literature [110, 111].

Signal Conditioning

In cyclic voltammetry a varying dc potential is applied to the chemical cell, and the output of interest is a dc current. The measurement of such currents is discussed in the section on potentiostat systems.

In ac polarography the signal of interest from the potentiostat is a periodically varying potential. This signal should be filtered to remove the higher harmonic components generated by the chemical cell. The next step is full-wave rectification of the ac signal to generate a dc component for convenient readout. The last step is removal of the ac ripple on the rectified signal using a low-pass filter.

Figure 6.29 shows a circuit for conveniently performing the above operations. Amplifier 1 and its associated feedback components is a twin-T tuned

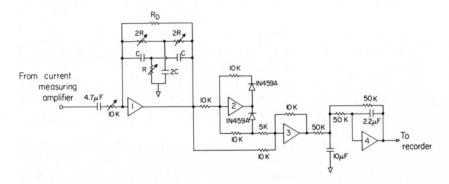

Fig. 6.29 Signal conditioning circuit for ac polarography. Frequency of tuned amplifier (No. 1) is $(4\pi RC)^{-1}$.

amplifier [5, 101, 112, 113] for initially filtering the current signal from the potentiostat. This configuration has been used successfully in ac polarography [114, 115] and particularly for second harmonic ac polarography [113, 115–118]. The frequency response is changed by varying the resistor values in the twin-T feedback network.

The output of the tuned amplifier now goes to a full-wave rectifier circuit represented by amplifiers 2 and 3. This circuit has been described elsewhere [112, 119, 120] and is used to accurately rectify an ac signal up to about 5 kHz, depending on the amplifiers used. The dc output is linear with ac input in this frequency range.

The output of the full-wave rectifier contains a large ac ripple which should be filtered before presentation to a recorder. One way of doing this is to put a large capacitor (1 to 20 μF) in the feedback of amplifier 3 [112]. However, better filtering is accomplished, particularly at low frequencies, by a higher-order active filter. Amplifier 4 is wired as a second-order filter with a Butterworth response [121]. The cutoff frequency of this filter, determined from the components used [121], is about 1.5 Hz, which effectively attenuates all ac ripple above 10 Hz yet does not attenuate the signal resulting from growth and fall of the DME [118], which has a fundamental frequency near 0.2 Hz (5-sec drop life).

In addition to the total current, a second parameter of importance in ac polarography is the phase angle, the significance of which is brought out in Section 14.

The phase angle can be measured point-by-point through observing the Lissajou pattern on an oscilloscope [113, 122, 123] or continuously [115, 124]. However, the phase angle need not be measured directly. Instead, the current component in phase with the applied potential can be measured by using a phase-sensitive rectifier. This is a full-wave rectifier in which a switch chops the incoming signal at the frequency of interest. This amounts to multiplying the sine wave by a square wave, and a Fourier analysis shows that the resulting dc component is proportional to the phase angle between the sine wave and the reference square wave [5].

Commercial instruments such as the Univector polarograph [97] use this technique for phase-sensitive and frequency-selective measurement [5, 125]. A circuit of this type has been used by Smith and co-workers in phase-sensitive ac polarography [5, 118].

Direct analog multiplication of the reference (potential input) signal and the current signal in ac polarography also produces a dc component proportional to the phase angle between the two signals [115]. This method has been successfully employed by Hayes and Reilley [106] to measure both the resistive and quadrature component of the ac current. Princeton Applied Research (PAR) lock-in amplifiers have been designed using the same concept [125]. Recently, an ac polarograph using a PAR lock-in amplifier as a phase-sensitive detector was described [126].

Operational amplifier manufacturers have recently developed multiplying modules with various degrees of accuracy, phase shift, and band pass [127–129]. Presumably these modules can be used with the same results, and perhaps more convenience, than some of the systems cited above for phase-sensitive ac current measurement.

Potentiostat Instrumentation

Many potentiostats designed for use with a three-electrode cell configuration have been described in the literature for ac polarography [106, 115, 130] and for cyclic voltammetry [130]. More recently, potentiostats with instrumental compensation of nonfaradaic effects have been described [107, 118, 131, 132] for use with several electrochemical techniques in addition to those under consideration here. One nonfaradaic effect to be corrected for is the iR potential drop between the reference electrode and the working electrode in the chemical cell resulting from solution and electrode resistance. The potential drop causes a loss in accurate potential control at the working electrode and can become extremely important in electroanalytical studies [5, 133–135].

In ac polarography, the raw data can be corrected for the iR potential losses by taking phase relations into account, but it is a tedious, point-by-point process. The problem of iR potential losses in cyclic voltammetry is even more insidious. In this case the potential drop must be considered in the boundary condition of applied cell potential [136]. In the region of a large current, the iR drop can become large enough to significantly alter the linearity of the applied potential scan which is an important parameter in the quantitative analysis of data. Nicholson has treated this problem for a reversible electrochemical reaction [136] and has shown that the effects of uncompensated iR drop are similar to those of a slow charge transfer reaction.

A second nonfaradaic effect that can be compensated for instrumentally is the charging current resulting from charging and discharging the double-layer capacities associated with the working electrode. Again, this nonfaradaic current can be subtracted manually from the raw data.

Another reason for compensation for the iR drop in the cell is to decrease the time constant of the cell and potentiostat system. This means that potentials can be applied more rapidly to the cell without the lag associated with charging a capacitor (double layer) through a resistor (cell resistance). Booman [137] has discussed the relationship between cell time constants and potentiostat time constants with respect to minimizing, but not completely eliminating, cell resistance. More recently, Brown et al. [138] have shown how complete iR compensation can improve the frequency response of a potentiostat cell system for ac polarography. Koopman [139] and Schroeder [132] have shown the improvement that can be gained in response to potential step inputs by complete iR compensation.

Figure 6.30 shows a potentiostat for use with ac polarography or cyclic voltammetry to measure faradaic current by instrumentally correcting for the iR potential drop and instrumentally subtracting the charging current. A signal proportional to the cell current is fed back into the summing function of the potential control amplifier C. This signal is generated by the current measuring amplifier I and a portion of its output equal to the cell resistance is fed back. The only independent experiment required is measurement of the cell resistance contributed by the electrodes and the solution.

Cell B contains only the supporting electrolyte and the output of the current-measuring amplifier is subtracted from the output of the current-measuring amplifier in cell A, which contains the system of interest. The output of the subtractor is equal to only the faradaic current, assuming: (1) iR compensation is complete in both cells, (2) the addition of faradaic species in cell A has no effect on the double-layer structure, (3) the working electrode areas are the same, and (4) no adsorption occurs. A recent study of

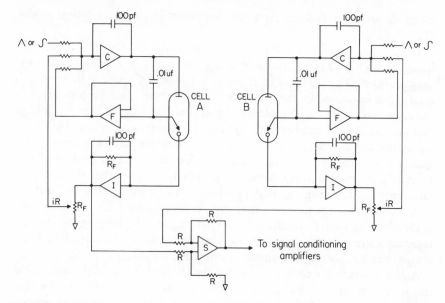

Fig. 6.30 Potentiostat with compensation for non-faradaic effects. Amplifier *C*—potential controlling amplifier. Amplifier *F*—follower amplifier. Amplifier *I*—current measuring amplifier. Amplifier *S*—subtractor amplifier. Cell *A*—reactant and supporting electrolyte. Cell *B*—supporting electrolyte only.

this total compensation of nonfaradaic effects [118] and studies with partial [140] or complete [132] iR compensation have shown the feasibility of the method.

When dropping mercury electrodes are used, it is necessary to synchronize their growth and fall in some way. This can be done by a simple timing circuit that drives two solenoids to knock the drops off mechanically. The column height of each mercury reservoir is adjusted to give equal flow rates in the two cells.

Electrodes and Cell Design

Much study in polarography has been directed toward designing electrodes and cells to yield current measurements that conform to the Ilkovic equation. These studies have attempted to minimize such effects as shielding (uneven current distribution resulting from geometric hindrance of mass transport), and creeping of solution into the mercury capillary. Other workers have extended these studies to techniques such as coulometry at mercury pool electrodes and stripping analysis at hanging mercury drop electrodes. Similar factors are important in ac polarography and cyclic voltammetry.

For fast, accurate response to potential changes applied to a chemical cell, the cell should be resistive in impedance and have no phase shift. In addition, the total cell impedance should be small. These two criteria are impossible to meet in a real cell, but they can be approached with good cell design. The effect of minimizing the double-layer capacitance has been emphasized recently by Koopman [139] and Brown et al. [138] in two separate theoretical studies on accuracy and stability of potentiostat-cell systems. The double-layer capacitance is proportional to electrode area, which is made as small as possible yet large enough to obtain easily measurable current signals.

The cell resistance should also be small to increase system stability [138] as well as increase potential control accuracy. Minimizing cell resistance is essential if instrumental iR compensation is not employed and is desirable even with instrumental compensation because of increased instrument stability when iR compensation is added [138, 140].

Solution resistance is decreased by placing the reference electrode tip as close as possible to the working electrode, without shielding the electrode [133]. The effects of iR drop in solution with various electrode placements and geometries has been studied by several workers [141–144]. In general. a fairly large potential gradient exists across the working electrode if the solution resistance is very large (nonaqueous media) or the current is very high (coulometry). These effects can be minimized by making the electrode arrangement as symmetrical as possible. The counter electrode should be large and surround the working electrode, if possible. This maintains the current lines and potential gradients in the solution as symmetrically as possible, even for large currents [143, 144]. In addition, the reference electrode should be placed near the working electrode surface where the current density is highest if the current lines are not symmetrical [143]. This minimizes the errors in potential at other points on the electrode.

The resistance of the working electrode itself is important, particularly with a DME, where the resistance of the mercury thread in the capillary may be as high as 100 Ω. One good way to reduce this resistance is to put a bulb at the bottom of the capillary with a small opening in the bottom [145]. A platinum wire is connected through the glass to the mercury in the bulb so that the electrical connection to the working electrode bypasses the capillary. The mercury drop then forms at a small hole in the bottom of the bulb, while the capillary controls the flow rate.

Other effects such as depletion, shielding, and "capillary response" should also be minimized. The term capillary response as used by von Stackelberg [145] refers to the spurious current signals resulting from solution entering the capillary after the mercury drop falls off and the capillary thread snaps back. This problem can be effectively eliminated by coating the electrode with an agent such as Beckman Desicote No. 18772, which when dry prevents the solution from wetting the inside of the capillary bore before the new mercury drop begins to grow.

Depletion is another important effect. Depletion occurs when the diffusion gradient formed around one mercury drop is not completely broken up by the time a new drop begins to grow. A blunt-tipped, vertical capillary from which the drop falls off does not eliminate the problem. In addition, this type of capillary does not allow the diffusion gradient to form either by planar or spherical diffusion [146]. Smoler [146] solved the problem of depletion effects by bending the tip of the capillary to a 45° angle. When the drop falls off, there is sufficient stirring at the electrode face to wash away the products of reaction occurring on that drop and return the concentration of the reacting species to essentially bulk concentration.

In another study of a DME, Triebel and Berg [147] found that grinding the capillary tip down similar to a pencil tip cut down the shielding of current lines and reduced the depletion effects as well. This technique was used by Underkofler and Shain [148] for studies with a hanging mercury drop. Recently, in faradaic impedance studies, the tapered electrode was found necessary to minimize the frequency dependence of measured double-layer capacitance at high frequency [149]. This indicates that a tapered electrode might be necessary for accurate measurements with ac polarography, particularly with subtraction, at high frequencies.

Figure 6.31 shows an electrochemical cell with electrode placement and geometry that can be used with ac polarography or cyclic voltammetry [150].

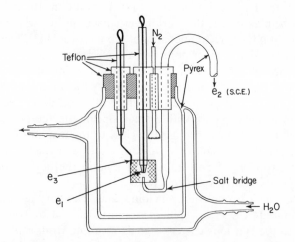

Fig. 6.31 Cell configuration for ac polarography and cyclic voltammetry (Courtesy Elsevier Publishing Co.).

The working electrode can be either a dropping mercury, a hanging mercury drop, or a solid electrode, as shown in the figure.

The above discussion on electrodes has referred implicitly to the DME, but all of the discussions apply to stationary electrodes also. One advantage of stationary electrodes is their much lower resistance. Even commercial electrodes which extrude a drop through a capillary with a micrometer dial use a bore large enough to keep the resistance down to less than 20 Ω.

14 RELATED TECHNIQUES

Ac Polarography and cyclic voltammetry are not the only electrochemical techniques that employ periodically varying potentials. Some of the other common techniques are briefly discussed in this section.

Faradaic impedance experiments preceded the use of ac polarography by many years, primarily because of the simple instrumentation required. The information obtained by both techniques is the same.

Techniques such as second harmonic polarography are a manifestation of the nonlinear potential-current behavior of the chemical cell. Other techniques based on this effect are "double-tone" or intermodulation ac polarography and faradaic rectification or "redoxokinetic" measurements. Neither of these last-mentioned two techniques is discussed here.

Other large-amplitude techniques such as oscillographic polarography are briefly mentioned to describe the scope of the method. References to more detailed discussions of these techniques are given in each section.

Faradaic Impedance

The expression relating the sine wave current to the applied alternating potential contains terms that can be equated with an impedance associated with the electrode reaction. This concept of faradaic impedance, a term first used by Grahame [151], has been known for many years. In fact, expressions for the faradaic impedance for the diffusion-controlled case were first derived by Warburg [152] at the turn of the century, hence the term Warburg impedance.

Under these conditions the impedance for a reversible redox couple can be considered to have a resistive and a capacitive component which are equal:

$$R_f = \frac{1}{\omega C_f} = \frac{RT}{n^2 F^2 A (2\omega)^{1/2}} \left[\frac{1}{C_O D_O^{1/2}} + \frac{1}{C_R D_R^{1/2}} \right]. \tag{6.123}$$

The Warburg impedance is then the vectorial combination of these two components:

$$Z_f = \left\{ R_f^2 + \left[\frac{1}{\omega C_f} \right]^2 \right\}^{1/2} = \frac{RT}{n^2 F^2 A \omega^{1/2}} \left[\frac{1}{C_O D_O^{1/2}} + \frac{1}{C_R D_R^{1/2}} \right]. \tag{6.124}$$

This expression has been obtained by Warburg [152], Grahame [151], Randles [153], Ershler [154], and Gerischer [155].

Under polarographic conditions in which the reactants are generated *in situ* at the electrode surface by the applied dc potential, the concentration terms C_O and C_R become potential dependent. If the dc concentration terms from (6.10a) and (6.10b) are substituted in (6.123) and (6.124), assuming C_R^* is zero, we have for the reversible case

$$R_f = \frac{1}{\omega C_f} = \frac{RT}{n^2 F^2 A (2\omega)^{1/2}} \left(\frac{e^{-j} + 2 + e^j}{C_O^* D_O^{1/2}} \right) = \frac{4RT \cosh^2 (j/2)}{n^2 F^2 A C_O^* (2\omega D_O)^{1/2}}. \quad (6.125)$$

This same expression can be obtained from (6.36), the expression for the fundamental harmonic current, by dividing $\Delta E \sin \omega t$ by $I(\omega t)$ [10].

From (6.125) it can be seen that the diffusion-controlled faradaic impedance is very large at negative potentials (large e^{-j}) and at positive potentials (large e^j), with a minimum impedance at the half-wave potential ($e^j = e^{-j} = 0$). This is reasonable if the meaning of the equivalent resistance and capacitance is considered. A resistor is a power dissipator and represents the loss of the product of the electrode reaction at the electrode surface, by diffusion in the reversible case. The capacitor is a storage device and the electrode reaction can be visualized as storing energy in the reaction product at the electrode surface because this material is available for the reverse electrolysis step. Both the "loss" impedance (resistive) and the storage impedance (capacitive) are inversely proportional to concentration of both species at the electrode surface (6.123), and at the half-wave potential these concentrations are equal. The impedance is then at a minimum. As the potential goes negative of $E_{1/2}$, the concentration of the oxidized form diminishes and the impedance increases; as the potential goes positive of $E_{1/2}$, the concentration of the reduced form diminishes and again the impedance increases. The fact that the impedance contains both a resistive and capacitive component means that there will be a phase shift between the potential applied to the cell and the resulting current. Because the values of the resistive and capacitive components change for different mechanisms, the phase shift varies also. This represents a valuable diagnostic tool as previous sections have shown.

In the reversible case in which both resistive and capacitive impedances are equal at any potential, the phase angle has a value of $\pi/4$ radians or $45°$. The cell current leads the applied potential by this value. For the case in which the charge transfer rate of the reaction is slow, the phase angle is always less than or equal to $45°$ at any frequency.

The faradaic impedance for the quasi-reversible case can be represented as having an additional resistive component in series with the Warburg (diffusion) impedance. As the charge transfer rate becomes slower, this resistance increases until it is the largest component of the faradaic reaction (irreversible reaction). Under these conditions the phase angle approaches zero. For intermediate values of the charge transfer rate (quasi-reversible), the faradaic impedance at the equilibrium potential is given by (6.126) and (6.127) [151–155]:

$$R_Q = \frac{RT}{n^2F^2A(2\omega)^{1/2}}\left[\frac{(2\omega)^{1/2}}{k_s} + \frac{1}{C_O D_O^{1/2}} + \frac{1}{C_R D_R^{1/2}}\right] \qquad (6.126)$$

$$\frac{1}{\omega C_Q} = \frac{RT}{n^2F^2A(2\omega)^{1/2}}\left[\frac{1}{C_O D_O^{1/2}} + \frac{1}{C_R D_R^{1/2}}\right]. \qquad (6.127)$$

The difference between R_Q and $1/C_Q$ is a measure of the charge transfer rate constant, just as in ac polarography the phase angle is used:

$$R_Q = \frac{1}{\omega C_Q} = \frac{RT}{n^2F^2Ak_s}. \qquad (6.128)$$

This equation has been used by Randles to measure the rates of some charge transfer reactions of metal ions [153, 156] and some organic oxidations, notably p-phenylenediamine and hydroquinone [157].

In line with the above reasoning, it might be predicted that coupled chemical reactions would affect both the resistive and capacitive components of the faradaic impedance. If the product of the electrode reaction reacts chemically, this is another pathway for loss of material at the electrode and represents another source of resistance. By the same reasoning a catalytic regeneration of the oxidized species from the reduced species might enhance the capacitive component. This is in general the case, and Gerischer [155] has calculated the faradaic impedance for the preceding chemical reaction, showing the frequency dependance of both the capacitive and resistive components at the equilibrium potential.

An equivalent circuit for a chemical cell including the nonfaradaic cell resistance and double-layer capacitance is shown in Fig. 6.32. One weakness

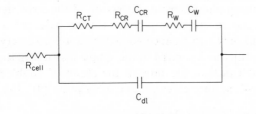

Fig. 6.32 Diagram of components of faradaic impedance. R_{cell}—cell resistance due to solution and electrodes. C_{dl}—double layer capacitance. R_w and C_w—Warburg impedance components. R_{CT}—charge transfer resistance. R_{CR} and C_{CR}—impedance due to coupled chemical reactions.

of this representation is that the various faradaic impedances change with $\omega^{1/2}$ while a real resistance is constant and capacitance changes with ω. A better representation of these impedances is a transmission line, as pointed out by Barker [158]. Delahay [159] also notes that the transient response of a transmission line is the same as for a faradaic process and makes this type of equivalent circuit a more relevant one.

The classic conditions for measuring faradaic impedance include having equal amounts of both forms of the redox couple present in solution. In this way the system is at equilibrium (no dc current flows) and the potential is the thermodynamic equilibrium potential (assuming equal diffusion coefficients). A small ac potential is applied to the cell, which is placed in the arm of a Wien bridge. The bridge is balanced to obtain the resistive and capacitive components of the cell impedance.

To make accurate measurements of the faradaic impedance, the effects of nonfaradaic impedances C_{dl} and R_{cell} shown in Fig. 6.32 must be minimized or corrected for. Delahay [160] has given a method of vectorial correction of the total cell impedance provided that the double-layer capacitance and cell resistance are measured, generally in the absence of reacting species. These terms can be measured in a cell containing reactant by extrapolating to high frequency [161]. More recently, Elliott and Buchanan [162] have corrected for these effects by using two cells in the bridge arms. One contains only supporting electrolyte and the other contains the reactants and supporting electrolyte. The balancing network is set up so that with the blank cell in the bridge, C_{dl} and R_{cell} are balanced. The reactant cell is put in the circuit and the faradaic impedances balanced. Provision is also made to balance the circuit if there is adsorption of the supporting electrolyte, for instance, specific adsorption of perchlorate ion [156, 162]. In this manner there is no tedious calculation needed to obtain the faradaic impedance values.

Bridge balance is generally accomplished by observing the amplified error signal (sine wave) on an oscilloscope and minimizing this signal by first adjusting capacitance, then resistance, then readjusting capacitance, and so on [163]. More recently, Thirsk and co-workers [164] have pointed out the advantages of using phase-sensitive detection of the error signal. In this way the capacitive and resistive components can be balanced independently of one another and only two balance adjustments are necessary.

The bridge technique has merit in the simple instrumentation required and is finding a good deal of use now in the study of double-layer relaxation phenomena [164], surface excesses, and adsorption [91, 162, 165].

Second Harmonic Ac Polarography

Harmonic current signals occur in a faradaic cell owing to the nonlinearity of the faradaic impedance; that is, the dc current is not linear with dc potential except near the equilibrium potential where a linear approximation can be used. The magnitude of these harmonic currents and associated dc rectification current [158, 159] can be calculated from (6.31) for higher values of p for the reversible case.

For $p = 2$ the second harmonic current is found in the same manner as for the fundamental case and is given by (6.129) [23, 166]:

$$I(2\omega t) = Z(2\omega)W(\omega) \sin (2\omega t - \pi/4), \qquad (6.129)$$

where

$$Z(2\omega) = \frac{n^2 F^2 A C_O{}^*(2\omega D_O)^{1/2} \Delta E}{4RT \cosh^2 (j/2)} \qquad (6.130)$$

$$W(\omega) = \frac{nF \Delta E \sinh (j/2)}{4RT \cosh (j/2)} \qquad (6.131)$$

Smith and Reinmuth [23] have shown experimentally that the second harmonic current is proportional to $\omega^{1/2}$ and to $(\Delta E)^2$ for the reversible reduction of ferric ion in oxalate media [43]. Smith [115] and Kooyman and Sluyters [167] have shown that the phase angle of the current lags the fundamental applied potential by $45°$ $(\pi/4)$ and the current changes sign ($180°$ phase shift) at the half-wave potential. The current magnitude at $E_{1/2}$ becomes zero, a manifestation of the linearity of the cell at this potential.

Figure 6.33 shows a second harmonic polarogram of the reversible reduction of nitrobenzene in acetonitrile to form the free radical [72]. The peak heights are equal as predicted by (6.129) [23], and the current goes to zero

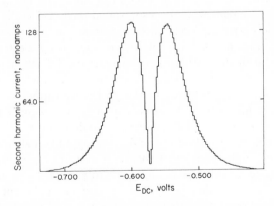

Fig. 6.33 Second harmonic ac polarogram of nitrobenzene in acetonitrile; 3.5 mM TEAP, 0.38 mM nitrobenzene, $\triangle E$ applied = 10.0 mV p-p at 23 Hz. Current measured at 46 Hz with complete iR compensation and no charging current subtraction. (Courtesy of Anal. Chem., Ref. 24.)

at the half-wave potential. The theoretical peak separation of $68/n$ mV [5] is not quite obeyed here, probably because of the extremely low supporting electrolyte concentration [24].

The theory for a quasi-reversible charge transfer mechanism has been derived by Paynter [168] and discussed by McCord and Smith [169]. A similar result is obtained by Bauer et al. [170], assuming that the dc process is reversible.

Recently, McCord and Smith [44] have presented an experimental study of two systems, $Cr(CN)_6^{3-}/Cr(CN)_6^{2-}$ in 1 M KCN and $Cd(II)/Cd(Hg)$ in 1 M Na_2SO_4, using second harmonic ac polarography. Their results for k_s and α agree well with previous studies of these systems. They also show that the experimental curves obtained are in excellent agreement with theory over an extended range of frequencies.

Smith has also described briefly some equations for second harmonic currents in systems with coupled chemical reactions [5], mainly to show their form and complexity. Recently, a general theory for these types of mechanisms utilizing the expanding plane model has been presented [171].

Higher harmonic current signals have been measured in the chemical cell [120, 168], but the experimental difficulty associated with such measurements has precluded any utility of this phenomenon.

Oscillographic Polarography

This technique, first introduced by Heyrovsky [172] in 1941, is similar to cyclic voltammetry in that a large-amplitude current function is used to polarize the cell instead of a large-amplitude potential function. The resulting output is potential measured as a function of time. The function may be a square wave, a triangular wave, or a sinusoidal wave. Generally, a sine wave is used and the current is large enough that the potential excursions are between the limits imposed by solvent or supporting electrolyte on the cathodic side and by the dissolution of mercury on the anodic side.

When a reactive species is placed in solution, the potential holds briefly while this species is electrolyzed, similar to the transition time in conventional chronopotentiometry. The sensitivity of the technique is enhanced by taking the derivative of potential with respect to time [173]. The minimum of the derivative occurs very close to the dc polarographic half-wave potential for a reversible couple, so it can be used to measure reversibility. A more convenient method for this type of analysis is to plot dE/dt against the potential of the cell at any instant [174]. Thus potentials of the minimum dE/dt can be determined quite accurately. The three types of experimental curves, E versus t, dE/dt versus t, and dE/dt versus E, that are obtained are shown in Fig. 6.34. This technique has found widespread acceptance in Czechoslovakia, where it originated, and there are several commercial instruments available

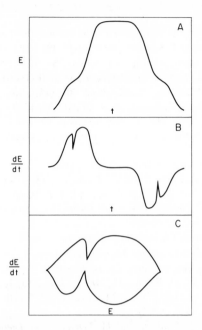

Fig. 6.34 Generalized experimental oscillographic curves. A, cell potential as a function of time. B, derivative of Curve *A* as a function of time. C, derivative of Curve *A* as a function of cell potential.

for studies employing the method. Recent monographs by Kalvoda and Heyrovsky [175, 176] summarize the theory and practice of this electrochemical tool.

References

1. A. A. Vlcek, *Progress in Polarography*, Vol. I, Interscience, New York, 1962, p. 269.
2. A. A. Vlcek, *Progress in Inorganic Chemistry*, Vol. 5, Interscience, New York, 1963, p. 211.
3. H. S. Carslaw and J. C. Jaeger, *Conduction of Heat in Solids*, Oxford University Press, London, 1947.
4. I. M. Kolthoff and J. J. Lingane, *Polarography*, 2nd ed., Interscience, New York and London, 1952.
5. D. E. Smith, *Electroanalytical Chemistry*, A. J. Bard, Ed., Vol. 1, Marcel Dekker, New York, 1966, Chapter 1.
6. P. Delahay, *Double Layer and Electrode Kinetics*, Interscience, New York, 1965.
7. D. M. Mohilner, *Electroanalytical Chemistry*, A. J. Bard, Ed., Vol. 1, Marcel Dekker, New York, 1966, p. 241.

8. R. S. Nicholson and I. Shain, *Anal. Chem.*, **36**, 706 (1964).
9. R. S. Nicholson and I. Shain, *Anal. Chem.*, **37**, 178 (1965).
10. R. S. Nicholson and I. Shain, *Anal. Chem.*, **37**, 190 (1965).
11. D. S. Polcyn and I. Shain, *Anal. Chem.*, **38**, 370 (1966).
12. D. S. Polcyn and I. Shain, *Anal. Chem.*, **38**, 376 (1966).
13. R. H. Wopschall and I. Shain, *Anal. Chem.*, **39**, 1514 (1967).
14. H. Matsuda and Y. Ayabe, *Z. Elektrochem.*, **59**, 494 (1955).
15. A. Y. Gokhshtein and Y. P. Gokhshtein, *Dokl. Akad. Nauk, SSSR*, **131**, 601 (1960).
16. W. H. Reinmuth, *Anal. Chem.*, **32**, 1891 (1960).
17. A. Sevcik, *Collection Czech. Chem. Commun.*, **13**, 349 (1948).
18. J. E. B. Randles, *Trans. Faraday Soc.*, **44**, 327 (1948).
19. R. Adams, *Electrochemistry at Solid Electrodes*, Marcel Dekker, New York, 1969.
20. M. S. Shuman, *Anal Chem.*, **41**, 142 (1969).
21. H. Matsuda, *Z. Elektrochem.*, **62**, 977 (1958).
22. W. H. Reinmuth, *Anal. Chem.*, **34**, 1446 (1962).
23. D. E. Smith and W. H. Reinmuth, *Anal. Chem.*, **33**, 482 (1961).
24. E. R. Brown, H. L. Hung, T. G. McCord, D. E. Smith, and G. L. Booman, *Anal. Chem.*, **40**, 1424 (1968).
25. S. Glasstone, K. J. Laidler, and H. Eyring, *The Theory of Rate Processes*, McGraw-Hill, New York, 1941, p. 575.
26. P. Delahay, *New Instrumental Methods in Electrochemistry*, Interscience, New York, 1954.
27. H. H. Bauer, *J. Electroanal. Chem.*, **16**, 419 (1968).
28. R. S. Nicholson, *Anal. Chem.*, **37**, 1351 (1965).
29. S. P. Perone, *Anal. Chem.*, **38**, 1158 (1966).
30. N. Tanaka and R. Tamamushi, *Electrochem. Acta*, **9**, 963 (1964).
31. H. L. Hung and D. E. Smith, *Anal. Chem.*, **36**, 922 (1964).
32. H. Matsuda, *Z. Elektrochem.*, **62**, 977 (1958).
33. D. E. Smith, T. G. McCord, and H. L. Hung, *Anal. Chem.*, **39**, 1149 (1967).
34. R. Tamamushi and N. Tanaka, *Z. Physik. Chem.*, **21**, 89 (1959).
35. B. Breyer and H. H. Bauer, *Alternating Current Polarography and Tensammetry in Chemical Analysis*, P. S. Elving and I. M. Kolthoff, Eds., Vol. 13, Interscience, New York, 1963.
36. P. Delahay, *Advances in Electrochemistry and Electrochemical Engineering*, P. Delahay and C. W. Tobias, Eds., Vol. 1, Interscience, New York, 1961, Chapter 5.
37. T. Berzins and P. Delahay, *Z. Elektrochem.*, **59**, 792 (1955).
38. D. E. Smith, *Anal. Chem.*, **35**, 610 (1963).
39. J. K. Frischmann and A. Timnick, *Anal. Chem.*, **39**, 507 (1967); J. D. McLean and A. Timnick, *Anal. Chem.*, **39**, 1669 (1967).
40. D. E. Smith and H. L. Hung, private communication, 1968.
41. D. E. Smith, Ph.D. Thesis, Columbia University, New York, 1961.
42. E. R. Brown, H. L. Hung, T. G. McCord, D. E. Smith, and G. L. Booman, *Anal. Chem.*, **40**, 1424 (1968).

43. J. E. B. Randles and K. W. Somerton, *Trans. Faraday Soc.*, **48**, 937 (1952).
44. T. G. McCord and D. E. Smith, *Anal. Chem.*, **41**, 131 (1969).
45. J. E. B. Randles, *Canadian J. Chem.*, **37**, 238 (1959).
46. P. Delahay, *J. Am. Chem. Soc.*, **75**, 1190 (1953).
47. G. Sundholm, *Electrochim. Acta*, **13**, 2111 (1968).
48. J. R. Delmastro and D. E. Smith, *J. Electroanal. Chem.*, **9**, 192 (1965).
49. B. Timmer, M. Sluyters-Rehback, and J. H. Sluyters, *J. Electroanal. Chem.*, **14**, 169 (1967).
50. B. Timmer, M. Sluyters-Rehbach, and J. H. Sluyters, *J. Electroanal. Chem.*, **14**, 181 (1967).
51. D. E. Smith and T. G. McCord, *Anal. Chem.*, **40**, 474 (1968).
52. L. Gierst and R. Cornelissen, *Collection Czech. Chem. Commun.*, **25**, 3004 (1960).
53. J. M. Saveant and E. Vianello, *Electrochim. Acta*, **10**, 905 (1965).
54. T. J. McCord and D. E. Smith, *Anal. Chem.*, **40**, 1959 (1968).
55. H. L. Hung, J. R. Delmastro, and D. E. Smith, *J. Electroanal. Chem.*, **7**, 1 (1964).
56. T. G. McCord, H. L. Hung, and D. E. Smith, *J. Electroanal. Chem.*, **21**, 5 (1969).
57. D. E. Smith, *Anal. Chem.*, **35**, 602 (1963).
58. J. Heyrovsky and J. Kuta, *Principles of Polarography*, Academic Press, New York, 1966.
59. T. G. McCord and D. E. Smith, *Anal. Chem.*, **41**, 116 (1969).
60. J. R. Kuempel and W. B. Schaap, *Inorg. Chem.*, **7**, 2435 (1968).
61. G. H. Aylward and J. W. Hayes, *Anal. Chem.*, **37**, 195 (1965).
62. K. Matsuda and R. Tamamushi, *Bull. Chem. Soc. Japan*, **41**, 1563 (1968).
63. D. G. Davis and D. T. Orleron, *Anal. Chem.*, **38**, 179 (1966).
64. H. E. Stapelfeldt and S. P. Perone, *Anal. Chem.*, **41**, 623 (1969).
65. S. P. Perone and W. T. Kretlow, *Anal. Chem.*, **38**, 1761 (1966).
66. M. L. Olmstead, R. G. Hamilton, and R. S. Nicholson, *Anal. Chem.*, **41**, 260 (1969).
67. M. Mastragostino, L. Nadjo, and J. M. Saveant, *Electrochimica Acta*, **13**, 721 (1968).
68. G. H. Aylward, J. W. Hayes, and R. Tamamushi, *Proceedings of the First Australian Conference on Electrochemistry 1963*, J. A. Friend and F. Gutman, Eds., Pergamon Press, Oxford, 1964, pp. 323–331.
69. D. N. Bailey, D. M. Hercules, and D. K. Roe, *J. Electrochem. Soc.*, **116**, 190 (1969).
70. T. G. McCord and D. E. Smith, *Anal. Chem.*, **41**, 116 (1969).
71. H. L. Hung, J. R. Delmastro, and D. E. Smith, *J. Electroanal. Chem.*, **7**, 1 (1964).
72. M. E. Peover, *Electroanalytical Chemistry*, A. J. Bard, Ed., Vol. 2, Marcel Dekker, New York, 1967, Chapter 1.
73. J. M. Saveant and E. Vianello, in *Advances in Polarography*, I. S. Longmuir, Ed., Vol. I, Pergamon Press, New York, 1960.
74. A. Blazek and J. Koryta, *Collection Czech. Chem. Commun.*, **18**, 326 (1953).

75. P. Delahay, C. C. Mattax, and T. Berzins, *J. Am. Chem. Soc.*, **76**, 5319 (1954).
76. O. Fisher, O. Dracha, and E. Visherova, *Collection Czech. Chem. Commun.*, **26**, 1505 (1961).
77. J. M. Saveant, *Electrochim. Acta*, **12**, 753 (1967).
78. E. T. Seo, R. F. Nelson, J. M. Fritsch, L. S. Marcoux, D. W. Leedy, and R. N. Adams, *J. Am. Chem. Soc.*, **88**, 3498 (1966).
79. M. D. Hawley and S. W. Feldberg, *J. Phys. Chem.*, **70**, 3459 (1966); R. N. Adams, M. D. Hawley, and S. W. Feldberg, *J. Phys. Chem.*, **71**, 851 (1967).
80. M. Mastragostino and J. M. Saveant, *Electrochim. Acta*, **13**, 751 (1968).
81. R. F. Nelson, *J. Electroanal. Chem.*, **18**, 329 (1968).
82. R. Parsons, *J. Electroanal. Chem.*, **5**, 397 (1963); **7**, 136 (1964).
83. R. H. Wopschall and I. Shain, *Anal. Chem.*, **39**, 1527 (1967).
84. R. H. Wopschall and I. Shain, *Anal. Chem.*, **39**, 1535 (1967).
85. B. Breyer and H. H. Bauer, *Australian J. Chem.*, **8**, 467, 472, 480 (1955).
86. H. H. Bauer, Doctoral Thesis, Univ. of Sidney, Sidney, Australia, 1955.
87. S. L. Gupta, *Kolloid-Z.*, **160**, 30 (1958); S. L. Gupta and S. K. D. Agarwal, *Kolloid-Z.*, **163**, 136 (1959).
88. W. Lorenz and E. O. Schmalz, *Z. Elektrochem.*, **62**, 301 (1958).
89. B. Breyer and H. H. Bauer, *Australian J. Chem.*, **9**, 437 (1956).
90. B. Breyer and S. Hacobian, *Australian J. Sci. Res.*, A5, 500 (1952).
91. B. Breyer and H. H. Bauer, *Alternating Current Polarography and Tensammetry*, Interscience, New York, 1963.
92. Beckman Instruments, Inc., Bulletin 7076-A, "Electroscan 30 Electroanalytical System," and Bulletin 7079, "An Introduction to Electroanalysis," 1966, Fullerton, California.
93. PAR Electrochemistry Instrument Model 170 and Model 171.
94. Heath Company, "Polarographic Module EUA-19-2," Benton Harbor, Michigan.
95. McKee-Pedersen Instruments, "MP-System 1000, Operation and Applications," 2nd ed., Danville, California, 1966.
96. A. R. F. Products, Inc., "Analytical Instruments," Raton, New Mexico.
97. G. A. Philbrick Researches, "Model RP Operational Manifold," Dedham, Massachusetts, 1967.
98. G. Jessop, Brit. Patent 640,768 (1950); Brit. Patent 776,543 (1957).
99. J. W. Hayes and H. H. Bauer, *J. Electroanal. Chem.*, **3**, 336 (1962).
100. R. C. Probst, "A Multipurpose Instrument for Electrochemical Studies," E. I. du Pont de Nemours and Co., Savannah River Laboratory, Aiken, South Carolina, USAEC Report DP-903 (1964).
101. C. F. Morrison, "Generalized Instrumentation for Research and Teaching," Washington State Univ., Pullman, Washington, 1964.
102. Operational Amplifiers Symposium, *Anal. Chem.*, **35**, 1770 (1963).
103. E. B. Buchanan and B. McCarten, *Anal. Chem.*, **37**, 29 (1965).
104. Symposium on Electroanalytical Instrumentation, *Anal. Chem.*, **38**, 1106 (1966).
105. D. M. Oglesby, S. H. Omang, and C. N. Reilley, *Anal. Chem.*, **37**, 1312 (1965).

106. J. W. Hayes and C. N. Reilley, *Anal. Chem.*, **37**, 1322 (1965).
107. D. Pouli, J. R. Huff, and J. C. Pearson, *Anal. Chem.*, **38**, 382 (1966).
108. M. T. Kelley, D. J. Fischer, and H. C. Jones, *Anal. Chem.*, **31**, 1475 (1959).
109. W. D. Weir and C. G. Enke, *Rev. Sci. Instr.*, **35**, 833 (1964).
110. Krohn-Hite Corporation Catalog, Section II, Cambridge, Massachusetts, 1967.
111. Hewlett Packard, 1968 Instrumentation Catalog, Palo Alto, California, p. 332.
112. "Handbook of Operational Amplifier Applications," Burr-Brown Research Corporation, Tuscon, Arizona, 1963.
113. D. E. Smith, *Anal. Chem.*, **35**, 1811 (1963).
114. D. E. Smith, *Anal. Chem.*, **35**, 610 (1963).
115. D. E. Smith and W. H. Reinmuth, *Anal. Chem.*, **33**, 482 (1961).
116. H. H. Bauer, *Australian J. Chem.*, **17**, 715 (1964).
117. R. Neeb, *Z. Anal. Chem.*, **186**, 53 (1962).
118. E. R. Brown, T. G. McCord, D. E. Smith, and D. D. DeFord, *Anal. Chem.*, **38**, 1119 (1967).
119. R. Howe, *Instr. Control Systems*, **34**, 1482 (1961).
120. E. R. Brown, Doctoral Thesis, Northwestern University, Evanston, Illinois, 1967.
121. G. A. Philbrich Researches, Inc., "The Lightning Empiricist," Vol. 13, Nos. 1 and 2, Dedham, Massachusetts, 1965.
122. H. H. Bauer and P. J. Elving, *J. Am. Chem. Soc.*, **82**, 2091 (1960).
123. D. E. Smith, presented at the 140th Meeting, Division of Analytical Chemistry, ACS, Chicago, September 1961.
124. Y. P. Yu, *Electronics*, **31**, 99 (Sept, 12, 1958).
125. R. D. Moore, *Electronics*, **35**, 40 (June 8, 1962).
126. R. F. Evilio and A. J. Diefenderfer, *Anal. Chem.*, **39**, 1885 (1967).
127. Analog Devices, "A Comprehensive Catalog and Guide to Operational Amplifiers," Cambridge, Massachusetts, January 1968.
128. Transmagnetics, Inc., "Short Form Catalog," Flushing, New York, 1967.
129. GPS Instrument Company, Inc., "New 400 Series Multipliers," Newton, Massachusetts, 1968.
130. W. L. Underkofler and I. Shain, *Anal. Chem.*, **35**, 1778 (1963).
131. C. N. Reilley, private communication, 1967.
132. R. R. Schroeder, Ph.D. Thesis, Univ. of Wisconsin, Madison, Wisconsin, 1967.
133. P. Delahay, *New Instrumental Methods in Electrochemistry*, Interscience, New York, 1954, pp. 132–134.
134. W. T. DeVries and E. von Dalen, *J. Electroanal. Chem.*, **10**, 183 (1965).
135. W. H. Reinmuth, *Anal. Chem.*, **36**, 211R (1964).
136. R. S. Nicholson, *Anal. Chem.*, **37**, 667 (1965).
137. G. L. Booman and W. B. Holbrook, *Anal. Chem.*, **37**, 795 (1965).
138. E. R. Brown, D. E. Smith, and G. L. Booman, *Anal. Chem.*, **40**, 1411 (1968).
139. R. Koopman, *Ber. Bunsenges, Phys. Chem*, **72**, 43 (1968).
140. G. Sauer and R. A. Osteryoung, *Anal. Chem.*, **38**, 1106 (1966).

141. R. deLevie, *J. Electroanal. Chem.*, **9**, 311 (1965).
142. W. B. Schaap and P. S. McKinney, *Anal. Chem.*, **36**, 29, 1251 (1964).
143. J. E. Harrar and I. Shain, *Anal Chem.*, **38**, 1148 (1966).
144. G. L. Booman and W. B. Holbrook, *Anal. Chem.*, **35**, 1793 (1963).
145. H. W. Nurnberg and M. von Stackelberg, *J. Electroanal. Chem.*, **2**, 181 (1961).
146. I. Smoler, *J. Electroanal. Chem.*, **6**, 465 (1963).
147. H. Triebel and H. Berg, *J. Electroanal. Chem.*, **2**, 467 (1961).
148. W. L. Underkofler and I. Shain, *Anal. Chem.*, **37**, 218 (1965).
149. G. Tessari, P. Delahay, and K. Holub, *J. Electroanal. Chem.*, **17**, 69 (1968).
150. W. R. Ruby and C. G. Tremmel, *J. Electroanal. Chem.*, **12**, 216 (1966).
151. D. C. Grahame, *J. Electrochem. Soc.*, **99**, 370C (1952).
152. E. Warburg, *Ann. Phys.*, **67**, 493 (1899); **69**, 125 (1901).
153. J. E. B. Randles, *Discussions Faraday Soc.*, **1**, 11 (1947).
154. B. V. Ershler, *Discussions Faraday Soc.*, **1**, 269 (1947); *Zh. Fiz. Khim.*, **22**, 683 (1948).
155. H. Gerischer, *Z. Physik. Chem.*, **198**, 286 (1951).
156. J. E. B. Randles and K. W. Somerton, *Trans. Faraday Soc.*, **48**, 937 (1952).
157. J. E. B. Randles and K. W. Somerton, *Trans. Faraday Soc.*, **48**, 951 (1952).
158. G. C. Barker, *Transactions of the Symposium on Electrode Processes, Phila.*, E. Yeager, Ed., Wiley, New York, 1961.
159. P. Delahay, *Advances in Electrochemistry and Electrochemical Engineering*, Vol. 1, P. Delahay, Ed., Interscience, New York, 1961, Chapter 5.
160. P. Delahay, *New Instrumental Methods in Electrochemistry*, Interscience, New York, 1954, Chapter 7.
161. R. deLevie, *Electrochim. Acta*, **10**, 395 (1965).
162. D. Elliott and G. S. Buchanan, *Anal. Chem.*, **39**, 1245 (1967).
163. D. C. Grahame, *J. Am. Chem. Soc.*, **71**, 2975 (1949).
164. R. D. Armstrong, W. P. Race, and H. R. Thirsk, *Electrochim. Acta*, **13**, 215 (1968).
165. H. A. Laitinen and J. E. B. Randles, *Trans. Faraday Soc.*, **51**, 54 (1955).
166. M. Senda and I. Tachi, *Bull. Chem. Soc. Japan*, **28**, 632 (1955).
167. D. J. Kooyman and J. H. Sluyters, *Rec. Trav. Chim.*, **83**, 587 (1964).
168. J. Paynter, Ph.D. Thesis, Columbia Univ., New York, 1964.
179. T. G. McCord and D. E. Smith, *Anal. Chem.*, **40**, 289 (1968).
170. H. H. Bauer and D. C. S. Foo, *Australian J. Chem.*, **19**, 1103 (1966); H. H. Bauer, *J. Sci. Res. (India)*, **24**, 372 (1965); H. H. Bauer, *Australian J. Chem.*, **17**, 715 (1964).
171. T. G. McCord and D. E. Smith, *Anal. Chem.*, **40**, 1967 (1968).
172. J. Heyrovsky, *Chem. Listy.*, **35**, 155 (1941).
173. J. Heyrovsky and J. Forejt, *Z. Physik. Chem.*, **193**, 77 (1943).
174. J. Heyrovsky, *Anal. Chim. Acta*, **2**, 533 (1948).
175. R. Kalvoda, *Techniques of Oscillopolarographic Measurements*, Elsevier, Amsterdam, 1964.
176. J. Heyrovsky and R. Kalvoda, *Oszillographische Polarographie mit Wechselstrom*, Akademie-Verlag, Berlin, 1960.

Chapter **VII**

VOLTAMMETRY WITH STATIONARY
AND ROTATED ELECTRODES

Stanley Piekarski and Ralph N. Adams

I INTRODUCTION

During the past two decades, there has been rapid growth in the theory and practice of electrochemistry at stationary electrodes. New electrode materials, such as wax-impregnated graphite, pyrolytic graphite, carbon paste, and boron carbide have been developed. Old electrode materials, platinum and gold, for example, have also been applied with good success. Nonaqueous solvents, primarily acetonitrile and dimethylformamide, have received extensive study. New techniques such as linear potential sweep voltammetry, cyclic voltammetry, and chronoamperometry have been applied to a wide variety of new problems. Parallel to this growth in practice has been the development of the theoretical understanding of these various techniques.

Voltammetry at rotating wire electrodes has been used for many years in analytical chemistry, especially for amperometric titrations. Largely as a result of the development of the theory of rotating disk electrodes by Levich and his school in the Soviet Union, voltammetry at rotating electrodes can now be applied to the study of electrode processes in a quantitative way.

Voltammetry is the study of current-potential relationships at micro-electrodes. Polarography is then a special case of voltammetry in which a dropping mercury electrode is the microelectrode. The dropping mercury electrode is excluded from this section since Müller discusses it in detail [1]. However, since the hanging mercury drop electrode is a stationary electrode, it is included in this chapter.

The limited anodic potential range of mercury has caused the renewed interest in solid electrodes. The oxidation of mercury occurs at $+0.4$ V versus SCE under the most favorable conditions [1, 2]. In supporting electrolytes that form insoluble salts or strong complexes with mercuric ion, the potential limit is considerably less anodic. This restriction is important since the oxidation of many organic compounds occurs at potentials more positive than $+0.4$ V.

Although several electrode materials can be polarized to potentials beyond +0.4 V none compares to the mercury electrode in the reproducibility of its surface. This shortcoming greatly restricts the use of solid electrodes. For example, while studying heterogeneous kinetics at a solid electrode, extreme care must be taken in the preparation of the electrode surface so that meaningful data result. In areas such as quantitative analysis and the study of chemical reactions coupled with the electron transfer, significant results have been obtained in recent years since the microscopic reproducibility is not crucial in most cases.

In discussing the theory of voltammetry, contrary to most treatments, oxidation processes are considered. Thus only the reductant (Red) is present in the solution at the start of the experiment. The extension of the treatment of reduction processes is obvious and in most cases requires only a change of sign. It is also assumed that both Ox and Red are soluble in the solution. The current-potential curves are plotted according to the usual polarographic conventions [1, 2]. The potential is plotted along the abscissa with the potential becoming more negative from left to right. Cathodic (reduction) current is positive by convention, while anodic (oxidation) current is negative. All potentials are given with respect to the saturated calomel electrode (SCE), which has a potential of +0.245 V versus the normal hydrogen electrode at 25.0°C [3].

2 STATIONARY ELECTRODES: THEORY

Mass Transport Processes

The voltammetric experiment is designed so that the mode of transport of the electroactive species to the electrode surface is well defined. The three important mass transport processes are migration, convection, and diffusion.

Migration is the movement of ions under the influence of an electrical field. The presence of migration serves only to complicate the results of the experiment. Migration is effectively eliminated by the addition of a 50- to 100-fold excess of an inert supporting electrolyte which serves to carry the current through the solution. Convection arises from accidental stirring of the solution and from density gradients. Convection can be minimized by using shielded electrodes.

Diffusion is the movement of ions or molecules caused by a concentration (activity) gradient. The movement is from a region of high concentration to one of low concentration. A consideration of the general reaction (7.1) shows how concentration gradients arise at stationary electrodes.

$$\text{Red} \rightleftharpoons \text{Ox} + ne^- \tag{7.1}$$

Prior to the initiation of the electrolysis, the concentration of Red is uniform throughout the solution. To eliminate convection, the solution is protected from vibrations for a period of at least 1 min prior to the electrolysis. When current flows, the concentration of Red at the electrode surface becomes less than that in the bulk of the solution. A concentration gradient for Ox is in the opposite direction; consequently, Ox diffuses away from the electrode surface.

The nature of the diffusion process depends on the shape of the electrode. Diffusion to a planar electrode immersed in a large amount of solution is called semiinfinite linear diffusion [4]. The flux of Red at time t at a distance x from the electrode surface is proportional to the concentration gradient. The proportionality constant D_R is called the diffusion coefficient. The flux of Red is given by Fick's first law of diffusion [4, 5]:

$$f(x, t) = D_R \frac{\partial [C_R(x, t)]}{\partial x}. \qquad (7.2)$$

The current flowing is dependent upon the flux at the electrode surface, that is, at $x = 0$. The current is negative because an oxidation process is being considered:

$$i = -nFAf(0, t). \qquad (7.3)$$

The number of electrons transferred for each mole of Red is n, F is Faraday's constant, and A is the area of the electrode. Thus if the flux at the electrode surface can be calculated, the current can also be calculated from (7.3). The time dependence of the concentration of Red is given by Fick's second law [4, 5]:

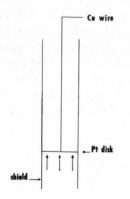

Fig. 7.1   Shielded electrode.

$$\frac{\partial [C_R(x, t)]}{\partial t} = D_R \frac{\partial^2 [C_R(x, t)]}{\partial x^2}. \qquad (7.4)$$

All voltammetric equations at planar, stationary electrodes are derived from these basic equations using the proper initial and boundary conditions characteristic of the particular technique.

Semiinfinite linear diffusion holds only if Red diffuses to the electrode in one direction (Fig. 7.1). Experimentally, this can be accomplished by using shielded electrodes such as the one shown in Fig. 7.1. With the use of shielded electrodes, semiinfinite linear diffusion has been observed for electrolysis times of greater than 10^3 sec [6–8].

Shielded electrodes are difficult to fabricate; therefore unshielded planar electrodes are widely used [8]. In aqueous solutions unshielded electrodes conform to semiinfinite linear diffusion for electrolysis times of less than 30 to 40 sec [7]. A particular system can be tested for semiinfinite linear diffusion by chronoamperometry [9]. A constant potential is applied to the electrode so the Red is oxidized immediately upon its arrival at the electrode surface. Under these conditions the current as a function of time is given by [7.5] [4, 10]:

$$i = - \frac{nFAD_R^{1/2}C_R}{\pi^{1/2}t^{1/2}}.$$ (7.5)

Table 7.1 The Influence of Convection on Current-Time Curves

t	$it^{1/2}$ (μA-sec$^{1/2}$)					Average Deviation	Average Deviation, (%)
	1	2	3	4	Average		
3.75	611	608	609	604	608	2	0.3
7.50	611	608	606	598	606	4	0.7
11.25	610	610	610	600	608	4	0.7
15.00	613	615	615	607	612	3	0.5
18.75	615	625	625	611	619	6	1.0
22.50	630	646	640	626	635	7	1.1
26.25	644	665	659	634	651	12	1.8
30.00	646	680	668	646	660	14	2.1
33.75	662	699	680	666	677	13	1.9
37.50	674	722	698	686	704	25	3.6

For a given experiment, $it^{1/2}$ is constant when semiinfinite linear diffusion holds. Table 7.1 shows a typical series of four chronoamperometric experiments for the oxidation of a 5×10^{-3} M solution of 4,5-dichlorocatechol in 2 M HClO$_4$ at a planar carbon paste electrode [11]. For 64 points taken between 3.75 to 15.0 sec, $it^{1/2}$ was equal to $609 \pm 4\mu$ A-sec$^{1/2}$. The average deviation is $\pm 0.7\%$. For electrolysis times beyond 15 sec, however, two effects can be observed: the $it^{1/2}$ constant increases rapidly, and the reproducibility among experiments is poorer. The electrolysis times beyond which semiinfinite linear diffusion no longer holds depend upon the density differences between the oxidized and reduced forms [6, 7] and the ability to eliminate external vibrations. Thus each system should be checked experimentally.

The diffusion equations for spherical and cylindrical electrodes are more complicated than those for planar electrodes [4]. For very short electrolysis times, the semiinfinite linear diffusion equations hold. With the availability of planar electrodes, wire electrodes are no longer widely used as stationary electrodes; consequently, cylindrical diffusion processes are ignored in this chapter. However, the hanging mercury drop electrode remains a very valuable one. Thus references to theoretical treatments of voltammetry at spherical electrodes are given. All equations given are for planar electrodes.

Linear Potential-Sweep Voltammetry for Uncomplicated Electron Transfers

Voltammetry with linear potential sweep can be used as a single- or multisweep technique (cyclic voltammetry). Stationary electrode voltammetry possesses several advantages over conventional polarography. The

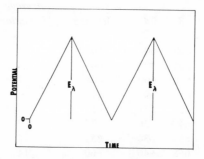

Fig. 7.2   Potential-time variation in cyclic voltammetry.

major advantage lies in the wide range of potential sweep rates that can be used. With pen and ink recording, sweep rates of 0.2–25 V/min can be used. With an oscilloscope sweep rates as high as 300,000 V/min have been used [12]. An experiment can be completed in a few seconds rather than the 5 to 10 min required in polarography. Stationary electrode voltammetry is also more sensitive than polarography [13, 14]. However, the analysis of mixtures is more difficult than in polarography because of uncertainties in the determination of the base line for the second process [14].

Cyclic voltammetry is a multisweep technique in which the potential is varied in the form of an isosceles triangle (Fig. 7.2). At $t = 0$ the potential sweep is begun at 0 V. The potential is swept at a constant rate until the switching potential E_λ is reached. At this point the direction is reversed, and the potential is swept back to 0 V. The potential sweep may be continued for as many cycles as desired. Cyclic voltammetry has become a most important tool in the investigation of organic electrode processes because it is capable of detecting electroactive intermediates [15–17]. These intermediates can be observed since a significant fraction of the product or products of the electrode process remains near the stationary electrode surface. These products can then be oxidized or reduced. Because of the wide range of sweep rates available, intermediates of high reactivity can be observed and their half-lives readily estimated.

Reversible Electron Transfers

The theoretical treatment for a reversible electron transfer at a stationary planar electrode has been developed by several workers [13, 18–23]. This

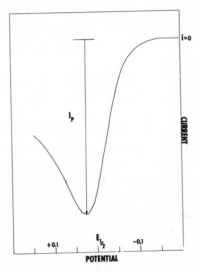

Fig. 7.3 Theoretical stationary electrode voltammogram for reversible system.

case has also been considered for spherical [20, 22–25] and cylindrical electrodes [26]. In the excellent study of Nicholson and Shain [23], extensive tables are given for the construction of theoretical voltammograms for uncomplicated electron transfers as well as for those coupled with homogeneous chemical reactions.

A voltammogram for a reversible oxidation process has the shape shown in Fig. 7.3. The potential sweep is begun cathodic of the formal potential ($E^{0'}$) of the couple, and the potential is swept in an anodic (positive) direction. As the formal potential is approached, the current increases rapidly until the peak (I_p) is reached and then the current decays. The rate of the electron transfer is exponentially related to the applied potential. Reductant diffuses toward the electrode surface because a concentration gradient is established.

Eventually, the rate of the electron transfer will be more rapid than the rate of diffusion of Red to the electrode surface; then the peak current is observed. Beyond the peak the current decays because the concentration gradient decreases as the diffusion layer becomes thicker. The diffusion layer is the distance from the electrode surface to the plane where the concentration of Red is equal to that of the bulk of the solution.

Equation (7.6) is the Randles-Sevcik equation which relates the peak current to the important experimental parameters[13, 18]:

$$i_p = -2.69 \times 10^5 n^{3/2} A D_R C_R V^{1/2}. \tag{7.6}$$

The constant given is from Nicholson and Shain [23] and has been verified experimentally [27, 28]. The number of electrons for each mole of Red is n, A is the area of the electrode in square centimeters, D_R is the diffusion coefficient of Red in square centimeters per second, C_R is the concentration in moles per cubic centimeter, V is the potential sweep rate in volts per second and i_p is the peak current in amperes which is negative because the process is an oxidation. In the more usual units of i_p in microamperes and C_R in moles per liter, (7.6) is:

$$i_p = -2.69 \times 10^8 n^{3/2} A D_R^{1/2} C_R V^{1/2}. \tag{7.7}$$

For a given sweep rate, the peak current is directly proportional to the concentration of Red in the solution. This fact can be used in analytical applications with a calibration curve of peak current versus concentration at constant sweep rate [13]. For a series of sweep rates and concentrations, $i_p/V^{1/2}C_R$ is a constant. This allows analytical work without the construction of a calibration curve.

Because the peak current increases with the one-half power of sweep rate, low concentrations of Red can be determined by increasing the sweep rate to high values. Stationary electrode voltammetry is more versatile than polarography in this respect. In polarography the mercury flow rate and the drop time cannot be varied widely. However, in stationary electrode voltammetry the current that results from the charging of the double layer increases more rapidly than the faradaic current. Calculations using reasonable values for the double-layer capacity, diffusion coefficient, and sweep rate have shown that the capacity current is important at concentrations of $10^{-5}\ M$ [29].

The formal reduction potential for the redox couple can be readily calculated from the current-potential curve [23]:

$$E_p = E^{0'} - \frac{RT}{nF} \ln \frac{D_O^{1/2}}{D_R^{1/2}} + 1.109 \frac{RT}{nF} \tag{7.8}$$

or

$$E_p = E_{1/2} + 1.109 \frac{RT}{nF}. \tag{7.9}$$

The $E_{1/2}$ that would be observed at a rotating disk electrode constructed from the same electrode material as the stationary electrode can be calculated from (7.9). If a planar mercury-plated or mercury pool electrode is used as the stationary electrode, the $E_{1/2}$ calculated from (7.9) is identical to that measured polarographically. The $E_{1/2}$ occurs at a potential where $i = 0.851i_p$ [23]. At 25.0°C, (7.9) is:

$$E_p = E_{1/2} + \frac{0.0285}{n}. \tag{7.10}$$

The precise determination of E_p is difficult since the observed peak is rather broad. However, $E_{p/2}$, the potential where $i = i_{p/2}$, can be used for the determination of $E_{1/2}$ [31]:

$$E_{p/2} = E_{1/2} - 1.09 \frac{RT}{nF}. \tag{7.11}$$

At 25.0°C, (7.11) is:

$$E_{p/2} = E_{1/2} - \frac{0.0280}{n}. \tag{7.12}$$

The difference between E_p and $E_{p/2}$ can be used as a criterion of reversibility for the electron transfer process. If n calculated from (7.13) (allowing for the uncertainties in the measurement of E_p) corresponds to the number of electrons transferred, then the electron transfer can be considered to be reversible:

$$E_p - E_{p/2} = \frac{0.0565}{n}. \tag{7.13}$$

The value of n has a large effect upon the shape of the current-potential curve. As n increases, the current-potential curve becomes sharper [see (7.13)]. Also, for two substances with identical diffusion coefficients, the value of the peak for a two-electron reversible process is 2.828 times that for a similar one-electron process (7.6).

Cyclic voltammetry can also be used to characterize reversible processes. The theory for cyclic voltammetry was originally derived for the steady-state conditions observed after many potential cycles [18, 32, 33]. Most cyclic voltammetric experiments are carried out for two to three cycles. The theory of cyclic voltammetry for the important first cycle has been given by Nicholson and Shain [23].

The first anodic sweep yields a peak $E_{p,a}$ which conforms to (7.6) through (7.13). However, the position of the cathodic peak for the reduction of Ox depends upon the switching potential E_λ. The position of $E_{p,c}$ as a function of $E_\lambda - E_{1/2}$ has been given [23]. If E_λ is more than $100/n$ mV anodic of $E_{1/2}$, the position of the cathodic peak becomes nearly independent of the switching potential. When $E_\lambda - E_{1/2}$ becomes much larger than $100/n$ mV $E_{p,a} - E_{p,c} = 58/n$ mV. Over the large range of switching potentials considered in the calculations, the position of $E_{p,c}$ varies only a total of 5 mV. Thus for reversible electron transfers $E_{p,a} - E_{p,c} = 60 \pm 2$ mV [23]. The deviation is usually within the error of measurement of the rounded peaks.

Measurement of Peak Heights

Figure 7.4 is a cyclic voltammogram for a reversible oxidation as recorded on an X-Y recorder. The measurement of the anodic peak current $i_{p,a}$ is

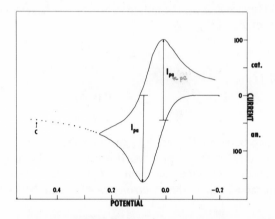

Fig. 7.4 Cyclic voltammogram for reversible electron transfer. Solution, 2.3 mM N,N,N',N'-tetramethyl-p-phenylenediamine; pH, 7.0; carbon paste electrode; sweep rate, 1.35 V/min.

simple. The current is measured from the residual current obtained from the solution containing only the supporting electrolyte. For a reversible process it is predicted that the ratio of the cathodic peak to the anodic peak current is 1.00 [23]. However, difficulties arise in the measurement of this ratio. The anodic current observed at the switching potential is not constant but is in fact decaying with time. Thus a simple base line from which to measure $i_{p,c}$ is not available. The cathodic peak current can be calculated in various ways.

The first method can be used successfully with a strip chart recorder with constant chart speed [23, 34]. The principle is illustrated in Fig. 7.5. A single-sweep voltammogram is run in an anodic direction. At the desired switching potential E, the potential sweep is stopped. The current continues to decay as a function of time, and this decay is recorded. A second anodic voltammogram is recorded. When E_λ is reached, the direction of the potential sweep is reversed, and the cathodic peak is recorded. The cathodic peak current is then measured from the decaying base line recorded during the single-sweep experiment.

A second method for measurement of the cathodic peak height is shown in Fig. 7.4. In this method it is assumed that the current beyond the peak decays as a function of $t^{-1/2}$. The descending portion of the anodic stationary electrode voltammogram can be considered to be a current-time curve at constant potential [see (7.5)] with $t = 0$ at a point between $E_{1/2}$ and E_p

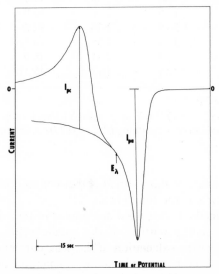

Fig. 7.5 Cyclic voltammogram recorded on strip chart recorder. Solution, 5 mM 4,5-dichlorocatechol in 2 M HClO$_4$; carbon paste electrode; sweep rate, 1.26 V/min.

[35–37]. The selection of $t = 0$ has been discussed by Polcyn and Shain [37]. Table 7.2 shows that $it^{1/2}$ is a constant to better than 1 % for the voltammogram in Fig. 7.4 [11]. The dotted line was calculated assuming that the $it^{1/2}$ constant held beyond the switching potential. The anodic peak current was -155 μA, and the current at point C (Table 7.2) was -45.3 μA. The cathodic peak current measured from the zero current base line was $+100$ μA. The actual peak current is $100-(-45.3)$ μA. This method is similar to the first method given except that the base line is calculated rather than measured.

A third method involves stopping the anodic potential sweep at E_λ and allowing the current to decay to a constant value [38]. The sweep is then reinitiated in the cathodic direction, and the cathodic peak is recorded. The cathodic peak is then measured from the constant-current base line obtained during the quiescent period. This method can be seen to be valid from Figs. 7.4 and 7.5; it is obvious that the current decays very slowly after a period of

Table 7.2   The Determination of $it^{1/2}$ from a Stationary Electrode
Voltammogram

E	t (sec)*	$t^{1/2}$	i (μA)	$it^{1/2}$
0.160	3.34	1.83	102.0	186.7
0.170	3.78	1.94	97.5	189.2
0.180	4.22	2.05	91.5	187.6
0.190	4.66	2.16	87.0	187.9
0.200	5.10	2.26	82.5	186.5
0.210	5.54	2.35	78.5	184.5
0.220	5.98	2.45	76.0	186.2
0.474	17.16	4.13	45.3†	

$it^{1/2} = 187.0 \pm 1.1$

* $t = 0$ at $E_{1/2} + 17.3$ mV $= +0.084$ V; sweep rate $= 1.35$ V/min.

† Calculated current assuming $it^{1/2} = 187.0$ corresponds to point C of Fig. 7.4.

time. The disadvantage of this method is that convection can lead to errors if the electrolysis time is too long (Table 7.1).

The first three methods discussed are similar in that they use a base line based on the descending portion of the anodic voltammogram. A semiempirical method for the calculation of the peak current ratio has also been given [39]. This method is based on the calculation of many theoretical cyclic voltammograms. Figure 7.6, the same voltammogram as shown in Fig. 7.4, shows the Nicholson method for the measurement of the peak

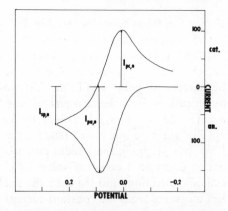

Fig. 7.6   Nicholson's method of peak height measurement (reversible system). Solution, 2.3 mM N,N,N',N'-tetramethyl-p-phenylenediamine; pH, 7.0; sweep rate, 1.35 V/min.

current ratio. By using (7.14), the peak ratio was calculated to be -0.952, thus $i_{p,c} = 148 \ \mu A$. The cathodic currents are positive, while the anodic currents are negative:

$$\frac{i_{p.c}}{i_{p.a}} = \frac{(i_{p.c})_O}{(i_{p.a})_O} - \frac{0.485(i_{SP})_O}{(i_{p_a})_O} - 0.086. \qquad (7.14)$$

Irreversible Electron Transfers

The reversibility of a heterogeneous electron transfer is dependent upon the heterogeneous rate constant and the rate of potential change. The electron transfer process can be written (7.15):

$$\text{Red} \underset{k_c}{\overset{k_a}{\rightleftharpoons}} \text{Ox} + ne \qquad (7.15)$$

The rate for the oxidation process at a given potential is characterized by a heterogeneous rate constant k_a, and the reduction process by k_c. The units for k_a and k_c are centimeters per second. The potential dependence of the heterogeneous rate constants is given by (7.16) and (7.17) [40, 41]:

$$k_a = k_s \exp\left[\frac{\beta n_a F}{RT}(E - E^0)\right] \qquad (7.16)$$

$$k_c = k_s \exp\left[\frac{-\alpha n_a F}{RT}(E - E^0)\right]. \qquad (7.17)$$

The heterogeneous rate constant at the standard potential of the couple E^0 is k_s, α and β are the transfer coefficients for the reduction and oxidation processes, respectively ($\alpha + \beta = 1$), and n_a is the number of electrons involved in the activation step [40, 41].

The rate of the electron transfer for an oxidation can be seen to increase as the potential is increased in an anodic (positive) direction. If the rate of the potential sweep is very slow and k_s is sufficiently large, (7.15) is in equilibrium at each potential [40]. The electrode process at the stationary electrode is then called reversible, and (7.6) through (7.13) describe the system.

As the sweep rate is increased, eventually the rate of the heterogeneous process will not be rapid enough for equilibrium to be established at each potential. Under such conditions the heterogeneous kinetics become an important factor in the current-potential curve. If k_s is very large, very rapid sweep rates are required to see the effect. If k_s is very small, then the heterogeneous kinetics are important at all practical sweep rates [42].

If the relative magnitude of k_c and k_a are such that $k_a \gg k_c$, and k_s is small, then the electron transfer is called totally irreversible. The theory for totally

$$\text{Red} \xrightarrow{k_a} \text{Ox} + ne \qquad (7.18)$$

irreversible electron transfer at stationary electrodes has been given for planar [19, 21–24, 29, 43] and spherical [22–23, 44] electrodes.

The peak current for an irreversible system at a planar electrode is given by (7.19) [43]:

$$i_p = -2.98 \times 10^5 (\beta n_a)^{1/2} n A D_R^{1/2} C_R V^{1/2}. \tag{7.19}$$

The magnitude of the peak current is dependent upon βn_a. Since $0.3 < \beta < 0.7$ and $n_a \leq n$, the peak current observed is less than that predicted for a reversible couple with an identical diffusion coefficient under the same experimental conditions.

The positions of E_p and $E_{p/2}$ with respect to the standard potential of the couple are given by (7.20) and (7.21) [19, 23, 43]. The position of the peak potential is a function of k_s and becomes more positive as k_s becomes smaller. Equation (7.22) relates $E_p - E_{p/2}$ to the value of βn_a [19, 23]. Since βn_a is usually less than n, the effect of β is to broaden the peak. The shape of the peak is independent of k_s:

$$E_p = E^0 + \frac{RT}{\beta n_a F}\left[0.780 + \tfrac{1}{2}\ln\left(\frac{D_R \beta n_a V F}{RT} \right) - \ln k_s \right] \tag{7.20}$$

$$E_{p/2} = E^0 + \frac{RT}{\beta n_a F}\left[\tfrac{1}{2}\ln\left(\frac{D_R \beta n_a V F}{RT} \right) - \ln k_s - 2.222 \right] \tag{7.21}$$

$$E_p - E_{p/2} = \frac{1.857RT}{\beta n_a F} \tag{7.22}$$

$$E_p - E_{p/2} = \frac{0.048}{\beta n_a} \qquad \text{at 25°C.} \tag{7.23}$$

The peak potential is also a function of the sweep rate as given by (7.24) [23]. The peak potential is seen to shift in an anodic direction by $0.030/\beta n_a$ V for a 10-fold increase in sweep rate. This fact can be used to distinguish an irreversible process from a reversible process, since the peak potential for a reversible process is independent of sweep rate [23, 43]:

$$(E_p)_2 - (E_p)_1 = (E_{p/2})_2 - (E_{p/2})_1 = \frac{RT}{\beta n_a F}\ln\left(\frac{V}{V_1} \right)^{1/2}. \tag{7.24}$$

The peak current is also a function of k_s as given by (7.25) [21, 23]. A plot of $\ln i_p$ versus $E_p - E^0$ or $E_{p/2} - E^0$ using a series of sweep rates can yield βn_a from the slope and k_s from the intercept [21, 23]. The heterogeneous kinetics of an irreversible electron transfer process can be studied by stationary electrode voltammetry provided the standard potential is known:

$$i_p = -0.227nFAC_R k_s \exp\left[\frac{\beta n_a F}{RT}(E_p - E^0) \right]. \tag{7.25}$$

Figure 7.7 compares the shapes of a reversible and an irreversible voltammogram. The curves were calculated from the tables of Nicholson and Shain [23]. The curves were calculated assuming that $n = 1$, and $\beta n_a = 0.5$. The potential axis is arbitrary (no k_s was assumed) since only the shapes of curves are of interest at present. The peak current for the irreversible case is lower than that for the reversible case, and βn_a acts to broaden the current-potential curve.

Cyclic voltammetry can also be used to distinguish between reversible and irreversible electron transfers. As can be seen from (7.18), no cathodic

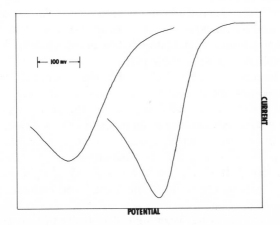

Fig. 7.7 Theoretical peak voltammograms for reversible and irreversible electron transfer.

current is observed for the reduction of Ox in a cyclic voltammogram of a totally irreversible process. In practical systems the reduction of Ox can be frequently observed, but its peak potential is very much more cathodic than that which would be observed for a reversible process. Great care must be taken in interpreting a cyclic voltammogram in which no reduction current is observed. When oxidized at an electrode, many organic compounds undergo very rapid homogeneous chemical reactions [17]. These reactions can effectively eliminate Ox from near the electrode surface, and no reduction current is observed in the cyclic voltammogram unless the product of the chemical reaction is electroactive [23].

Quasi-reversible Electron Transfers

$$\text{Red} \underset{k_c}{\overset{k_a}{\rightleftharpoons}} \text{Ox} + ne \tag{7.15}$$

Electron transfer reactions in which the heterogeneous process is neither reversible nor irreversible are common in organic systems. The degree of apparent reversibility is a function of the sweep rate. The quasi-reversible case is difficult to treat theoretically because of the many variables. However, this case has been treated for many cycles (steady state) [19] and for a single cycle [42].

Equation 7.26 defines the parameter ψ used by Nicholson to treat the degree of reversibility of the electron transfer process [42]:

$$\psi = \frac{(D_R/D_O)^{\beta/2} k_s (RT)^{1/2}}{\pi^{1/2} n^{1/2} V^{1/2} D_R^{1/2}}.$$ (7.26)

When $\psi > 7$, the electron transfer is reversible within experimental error. This can be true if k_s is large or if V is small. When $\psi < 10^{-3}$, the electron transfer can be considered to be totally irreversible. For $10^{-3} < \psi < 7$, the electron transfer is quasi-reversible [42].

By using the treatment of Nicholson [42], the heterogeneous kinetics of an electrode process can be studied by cyclic voltammetry. The value of $E_{p,a} - E_{p,c} (\Delta E_p)$ as a function of the sweep rate is the experimentally measured variable. As ψ decreases because of an increase in sweep rate, ΔE_p becomes larger since the electron transfer process is more irreversible. The shape of the current-potential curve and its position on the potential axis is also dependent upon β. However, for $0.3 < \beta < 0.7$, ΔE_p is essentially independent of β because the shifts in the anodic and cathodic peaks tend to cancel [42].

Nicholson's treatment has been applied to the reduction of Cd^{2+} [12, 42] and to the oxidation of diphenylanthracene [45]. Care must be taken to eliminate uncompensated resistance in the electrochemical cell and associated circuitry when high sweep rates are used [12, 42, 46]. Uncompensated iR losses cause effects which are very similar to those that result from quasi reversibility [46]. Modern potentiostatic circuits used with three electrodes act to eliminate most of the iR losses [47, 48]. However, a small amount of uncompensated resistance is necessary for these circuits to function properly [49–51]. In ordinary work uncompensated iR losses are small because the amount of current passed is usually small.

Consecutive Electron Transfers

$$Red_1 \rightleftharpoons Ox_1 + n_1 e$$

$$Red_2 \rightleftharpoons Ox_2 + n_2 e$$ (7.27)

A mixture of two oxidizable species, Red_1 and Red_2, can be determined rather easily at a dropping mercury or a rotating electrode. The limiting current that results from the oxidation of Red_2 can be measured from the base line defined by the limiting current for the oxidation of Red_1. At stationary electrodes a limiting current is not observed, and thus no such linear extrapolation can be made.

The peak current for the oxidation of Red_1 can be measured without difficulty from the residual current base line. Under favorable conditions the peak current for the oxidation of Red_2 can be measured as shown in Fig. 7.8 [29]. The peak current for the oxidation of Red_2 must be measured from the base line defined by the descending portion from the oxidation of Red_1.

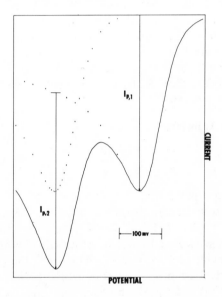

Fig. 7.8 Voltammogram of consecutive electron transfers.

Methods for the determination of the decaying base line have been given previously.

The presence of Red_2 in solution also affects the accuracy to which $i_{p,1}$ can be determined. If the potentials for the oxidation of Red_1 and Red_2 are very close together, the two peaks will not be resolved. The ability of stationary electrode voltammetry to resolve redox couples completely is slightly better than in polarography or in rotating electrode voltammetry. This is because in stationary electrode voltammetry E_p occurs only $28/n$ mV anodic of $E_{1/2}$ for a reversible system [see (7.10)]. In polarography at $28/n$ mV anodic of $E_{1/2}$, the current is only $0.75\ i_{\text{lim}}$.

Equation (7.28) describes the current-potential curve for a reversible oxidation at a dropping mercury electrode or rotating electrode [2]:

$$E = E_{1/2} - \frac{0.059}{n} \log \frac{i_L - i}{i}. \tag{7.28}$$

Using (7.28), the differences in $E_{1/2}$ necessary for the full resolution of the limiting current for the first couple can be estimated. The limiting current (to 2%) occurs $100/n$ mV anodic of $(E_{1/2})_1$. Since the current-potential curve at a dropping mercury or rotating electrode is symmetrical about $E_{1/2}$, the necessary difference between $(E_{1/2})_2$ and $(E_{1/2})_1$ is $2 \times 100/n = 200/n$ mV.

A similar calculation for stationary electrode voltammetry can be made using the tables of Nicholson and Shain [23]. In order for the current from the oxidation of Red_2 to contribute only 2% to the peak current for Red_1, $(E_{1/2})_2 - (E_{1/2})_1$ must be $148/n$ mV. These calculations assume that there are equal concentrations of Red_1 and Red_2 and that $n_1 = n_2$. If the concentration of Red_1 is greater than Red_2, then the difference in $E_{1/2}$ values need not be greater than $148/n$ mV for the accurate determination of $i_{p,1}$. If the concentration of Red_1 is less than Red_2, then a greater difference in half-wave potentials will be necessary.

Multistep Electron Transfers

$$Red \rightleftharpoons Ox_1 + n_1 e \qquad at\ E_1^{0'}$$

$$Ox_1 \rightleftharpoons Ox_2 + n_2 e \qquad at\ E_2^{0'} \qquad (7.29)$$

The shapes of the peaks and the magnitudes of the peak currents for multistep electron transfers are dependent upon the differences in the formal potentials $E_2^{0'} - E_1^{0'}$ [36, 37, 52]. If Red is much more easily oxidized than $Ox_1(E_2^{0'} > E_1^{0'})$, then two peaks will be observed. If both electron transfer reactions are reversible and $E_2^{0'} - E_1^{0'} > 118/n$ mV, the peak height for the second oxidation will be proportional to $(n_2)^{3/2}$ [37]. For cases in which $E_2^{0'} - E_1^{0'} < 118/n$ mV, the peaks broaden and eventually merge into one peak. When two peaks are still observed, the peak heights must be calculated theoretically for each value of $E_2^{0'} - E_1^{0'}$ [37].

Two processes contribute to the observed current beyond the first peak. Red is still diffusing to the electrode surface and is being oxidized by an $(n_1 + n_2)$-electron process to Ox_2. Also, Ox_1 formed by the n_1-electron oxidation of Red and which has not diffused away from near the electrode surface is oxidized by a n_2-electron process to Ox_2. When $E_2^{0'} - E_1^{0'} > 118/n$ mV, only the second process is important and two independent peaks are observed. When $E_2^{0'} - E_1^{0'} < 118/n$ mV, both processes are important [37].

For the special case in which $E_2^{0'} = E_1^{0'}$, only one wave is observed, but the height of the peak current is not proportional to $(n_1 + n_2)^{3/2}$. The peak height is intermediate between that for an n_1- and an $(n_1 + n_2)$-electron transfer. As $E_2^{0'} - E_1^{0'}$ becomes much less than zero, however, the peak does become proportional to $(n_1 + n_2)^{3/2}$ [37]. The theory of multistep electron transfers for other combinations of reversible and irreversible electron transfers has also been given [36, 37, 52].

An example of a cyclic voltammogram for two reversible one-electron transfers is shown in Fig. 7.9. The system is the oxidation of 5,10-dihydro-5,10-dimethylphenazine in propylene carbonate [11, 38]. The peak current for each peak is measured from the base line determined by the decaying portion of the preceding peak. For example, $i_{p,c}$ is measured from the base line determined from the descending portion of B [37, 38]. The peak currents measured in this way were $A = 155$, $B = 154$, $C = 158$, $D = 155 \mu$A [11].

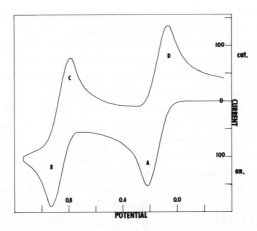

Fig. 7.9 Cyclic voltammogram for oxidation of 5,10-dihydro-5,10-dimethylphenazine. Solvent, propylene carbonate; platinum button electrode; sweep rate, 3.8 V/min.

Model Organic Systems

In order to make meaningful theoretical calculations using (7.6) and (7.19), it is necessary to know the value of the diffusion coefficient. The diffusion coefficient can be calculated from stationary electrode data using (7.6) and (7.19), provided the other parameters are known. It is always necessary to know the number of electrons transferred n if the process is reversible, and βn_a if the process is irreversible. These two quantities are, in many cases, sought in the experiment, and thus the diffusion coefficient must be known. Diffusion coefficients can be determined from other electrochemical techniques, but each technique has limitations. The value of the transition time in chronopotentiometry [see (7.30)], $it^{1/2}$ in chronoamperometry [see (7.5)], and the limiting current at rotating disk electrodes [see (7.64)] have the advantage that they are all independent of the reversibility of the electron transfer.

The transition time τ in chronopotentiometry is given by (7.30) for planar electrodes [9]:

$$\tau^{1/2} = \frac{\pi^{1/2} n F A D_R^{1/2} C_R}{2i}. \tag{7.30}$$

The value of the transition time is dependent upon the method by which it is measured [53], which leads to uncertainty in the calculated diffusion coefficient. Diffusion coefficients obtained from rotating disk electrodes show considerable promise [54]. Care must be taken in the construction of the electrode because any eccentricity in the rotation of the disk may lead to errors. The measurement of diffusion coefficients by chronoamperometry [see (7.5)] can yield excellent diffusion coefficients provided that natural convection can be avoided [6–8, 55, 56].

If the geometrical area of the electrode is accurately known, it can be used in these calculations. The area of the electrode can also be calibrated using a compound having a known diffusion coefficient [56]. Reliable diffusion coefficients, such as those of ferrocyanide in 2.0 M KCl [$D = (6.29 \pm 0.03) \times 10^{-6}$ cm^2/sec], can be used for calibration purposes [56, 57]. Only geometric areas should be used with the rotating disk electrode.

The choice of n is a very important factor in the calculation of diffusion coefficients. For example, an incorrect choice of n in the oxidation of anthracene led to the incorrect diffusion coefficient for anthracene [58–60]. Controlled-potential coulometry can be used to determine n in favorable cases [61]. However, in organic systems this can lead to erroneous results when chemical reactions that yield electroactive products interfere.

The diffusion coefficient of o-dianisidine (3,3′-dimethoxybenzidine) in aqueous media has been determined by several techniques [8]. The oxidation of o-dianisidine in 1 M H$_2$SO$_4$ satisfies all the electrochemical criteria for a reversible two-electron process [8]. The diffusion coefficient has been determined by tracer studies (4.7 × 10^{-5}), chronoamperometry (4.4 × 10^{-5}), stationary electrode voltammetry (4.1 × 10^{-5}), and rotating disk electrode voltammetry (4.1 × 10^{-5}) [8, 62]. The diffusion coefficients can be seen to agree reasonably well, but not well enough for precise work. A more thorough discussion of diffusion coefficients and a listing of known organic diffusion coefficients have been given [56].

In the absence of reliable diffusion coefficients for most organic compounds, model compounds of similar size (hence similar D) can be used to estimate n for the new system [2]. This is especially important when chemical reactions that produce electroactive species preclude the use of controlled-potential coulometry to determine n.

Table 7.3 lists several compounds that have been shown in this laboratory or in the literature to undergo totally reversible electron transfer processes [8, 63]. The conditions listed in Table 7.3 are only those that have been experimentally confirmed, however, this does not imply that these systems are not reversible under other conditions [63]. The references given in Table 7.3 are examples of voltammetric studies for a particular compound; however, in many cases these studies were not undertaken to show that the

Table 7.3 Model Organic Systems*

Compound	Solvent	Electrode	Reduction, Oxidation, n	Ref.
CH_3–N–CH_3 phenyl with OCH_3	CH_3CN,	Pt	$1e^-$, Ox	64, 65, 66
	pH 2–6	Pt, CPE	$1e^-$, Ox	63
CH_3–N–CH_3 phenyl with N–CH_3, CH_3	CH_3CN	Pt	$2 \times 1e^-$, Ox	64, 66
	pH 5–8	CPE	$1e^-$, Ox	63, 67
N(C_6H_4–OCH_3)$_3$	CH_3CN	Pt	$1e^-$, Ox	65
	1:1 Acetone–water, "pH" 2–6	Pt	$1e^-$, Ox	65
phenazine with CH_3, CH_3	CH_3CN,	Pt	$2 \times 1e^-$, Ox	38
	Propylene carbonate	Pt	$2 \times 1e^-$, Ox	38
$(C_6H_5)_2$N–N$(C_6H_5)_2$	CH_3CN	Pt	$1e^-$, Ox	68
	CH_3CN	Pt	$1e^-$, Red	68
9,10-diphenylanthracene	CH_3CN, CH_2Cl_2	Pt	$1e^-$, Ox	45, 69
	Nitrobenzene	Pt	$1e^-$, Ox	70
	DMF	Pt, HDE	$1e^-$, Red	71, 72

* DMF, dimethylformamide; Pt, platinum; CPE, carbon paste electrode; HDE, hanging mercury drop electrode.

Table 7.3   *(contd.)*

Compound	Solvent	Electrode	Reduction, Oxidation, n	Ref.
NH_2–(OCH$_3$ / OCH$_3$ biphenyl)–NH_2	1 M H$_2$SO$_4$	CPE, B$_4$C	$2e^-$, Ox	31, 73
NH_2–NH–NO_2	pH 2–4	CPE	$2e^-$, Ox	63
nitrobenzene (NO$_2$)	CH$_3$CN	Pt	$1e^-$, Red	63
benzoquinone (O=...=O)	DMF, CH$_3$CN	Pt	$1e^-$, Red	63, 74
bipyridyl N,N'-dioxide azo compound	DMF	Pt, HDE	$1e^-$, Red	75

process is reversible. Since reversibility depends upon the sweep rate, the criterion used for all the compounds listed in Table 7.3 is that they behave reversibly by cyclic voltammetry using sweep rates up to 25 V/min.

An interesting criterion for reversibility was first suggested by Mueller [31]. The $it^{1/2}$ constant obtained by chronoamperometry [see (7.5)] is valid regardless of the reversibility of the electron transfer, provided $k_a t^{1/2}/D_r^{1/2} > 4$ [40]. However, the Randles-Sevcik equation [see (7.6)] for stationary electrode voltammetry is valid only for reversible systems. Equations (7.5) and (7.6) can be combined to give (7.31) which is valid only for reversible electron transfer processes:

$$i_p = \frac{2.69 \times 10^5 n^{1/2}(it^{1/2})\pi^{1/2}V^{1/2}}{F}. \tag{7.31}$$

If one records a current-time curve on a particular solution of Red, $it^{1/2}$ can be readily calculated. A stationary electrode voltammogram can then be run on the same solution and the observed peak current compared with that calculated from (7.31). In this experiment the diffusion coefficient, the area of the electrode, and the concentration of Red need not be known. For a reversible electron transfer, the number of electrons transferred can be calculated without a knowledge of the diffusion coefficient.

Similar equations that relate stationary electrode voltammetry to chronopotentiometry [see (7.32)] and chronoamperometry to chronopotentiometry [see (7.33)] can also be derived [31]. Equation (7.33) can be used to check the reliability of the results since it holds for reversible, quasi-reversible, and irreversible electron transfers [31]:

$$i_p = \frac{2 \cdot 2.69 \times 10^5 n^{1/2}(i_0\tau^{1/2})V^{1/2}}{\pi^{1/2}F} \tag{7.32}$$

$$(i\tau^{1/2}) = \frac{\pi(it^{1/2})}{2}. \tag{7.33}$$

An example of the application of (7.31) is the verification of the one-electron reduction of nitrobenzene in acetonitrile at a planar platinum electrode [11]. For a $8.2 \times 10^{-4}\ M$ nitrobenzene solution, $it^{1/2}$ was determined by chronoamperometry to be $55.4\ \mu A\text{-sec}^{1/2}$. The peak current calculated from (7.31) was $109\ \mu A$ for a one-electron reduction and $154\ \mu A$ for a two-electron reduction for a sweep rate of 9.63 V/min. The observed peak current was $109\ \mu A$, which indicated a reversible one-electron reduction. A similar calculation was carried out for the oxidation of N,N,N',N'-tetramethyl-p-phenylenediamine in pH 7.0 phosphate buffer at a carbon paste electrode [11]. For a $2.3 \times 10^{-3}\ M$ solution, $it^{1/2}$ was $84.7\ \mu A\text{-sec}^{1/2}$ and i_p was $378\ \mu A$ at 9.00 V/min. The predicted peak current from (7.31) for a one-electron oxidation is $364\ \mu A$, and $515\ \mu A$ for a two-electron oxidation.

When the electron transfer process is quasi-reversible, (7.31) is not valid. The oxidation of 4,5-dichlorocatechol in $2\ M$ $HClO_4$ is a quasi-reversible two-electron process at a carbon paste electrode [76]. Table 7.1 shows the data obtained from current-time curves for this system, and $it^{1/2} = 609 \pm 4\ \mu A\text{-sec}^{1/2}$ [11]. The chronopotentiometric $it^{1/2}$ was observed to be $957 \pm 6\ \mu A\text{-sec}^{1/2}$, while (7.33) predicts $954\ \mu A\text{-sec}^{1/2}$. When a series of stationary electrode voltammograms were recorded, however, $i_p/V^{1/2}$ was $(3.76 \pm 0.02) \times 10^3\ \mu A\text{-sec}^{1/2}\text{-V}^{-1/2}$ [11]. Substitution of these numbers into (7.31) yielded $n = 1.54$, showing that the process is not reversible.

Linear Potential Sweep Voltammetry for Electron Transfers Coupled to Chemical Reactions

The interesting feature of the oxidation of reduction of organic compounds is that most often chemical reactions are associated with the electron transfer process. This is especially true of oxidations in aqueous media. The initial product of the oxidation is an electron-deficient species and hence can react with the nucleophiles H_2O and OH^-.

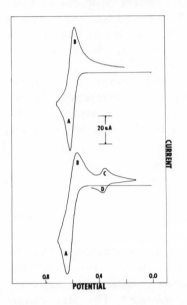

Cyclic voltammetry is the single most valuable tool available to the organic chemist studying the total electrode process. The potential range of the solvent can be swept with a wide range of sweep rates, allowing the observation of electroactive intermediates of widely varying stabilities [15–17]. The degree of reversibility of the initial electron transfer process and of any products observed can in many cases be immediately estimated. Important studies have been made recently by Nicholson, Shain [23, 46, 77–79], and Saveant and Vianello [80–88] in the quantitative treatment of stationary electrode voltammetry complicated by homogeneous chemical reactions. In Chapter VI of this volume, Large and Brown have given a detailed discussion of some of this work. They present experimental criteria and work relating to several types of mechanisms from which kinetic

Fig. 7.10 Oxidation of 4,5-dichloro-catechol. Top: 4,5-dichlorocatechol alone; bottom: in presence of aniline. Carbon paste electrode; sweep rate, 5.0 V/min.

parameters can be obtained. Many important systems have yet to be treated quantitatively; of special interest are chemical reactions associated with quasi-reversible electron transfers. However, the general picture of the electrode process can be readily obtained in the absence of a quantitative theory [15–17, 89–92, 95].

Figure 7.10 (top) is a cyclic voltammogram of 4,5-dichlorocatechol in 1.2 M $HClO_4$ at a carbon paste electrode [76]. The anodic potential sweep is initiated at 0.0 V, and the sweep rate is 5.0 V/min. On the first anodic sweep, the oxidation (A) of 4,5-dichlorocatechol [see (7.34)] to 4,5-dichloro-o-benzoquinone is observed at $E_p = +0.61$ V. The direction of the potential sweep is reversed at $+0.70$ V, and the reduction (B) of 4,5-dichloro-o-benzoquinone is observed at $+0.57$ V. The value of $E_{p,a} - E_{p,c} = 40$ mV, which indicates a quasi-reversible two-electron transfer. Subsequent potential cycles of this system yield no new couples.

Figure 7.10 (bottom) is the cyclic voltammogram of the same solution of 4,5-dichlorocatechol with $5.5 \times 10^{-2}\,M$ aniline added. On the first anodic potential sweep, only the oxidation (A) of 4,5-dichlorocatechol is observed. After reversal of the potential sweep at $+0.70$ V, the peak current (B) for the reduction of 4,5-dichloro-o-benzoquinone is diminished and a new cathodic peak (C) is observed at $+0.35$ V. The new peak corresponds to the reduction of 4-anilino-5-chloro-o-benzoquinone to 4-anilino-5-chlorocatechol [see 7.36]. The 4-anilino-5-chloro-o-benzoquinone is the product of a chemical reaction [see (7.35)]. On the second anodic sweep, a new anodic peak (D) at $+0.38$ V corresponding to the oxidation of 4-anilino-5-chlorocatechol is observed. The new couple [see (7.36)] is a reversible two-electron process [76]:

$$\text{(structure)} \rightleftharpoons \text{(structure)} + 2H^+ + 2e \qquad (7.34)$$

$$\text{(structure)} + \text{(structure)} \xrightarrow{k} \text{(structure)} + H^+ + Cl^- \qquad (7.35)$$

$$\text{(structure)} + 2H^+ + 2e \rightleftharpoons \text{(structure)} \qquad (7.36)$$

If a much faster sweep rate is selected to record the cyclic voltammogram in the presence of aniline, the product of the chemical reaction is not observed. The chemical reaction does not have sufficient time to take place before the 4,5-dichloro-o-benzoquinone is reduced to the unreactive 4,5-dichloro-catechol. If a much slower sweep rate is selected, then the reduction (B) of 4,5-dichloro-o-benzoquinone is not observed at all. The chemical reaction in this case can go to completion before the potential is sufficient to reduce the 4,5-dichloro-o-benzoquinone. Thus the sweep rate is the important experimental variable to consider when the electron transfer process is coupled to a chemical reaction.

The product of the chemical reaction can in many cases be identified by cyclic voltammetry [15–17, 65, 91–94]. Cyclic voltammetric experiments are carried out on all the possible products of the chemical reaction under conditions identical to the original experiment. If the peak potentials and electron transfer characteristics (reversibility) of the known compound are identical to that of the unknown product, a tentative identification can be made [95].

The homogeneous chemical reaction has an effect upon various features of the current-potential curve as the sweep rate is varied. The position of E_p or $E_{p/2}$ usually changes as the sweep rate is varied. These effects are small, and thus are not useful if the rate constant for the chemical reaction is sought. However, the shift at E_p is valuable as a diagnostic criterion for the type of chemical reaction [23]. Of more importance in the determination of the chemical rate constant are the anodic peak for a single-sweep experiment and the ratio of cathodic to anodic peak currents in cyclic voltammetry. When the rate of the chemical reaction is significant, one or both of these quantities varies as a function of sweep rate.

Chemical Reactions Following Electron Transfer

IRREVERSIBLE CHEMICAL REACTIONS

$$\text{Red} \rightleftharpoons \text{Ox} + ne$$
$$\text{Ox} \xrightarrow{k_f} Z \tag{7.37}$$

The general scheme for a chemical reaction following electron transfer is shown in (7.37). The compound Red is oxidized by a reversible n-electron process to give Ox. The species Ox is unstable and reacts by a first-order or pseudo-first-order process to give Z which is not electroactive. In many cases Z is in fact reducible at more negative (cathodic) potentials than Ox, but this does not complicate the theoretical treatment. This type of a chemical complication is also widely known as an EC (electron transfer–chemical reaction) [96].

The theory for single-sweep and cyclic voltammetry for this chemical complication has been given for planar [22, 23] and spherical [97] electrodes. It is convenient to use a parameter, with the units $\sec^{-1}$, which is proportional to the sweep rate. Nicholson and Shain define a as nFV/RT [23]. At 25.0°C, $a = 38.9nV$ where V is in volts per second.

The characteristics of the current-potential curve for the oxidation of Red depend upon the magnitude of k_f/a, a dimensionless parameter. If k_f/a is very small (k_f small and/or a large), then the current-potential curve is identical to that for an uncomplicated reversible electron transfer. The peak potential is independent of sweep rate, and $i_p/V^{1/2}$ is constant as a function of sweep rate. If k_f/a is very large, the peak potential is no longer independent of sweep rate. The peak potential is given by (7.38) [23, 80]:

$$E_p = E_{1/2} + \frac{RT}{nF}[0.78 - \tfrac{1}{2}\ln(k_f/a)]. \tag{7.38}$$

The peak potential now shifts in a cathodic direction by $0.030/n$ V for each 10-fold increase in k_f/a [23, 80]. Experimentally, it is more convenient to vary the sweep rate since at a particular temperature, pH, and so on, k_f will be a constant. Thus for a rapid irreversible chemical reaction, a 10-fold increase in sweep rate leads to a $30/n$ mV anodic shift of E_p. When k_f/a is very large, the peak current is greater than that for an uncomplicated reversible electron transfer at the same sweep rate. The constant 2.98×10^5 is substituted for 2.69×10^5 in (7.6) [23]. It is obvious that the single-sweep method is not very useful for the quantitative measurement of k_f because a 10-fold variation in k_f causes only a $30/n$ mV shift in E_p, and i_p increases by only 10% for an increase in k_f from zero to infinity [23].

Cyclic voltammetry is very useful for the measurement of k_f because the ratio $i_{p,c}/i_{p,a}$ is very dependent upon k_f/a. When k_f/a is very small, the ratio $i_{p,c}/i_{p,a}$ is equal to -1.00. The rate of the chemical reaction is slow relative to the sweep rate. When k_f/a is very large, no current for reduction of Ox is observed, and $i_{p,c}/i_{p,a}$ equals 0 [23].

The value of $i_{p,c}/i_{p,a}$ as a function of sweep rate cannot be obtained in an explicit form. In many cases, however, working curves can be constructed which relate a dimensionless experimental parameter such as $i_{p,c}/i_{p,a}$ to a dimensionless parameter containing the rate constant and a time function (k_f/a) [98]. Nicholson and Shain [23] have tabulated the data necessary for the construction of such a working curve for the cyclic voltammetric determination of k_f. The ratio $i_{p,c}/i_{p,a}$ is given as a function of $k_f\tau$, where τ is the time necessary to sweep the potential from $E_{1/2}$ to the switching potential E_λ [23].

Only reversible electron transfer processes have been treated theoretically. The chemical reaction has no effect upon the single-sweep behavior of an irreversible process [23]. If the electron transfer process is so irreversible that no cathodic current is observed by cyclic voltammetry, the presence of an EC process will be difficult to establish. If Z is electroactive, then the EC process can be identified. The important case of a chemical reaction following a quasi-reversible electron transfer has not been treated theoretically [see (7.34) and (7.35)].

The cyclic voltammogram for a totally irreversible electron transfer is qualitatively similar to that for a reversible electron transfer followed by a rapid, irreversible chemical reaction if Z is not electroactive. No cathodic current is observed in either case. The two electrode processes can be distinguished by the single-sweep method. The value of E_p shifts in an anodic direction with increasing sweep rate in both cases. The value of E_p shifts by $30/\beta n_a$ mV for an irreversible electron transfer, and $30/n$ mV for a rapid chemical reaction when V is increased 10-fold. The value of $E_p - E_{p/2} = 0.048/\beta n_a$ V for a totally irreversible electron transfer. Since βn_a is usually less than n, the value of $E_p - E_{p/2}$ is greater than $48/n$ mV. When a reversible electron transfer is followed by a rapid, irreversible chemical reaction, $E_p - E_{p/2} = 0.048/n$ V [23, 46].

The two processes can also be differentiated by using (7.31) and (7.32). If the electron transfer is reversible but followed by a rapid chemical reaction, the peak current observed by stationary electrode voltammetry can be calculated from (7.31) or (7.32) by substituting 2.98×10^5 for 2.69×10^5. If the electron transfer is totally irreversible, however, the observed peak current will be much smaller than that calculated. None of these criteria are of value if $\beta n_a = n$.

$$\text{Red} \rightleftharpoons \text{Ox} + ne$$
$$2\text{Ox} \xrightarrow{k_f} \text{Z} \tag{7.39}$$

An irreversible dimerization reaction following a reversible electron transfer has received some attention as well [46, 80]. The only case that has been considered is the one in which the rate of the chemical reaction is very rapid compared to the sweep rate. Under these conditions the peak potential is given by (7.40) [46, 80]:

$$E_p = E^0 + \frac{RT}{3nF} \ln \left(\frac{4.78\pi 3 D_R}{2D_O} \right) + \frac{RT}{3nF} \ln \frac{a}{k_f C_R}. \tag{7.40}$$

The peak potential is not only dependent upon k_f and V but also on the concentration of Red. As the sweep rate is increased by a factor of 10, the peak potential shifts in an anodic direction by $20/n$ mV. If the concentration of Red is increased 10-fold, the peak potential will shift cathodically by $20/n$ mV. Thus a first-order chemical reaction can be distinguished from a second-order reaction easily. Again the current-potential curve is sharper than for an uncomplicated reversible electron transfer since $E_p - E_{p/2} = 39/n$ mV [46].

The peak current for a rapid dimerization reaction following a reversible electron transfer is given by (7.41) at 25.0°C [46, 88]:

$$i_p = -3.25 \times 10^5 n^{3/2} A D_R^{1/2} C_R V^{1/2} \tag{7.41}$$

Thus the peak current for the oxidation of Red increases only 20% for an increase in k_f from 0 to ∞. A treatment of dimerization reactions in cyclic voltammetry has just appeared [242].

REVERSIBLE CHEMICAL REACTIONS

$$\text{Red} \rightleftharpoons \text{Ox} + ne$$
$$\text{Ox} \underset{k_b}{\overset{k_f}{\rightleftharpoons}} \text{Z} \tag{7.42}$$

The voltammetric behavior of a reversible electron transfer followed by a reversible chemical reaction depends not only on a and k_f but also on the value of the equilibrium constant $K = k_f/k_b$ [23, 88]. If K is very large, then (7.42) will simplify to (7.37) which has been previously discussed. For other values of K, the voltammetric behavior of Red can be correlated to a dimensionless parameter $K(a/k_f + k_b)^{1/2}$ given by Nicholson and Shain [23].

If the rate of the chemical reaction is very large compared to the sweep rate ($k_f + k_b \gg a$), then the system will be at equilibrium at all times [23, 88]. The current-potential will be shifted by $-RT/nF \ln (1 + K)$ V with respect to its $E_{1/2}$ when the chemical reaction is not present. As K increases, the shift becomes more marked. The position of E_p is independent of the sweep rate [23, 88]. This condition is true when $K(a/k_f + k_b)^{1/2}$ is less than 0.05 [23].

If the parameter $K[a/(k_f + k_b)]^{1/2}$ is greater than 5 and $k_f + k_b \gg a$, the position of the peak potential depends upon the sweep rate as well as in the equilibrium constant. The peak potential is given by (7.43) [23, 88]:

$$E_p = E_{1/2} + \frac{RT}{nF}\left[0.780 + \tfrac{1}{2}\ln\left(\frac{a}{k_f + k_b}\right) + \ln\left(\frac{K}{1 + K}\right)\right]. \quad (7.43)$$

When K becomes very large, (7.43) reduces to (7.38) as expected. The value of E_p is seen to shift anodically by $0.030/n$ V for each 10-fold increase in the sweep rate [23].

As in the case of an irreversible chemical reaction, $i_p/V^{1/2}$ increases by only 10% for the oxidation of Red when $K(a/k_f + k_b)^{1/2}$ increases from 0 to ∞. Nicholson and Shain [23] have calculated working curves for the cyclic voltammetric determination of k_f and k_b. The ratio of $i_{p,c}/i_{p,a}$ is plotted in the working curves. A different working curve must be constructed for each value of the switching potential [23].

$$\text{Red} \rightleftharpoons \text{Ox} + ne$$
$$2\text{Ox} \underset{\overrightarrow{k_b}}{\rightleftharpoons} \text{Z} \quad (7.44)$$

A reversible dimerization reaction following a reversible electron transfer has been considered by Saveant and Vianello [80, 85, 88]. The kinetics of the chemical reaction are considered to be so fast compared to the sweep rate that the system is in equilibrium at all times. The results are similar to those described for first-order reactions except that the initial concentration of Red is important. When K is very large, (7.40) holds. When K is small, the peak potential is independent of sweep but is dependent on the equilibrium constant and the concentration of Red by (7.45) [88]. Thus a 10-fold increase in the concentration of Red causes a $30/n$ mV cathodic shift in the peak potential:

$$E_p = E^0 + \frac{RT}{nF}\left[0.70 + \ln\frac{D_R}{D_Z} + \tfrac{1}{2}\ln K - \tfrac{1}{2}\ln C_R\right]. \quad (7.45)$$

ORGANIC SYSTEMS

Chemical reactions following an electron transfer have been shown to be very common in organic systems [15–17, 76, 79, 83, 91, 94, 98–105]. Only three organic reactions have been studied quantitatively by cyclic voltammetry [79, 102, 104] because of limitations caused by the reversibility of the electron transfer. In many cases, however, cyclic voltammetry was the major technique used to establish the type of electrode process, while the kinetics were studied by other electrochemical techniques.

The oxidation of p-aminophenol in aqueous media occurs according to (7.46) [98, 100, 102].

$$+ 2H^+ + 2e \tag{7.46}$$

The p-aminophenol is oxidized by a two-electron process to the quinoneimine. Hawley [102] has studied the kinetics of this chemical reaction as a function of pH by cyclic voltammetry. The results agreed well with rate constants obtained by other electrochemical techniques [98, 100, 102, 106–108]. The pH profile for the hydrolysis of various substituted p-benzoquinoneimines is bell-shaped [108], which is typical of the hydrolysis of Schiff bases [109].

The benzidine rearrangement has been studied by cyclic voltammetry [79] as well as by other electrochemical techniques [101, 103, 110, 111]. Hydrazobenzene is generated at the electrode surface by the two-electron reduction of azobenzene. The hydrazobenzene undergoes an acid-catalyzed rearrangement to benzidine [see (7.47)]. Good agreement was obtained among the electrochemical techniques [79].

$$+ 2H^+ + 2e \rightleftharpoons \tag{7.47}$$

The irreversible hydration of oxidized ascorbic acid [see (7.48)] has been studied by cyclic voltammetry [104] as well as by other electrochemical techniques [104, 113]. At pH 7.2, $k_f = 1.39 \times 10^3 \, \text{sec}^{-1}$ by cyclic voltammetry, and $k_f = 1.31 \times 10^3 \, \text{sec}^{-1}$ by step-functional chronoamperometry [104].

$$(7.48)$$

Chemical Reactions Preceding Electron Transfer

REVERSIBLE ELECTRON TRANSFERS

$$Z \underset{k_b}{\overset{k_f}{\rightleftharpoons}} \text{Red}$$

$$(7.49)$$

$$\text{Red} \rightleftharpoons \text{Ox} + ne$$

The preceding chemical reaction is another important chemical complication associated with the oxidation and reduction of organic compounds. In most cases it involves acid-base reactions. The voltammetric behavior depends upon the relative values of $K = k_f/k_b$, $k_f + k_b$ and $a = nFV/RT$ [23, 82, 85]. A dimensionless parameter $a^{1/2}/K(k_f + k_b)^{1/2}$ is again useful in the discussion of this process [23, 82].

Several limiting cases must be considered to fully treat the preceding chemical reaction. If the rate of the chemical reaction is very small compared to the sweep rate $[(k_f + k_b)/a$ is very small], then the experiment will be complete before any chemical reaction occurs. The peak current is proportional to the equilibrium concentration of Red in the solution as given by (7.50) [23, 82]:

$$i_p = -2.69 \times 10^5 n^{3/2} A D_R^{1/2} V^{1/2} C_R \left(\frac{K}{1+K}\right).$$

$$(7.50)$$

When K is large, the peak current is the same as that observed for an uncomplicated reversible electron transfer, while for small values of K the current is proportional to K [28, 82]. Equation (7.50) is valid for sweep rates such that $(k_f + k_b)/a < 0.025$, and $K < 0.5$ [82]. Equation (7.50) can be used to determine the equilibrium constant provided that the diffusion coefficient is known or can be determined [82, 112]. The peak potential is identical to that for an uncomplicated electron transfer [23, 82].

The other limiting cases are obtained when $k_f + k_b \gg a$ [23, 82]. When K is large, this coincides exactly with that observed for an uncomplicated case, and E_p or $E_{p/2}$ are independent of sweep rate [23, 82]. When $K < 0.5$, $(k_f + k_b)/a$ must be greater than $100/K^2$ [82].

When $k_f + k_b \gg a$ and K is very small, the characteristic peak is no longer observed with linear potential sweep voltammetry. Instead, the current-potential response is in the form of an S-shaped curve such as a polarogram. The limiting current is proportional to $k_b^{1/2}$ if $(k_f \ll k_b)$ and the equilibrium constant. The current is also independent of sweep rate [23, 82]:

$$i_L = -nFAD_R^{1/2}C_RKk_b^{1/2}. \tag{7.51}$$

If K is known, the rate constant k_b can be determined. The value of $E_{1/2}$ shifts in a cathodic direction by $30/n$ mV for a 10-fold increase in sweep rate [23, 89].

For intermediate values of K, and where $k_f + k_b \gg a$, working curves have been constructed relating i_k/i_d to $a^{1/2}/K(k_f + k_b)^{1/2}$ [23]. The current i_k observed when there are kinetic complications is related to i_d, the peak current observed at the same sweep rate in the absence of chemical complications. An empirical equation has also been calculated to simplify the use of such a working curve [23]. Equation (7.52) is valid to better than 1.5% except for small values of $a^{1/2}/K(k_f + k_b)^{1/2}$ [23]. When the preceding reaction takes place, the ratio of i_k/i_d decreases with increasing sweep rate:

$$\frac{i_k}{i_d} = \frac{1}{1.02 + \dfrac{0.471a^{1/2}}{K(k_f + k_b)^{1/2}}}. \tag{7.52}$$

For many systems i_d cannot be experimentally observed, and the diffusion coefficient may be unknown so that i_d cannot be calculated. In such cases cyclic voltammetry can be used to measure the kinetics. The ratio $i_{p,c}/i_{p,a}$ can be obtained from cyclic voltammetry. Working curves have been prepared relating $i_{p,c}/i_{p,a}$ to the parameter $a^{1/2}/K(k_f + k_b)^{1/2}$ [23].

IRREVERSIBLE ELECTRON TRANSFERS

$$Z \underset{k_b}{\overset{k_f}{\rightleftharpoons}} \text{Red}$$

$$\text{Red} \xrightarrow{k_a} \text{Ox} + ne \tag{7.53}$$

The effect of a preceding chemical reaction upon the stationary electrode voltammetric behavior of an irreversible electron transfer process is very similar to that for a reversible electron transfer [23, 82, 85]. When $k_f + k_b \ll b$ ($b = \beta n_a F / RT$), the experiment is complete before a significant amount of Z can be converted to Red. Under these conditions the oxidation of Red occurs at the same potential as when no chemical complications are present. The difference is that the peak current is proportional to the equilibrium concentration of Red [23].

When $k_f + k_b \gg b$, there are two limiting cases which depend upon the value of the parameter $b^{1/2} / K(k_f + k_b)^{1/2}$. When $b^{1/2} / K(k_f + k_b)^{1/2}$ is small (K is large), the current-potential curve is the same for the uncomplicated case except that E_p is shifted by $-RT/\beta n_a F \ln K/1 + K$ V [23]. The peak potential shifts in a cathodic direction by $30/\beta n_a$ mV for a 10-fold increase in sweep rate. When $b^{1/2}/K(k_f + k_b)^{1/2}$ is large (K is small), then a peak is no longer observed; instead there is an S-shaped curve whose limiting current is given by (7.51). The limiting current is independent of the sweep rate, as is $E_{1/2}$. For intermediate values of $b^{1/2}/K(k_f + k_b)^{1/2}$, (7.54) can be used to study the kinetics of the preceding reaction [23]:

$$\frac{i_k}{i_d} = \frac{1}{1.02 + \dfrac{0.531 b^{1/2}}{K(k_f + k_b)^{1/2}}} . \tag{7.54}$$

ORGANIC SYSTEMS

Thus far no organic systems that are complicated by preceding chemical reactions have been studied by linear potential sweep voltammetry. Many such reactions have been studied by other electrochemical methods [114–127], and these systems can be studies by stationary electrode voltammetry as well.

Electron Transfer with Cyclic Regeneration of the Reductant

REVERSIBLE AND IRREVERSIBLE ELECTRON TRANSFERS

$$\text{Red} \rightleftharpoons \text{Ox} + ne$$
$$\text{Ox} + \text{Z} \xrightarrow{k_f} \text{Red} + \text{P} \tag{7.55}$$

Equation (7.55) is the cyclic or catalytic mechanism in its simplest form. After oxidation of Red at the electrode surface by an n-electron process, a reducing agent Z reacts with Ox to regenerate Red. Both Z and P are not electroactive in the potential range of interest, and Z is present in sufficient excess so that pseudo-first-order kinetics are followed. The requirements for Z are rather stringent, since Z must be a powerful enough reducing agent so that it can reduce Ox irreversibly. Yet Z must not be oxidized at the electrode and thus there must be a high overpotential for the oxidation of Z at the electrode being used [128]. This kinetic case has been treated for reversible electron transfers at planar [23, 81, 83, 84] and spherical [84, 129] electrodes, for irreversible electron transfers at planar electrodes [23, 83, 84], and for multistep electron transfers [78].

When the sweep rate is very large such that k_f/a is very small, it is apparent that the voltammetric behavior of Red is identical to a reversible electron transfer process without kinetic complication [23, 84]. When k_f/a is larger than 1.0, a peak current is no longer observed; instead a limiting is observed which is independent of the sweep rate:

$$i_L = -nFAC_R D_R^{1/2} k_f^{1/2}. \tag{7.56}$$

If the diffusion coefficient is known, the rate constant can be determined by comparison of the current observed at a particular sweep rate i_k to that calculated for the sweep rate if there were no chemical complication, i_d [23]. The value of i_d can be obtained at high sweep rates where the chemical reaction is not important, or from solutions when Z is not present. For values of $k_f/a > 1$, the plot of i_k/i_d is linear and is given by (7.6), using the proper units:

$$\frac{i_k}{i_d} = 2.24 \left(\frac{k_f}{a}\right)^{1/2}. \tag{7.57}$$

When k_f/a is small, E_p and $E_{p/2}$ are independent of the sweep rate. When k_f/a is greater than 10, $E_{p/2}$ for the S-shaped curve is again independent of the sweep rate and equal to $E_{1/2}$ for the redox couple. For intermediate values of k_f/a, $E_{p/2}$ shifts in a cathodic direction as the sweep rate is increased but not at a constant rate [23, 84].

The effect of a cyclic process upon an irreversible electron transfer is very similar to that on a reversible electron transfer [23, 83, 84]. For small values of k_f/b, the current potential curve has all the properties of an uncomplicated irreversible electron transfer process. When k_f/b is large, a limiting current plateau is observed. The limiting current is independent of the sweep rate and is given by (7.56) [23, 84]. Working curves relating i_k/i_d to $(k_f/b)^{1/2}$ have also been constructed [23].

When k_f/b is small, E_p and $E_{p/2}$ shift in an anodic direction by $30/\beta n_a$ mV for each 10-fold increase in the sweep rate. This behavior is typical for an irreversible electron transfer. When k_f/b is large, $E_{1/2}$ for the wave is independent of the sweep rate [23, 84].

ORGANIC SYSTEMS

The theory for the catalytic process has been tested with good results on systems in which the redox couple is inorganic (Ti^{4+}, Fe^{3+}) and Z is organic (NH_2OH) [78, 84]. The redox reaction in both cases were reductions.

Recently, a catalytic mechanism has been found in which Red is an organic compound [130]. Tri-p-anisylamine and 9,10-diphenylanthracene are oxidized by a reversible one-electron process in acetonitrile at a platinum electrode. In the presence of the cyanide ion, the radical cation of the organic species is rapidly reduced to the starting material. The approximate rates of these processes have been measured by stationary electrode voltammetry [130].

Chemical Reactions Coupled between Two-Electron Transfers

$$\text{Red}_1 \rightleftharpoons \text{Ox}_1 + n_1 e \qquad E_1^0$$

$$\text{Ox}_1 \xrightarrow{k_f} \text{Red}_2 \tag{7.58}$$

$$\text{Red}_2 \rightleftharpoons \text{Ox}_2 + n_2 e \qquad E_2^0$$

The ECE (electron transfer–chemical reaction–electron transfer) process [96] has been shown to occur frequently in the oxidation and reduction of organic molecules [38, 45, 59, 65, 76, 77, 86, 90, 93, 96, 104, 131–153]. The detailed behavior of the stationary voltammogram depends upon many factors [77, 86, 93]. The reversibility of the chemical reaction, the relative values of E_1^0 and E_2^0, and the ratio of n_2/n_1 significantly affect the stationary electrode voltammogram. All these factors have been taken into account in the theoretical treatment [77]. The only case considered here is the one in which the electron transfers are reversible, the chemical reaction is irreversible, Red_2 is more easily oxidized than Red_1 ($E_2^0 < E_1^0$), and $n_1 = n_2$. This appears to be the most prevalent case observed experimentally [93]. Other important experimental cases in which the oxidation of Red is quasi-reversible [76, 105, 146, 153], or the chemical reaction is second order in Ox_1 [65, 89, 90, 152], have not been treated theoretically for linear potential sweep voltammetry.

When Red_2 is much more easily oxidized than Red_1, it can be seen that as the reactions go to completion the number of electrons transferred increases from n_1 to $n_1 + n_2$ [77, 93, 140]. Thus when k_f/a is very small, the chemical reaction has no effect upon the oxidation of Red_1, and the current-potential curve is given by (7.6). As the sweep rate is decreased, the chemical reaction becomes more important; and as k_f/a becomes very large, the peak current is given by (7.59) [77]:

$$i_p = -2.69 \times 10^5 (n_1 + n_2) n_1^{1/2} A D_R C_R V^{1/2}. \tag{7.59}$$

This behavior can be used to quantitatively study the kinetics of the chemical reaction [93, 145]. The kinetics can be studied by measuring the peak currents at various sweep rates and applying (7.60) [77]. The peak current i_k observed when the chemical reaction is important is compared to the peak current i_d at the same sweep rate that would have been observed if the chemical reaction had not been present. The value of i_d can be calculated from the value of $i_d/V^{1/2}$ observed at very fast sweep rates when the chemical reaction is unimportant [93]:

$$\frac{i_k}{i_d} = \frac{0.400 + k_f/a}{0.396 + 0.469k_f/a}.$$ (7.60)

Only two similar organic systems that follow the ECE mechanism have been studied by stationary electrode voltammetry [93, 145]. However, several other organic ECE processes have been studied quantitatively by other electrochemical techniques [76, 105, 138, 140–142, 144–148, 152, 153].

The reduction of p-nitrosophenol in aqueous alcohol solutions has been shown to proceed by the ECE mechanism shown in (7.61) [93, 135, 137, 140, 142]. The p-hydroxylaminophenol is rapidly dehydrated to p-benzoquinone-imine, which is then reduced by a two-electron process to p-aminophenol. The values of k_f obtained by stationary electrode voltammetry [93] agreed well with those obtained by chronoamperometry and chronopotentiometry [140]. Leedy [145] has studied an ECE process very similar to (7.61) except that the N,N-dimethyl-p-nitrosoaniline was reduced.

$$\text{(chemical reactions)} \tag{7.61}$$

N=O (on benzene ring with OH) $+ 2H^+ + 2e \rightleftharpoons$ HO–N–H (on benzene ring with OH)

HO–N–H (on benzene ring with OH) $\xrightarrow[-H_2O]{k_f}$ NH (benzoquinone-imine with O)

NH (benzoquinone-imine with O) $+ 2H^+ + 2e \rightleftharpoons$ NH$_2$ (on benzene ring with OH)

3 ROTATING ELECTRODES: THEORY

Whereas natural convection complicates the study of electrode processes at stationary electrodes, forced convection can be used to advantage as a mass transport process. Forced convection can be reproducibly obtained in two ways: the electrode can be moved in a reproducible fashion (vibration) [154–158], or the electrode can be fixed and the solution forced past the electrode [128, 159–187]. One very important advantage of convective mass transport is that a steady-state current is observed; that is, the current is independent of time.

The most widely used method of achieving convective mass transport is by rotating the electrode. The rotating wire electrode was introduced by Nernst [188, 189]. Limiting currents similar to those found in polarography are observed and are proportional to the concentration of the electroactive species. This fact is important in the analytical application of rotating electrodes [190].

Nernst introduced the concept that there is a thin layer called the diffusion layer δ_n very close to the electrode surface [188]. Within this diffusion layer the solution is considered to be motionless, and the only means of mass transport is diffusion. The forced convection caused by the rotation of the electrode serves to replenish the concentration of the electroactive material at the boundary of the diffusion layer. The concentration gradient with δ_n is considered to be linear, and the current for an oxidation given by (7.62). In (7.62), x is the distance from the electrode surface in centimeters and δ_n is also in centimeters:

$$i = \frac{-nFAD_R[C_R(x = \alpha) - C_R(x = 0)]}{\delta_n}. \tag{7.62}$$

At a potential sufficiently positive such that the limiting current is observed, (7.62) reduces to (7.63) [188]:

$$i_L = \frac{-nFAD_RC_R}{\delta_n}. \tag{7.63}$$

If experimental conditions such as rotation rate are held constant, (7.63) predicts that the limiting current is proportional to the concentration of the electroactive species. This fact has been confirmed [189, 190].

Problems in the interpretation of the limiting currents arise when the experimental conditions are varied, since (7.63) contains no explicit terms for rotation rate, viscosity of the solution, and so on. The defects of the Nernst theory have been pointed out by several workers [191]. The main problems are that the limiting current is not directly proportional to the diffusion coefficient [191, 192], the solution has motion much closer to the electrode than that predicted by (7.63) [191, 193], and the effect of rotation rate is dependent upon the system studied and the electrode shape used [191].

Two approaches have been used to obtain quantitative information from rotating electrodes: the first approach is to vary the experimental parameters in a controlled fashion and to empirically determine the effect of each parameter upon the limiting current [191, 194–200]; the second approach involves the solution of the mass transport problem by the application of hydrodynamics [54, 159, 161, 162, 201, 202]. The hydrodynamic equations are very complex and have been applied to only a few rotating electrode geometries. The rotating disk electrode has been studied in the most detail [54, 159, 160, 161, 162, 201].

Rotating Disk Electrodes

The limiting current at a rotating disk electrode [see (7.64)] was derived by Levich from the principles of hydrodynamics [161]:

$$i_L = -0.62nFAD^{2/3}C_R v^{-1/6}\omega^{1/2}. \tag{7.64}$$

The kinematic viscosity of the solution v is given in square centimeters per second; $\omega = 2\pi N$ is the angular velocity of the disk; N is the number of rotations per second, while the other quantities remain as before. The kinematic viscosity is defined as the viscosity of the solution divided by its density. When the viscosity is given in poise (grams per centimeter-second) and the density in grams per cubic centimeter, the kinematic viscosity is in square centimeters per second. The kinematic viscosity of most aqueous solutions is approximately 0.01 cm²/sec.

In mass transport by pure diffusion as is ideally observed at stationary electrodes, the current is independent of the solution parameters such as kinematic viscosity. At the rotating disk electrode, however, the observed current depends upon the diffusive properties of the electroactive species (D) as well as on the fluid flow properties of the solution (v). If one considers the bulk of the solution volume, convection is the predominate process. However, very near the electrode surface diffusion is also important and must be considered in any theoretical treatment. Thus at the rotating disk electrode, the mass transport process can be called convective diffusion.

The pattern of the fluid flow is easy to visualize qualitatively. As the electrode begins to rotate, the solution very near the electrode surface also begins to rotate because of its viscous properties. The solution also acquires a component of velocity away from the center of the disk toward the edges because of centrifugal force. The solution is replenished at the electrode by an upward motion of the solution toward the electrode.

The derivation of (7.64) assumes that the disk has infinite radius, that it is rotating at a constant rate in a solution of infinite volume, and that the flow is laminar [54, 159, 161]. It has been shown that laminar flow can be obtained at values of $r^2\omega/v$ less than 2×10^5 [54, 159]. At Reynolds numbers ($r^2\omega/v$) much above 10^5, turbulent flow contributes to the fluid flow. Turbulence can also be important at lower Reynolds numbers if the rotating disk electrode is not centered properly on the shaft or wobbles [54]. Experimentally, the onset of turbulent flow can be observed since the limiting current deviates significantly from the $\omega^{1/2}$ dependence given in (7.64). The radius r in the Reynolds number includes not only the radius of the actual electrode surface but also that of any nonconducting material surrounding the disk.

Various electrode shapes have been used as rotating disk electrodes [54, 203–205]. The shapes of practical rotating disk electrodes have been discussed from the point of view of their fluid flow patterns by Riddiford [54, 203]. Several electrodes can be designed that conform to the theoretical limiting current behavior [54, 203–205].

Gregory and Riddiford have reconsidered the derivation of the Levich equation [see (7.64)]; they found that the constant 0.620 should include another term as shown in (7.65) [201]:

$$i_L = - \frac{0.554}{0.8934 + 0.316\left(\dfrac{D}{\nu}\right)^{0.36}} (nFAD^{2/3}C_R\nu^{-1/6}\omega^{1/2}). \qquad (7.65)$$

Equation 7.65 is valid for D/ν values in the range 0 to 4×10^{-3}. If $D = 10^{-5}$ and $\nu = 0.01$, then the constant in (7.4) becomes 0.602. This constant is 3 % lower than the constant in the Levich equation. For a particular combination of solvent and electroactive compound, the ratio D/ν may be larger than the example above. For precise work the Gregory and Riddiford correction should be used. Their correction has been experimentally confirmed [201].

A more exact solution of the same equations has been carried out by Newman [206]. The limiting current is given by (7.66) and holds for D/ν values of 0 to 10^{-2} [206]:

$$i_L = - \frac{0.6205(nFAD^{2/3}C_R\nu^{-1/6}\omega^{1/2})}{1 + 0.2980\left(\dfrac{D}{\nu}\right)^{1/3} + 0.1451\left(\dfrac{D}{\nu}\right)^{2/3}}. \qquad (7.66)$$

According to Levich, the diffusion boundary layer at a rotating disk electrode is given by 7.67:

$$\delta = 1.61 D^{1/3}\nu^{1/6}\omega^{-1/2}. \qquad (7.67)$$

Substitution of (7.67) into the Nernst equation [see (7.63)] yields (7.64). An important feature of (7.67) is that the diffusion layer thickness is independent of the distance from the center of the disk. Thus the current density is also independent of the distance from the center of the disk; that is, the current density is uniform over the entire disk area [159]. Almost no other electrode shape has this feature, an important one when studying heterogeneous kinetics.

Recently, Newman [207–211] has shown that the current density is not uniform under all conditions. In solutions in which the concentration of electroactive species is high and/or the conductivity is low, the surface is no longer uniformly accessible. A theory has been presented to deal with these effects [210].

A current-potential curve at a rotating disk electrode is shown in Fig. 7.11. The limiting current is measured from the residual current observed for a solution of the supporting electrolyte without the electroactive species present. The half-wave potential $E_{1/2}$ is taken as the potential where the current is one-half of the limiting current. The current-potential curve for a reversible

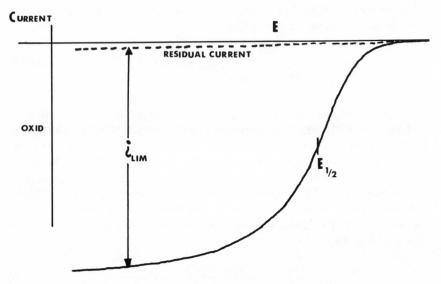

Fig. 7.11   Reversible oxidation at the rotating disk electrode.

electron transfer has a shape identical to that observed in dropping mercury electrode polarography; the curve is given by (7.68) [159]:

$$E = E_{1/2} - \frac{RT}{nF} \ln \frac{i_L - i}{i}. \tag{7.68}$$

For a reversible electron transfer, a plot of E versus $\log (i_L - i)/i$ at $25°C$ has a slope of $0.059/n$; n then can be determined from the slope. The intercept of such a plot yields $E_{1/2}$. Provided the flow is laminar, the slope and $E_{1/2}$ are independent of rotation rate for a reversible electron transfer process.

Irreversible Electron Transfers

The rotating disk electrode is a powerful tool in the study of the kinetics and mechanisms of electron transfer processes [54]. This is because the rate of mass transport can be widely varied by a change in the rotation comparable to the mass transport rate; the expression for the current at a rotating disk is given by (7.69) [159]:

$$i = \frac{-nFAD_RC_R}{\dfrac{1.61D^{1/3}\nu^{1/6}}{\omega^{1/2}} + \dfrac{D}{k_a}} . \qquad (7.69)$$

At a given potential the heterogeneous rate constant k_a is given by (7.16). When the rotation rate is small and the heterogeneous constant is large, (7.69) reduces to (7.64). A limiting current is observed which increases with $\omega^{1/2}$. In such a case the rate of the electron transfer is rapid compared to the rate of the mass transport. If the rotation rate can be increased so that the rate of mass transport is very rapid compared to the electron transfer rate, or if k_a is very small, (7.69) reduces to (7.78):

$$i = -nFAk_aC_R. \qquad (7.70)$$

The current observed is now independent of the rotation rate of the electrode. The current is limited entirely by the rate of the electron transfer. For intermediate values of k_a and/or ω, the current is given by (7.69) or one of its equivalent forms.

When the heterogeneous rate constant k_a is greater than 10^{-1} cm/sec, the electron transfer can be considered to be reversible at all practical rotation rates [212, 213]. Similarly, when $k < 10^{-4}$ cm/sec, the electron transfer is so low that (7.9) holds at all rotation rates [213]. Riddiford has given a thorough discussion of the possibilities for studying the mechanisms of heterogeneous electron transfer reactions using the rotating disk electrode [54].

Chemical Reactions

Only those chemical reactions that directly involve Red (preceding, catalytic) or form an electroactive product (ECE) have an influence upon the limiting current observed for the oxidation of Red. The reaction rate in the important case of a following chemical reaction [see (7.37)] cannot be studied quantitatively at the rotating disk electrode. The $E_{1/2}$ shifts as a function of k_f, however [214, 215]. The rotating ring-disk electrode can be used to obtain quantitative reaction rates.

CHEMICAL REACTION PRECEDING ELECTRON TRANSFER

$$\begin{aligned} Z &\underset{k_b}{\overset{k_f}{\rightleftharpoons}} Red \\ Red &\rightleftharpoons Ox + ne \end{aligned} \qquad (7.49)$$

The effect of a preceding chemical reaction on the limiting current for the oxidation of Red has been treated by the reaction layer concept [159, 216,

217] and numerically [218]. When k_f and k_b are very large, the limiting current is given by (7.71), where $(C_R)_T = C_R + C_Z$ [159, 216]:

$$i_L = -\frac{nFAD_R(C_R)_T}{\dfrac{1.61D^{1/3}\nu^{1/6}}{\omega^{1/2}} + \dfrac{D^{1/2}}{K(k_f + k_b)^{1/2}}}. \tag{7.71}$$

Any effect of the chemical reaction on the limiting current depends upon the value of $K = k_f/k_b$, since k_f and k_b are assumed to be large. When K is very large, (7.71) reduces to (7.64) and the chemical reaction has no effect upon the limiting current. When K is small, a limiting current independent of the rotation rate is observed. Note that (7.72) is identical to (7.51):

$$i_L = -nFAD^{1/2}k_b^{1/2}K(C_R)_T. \tag{7.72}$$

Equation (7.71) can be used to study the kinetics of the chemical reaction by observing the limiting current as a function of rotation rate. The behavior of the limiting current as a function of the rotation rate can be seen by rearranging (7.71) [120]:

$$\frac{i_L}{\omega^{1/2}} = -0.620nFAD_R^{2/3}\nu^{-1/6}(C_R)_T - \frac{D^{1/2}i_L}{1.61\nu^{1/6}K(k_f + k_b)^{1/2}}. \tag{7.73}$$

The slope of the plot of $i_L/\omega^{1/2}$ versus i_L gives k_f and k_b provided K is known. This technique has been used to determine the rate of proton transfer to organic bases [120, 121, 124, 219, 220].

ELECTRON TRANSFER FOLLOWED BY THE CYCLIC
REGENERATION OF REDUCTANT

$$\text{Red} \rightleftharpoons \text{Ox} + ne$$

$$\text{Ox} + \text{Z} \xrightarrow{k_f} \text{Red} + \text{P} \tag{7.55}$$

The reaction layer concept (k_f very large) was also used to derive an expression for the limiting current for cyclic kinetic complication [159, 216, 221]. When the chemical reaction is very fast, the limiting current is independent of the rotation rate [see (7.74)]. Equation (7.74) is identical to that derived for stationary electrodes [see (7.56)] when k_f is large:

$$i_L = -nFAC_RD_R^{1/2}k_f^{1/2}. \tag{7.74}$$

CHEMICAL REACTION COUPLED BETWEEN TWO-ELECTRON TRANSFERS

$$\text{Red}_1 \rightleftharpoons \text{Ox}_1 + n_1e$$

$$\text{Ox}_1 \overset{k_f}{\rightleftharpoons} \text{Red}_2 \tag{7.58}$$

$$\text{Red}_2 \rightleftharpoons \text{Ox}_2 + n_2e$$

The theoretical treatment of an ECE reaction at a rotating disk electrode has been given with varying degrees of approximation [215, 221, 222]. Of most importance is the case in which Red_2 is more easily oxidized than Red_1 because the value of k_f affects the limiting current. When k_f is very small, the chemical reaction has no effect upon the limiting current. The limiting current is given by (7.64). When k_f is very large, the limiting current is given by (7.64), but with $n_1 + n_2$ substituted for n_1. For intermediate values of k_f, the limiting current is between these two limits [215, 221, 222]. The rotating disk electrode can be used to study ECE reactions that are more rapid than can be measured by chronoamperometry [221].

The reduction of p-nitrosophenol [see (7.61)] has been studied using a rotating disk electrode. The rate constant agreed with those obtained by other electrochemical techniques [221].

A rotating disk electrode also was used to study the coupling rates of phenols with quinoneimines [see (7.75)]. The results agreed with a spectrophotometric flow technique but were not as precise [215]. A similar reaction was studied by chronopotentiometry with current reversal [223].

$$(7.75)$$

The corresponding second-order chemical reaction has been treated by Feldberg [224] and has been experimentally confirmed [224, 225]. The oxidation of triphenylamines is initially a one-electron process in acetonitrile [65]. When substituents are present at all three *para* positions, the cation radical is stable. However, when a *para* position is substituted, the cation radicals dimerize according to (7.76). The rate of the reaction increases as X becomes more electron withdrawing. The rate of the reaction has been measured by chronoamperometry [152] and at a rotating disk electrode; the results showed good agreement.

$$(7.76)$$

where XPh = X—⟨○⟩—

Rotating Ring-Disk Electrodes

An interesting extension of the rotating disk electrode is the rotating ring-disk electrode [226–231]. The usual disk is in the center of the electrode and is surrounded by a thin, insulated area concentric to it. A conducting ring surrounds the insulated ring and is also concentric to the disk. The flow pattern from the central disk is such that the material produced at the disk by the electrochemical reaction is swept out toward the ring where intermediates in the electrode process can be detected [226–231]. The instrumentation must be such that the potential of the ring is independent of that of the disk [232].

Consider that the simple electrochemical process given in (7.77) is occurring at the disk:

$$\text{Red} \rightleftharpoons \text{Ox} + n e \tag{7.77}$$

The potential applied to the disk corresponds to the limiting current plateau for (7.77), and potential applied to the ring is on the limiting current plateau for the reverse process:

$$\text{Ox} + n e \rightleftharpoons \text{Red} \tag{7.78}$$

The species Ox is produced by the electron transfer process at the disk and then is swept by the fluid flow to the ring where it can be reduced. Not all the Ox produced at the disk reaches the ring. A collection efficiency N, defined by (7.79), must be calculated or experimentally determined for each electrode:

$$N = - \frac{i_{\text{ring}}}{i_{\text{disk}}}. \tag{7.79}$$

The collection efficiency depends upon the dimensions of the insulating gap and the ring. Approximate equations have been derived to calculate N for various electrodes [228, 233], and exact equations have been derived for electrodes with thin insulating gaps and thin rings [234, 235]. The experimental values of N compare well with theory [235]. If a particular electrode's dimensions are not listed in the tables provided by Albery and Bruckenstein [235], an experimental N can be determined using an uncomplicated electron transfer process [108, 227].

Important practical applications of ring-disk electrodes have begun to appear, particularly from the groups of Bruckenstein, Albery, and Nekrasov. A variety of practical systems have been examined by Bruckenstein and co-workers [252–256]. Bruckenstein and Napp showed that the ring disk provided a remarkably simple yet powerful experimental technique for detecting adsorption processes [257]. Albery and co-workers have continued the study of homogeneous chemical reactions with the ring disk [258]. Nekrasov and Korsun have examined the reduction of carbonyls at the ring disk, including benzephenone and benzaldehyde among other compounds [259, 260]. Many more applications of the ring-disk electrode can be expected, especially since the electrodes are now commercially available.

Chemical Reactions Following Electron Transfer

If Ox in (7.78) is unstable, the rotating ring-disk electrode can be a powerful tool in the study of the decay reaction. This type of process has been treated quantitatively for both first-order [236] and second-order reactions [237]. The collection efficiency is a function of ω. When ω is very large, the chemical reaction has no time to occur before Ox reaches the ring and is reduced. The collection efficiency is that calculated for the absence of chemical complications. When ω is small, however, Ox can be consumed by the chemical reaction before it can reach the ring.

The theory was rigorously derived for cases in which the first-order reaction was very fast or very slow [236]. However, it was suggested that intermediate reaction rates could be determined empirically [236]. Malachesky et al. have used a rotating ring-disk electrode to study first-order reaction rates after constructing a calibration curve using known rate constants [108]. The calibration curve, which varies from electrode to electrode, correlated N with the dimensionless parameter $(k_f/\omega)^{1/2}$. Unknown rate constants can then be determined from the calibration curve with good results. The hydrolysis of various quinoneimines were studied by this method [see (7.46)]

[108]. Intermediates with half-lives of the order of milliseconds can be observed at a rotating ring-disk electrode [233].

Rotating Wire Electrodes

Since the introduction of the rotating wire electrode for analytical purposes by Laitinen and Kolthoff [190], this electrode frequently has been applied to problems in analytical chemistry, especially in amperometric titrations [238]. This electrode has also been used for the determination of $E_{1/2}$ values for a wide range of organic compounds. The $E_{1/2}$ values of the organic compounds are then correlated with Hammett substituent constants or molecular orbital parameters [239–242].

The limiting current expression has been obtained for a rotating wire [202]. The wire in this case is at an angle of 90° to the axis of rotation. It was found that the effect upon the limiting current of the rotation rate was variable. Between 200 and 1600 rpm, the limiting current was observed to be proportional to $\omega^{1/3}$. From 10 to 90 rpm, the limiting current was proportional to $\omega^{1/2}$ [202].

The exact effect of the rotation rate varies from electrode to electrode [191, 196, 197, 202]. Turbulent flow causes the difficulties found when using rotating wire electrodes [197, 202]. Although rotating wire electrodes are valuable in analytical applications and for $E_{1/2}$ measurements at constant rotation rate, they cannot be recommended for the study of electrode processes [56].

4 EXPERIMENTAL

Instrumentation

Anyone reasonably well versed in polarography might well conclude that any of the experiments outlined in the previous sections could be carried out with a good recording polarograph. The instrument should have a wide range of sweep rates for stationary electrode voltammetry. This sweep rate variation is not ordinarily found in instruments designed for dropping mercury polarography, but proper gear trains or variable speed motors would accomplish the modification easily. For cyclic voltammetry, the instrument should be rapidly reversible, that is, capable of producing a triangular sweep signal. In chronoamperometry, the polarograph needs only suddenly to apply (at $t = 0$) a preselected constant voltage and record the current-time curve. Finally, since the rotating disk electrode usually provides time-independent polarograms, any recording polarograph can be used.

The analysis above is basically correct—any recording polarograph can actually perform the experiments. The output information is entirely satisfactory for qualitative and, in most instances, semiquantitative studies.

However, for true adherence to theoretical equations, or for precise evaluation of kinetic parameters, the simple polarograph fails. This is because the heart of most ordinary polarographs consists of a motor-driven slide-wire. This slide-wire is a voltage divider which provides the applied voltage to a two-electrode system. Thus the polarograph can only record resulting current flow versus total applied voltage. The applied voltage is usually in terms of synchronization of the motor-driven slide-wire with the chart speed. This applied voltage is not the important parameter in voltammetry. What we need to know is the real *potential of the working electrode* at all times. It is true this value can be obtained arithmetically by various applications of Ohm's law whenever the current is known. However, the simple two-electrode polarograph cannot control this real potential and provide instantaneous information on its value. This is absolutely necessary for quantitative applications of voltammetry. To do this requires a three-electrode polarographic system and a potentiostat.

As its name implies, a potentiostat controls the potential—in this case, of the working electrode of interest. It does so by passing required currents through the working electrode and an auxiliary electrode, so as to maintain any programmed value of working electrode potential versus a reference half-cell. Thus in scanning operations it linearly sweeps the working electrode potential (for instance, versus SCE) at any predetermined rate. For chronoamperometry, regardless of the instantaneous current flow, it maintains the working electrode potential constant (within its electronic limitations). A potentiostat should be capable of maintaining a constant potential, suddenly jumping from one value to another, sweeping linearly and reversibly, or providing any other time variations of potential. All of these operations are easily accomplished by applying appropriate program signals to the input of the potentiostat. Ordinarily, in addition, the potentiostat monitors the voltammetric current. This information is normally supplied as output to a suitable recorder. For cyclic voltammetry, this is an X-Y recorder or, for fast information, an oscilloscope.

These three-electrode measurements, which produce current-potential curves as opposed to current–applied voltage polarograms, are not new. Manual measurements of the working electrode potential versus an independent, third reference electrode were advocated long ago by Kolthoff and Lingane [2] and Furness [221]. Manual control, however, is not satisfactory for the rapid operations of modern voltammetry; hence the need for essentially instantaneous electronic monitoring via the potentiostat.

Almost all modern potentiostatic polarographs are built around operational amplifier units. An abundance of circuits are discussed in the literature and no detailed descriptions are necessary here. Some circuits are multipurpose and highly specialized, while others are of relatively simple design. Individual experimenters choosing from this variety of designs are faced only with the problem of how versatile a circuit they wish to construct or purchase. The minimum performance characteristics should include both single- and cyclic sweep operations to about 100 V/min, chronoamperometric, and probably chronopotentiometric capabilities. Detailed discussions of circuits are given in reviews of at least two symposia devoted to operational amplifier instruments, and in individual references [243–245]. Hume's recent review summarizes some of the commercial instrumentation available [246].

Electrode Designs

Stationary Electrodes

Because of the large variety of stationary electrode designs, it is inappropriate to discuss the fabrication of each system. Construction details for quite a few solid electrodes have recently been summarized [95]. The purpose of the following section is to illustrate the general characteristics of selected electrode systems and to offer some critical comments on the choice of electrodes for particular studies.

Electrodes that adhere *rigorously* to semiinfinite linear diffusion conditions are few in number. The only planar electrode that obeys the Cottrell equation for long periods of time is the original design of von Stackelberg and co-workers [8, 57]. This is particularly impractical for routine work. It is difficult both to fabricate and to use. Fortunately, long-term experiments in chronoamperometry are seldom necessary. Simple, unshielded, planar electrodes usually work quite well for time <30 sec. (Almost all electrode types obey semiinfinite linear diffusion conditions for very short times, ca. <10 sec.) One of the best forms appears to be a horizontal, planar electrode surface set flush in a mounting shroud. Deviations of the electrode surface from the horizontal are frequently noticeable by inconstancy of the $(it^{1/2})$ product.

The actual electrode surfaces in practical designs may be platinum, gold, or a mercury-coated metal, or one of the various types of carbons. The carbon surfaces include carbon paste, pyrolytic graphite, glassy carbon, or even wax-impregnated carbon rods. In the latter case the shroud is really only the outer surface covering of the rod—usually an insoluble plastic coating. One of the most practical platinum electrodes of this type is a commercial product. This is the so-called button or platinum inlay electrode (Beckman No. 39273). They are rugged and reliable and have an area of about 0.2 cm² which is well suited for millimolar concentrations of electroactive species. With proper precautions, button electrodes provide excellent linear diffusion response for short periods of time (ca. <45 sec).

Stationary wire electrodes are easy to construct by merely sealing platinum wire in soft glass, or pressure-fitting platinum or gold wires to Teflon or other inert material. They are easily damaged by bending, which can alter any calibration characteristics. Glass seals readily develop leaks at the sealing point. The response of wire electrodes usually requires correction for cylindrical diffusion. With the ready availability of practical semiinfinite linear diffusion electrodes mentioned above, it hardly seems worthwhile to use stationary wires. Wires that do not extend vertically downward are to be avoided. Mass transfer to such surfaces includes nonreproducible natural convection.

There surely is little reason to employ metallic spheres in quiet solution although spherical diffusion calculations are readily available. The spherical mercury electrode surface (the hanging drop electrode) has very desirable properties—but only because it is a reproducibly renewable mercury surface— not because of its sphericity. Even in this case thin-film mercury-plated electrodes offer some distinct advantages as discussed in a recent review [246].

One of the serious difficulties of stationary mercury pool electrodes—the curved mercury surface as it contacts the pool retaining surface—has been recently eliminated by a design of Kuempel and Schaap [247]. A platinum ring, wetted by the mercury, defines a planar pool surface. Jordan and co-workers have designed a new type of mercury frit electrode which performs well for peak voltammetry and chronoamperometry. Because of the unusual area open to diffusion in this electrode, modified forms of the Randles-Sevcik and Cottrell equations apply. The electrode has the unique characteristic of giving reproducible peak polarograms in stirred solution and shows promise in monitoring operations [248]. The advantages of mercury-plated planar electrodes were mentioned previously.

Although one is reluctant to single out an electrode developed in his laboratory for special consideration, the carbon paste electrode has become widely used. It is, we believe, one of the most practical electrodes for anodic work. The carbon paste is simply a well-mixed mull of carbon (usually graphite) with any liquid that is nonmiscible with the test solution and nonelectroactive itself within the potential range of interest. Obviously, highly volatile liquids are undesirable as are those with electroactive impurities. Some of the most useful pasting liquids have been Nujol and bromoform.

The paste is mixed by hand (a typical formulation would be 15 gm of carbon and 9 ml of Nujol). We prefer Acheson Grade 38 graphite or equivalent, but the choice of carbon, provided it is not gritty—or too smooth as are some grades of spectroscopic graphite—is not critical. The paste has been described as having the consistency of peanut butter. It is perhaps better to consider it as slightly dried-out peanut butter. The drier paste packs and smooths more readily.

The paste is packed into a well-type electrode holder—frequently a depression drilled in a Teflon rod, or the female end of a glass ball joint. Electrical contact with the paste is made via a copper or platinum wire through the inside of the pool [95]. The wire should be within a few millimeters of the front surface of the paste to prevent excessive internal iR drop. After firmly packing the well with paste using a small spatula, the front surface is smoothed by repeated and uniform rubbing on a hard-finish paper (used computer program cards are ideal). It is this finishing technique that determines the reproducibility of the electrode surface. After about a week of practice, undergraduate students can prepare electrodes routinely that give ± 2 to 3% reproducibility in limiting or peak currents. Those with special finesse can do better.

Anodic residual currents with Nujol carbon pastes are, for practical purposes, nonexistent. The residual current is relatively unaffected by potential excursions—although high anodic potentials do sometimes alter the residual currents. While much more free of filming effects than platinum, many organic oxidations give film formation on carbon paste. There certainly are practically none of the surface oxidation-reduction residual current effects of platinum or gold, or the frequent large residual currents of improperly wax-impregnated graphite electrodes.

It is known that some neutral organic compounds actually extract into the Nujol medium. Chambers and Lee, using tritiated N,N-dimethylaniline showed this extraction tendency at high pH [261] Davis and Everhart suggested that bromine adsorbed on carbon pastes [262]. A detailed understanding of these effects is not yet available, but in general one should be aware of possible anomalous results with carbon paste electrodes in very basic (pH $>$ 11) or highly acidic media (4 M acid or greater).

The carbon paste electrode also offers the possibility of dissolving electroactive organic compounds in the paste itself. This technique has been used by Kuwana and co-workers with interesting results [263, 264]. At Eastman Kodak carbon pastes containing silver halides have been used to study details of interest to photographic chemistry [265]. There are many such intriguing possibilities with this type of electrode. The use of carbon pastes for cathodic reduction is possible although a small residual cathodic background current is almost impossible to remove. This residual current is fairly flat over a wide potential range and, if concentrations of electroactive material are 2 to 3 mM or greater, one can easily measure peak or limiting currents accurately above this background. Metal depositions followed by anodic stripping have been quite successful [266].

There is still a tremendous need for a reproducible, easily *renewable* stationary electrode for the anodic region. The carbon paste surface, which can be replenished after each run with relative ease, is only partially satisfactory in this respect. The renewing cycle often must be within the time interval of a single polarographic scan. The revolving wheel electrode of Lawrence and Chambers [249] and the extrusion technique of Linquist [250], as well as the electrode scraping methods of Eyring et al. [251], are incomplete answers to this problem. Hopefully, some ingenuous experiments will soon provide this most needed device.

Convective Mass Transport Electrodes

Electrodes with convective mass transport include rotating electrodes as well as those set in stirred or flowing solutions. Conventional magnetic stirring is not a very efficient means of mass transfer to a small electrode and is seldom used in voltammetry (although it is used in various voltammetric end-point detection methods in electrometric titrations). Hence it suffices to say that the response of small electrodes in stirred solution is similar in nature to that obtained at rotating electrodes, although it is generally speaking less reproducible.

The common rotating electrodes include the rotating wire and rotating disk electrodes. Rotating wires are built exactly as their stationary counterparts are except for some sort of brush or slip-ring electrical contact. It is recommended that experimenters replace rotating wires with rotating disk electrodes. The latter are just as easy to construct and their reproducibility and reliability far exceed that of wires.

The theory of the rotating disk electrode (e.g., the Levich equation) applies strictly to a thin disk rotating precisely about a central vertical shaft. Only the bottom surface of the disk is used as an electroactive surface. In practice, the disk is always housed in some sort of supporting shroud. A simple platinum rotating disk electrode can be made from the commercially available platinum inlay electrode mentioned previously. Since it already is built with a smooth glass shroud, one need only cut off the top with a glass saw, insert a central rotator shaft of brass or stainless steel, and cement it in place. Complete directions for preparing this and other rotating disk electrodes have been summarized [56].

Just recently, commercial instrumentation for rotating disk electrodes has become available. Beckman Instruments Division produces a fixed-speed rotator device (30 rps). The rotator drives a convenient flexible drive to which a variety of rotating disk electrodes can be attached. A similar unit, but with 1% speed control over the range 3 to 100 rps, with similar electrode attachments is also available. These rotating disk electrodes are excellent for all ordinary applications. As suggested previously, they should replace rotating wire electrodes for routine analytical applications. The fixed-rotation device above should serve ideally in this capacity.

Ring-disk electrodes are difficult to fabricate, especially with thin gaps between the disk and ring. This is a job for each individual designer and requires good glass-blowing and a machinist in general. Fortunately, platinum ring-disk electrodes are now available commercially through Pine Instrument Company, Grove City, Pennsylvania.

References

1. O. H. Müller, "Polarography," in *Physical Methods of Organic Chemistry*, A. Weissberger and B. W. Rossiter, Eds., 4th ed., Part II, Interscience, New York, 1969, Chapter V.
2. I. M. Kolthoff and J. J. Lingane, *Polarography*, 2nd ed., Vol. I, Interscience, New York, 1952.
3. D. J. G. Ives and G. J. Janz, *Reference Electrodes*, Academic Press, New York, 1961, p. 160.
4. P. Delahay, *New Instrumental Methods in Electrochemistry*, Interscience, New York, 1954, Chapter 3.
5. A. Fick, *Pogg. Ann.*, **94**, 59 (1855).
6. H. A. Laitinen, and I. M. Kolthoff, *J. Am. Chem. Soc.*, **61**, 3344 (1939).
7. A. J. Bard, *Anal. Chem.*, **33**, 11 (1961).
8. J. F. Zimmerman, Ph.D. Thesis, University of Kansas, Lawrence, Kansas, 1964.
9. R. W. Murray, "Chronopotentiometry, Chronoamperometry, and Chronocoulometry" in *Physical Methods of Organic Chemistry*, A. Weissberger and B. W. Rossiter, Eds., 4th ed., Part II, Interscience, New York, 1969, Chapter VIII.
10. F. G. Cottrell, *Z. Physik. Chem.*, **42**, 385 (1902).
11. S. Piekarski and R. N. Adams, unpublished experiments, 1967.
12. S. P. Perone, *Anal. Chem.*, **38**, 1158 (1966).
13. J. E. B. Randles, *Trans. Faraday Soc.*, **44**, 327 (1948).
14. J. W. Ross, R. D. DeMars, and I. Shain, *Anal. Chem.*, **28**, 1768 (1956).
15. W. Kemula and Z. Kublik, *Bull. Acad. Polon. Sci., Ser. Sci. Chim.*, **6**, 661 (1958).
16. Z. Galus and R. N. Adams, *J. Phys. Chem.*, **67**, 862 (1963).
17. Z. Galus, H. Y. Lee, and R. N. Adams, *J. Electroanal. Chem.*, **5**, 17 (1963).
18. A. Sevcik, *Collection Czech. Chem. Commun.*, **13**, 349 (1948).
19. H. Matsuda and Y. Ayabe, *Z. Elektrochem.*, **59**, 494 (1955).
20. W. H. Reinmuth, *J. Am. Chem. Soc.*, **79**, 6358 (1957).
21. Y. P. Gokhshtein, *Dokl. Akad. Nauk SSSR* **131**, 601 (1960).
22. W. H. Reinmuth, *Anal. Chem.*, **33**, 1793 (1961).
23. R. S. Nicholson and I. Shain, *Anal. Chem.*, **36**, 706 (1964).
24. R. P. Frankenthal and I. Shain, *J. Am. Chem. Soc.*, **78**, 2969 (1956).
25. W. H. Reinmuth, *Anal. Chem.*, **33**, 185 (1964).
26. M. M. Nicholson, *J. Am. Chem. Soc.*, **76**, 2539 (1954).
27. C. A. Streuli and W. P. Cooke, *Anal. Chem.*, **25**, 1691 (1953).
28. T. R. Mueller and R. N. Adams, *Anal. Chim. Acta*, **23**, 467 (1960).

29. P. Delahay, *New Instrumental Methods in Electrochemistry*, Interscience, New York, 1954, Chapter 6.

30. J. W. Loveland and P. J. Elving, *J. Phys. Chem.*, **56**, 250 (1952).

31. T. R. Mueller and R. N. Adams, *Anal. Chim. Acta*, **25**, 482 (1961).

32. H. Matsuda, *Z. Elektrochem.*, **61**, 489 (1957).

33. Y. P. Gokhshtein, *Dokl. Akad. Nauk SSSR*, **126**, 598 (1959).

34. W. H. Reinmuth, Columbia University, New York, unpublished data, 1964, cited in Ref. 23.

35. Y. P. Gokhshtein and A. Y. Gokhshtein, *Zh. Fiz. Khim.*, **34**, 1654 (1960).

36. Y. P. Gokhshtein and A. Y. Gokhshtein, in *Advances in Polarography*, I. S. Longmuir, Ed., Vol. 2, Pergamon, Oxford, 1960, p. 465.

37. D. S. Polcyn and I. Shain, *Anal. Chem.*, **38**, 370 (1966).

38. R. F. Nelson, D. W. Leedy, E. T. Seo, and R. N. Adams, *Z. Anal. Chem.*, **224**, 184 (1967).

39. R. S. Nicholson, *Anal. Chem.*, **38**, 1406 (1966).

40. P. Delahay, *New Instrumental Methods in Electrochemistry*, Interscience, New York, London, 1954, Chapters 2 and 4.

41. H. H. Bauer, *J. Electroanal. Chem. Interfacial Electrochem.*, **16**, 419 (1968).

42. R. S. Nicholson, *Anal. Chem.*, **37**, 1351 (1965).

43. P. Delahay, *J. Am. Chem. Soc.*, **75**, 1190 (1953).

44. R. D. DeMars and I. Shain, *J. Am. Chem. Soc.*, **81**, 2654 (1959).

45. M. E. Peover and R. S. White, *J. Electroanal. Chem. Interfacial Electrochem.*, **13**, 93 (1967).

46. R. S. Nicholson, *Anal. Chem.*, **37**, 667 (1965).

47. G. L. Booman, *Anal. Chem.*, **29**, 213 (1957).

48. D. D. Deford, Division of Analytical Chemistry, 133rd National Meeting, American Chemical Society, San Francisco, California, April 1958.

49. G. L. Booman and W. B. Holbrook, *Anal. Chem.*, **35**, 1793 (1963).

50. G. L. Booman and W. B. Holbrook, *Anal. Chem.*, **37**, 795 (1965).

51. W. M. Schwarz and I. Shain, *Anal. Chem.*, **35**, 1770 (1963).

52. Y. P. Gokhshtein and A. Y. Gokhshtein, *Dokl. Akad. Nauk SSSR*, **128**, 985 (1959).

53. C. D. Russell and J. M. Peterson, *J. Electroanal. Chem.*, **5**, 457 (1963).

54. A. C. Riddiford, "The Rotating Disk System," in *Advances in Electrochemistry and Electrochemical Engineering*, P. Delahay, Ed., Vol. 4, Interscience, New York, 1966, p. 47.

55. I. Shain and K. J. Martin, *J. Phys. Chem.*, **65**, 254 (1961).

56. R. N. Adams, *Electrochemistry at Solid Electrodes*, Marcel Dekker, New York, 1969.

57. M. Von Stackelberg, M. Pilgram, and V. Toome, *Z. Elektrochem.*, **57**, 342 (1953).

58. J. D. Voorhies and N. H. Furman, *Anal. Chem.*, **31**, 381 (1959).

59. T. A. Gough and M. E. Peover, in *Polarography, 1964*, G. J. Hills, Ed., Macmillan, London, 1966, p. 1017.

60. T. A. Miller, B. Prater, J. K. Lee, and R. N. Adams, *J. Am. Chem. Soc.*, **87**, 121 (1965).

61. L. Meites, "Controlled-Potential Electrolysis," in *Physical Methods of Organic Chemistry*, A. Weissberger and B. W. Rossiter, Eds., 4th ed., Part II, Interscience, New York, 1969, Chapter IX.

62. T. A. Miller, B. Lamb, K. Prater, J. K. Lee, and R. N. Adams, *Anal. Chem.*, **36**, 418 (1964).

63. A. J. Bacon, D. W. Leedy, L. S. Marcoux, R. F. Nelson, S. Piekarski, and R. N. Adams, University of Kansas, Lawrence, Kansas, unpublished experiments.

64. A. Zweig, J. E. Lancaster, M. T. Neglia, and W. H. Jura, *J. Am. Chem. Soc.*, **86**, 4130 (1964).

65. E. T. Seo, R. F. Nelson, J. M. Fritsch, L. S. Marcoux, D. M. Leedy, and R. N. Adams, *J. Am. Chem. Soc.*, **88**, 3498 (1966).

66. R. F. Nelson, Ph.D. Thesis, University of Kansas, Lawrence, Kansas, 1966.

67. J. A. Friend and N. K. Roberts, *Australian J. Chem.*, **11**, 104 (1958).

68. E. Solon and A. J. Bard, *J. Am. Chem. Soc.*, **86**, 1926 (1964).

69. J. Phelps, K. S. V. Santhanam, and A. J. Bard, *J. Am. Chem. Soc.*, **89**, 1752 (1967).

70. L. S. Marcoux, J. M. Fritsch, and R. N. Adams, *J. Am. Chem. Soc.*, **89**, 5766 (1967).

71. R. E. Visco and E. A. Chandross, *J. Am. Chem. Soc.*, **86**, 5350 (1964).

72. K. S. V. Santhanam and A. J. Bard, *J. Am. Chem. Soc.*, **88**, 2669 (1966).

73. C. Olson and R. N. Adams, *Anal. Chim. Acta*, **22**, 582 (1960).

74. M. E. Peover, *J. Chem. Soc.*, **1962**, 4540.

75. J. L. Sadler and A. J. Bard, *J. Electrochem. Soc.*, **115**, 343 (1968).

76. S. Piekarski, Ph.D. Thesis, University of Kansas, Lawrence, Kansas, 1967.

77. R. S. Nicholson and I. Shain, *Anal. Chem.*, **37**, 178 (1965).

78. D. S. Polcyn and I. Shain, *Anal. Chem.*, **38**, 376 (1966).

79. R. H. Wopschall and I. Shain, *Anal. Chem.*, **39**, 1535 (1967).

80. J. M. Saveant and E. Vianello, *Compt. Rend.*, **256**, 2597 (1963).

81. J. M. Saveant and E. Vianello, in *Advances in Polarography*, I. S. Longmuir, Ed., Vol. 1, Pergamon, Oxford, 1960, p. 367.

82. J. M. Saveant and E. Vianello, *Electrochim. Acta*, **8**, 905 (1963).

83. J. M. Saveant and E. Vianello, *Compt. Rend.*, **259**, 4017 (1964).

84. J. M. Saveant and E. Vianello, *Electrochim. Acta*, **10**, 905 (1965).

85. J. M. Saveant and E. Vianello, *Electrochim. Acta*, **12**, 629 (1967).

86. J. M. Saveant, *Electrochim. Acta*, **12**, 753 (1967).

87. J. M. Saveant, *Electrochim. Acta*, **12**, 999 (1967).

88. J. M. Saveant and E. Vianello, *Electrochim. Acta*, **12**, 1545 (1967).

89. T. Mizoguchi and R. N. Adams, *J. Am. Chem. Soc.*, **84**, 2058 (1962).

90. Z. Galus and R. N. Adams, *J. Am. Chem. Soc.*, **84**, 2061 (1962).

91. W. Kemula and Z. Kublik, *Nature*, **182**, 793 (1958).

92. W. Kemula, in *Advances in Polarography*, I. S. Longmuir, Ed., Vol. 10, Pergamon, New York, 1960, p. 105.

93. R. S. Nicholson and I. Shain, *Anal. Chem.*, **37**, 190 (1965).

94. J. G. Lawless and M. D. Hawley, *Anal. Chem.*, **40**, 948 (1968).

95. R. N. Adams, *Electrochemistry at Solid Electrodes*, Marcel Dekker, New York, 1969.
96. A. C. Testa and W. H. Reinmuth, *Anal. Chem.*, **33**, 1320 (1961).
97. M. L. Olmstead and R. S. Nicholson, *J. Electroanal. Chem. Interfacial Electrochem.*, **14**, 133 (1967).
98. A. C. Testa and W. H. Reinmuth, *Anal. Chem.*, **32**, 1512 (1960).
99. D. Hawley and R. N. Adams, *J. Electroanal. Chem.*, **8**, 163 (1964).
100. H. B. Herman and A. J. Bard, *Anal. Chem.*, **36**, 510 (1964).
101. W. M. Schwarz and I. Shain, *J. Phys. Chem.*, **69**, 30 (1965).
102. D. Hawley and R. N. Adams, *J. Electroanal. Chem.*, **10**, 376 (1965).
103. W. M. Schwarz and I. Shain, *J. Phys. Chem.*, **70**, 845 (1966).
104. S. P. Perone and W. J. Kretlow, *Anal. Chem.*, **38**, 1760 (1966).
105. S. V. Tatawawadi, S. Piekarski, M. D. Hawley, and R. N. Adams, *Chem. Listy*, **61**, 624 (1967).
106. P. A. Malachesky, Ph.D. Thesis, University of Kansas, Lawrence, Kansas, 1966.
107. C. R. Christensen and F. C. Anson, *Anal. Chem.*, **36**, 495 (1964).
108. P. A. Malachesky, K. B. Prater, G. Petrie, and R. N. Adams, *J. Electroanal. Chem. Interfacial Electrochem.*, **16**, 41 (1968).
109. W. P. Jencks, in *Progress in Physical Organic Chemistry*, S. G. Cohen, A. Streitweiser, and R. W. Taft, Eds., Vol. 2, Interscience, New York, 1964, p. 63.
110. D. M. Oglesby, J. D. Johnson, and C. N. Reilley, *Anal. Chem.*, **38**, 385 (1966).
111. J. T. Lundquist and R. S. Nicholson, *J. Electroanal. Chem. Interfacial Electrochem.*, **16**, 445 (1968).
112. P. Papoff, *J. Am. Chem. Soc.*, **81**, 3254 (1959).
113. W. Jaenicke and H. Hoffmann, *Z. Elektrochem.*, **66**, 814 (1962).
114. E. F. Caldin, *Fast Reactions in Solution*, Wiley, New York, 1964, Chapter 9.
115. W. J. Albery, *Ann. Rept. Progr. Chem.*, **60**, 40 (1963).
116. R. B. Buck, *Anal. Chem.*, **35**, 1853 (1963).
117. H. B. Mark and F. C. Anson, *Anal. Chem.*, **35**, 722 (1963).
118. R. Brdicka, *Z. Elektrochem.*, **64**, 16 (1960).
119. R. Brdicka, in *Advances in Polarography*, I. S. Longmuir, Ed., Vol. 2, Pergamon, New York, 1960, p. 655.
120. W. Vielstich and D. Jahn, *Z. Elektrochem.*, **64**, 43 (1960).
121. W. J. Albery and R. P. Bell, *Proc. Chem. Soc.*, **1963**, 169.
122. P. Delahay and S. Oka, *J. Am. Chem. Soc.*, **82**, 329 (1960).
123. P. Delahay and W. Vielstich, *J. Am. Chem. Soc.*, **77**, 4955 (1955).
124. J. Giner and W. Vielstich, *Z. Elektrochem.*, **64**, 128 (1960).
125. H. Hoffmann and W. Jaenicke, *Z. Elektrochem.*, **66**, 7 (1962).
126. M. D. Hawley, Ph.D. Thesis, University of Kansas, Lawrence, Kansas, 1965.
127. D. W. Leedy, Ph.D. Thesis, University of Kansas, Lawrence, Kansas, 1968.
128. L. N. Klatt and W. J. Blaedel, *Anal. Chem.*, **40**, 512 (1968).
129. J. Weber, *Collection Czech. Chem. Commun.*, **24**, 1770 (1959).

130. L. Papouchado, S. Feldberg, and R. N. Adams, *J. Electroanal. Chem.*, **21**, 408 (1969).
131. M. J. Astle and W. V. McConnell, *J. Am. Chem. Soc.*, **65**, 35 (1943).
132. J. E. Page, J. W. Smith, and J. G. Waller, *J. Phys. Chem.*, **53**, 545 (1949).
133. D. Stocesova, *Collection Czech. Chem. Commun.*, **14**, 615 (1949).
134. J. Koutecky, *Collection Czech. Chem. Commun.*, **18**, 183 (1953).
135. M. Suzuki, *Mem. Coll. Agr. Kyoto Univ.*, **67**, (1954).
136. I. Tachi and M. Senda, in *Advances in Polarography*, I. S. Longmuir, Ed., Vol. 2, Pergamon, Oxford, 1960, p. 454.
137. L. Holleck and R. Schindler, *Z. Elektrochem.*, **60**, 1138 (1960).
138. A. C. Testa and W. H. Reinmuth, *J. Am. Chem. Soc.*, **83**, 784 (1961).
139. W. Jaenicke and H. Hoffman, *Z. Elektrochem.*, **66**, 803 (1962).
140. G. S. Alberts and I. Shain, *Anal. Chem.*, **35**, 1859 (1963).
141. H. B. Herman and A. J. Bard, *J. Phys. Chem.*, **70**, 396 (1966).
142. R. S. Nicholson, J. M. Wilson, and M. L. Olmstead, *Anal. Chem.*, **38**, 542 (1966).
143. M. D. Hawley and S. W. Feldberg, *J. Phys. Chem.*, **70**, 3459 (1966).
144. P. A. Malachesky, L. S. Marcoux, and R. N. Adams, *J. Phys. Chem.*, **70**, 4068 (1966).
145. D. W. Leedy and R. N. Adams, *J. Electroanal. Chem. Interfacial Electrochem.*, **14**, 119 (1967).
146. M. D. Hawley, S. V. Tatawawadi, S. Piekarski, and R. N. Adams, *J. Am. Chem. Soc.*, **89**, 447 (1967).
147. L. K. J. Tong, K. Liang, and W. R. Ruby, *J. Electroanal. Chem. Interfacial Electrochem.*, **13**, 245 (1967).
148. R. N. Adams, M. D. Hawley, and S. W. Feldberg, *J. Phys. Chem.*, **71**, 851 (1967).
149. J. Phelps, K. S. V. Santhanam, and A. J. Bard, *J. Am. Chem. Soc.*, **89**, 1752 (1967).
150. L. S. Marcoux, J. M. Fritsch, and R. N. Adams, *J. Am. Chem. Soc.*, **89**, 5766 (1967).
151. M. F. Marcus and M. D. Hawley, *J. Electroanal. Chem. Interfacial Electrochem.*, **18**, 175 (1968).
152. R. F. Nelson and R. N. Adams, *J. Am. Chem. Soc.*, **90**, 3925 (1968).
153. S. Piekarski, Ph.D. Thesis, University of Kansas, Lawrence, Kansas, 1967.
154. E. D. Harris and A. J. Lindsey, *Nature*, **162**, 413 (1948).
155. E. D. Harris and A. M. Lindsey, *Analyst*, **76**, 647 (1951).
156. E. D. Harris and A. J. Lindsey, *Analyst*, **76**, 650 (1951).
157. E. R. Roberts and J. S. Meek, *Analyst*, **77**, 43 (1952).
158. A. J. Lindsey, *J. Phys. Chem.*, **56**, 439 (1952).
159. V. G. Levich, *Physicochemical Hydrodynamics*, Prentice-Hall, Englewood Cliffs, New Jersey, 1962.
160. V. G. Levich, *Discussions Faraday Soc.*, **1**, 37 (1947).
161. V. G. Levich, *Acta Physicochem. URSS*, **17**, 257 (1942).
162. O. H. Müller, *J. Am. Chem. Soc.*, **69**, 2992 (1947).
163. G. Trümpler and H. Zeller, *Helv. Chim. Acta*, **34**, 952 (1951).

164. P. Delahay, *New Instrumental Methods in Electrochemistry*, Interscience, New York, 1954, Chapter 9.
165. J. Jordan, *Anal. Chem.*, **27**, 1708 (1955).
166. J. Jordan and R. A. Javick, *J. Am. Chem. Soc.*, **80**, 1264 (1958).
167. J. Jordan, R. A. Javick, and W. E. Ranz, *J. Am. Chem. Soc.*, **80**, 3846 (1958).
168. F. Stráfelda, *Collection Czech. Chem. Commun.*, **25**, 862 (1960).
169. J. Jordan and R. A. Javick, *Electrochim. Acta*, **6**, 23 (1962).
170. G. Wranglen and O. Nilsson, *Electrochim. Acta*, **7**, 121 (1962).
171. F. Stráfelda and A. Kimla, *Collection Czech. Chem. Commun.*, **28**, 1516 (1963).
172. Y. G. Gurinov and S. V. Gorbachev, *Zh. Fiz. Khim.*, **37**, 1141 (1963).
173. W. J. Blaedel, C. L. Olson, and R. L. Sharma, *Anal. Chem.*, **35**, 2100 (1963).
174. J. Jordan and H. A. Catherino, *J. Phys. Chem.*, **67**, 2241 (1963).
175. J. C. Bazán and A. J. Arvia, *Electrochim. Acta*, **9**, 17 (1969).
176. T. K. Ross and A. A. Wragg, *Electrochim. Acta*, **10**, 1093 (1965).
177. F. Stráfelda and A. Kimla, *Collection Czech. Chem. Commun.*, **30**, 3606 (1965).
178. W. J. Blaedel and L. N. Klatt, *Anal. Chem.*, **38**, 879 (1966).
179. S. L. Marchiano and A. J. Arvia, *Electrochim. Acta*, **12**, 801 (1967).
180. J. S. W. Carrozza, S. L. Marchiano, J. J. Podesta, and A. J. Arvia, *Electrochim. Acta*, **12**, 809 (1967).
181. J. D. Bazán, S. L. Marchiano, and A. J. Arvia, *Electrochim. Acta*, **12**, 821 (1967).
182. L. N. Klatt and W. J. Blaedel, *Anal. Chem.*, **39**, 1065 (1967).
183. L. P. Kholpanov, *Russ. J. Phys. Chem.*, **40**, 930 (1966).
184. L. P. Kholpanov, *Russ. J. Phys. Chem.*, **40**, 1380 (1966).
185. H. Matsuda, *J. Electroanal. Chem. Interfacial Electrochem.*, **15**, 109 (1967).
186. H. Matsuda, *J. Electroanal. Chem. Interfacial Electrochem.*, **15**, 325 (1967).
187. T. O. Oesterling and C. L. Olson, *Anal. Chem.*, **39**, 1543 (1967).
188. W. Nernst, *Z. Physik. Chem.*, **47**, 52 (1904).
189. W. Nernst and E. S. Merriam, *Z. Physik. Chem.*, **53**, 235 (1905).
190. H. A. Laitinen and I. M. Kolthoff, *J. Phys. Chem.*, **45**, 1079 (1941).
191. L. L. Bircumshaw and A. C. Riddiford, *Quart. Rev.*, **6**, 157 (1952).
192. C. V. King and W. H. Cathcart, *J. Am. Chem. Soc.*, **59**, 63 (1937).
193. A. Fage and H. C. H. Townsend, *Proc. Roy. Soc.*, (*London*), **A135**, 656 (1932).
194. C. R. Wilke, M. Eisenberg, and C. W. Tobias, *J. Electrochem. Soc.*, **100**, 513 (1953).
195. J. N. Agar, *Discussions Faraday Soc.*, **1**, 26 (1947).
196. M. Eisenberg, C. W. Tobias, and C. R. Wilke, *J. Electrochem Soc.*, **101**, 306 (1954).
197. D. J. Ferrett and C. S. G. Phillips, *Trans. Faraday Soc.*, **51**, 390 (1955).
198. N. Ibl, *Chimia*, **9**, 135 (1955).
199. E. R. Nightengale, *Anal. Chim. Acta*, **16**, 495 (1957).
200. N. Ibl, *Electrochim. Acta*, **1**, 3 (1959).
201. D. P. Gregory and A. C. Riddiford, *J. Chem. Soc.*, **1956**, 3756.

202. T. Tsukamoto, T. Kambara, and I. Tachi, *Proc. Intern. Polarographic Congr.*, **1**, Prague, 1951, p. 525.
203. K. F. Blurtan and A. C. Riddiford, *J. Electroanal. Chem.*, **10**, 457 (1965).
204. S. Azim and A. C. Riddiford, *Anal. Chem.*, **34**, 1023 (1962).
205. K. B. Prater and R. N. Adams, *Anal. Chem.*, **38**, 153 (1966).
206. J. Newman, *J. Phys. Chem.*, **70**, 1327 (1966).
207. J. Newman, *J. Electrochem. Soc.*, **113**, 501 (1966).
208. J. Newman and L. Hsueh, *Electrochim. Acta*, **12**, 417 (1967).
209. L. Hsueh and J. Newman, *Electrochim. Acta*, **12**, 429 (1967).
210. J. Newman, *J. Electrochem. Soc.*, **113**, 1235 (1966).
211. W. J. Albery and J. Ulstrup, *Electrochim. Acta*, **13**, 281 (1968).
212. D. Jahn and W. Vielstich, *J. Electrochem. Soc.*, **109**, 849 (1962).
213. Z. Galus and R. N. Adams, *J. Phys. Chem.*, **67**, 866 (1963).
214. Z. Galus and R. N. Adams, *J. Electroanal. Chem.*, **4**, 248 (1962).
215. L. K. J. Tang, K. Liang, and W. R. Ruby, *J. Electroanal. Chem. Interfacial Electrochem.*, **13**, 245 (1967).
216. J. Kòutecky and V. G. Levich, *Zh. Fiz. Khim.*, **32**, 1565 (1958).
217. R. R. Dogonadze, *Zh. Fiz. Khim.*, **32**, 2437 (1958).
218. J. M. Hale, *J. Electroanal. Chem.*, **8**, 332 (1964).
219. W. Vielstich and D. Jahn, in *Advances in Polarography*, I. S. Longmuir, Ed., Vol. 1, Pergamon, Oxford, 1960, p. 281.
220. D. Jahn and W. Vielstich, *Z. Elektrochem.*, **64**, 43 (1960).
221. P. A. Malachesky, L. S. Marcoux, and R. N. Adams, *J. Phys. Chem.*, **70**, 4068 (1966).
222. S. Karp, *J. Phys. Chem.*, **72**, 1082 (1966).
223. W. Jaenicke and H. Hoffman, *Z. Elektrochem.*, **66**, 803 (1962).
224. L. S. Marcoux, S. Feldberg, and R. N. Adams, *J. Phys. Chem.*, **73**, 2611 (1969).
225. L. S. Marcoux, Ph.D. Thesis, University of Kansas, Lawrence, Kansas, 1967.
226. A. N. Frumkin and L. N. Nekrasov, *Dokl. Akad. Nauk SSSR*, **126**, 115 (1959).
227. A. Frumkin, L. Nekrasov, B. Levich, and Ju. Ivanov, *J. Electroanal. Chem.*, **1**, 84 (1959).
228. Yu. B. Ivanov and V. G. Levich, *Dokl. Akad. Nauk SSSR*, **126**, 1029 (1959).
229. Z. Galus, C. Olson, H. Y. Lee, and R. N. Adams, *Anal. Chem.*, **34**, 164 (1962).
230. L. N. Nekrasov and N. P. Berezina, *Dokl. Akad. Nauk SSSR*, **142**, 855 (1962).
231. Z. Galus and R. N. Adams, *J. Am. Chem. Soc.*, **84**, 2061 (1962).
232. D. T. Napp, D. C. Johnson, and S. Bruckenstein, *Anal. Chem.*, **39**, 481 (1967).
233. S. Bruckenstein and G. A. Feldman, *J. Electroanal. Chem.*, **9**, 395 (1965).
234. W. J. Albery, *Trans. Faraday Soc.*, **62**, 1915 (1966).
235. W. J. Albery and S. Bruckenstein, *Trans. Faraday Soc.*, **62**, 1920 (1966).
236. W. J. Albery and S. Bruckenstein, *Trans. Faraday Soc.*, **62**, 1946 (1966).
237. W. J. Albery and S. Bruckenstein, *Trans. Faraday Soc.*, **62**, 2584 (1966).
238. J. T. Stock, *Amperometric Titrations*, Interscience, New York, 1965.
239. A. Streitweiser, *Molecular Orbital Theory for Organic Chemists*, Wiley, New York, 1961, pp. 173–187.

240. D. J. Pietrzyk, *Anal. Chem.*, **38**, 278R (1966).
241. D. J. Pietrzyk, *Anal. Chem.*, **40**, 194R (1968).
242. M. L. Olmstead, R. G. Hamilton, and R. S. Nicholson, *Anal. Chem.*, **41**, 260 (1969).
243. Symposium on Electroanalytical Instrumentation, *Anal. Chem.*, **38**, 1106 (1966).
244. A. D. Goolsby and D. T. Sawyer, *Anal. Chem.*, **39**, 411 (1967).
245. E. R. Brown, D. E. Smith, and G. L. Booman, *Anal. Chem.*, **40**, 1411 (1968).
246. D. N. Hume, *Anal. Chem.*, **40**, 174R (1968).
247. J. R. Kuempel and W. B. Schaap, *Anal. Chem.*, **38**, 664 (1966).
248. J. H. Clausen, G. B. Moss, and J. Jordan, *Anal. Chem.*, **38**, 1398 (1966).
249. R. J. Lawrence and J. Q. Chambers, *Anal. Chem.*, **39**, 134 (1967).
250. J. Lindquist, *J. Electroanal. Chem.*, **18**, 204 (1967).
251. R. S. Perkins, R. C. Livingston, T. N. Anderson, and H. Eyring, *J. Phys. Chem.*, **69**, 3329 (1965).
252. D. C. Johnson, D. T. Napp, and S. Bruckenstein, *Anal. Chem.*, **40**, 482 (1968).
253. P. Beran and S. Bruckenstein, *Anal. Chem.*, **40**, 1044 (1968).
254. G. W. Tindall and S. Bruckenstein, *Anal. Chem.*, **40**, 1051 (1968).
255. G. W. Tindall and S. Bruckenstein, *Anal. Chem.*, **40**, 1402 (1968).
256. G. W. Tindall and S. Bruckenstein, *Anal. Chem.*, **40**, 1637 (1968).
257. D. T. Napp and S. Bruckenstein, *Anal. Chem.*, **40**, 1036 (1968).
258. W. J. Albery, M. L. Hitchman, and J. Ulstrup, *Trans. Faraday Soc.*, **64**, 2831 (1968).
259. L. N. Nekrasov and A. D. Korsun, *Elektrokhimiya*, **4**, 539, 996 (1968).
260. A. D. Korsun and L. N. Nekrasov, *Elektrokhimiya*, **4**, 1501 (1968).
261. C. A. H. Chambers and J. K. Lee, *J. Electroanal. Chem.*, **14**, 309 (1967).
262. D. G. Davis and M. E. Everhart, *Anal. Chem.*, **36**, 38 (1964).
263. T. Kuwana and W. G. French, *Anal. Chem.*, **36**, 241 (1964).
264. F. A. Schultz and T. Kuwana, *J. Electroanal. Chem.*, **10**, 95 (1965).
265. W. B. Ruby and C. G. Tremmel, *J. Electroanal. Chem.*, **18**, 231 (1968).
266. E. S. Jacobs, *Anal. Chem.*, **35**, 2112 (1963).

Chapter **VIII**

CHRONOAMPEROMETRY, CHRONOCOULOMETRY, AND CHRONOPOTENTIOMETRY

Royce W. Murray

I INTRODUCTION

We consider in this chapter three electrochemical techniques involving application of either a current or potential step excitation to an electrochemical cell and measurement of some transient response related to electrolysis of a component of that cell. These techniques, chronopotentiometry, chronoamperometry, and chronocoulometry, are collectively herein referred to as the "chronomethods." In each, the solution in the cell compartment containing the *working electrode* is *unstirred* and contains, in addition to the electroactive sample component, a large excess of an electroinactive salt, the *supporting* electrolyte. These conditions insure that mass transfer of sample component to the working electrode surface occurs through *diffusional* motion.

The sample component in our discussions is represented as an oxidized substance O, and its electrolysis, consuming *n* electrons per mole, as

$$O + ne \to R \tag{8.1}$$

Species O and product R are solution-soluble unless otherwise noted. Their initial concentrations in the cell are denoted as $C_O{}^b$ and $C_R{}^b$, their instantaneous concentrations at distance x from the electrode surface as $C_{O(x,t)}$ and $C_{R(x,t)}$, and their diffusion coefficients as D_O and D_R, respectively. Our discussions of this reduction process are transposable to the reverse (oxidation) reaction with no essential differences.

Elements of Chronopotentiometric, Chronoamperometric, and Chronocoulometric Experiments

The most common chronopotentiometric experiment involves application of a step current excitation to the working electrode (*A* in Fig. 8.1), with subsequent measurement of its time-dependent potential relative to the potential of a reference electrode in the cell (*B* in Fig. 8.1). A cathodic current step causes a change in working electrode potential (at a rate limited by the capacitive charging of the working electrode's double layer) to a value sufficiently negative for reaction (8.1) to commence. The potential then changes more slowly as O is reduced until the supply of O at the electrode surface is exhausted $(C_{O(0,t)} \rightarrow 0)$, at which time, the *transition time* (τ), the current efficiency for reaction (8.1) drops below unity and the potential rises to that of another electrode reaction (which we shall for the moment assume is uninteresting). The parameters of interest in the potential-time response curve (the *chronopotentiogram*) for reaction (8.1) are the value of τ and the

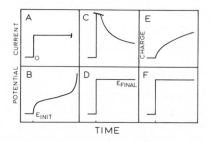

Fig. 8.1 Excitations and responses of chronomethods: Chronopotentiometry, curves *A* (excitation) and *B* (response); chronoamperometry, curves *C* (response) and *D* (excitation); chronocoulometry, curves *E* (response) and *F* (excitation).

shape of the *E-t* curve, and their dependencies on the value of the impressed cathodic current *i*.

In a chronoamperometric experiment a potential step is applied to the working electrode (*D* in Fig. 8.1) and one measures the resulting current-time response (*C* in Fig. 8.1). In the simplest case the initial working electrode potential (E_{init}) is sufficiently positive that reaction (8.1) does not proceed, and the potential attained by the step E_{final} is sufficiently negative to immediately drive the surface concentration of reactant O to zero ($C_{O(0,t)} \rightarrow 0$). The initially large cathodic current rapidly decays as the solution volume around the working electrode becomes depleted of O. The current-time ($i - t$) response, or *chronoamperogram*, is of interest with respect to its shape and magnitude of current at any given time, and its dependency on the value of E_{final} (in cases insufficiently negative to drive $C_{O(0,t)}$ to zero).

The chronocoulometric experiment is operationally identical to the chronoamperometric one except that one measures the charge flow attributable to reaction (8.1) as a function of time (*E* and *F* in Fig. 8.1). Such integral measurement is in principle no more informative than the current-time measurement, but in practice the charge-time (*Q-t*) data are sometimes more readily interpretable. Parameters of interest in the *Q-t* curve, or *chronocoulogram*, are, as in chronoamperometry, shape, amplitude, and dependency on E_{final}.

Useful variants of these experiments exist, notably ones in which a *reverse step excitation* follows the initial "forward" one. In such cases reaction (8.1) is driven in the oxidation direction following application of the reverse step. For chronopotentiometry, chronoamperometry, and chronocoulometry, the reverse *E-t* curve and transition time (τ_b), reverse *i-t* curve shape and amplitude, and reverse *Q-t* curve shape and amplitude, respectively, are determined by (1) the quantity of species R generated during the forward electrolysis time and (2) any factors tending to diminish the recoverable part of this amount of R, such as diffusion of R away from the electrode surface and any chemical instability of R. The latter factor is the one of prime interest in the reverse step experiments.

Elements of Chronomethods Theory

Three important kinetic events occur in electrochemical processes: *diffusion rates* of electrode reactants and products (O and R), the *rate of the heterogeneous electron transfer process* itself (8.1), and the *rates of any chemical reactions coupled to the electron transfer* through their generation or consumption of O or R near the electrode surface. In the simplest electrode processes and experiments, only the diffusional event need be considered; such instances yield straightforward and analytically useful relations between the measured chronomethod response and bulk reactant concentration ($C_O{}^b$). The main virtues of the chronomethods, however, lie in their transient experimental capabilities for study of the physical phenomena of heterogeneous electron transfer and coupled chemical reactions. Also, we shall see application to study of another important electrochemical event, the *adsorption* of species of O or R on the working electrode surface.

Let us consider some basic theoretical relations useful for description of the three above-mentioned kinetic events and of adsorption.

Theoretical characterization of the diffusion rate dependency of a chronomethod requires specification of the physical geometry of the working electrode and the surrounding body of solution. Diffusion conditions are classed as either *semiinfinite* or *finite* (bounded). The former are those in which the layer of solution within which the electrode reaction alters the concentrations of O or R (the *diffusion layer*) is thin in comparison to the electrode-to-cell wall distance. In finite diffusion, the cell dimensions are much smaller, and the diffusion layer impinges more or less severely on the cell wall; cases of this diffusional condition fall within the realm of "thin-layer electrochemical" experiments. Chronomethods experiments are usually performed under semiinfinite diffusion conditions. Bounded diffusion experiments have been recently reviewed [1].

Semiinfinite diffusion conditions are subdivided into cases of *planar*, *spherical*, and *cylindrical* working-electrode geometries (Fig. 8.2). Planar, or linear, diffusion is the mathematically simplest condition and is generally striven for in practical experiments. Fick's second law is the basic equation of linear diffusional motion:

$$\frac{\partial C_{(x,t)}}{\partial t} = D \frac{\partial^2 C_{(x,t)}}{\partial x^2}, \tag{8.2}$$

where D (square centimeters per second) is the diffusion coefficient. Theoretical description of any electrochemical experiment involving linear mass

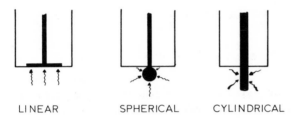

LINEAR   SPHERICAL   CYLINDRICAL

Fig. 8.2 Illustration of linear, spherical, and cylindrical diffusions. Arrows represent average diffusional paths.

transport of O or R requires solution of this differential equation under initial and boundary conditions descriptive of the experiment. For a solution for O diffusion, the usual initial condition is that of a homogeneous solution at $t = 0$:

$$C_{O(x,0)} = C_O{}^b. \tag{8.3}$$

The boundary condition statement of semiinfinite diffusion is

$$C_{O(\infty,t)} = C_O{}^b. \tag{8.4}$$

The second boundary condition describes the electrochemical excitation signal (current or potential step for the chronomethods). The resultant solution of Fick's law is an equation for $C_{O(x,t)}$ describing the concentration-distance-time profiles during the electrochemical experiment. This equation is then manipulated to produce a relation for the desired response term in the experiment.

Many interesting electrode processes are those involving coupling of a chemical process to the electrode reaction. As an example, consider an initial equilibrium of reactant O with an electroinactive species Z:

$$Z \underset{k_b}{\overset{k_f}{\rightleftharpoons}} O; \qquad O + ne \rightarrow R \tag{8.5}$$

It is evident that for $Z \rightarrow 0$ conversion rates fast in comparison to the rate of electrolytic consumption of O, the effective concentration of electrode reactant becomes $C_Z{}^b + C_O{}^b$. Electrolytic consumption rates comparable to the chemical $Z \rightarrow 0$ rate, however, can reveal control of the supply of O by k_f and k_b and by the diffusion rate of Z. Study and evaluation of the reaction kinetics then becomes possible.

Theoretical description of the "prekinetic" reaction (8.5) (also called a CE case, for chemical-electrochemical reaction sequence) requires simultaneous solution of Fick's law equations for O and Z, each modified to include the chemical reaction rate term:

$$\frac{\partial C_{O(x,t)}}{\partial t} = D_O \frac{\partial^2 C_{O(x,t)}}{\partial x^2} + k_f C_{Z(x,t)} - k_b C_{O(x,t)} \tag{8.6}$$

$$\frac{\partial C_{Z(x,t)}}{\partial t} = D_Z \frac{\partial^2 C_{Z(x,t)}}{\partial x^2} - k_f C_{Z(x,t)} + k_b C_{O(x,t)}. \tag{8.7}$$

Numerous other types of chemical reaction coupling are known, important among which are the postkinetic EC and ECE cases:

$$O + ne \rightarrow R; \qquad R \rightleftarrows Y \tag{8.8}$$

$$O + n_1 e \rightarrow R; \qquad R \rightleftarrows Y; \qquad Y + n_2 e \rightarrow X \tag{8.9}$$

Study of these forms is often accomplished in the chronomethods by application of reverse excitations. Investigation of chemical processes through observation of their influence on an electrochemical response constitutes an important branch of electrochemical research, inasmuch as rather labile and often unique solution chemistry can be examined and quantitatively characterized. Reviews of organic electrochemical reactions particularly illustrative of this point have been prepared by Peover [2] and Adams [3].

The third important kinetic event in electrode processes concerns the rate of the electron transfer reaction (8.1) itself. The rate constant for charge transfer is denoted k_s when the potential scale is referenced to the standard potential E^0 of reaction (8.1). The net current flow (amperes per square centimeter) at the electrode is the sum of a cathodic (positive) and anodic (negative sign) current component, or

$$i = i_c + i_a = nFk_s \Bigg\{ C_{O(0,t)} \exp\left[\frac{-\alpha n_a F}{RT}(E - E^0)\right]$$

$$- C_{R(0,t)} \exp\left[\frac{(1 - \alpha)n_a F}{RT}(E - E^0)\right] \Bigg\}, \tag{8.10}$$

where F is the Faraday, α the *transfer coefficient* (values of 0 to 1, a measure of the electron transfer energy barrier's symmetry), and n_a the electrons per mole for the rate-determining step in the overall electrode reaction (recognizing that microscopically a sequence of chemical events and charge transfers may occur during reaction (8.1) at the electrode surface, n_a may be less than n), R the gas constant (joules per mole-degree), and T the absolute temperature. An alternative statement of (8.10) is

$$i = i_0\left\{\exp\left[\frac{-\alpha n_a F}{RT}(E - E_{eq})\right] - \exp\left[\frac{(1 - \alpha)n_a F}{RT}(E - E_{eq})\right]\right\}, \quad (8.11)$$

where i_0 is the *exchange current*. The exchange current is the current that flows equally in the cathodic (i_c) and anodic (i_a) directions when the *net* current is zero ($i_c = -i_a$, at $E = E_{eq}$) and is a reflection of the dynamic nature of a heterogeneous charge transfer process (compare to a chemical reaction at equilibrium in a bulk solution). Exchange current is concentration dependent and related to the more fundamental rate parameter k_s by

$$i_0 = nFk_s[C_{O(0,t)}]^{1-\alpha}[C_{R(0,t)}]^{\alpha}. \quad (8.12)$$

Under the condition $i = 0$, (8.10) or (8.11) reduces to the familiar *Nernst equation*:

$$E = E^0 - 2.303\frac{RT}{nF}\log\frac{C_{R(0,t)}}{C_{O(0,t)}} \quad (8.13)$$

Note that the concentrations in the above equations are at $x = 0$. The term $2.303RT/nF = 0.0591/n$ at 25°C.

If during an actual electrolysis the net $i \ll i_0$, then (8.11) can again be reduced to the Nernst equation (8.13) with reasonable accuracy. That is, the electron transfer rate is sufficiently rapid that its displacement from an equilibrium condition is small and the surface concentrations $C_{O(0,t)}$ and $C_{R(0,t)}$ extant during electrolysis are describable by the equilibrium equation (8.13). The electrode process is in this case said to be *reversible*. A value of i significant in comparison to i_0, however, produces a displacement of concentrations from those given by (8.13), and the electrode process is then said to be *irreversible*. In cases in which $i_c \gg -i_a$ also, the electrode process is said to be "totally irreversible." It should be clear that in the electrochemical context the words reversible and irreversible are entirely qualitative. Reversibility can depend on the conditions of electrolysis; an electrode reaction reversible at low current flow may exhibit substantial irreversibility under experimental conditions demanding much higher currents. The study of electron transfer rates has long been an important feature of attempts to understand the detailed charge transfer process, and the chronomethods find application in such kinetic measurements.

Another important electrochemical event concerns the occasional tendency of electrode reactant O and/or product R to become adsorbed on the working electrode surface. Study of the fundamental factors governing such adsorption is of interest; adsorption can play a role in the detailed charge transfer process and in charge transfer-coupled chemical reactions (some of which may actually be heterogeneous or surface reactions). The presence of adsorbed reactant or product becomes reflected in most transient electrochemical experiments, including the chronomethods, through alteration of the electrochemical response from that anticipated for purely diffusion-controlled mass transport. For instance, if the reactant O is adsorbed and both solution and adsorbed reactant are electroactive (the usual case),

$$
\begin{array}{c}
O_{ads} \\
\Big\updownarrow \;\searrow^{ne} \\
\;\;\;\;\;\;\;\;\;\;\searrow R \\
\Big\updownarrow \;\nearrow_{ne} \\
O_{soln}
\end{array}
\tag{8.14}
$$

and the adsorption equilibrium is established prior to $t = 0$ (so that $C_{O(x,0)soln} = C_O{}^b$), then the charge (coulombs per square centimeter) passed to reduce both forms of reactant in a chronocoulometric experiment exceeds that expected for reaction only of diffusing reactant (Q_t) by

$$
Q_{ads} = nF\Gamma_0, \tag{8.15}
$$

where Γ_0 is the *surface excess* (moles per square centimeter) of adsorbed reactant. Measurement of Γ_0 depends on separation of the total measured charge into its Q_{ads} and time-dependent Q_t parts.

2 CHRONOAMPEROMETRY AND CHRONOCOULOMETRY

As noted earlier, the electrochemical excitation in both the chronoamperometric and chronocoulometric experiments is a potential step applied suddenly to an electrode. The response of the ensuing electrode reaction is a current decaying with time; in chronoamperometry this current-time curve is the measured response, but in chronocoulometry one measures the integral of this response as a charge-time curve. Because of the basic similarity of the two methods, it is logical here to consider them together.

The chronoamperometric experiment (also known as the potentiostatic or potential step method) has been known for many years, its most basic equation, the Cottrell equation [4], having been subjected to experimental scrutiny nearly three decades ago [5, 6]. The potential step method saw moderate use thereafter; since the late 1950s, and particularly since the application of operational amplifiers revolutionized electrochemical instrumentation (producing, among other things, increasingly excellent potentiostatic instruments), its popularity has steadily risen. Today it can be speculated that chronoamperometry, and its relatively youthful offspring, chrono-coulometry [7], may be widely accepted as the method of choice for many transient electrochemical experiments concerned with charge transfer and coupled chemical reaction rates and adsorption.

Mass Transfer Control by Diffusion

Single Potential Step Experiment

Consider application to a working electrode of a potential step sufficiently cathodic that the concentration of reactant species O is forced immediately to an essentially zero value:

$$C_{O(0,t)} = 0. \tag{8.16}$$

Application of this and the boundary conditions of (8.3) and (8.4) to the solution of Fick's law [see (8.2)] for O yields an expression for the concentration of O as a function of time and distance from the electrode surface

$$C_{O(x,t)} = C_O^b \, \text{erf} \left[\frac{x}{2D_O^{1/2}t^{1/2}} \right], \tag{8.17}$$

where erf denotes the *error function*. Some concentration profiles are plotted in Fig. 8.3; we note that the depth of the depleted region around the electrode surface, the diffusion layer, is actually quite small. It is crudely approximated by the term $(Dt)^{1/2}$ cm. The current flow is proportional, through Fick's first law, to the instantaneous concentration gradient of O at the electrode surface:

$$i = nFD_O \frac{\partial C_{O(0,t)}}{\partial x}, \tag{8.18}$$

which, when applied by differentiation to (8.17) gives

$$i = nFC_O^b \left(\frac{D_O}{\pi t} \right)^{1/2}. \tag{8.19}$$

This equation, known as the *Cottrell equation*, shows that a diffusion-controlled electrolysis of reactant at a planar, stationary electrode gives a current decaying as a $t^{-1/2}$ function which is proportional at any given time to the bulk concentration of electrode reactant.

If now the applied potential step is sufficiently negative to cause electrolysis of O but insufficiently negative to yield the condition of (8.16), that is, yields $C_O^b > C_{O(0,t)} > 0$, the condition of (8.16) must be replaced with one describing the exact nature of the E_{final}—$C_{O(0,t)}$ dependency. Two situations arise: (1) the $O + ne \rightarrow R$ charge transfer is very fast (reversible reaction) and E_{final} is related to $C_{O(0,t)}$, and $C_{R(0,t)}$, through the Nernst equation [see (8.13)] or (2) the reaction involves a low k_s, and we apply the statement of

(8.10). The latter, irreversible case is considered later in this Section. For reversible charge transfer, simultaneous solution of Fick's law for O and R and use of (8.13) and other appropriate boundary conditions [8] gives

$$i = nFC_O{}^b \left(\frac{D_O}{\pi t}\right)^{1/2} \left\{1 + \left(\frac{D_O}{D_R}\right)^{1/2} \exp\left[\frac{nF}{RT}(E - E^0)\right]\right\}^{-1}. \quad (8.20)$$

In the limit of sufficiently negative potential, this expression reduces to (8.19). The effect of E_{final} in a reversible reaction is seen to simply scale the current-time response curve.

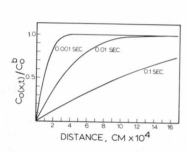

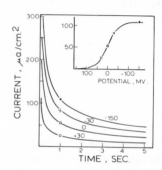

Fig. 8.3 Potential step concentration-distance profiles at various times for reactant 0; $D_O = 1.0 \times 10^{-5}$ cm²/sec.

Fig. 8.4 Chronoamperometric current-time curves calculated from (8.20) for various $E - E^0$ (numbers on curves are $E - E^0$ in mV) and $n = 1$, $D_O = D_R = 4.0 \times 10^{-6}$ cm²/sec, $C_O{}^b = 1.0 \times 10^{-6}$M/cm³, $T = 25°C$. Inset is current at 1 sec.

A series of chronoamperometric i-t curves are plotted in Fig. 8.4. From these we note that if currents at any given time t_x are plotted against the value of applied potential, a current-potential "wave" for the reversible reaction results which has as its $E_{1/2}$ (value of potential halfway up the wave)

$$E_{1/2} = E^0 - \frac{RT}{nF} \ln \left(\frac{D_O}{D_R}\right)^{1/2}. \quad (8.21)$$

We now recognize the relation of the above equations to the well-known polarographic experiment in which a potential (effectively a step potential) is applied to a growing mercury drop. The resulting current-time curve is a composite of the $t^{2/3}$ area dependency of the growing mercury drop and the $t^{-1/2}$ dependency of (8.19) and (8.20) (giving a $t^{1/6}$ result for current). (A correction for spherical diffusion effects is applied in exact polarographic theory.) A fuller description of the polarographic experiment is given in Chapter 5.

The chronoamperometric experiment finds its most frequent application in its polarographic modification. Although chronoamperometric current-time curves at stationary electrodes also provide concentration-proportional currents suitable for analytical applications, few advantages accrue as compared to the polarographic experiment, and it is rarely thus applied. The stationary electrode chronoamperometric experiment does, as we shall see below, find highly useful application to the study of charge transfer and coupled chemical reaction rates and adsorption, and experimental data are often compared to (8.19) and (8.20) for the purpose of detecting the presence of such events.

The chronocoulometric, or charge-time, response of the potential step experiment is elicited from (8.19) and (8.20) simply by current-time integration. For the limiting case of sufficiently negative E_{final}, this gives

$$Q = 2nFC_O{}^b\left(\frac{D_O t}{\pi}\right)^{1/2}. \tag{8.22}$$

Thus a diffusion-controlled chronocoulometric response exhibits a Q—$t^{1/2}$ proportionality. Actually, when "raw" data are plotted against $t^{1/2}$, the resulting straight-line plot for a diffusion-controlled system exhibits a positive zero-time intercept. This charge intercept term reflects the difference in the double-layer charge on the working electrode at E_{init} and E_{final}, Q_{dl}. The double-layer charge can also be measured in a blank experiment carried out in the absence of sample O. A double-layer charging current appears in the chronoamperometric response as a very short time current "spike" which is often not clearly recorded.

Double Potential Step Experiment

This experiment, of recent vintage, involves application of first a cathodic potential step, causing reduction of O and the condition $C_{O(0,t)} = 0$, followed at some time τ by an anodic potential step causing reoxidation of the product R (generated during $t < \tau$) and the condition $C_{R(0,t)} = 0$. The shape of the chronoamperometric and chronocoulometric response curves are illustrated in Fig. 8.5. At times $t < \tau$, the current and charge-time curves are described by the previous equations (8.19) and (8.22). At times $t > \tau$, the oxidation of R occurs from the diffusion layer of R created during the cathodic step. The diffusion profile description of R is thus in effect required for the initial condition of the reverse potential step; the R diffusion profile is connected to that of O by the expression

$$D_O\frac{\partial C_{O(0,t)}}{\partial x} = -D_R\frac{\partial C_{R(0,t)}}{\partial x} \tag{8.23}$$

and application of this to a Fickian solution for R yields for $t \leq \tau$

$$C_{R(x,t)} = C_O{}^b\left(\frac{D_O}{D_R}\right)^{1/2}\text{erfc}\left(\frac{x}{2D_R^{1/2}t^{1/2}}\right), \tag{8.24}$$

where erfc is the complementary error function $[1 - \text{erf}(Z)]$. (In the event that $D_O = D_R$, the diffusion profile of R is an exact mirror image of that of O.)

The solution of the response to the reverse potential step was obtained for the chronoamperometric case by Schwarz and Shain [9], and for the chronocoulometric case by Christie [10]:

$$i_a = \frac{nFD_O^{1/2}C_O^{\ b}}{\pi^{1/2}}\left[\frac{1}{(t-\tau)^{1/2}} - \frac{1}{t^{1/2}}\right] \tag{8.25}$$

$$Q_a = \frac{2nFD_O^{1/2}C_O^{\ '}}{\pi^{1/2}}[t^{1/2} - (t-\tau)^{1/2}]. \tag{8.26}$$

Comparison of experimental data to these expressions is accomplished by the obvious i_a—$f(t)$ or Q_a—$f(t)$ plots, or by examination of i_a/i_c and Q_a/Q_c ratios. In the latter it is convenient to measure the forward and backward currents and charges at the same time following their respective potential steps (times t and $t - \tau$), and to take for the reverse charge the quantity $Q_\tau - Q_a$. Taking ratios of (8.19) and (8.25) and (8.22) and (8.26):

$$\left|\frac{i_a}{i_c}\right| = 1 - \left(\frac{\theta}{1+\theta}\right)^{1/2} \tag{8.27}$$

$$\frac{Q_\tau - Q_a}{Q_c} = 1 - \left(\frac{1+\theta}{\theta}\right)^{1/2} + \left(\frac{1}{\theta}\right)^{1/2}, \tag{8.28}$$

where $\theta = (t - \tau)/\tau$ (an intuitively significant quantity expressing the time scale of the forward and backward measurement relative to that of τ) and

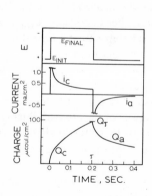

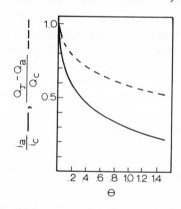

Fig. 8.5 Chronoamperometric and chronocoulometric double potential step responses calculated [see (8.19), (8.22), (8.25), and (8.26)] for $n = 1$, $D = 4.0 \times 10^{-6}$ cm/sec, $C_O^b = 1.0 \times 10^{-6} M/cm^3$, $\tau = 0.200$ sec.

Fig. 8.6 Variation of double potential step current and charge ratios with θ [see (8.27) and (8.28)].

the absolute value recalls the opposite signing of i_a and i_c. Figure 8.6 shows how the ratios depend on θ. Adherence of experimental data to the predictions of (8.27) and (8.28) requires that both reactant O and product R react in a diffusion-controlled manner. The discussion following will show that ratios of Fig. 8.6 can be quite sensitive to adsorption of reactant O and to chemical reactions generating O or consuming R.

Mass Transfer Control by Coupled Chemical Reactions

If chemical reactions generating O or consuming R occur with reaction half-lives comparable to the observation time of a chronoamperometric or chronocoulometric response curve, their presence can become reflected in an alteration of the responses from the diffusion-controlled shapes. Chrono-amperometric and chronocoulometric theory have been examined for a variety of reaction types; we examine here four important cases.

Preceding Chemical Reactions (Prekinetics)

Consider the case of reactant O lying at initial equilibrium with an electroinactive substance Z

$$Z \underset{k_b}{\overset{k_f}{\rightleftharpoons}} O; \qquad O + ne \to R \tag{8.5}$$

with the rates and initial concentrations $C_O{}^b$ and $C_Z{}^b$ being connected by the equilibrium constant,

$$K = k_f/k_b = C_O{}^b/C_Z{}^b \tag{8.29}$$

For very small k_f, the current-time potential step response is controlled solely by the initial concentration of O through the Cottrell equation (8.19). Larger k_f causes significant Z $\to$ O conversion, however, and the Fickian solution of diffusion equations for Z and O [see (8.6) and (8.7)] must be obtained, using the boundary conditions equations (8.3) and (8.4) (written for both O and Z), (8.16), (8.29), and

$$\frac{\partial C_{Z(0,t)}}{\partial x} = 0, \tag{8.30}$$

which is a statement of electroinactivity of Z at the potential for O reduction. There results [8], under the simplifying assumption that $D_O = D_Z$,

$$i = nF(C_O{}^b + C_Z{}^b)(Dk_f K)^{1/2} \exp [k_f Kt] \operatorname{erfc} [k_f Kt]^{1/2}. \tag{8.31}$$

The behavior of this kinetically controlled current is assessed by comparison to the Z and O diffusion-controlled current obtained for very rapid Z $\to$ O conversion, which is found from (8.31) at large k_f to be

$$i_\infty = nF(C_O{}^b + C_Z{}^b)(D/\pi t)^{1/2}. \tag{8.32}$$

Dependency of the ratio of (8.31) and (8.32)

$$\frac{i}{i_\infty} = (\pi k_f K t)^{1/2} \exp\,[k_f K t]\,\text{erfc}\,[k_f K t]^{1/2} \tag{8.33}$$

on the parameter $(k_f K t)^{1/2}$ is illustrated in Fig. 8.7. Thus at sufficiently large k_f, K, or time, the observed current approaches a diffusion-controlled value. Data taken from Fig. 8.7 using experimental i/i_∞ values at various times should yield a constant $k_f K$ to confirm obedience of the experimental case to reaction (8.5). A separate measurement of the equilibrium constant by any appropriate method then allows evaluation of the rate constants k_f and k_b.

The chronocoulometric experiment employs an integrated form of (8.33) and working curves of Q/Q_∞ similar to those of Fig. (8.7). It has recently been applied to study of the kinetics of the monomer–dimer equilibrium between Fe(III)–EDTA complexes [11]; the problem assumes a somewhat different theoretical form in this case owing to the second-order character of this type of reaction.

Following Chemical Reactions (Postkinetics)

Study of much interesting chemistry can become possible via electrochemical generation of an unstable species R, and investigations in this area are an important facet of electrochemical research. Three important reaction types following the $O + ne \rightarrow R$ electrode reaction are the simple postkinetic (EC) case, the *catalytic* case, and the case of a postkinetic reaction leading to an additional electroactive species (ECE). The postkinetic EC and ECE cases may be either chemically reversible or irreversible ($k_b = 0$); the catalytic case, regenerating reactant O, is typically irreversible (else the cell solution decays spontaneously). The available theory for first-order decay reactions applies equally well of course to pseudo-first-order reactions in which the other reaction components are in large excess or are buffered.

EC CASE

$$O + ne \rightarrow R$$
$$R \underset{k_b}{\overset{k_f}{\rightleftharpoons}} Y \tag{8.34}$$

A single potential step reducing O and producing the condition $C_{O(0,t)} = 0$ yields only the O-diffusion-controlled responses of (8.19) and (8.22). The double potential step experiment must be applied for detection and kinetic characterization of a following decomposition of R; obviously, the reverse, anodic currents or charges for R oxidation are less than the diffusion-only values according to the value of k_f in (8.34). For the irreversible chemical case ($k_b = 0$), simultaneous solution of the diffusion equation for O (8.2) and R,

$$\frac{\partial C_{R(x,t)}}{\partial t} = D_R \frac{\partial^2 C_{R(x,t)}}{\partial x^2} - k_f C_{R(x,t)}, \tag{8.35}$$

under appropriate boundary conditions was accomplished by Schwarz and Shain [9]. The chronoamperometric $i_c - t$ behavior prior to reverse potential stepping at time τ is given by (8.19); that at $t > \tau$ is

$$i_a = nFC_O{}^b \left(\frac{D_O}{\pi}\right)^{1/2} \left[\frac{\phi}{(t-\tau)^{1/2}} - \frac{1}{t^{1/2}}\right], \tag{8.36}$$

where ϕ is a complex but numerically evaluable function of k_f, τ, and t. This expression is identified with the diffusion-only equation (8.25) when $k_f = 0$; that is, $\phi = 1.00$ in the absence of any R decomposition.

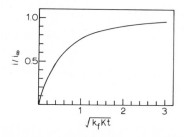

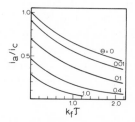

Fig. 8.7 Working plot of (8.33) for chrono-amperometric evaluation of prekinetic rate.

Fig. 8.8 Working plot of (8.37), for selected θ, for double potential step chronoamperometric evaluation of EC postkinetics.

As in the diffusion-controlled double potential step case, it is convenient to examine the ratio i_a/i_c where i_a and i_c are measured on the current-time response curve (see Fig. 8.5) at times t and $t - \tau$, respectively. The ratio of (8.19) and (8.36) written at these times is

$$\left|\frac{i_a}{i_c}\right| = \phi - \left[\frac{\theta}{1+\theta}\right]^{1/2}, \tag{8.37}$$

where θ is again $(t - \tau)/\tau$. Figure 8.8 shows working curves prepared [9] for dependency of the current ratio on $k_f\tau$ at various selected measurement times θ. Values of k_f result directly from use of these curves and experimental i_a/i_c measured at appropriate θ. At short τ the ratios tend to their diffusion-only values (Fig. 8.6); at long τ, toward zero as R becomes more or less completely consumed before reverse potential stepping. It is evident that optimum sensitivity of i_a/i_c data to the decomposition of R is obtained when $\tau = 0.5$ to $1.5k$; that is, the reverse potential step time τ employed should be of magnitude comparable to the reaction half-life of the R $\rightarrow$ Y reaction.

The double potential step chronoamperometric experiment was applied by Schwarz and Shain [9] to the rearrangement reaction of hydrazobenzene (to benzidine) produced by reduction of azobenzene in acid solution. Rate constants measured ranged from 0.6 to 86 sec^{-1}, depending on pH. The capabilities of the method have been estimated to encompass $0.04 < k_f < 10^3 sec^{-1}$ by Koopman [12, 13], who has applied the method to a reaction following electroreduction of nitrophenol. Ascorbic acid undergoes a rapid hydration reaction following two-electron oxidation, and Perone and Kretlow [14] have successfully employed reverse potential stepping times $\tau = 0.5$ to 1.0 msec to evaluate a $k_f = 1.3 \times 10^3 sec^{-1}$. (The upper limit of accessible k_f estimated by Koopman, which itself labels the double potential step chronoamperometric method as a powerful fast kinetics technique, may in fact be too low.) An intermetallic compound formation between Zn and $Co_2(Hg)_x$ occurs upon reduction of $Zn(II)$ into a $Co_2(Hg)_x$ amalgam electrode, and Hovsepian and Shain [15] have employed the double potential step method for its kinetic characterization.

The chronocoulometric response to the EC case in the double potential step experiment was obtained by Christie [10]; the Q_c-t curve at $t \leq \tau$ is given by (8.22) and the Q_a-t curve $(t > \tau)$ by

$$Q_a = 2nFC_O^b \left(\frac{D_O}{\pi}\right)^{1/2} \left[t^{1/2} - (t - \tau)^{1/2} \int \phi \, dt \right], \qquad (8.38)$$

where $\int \phi \, dt$ is again a complex but numerically evaluable quantity which reduces to 1.00 at $k_f = 0$ [see (8.26)]. As in the diffusion-only case, we examine the $(Q_\tau - Q_a)/Q_c$ ratio; Fig. 8.9 shows the ratio as a function of $(k_f\tau)^{1/2}$ for the particular case of $\theta = 1.00$ (Q_c measured at $t = \tau$, Q_a at $t = 2\tau$). While the results of reference [10] and Fig. 8.9 are qualitatively correct, a recent report by Van Duyne [15 a] should be consulted for a quantitative correction of the theory for the EC double potential step chronocoulometric case. For k_f or $\tau \to 0$, the ratio tends to the diffusional value of 0.586.

CATALYTIC CASE

$$O + ne \to R$$
$$V + R \xrightarrow{k'} O + V' \qquad (8.39)$$

The special circumstance in which the following chemical reaction is a redox process regenerating the original reactant O, produces a kinetic control of both the cathodic current (or charge) in a single potential step experiment (O reduction) and of the anodic current (or charge) in a double potential step experiment (R reoxidation). The cathodic response is larger, and the anodic smaller, as compared to the diffusion-only predictions of (8.19), (8.22), (8.25), and (8.26).

The pertinent Fickian equations are (8.35) for R, and for O:

$$\frac{\partial C_{O(x,t)}}{\partial t} = D_O \frac{\partial^2 C_{O(x,t)}}{\partial x^2} + kC_{R(x,t)}. \tag{8.40}$$

Solution of these, under the potential step condition $C_{O(0,t)} = 0$ and other appropriate conditions, was obtained by Delahay and Stiehl [16], Miller [17], and Pospisil [18]:

$$i_{\text{cat}} = nFC_O^b D^{1/2} \left[k^{1/2} \operatorname{erf} (kt)^{1/2} + \frac{\exp(-kt)}{(\pi t)^{1/2}} \right]. \tag{8.41}$$

The expected behavior of the *catalytic current* is examined by comparison to the diffusion-only current (8.19) and preparation of a working plot of $i_{\text{cat}}/i_{k=0}$

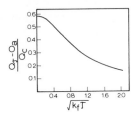

Fig. 8.9 Working plot for double potential step chronocoulometric evaluation of EC postkinetics; $\theta = 1.00$.

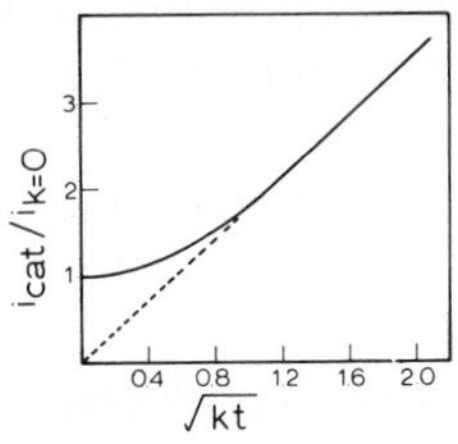

Fig. 8.10 Working plot of (8.42) for chronoamperometric evaluation of catalytic kinetics.

against the parameter $(kt)^{1/2}$ (Fig. 8.10). We note from the figure and equation for $i_{\text{cat}}/i_{k=0}$,

$$\frac{i_{\text{cat}}}{i_{k=0}} = (kt)^{1/2} \left[\pi^{1/2} \operatorname{erf} (kt)^{1/2} + \frac{\exp(-kt)}{(kt)^{1/2}} \right], \tag{8.42}$$

that at large $(kt)^{1/2}$, (>2), a linear variation results with slope $\pi^{1/2}$. The rate constant k in these expressions is actually $k'[V]$. It is apparent that the magnifying effect of the catalytic process on the current grows with increasing time; one must recall during longer time measurements that the theory holds only for the condition that V is not seriously depleted. This is readily shown [8] to be

$$C_V^b/C_O^b \gg (\pi kt)^{1/2}. \tag{8.43}$$

Delahay and Stiehl [16] and Miller [17] both applied the above theory to the catalytic reaction between Fe(II) and hydrogen peroxide following Fe(III) reduction, but in the polarographic chronoamperometric context. Guidelli and Cozzi [19] have developed the theory further. Booman and Pence [20, 21] have considered the theory for a disproportionation reaction of R $(2R \to O + Y$, a second-order process), and applied the chronoampero-metric experiment to the case of $U(V)$. Sometimes coupled chemical reactions occur heterogeneously (*on* the electrode surface) rather than following diffusion into the solution surrounding the electrode, and McIntyre [22] and Guidelli [23] have recently made theoretical inroads into this important problem in the catalytic and other reaction cases.

The single potential step chronocoulometric response for the catalytic case, according to Christie [10], is

$$Q_{cat} = nFC_O{}^b D^{1/2} \left\{ \left[k^{1/2}t + \frac{1}{2k^{1/2}} \right] \mathrm{erf}\,(kt)^{1/2} + \left(\frac{t}{\pi} \right)^{1/2} \exp\,(-kt) \right\}. \quad (8.44)$$

This somewhat more complex expression can be examined and compared to experimental data in the same basic manner as (8.43) (a plot of $Q_{cat}/Q_{k=0}$ against kt) except that a family of curves representing different values of k is required.

In a double potential step experiment, the current and charge-time responses for $t \leq \tau$ are given by (8.41) and (8.44); those at $t > \tau$ are [10]

$$i_{a(cat)} = nFC_O{}^b(kD)^{1/2} \left\{ \mathrm{erf}\,(kt)^{1/2} + \frac{\exp\,(-kt)}{(\pi kt)^{1/2}} \right.$$

$$\left. - \mathrm{erf}\,[k(t-\tau)]^{1/2} - \frac{\exp\,[-k(t-\tau)]}{[k\pi(t-\tau)]^{1/2}} \right\} \quad (8.45)$$

$$Q_{a(cat)} = nFC_O{}^b \left(\frac{D}{k} \right)^{1/2} \left\{ (kt + \tfrac{1}{2}) \mathrm{erf}\,(kt)^{1/2} + \left(\frac{kt}{\pi} \right)^{1/2} \exp\,(-kt) \right.$$

$$\left. - [k(t-\tau) + \tfrac{1}{2}] \mathrm{erf}\,[k(t-\tau)]^{1/2} - \left[\frac{k(t-\tau)}{\pi} \right]^{1/2} \exp\,[-k(t-\tau)] \right\}. \quad (8.46)$$

The chronocoulometric response was evaluated [10] in the same way as the EC equation (8.38)—a plot of $(Q_{\tau(cat)} - Q_{a(cat)})/Q_{c(cat)}$ against $(k\tau)^{1/2}$ for $\theta = 1.00$—and the form of the plot is qualitatively similar to that of Fig. 8.9. The charge ratio tends to the diffusional value 0.586 at very short times of potential reversal and to zero at long τ or large k. Lingane and Christie [24] applied the double potential step chronocoulometric method to the reduction of Ti(IV) to Ti(III) with catalytic regeneration of Ti(IV) produced by reaction of Ti(III) with hydroxylamine; a $k' = 43.4$ liter mole^{-1} sec^{-1} results. It should be observed, however, that the advantages of the integral response cited by the authors could perhaps have been more simply procured by use of the single potential step method (8.44).

ECE CASE

$$O + ne \to R; \qquad R \underset{k_b}{\overset{k_f}{\rightleftharpoons}} Y; \qquad Y + n'e \to X \qquad (8.47)$$

A postkinetic chemical process which leads to a product Y, which is more readily reducible than the original electrode reactant O, is an especially interesting circumstance in electrode process, occurring in a variety of organic electrode reactions. (An example is the one-electron reduction of an aromatic hydrocarbon followed by protonation of the radical anion to yield a further reducible neutral radical [2]). The cathodic current in a single potential step experiment is enhanced over the diffusion-only value for O, [see (8.19)], according to the values of the rate constants and the relative values of n and n'.

Single potential-step chronoamperometric ECE theory, first given by Koutecky [25] and by Tachi and Senda [26], has been presented by Alberts and Shain [27] in a more intelligible form. Under the potential step condition that $C_{O(0,t)} = C_{Y(0,t)} = 0$, and for the irreversible chemical reaction case ($k_b = 0$), the chronoamperometric result is

$$i = FC_O{}^b\left(\frac{D}{\pi t}\right)^{1/2}\{n + n'(1 - \exp[-k_f t])\}. \tag{8.48}$$

The effect of the ECE process on the current-time response is illustrated in Fig. 8.11 as a i-$t^{-1/2}$ plot. At short times the i-t response is mainly controlled by electrolysis of O and the slope of the i-$t^{-1/2}$ plot is proportional to n. At long times, the chemical process "keeps pace" with the electrochemistry, the i-t response is governed by joint O + Y electrolysis, and the i-$t^{-1/2}$ slope is proportional to $n + n'$. At intermediate times the rate k_f is also a current-controlling and thus evaluable factor.

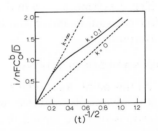

Experimental data is compared to (8.48) in two ways: first, a series of working curves similar to Fig. 8.11 is prepared, varying k_f to find the working curve of "best fit" to experimental data, and (2) a working plot

Fig. 8.11 Current-time responses for chronoamperometry with ECE kinetics for $k_f = \infty$, 0.1, and 0, and $n = n' = 1$.

of the ratio of (8.48) to the expression for $k_f = \infty$ [see (8.19)] written for $n = n + n'$ plotted against $\log(k_f t)$ is prepared and experimental values of i/i_∞ for various times are used to read off values of $k_f t$. Various approaches can be used to obtain data for i_∞ for the system under study [27].

The case for $k_b \neq 0$ was also treated by Alberts and Shain [27] but, for reasons they discuss, nonzero values of k_b have little effect on the working curves. The chronoamperometric theory was applied to the reduction of p-nitrosophenol to p-hydroxylaminophenol, which dehydrates with $k_f = 1.4\ \text{sec}^{-1}$ (pH = 4.8) to further reducible p-benzoquinoneimine. More recently, Nelson and Adams [27a] have applied chronoamperometry in an elegant study of the oxidation of triphenylamines, which undergo dimerization to further oxidizable benzidines in acetonitrile solvent.

A possibility overlooked in the early chronoamperometric ECE theory [27] was the electron transfer process

$$O + X \underset{k_2}{\overset{k_1}{\rightleftharpoons}} R + Y \qquad (8.49)$$

which could occur in the diffusion layer mixture of these species. Attention was drawn to this "nuance" in the ECE reaction by Hawley and Feldberg [28], who showed that results obtained neglecting this process could be biased high or low (in k_f), depending on the value of k_1/k_2. Theoretical i/i_∞-$k_f t$ working curves for various ratios k_1/k_2 were calculated for the reasonable assumption that (8.49) maintained an equilibrium position throughout the electrolysis. These were applied by Adams et al. [29] to an experimental case, the 1,4-addition of amines or hydrochloric acid to o-benzoquinones electrogenerated by catechol reductions, to demonstrate the importance of the nuance reaction.

Koopman [29a] has presented an elegant experimental and theoretical treatment of the double potential step chronocoulometric experiment in which the chemical step in the ECE sequence occurs heterogeneously on the electrode surface and reactants and products are involved in a linear isotherm adsorption.

SECOND-ORDER REACTIONS

At this point, we should recognize an important general problem in theoretical treatments of electrode-coupled chemical reactions. This concerns reactions that are second order. In the ECE case, for instance, one of the Fickian expressions that must be solved, that for O, becomes in the presence of the nuance equation (8.49)]:

$$\frac{\partial C_{O(x,t)}}{\partial t} = D_O \frac{\partial^2 C_{O(x,t)}}{\partial x^2} - k_1 C_{O(x,t)} C_{X(x,t)} + k_2 C_{R(x,t)} C_{Y(x,t)}. \quad (8.50)$$

Explicit solution of expressions such as this are, with a few exceptions, made impossible because of the presence of cross products of concentration terms. For many years the usual approach to this problem lay in application of an approximate concept known as the reaction layer [8]. Only recently have more exact results for second-order kinetic cases become available through application of a calculation technique introduced by Feldberg and Auerbach [30]. Diffusion is treated as transport from one thin volume layer to another; an explicit equation is not obtained but rather numerical calculations (computer) of diffusion profiles, current-time curves, and so on. Such results, esthetically less pleasing than explicit formulas but nevertheless highly useful, have opened numerous new avenues to study of coupled chemical processes that were previously avoided because of lack of suitable theoretical description. Feldberg [31] has recently described the calculation in some detail.

Adsorption of Electrode Reactant

Single Potential Step Chronocoulometry

Consider the case in which the reactant O lies at $t = 0$ in an equilibrium between adsorbed and solution forms

$$
\begin{array}{c}
O_{ads} \\
\Big\updownarrow \searrow^{ne} \\
\quad\quad R \\
\Big\updownarrow \nearrow_{ne} \\
O_{soln}
\end{array}
\qquad (8.14)
$$

Both O_{ads} and O_{soln} are presumed electroactive. Application of a potential step sufficiently negative to produce $C_{O(soln)(0,t)} = C_{O(ads)(0,t)} = 0$ yields the normal diffusional current-time response [see (8.19)] for electrolysis of O_{soln}, but the adsorbed species O_{ads}, not requiring a mass transfer step prior to reduction and present in a limited amount (determined by the surface excess of O_{ads}, Γ_0 at $t = 0$), reacts in a sharp current spike at exceedingly short time. Although the presence of the current spike for reaction of O_{ads} can sometimes be discerned, it is difficult to measure this quantitatively. If, however, we record the chronocoulometric (Q-t) response of this experiment, the charge attributable to reaction of O_{ads}

$$
Q_{ads} = nF\Gamma_0 \quad \text{(coulombs/cm}^2) \qquad (8.15)
$$

is "remembered" at longer times and in fact simply adds at any time to the time-dependent charge accumulated by reaction of O_{soln} (8.22).

The expression for the chronocoulometric single potential step response in the presence of reactant adsorption is then

$$
Q = 2nFC_O{}^b \left(\frac{D_O t}{\pi}\right)^{1/2} + Q_{ads} + Q_{dl}, \qquad (8.51)
$$

where $C_O{}^b$ is the bulk concentration of O_{soln}, Q_{dl} is the difference in the charge on the electrical double layer between E_{init} and E_{final} of the potential step, and Q_{ads} represents $nF\Gamma_0$ at E_{init}. The experimental Q-t data is analyzed through a Q-$t^{1/2}$ plot, of which Fig. 8.12 is an example. We see that the zero-time intercept provides $Q_{ads} + Q_{dl}$, from which, after correction for Q_{dl}, we obtain the surface excess Γ_0 with (8.15).

The correction Q_{dl} is obtained from a blank potential step from E_{init} to E_{final} in supporting electrolyte solution alone. As the electrode double layer is in effect a capacitor, this is a time-independent charge, Q_{dl}^b. We must also, however, account for any change in the double-layer charge at E_{init} caused by the presence of O_{ads}. This can be evaluated for mercury electrodes by integrating the charge accumulated on a growing DME drop, at E_{init}, in the presence and absence of O_{ads} [32]. The difference in these charges, ΔQ_{dl}, when subtracted from Q_{dl}^b, yields the appropriate Q_{dl} for application to (8.51) and evaluation of Γ.

The chronocoulometric method of adsorption measurement was introduced by Christie and co-workers [33]. Because of its obvious superiority to the chronopotentiometric method then in use for adsorption measurements (see Section 3), which had some shortcomings, chrono-coulometry quickly supplanted this method and now, with its double potential step partner (*vide infra*), is the present "Fara-daic method of choice" for accurate and sensitive surface excess studies. The single potential step method has been applied to adsorption of metal complexes on mercury electrodes, including studies of adsorption of Co(III)–EDTA complexes by Anson [34], of Pb(II) iodide, bromide, and thiocyanate by Murray and Gross [32], of Hg(II) thiocyanate by Anson and Payne [35], of Cd(II) iodide and In(III) thiocyanate by O'Dom and Murray [36, 37], and of Zn(II) thiocyanate by Osteryoung and Christie [38].

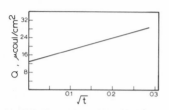

Fig. 8.12 Analysis of chrono-coulometric data for reactant adsorption by (8.51). $n = 1$, $D = 4.0 \times 10^{-6}$ cm²/sec, $C_O^b = 0.5$ M/cm³, $\Gamma_0 = 1.0 \times 10^{-10}$ M/cm², $Q_{dl} = 8\ \mu C/cm^2$.

Double Potential Step Chronocoulometry

Inasmuch as Γ_0, and consequently Q_{ads}, can often be an exceedingly small term (Γ_0 values measured have ranged from 3×10^{-12} to 8×10^{-10} mole/cm²), Q_{dl} frequently exceeds Q_{ads}. The necessity for accurate evaluation of Q_{dl} can then become extreme, and the accumulation of errors in the total of four experiments required in the single potential step method can limit the sensitivity of detection and quantitative evaluation of Q_{ads}. The double potential-step chronocoulometric experiment, introduced by Anson [39] and theoretically characterized by Christie et al. [40] eliminates much of the experimental error incurred in the Q_{dl} correction by acquiring the $Q_{ads} + Q_{dl}$ and Q_{dl} data in a single experiment.

In the double potential step chronocoulometric experiment with reactant adsorption, the charge during $t \leq \tau$ is again given by (8.51) and the charge during $t > \tau$ by [40]

$$Q_a = 2nFC_O^b \left(\frac{D_O}{\pi}\right)^{1/2} [t^{1/2} - (t - \tau)^{1/2}] + Q_{ads}\left[\frac{2}{\pi}\sin^{-1}\left(\frac{\tau}{t}\right)^{1/2}\right]. \quad (8.52)$$

This expression, describing charge for oxidation of R, differs from the diffusion-only equation (8.26) only in the last term, which accounts for the extra R (in the R diffusion profile) arising from previous reduction of the adsorbed O component. This term is small and well approximated by a simpler expression; making such approximation and now measuring Q_a as a difference from the cathodic charge at τ [Q_τ, (8.51) written for $t = \tau$] gives

$$Q_\tau - Q_a = 2nFC_O{}^b \left(\frac{D_O}{\pi}\right)^{1/2}\left[1 + \frac{a_1\Gamma_0\pi^{1/2}}{2C_O{}^b D_O^{1/2}\tau^{1/2}}\right]$$

$$\times\ [(t - \tau)^{1/2} + \tau^{1/2} - t^{1/2}] + a_0Q_{ads} + Q_{dl}. \quad (8.53)$$

The terms a_0 and a_1 are constants resulting from the above approximation over a specified range of t and τ; typical values [40] are $a_0 = -0.0688$ and $a_1 = 0.970$.

Experimental data are treated according to (8.51) and (8.53) by plots of Q_c against $t^{1/2}$ and $Q_\tau - Q_a$ against $(t - \tau)^{1/2} + \tau^{1/2} - t^{1/2}$; an example is shown in Fig. 8.13. Comparison of (8.51) and (8.53) shows that the intercepts of these plots are $Q_{ads} + Q_{dl}$ and $a_0Q_{ads} + Q_{dl}$; the difference between these intercepts eliminates Q_{dl} and is $Q_{ads}\ (1 - a_0)$, allowing calculation of Γ_0.

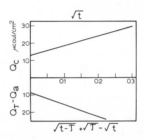

Fig. 8.13 Analysis of double potential step chronocoulometric data for reactant adsorption. Parameters same as in Fig. 8.12; $\tau = 0.100$ sec.

Evaluation of reactant surface excesses by the double potential step method yields experimental uncertainties in Γ_0 from 2 to 5 times smaller than those typical ($\pm 1 \times 10^{-11}$ mole/cm²) [36] for the single potential step experiment. A further, extremely significant, advance has been the application [41] of on-line computerized data acquisition methods to the double potential step experiment. The considerably greater accuracy of data recording thus possible has lowered the uncertainty in Γ_0 adsorption measurements further to about $\pm 1 \times 10^{-12}$ mole/cm² [42].

Anson and co-workers have applied this experiment to adsorptions, on mercury working electrodes, of Cd(II) from thiocyanate [43], thiosulfate [44], chloride [41], iodide and bromide [42] solutions, and of anthraquinone monosulfonate ion [45].

Charge Transfer Kinetics

As shown earlier by (8.10) and (8.11), the rate of a charge transfer reaction $O + ne \rightarrow R$ depends on both its rate constant k_s and the value of the applied potential. The influence of k_s on current is removed at sufficiently negative applied potential, as then the charge transfer rate process is made fast enough that mass transport of reactant becomes the current-limiting factor. Such was the case in the chronoamperometric and chronocoulometric experiments discussed above; the relations derived there had no dependence on the value of k_s, or the reversibility of the charge transfer reaction. We now consider use of much smaller potential steps and the resulting charge transfer rate connotations in chronoamperometric and chronocoulometric experiments.

The chronoamperometric, or potentiostatic (as it is commonly called in this area) experiment was developed for charge transfer kinetic studies by Gerischer and Vielstich [46, 47]. Consider first the experiment as conducted in a solution containing both O and R; E_{init} is thus the equilibrium ($i = 0$) potential E_{eq} of this mixture and E_{final} is a potential somewhat more negative than E_{eq}. Solution of the Fickian equations for O and R using (8.11) as a boundary condition yields

$$i = i_0 \left\{ \exp \left[\frac{-\alpha n F}{RT} \eta \right] - \exp \left[\frac{(1 - \alpha) n F}{RT} \eta \right] \right\} \exp (\lambda^2 t) \, \text{erfc} \, (\lambda t^{1/2}), \quad (8.54)$$

where

$$\lambda = \frac{i_0}{nF} \left(\frac{\exp \left[\dfrac{-\alpha n F}{RT} \eta \right]}{C_O^b D_O^{1/2}} + \frac{\exp \left[\dfrac{(1 - \alpha) n F}{RT} \eta \right]}{C_R^b D_R^{1/2}} \right) \quad (8.55)$$

and $\eta = E_{\text{final}} - E_{\text{eq}}$, the *charge transfer overvoltage*. The first term of (8.54) is recognized as identical to (8.11). The second, time-dependent term represents the *mass transfer overvoltage*, describing the decay of current resulting from electrolytic depletion of O, and enhancement of R, in a diffusion layer around the working electrode. The time dependency of the mass transfer term is illustrated in Fig. 8.14. At sufficiently long time ($\lambda t^{1/2} > 5$), the erfc $(\lambda t^{1/2})$ term is $\approx \exp [-\lambda^2 t]/\pi^{1/2} \lambda t^{1/2}$, λ and i_0 vanish from (8.54) and the current becomes dominated by a i-$t^{-1/2}$ mass transfer similar to that of the Cottrell equation (8.19). The most pronounced dependency on the charge transfer kinetic term i_0 occurs at short electrolysis times, and the current-time data are taken with this in mind.

Two approaches are useful for dealing with (8.54) at short electrolysis times. In one, measurement at times sufficiently small that $1 \gg \lambda t^{1/2}$ allows restatement of the mass transfer term as $1 - 2\lambda t^{1/2}/\pi^{1/2}$. Thus a plot of current against $t^{1/2}$ is linear with zero-time intercept current equal to (8.11). If also the potential step is sufficiently small ($|\eta| < \alpha n F/RT$, a few millivolts), then (8.11) simplifies to

$$i_{\eta \to 0} = - \frac{n F i_0 \eta}{RT}, \quad (8.56)$$

from which the exchange current i_0 is calculable. Application of (8.12) then yields the charge transfer rate k_s. In the second approach [48, 49], current data over a span of time when $0.14 < \lambda t^{1/2} < 1.0$ are employed with an alternative mathematical expression for the mass transfer term to again obtain the current of (8.11). Further details can be found in a review of methods for charge transfer study by Delahay [50]. It is estimated that the potentiostatic method is applicable to measurement of charge transfer rates $k_s < 1$ cm/sec.

The potentiostatic experiment can also be conducted on solutions containing only reactant O $(C_R{}^b = 0)$, a convenience when solutions of R are

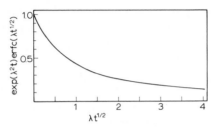

Fig. 8.14 Variation of mass transfer overvoltage term of (8.54) with its argument $\lambda t^{1/2}$.

difficult to maintain or prepare. In this case the Fickian solution for mass transport of O and R employs (8.10) as a boundary condition, and there results [8]

$$i = nFk_sC_O{}^b\left\{\exp\left[\frac{-\alpha nF(E - E^0)}{RT}\right]\right\} \exp(\gamma^2 t)\,\text{erfc}\,(\gamma t^{1/2}), \quad (8.57)$$

where

$$\gamma = k_s\left\{\frac{\exp\left[\dfrac{-\alpha nF(E - E^0)}{RT}\right]}{D_O^{1/2}} + \frac{\exp\left[\dfrac{(1 - \alpha)nF(E - E^0)}{RT}\right]}{D_R^{1/2}}\right\} \quad (8.58)$$

and E is E_{final} in the potential step. Again, at long times (large $\gamma t^{1/2}$), (8.57) reduces to a largely mass transfer-controlled condition. At shorter times (8.57) is handled in the same manner as (8.54); a plot of current against $t^{1/2}$ yields

$$i_{t=0} = nFk_sC_O{}^b\left\{\exp\left[\frac{-\alpha nF(E - E^0)}{RT}\right]\right\} \quad (8.59)$$

from a logarithmic form of which, with a series of experiments at different E, data for k_s results.

Chronocoulometric theory for the case of $C_R{}^b = 0$ has been given by

Christie et al. [51]; the integrated form of (8.57) is:

$$Q = \frac{nFk_sC_O{}^b}{\gamma^2}\left\{\exp\left[\frac{-\alpha nF(E-E^0)}{RT}\right]\right\}\left\{\exp(\gamma^2 t)\,\mathrm{erfc}\,(\gamma t^{1/2}) + \frac{2\gamma t^{1/2}}{\pi^{1/2}} - 1\right\}.$$

(8.60)

At sufficiently long times ($\gamma t^{1/2} > 5$), the term $\exp(\gamma^2 t)\,\mathrm{erfc}\,(\gamma t^{1/2})$ becomes negligible; thus a plot of Q against $t^{1/2}$ becomes linear with extrapolated intercept on the charge axis of

$$t_{\mathrm{int}}^{1/2} = \frac{\pi^{1/2}}{2\gamma}.$$

(8.61)

An example of such a plot is shown in Fig. 8.15. Values of γ obtained for a series of potential steps at different E pass through a minimum at $E \approx E^0$, as is apparent from (8.58). For $\alpha = 0.5$ and $D_O = D_R$, this occurs at $E = E^0$, and $\gamma_{\min} = 2k_s/D^{1/2}$. (See Ref. 51 for more general equations for $\gamma_{\min}$.)

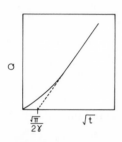

Fig. 8.15 Analysis of charge transfer-controlled chronocoulometric data.

It is interesting that in the chronocoulometric method data taken at long times are useful for extraction of k_s; just the opposite is true in chronoamperometry. This is a direct consequence of the integral (charge) mode of data-recording (effecting a "memorization" of preceding kinetic influences on the flow of current) and is a practical convenience of this method. Difficulties can arise, however [51], for slower charge transfer rates, inasmuch as the electrolysis times necessary to achieve linearity in the Q-$t^{1/2}$ plot (to obtain $\gamma t^{1/2} > 5$) may be sufficiently long to be susceptible to convective and other deleterious effects on the diffusion mass transport process. Lingane and Christie [52], by combining chronoamperometric and chronocoulometric data, seek to eliminate short and long time measurement problems, respectively, of these methods.

3 CHRONOPOTENTIOMETRY

The elementary theory for current step, or constant-current chronopotentiometry was set forth in the early twentieth century [53–56]. The term "transition time" was coined in later work [57, 58]. Although some (albeit inaccurate) experiments were described in the early reports, it was not until the work of Gierst and Juliard [59], followed closely by a timely chapter on the subject [8] and coincident papers by Delahay and Mamantov [60] and Reilley et al. [61], that modern interest in chronopotentiometry was aroused. Activity quickly became intense, and today most aspects of chronopotentiometric theory and experiment, areas of applicability, advantages, and limitations are understood.

Mass Transfer Control by Diffusion

Constant-Current Experiment

The boundary condition describing application of a current step causing reduction of electrode reactant O is expressed by Fick's first law:

$$\frac{i}{nFD_O} = \frac{\partial C_{O(0,t)}}{\partial x}. \tag{8.18}$$

Solution of the linear diffusion equation (8.2) under this and the conditions of (8.3) and (8.4) yields the concentration-time-distance profile of species O as

$$C_{O(x,t)} = C_O{}^b - \frac{2it^{1/2}}{nFD_O^{1/2}\pi^{1/2}} \exp\left[-\frac{x^2}{4D_Ot}\right] + \frac{ix}{nFD_O} \operatorname{erfc}\left[\frac{x}{2D_O^{1/2}t^{1/2}}\right]. \tag{8.62}$$

The analogous Fickian solution for product R, but with initial condition $C_{R(x,0)} = 0$, yields for the profile of R

$$C_{R(x,t)} = \frac{2it^{1/2}}{nFD_R^{1/2}\pi^{1/2}} \exp\left[-\frac{x^2}{4D_Rt}\right] - \frac{ix}{nFD_R} \operatorname{erfc}\left[\frac{x}{2D_R^{1/2}t^{1/2}}\right]. \tag{8.63}$$

Concentration-distance profiles calculated from (8.62) are shown in Fig. 8.16. The transition time for this chronopotentiometric electrolysis (which occurs when $C_{O(0,t)} = 0$) is seen to be 1.9 sec, and a break in the potential-time curve is found at this point. Setting $x = 0$ and $t = \tau$ in (8.62) yields the *Sand equation*, the basic expression for many chronopotentiometric applications:

$$\frac{i\tau^{1/2}}{C_O{}^b} = \frac{nFD_O^{1/2}\pi^{1/2}}{2}. \tag{8.64}$$

Several features of the Sand equation require comment. First, the $\tau^{1/2}$-$C_O{}^b$ relationship is the basis for quantitative chronopotentiometric determinations of concentration. (A thorough summary of analytical applications of chronopotentiometry has been given by Davis [62].) For known n and $C_O{}^b$, evaluation of the diffusion coefficient D_O becomes possible. Next, although transition time τ decreases at increased applied current i, the product $i\tau^{1/2}$ should remain constant; experimental constancy of $i\tau^{1/2}$ provides a criterion that mass transport of reactant O is diffusion-controlled over the range of times inspected. Finally, the factor $i\tau^{1/2}/C_O{}^b$, the *chronopotentiometric constant*, is characteristic of the species O. Equation (8.64) has been verified by numerous workers; see Ref. 61 for an early example.

Equation (8.64) is derived from mass transfer considerations alone; thus the chronopotentiometric transition time is not altered by charge transfer reversibility considerations. The charge transfer rate constant does, however, govern the shape of the chronopotentiometric E-t curve and thus indirectly affects the ease of assessment of τ in experiments.

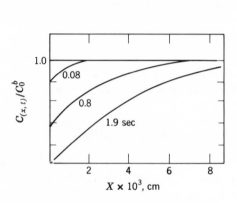

Fig. 8.16 Chronopotentiometric concentration-distance curves for electrolysis with transition time 1.9 sec and $n = 1$, $D = 1.0 \times 10^{-5}$ cm²/sec, $C_o{}^b = 1.0 \times 10^{-6}$ M/cm³, and $i = 2.0 \times 10^{-3}$ A/cm².

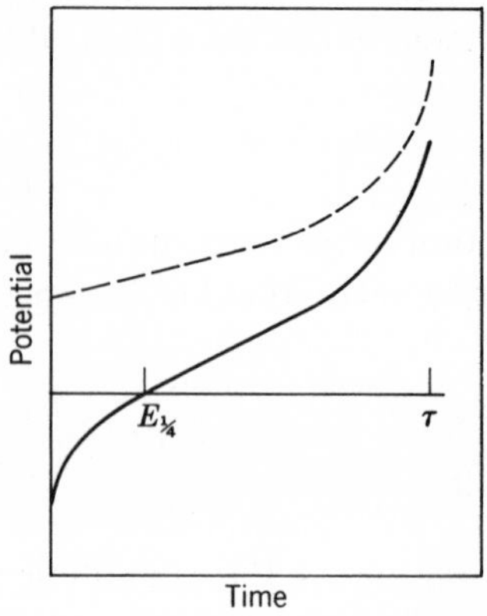

Fig. 8.17 Reversible (———) and irreversible (– – – –) chronopotentiograms.

For a reversible electrode reaction, insertion of (8.62) and (8.63) (at $x = 0$) into the Nernst equation yields, after rearrangement and use of (8.64),

$$E = E^0 + \frac{0.0591}{n} \log \frac{D_R^{1/2}}{D_O^{1/2}} + \frac{0.0591}{n} \log \frac{\tau^{1/2} - t^{1/2}}{t^{1/2}}. \qquad (8.65)$$

The shape of the reversible chronopotentiogram is illustrated in Fig. 8.17. The criterion for chronopotentiometric reversibility is a linear plot of E versus $\log [(\tau^{1/2} - t^{1/2})/t^{1/2}]$ with slope $0.0591/n$; an irreversible reaction gives a slope larger than $0.0591/n$, or a curved plot. At $t = \tau/4$, we see that a characteristic potential, the *quarter-wave potential* $E_{1/4}$, results which can be identified with the polarographic half-wave potential $E_{1/2}$.

The potential-time relation for the totally irreversible case is found by insertion of (8.62) (at $x = 0$) into (8.10) (retaining only the i_c term), which yields [63]

$$E = E^0 + \frac{0.0591}{\alpha n_a} \log \frac{2k_s}{\pi^{1/2} D_O^{1/2}} + \frac{0.0591}{\alpha n_a} \log (\tau^{1/2} - t^{1/2}). \qquad (8.66)$$

The totally irreversible chronopotentiometric wave is cathodically displaced from the reversible one according to both the value of k_s and applied i (through τ), is of different shape, and exhibits a less distinct transition time break. It is evident that (8.66), and the more complex one that results from retention of both terms of (8.10) (quasi-reversible case), provide a route to chronopotentiometric evaluation of the charge transfer rate k_s.

At this point we should mention another experiment, the galvanostatic (or amperometric) experiment, which is useful for studies at higher charge transfer rates and, similar to chronopotentiometry, employs a current step excitation. In the galvanostatic experiment, however, the potential-time response of interest is that in the very short time portion of the ensuing chronopotentiogram. In a solution containing O and R, a current step is applied such as to produce only a small potential change; (η, recall Section 2) after a very short time, an η-$t^{1/2}$ response is obtained which is extrapolated to zero time to yield an η related to the applied current by (8.11). Further details on this and a related (double current pulse) experiment can be found in the review by Delahay [50]; an important critique of the double-pulse technique has been given by Kooijman and Sluyters [64].

In chronopotentiometry of solutions containing more than one electroactive component,

$$O_1 + n_1 e \rightarrow R_1; \qquad O_2 + n_2 e \rightarrow R_2 \tag{8.67}$$

multiple waves are observed when the reactions are spaced by >0.1 V. While the transition time for the more easily reduced component O_1 is again described by (8.64), that of the second is elongated by the continued influx and reduction of O_1. Correction of this "residual diffusion" effect gives, for the transition time of O_2 [61, 65]

$$(\tau_1 + \tau_2)^{1/2} - (\tau_1)^{1/2} = \frac{n_2 F D_2^{1/2} \pi^{1/2} C_2^{b}}{2i}, \tag{8.68}$$

where τ_2 is measured from τ_1. For the general case of m components, the expression for the mth wave is [61]

$$\left(\sum_{j=1,2}^{m} \tau_j \right)^{1/2} - \left(\sum_{j=1,2}^{m-1} \tau_j \right)^{1/2} = \frac{n_m F D_m^{1/2} \pi^{1/2} C_m^{b}}{2i}. \tag{8.69}$$

These equations permit translation of transition time data into concentration for analysis of multicomponent systems.

Multiple chronopotentiometric waves can also result when a single electroactive species is reduced in several stages:

$$O + n_1 e \rightarrow R_1; \qquad R_1 + n_2 e \rightarrow R_2 \tag{8.70}$$

Examples of such behavior are found in oxygen, vanadium(V $\rightarrow$ III $\rightarrow$ II), and copper(II $\rightarrow$ I $\rightarrow$ O). A concentration-independent relation [61, 65] exists between τ_1 and τ_2 in such cases:

$$\tau_1\left(\frac{n_1 + n_2}{n_1}\right)^2 - \tau_1 = \tau_2. \tag{8.71}$$

For $n_1/n_2 = 1$, 0.5, and 2, $\tau_2/\tau_1 = 3.0$, 8.0, and 1.25, respectively. Actual values of τ_1 and $\tau_1 + \tau_2$ are also given by the Sand equation written for $n = n_1$ and $n = n_1 + n_2$, the latter expression applying whether or not one can accurately discern the individual value of τ_1 in an experimental case.

Current Reversal Experiment

The current reversal chronopotentiometric experiment involves replacement of the applied cathodic current step excitation, at a time $t_r < \tau$, with an anodic step excitation, causing the product obtained during the forward step R to be reoxidized. The resulting reverse wave terminates in a transition time τ_b when the supply of R at the electrode surface is exhausted ($C_{R(0,t)} = 0$). The overall forward and reverse response curve is illustrated in Fig. (8.18) for reversible and irreversible electrode processes. Description of the reverse transition time τ_b was first given by Berzins and Delahay [65]. For the case in which equal forward and reverse current steps are employed ($i_c = -i_a$), one obtains the remarkably simple and $C_O{}^b$-, D_O-, D_R-, and i_c-independent result:

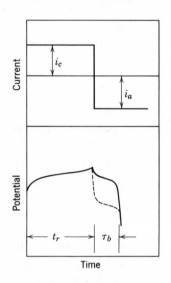

$$t_r/\tau_b = 3.00. \tag{8.72}$$

In the more general case for any relative value of i_c and i_a,

Fig. 8.18 Reversible (———) and irreversible (– – – –) current reversal chronopotentiograms.

$$\frac{i_a{}^2 - 2i_a i_c}{i_c{}^2} = \frac{t_r}{\tau_b}. \tag{8.73}$$

Equation (8.72) provides an experimental criterion to assess the diffusion-controlled behavior of species R. As long as R diffuses freely in the solution (or in the electrode, as in the case of a metal amalgam produced in a mercury working electrode), one-third of the R generated during the forward current step is recovered in the reverse step (the remainder of R being "lost" by diffusion away from the electrode surface). If R is very insoluble and thus quantitatively retained at the electrode surface, $\tau_b = t_r$. An example is the anodic (forward) generation of silver chloride on a silver working electrode in a chloride-containing solution. Values of t_r/τ_b intermediate between 1.00 and 3.00 are found in instances of specific adsorption of R on the electrode surface, as then, depending on the strength of the adsorption, a mixture of diffusing R (approximately one-third recoverable during τ_b) and adsorbed R (quantitatively recoverable, usually during τ_b) is generated during t_r. For a chemical decomposition of R into a less readily reoxidized substance, less R becomes recoverable during τ_b, and t_r/τ_b exceeds 3.00. This case is considered later in Section 3.

Although the mass transfer equations (8.72) and (8.73) are independent of charge transfer kinetics, the relative placement of the forward and reverse chronopotentiometric waves on the potential scale (see Fig. 8.18) and their shapes are functions of k_s. If k_s is sufficiently large (the reversible case), the potential-time behavior of the reverse wave is given by [65]

$$E = E^0 + \frac{0.0591}{n} \log \frac{D_R^{1/2}}{D_O^{1/2}} + \frac{0.0591}{n} \log \frac{t_r^{1/2} - t^{1/2} + 2(t - t_r)^{1/2}}{t^{1/2} - 2(t - t_r)^{1/2}} . \quad (8.74)$$

The potential analogous to $E_{1/4}$ during the forward wave (that at which the extreme right-hand term vanishes) occurs at $t - t_r = 0.215\tau_b$. The equality (forward)$E_{1/4} = $ (reverse)$E_{0.215}$ constitutes an additional reversibility criterion.

Smaller values of k_s result in eventual irreversibility, and $E_{1/4}$ becomes more negative than $E_{0.215}$ (for a cathodic-anodic reversal sequence). The degree of this splitting is a function of k_s and the applied currents, and can be employed for measurement of k_s. Theory has been presented for such charge transfer rate applications [8, 66, 67]; it appears that, except for highly irreversible reactions, use of the current reversal method should enjoy some superiority over the forward current experiment [see (8.66) and quasi-reversible modification] in terms of sensitivity of experimental data to k_s. (Some of this advantage, however, is removed by charging current limitations.)

Mass Transfer Control by Coupled Chemical Reactions

As in the chronoamperometric and chronocoulometric experiments, the existence of a chemical reaction coupled to O or R can produce an alteration of the constant-current and current reversal chronopotentiometric responses from their diffusion-only values. We examine four important types of coupled chemical reactions whose theory has been chronopotentiometrically characterized.

Preceding Chemical Reactions (Prekinetics)

Solution of the boundary value problem for reactant O lying at initial equilibrium with an electroinactive species Z [see (8.5)] is accomplished, for chronopotentiometry, by application of conditions (8.3), (8.4), (8.18), (8.29), and (8.30) to (8.6) and (8.7). The result, obtained by Delahay and Berzins [68], is

$$i\tau_k^{1/2} = \frac{nFD^{1/2}\pi^{1/2}[C_O^b + C_Z^b]}{2} - \frac{i\pi^{1/2}}{2K(k_f + k_b)^{1/2}}\,\mathrm{erf}\,[(k_f + k_b)^{1/2}\tau_k^{1/2}]. \quad (8.75)$$

We see that the chronopotentiometric $i\tau^{1/2}$ product exhibits a dependency on i for prekinetics; Fig. 8.19 illustrates this for several degrees of kinetic control.

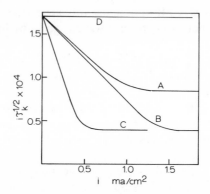

Fig. 8.19 Effect of prekinetics on chronopotentiometric response. $n = 1$, $D = 4.0 \times 10^{-6}$ cm^2/sec, $C_O^b + C_Z^b = 1.0 \times 10^{-6}$ M/cm^3; curve A: $K = 1.0$, $k_f + k_b = 100$; curve B: $K = 0.3$, $k_f + k_b = 900$; curve C: $K = 0.3$, $k_f + k_b = 100$; curve D: very large k_f or K.

For very small chemical rates or transition time (conversion of Z to O is insignificant during the chronopotentiogram), (8.75) reduces, through expansion of the error function at small argument to erf $(\beta) = 2\beta/\pi^{1/2}$, to the Sand equation (8.64) written for O, or

$$i\tau^{1/2} = \frac{nFD^{1/2}\pi^{1/2}[C_O^b + C_Z^b]}{2 + (2/K)}. \quad (8.76)$$

Attainment of this limiting condition is of course determined by experimental accessibility of τ sufficiently short to "beat" the chemical conversion rate; attainment of the limit allows evaluation of the equilibrium constant K if it is otherwise unknown.

At the other extreme (large chemical rate, large τ, small current), erf $(\beta) = 1$ for $\beta > 2$, and we have from (8.75)

$$i\tau_k^{1/2} = \frac{nFD^{1/2}\pi^{1/2}[C_O{}^b + C_Z{}^b]}{2} - \frac{i\pi^{1/2}}{2K(k_f + k_b)^{1/2}} . \tag{8.77}$$

Thus a linear relation between $i\tau_k^{1/2}$ and i prevails; extrapolation of the curve to zero current yields the $i\tau^{1/2}$ limit for "infinitely fast" prekinetics, and the value of $K(k_f + k_b)^{1/2}$ can be obtained from the slope. Given that a variation of 10% in $i\tau^{1/2}$ can be observed over the range of useful chronopotentiometric transition times (*vide infra*), it has been estimated [8] that values of $K(k_f + k_b)^{1/2} < 500$ sec^{-1} could be characterized.

A number of chronopotentiometric applications to the study of chemical prekinetics has been reported, mainly in cases in which the prekinetic reaction involves either a deprotonation reaction of a weak acid [69–72] or loss of a ligand from a metal complex to yield a more readily reducible coordination state [59]. In the latter area, the absence of the chronopotentiometric prekinetic effect has also provided a criterion for judging whether or not a metal complex is reduced directly without predissociation of ligand [8, 63]. Theory for chronopotentiometric prekinetic reactions of the form $pZ \rightleftharpoons$ and $Z \rightleftharpoons pO$ has also been presented and discussed [73].

Following Chemical Reactions (Postkinetics)

EC CASE

A cathodic chronopotentiogram leading to a decomposing species R according to (8.34) yields a transition time no different from that observed in the absence of the chemical reaction, inasmuch as τ is controlled only by mass transport of the reactant O. The shape and placement of the wave on the potential axis is influenced by the chemical process, the general effect being a positive shift of the value of $E_{1/4}$ of magnitude controlled by values of k_f/k_b, k_f, and the applied current. The theory for this effect has been derived and applied to a chemical example [74–76]. A severe limitation of use of the potential shift effect, however, is the requirement of a reversible charge transfer process. Current reversal chronopotentiometry does not suffer from this reversibility requirement.

The qualitative effect of (8.34) on the chronopotentiometric current reversal ratio t_r/τ_b is to lower the quantity of recoverable charge transfer product R and to increase the ratio to values exceeding 3.00. The theory describing this effect, for an irreversible chemical reaction ($k_b = 0$), is [77–79]

$$2 \text{ erf } (k_f\tau_b)^{1/2} = \text{erf } [k_f(t_r + \tau_b)^{1/2}]. \tag{8.78}$$

The analytic behavior of the reversal ratio τ_b/t_r is described, for purposes of evaluation of k_f, as a working curve (Fig. 8.20). The ratio approaches the diffusional value for short t_r or small k_f and dwindles to zero at sufficiently long t_r; obviously, values of $t_r k_f = 1$ to 3 provide the optimum experimental situation for measurement of k_f. It is estimated that first-order rates of about 10^2 sec^{-1} can be evaluated using the current reversal method [77]. It has been applied to study of the hydrolysis reaction of electrogenerated benzoquinone-imine by Testa and Reinmuth [79].

CATALYTIC CASE

Regeneration of the original reactant by a reaction of R [see (8.39)], can cause kinetic control of both O and R mass transport. The cathodic transition

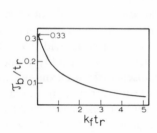

Fig. 8.20 Working plot of (8.78) for current reversal chronopotentiometric evaluation of EC postkinetics.

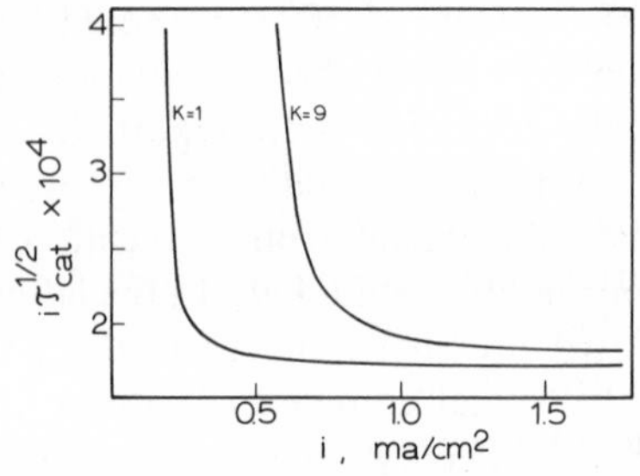

Fig. 8.21 Effect of catalytic kinetics on chronopotentiometric response. $n = 1$, $D = 4.0 \times 10^{-6}$ cm^2/sec, $C_o{}^b = 1.0 \times 10^{-6}$ M/cm^3.

time in constant-current chronopotentiometry becomes elongated by the extra supply of O over its diffusion-controlled value, and the reverse transition time τ_b in current reversal chronopotentiometry by the same token becomes abbreviated. The ratio t_r/τ_b in the current reversal experiment is again described [78] by (8.78), and a kinetic study is conducted in that instance in the same manner as for the simple EC postkinetics case.

The solution for the catalytically controlled forward transition time for species O was derived and experimentally tested by Delahay et al. [74] and is

$$i\tau_{\text{cat}}^{1/2} = \frac{nFD^{1/2}C_O{}^b(k\tau_{\text{cat}})^{1/2}}{\text{erf}\,(k\tau_{\text{cat}})^{1/2}}. \tag{8.79}$$

The behavior of (8.79) is illustrated in Fig. 8.21. At sufficiently small rate or transition time (large current), the error function reduces to $2(k\tau_{\text{cat}})^{1/2}/\pi^{1/2}$, and the system displays Sand equation behavior (current-independent $i\tau^{1/2}$). At larger rate or transition time, the value of $i\tau^{1/2}$ rises sharply. The applications of chronopotentiometry to catalytic kinetics study have emphasized the chemical regeneration of iron(III) or titanium(IV) following their electroreduction [74, 78, 80, 81].

ECE CASE

Constant-current chronopotentiometric theory for a postkinetic chemical process leading to a product more readily reducible than the original electrode reactant [see (8.47)] was presented by Testa and Reinmuth [82] for the case $k_b = 0$. The solution for $i\tau_k^{1/2}$ is complex and best examined and compared with experimental data through use of a computed working curve [83, 84] such as that shown in Fig. 8.22. The vertical axis is presented in a normalized fashion as the ratio of the $i\tau_\infty^{1/2}$ product that would prevail for infinitely fast kinetics ($k_f \to \infty$) to the kinetically controlled $i\tau_k^{1/2}$ product. The former product is given by the Sand equation (8.64) written for $n + n'$ electrons. The figure shows that the limit of $i\tau_\infty^{1/2}$ is approached at large transition times (low current), and the other extreme limit corresponds to transition times so short as to "beat" the kinetic process and attain control of the experimental $i\tau^{1/2}$ by solely the $O + ne \to$ reaction. At intermediate times experimental $i\tau_\infty^{1/2}/i\tau_k^{1/2}$ ratios can be

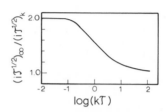

Fig. 8.22 Working plot for chronopotentiometric evaluation of ECE kinetics. $n = n'$.

used with the working curve to read off values of $k_f\tau_k$. A difficulty in the experiment can be evaluation of $i\tau_\infty^{1/2}$.

Chronopotentiometric studies of ECE cases include the reductions and subsequent reactions of o-nitrophenol [83] and p-nitrosophenol [84].

Adsorption of Electrode Reactant

As noted earlier (Section 1), the presence of adsorbed reactant or product on the working electrode surface causes alteration of the electrochemical response signal from its anticipated diffusion-controlled value. In the chrono-potentiometric experiment this alteration appears, for the case of adsorbed reactant, as an increase in the chronopotentiometric product $i\tau^{1/2}$ with increasing applied current. A qualitative rationale for this effect, first recognized and applied by Lorenz [85–87], results from recognizing that as the applied current is increased the value of τ_s (transition time for a solution reactant) decreases according to i^2 while τ_a (transition time for an adsorbed reactant) decreases linearly with i. Since the products $i\tau_a$ and $i\tau_s^{1/2}$ are constants, the measured $i(\tau_a + \tau_s)^{1/2}$ must increase at short times or high applied currents.

The percentage increase in $i\tau^{1/2}$ at large currents depends on the relative values of Γ_0 (surface excess of O) and $C_O{}^b$. The precise shape of the $i\tau^{1/2}$-t curve is, however, also a function of the order in which adsorbed reactant (hereafter denoted as AR) and solution reactant (denoted SR) are reduced during the chronopotentiogram, an unfortunate functionality in that assessment of this order of reaction must be an implicit part of the Γ_0 measurement. Figure 8.23 illustrates four situations that can logically arise. The SR,AR, and AR,SR cases refer to attainment of surface concentration depletion of SR prior to reduction of any AR, and the reverse of this, respectively. This leads to split waves, as shown. In the SAR case, SR and AR react at similar potentials so that there is no clear wave splitting. Once the initial equilibrium between SR and AR is established and the chronopotentiometric experiment started, the species SR and AR react in these cases as individual electroactive chemicals, one a solution component and one an adsorbed component. In the EQUIL case, an equilibrating interconversion AR $\rightleftharpoons$ SR occurs throughout the chronopotentiogram with sufficient lability that the surface concentrations of both attain zero simultaneously at a common transition time.

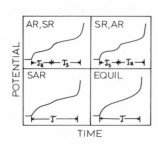

Fig. 8.23 Chronopotentiograms typical of the various adsorption models.

Differentiation between these cases and others which might be imagined depends on establishing the theory of their $i\tau^{1/2}$-i functionalities and comparison of experimental data thereto. Clear adherence of experimental data to one of the models, and unanimous failure of all others, then yields a value for Γ_0.

SR,AR Case

A clearly split wave in this case should produce for the first step τ_s a constant $i\tau_s^{1/2}$ product conforming to the Sand equation (8.64). As the chronopotentiogram is continued to the second transition time τ_a, we find that $\tau_a > nF\Gamma_0/i$ owing to the continued influx of SR during τ_a. This transition time was first described (approximately) by Lorenz [85] and subsequently rigorously by Reinmuth [88] and Anson [89]:

$$\frac{nF\pi\Gamma_0}{i} = (\tau_s + \tau_a) \arccos\left[\frac{\tau_s - \tau_a}{\tau_s + \tau_a}\right] - 2(\tau_s\tau_a)^{1/2}. \tag{8.80}$$

For comparison to experimental results, individual values of τ_s and τ_a are inserted to yield a current-independent value of Γ_0. When τ_s is indistinct, a computer solution for Γ_0 can be sought from (8.80) using values of $\tau_s + \tau_a$ at several currents [90], or alternatively an operational test using current sweep chronopotentiometry can be employed [91].

The SR,AR case is the one thermodynamically expected for strong adsorption of reactant. According to Brdička [92], the adsorbed reactant, being in a lower free energy state and having a more negative standard potential (E^0) than solution reactant SR, yields a wave cathodically displaced from that of SR. The validity of the Brdička argument of course depends on charge transfer reversibility for both SR and AR. Documented cases of SR,AR behavior in chronopotentiometric reactions are actually rather rare, the closest approximation being the adsorption of lead iodide complex from iodide supporting electrolyte [32].

EQUIL Case

This case can result from a combination of weak to moderately strong adsorption and a labile adsorption equilibrium between SR and AR. For a linear adsorption isotherm, Lorenz [85] gives the following theoretical description:

$$\tau^{1/2} + \frac{\pi^{1/2}\Gamma_0}{2D^{1/2}C_O{}^b}\,\phi = \frac{nFC_O{}^b(D\pi)^{1/2}}{2i} + \frac{\pi^{1/2}\Gamma_0}{2D^{1/2}C_O{}^b}, \qquad (8.81)$$

where $\phi = \exp(C_O{}^{b^2}D\tau/\Gamma_0{}^2)\,\mathrm{erfc}\,(C_O{}^bD^{1/2}\tau^{1/2}/\Gamma_0)$. Comparison of experimental data to this expression can be accomplished through a computer fitting or through an iterative graphical approach [32]. The adsorptions of several mercury(II) complexes on mercury electrodes [32] adhere to (8.81).

AR,SR Case

In this case, the first transition time is governed by $i\tau_a = nF\Gamma_0$, the second by the Sand equation (8.64) for component SR; thus for distinct wave-splitting, constant $i\tau_a$ and $i\tau_s{}^{1/2}$ products must result. Indistinct splitting is treated by the combined expression given by Lorenz [85]

$$i\tau = nF\Gamma_0 + \frac{D\pi(nFC_O{}^b)^2}{4i}, \qquad (8.82)$$

where $\tau = \tau_a + \tau_s$. Data comparison is accomplished by a plot of $i\tau$ against $1/i$. This case can result for irreversible charge transfer with the charge transfer rate for SR being lower than that for AR. Examples adhering exactly to this case are unknown; it is approximated in the adsorption of phenylmercuric ion [93] on mercury electrodes.

SAR Case

This case is an approximation for the situation in which equilibration between SR and AR does not proceed rapidly and the reduction potentials for AR and SR are similar. A fixed division between current reducing SR and that reducing AR is presumed, the division being such as to procure a common transition time. The theory [85] for this model is

$$i\tau = nF\Gamma_0 + \frac{nFC_O{}^b(\pi D\tau)^{1/2}}{2}. \tag{8.83}$$

Experimental data are examined by a plot of $i\tau$ against $\tau^{1/2}$.

Following the initial work by Lorenz [85–87], a variety of adsorption measurements were conducted: adsorption of riboflavin on mercury by Tatwawadi and Bard [90], of Alizarin Red S and cobalt chloride–ethylene-diamine on mercury by Laitinen and Chambers [94], of carbon monoxide on platinum by Munson [95, 96], and others mentioned above. This considerable activity in adsorption measurements (suitable methods had been previously lacking) was short-lived owing to the existence of certain important short-comings of the chronopotentiometric experiment, plus development of a second, superior technique (chronocoulometry). The necessary "single-fit" comparison of data to the various theoretical expressions suffers from the basic problem that, although they appear mathematically diverse, in reality their $i\tau^{1/2}$-i characteristics are quite similar. Unless the adsorption is rather strong or a distinctive wave morphology appears, it is often difficult, if not impossible [97], to obtain a clear-cut, single-fit identification of the proper model for assessment of precise surface excess values. Factors optimizing possibilities of the correct model identification have been discussed [32]. Also, experimental bias can occur, notably through the positive contribution of double layer charging [98] to τ at short times and a possible potential dependency of the adsorption [99]. Finally, the fact that a series of experiments (20 or more) is required to describe a system's $i\tau^{1/2}$-i behavior adequately for model comparison makes the experimental labor per value of Γ_0 considerable.

Other Chronopotentiometric Techniques

The most commonly employed chronopotentiometric experiments are those discussed above—the current step and current reversal experiments. It is possible, however, both in terms of required instrumentation and mass transport theory, to employ a variety of current excitation shapes, that is, to use various current programs. Some examples are currents that increase with any power of time [100, 101] (applied $i = \beta t^r$) or exponentially with time [101]. The case of current increasing as the square root of time ($r = \frac{1}{2}$) is interesting as it yields chronopotentiograms exhibiting a transition time τ directly proportional to concentration and inversely proportional to current ($\beta\tau/C^b = $ constant). This case has certain advantages in applications to analysis as compared to the constant-current experiment.

The form of programmed current chronopotentiometry most useful in study of coupled chemical reactions is that termed cyclic chronopotentiometry. This experiment, illustrated in Fig. 8.24, consists of a series of current steps with current reversals occurring at each successive cathodic and anodic transition time. The theory for cyclic chronopotentiometry was developed by Herman and Bard [104–107] for diffusion control of O and R and also for various kinetic cases, such as prekinetics and EC, catalytic, and ECE post-kinetics. For the diffusion-controlled case, the ratio of transition times for the first ($n = 1$) and second ($n = 2$) current reversal τ_1/τ_2 has the familiar value of 3.00 [see (8.72)]. Inasmuch as the reactant O consumed during τ_1 is not completely recovered during τ_2, the second cathodic transition time τ_3 is less than τ_1 and also $\tau_3/\tau_4 < 3.00$. In fact, at large n, $\tau_n/\tau_{n+1} \rightarrow 1.00$. Table 8.1 gives computed values for transition times relative to τ_1 for the diffusion-controlled, EC, and catalytic cases (for a selected decomposition rate of R, $k = 1.00$). Such theoretical data (for various k) can be compared with experimental τ_n values through dependency of either τ_n/τ_{n+1} (a cathodic-anodic pair) or τ_n/τ_{n+2} (successive cathodic transitions) ratios on the value of n; the functionality of these ratios with n is obviously quite different and is also sensitive to k.

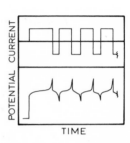

Fig. 8.24 Cyclic chronopotentiometry: excitation, and response. Current reversed at preselected potential at transition time.

In principle, the cyclic chronopotentiometric experiment yields no extra kinetic information beyond that available solely from $i\tau^{1/2}$ or τ_1/τ_2. For the EC case, we see in fact that τ_n/τ_{n+1} is less affected by the chemical process (compare to the diffusion-only value) as n increases. The merit of the cyclic experiment lies primarily in its collection in one experiment of a series of data points, all of which contain some kinetic information; random errors are to some extent averaged when one relies on the entire series to evaluate a rate k. Also, the series of data points, collectively, can reveal deviations from ideal adherence of a chemical system to a supposed kinetic scheme with greater sensitivity, as such deviations have a cumulatively larger effect on transition times at larger n.

Another form of chronopotentiometry which should be mentioned is derivative chronopotentiometry, in which one electronically measures and records dE/dt against time in the chronopotentiogram (rather than potential E against time). As pointed out by Peters and Burden [108], and as is evident from the reversible chronopotentiogram of Fig. 8.17, dE/dt passes through a minimum at $t < \tau$. For a reversible reaction, $t_{\min} = 4\tau/9$. This can permit a somewhat more reliable measurement of τ but unfortunately suffers from the dependency of t_{min} on the reversibility of the electrode reaction (for a completely irreversible reaction $t_{min} = \tau/4$). Sturrock et al. have [109, 110] extended the theoretical relations for dE/dt to multicomponent systems and to low-concentration analysis.

Sturrock [111] had actually earlier, but without commensurate theory, introduced the derivative chronopotentiometric readout in application to the cyclic chronopotentiometric experiment, recording in this case dE/dt against

Table 8.1   Relative Cyclic Chronopotentiometric Transition
Times [104, 105]

n	τ_n Diffusion Only	EC Case $k = 1.00$	Catalytic Case $k = 1.00$
1	1.000	1.000	1.000
2	0.333	0.167	0.167
3	0.588	0 384	0 673
4	0 355	0 138	0.169
5	0.546	0.292	0.658
6	0.366	0.120	0.169
7	0.525	0.243	0.655
8	0.373	0.108	0.169
9	0.513	0.211	0.654
10	0.378	0.099	0.169

E. This experiment is a quantitation of the oscillographic polarographic method [112] popular among European electrochemists.

4 THE PRACTICE OF CHRONOAMPEROMETRY, CHRONOCOULOMETRY, AND CHRONOPOTENTIOMETRY

Response Quality

The sensitivity of the chronomethods to the phenomena of coupled chemical reaction and charge transfer rates and adsorption has been detailed in the preceding Sections 2 and 3. It is now appropriate to examine the reduction of this theoretical sensitivity to practice for quantitative, experimental studies We are concerned with the *response quality* in a chronomethods experiment or how faithfully experimental current-time data in chronoamperometry, charge-time data in chronocoulometry, and transition time data in chrono-potentiometry reflect the rigorous theoretical predictions appropriate to the chemical system at hand. Deviation of experiment from theory, or imperfection in response quality, can arise from sources enumerated as follows. Any chronomethods theoretical relation is grounded on a certain, defined set of mass transport conditions (linear diffusion for theory considered in this chapter). Response quality is lowered by any perturbation of the diffusion conditions ideally present around the working electrode employed. Chronomethods theory is also predicated on application to the working electrode of current or potential steps of "infinitely fast" risetime and of precisely prescribed magnitude. Limitations are imposed on the exactness of both current and potential step excitations by a combination of the electrode double-layer capacitance, the ohmic resistance of the cell solution, and the quality of the electrochemical instrument employed. The accuracy of the data recording system is also important; at fast electrolysis times in chronoamperometry, for instance, one finds that the compressed data display of an oscilloscope screen is considerably less precise than that of the modern strip chart recorder useful on slower time scales. Finally, in the case of chronopotentiometry, a limitation on response quality occurs through the uncertainty of how best to extract a numerical value for transition time τ from the recorded chronopotentiogram.

The relative importance of the above sources of response quality imperfection largely depends of course on the particular circumstances of the chronomethod application. A generally influential circumstance, however, is the time scale on which data are desired. The chronomethods in practice provide an experimental "time window"; beyond an upper and lower time value, response quality steadily diminishes. The long time limit is most often imposed by effects associated with maintenance of ideal diffusion conditions. The short time limit is usually a function of the felicity with which the desired current or potential excitation signal can be forced on the working electrode.

Further comments on the working electrodes and instrumentation employed in the chronomethods, and the relation of these to response quality, are given in the following sections.

Working Electrodes and Diffusional Imperfections

Three working electrode geometries usable for chronomethods experiments were illustrated earlier in Fig. 8.2. The disk, spherical, or cylindrical electrode can be fabricated from a variety of materials, among which mercury, platinum, and gold are common choices. The spherical mercury electrode is a pendant or hanging mercury drop electrode (HMDE); a disk or cylindrical mercury electrode surface is provided by coating mercury onto an underlying platinum disk or wire. Mercury can also be used in the form of a pool electrode.

Strictly speaking, only a planar working electrode surface provides data to which the linear diffusion theory presented in Sections 2 and 3 is applicable. In practice, however, any nonplanar working electrode yields experimental chronomethod results conforming to linear diffusion theory if the results are obtained at electrolysis times sufficiently short that the diffusion layer remains thin in comparison to the radius of curvature of the nonplanar electrode surface. Thus the spherical and cylindrical electrode geometries of Fig. 8.2 are useful for short-time chronomethod experiments; the spherical HMDE is, in fact, one of the most popular working electrodes for all three chronomethod techniques. This electrode was developed by DeMars and Shain [113] and Kemula and Kublik [114], who employ different procedures for preparing the pendant mercury drop. An HMDE of area 0.05 cm², for example, provides chronoamperometric data adhering within 1 to 2% to the linear diffusion Cottrell equation [see (8.19)] out to times of about 1 sec. For HMDE chronomethod data at longer electrolysis times, the appropriate spherical diffusion boundary value problem must be solved to allow interpretation of the results. An alternative procedure for long electrolysis times, and preferable for very long times or when the bulk of experimentation is to be done in this time range, is of course to select a more planar working electrode surface.

In any event, a typical preamble to chronomethod experimentation with an unfamiliar electrode is a careful assessment, over the electrolysis time interval of interest, of the diffusion conditions prevailing at that electrode. This is done using a model, diffusion-controlled electrode reaction and, for linear diffusion geometry, is based on the time independency of the chronoamperometric $it^{1/2}$, chronocoulometric $Q/t^{1/2}$, and chronopotentiometric $i\tau^{1/2}$ parameters of (8.19), (8.22), and (8.64), respectively. Onset of an appreciable spherical diffusion component when using the HMDE and a solution soluble model reactant is signaled in this experimental test by an increase in these parameters with increasing electrolysis time.

The model reaction test of the working electrode can also on occasion reveal by a time dependency of the diffusion parameters, the presence of other diffusional nonidealities at the working electrode. At long electrolysis times, with any given working electrode, convective disturbances can introduce an additional mass transport mode and enlarge the chronomethod diffusion parameters. Such disturbances arise from vibrational motions (a function of rigidity of the cell mounting), or from solution motion induced by a density gradient within the diffusion layer. With a disk electrode proper spatial orientation of the working electrode surface and shielding of its perimeter (with a skirt) can, as shown by Bard [115] in a chronopotentiometric study, substantially minimize this problem. In general, however, complete freedom from convective disturbances for electrolysis times exceeding 100 sec is rarely attained. Another type of convective disturbance sometimes results, in the short-time domain, from surface tension changes at mercury electrodes (and resulting mercury pool or drop shape changes) accompanying abrupt alterations in the electrode potential. The data of Shain and Martin [116] provide a relatively severe example of this. A third type of solution motion at mercury electrodes, convective streaming (analogous to that causing "maxima" in polarography), can occur unpredictably; this effect might be absent in the model electrode reaction but appear in a chemical system subsequently examined using the same working electrode. Attempts to eliminate this effect with maximum suppressors (added surfactants) require considerable caution, as interference with the kinetic or adsorption process under study can result. Finally, it is evident that the region around the "neck" of an HMDE (see Fig. 8.2) can encounter a shielding from true semiinfinite diffusion conditions by the lower surface of the capillary from which the mercury drop is hung. This shielding is most severe at long electrolysis times, where more of the thickening diffusion layer impinges on the obstructing surface, and causes a diminution of the chronomethod diffusion parameter [116, 117]. An HMDE shielding effect can also be experienced in chronomethod reversal experiments; in current reversal chronopotentiometry, for example, less of a solution-soluble species generated during t_r is diffusionally "lost" (because of diffusion layer reflection off the shielding surface) and τ_b exceeds its theoretical value ($t_r/\tau_b < 3$). Shielding is less noticeable in the double potential step experiments, where shielding effects during forward and reverse electrolysis times tend to cancel in the current or charge ratios.

Since most of the above diffusional nonidealities are more likely to produce difficulties when using an HMDE electrode as opposed to, for example, a platinum disk electrode, it is pertinent to comment on why, given a reaction occurring in a potential region where either electrode could, in principle, be employed, the HMDE is generally still the electrode of choice. In a kinetics or adsorption-directed study, the preference for an mercury electrode rests primarily with its higher assurance of a clean and reproducible surface. A platinum or other solid electrode is kept free from surface oxides or prior-adsorbed impurities only with extreme difficulty; even then, its surface is heterogeneous in terms of an array of exposed crystal planes and surface defects. The problem of understanding an adsorption process in its most basic terms is greatly complicated by these factors and is further compounded by the necessity for assessing the true surface area of the solid metal. Charge transfer kinetic rates can be quite sensitive to the condition of the working electrode surface; an excellent illustration of this has been given by Daum and Enke [118]. Study of coupled chemical reaction rates is less affected by use of a solid electrode, but even there a heterogeneous component of the chemical reaction is possible, and determination of double-layer charge corrections is less accurate on the less reproducible solid surface. In general, then, all other things being equal, the usual preference is for an mercury surface in chronomethod studies. All other things are of course frequently not equal; an intractable diffusional disturbance at an HMDE or a necessity for working at potentials anodic of the mercury dissolution potential can, for example, dictate choice of a solid working electrode.

Control and Measurement—Chronopotentiometry

Of the several instrumental approaches to the application and control of potential or current at a working electrode, that utilizing operational amplifiers is versatile and generally satisfactory and is illustrated here and in the following section. Figure 8.25 shows a simple operational amplifier chronopotentiometric circuit. The experiment is initiated by sudden application of the potential V, which, dropping across resistance R to the virtual ground of the operational amplifier CON summing point (*), determines the current flowing between the working (——o) and auxiliary (——|) electrodes in the amplifier feedback loop. This circuit is preferable to the "two-electrode" version in which the auxiliary electrode shown is replaced with a poised reference electrode. The cell potential, measured at the output of the CON amplifier, in that case includes the ohmic resistance drop across the solution (the product of the applied current and solution resistance). The "three-electrode" circuit of Fig. 8.25 avoids much of this ohmic component by monitoring the potential of a third, reference electrode (——→) against the working electrode with a high input impedance voltage follower. The ohmic drop, or uncompensated resistance still included in the output cell potential depends on the working-electrode geometry and the working–reference electrode spacing (proximity is desirable). The ohmic component eliminated by use of the three-electrode circuit is termed the compensated resistance. For current reversal chronopotentiometry, the sign of the constant potential V (also typically generated by an operational amplifier device) is altered at the desired time t_r. The equipment used to record the chronopotentiometric potential-time output depends on the particular application at hand. For charge transfer kinetic studies, accuracy of the potential scale is crucial. Minimization of the uncompensated resistance then receives much attention in the cell design, and the potential output is zero suppressed and scale expanded. For other work a relatively coarse potential scale expansion is usually sufficient.

This brings us to the measurement of transition times from recorded chronopotentiograms, upon whose values applications in analysis, adsorption, and coupled chemical reaction kinetics are founded. Unfortunately, transition time measurements are sometimes rendered uncertain by contributions from processes other than electrolysis of the cell reactant. The main villain in this respect is the double-layer charging process (another source is the reaction of surface oxides or extraneous adsorbed impurities). The current required to charge the double layer is related to the rate of potential change and the double layer capacitance C_{dl} by

$$i_{dl} = C_{dl} \frac{dE}{dt}, \qquad (8.84)$$

where C_{dl} is some function of the value of potential E. Thus the current actually driving the desired cell reaction is less than that applied according

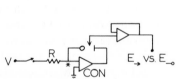

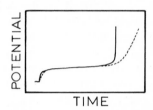

Fig. 8.25 Elementary operational amplifier apparatus for chronopotentiometry. $i_{cell} = -V/R$.

Fig. 8.26 Illustration of effect of double-layer charging on chronopotentiograms. Ideal reversible curve (———); with double-layer effect (– – – –).

to C_{dl} and the instantaneous value of dE/dt; the latter term varies widely and throughout the chronopotentiogram, being largest at early times and around τ. A double-layer charging current of course flows in any electrochemical experiment in which the working electrode potential is altered; the point in interpretation of experimental results is whether or not correction for it can be easily and accurately accomplished. In chronopotentiometry, the complex variation of i_{dl} and its persistence throughout the experimental curve makes such correction quite difficult. The influence of double-layer charging is most pronounced at short transition times (large dE/dt) and at low reactant concentrations (small applied current) and is reflected by a distortion of the potential-time response. The qualitative illustration of Fig. 8.26 shows that the potential breaks are smeared and the time elapsing between them is elongated.

The approaches dealing with double-layer charging, when this constitutes a problem in obtaining reliable data, are several. One is instrumental and attempts to electronically supply and add the current of (8.84) to the desired applied current. Two methods for this can be envisaged: In one, two identical cells are employed, one containing sample and the other (the blank) only supporting electrolyte [119]. A chronopotentiometric circuit applies the desired constant current to the sample cell; its potential output simultaneously controls the working electrode potential in the blank cell. The blank cell current is thus i_{dl} (if the working electrodes are perfectly matched), which is added to the constant sample cell current. In the second method, a single cell is used; the output of the reference electrode is electronically differentiated and returned to the cell as a current through a second input resistance to the CON amplifier [120, 121]. A constant value of C_{dl} is assumed. While effective for long τ values, both of these methods rely on positive electronic feedback and thus experience instrument instability at short electrolysis times.

Another approach to double-layer charging distortion relies on various graphical procedures for measuring transition time which are empirically judged to compensate to some extent for the distortion. Several such procedures have seen use [60, 72, 122–124], but none seems to have gained uniform acceptance as reliably removing a double-layer contribution to the τ value. Given transition time data measured by some graphical method, a further treatment is proposed by Bard [125] and Lingane [126], in which the term i_{dl} is assumed to be a constant fraction of the applied current throughout the chronopotentiogram. This amounts to treating the double-layer charge as an SAR-adsorbed component; a plot of $i\tau$ against $\tau^{1/2}$ yields an intercept term representative of the double layer which is applied as a correction to each individual transition time data point. This method, although involving obvious approximation, provides significantly improved data; Evans [127] has illustrated its efficacy in application to some chronopotentiometric data for an ECE chemical reaction system. Finally, we should note that careful scrutiny of any procedure for dealing with the double-layer effect on transition time is greatly facilitated by a knowledge of the theoretical form of the double-layer-distorted potential-time curve. Such theoretical curves, obtained by numerical computation methods by DeVries [128] and Rodgers and Meites [129], themselves involve assumptions of a potential-independent C_{dl} and of a specified reversibility of the ideal chronopotentiogram but are very revealing of the seriousness of the double-layer influence on τ and of the inefficiency of the purely graphical transition time measurement schemes for dealing with the problem. This theory does not in itself, however, provide an adequate solution to the problem and is best employed as an evaluative method.

In summary of the problem of chronopotentiometric transition time measurement, we should note that accurate work can be accomplished under circumstances (long τ and higher concentration levels) in which double-layer distortion is minimal. For work in the short-time domain (about 10 msec, for example), however, the measurement of transition time is attended by sufficient uncertainties as to make this technique poorly competitive, for quantitative studies, with the other chronomethods in which double-layer correction can be more effectively accomplished.

Control and Measurement—Chronoamperometry and Chronocoulometry

The chronoamperometric and chronocoulometric experiments approach in practice the ideal boundary value form of the potential step excitation much more closely than is the typical case in chronopotentiometry. This of course is a consequence of the localization of the working electrode potential change, and concurrent double-layer charging process, to a single (short-time) portion of the experiment. Thus the lower time limit for good response quality in a potential step experiment is a direct function of quality of the potentiostat instrument employed. Potentiostat design is subject to continual improvement; many advances have been made in recent years, and reliable current and charge measurements at times of a few microseconds will undoubtedly be widely achieved as progress continues. Measurements at 100 μsec are already commonplace.

An elementary three-electrode operational amplifier potentiostat for chronoamperometry is shown in Fig. 8.27. The chronoamperometric experiment is initiated by sudden addition of the desired step potential E_{step} to the initial potential E_{init}. Potential control of the relative working (——O) and reference (——→) electrode potentials is accomplished in this circuit by amplifier CON. This amplifier senses any imbalance between the reference electrode potential input and E_{init} (or $E_{init} + E_{step}$), providing a voltage output and resultant current flow at the working electrode to maintain equality of the reference-working and E_{init} (or $E_{init} + E_{step}$) potentials. The current flow at the working electrode is monitored by the current follower amplifier CUR, whose output E_1 is related to the cell current as shown. The circuit for chronocoulometry is a simple extension of this; the voltage E_1 is integrated by amplifier Q to yield a voltage E_2 proportional to the total charge passed in the cell. A double potential step experiment is accomplished by, following the above, removal of E_{step} at the desired time.

The current at times early in a chronoamperometric experiment is quite large as the double layer becomes charged to its new potential and the faradaic electrode process commences. Output current capability of the operational amplifiers employed in the potentiostat circuit is of obvious importance. A more useful measure of potentiostat performance, however, is its risetime, or the time required for the relative working and reference electrode potentials to achieve precisely the given value of $E_{init} + E_{step}$. Potentiostat risetime is a function not only of the amplifiers employed but is also governed by the electrical properties of the electrochemical cell. Thus overall performance must be stated with reference to the potentiostat circuit in combination with the array of resistance (compensated and uncompensated solution and faradaic resistance) and capacitance (double layer) which constitutes the particular electrochemical cell used. The electrochemical cell can in fact be represented by a physical analog resistance-capacitance network; such "dummy cells" are convenient for testing the risetime of a given

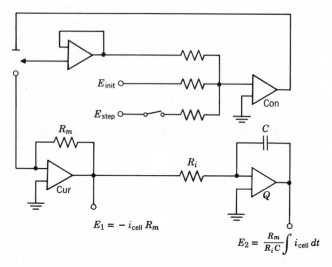

$$E_1 = -i_{cell} R_m$$

$$E_2 = \frac{R_m}{R_i C} \int i_{cell} \, dt$$

Fig. 8.27 Elementary operational amplifier potentiostat for chronoamperometry and chronocoulometry.

potentiostat under various potential step conditions. Evaluation of the potentiostat risetime is important not only from the standpoint of attempting circuit optimization for minimal risetime but also to define the lower time limit on which current or charge-time data of acceptable quality can be expected. In general, deviation from theory exceeding several percent is encountered at times less than about 10 times risetime.

A more fundamental approach can be taken to optimization of potentiostat performance by writing the transfer function of the potentiostat-cell system [21, 130–132]. The transfer function theory, a mathematical representation of the frequency response of the system to a specified potential excitation, gives insight into how to adjust best the potentiostat parameters (and the cell parameters where possible) to achieve minimal risetime. The transfer function of the actual experimental system can also be measured to fully define its risetime capabilities and limitations.

One of the most significant cell parameters, in terms of influence on the risetime of the potentiostat-cell system, is the portion of the cell ohmic drop sensed by the reference electrode: the uncompensated cell resistance. The potential value actually operative at the working electrode is in effect less than that applied by the product of uncompensated resistance and the instantaneous cell current. At short electrolysis times, at which the cell current is large, this deviation from the control potential can be large and provide a major fraction of the overall system risetime [133]. In the cell design, therefore, the reference electrode junction with the cell solution, usually in the form of a capillary salt bridge (a Luggin capillary), is positioned as close to the working electrode as is practical. Theory relating the interelectrode distance to the uncompensated term has been given by Nemec [134]; this theory shows that the spacing must be quite close to achieve a major reduction. (For instance, only a 50% reduction results for a spacing of one electrode radius from a spherical working electrode [133].) Further relief from the uncompensated resistance effect on risetime can be obtained through feedback and addition of a voltage proportional to the output of amplifier CUR to the control potential, which cancels a portion of the uncompensated potential drop [133, 135]. This involves a positive feedback loop, so full compensation is not consistent with maintaining potentiostat stability, but nevertheless valuable lowering of risetime can be attained.

Other features of the cell and experiment design that influence the potential step risetime include the working electrode area and the sample concentration, to which the cell current and uncompensated resistance are proportional. One thus avoids use of sample concentration significantly larger than is necessary to yield a sample current to which background current correction (obtained in a blank experiment) can be accurately applied. The working electrode area is kept as small as is consistent with obtaining conformity to the desired diffusion conditions. Large supporting electrolyte concentrations lower the solution and uncompensated resistances; studies in nonaqueous media generally experience poorer risetimes because of the higher solution resistances typical of such media. Last, there can be significant differences in the requisite potentiostat risetimes for chronoamperometric or chrono-coulometric experiments directed at charge transfer measurements and those elsewhere directed. In the former, charge transfer experiment much smaller potential steps are used, and small losses in potential control at short electrolysis times produce a relatively greater uncertainty in the transient data.

Measurement of the current and charge-time transients in the potential step experiment can be accomplished by oscilloscope recording when data at short times are required, or strip chart recording when data are taken at times of about 1 sec and longer. Recent years have seen, however, the advent of modern data acquisition systems, which are exerting a considerable impact on electrochemical experimentation, in particular that in the chronoampero-metric and chronocoulometric areas. Such systems consist of the potentiostat-cell system, a device for automatic, real-time conversion of analog output signals from the potentiostat to a digital data form, a digital memory device for storage of this data, and appropriate "interfacing" apparatus for inter-connecting the analog and digital sections of the system. In the earliest data acquisition systems, after completion of the experiment, the data is trans-ferred manually (through paper or magnetic tape) from the system's memory device to the memory of a digital computer for processing [136–139]. A more powerful and convenient approach involves the use of the memory of the computer itself (typically a small, captive version) as the on-line digital data storage device during the potential step experiment, so that the data analysis can be initiated immediately following termination of the experiment [41, 140]. Both types of systems produce enormous savings in the labor of data analysis; the latter is considerably more effective in this respect and can reduce the total time required for a transient experiment, including the entire sequence of computer data collection, computer data interpretation, and teletypewriter printout of 'answers' to 1 to 2 min. Such immediate data interpretation produces efficiency in allowing a series of potential step experiments on a given sample solution to be tailored and adjusted as the series progresses to yield a more optimum experimental characterization. Two other important features of the computerized data acquisition system are the much higher levels of recording accuracy possible at short times (as compared to data taken oscilloscopically), and the use of the computer system itself to supply the requisite waveform or timing of the potential step excitation signal.

ACKNOWLEDGMENT

The preparation of this chapter was facilitated by support of the author's research in several of the chrono-methods by the Air Force Office of Scientific Research under Grant AFOSR-69-1625.

References

1. C. N. Reilley, *Rev. Pure Appl. Chem.*, **18**, 137 (1968).
2. M. E. Peover, "Electrochemistry of Aromatic Hydrocarbons and Related Substances," in *Electroanalytical Chemistry*, Vol. 2, A. J. Bard, Ed., Marcel Dekker, New York, 1967.
3. R. N. Adams, *Accounts Chem. Res.*, **2**, 175 (1969).
4. K. F. Herzfeld, *Physik. Z.*, **14**, 29 (1913).
5. H. A. Laitinen and I. M. Kolthoff, *J. Am. Chem. Soc.*, **61**, 3344 (1939).

6. H. A. Laitinen, *Trans. Electrochem. Soc.*, **82**, 289 (1942).
7. J. H. Christie, G. Lauer, R. A. Osteryoung, and F. C. Anson, *Anal. Chem.*, **35**, 1979 (1963).
8. P. Delahay, *New Instrumental Methods in Electrochemistry*, Interscience, New York, 1954.
9. W. M. Schwarz and I. Shain, *J. Phys. Chem.*, **69**, 30 (1965).
10. J. H. Christie, *J. Electroanal. Chem.*, **13**, 79 (1967).
11. H. J. Schugar, A. T. Hubbard, F. C. Anson, and H. B. Gray, *J. Am. Chem. Soc.*, **91**, 71 (1969).
12. Von R. Koopman, *Ber. Bunsenges. Physik. Chem.*, **70**, 121 (1967).
13. Von R. Koopman and H. Gerischer, *Ber. Bunsenges. Physik. Chem.*, **70**, 127 (1967).
14. S. P. Perone and W. J. Kretlow, *Anal. Chem.*, **38**, 1760 (1966).
15. B. K. Hovsepian and I. Shain, *J. Electroanal. Chem.*, **14**, 1 (1967).
15a. R. P. Van Duyne, Ph. D Thesis, University of North Carolina, Chapel Hill, N. C., 1961.
16. P. Delahay and G. L. Stiehl, *J. Am. Chem. Soc.*, **74**, 3500 (1952).
17. S. L. Miller, *J. Am. Chem. Soc.*, **74**, 4130 (1952).
18. Z. Pospisil, *Collection Czech. Chem. Commun.*, **18**, 337 (1953).
19. R. Guidelli and D. Cozzi, *J. Electroanal. Chem.*, **14**, 245 (1967).
20. G. L. Booman and D. T. Pence, *Anal. Chem.*, **37**, 1366 (1965).
21. D. T. Pence and G. L. Booman, *Anal. Chem.*, **38**, 1112 (1966).
22. J. D. E. McIntyre, *J. Phys. Chem.*, **71**, 1196 (1967).
23. R. Guidelli, *J. Phys. Chem.*, **72**, 3535 (1969).
24. P. J. Lingane and J. H. Christie, *J. Electroanal. Chem.*, **13**, 227 (1967).
25. J. Koutecky, *Collection Czech. Chem. Commun.*, **18**, 183 (1953).
26. I. Tachi and M. Senda, *Advances in Polarography*, I. Longmuir, Ed., Pergamon Press, New York, 1960, p. 454.
27. G. S. Alberts and I. Shain, *Anal. Chem.*, **35**, 1859 (1963).
27a. R. F. Nelson and R. N. Adams, *J. Am. Chem. Soc.*, **90**, 3925 (1968).
28. M. D. Hawley and S. W. Feldberg, *J. Phys. Chem.*, **70**, 3459 (1966).
29. R. N. Adams, M. D. Hawley, and S. W. Feldberg, *J. Phys. Chem.*, **71**, 851 (1967).
28a. R. Koopman *Ber. Bunsengesellsch. phxsik. Chem.*, **72**, 32 (1938).
30. S. W. Feldberg and C. Auerbach, *Anal. Chem.*, **36**, 505 (1964).
31. S. W. Feldberg, "Digital Simulation: A General Method for Solving Electrochemical Diffusion-Kinetic Problems," in *Electroanalytical Chemistry*, Vol. 3, A. J. Bard, Ed., Marcel Dekker, New York, 1969.
32. R. W. Murray and D. J. Gross, *Anal. Chem.*, **38**, 393 (1966).
33. J. H. Christie, G. Lauer, R. A. Osteryoung, and F. C. Anson, *Anal. Chem.*, **35**, 1979 (1963).
34. F. C. Anson, *Anal. Chem.*, **36**, 932 (1964).
35. F. C. Anson and D. A. Payne, *J. Electroanal. Chem.*, **13**, 35 (1967).
36. G. W. O'Dom and R. W. Murray, *Anal. Chem.*, **39**, 51 (1967).
37. G. W. O'Dom and R. W. Murray, *J. Electroanal. Chem.*, **16**, 327 (1968).
38. R. A. Osteryoung and J. H. Christie, *J. Phys. Chem.*, **71**, 1348 (1967).

39. F. C. Anson, *Anal. Chem.*, **38**, 55 (1966).
40. J. H. Christie, R. A. Osteryoung, and F. C. Anson, *J. Electroanal. Chem.*, **13**, 236 (1967).
41. G. Lauer, R. Abel, and F. C. Anson, *Anal. Chem.*, **39**, 765 (1967).
42. F. C. Anson and D. J. Barclay, *Anal. Chem.*, **40**, 1791 (1968).
43. F. C. Anson, J. H. Christie, and R. A. Osteryoung, *J. Electroanal. Chem.*, **13**, 343 (1967).
44. D. J. Barclay and F. C. Anson, *J. Electrochem. Soc.*, **116**, 438 (1969).
45. F. C. Anson and B. Epstein, *J. Electrochem. Soc.*, **115**, 1155 (1968).
46. H. Gerischer and W. Vielstich, *Z. Physik. Chem. (Frankfurt)*, **3**, 16 (1955).
47. W. Vielstich and H. Gerischer, *Z. Physik. Chem. (Frankfurt)*, **4**, 10 (1955).
48. C. A. Johnson and S. Barnartt, *J. Electrochem. Soc.*, **114**, 1256 (1967).
49. C. A. Johnson and S. Barnartt, *J. Phys. Chem.*, **71**, 1637 (1967).
50. P. Delahay, "Study of Fast Electrode Processes," in *Advances in Electrochemistry and Electrochemical Engineering*, Vol. 1, P. Delahay, Ed., Interscience, New York, 1961.
51. J. H. Christie, G. Lauer, and R. A. Osteryoung, *J. Electroanal. Chem.*, **7**, 60 (1964).
52. P. J. Lingane and J. H. Christie, *J. Electroanal. Chem.*, **10**, 284 (1965).
53. H. F. Weber, *Wied. Ann.*, **7**, 536 (1879).
54. H. J. S. Sand, *Phil. Mag.*, **1**, 45 (1901).
55. T. R. Rosebrugh and W. L. Miller, *J. Phys. Chem.*, **14**, 816 (1910).
56. Z. Karaoglanoff, *Z. Elektrochem.*, **12**, 5 (1906).
57. J. A. V. Butler and G. Armstrong, *Proc. Roy. Soc. (London)*, **A139**, 406 (1933).
58. J. A. V. Butler and G. Armstrong, *Trans. Far. Soc.*, **30**, 1173 (1934).
59. L. Gierst and A. Juliard, *J. Phys. Chem.*, **57**, 701 (1953).
60. P. Delahay and G. Mamantov, *Anal. Chem.*, **27**, 478 (1955).
61. C. N. Reilley, G. W. Everett, and R. H. Johns, *Anal. Chem.*, **27**, 483 (1955).
62. D. G. Davis, "Applications of Chronopotentiometry to Problems in Analytical Chemistry," in *Electroanalytical Chemistry*, Vol. 1, A. J. Bard, Ed., Marcel Dekker, New York, 1966.
63. P. Delahay and T. Berzins, *J. Am. Chem. Soc.*, **75**, 2486 (1953).
64. D. J. Kooijman and J. H. Sluyters, *J. Electroanal. Chem.*, **13**, 152 (1967).
65. T. Berzins and P. Delahay, *J. Am. Chem. Soc.*, **75**, 4205 (1953).
66. L. B. Anderson and D. J. Macero, *Anal. Chem.*, **37**, 322 (1965).
67. F. H. Beyerlein and R. S. Nicholson, *Anal. Chem.*, **40**, 286 (1968).
68. P. Delahay and T. Berzins, *J. Am. Chem. Soc.*, **75**, 2486 (1953).
69. N. Tanaka and T. Murayama, *Z. Physik. Chem. (Frankfurt)*, **21**, 146 (1959).
70. R. P. Buck and L. R. Griffith, *J. Electrochem. Soc.*, **109**, 1005 (1962).
71. H. B. Mark and F. C. Anson, *Anal. Chem.*, **35**, 722 (1963).
72. R. P. Buck, *Anal. Chem.*, **35**, 1853 (1963).
73. W. H. Reinmuth, *Anal. Chem.*, **33**, 322 (1961).
74. P. Delahay, C. C. Mattax, and T. Berzins, *J. Am. Chem. Soc.*, **76**, 5319 (1954).
75. W. K. Snead and A. E. Remick, *J. Am. Chem. Soc.*, **79**, 6121 (1957).
76. A. C. Testa and W. H. Reinmuth, *Anal. Chem.*, **32**, 1518 (1960).
77. O. Dracka, *Collection Czech. Chem. Commun.*, **25**, 338 (1960).

78. C. Furlani and G. Morpurgo, *J. Electroanal. Chem.*, **1**, 351 (1960).
79. A. C. Testa and W. H. Reinmuth, *Anal. Chem.*, **32**, 1512 (1960).
80. H. B. Herman and A. J. Bard, *Anal. Chem.*, **36**, 510 (1964).
81. O. Fischer, O. Dracka, and E. Fischerova, *Collection Czech. Chem. Commun.*, **25**, 323 (1960).
82. A. C. Testa and W. H. Reinmuth, *Anal. Chem.*, **33**, 1320 (1961).
83. A. C. Testa and W. H. Reinmuth, *J. Am. Chem. Soc.*, **83**, 784 (1961).
84. G. S. Alberts and I. Shain, *Anal. Chem.*, **35**, 1859 (1963).
85. W. Lorenz, *Z. Elektrochem.*, **59**, 730 (1955).
86. W. Lorenz and H. Mühlberg, *Z. Elektrochem.*, **59**, 736 (1955).
87. W. Lorenz and H. Mühlberg, *Z. Physik. Chem. (Frankfurt)*, **17**, 129 (1958).
88. W. H. Reinmuth, *Anal. Chem.*, **33**, 322 (1961).
89. F. C. Anson, *Anal. Chem.*, **33**, 1123 (1961).
90. S. V. Tatwawadi and A. J. Bard, *Anal. Chem.*, **36**, 2 (1964).
91. R. W. Murray, *J. Electroanal. Chem.*, **7**, 242 (1964).
92. R. Brdička, *Z. Elektrochem.*, **48**, 278 (1942).
93. R. F. Broman and R. W. Murray, *Anal. Chem.*, **37**, 1408 (1965).
94. H. A. Laitinen and L. M. Chambers, *Anal. Chem.*, **36**, 5 (1964).
95. R. A. Munson, *J. Electroanal. Chem.*, **5**, 292 (1963).
96. R. A. Munson, *J. Phys. Chem.*, **66**, 727 (1962).
97. P. J. Lingane, *Anal. Chem.*, **39**, 485 (1967).
98. A. J. Bard, *Anal. Chem.*, **35**, 340 (1963).
99. J. H. Christie and R. A. Osteryoung, *Anal. Chem.*, **38**, 1620 (1966).
100. R. W. Murray and C. N. Reilley, *J. Electroanal. Chem.*, **3**, 64 (1962).
101. R. W. Murray, *Anal. Chem.*, **35**, 1784 (1963).
102. H. Hurwitz and L. Gierst, *J. Electroanal. Chem.*, **2**, 128 (1961).
103. H. Hurwitz, *J. Electroanal. Chem.*, **2**, 142, 328 (1961).
104. H. B. Herman and A. J. Bard, *Anal. Chem.*, **35**, 1121 (1963).
105. H. B. Herman and A. J. Bard, *Anal. Chem.*, **36**, 510, 971 (1964).
106. H. B. Herman and A. J. Bard, *J. Phys. Chem.*, **70**, 396 (1966).
107. H. B. Herman and A. J. Bard, *J. Electrochem. Soc.*, **115**, 1028 (1968).
108. D. G. Peters and S. L. Burden, *Anal. Chem.*, **38**, 530 (1966).
109. P. E. Sturrock, W. D. Anstine, and R. H. Gibson, *Anal. Chem.*, **40**, 505 (1968).
110. P. E. Sturrock, G. Privett, and A. R. Tarpley, *J. Electroanal. Chem.*, **14**, 303 (1967).
111. P. E. Sturrock, *J. Electroanal. Chem.*, **8**, 425 (1964).
112. R. Kalvoda, *Techniques of Oscillographic Polarography*, Elsevier, New York, 1965.
113. R. D. DeMars and I. Shain, *Anal. Chem.*, **29**, 1825 (1957).
114. W. Kemula and Z. Kublik, *Roczniki Chem.*, **33**, 1431 (1959).
115. A. J. Bard, *Anal. Chem.*, **33**, 11 (1961).
116. I. Shain and K. J. Martin, *J. Phys. Chem.*, **65**, 254 (1961).
117. R. W. Murray and D. J. Gross, *J. Electroanal. Chem.*, **13**, 132 (1967).
118. P. H. Daum and C. G. Enke, *Anal. Chem.*, **41**, 653 (1969).
119. W. D. Shults, F. E. Haga, T. R. Mueller, and H. C. Jones, *Anal. Chem.*, **37**, 1415 (1965).

120. C. G. Enke, Princeton University, private communication, 1966, referenced in W. H. Reinmuth, *Anal. Chem.*, **38**, 270R (1966).
121. R. W. Murray, University of North Carolina, unpublished results, 1963.
122. P. Delahay and C. C. Mattax, *J. Am. Chem. Soc.*, **76**, 874 (1954).
123. R. W. Laity and J. D. E. McIntyre, *J. Am. Chem. Soc.*, **87**, 3306 (1965).
124. W. H. Reinmuth, *Anal. Chem.*, **33**, 485 (1961).
125. A. J. Bard, *Anal. Chem.*, **35**, 340 (1963).
126. J. J. Lingane, *J. Electroanal. Chem.*, **1**, 379 (1960).
127. D. H. Evans, *Anal. Chem.*, **36**, 2027 (1964).
128. W. T. DeVries, *J. Electroanal. Interfacial Chem.*, **17**, 31 (1968).
129. R. S. Rodgers and L. Meites, *J. Electroanal. Interfacial Chem.*, **16**, 1 (1968).
130. G. L. Booman and W. B. Holbrook, *Anal. Chem.*, **35**, 1793 (1963).
131. G. L. Booman and W. B. Holbrook, *Anal. Chem.*, **37**, 795 (1965).
132. I. Shain, J. E. Harrar, and G. L. Booman, *Anal. Chem.*, **37**, 1768 (1965).
133. G. Lauer and R. A. Osteryoung, *Anal. Chem.*, **38**, 1106 (1966).
134. L. Nemec, *J. Electroanal. Chem.*, **8**, 166 (1964).
135. E. R. Brown, T. G. McCord, D. E. Smith, and D. D. DeFord, *Anal. Chem.*, **38**, 1119 (1966).
136. M. W. Breiter, *J. Electrochem. Soc.*, **112**, 845 (1965).
137. M. W. Breiter, *J. Electrochem. Soc.*, **113**, 1071 (1966).
138. E. R. Brown, D. E. Smith, and D. D. DeFord, *Anal. Chem.*, **38**, 1130 (1966).
139. G. Lauer and R. A. Osteryoung, *Anal. Chem.*, **38**, 1137 (1966).
140. G. Lauer and R. A. Osteryoung, *Anal. Chem.*, **40** (10), 40A (1968).

Chapter **IX**

CONTROLLED-POTENTIAL ELECTROLYSIS

Louis Meites

I   INTRODUCTION

There have been many studies of the reactions that occur when solutions of organic substances are electrolyzed, of the products they yield, and of the mechanisms by which they take place. Some of these have been made to elucidate the ways in which electrons are transferred across the interface between an electrode and a solution. Others have been made to obtain information that can be used in devising new procedures of separation and analysis, or in searching for simple and convenient ways of preparing compounds otherwise difficult to obtain. Still others have been made in attempts to correlate the electrochemical behaviors of different substances with their chemical constitutions or molecular structures, or to reveal similarities or differences between the mechanisms of related chemical and electrochemical processes.

Partly because of this diversity of aims, and partly because of the number of decades over which these studies have extended, they have employed three related but distinctly different techniques. The oldest is constant-voltage electrolysis, reviewed by Sherlock Swann, "Catalytic, Photochemical, and Electrolytic Reactions," in Volume II of this series. A nearly constant and usually fairly high voltage is applied to a cell in which the electrodes have large areas; a large current flows and substantial amounts of the reaction products are formed in short periods of time. Most large-scale industrial electrosyntheses are still performed in this fashion because only relatively simple and inexpensive apparatus is required. However, the technique is not selective. Different half-reactions often occur simultaneously at the electrode of interest, giving rise to mixtures of products, and the relative rates of these half-reactions often vary greatly as the electrolysis proceeds, so that fundamental information and even reproducible results are difficult to secure.

A second and newer approach is voltammetry, described by Müller in Chapter V and by Piekarski and Adams in Chapter VII. This usually involves the use of a cell in which the electrode of interest is very small. Only small currents can flow through such a cell, and it becomes a simple matter to measure or control the potential of this electrode. It is the electrode potential that governs the rates and extents of the half-reactions that take place. By proper selection of the potential, it is often possible to isolate and study individual half-reactions to the exclusion of others: to reduce or oxidize one constituent of a mixture while leaving others unaffected, or to decipher the details of a stepwise reduction or oxidation. This selectivity makes voltammetry very valuable, but the smallness of the current makes it very difficult to prepare a significant amount of product within a reasonable length of time.

The third approach, which is the subject of this chapter, combines the valuable features of the others. It involves the use of a large electrode, which permits the flow of relatively high currents and makes rapid quantitative conversions of starting materials into products possible. It also involves the use of special equipment and techniques for controlling the potential of that electrode, thus preserving the selectivity of voltammetry on a larger scale. This combination of preparative utility with high selectivity entitles controlled-potential electrolysis to a place among the most versatile and useful techniques of electroorganic chemistry.

It is often possible to find conditions under which only a single substance is reduced or oxidized at the electrode whose potential is being controlled, and under which only a single product is obtained. Then the total quantity of electricity consumed in the complete reduction or oxidation of N^0 moles of starting material is given by Faraday's law:

$$Q_\infty = \int_0^\infty i\, dt = nFN^0, \tag{9.1}$$

where Q_∞ is the quantity of electricity in coulombs, i is the current (in amperes) t seconds after the start of the electrolysis, n is the number of electrons involved in the half-reaction in question, and F is the number of coulombs per faraday. Integrating the current under such conditions permits the calculation of n if N^0 is known, or of N^0 if n is known. The former possibility is helpful in elucidating the course and products of the electrode reaction; the latter is of much interest in chemical analysis.

As (9.1) implies, the current that flows in a controlled-potential electrolysis is time-dependent. This is because substances, whether present initially or formed as intermediates, that can undergo reduction or oxidation under the conditions employed are converted as time goes on into other substances that cannot. For a simple half-reaction, in which an electroactive reactant is converted into an inert product in a single step (or a single series of fast steps), the current decays exponentially with time. Other kinds of time dependences arise from the occurrence of slow steps, which may be either chemical or electrochemical. They are of two kinds: those that occur in consecutive mechanisms and those that occur in branched ones. A consecutive mechanism, in which each substance can react in only one way, yields a single product and gives an integral n-value. A branched one generally yields a mixture of products and gives an n-value that is non-integral and, if a pseudo-second-order competing step is involved, also dependent on the initial concentration of the starting material. By combining coulometric data, current-time curves, and identification and determination of the products obtained, it is possible to obtain a great deal of information about the mechanism of the process and often even to evaluate the rate constants of the slow steps.

2   PRINCIPLES OF ELECTROLYSIS. THE BEHAVIORS OF SIMPLE HALF-REACTIONS

An electrolytic reaction is brought about by applying a dc voltage to a cell consisting of two electrodes in contact with a solution. This causes electrons to flow through the external circuit from the anode to the cathode. At the interface between the cathode and the solution, electrons are transferred from the cathode to some constituent of the solution, which is thereby reduced. Sometimes this constituent may be the solvent itself, and the solvated electron formed initially may diffuse some distance away from the electrode surface before reacting further, but we need not be concerned here with the details of the electron-transfer process. Conversely, at the surface of the anode, some constituent of the solution loses electrons to the anode and is thereby oxidized. The flow of electricity through the body of the solution is accomplished by the motion of dissolved ions; cations move toward the negative electrode, while anions move toward the positive one. It may be mentioned that, contrary to a popular misconception, the cathode is not necessarily the negative electrode and the anode is not necessarily the positive electrode.

Only when two identical electrodes are immersed in the same solution is the common belief correct.

The resistance R of the solution depends on the concentration and mobility (which in turn is affected by the temperature and viscosity of the solution) of each of the ionic species present, by the areas of the electrodes and the distance between them, and by the thickness and porosity of any diaphragm used to prevent mixing of the solutions around the anode and cathode. A high concentration of an acid, base, buffer mixture, or neutral salt is usually employed to minimize the value of R.

The total voltage V impressed across the cell is distributed in the following manner:

$$V = E_a - E_c + iR. \tag{9.2}$$

Here E_a and E_c are the potentials of the anode and cathode, respectively, referred to the same reference electrode. Although it does not matter what reference electrode is chosen, it is conventional in voltammetry to refer all electrode potentials to the saturated calomel electrode (SCE), whose potential is $+0.241$ V versus the normal hydrogen electrode (NHE), and this convention is followed throughout this chapter. The symbols i and R denote, respectively, the current in amperes and the cell resistance in ohms.

Each of the electrode potentials may be regarded as the sum of a reversible potential and an overpotential. The former depends on the standard potential of the half-reaction taking place at the electrode in question and on the activities, at the surface of the electrode, of the substances involved in that half-reaction. It is the potential that the electrode would have if electrochemical equilibrium at its surface were attained instantaneously. There are some organic couples for which this is the case, but the great majority of organic electrode processes are irreversible in the sense of involving an appreciable overpotential. The overpotential reflects the occurrence of a slow step in the electron-transfer process, and its value is affected by a variation of any experimental parameter that influences the rate of the slow step, such as the current density, the temperature, the composition (and especially the pH and ionic strength) of the solution, and the nature and sometimes even the prior history of the electrode surface. Under identical experimental conditions, different half-reactions are characterized by different overpotentials, and these vary in different and often unpredictable ways as the experimental conditions are changed. This is one reason why even apparently minor changes in experimental conditions often have profound effects on the yields and even the products of electrolyses of organic substances. Many striking examples of these effects are cited by Swann [1] and Allen [2].

The significance of the electrode potential can be described by considering the electrolytic reduction of p-benzoquinone. Figure 9.1 is a polarogram of a solution containing approximately 0.001 M p-benzoquinone in a supporting electrolyte containing 0.1 F potassium dihydrogen phosphate, 0.1 F potassium monohydrogen phosphate, and 1 F potassium nitrate, and having a pH of 6.8. It was obtained with a conventional dropping mercury electrode [3, 4] as the indicator electrode and a saturated calomel electrode as the reference electrode. Both the current and the cell resistance were so small that the iR

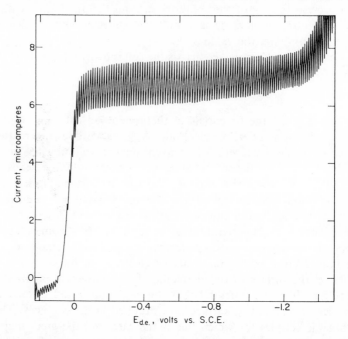

Fig. 9.1 Polarogram of 0.001 M p-benzoquinone in 0.1 F potassium dihydrogen phosphate—0.1 F potassium monohydrogen phosphate—1 F potassium nitrate, pH 6.8, secured with a conventional dropping mercury electrode.

term in (9.2) was always negligible, so that the voltage applied to the cell at each point was equal to the difference between the electrode potentials.

The polarogram may be divided into five regions. In the first, at potentials more positive than about +0.2 V versus SCE, the potential of the dropping electrode is so positive that mercury is oxidized. Under these conditions the product is an insoluble mercury(I) phosphate; under some other conditions it may be some constituent of the supporting electrolyte that is oxidized, but in any case this region is uninteresting in controlled-potential electrolysis. In the second region, between about +0.2 and +0.1 V versus SCE, the potential of the dropping electrode is too negative to permit the oxidation of mercury to occur to any appreciable extent but is still not negative enough to effect any appreciable reduction of the quinone. The third region extends from about $+0.1$ to -0.1 V versus SCE and contains the rising portion of the quinone wave. This corresponds to the half-reaction

$$\text{(quinone, O=C}_6\text{H}_4\text{=O)} + 2\text{H}^+ + 2e = \text{(hydroquinone, HO-C}_6\text{H}_4\text{-OH)}$$

As the potential of the dropping electrode is varied by the action of the external electrical circuit, the concentrations of the quinone and hydroquinone in the layer of solution immediately adjacent to the surface of the electrode vary in accordance with the Nernst equation [5]:

$$E = E^0 - \frac{2.303RT}{nF} \log \frac{[\text{H}_2\text{Q}]}{[\text{H}^+]^2[\text{Q}]} = E^0 - \frac{0.05916}{2} \log \frac{[\text{H}_2\text{Q}]}{[\text{H}^+]^2[\text{Q}]}. \quad (9.3)$$

The second equality is written for $T = 25°\text{C}$. The quantities $[\text{H}_2\text{Q}]$, $[\text{H}^+]$, and $[\text{Q}]$ represent the concentrations (or, more accurately, the activities) of the hydroquinone, hydrogen ion, and the quinone, respectively, at the electrode surface, and E^0 is the standard potential of the quinone-hydroquinone half-reaction. Since both E^0 and, because the solution is well buffered, $[\text{H}^+]$ are constant, the ratio $[\text{H}_2\text{Q}]/[\text{Q}]$ must vary as the potential changes. As the potential becomes more negative the ratio increases. At a potential around $+0.1$ V, for example, where the ratio is very small, only a very small current suffices to produce the required small concentration of the hydroquinone. But at a potential around -0.1 V, for example, where the ratio is very large, almost every molecule of the quinone that reaches the electrode surface must be reduced, and a comparatively large current is needed to effect the reduction. It can be shown that the dependence of the current on the potential of the dropping electrode is closely approximated by the equation

$$E = E^0 - \frac{0.05916}{2} \text{pH} - \frac{0.05916}{2} \log \frac{i}{i_d - i}, \quad (9.4)$$

where i is the measured current at the potential E and i_d is the diffusion current or overall height of the wave. The diffusion current is the current that flows in the fourth region, which extends from about -0.1 to -1.2 V versus SCE, where molecules of the quinone are reduced as rapidly as they diffuse to the electrode surface. The fifth and final region begins at about -1.2 V; here a second half-reaction, in this case the reduction of dihydrogen phosphate ion,

$$\text{H}_2\text{PO}_4^- + e \rightarrow \tfrac{1}{2}\text{H}_2 + \text{HPO}_4^=$$

begins to take place.

A polarogram of very similar shape would have been obtained with an irreversible couple, but it would have a slightly different significance. The irreversible half-reaction $O + ne \rightarrow R$ may be characterized by a heterogeneous rate constant $k_{s,h}$ that is the common rate constant for the forward (reduction) and reverse (oxidation) processes at the standard potential of the couple. If $k_{s,h}$ is small, which is what the term "irreversible" implies, only a very small current results from the reduction of O at the standard potential, but as the potential of the electrode changes the current changes in accordance with the equation [6, 7]

$$i = nFAC_O^0 k_{s,h} \exp\left[-\alpha n_a F(E - E^0)/RT\right], \qquad (9.5)$$

where i is the current in amperes, n is the number of faradays consumed in the reduction of 1 mole of O, A is the area of the electrode in square centimeters, C_O^0 is the concentration of O at the surface of the electrode in mole-cm^{-3}, R is the gas constant (8.316 volt-coulombs per degree), and T is the absolute temperature. The transfer coefficient α is a parameter characteristic of the half-reaction under fixed experimental conditions; its value is potential-dependent but usually lies between about 0.3 and 0.7. The parameter n_a is the number of electrons involved in the rate-controlling step (of which there is assumed to be only one in deriving this simple expression); its value cannot exceed that of n, and is usually equal to either 1 or 2.

Although i is small for an irreversible half-reaction when $E = E^0$, the increase of the exponential term in (9.5) as the potential becomes more negative eventually causes i to assume finite values. However, as i increases, the concentration of O at the electrode surface decreases and these two opposing effects finally lead, at a potential much more negative than E^0, to a situation in which molecules of O are reduced as rapidly as they reach the surface of the electrode.

Thus the general form of the current-potential curve does not depend on whether the half-reaction is reversible or not. However, reversible and irreversible couples do behave quite differently in electrolyses carried out in the third portion of a current-potential curve. The difference is described in a later section.

Figure 9.2 is another current-potential curve for the p-benzoquinone solution whose polarogram was shown in Fig. 9.1. It was obtained on replacing the dropping mercury electrode with a large mercury-pool electrode and using an efficient propeller-type stirrer to agitate the interface between the electrode and the solution. The general shapes of the two curves are very much the same: the same five regions can be identified without difficulty in Fig. 9.2, and they occur at very nearly the same potentials as they do in Fig. 9.1. There is one very important difference between them: whereas the height of the wave in Fig. 9.1 is only about 6 μA, that of the wave in Fig. 9.2 is about

60 mA. This is attributable partly to the much larger area of the pool electrode and partly to the effect of stirring on the rate at which molecules of the quinone arrive at the electrode surface. It is the fact that large stirred electrodes yield high limiting currents that makes controlled-potential electrolysis practical and useful.

Both analytical and synthetic controlled-potential electrolyses are most often performed in the fourth region, on the plateau of the wave, because this is where they proceed most rapidly. One then attempts to avoid encroaching on either the third region or the fifth. On the one hand, an electrolysis

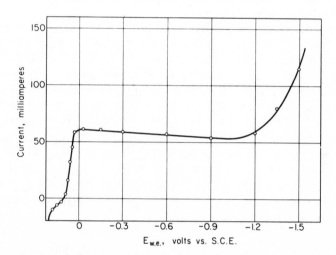

Fig. 9.2 Current-potential curve of the solution of Fig. 9.1 at a stirred mercury-pool electrode having an area of 40 cm². The small decrease of the current between 0 and −0.9 V versus SCE is attributable to the partial electrolytic depletion of the solution while the measurements were being made.

performed on the rising part of the wave is either slower or less complete than one performed on the plateau; on the other hand, the product of the foreign half-reaction that occurs in the fifth region will not always be as innocuous as the hydrogen and monohydrogen phosphate ion here, and (9.1) becomes irrelevant if two half-reactions occur at once.

Other regions are useful for other purposes. A solution containing *p*-benzoquinone might be electrolyzed in the second region, below the foot of the wave, to rid it of more easily reducible substances before determining the quinone polarographically or in some other way, to avoid reducing any appreciable fraction of the quinone while determining a more easily reducible substance coulometrically, or to oxidize any of the hydroquinone that might be present as an impurity. Electrolyses in the third region, on the rising part of the wave, show how the fraction reduced depends on potential and provide values of the formal potential if the couple is reversible; the fact that the chemical branch of a mechanism similar to

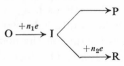

is more prominent, because the rate constant for the reduction of I is smaller, on the rising part of the wave than on the plateau is advantageous in mechanistic studies and some synthetic work.

In any event, it is hardly possible to select the potential to be used unless the boundaries of the region of interest are known. As Figs. 9.1 and 9.2 suggest, polarographic data, which are readily available [8–10] in modest profusion, provide a useful starting point. For example, a chemist who wants to determine p-benzoquinone in this medium by controlled-potential coulometry could tell from Fig. 9.1 that a preliminary electrolysis with a stirred mercury pool maintained at about $+0.15$ V versus SCE would remove more easily reducible constituents without affecting the quinone, and that the quinone could then be rapidly and quantitatively reduced at any potential between about -0.1 and -1.2 V versus SCE. To avoid both the devil (incomplete reduction of the quinone) and the deep blue sea (consumption of excess current by the reduction of dihydrogen phosphate ion), he might select a potential halfway between these two extremes, perhaps -0.6 V versus SCE.

With irreversible couples the matter is not as simple. The polarographic half-wave potential then depends on the drop time [11, 12], and the rapidity of mass transfer to the stirred pool exaggerates the slowness of the rate-determining electron-transfer step. The difference between the polarographic half-wave potential $E_{1/2,\mathrm{DME}}$, where the maximum current attributable to the reduction of the substance responsible for the wave is half as large as on the plateau, and the voltammetric half-wave potential $E_{1/2,V}$ at a large stirred mercury-pool electrode is given by [7, 13]

$$E_{1/2,\mathrm{DME}} - E_{1/2,V} = \frac{0.05916}{\alpha n_a} \log \frac{1.349 s_O V t^{1/2}}{D_O^{1/2} A} \tag{9.6}$$

at 25°C, where s_O is the mass-transfer constant for O, defined by 9.9, D_O is the diffusion coefficient of O in cm²-sec⁻¹, V is the volume of solution in cubic centimeters employed in the voltammetric measurement, and t is the drop time at the polarographic half-wave potential. The difference may be very large. Streuli and Cooke [14] reported that the half-wave potential of 1-nitroso-2-naphthol in 0.1 F ammonia–0.1 F ammonium chloride was

−0.80 V versus SCE at a mercury pool electrode but −0.27 V versus SCE at a dropping mercury electrode. Karp and Meites [13] found that the half-wave potential of hydrogen peroxide in a neutral phosphate buffer was —1.31 V versus SCE at a large stirred pool electrode but —0.75 V versus SCE at a dropping electrode. Hence, as has been pointed out by Charlot, Badoz-Lambling, and Trémillon [15] and by Rechnitz [16], although polarographic data provide useful indications of what can be done and how they should not be relied on too implicitly. It is more useful to obtain current-potential curves with a large stirred pool electrode under exactly the same conditions, in which the stirring efficiency is by far the most important, that will be used in the controlled-potential electrolysis. Such curves can be obtained manually by adjusting the potentiostat to provide each of a number of preselected potentials successively and measuring the current at each; speed is essential to avoid excessive electrolytic depletion of the solution during the measurements. Some potentiostats are equipped with synchronous-motor-driven bridges to permit automatic recording.

Mercury pool electrodes have been more widely used than any others in controlled-potential electrolysis, partly for the same reasons that have made the dropping mercury electrode the most widely used voltammetric indicator electrode and partly because the ready availability of polarographic data makes it easier to decide whether a desired result is feasible and to select conditions under which it may be achieved than would be the case if electrodes of other materials were used. This is why mercury pools are stressed in the present discussion. What is said about them can be applied, *mutatis mutandis*, to electrodes made of platinum, gold, graphite, boron carbide, or any other material.

The rate of a controlled-potential electrolysis is interesting in analytical and synthetic work and is crucial in mechanistic studies. Assuming that the half-reaction is simply $O + ne \rightarrow R$ and that the rate of reoxidation of R is negligible—which is true on the plateau of any wave and is also true on the rising part of a totally irreversible one—Karp and Meites [13] obtained the equation*

$$i = \frac{s_O nFVAk_O C_O^{b,0}}{Ak_O + Vs_O} \exp\left(-\frac{s_O Ak_O}{Ak_O + Vs_O} t\right). \tag{9.7}$$

Here $C_O^{b,0}$ is the concentration of O (in mole-cm^{-3}) in the bulk of the solution at the start of the electrolysis; k_O is the heterogeneous rate constant (in cm-sec^{-1}) of the electron-transfer process, given in view of (9.5) by

$$k_O = k_{s,h} \exp\left[-\alpha n_a F(E - E^0)/RT\right], \tag{9.8}$$

* The quantity here denoted by s was formerly denoted by β by the writer and by p by Bard and his associates; the symbol s is being recommended by Commission V. 5 on Electroanalytical Chemistry of the International Union of Pure and Applied Chemistry.

and s_O is a mass transfer constant (in sec^{-1}) defined by

$$s_O = D_O A/V\delta,\tag{9.9}$$

in which δ is a distance (in centimeters) equivalent to the thickness of the diffusion–turbulence-damping region in the layer of solution adjacent to the electrode surface, familiarly known as the "thickness of the Nernst diffusion layer." It is convenient to define a potential-dependent rate constant s_O^* by writing

$$s_O^* = \frac{s_O A k_O}{A k_O + V s_O}.\tag{9.10}$$

Combining this with (9.7) yields

$$i = s_O^* n F V C_O^{b;0} \exp(-s_O^* t)\tag{9.11}$$

as the equation for the current-time curve obtained in this simple case.

There are two extreme situations. On the plateau of the wave, k_O is very much larger than $V s_O/A$; then s_O^* becomes equal to s_O and (9.11) becomes

$$i = s_O n F V C_O^{b;0} \exp(-s_O t),\tag{9.12}$$

which is equivalent to an expression obtained by Lingane [17]. Near the foot of a totally irreversible wave, however, k_O is much smaller than $V s_O/A$; then s_O^* becomes equal to $A k_O/V$ and (9.11) becomes

$$i = n F A k_O C_O^{b;0} \exp\left(-\frac{A k_O}{V} t\right).\tag{9.13}$$

Values of k_O obtained from measurements at different potentials near the foot of such a wave may be combined with (9.8) to evaluate the important parameter αn_a. The dependence of s_O^* on potential between the foot and plateau of a totally irreversible wave is given by [13, 18]

$$E = E^0 + \frac{RT}{\alpha n_a F} \ln \frac{A k_{s,h}}{V s_O} - \frac{RT}{\alpha n_a F} \ln \frac{s_O^*}{s_O - s_O^*}.\tag{9.14}$$

According to (9.11), the current decays exponentially with time; this is one of the characteristic features of the one-step mechanism contemplated. It is convenient to write

$$\log i = \log i^0 - 0.434 s_O^* t,\tag{9.15}$$

where i^0 is the current at the start of the electrolysis, when $t = 0$. A typical current-time curve, together with the corresponding plot of $\log i$ versus t, is shown in Fig. 9.3.

This variation of current has important consequences. One is that in view of (9.2) it necessitates continuous readjustment of the voltage V applied to the cell in order to keep the potential of one electrode constant. Whereas the difference between the two electrode potentials on the right-hand side of (9.2) is unlikely to exceed 5 V and is usually much smaller, the iR term may be as large as 100 V or even higher at the start of an electrolysis. As the current decays, V must be changed to prevent the potential of the electrode of interest from straying outside the permissible limits. This more or less continuous variation of the applied voltage is the experimental heart of controlled-potential

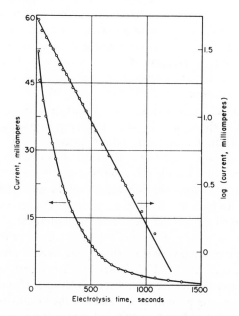

Fig. 9.3 Plots of current and of the logarithm of the current against time during an electrolysis of 100 ml of the solution of Figs. 9.1 and 9.2 with a stirred mercury-pool cathode having an area of 40 cm² and maintained at a potential of -0.5 V versus SCE.

electrolysis. Another consequence is that a measurement of the quantity of electricity consumed during a controlled-potential electrolysis involves the evaluation of the time integral of a current that ranges over three or four orders of magnitude in a typical case. Until fairly recently, these problems occupied most of the attention of those active in the field; it is probable that for some years more potentiostats (which effect the variation of applied voltage automatically) and current integrators were being designed and tested than were being applied to chemical problems. These distractions are now fortunately things of the past. The currently popular solutions to them are dealt with summarily in Section 3.

Equation (9.11) provides estimates of the time required to achieve any desired degree of completion when the half-reaction proceeds by the simple mechanism under consideration. Suppose, for example, that the p-benzo-quinone solution of Figs. 9.1 and 9.2 is electrolyzed with a stirred mercury pool at a potential of -0.10 V versus SCE under such conditions that $s_O = 5 \times 10^{-3} \sec^{-1}$. As this potential lies on the plateau of the stirred-pool voltammogram, $s_O{}^*$ and s_O are identical. It is easy to show that the current, and consequently the concentration of the quinone remaining, decreases to one-tenth of the original value after 461 sec, to one-hundredth of the original value after 921 sec, and so on. One-half hour suffices to reduce nearly 99.99% of the starting material. As is true of any other first-order process—and this electrolysis is a first-order process because the current at any instant is proportional both to the concentration of the quinone and to its rate of change—the time required to reach any particular degree of completion is independent of the initial concentration.

Controlled-potential electrolyses have sometimes been stigmatized as time-consuming. On the basis of the preceding paragraph, the reader can evaluate this assertion for himself, taking into account the fact that a potentiostat frees the operator from the necessity of paying any attention whatever to the electrolysis while it is in progress. It need only be added that the value of s_O given above is conservative; values an order of magnitude higher have been attained by ultrasonic agitation [19] or extremely efficient mechanical stirring [20] of the pool-solution interface. Very prolonged electrolysis is necessary only if insufficient attention is paid to the stirring efficiency or if there is some very slow step in the overall process.

There is of course a limit, which is set by the Nernst equation, to the degree of completion that can be achieved. Letting $N_O{}^0$ be the number of moles of p-benzoquinone present at the start of the electrolysis cited above, and f the fraction of it reduced to the hydroquinone at equilibrium, (9.3) becomes, at 25°C,

$$E = E^0 - \frac{0.05916}{2} \log \frac{fN_O{}^0/V}{[H^+]^2[(1-f)N_O{}^0/V]}$$

$$= E^0 - 0.05916 \, pH - \frac{0.05916}{2} \log \frac{f}{1-f}. \qquad (9.16)$$

The standard potential of the couple is $+0.699$ V versus NHE, or $+0.458$ V versus SCE, at 25°C; introducing this value together with $E = -0.10$ V versus SCE and pH $= 6.8$ into (9.16) yields log $[f/(1-f)] = 5.24$. Hence $(1-f)$, the fraction remaining unreduced, is 6×10^{-6}. If 99.9994% complete reduction did not suffice (as might be the case if an enormously large excess of the quinone had to be reduced in order to detect or determine some more difficultly reducible trace constituent of the solution), it would be necessary to perform the electrolysis at a more negative potential.

Equation (9.16) is the basis of a technique for evaluating the standard (or, more properly, the formal) potential of a reversible couple [21–24]. Two alternative schemes are available. One consists of electrolyzing a solution of the oxidized form O at a potential $E_{w.e.}$ on the rising portion of the wave and measuring the quantity of electricity $Q_{\infty,r}$ consumed. In the resulting equilibrium mixture

$$[O] = C_O{}^0 - (Q_{\infty,r}/nFV)$$

and

$$[R] = Q_{\infty,r}/nFV,$$

so that

$$E^{0\prime} = E_{w.e.} + \frac{2.303RT}{nF} \log \frac{Q_{\infty,r}}{nFVC_O{}^0 - Q_{\infty,r}}, \tag{9.17a}$$

where $C_O{}^0$ is the initial concentration of O and is equal to $N_O{}^0/V$. The formal potential pertains to the particular supporting electrolyte employed and includes the effect of pH as well as activity coefficients omitted from (9.3). The product $nFVC_O{}^0$ corresponds to the right-hand side of (9.1) and is simply the quantity of electricity $Q_{\infty,p}$ that would be consumed in the exhaustive electrolysis of the same solution at a potential on the plateau of the wave. Very high precision is required of the potentiostat because $Q_{\infty,r}$ is very sensitive to fluctuations of the potential. An easier method is to perform the electrolysis at a potential on the plateau throughout. When some suitable fraction—perhaps one-tenth—of the O has been reduced, the potentiostat is disconnected from the cell, the quantity of electricity Q that has accumulated is recorded, the zero-current potential E between the mercury pool and the reference electrode is measured with a potentiometer, and the electrolysis is then resumed and the measurements repeated after it has proceeded further. After the desired number of points have been obtained, the electrolysis is allowed to proceed to completion to provide a value of $Q_{\infty,p}$. The values of Q and E at each point are combined with an equation exactly analogous to (9.17a):

$$E^{0\prime} = E + \frac{2.303RT}{nF} \log \frac{Q}{Q_{\infty,p} - Q} \tag{9.17b}$$

to provide a value of $E^{0\prime}$.

If the reduction of O is irreversible, the potential required to bring it about at an appreciable rate is considerably more negative than the formal potential. Because the Nernst equation still governs the equilibrium that is eventually established, the reduction may proceed to completion even at a potential very near the foot of the wave on the stirred-pool voltammogram. Figure 9.4 shows two different cases. Curve a represents the reduction of hexaaquo-chromium(III) ion, and curve b represents the oxidation of the hexaaquo-chromium(II) ion formed by this reduction. According to these curves, it is possible to reduce hexaaquochromium(III) quantitatively even at a potential as positive as −0.8 V versus SCE. An electrolysis at that potential would be very slow because the current is only a small fraction of the limiting current, but it would approach completion because the rate of reoxidation of chromium(II) is entirely insignificant. The electrolytic oxidation of hexaaquochromium(II) yields the chloropentaaquochromium(III) ion if only a small

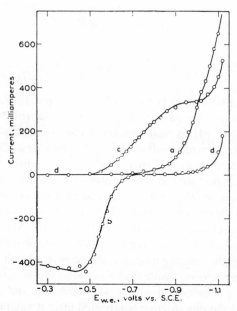

Fig. 9.4 Current-potential curves obtained with a stirred mercury pool electrode having an area of 40 cm² in 6 F hydrochloric acid solutions containing (a) 14 mM hexaaquochromium(III); (b) 14 mM chromium(II), prepared by the controlled-potential electroreduction of solution (a) at −1.10 V versus SCE; and (c) 14 mM dichlorotetraaquochromium(III), prepared by the controlled-potential electrooxidation of solution (b) at −0.40 V versus SCE. Curve (d) is the residual current curve of 6 F hydrochloric acid alone.

excess of chloride is present [25], but under the conditions of Fig. 9.4 the dichlorotetraaquochromium(III) ion predominates [26], and curves b and c therefore represent the behavior of the couple

$$CrCl_2(H_2O)_4^+ + 2H_2O + 2e = Cr(H_2O)_6^{2+} + 2Cl^-$$

which is much more nearly reversible. Electrolyzing a solution of dichlorotetraaquochromium(III) at a potential near the foot of its wave, at −0.55 V versus SCE, for example, reduces only a fraction of it because the rate of reoxidation of chromium(II) is appreciable at this potential. To effect quantitative reduction it would again be necessary to employ a potential at least as negative as -0.8 V versus SCE, thus preventing reoxidation of the chromium(II); this is still not quite on the plateau of the cathodic wave. Electrolysis on the rising part of an irreversible wave, when properly employed, is a useful way of bringing about a desired half-reaction while minimizing the extent of another that proceeds with nearly equal facility. The cloud that accompanies this silver lining is that an irreversible reduction takes place slowly even at a potential so positive that hardly any current can be seen on the voltammogram. Great care is therefore needed to avoid the troubles that may result from the occurrence of such a reduction when it is not desired.

3 APPARATUS FOR CONTROLLED-POTENTIAL ELECTROLYSIS

Electrical Apparatus

We have seen that controlled-potential electrolyses involve measuring the difference of potential between the electrode of interest [27] and a reference electrode, and adjusting the voltage applied across the working electrode and the other current-carrying electrode [28] in such a way that this difference of potential remains virtually constant. The variation of potential that can be tolerated depends on the characteristics of the current-potential curve and on the purpose of the electrolysis. Under the conditions of Figs. 9.1 and 9.2, fluctuations of ±0.8 V around a nominal potential of -0.75 V versus SCE could be tolerated if the desideratum were simply to obtain a quantitative yield of the hydroquinone. At an instant when the potential lies at the positive extreme of this range, the rate of reduction decreases, and if this instant occurs late in the electrolysis some reoxidation of the hydroquinone already formed would even take place, but these things would merely prolong the electrolysis a little. At an instant when the potential lies at the negative extreme, some hydrogen is formed, but this would merely increase the electric bill a little. If, however, the same electrolysis were being performed for the sake of a precise coulometric determination of the amount of the quinone present, the potential would have to be kept within about ±0.45 V of a nominal value of -0.45 V versus SCE, for even momentary excursions to potentials more negative than -0.9 V versus SCE would lead to erroneously high results because of the consumption of electricity by the reduction of dihydrogen phosphate ion. In either of these situations, the allowable range would be narrower if the plateau were shorter. Sometimes the waves for different constituents of a solution, or for successive reductions or oxidations of a single substance, are so close together that control to ±0.05 V may be required in an electrolysis on the plateau of the first wave. In electrolyses performed on the rising parts of waves, especially reversible ones, the requirements may be very much more stringent; the user of (9.17a) would certainly wish to have the potential controlled within 1 or 2 mV.

A simple manual circuit for controlled-potential electrolysis is shown in Fig. 9.5. The potential of the working electrode is observed with a vacuum-tube or digital voltmeter or, less conveniently, with a potentiometer, and the

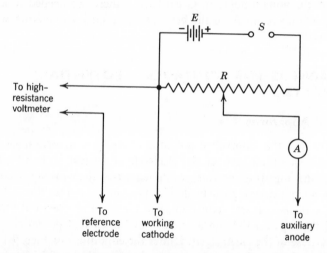

Fig. 9.5 Simple manual circuit for controlled-potential electrolysis. E is a storage battery or other dc source free from appreciable ac ripple, S is a single-pole-single-throw on-off switch, A is a milliammeter, and R is a heavy-duty rheostat having a resistance of about $10 \, \Omega$.

applied voltage is adjusted by means of a rheostat acting as a voltage divider. A milliammeter in series with the auxiliary electrode serves to indicate the progress of the electrolysis.

Manual apparatus is not only tedious but sometimes even impossible to use. The iR drop is often so large and decreases so rapidly near the start of an electrolysis that the required rate of change in voltage may be as high as 5–20 V/min. It is not within human capability to effect so rapid a change while keeping the value within even 0.1 V of the one that would produce exactly the control potential desired. Controlled-potential electrolysis was therefore but little used until Hickling [29] built the first potentiostat in 1942. A potentiostat is an automatic device that continuously senses the difference of potential between the working and reference electrodes, compares this with a preset voltage whose numerical value is equal to that of the desired control potential, and adjusts the voltage applied across the cell so as to restore any difference to zero.

Potentiostats are of two types: electromechanical and electronic. In an electromechanical instrument the dc error signal is chopped, amplified, and then used to actuate a control motor that drives the shaft of either a rheostat acting as a dc voltage divider or an autotransformer whose ac output is rectified and heavily filtered; in either case the dc output voltage is applied across the working and auxiliary electrodes. In an electronic instrument, however, the control loop is fully electronic. Electromechanical instruments respond much more slowly than electronic ones but have very much larger power outputs.

The required power output depends on the initial concentration of the solution being electrolyzed and the resistance of the cell. At the very beginning of an electrolysis on the plateau of the wave of O, the current i^0 is given by

$$i^0 = s_O nFVC_O^{b,0}. \tag{9.18a}$$

Typically, $s_O = 5 \times 10^{-3} \text{ sec}^{-1}$ if $V = 0.06$ l, so that

$$i^0 \simeq 30nC_O^{b,0}. \tag{9.18b}$$

In a large-scale synthesis or in the electroseparation of a major constituent, $C_O^{b,0}$ may be as large as 0.1 M, and if $n = 2$ currents as large as 6 A are required. The resistance of such a solution is unlikely to exceed 10 Ω even in a double-diaphragm cell, but even this corresponds to a required voltage output of over 60 V. Electronic potentiostats are generally incapable of supplying such currents at such voltages and are also generally unable to cope with the other extreme situation, in which $C_O^{b,0}$ is more moderate but the cell resistance is high. This is often encountered in work with nonaqueous solutions. If, for example, the cell resistance is 1000 Ω, which is not at all unlikely in such work, a voltage output of 60 V is needed in the electrolysis of even a 1 mM solution if $n = 2$.

The danger is that the potentiostat may be unable to maintain the desired control potential throughout the electrolysis. Suppose that one wants to perform a reduction on the plateau of a wave but that the limiting current is initially so large that the value of V in (9.2) exceeds the maximum voltage available. Then the working cathode assumes a potential on the rising portion of the wave. As time goes on, the concentration and limiting current decrease, the current that can be forced through the cell becomes a larger fraction of the limiting current, and the potential of the working cathode drifts toward the desired value on the plateau. The current is nearly constant until the desired value is obtained; only then does it begin to decrease in the expected fashion. Typical current-time curves obtained under such conditions are shown in Fig. 9.6.

Although there are cases in which this behavior is innocuous, there are others in which it is not. The reduction of benzaldehyde at -1.5 V versus SCE at pH 3 gives a quantitative yield of benzyl alcohol by a mechanism that can be represented by the equations

$$\phi CHO \xrightarrow{+H^+ + e} \phi\dot{C}HOH \xrightarrow[\text{very fast}]{+H^+ + e} \phi CH_2OH$$

If part of the electrolysis is conducted at less negative potentials, the radical intermediate can no longer be reduced as rapidly as it is formed at the electrode surface; the decreased rate of the second electron-transfer step results in the formation of more or less hydrobenzoin by the side reaction

$$2\phi\dot{C}HOH \longrightarrow \begin{array}{c} \phi CHOH \\ | \\ \phi CHOH \end{array}$$

The yield of benzyl alcohol and the value of n obtained coulometrically will be decreased, and so may the experimenter's understanding of the process.

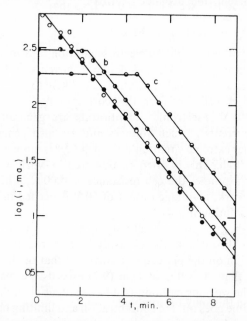

Fig. 9.6 Effect of a limited voltage output on the current-time curve in controlled-potential electrolysis. The data were secured in oxidations of vanadium(II) to vanadium(III) at a mercury anode in 4 F hydrochloric acid. The cell resistance was 18.8 Ω; higher values were secured by connecting a resistance box in series with the cell. The nominal control potential was -0.30 V versus SCE, but the potentiostat was adjusted to provide a maximum output voltage of only 3.5 V. The total resistance was (a) 18.8 (open circles) or 25 Ω (solid circles), (b) 45 Ω, and (c) 75 Ω.

It is relatively easy to increase the output voltage that can be obtained from an electromechanical potentiostat. One need only connect a bank of dry cells or storage batteries in series with its output through a motor-driven stepping switch actuated by limit switches attached to the rheostat or autotransformer. Then, when the slider approaches one extreme of its travel, the voltage of one dry cell or battery is automatically added to or subtracted from the voltage output of the potentiostat itself. Automatic operation is preserved over a range of as many tens or hundreds of volts as needed.

Most of the instruments now available commercially are electronic, and the characteristics of most of them have been summarized by Rechnitz [16] and Lott [30]. Response times as fast as 0.2 μsec are available but are intended for studies of fast reactions at microelectrodes rather than for controlled-potential electrolysis, where time constants of the order of milliseconds are more appropriate. The transient responses of potentiostats and techniques for matching them to electrolysis cells have been discussed by Booman et al. [31] and by Harrar and Behrin [32].

Data on the current-time dependence are often of great value because they reveal the occurrence of side reactions and other complications in the mechanism of the half-reaction at the working electrode. Plots of log i versus t are more convenient than those of i versus t. For the simple case described in connection with (9.7), a plot of log i versus t is linear, while one of i versus t is exponential, and the deviations produced by mechanistic complications are easier to detect on plots of log i versus t. These are most easily obtained by connecting a logarithmic recording potentiometer across a standard resistor in series with the working electrode. The recorder should be fairly heavily damped to minimize the random excursions that result from momentary fluctuations in stirring efficiency.

These fluctuations make it difficult to obtain high precision in measuring the current or evaluating a rate constant, and make many mechanistic complications harder to detect. Data much more useful for these purposes can be obtained by measuring Q_t, the quantity of electricity consumed between the start of an electrolysis and an instant t seconds later:

$$Q_t = \int_0^t i \, dt \tag{9.19}$$

In the simple case this may be combined with (9.1) and (9.11) to yield

$$Q_t = Q_\infty (1 - e^{-s_0 * t}) \tag{9.20}$$

Values of Q_t at predetermined values of t may be obtained from the register of an electromechanical current integrator or from a digital voltmeter connected across the integrating capacitor of an electronic one, and in either case may be printed out by conventional techniques of digital data acquisition.

If these values of t occur at uniform intervals $2\Delta t$, estimates of i that are virtually immune to fluctuations in stirring efficiency may be obtained from the equation [33]

$$i_{t,\text{est}} = \frac{Q_{t+\Delta t} - Q_{t-\Delta t}}{2\,\Delta t}. \tag{9.21}$$

It is easily shown [34] that the ratio of this estimate to the actual current at the midpoint of the interval is given by

$$\frac{i_{t,\text{est}}}{i_{t,\text{true}}} = \frac{\sinh s_O{}^* \Delta t}{s_O{}^* \Delta t} \tag{9.22}$$

in the simple case, and thus that the error is less than 0.2% if $\Delta t \leq 0.1/s_O{}^*$. The linearity of a plot of $\log i_{t,\text{est}}$ versus t provides a fairly sensitive criterion of adherence to the simple mechanism.

A still more sensitive one is based on the behavior of the quantity Q_R defined by the equation [35]

$$Q_R = Q_\infty - Q_t \tag{9.23}$$

This is the quantity of electricity that remains to be accumulated when t seconds have already elapsed since the start of the electrolysis. Values of Q_R can of course be computed from those of Q_t and Q_∞, but it is more convenient to measure them directly and this can be done in either of two ways [36]. One involves connecting a helical potentiometer to the gear train of an electromechanical integrator and to an appropriate dc voltage source, and recording the output voltage of the potentiometer as the electrolysis proceeds. The other involves connecting an adjustable dc voltage source in series with the integrating capacitor of an electronic current integrator, and recording the algebraic sum of this voltage and the voltage stored in the capacitor. The initial position of the slider in the first of these, or the voltage in series with the capacitor in the second, is adjusted so that the quantity recorded just becomes equal to zero when the electrolysis is complete. A preliminary electrolysis under identical conditions provides the datum needed in making this adjustment.

Combining (9.20) and (9.23) yields

$$s_O{}^* = -\frac{1}{t} \ln \frac{Q_R}{Q_\infty} \tag{9.24a}$$

or, with even higher sensitivity

$$s_O{}^* = -\frac{1}{t_2 - t_1} \ln \frac{Q_{R_1}}{Q_{R_2}} \tag{9.24b}$$

where Q_{R_1} and Q_{R_2} are the values of Q_R at the times t_1 and t_2, respectively. In the simple case the value of s_O^* thus computed is of course constant, but if any moderately slow coupled chemical or electron-transfer step is involved the value of s_O^* will vary systematically with time. Typical behavior is illustrated by Figs. 9.7 and 9.8. The reduction of cadmium ion is a single-step process devoid of complications under these conditions, and it therefore

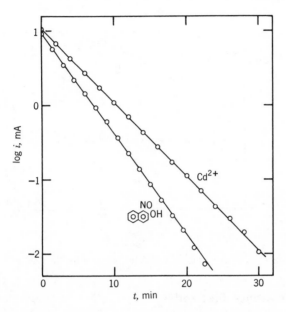

Fig. 9.7 Plots of $\log i$ versus t for the controlled-potential electroreductions of cadmium(II) and 1-nitroso-2-naphthol at mercury electrodes in acetate buffers at potentials on the plateaus of their respective waves.

gives an accurately linear plot of $\log i$ versus t in Fig. 9.6 and an essentially constant value of s_O^* in Fig. 9.8. The drift in the values of s_O^* toward the end of the electrolysis corresponds to an uncertainty of a few hundredths of a percent in the value of Q_∞. The reduction of 1-nitroso-2-naphthol, however, probably occurs by the ECE mechanism which is analogous to the one involved in the reduction of p-nitrosophenol [37–39]. The loss of water is too rapid to produce any evident distortion of the plot of $\log i$ versus t in Fig. 9.6, but the fact that this plot is not strictly linear is made clearly apparent by the corresponding plot of $s_O{}^*$ versus t in Fig. 9.7. Both analog- and digital-computer circuitry yielding plots of $-(1/t)$ $\ln Q_R$ versus t directly, with the aid of an ordinary Y-T recorder, were described by Harrar et al. [39a].

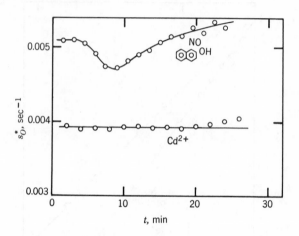

Fig. 9.8   Plots of $s_O{}^*$, computed from (9.24b), versus t for the data of Fig. 9.7.

Cells and Reference Electrodes

Almost everyone who uses controlled-potential electrolysis develops his own ideas about cell design, and descriptions of all these ideas would consume more space than is available here. This section is therefore limited to a review of the functions that controlled-potential electrolytic cells must perform and descriptions of a few typical cells that have served different purposes successfully. Figure 9.9 shows the very simple cell employed by Lingane [40] for controlled-potential electroseparations and electrogravimetric determinations of metals by deposition onto a platinum cathode. Concentric cylinders of platinum gauze are employed as the working and auxiliary electrodes, a saturated calomel electrode serves as the reference electrode, and the solution is stirred by means of a magnetic stirrer. Generally similar cells have been employed by Diehl [41] and many others.

Because they are so simple and can be handled so easily, such cells are very convenient for the purposes for which they were intended, but they have two interrelated drawbacks that render them unfit for most other applications. One is that the product of the desired half-reaction at the working electrode may take part in some other half-reaction when it reaches the auxiliary electrode. If this second half-reaction is simply the reverse of the desired one, a cyclic process will be established; a steady state will eventually be reached when the starting material is regenerated at the auxiliary electrode just as rapidly as it is being consumed at the working electrode. The yield of product

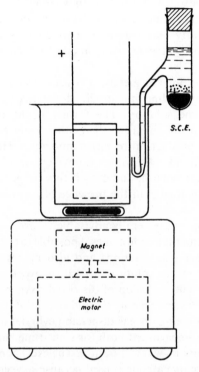

Fig. 9.9   Cell for controlled-potential separations and determinations of metals with a platinum working electrode.

will be relatively low and its isolation will be complicated by the necessity of separating it from the starting material that remains; moreover, the current-time curve will be nearly impossible to interpret and coulometric measurements will be meaningless. This situation would arise in attempting to reduce a quinone to the hydroquinone. Another situation arises if the half-reaction at the auxiliary electrode does not regenerate the starting material, as would be the case in the reduction of nitrobenzene under certain conditions; the phenylhydroxylamine formed at the cathode can be reoxidized to nitrosobenzene at the anode, and the less said about the ensuing complications the better.

Certainly there will not be much of the expected product (*p*-aminophenol) in the solution when the dust has settled. The other drawback is that the product of the half-reaction at the auxiliary electrode may react, either electrochemically at the surface of the working electrode or chemically in the bulk of the solution, with the product of the desired reaction or with some intermediate. In an aqueous solution the most likely product at the auxiliary electrode is either oxygen or hydrogen, depending on whether this electrode is the anode or the cathode. In the former case the oxygen might react with a free-radical intermediate; similarly, electrolysis in a moderately concentrated chloride solution yields chlorine at an auxiliary anode, and chlorination of the desired product is all too likely.

In cathodic depositions of metals, it is usual to circumvent the second of these drawbacks by adding an anodic depolarizer such as hydrazine or hydroxylamine to the solution. The former is the more efficacious [42]. Hydrazine is more readily oxidized than either water or chloride ion, and its principal oxidation product (nitrogen) is innocuous. However, the oxidation proceeds through diimide [43], which is nowhere near as certainly unreactive toward organic intermediates, and even if this danger did not exist it would be reckless to use hydrazine in the presence of any compound containing a carbonyl group.

For these reasons it is generally advisable to use a diaphragm cell. Separating the cathode and the anode makes it possible to eliminate cyclic processes and other extraneous side reactions and also makes it feasible to exclude dissolved oxygen from the solution around the working electrode whenever this is the cathode. The exclusion of dissolved oxygen is discussed below. It is usually accomplished by passing a rapid stream of oxygen-free nitrogen or argon through a sintered-glass gas-dispersion cylinder into the solution in the working-electrode compartment both for some time before the electrolysis is begun and throughout its whole duration. The exclusion of oxygen is customary whenever a mercury cathode is used because oxygen can be reduced over much of the range of potentials attainable with mercury electrodes, and it is of course essential if current-time or coulometric data are wanted.

One widely useful double-diaphragm cell is shown in Fig. 9.10 [44]. The solution in the working-electrode compartment contains the substance of interest; the other two compartments usually contain the supporting electrolyte alone, although in chloride media it is advisable to add a little hydrazine to the auxiliary-electrode compartment to avoid anodic attack on a platinum auxiliary anode. The reference electrode may be a commercial saturated calomel electrode, such as is sold for use with pH meters, provided that the solution being electrolyzed does not cause potassium chloride or any other substance to precipitate at the liquid junction. The danger is least with asbestos fiber-junction electrodes; it is so great with palladium-junction electrodes that these are useless for the purpose. With solutions containing high concentrations of non-aqueous solvents, hydrochloric acid, or perchlorate, it is better to use a silver–silver chloride electrode of the type shown in Fig. 9.11 [45], which makes it possible to select a solvent–electrolyte system that will not give rise to precipitation. For example, sodium chloride might be used in place of potassium chloride, both in the reference electrode itself and

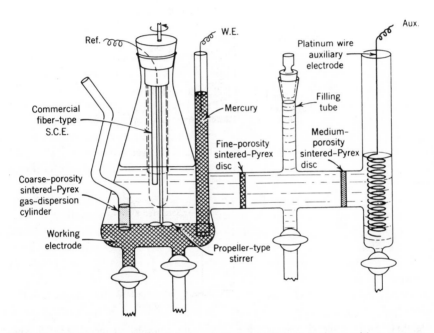

Fig. 9.10 Double-diaphragm cell for controlled-potential electrolyses with a mercury working electrode.

in the bridge that intervenes between this and the solution being electrolyzed, in work with perchlorate solutions, or 50% aqueous ethanol saturated with lithium chloride might be used in work with solutions containing 50% ethanol. Whatever reference electrode is used should be so placed in the cell that the liquid junction is made as near the surface of the mercury as possible, for otherwise there may be an appreciable iR drop between the reference and working electrodes, especially if the specific conductance of the solution being electrolyzed is not very high.

The two-bladed propeller-type stirrer shown in Fig. 9.9 should be about half immersed in the mercury pool to provide the greatest stirring efficiency and minimize the time required. The stirring motor should usually be adjusted to a speed that is just high enough to throw an occasional droplet of mercury up into the solution. However, if the solution contains a strong oxidizing agent and an anion that yields an insoluble mercury(I) salt, as does a solution of picric acid in hydrochloric acid as the supporting electrolyte, the drops that are detached will become covered with a film as they fly through the

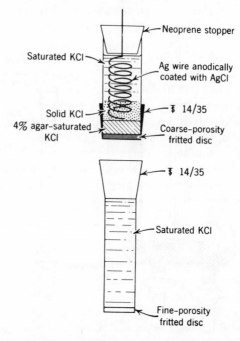

Fig. 9.11   Silver-silver chloride-saturated potassium chloride reference electrode and bridge.

solution and then will not coalesce with the main body of the pool, which rapidly turns into a nonconducting emulsion. This is easily solved by adjusting the propeller so that it is a few millimeters above the surface of the mercury at the start of the electrolysis and lowering it into the mercury only when the electrolysis is nearly complete.

The efficiency of stirring is important because it governs the value of s_O, which by virtue of (9.12) affects the rate of the overall process. The arrangement just described usually yields values of s_O in the vicinity of 5×10^{-3} sec^{-1}, so that 99.9% completion is attained in about 20 to 25 min if about 75 ml of solution is used. In any one cell the product $s_O V$ is accurately constant under fixed experimental conditions [46], so that s_O is about

0.01 sec^{-1} with 35 to 40 ml of solution. Somewhat larger values are secured with a large four-bladed paddle-type stirrer rapidly rotated in the pool-solution interface [20], while values as high as 0.1 sec^{-1} should be attainable with ultrasonic agitation [19].

Although the cell in Fig. 9.9 is suitable for most purposes, it has two drawbacks in work with current-time curves. One arises from the asymmetrical placement of the auxiliary electrode and the other from the fact that the efficiency of stirring is far from constant over the whole area of the working electrode, being highest in the center of the cell and lowest near its periphery.

The second of these is rather exotic and is not discussed here; the first gives rise to an iR drop through the solution across the surface of the working electrode. As the electrode has an equipotential surface, this gives rise to a gradient of the difference of potential between the electrode and the solution. If the working electrode is the cathode, the sign of the gradient is such that the left-hand side of the mercury pool in Fig. 9.9 appears to be less cathodic, while the right-hand side appears to be more so, than the point about halfway between them, where the tip of the reference electrode is located. The problem has been discussed by Booman and Holbrook [47] and by Harrar and Shain [48]. The variation of potential increases as the current increases or as the specific conductance of the solution decreases; its effect becomes more important as the range of permissible potentials becomes narrower. If the nominal control potential lies very near the final current rise, an appreciable quantity of electricity may be consumed in reducing the supporting electrolyte at the most cathodic part of the working electrode. This gives rise to an error in coulometric data and, because the rate of supporting-electrolyte reduction decreases rapidly as the overall current decreases, may cause a plot of log i versus t to be visibly concave upward. Conversely, if the nominal control potential is on the plateau but close to the rising portion of the wave, the least cathodic part of the working electrode will be on this rising portion at the start of the electrolysis. If the overall half-reaction follows only a single course, this will have no effect on either the yield or the coulometric n-value, but it will cause a plot of log i versus t to be concave downward. Electrolysis on the rising portion of a wave does not always give the same product as electrolysis on the plateau, however, and when it does not the consequences of this variation of potential may be severe. The least favorable case is of course that in which one is attempting to perform an electrolysis on a plateau so short that neither the devil nor the deep blue sea can be avoided.

These problems can be avoided only by having the working electrode everywhere equidistant from the auxiliary electrode. Booman and Holbrook [47] designed a cell in which the center of the pool is occupied by a solid stirrer so that electrolysis proceeds only at the surrounding ring-shaped area;

this arrangement gives less noisy current-time curves, because it produces much less fluctuation in the electrode area, than an ordinary propeller-type stirrer. Above this ring-shaped area, a flat circle of platinum wire enclosed in a toroidal auxiliary-electrode compartment was placed. Jones et al. [49] placed a platinum-wire auxiliary electrode in an unfired Vycor tube above the mercury pool but a little distance from its center, which was occupied by the stirrer shaft. Karp [20] designed the cell shown in Fig. 9.12, in which

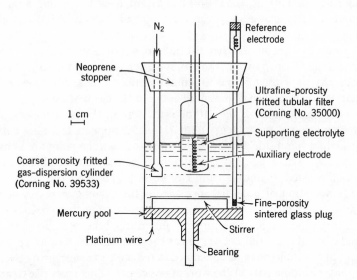

Fig. 9.12 Diaphragm cell for minimizing point-to-point variations of potential at the surface of the pool.

the stirrer shaft is brought up through the bottom of the cell to permit the placing of the auxiliary electrode above the center of the working electrode. Under typical conditions no significant variation of potential can be discerned with any of these cells.

Auxiliary Equipment

For the sake of simplicity, many controlled-potential electrolyses have been performed in cells open to the air. Even at a potential where oxygen can be reduced at the working electrode, its presence may have no effect on the completeness of a separation or on the course and yield of an electrosynthesis, and when this is true and when the effects of its reduction on coulometric data and current-time curves are unimportant there is no need to exclude it. However, it cannot be presumed to be innocuous at potentials where it is not reduced, for it is thermodynamically a much more powerful oxidizing agent than its voltammetric properties suggest. Hence it may be reduced by an appreciable rate by an intermediate or product even though its direct electrolytic reduction is immeasurably slow. For example, in the oxidation of anthracene in acetonitrile [50], coulometric data indicate that two faradays are consumed by each mole of anthracene, but bianthrone, whose formation corresponds to a three-electron process, is obtained as the product, apparently as the result of air-oxidation of the substance formed by the two-electron oxidation. In general, it is unsafe to allow oxygen to come into contact with the solution, either during electrolysis itself or during the subsequent work-up, unless its presence is shown to have no deleterious effects.

Probably the most common expedient is to bubble tank "prepurified" or "Seaford-grade" nitrogen through the solution, on the ground that the resulting concentration of dissolved oxygen is too low to have any significant effect. This may or may not be true; as usual, it depends on the meaning of "significant." If the gas contains 10 ppm of oxygen, the equilibrium concentration of dissolved oxygen will be about 10^{-8} M. Assume that 60 ml of a solution containing this concentration of dissolved oxygen is electrolyzed with a mercury cathode at a potential on the plateau of the second oxygen wave, where $n = 4$ for the reduction of oxygen to water, and that s_{O_2} is 5×10^{-3} sec^{-1}. In accordance with (9.18b), a steady current of 1.2 μA is consumed in the reduction of oxygen; if the electrolysis proceeds for a total of 2000 sec, the quantity of electricity thus consumed will be 25 nF. If the original solution were 1 mM in a substance Ox which undergoes the half-reaction Ox $+ 2e =$ Red with $s_O = 5 \times 10^{-3}$ sec^{-1}, the total initial current would be 60 mA and the total quantity of electricity consumed would be 120 μF. The current consumed by oxygen would produce a noticeable upward concavity in a plot of log i versus t after it had decayed through several decades, and the consumption of electricity would lead to an error of 0.02% in a coulometric measurement; these are insignificant effects by any reasonable standard. If the initial concentration of Ox were 1 μM, the corresponding coulometric error would be 20%; this is not insignificant, but the expenditure of a little more time would make it possible to apply a correction accurate enough to avert tragedy. The real danger is that the reaction between oxygen and the reduced form Red may be rapid enough to consume most of the dissolved oxygen present; when the rate of flow of gas is, for example, 300 ml/min in the above electrolysis, there is a total of 40 μmoles of oxygen available for this reaction. If the stoichiometry of the reaction is 2 Red $+$ $O_2 = 2$ Ox $+ 2H_2O$, the coulometric error would be about 60% with the 1 mM solution and several orders of magnitude with the 1 μM one if the above assumptions could be taken at face value. As the current decays only to a fairly substantial value (which is nearly the same as the initial current for the 1 μM solution), the situation is even worse than these calculations indicate.

In such circumstances it is necessary to remove even the trace of oxygen present in prepurified nitrogen. This may be done by passing the gas either over hot copper or through chromous chloride. Passage through a 50-cm Vycor or quartz tube, electrically heated to 450 to 500°C and filled with copper turnings, decreases the oxygen content of a stream of nitrogen to an undetectably low value. The surface of the copper must be regenerated occasionally by passing tank hydrogen through the column for about 10 min; convenience and safety are best combined by doing this whenever the tank of nitrogen is changed.

Chromous chloride solutions for removing oxygen from gas streams may be prepared by placing about 25 gm of heavily amalgamated 20-mesh zinc in a 250-ml Corning 31760 gas-washing bottle with a coarse- or fine-porosity fritted disk, adding about 150 ml of 2 F hydrochloric acid containing 0.05 F potassium dichromate, stoppering tightly, and letting the mixture stand for 1 or 2 days. During this time the dichromate is reduced to chromous ion by the zinc, and the solution is ready for use as soon as the characteristic blue color of chromous ion has appeared. A single wash bottle of this sort suffices for all purposes except coulometry on the microgram or submicrogram scale, for which two or three should be used. A similar wash bottle filled with water should follow the chromous chloride train to trap droplets of chromous chloride solution that might otherwise be sprayed over into the cell. Chromous chloride solutions deteriorate slowly on standing, turning brownish and then depositing precipitates, but may easily be regenerated by adding 10 to 20 ml of concentrated hydrochloric acid.

The gas train should be constructed entirely from Tygon and glass; rubber, polyethylene, and other porous materials are unsuitable.

Procedures for purifying mercury for use in polarography and other electrochemical techniques have been described by several authors [3, 51, 52] and are generally intended to remove metallic contaminants. Used mercury may be returned to the supplier for reprocessing or may be distilled to remove metals such as platinum and gold, which defy chemical separation, but a laboratory doing much controlled-potential electrolysis and consequently using as much as 10 to 20 lb of mercury each day is likely to deem the first of these expensive and the second both tedious and hazardous. Of the simpler procedures that have been suggested, the writer naturally prefers his own [3]. For most organic applications, however, it probably suffices to wash used mercury with solvents for the organic compounds and the inorganic constituents of the supporting electrolyte involved, dry it thoroughly, and pinhole it once or twice. Occasional more thorough purification is advisable to prevent the accumulation of excessive amounts of heavy metals, such as lead and zinc, whose ions are present as trace impurities in most supporting electrolytes.

Most controlled-potential electrolyses are performed at ambient temperature. However, for work at other temperatures, for large-scale synthetic work with a diaphragm cell (where there is some danger of excessive electrical heating), or for a kinetic study requiring close control of the temperature, it is advantageous to use a water-jacketed cell in conjunction with an external thermostat. A cell suitable for these purposes was designed by Rechnitz [16].

One often wants to detect or identify an intermediate formed during an electrolysis or to observe the growth and decay of its concentration. If the intermediate in question is stable, it is appropriate to interrupt the electrolysis, remove a portion of the solution from the working-electrode compartment to an external spectrophotometric or polarographic cell, obtain the desired information, return the solution to the electrolytic cell, and resume the electrolysis. This was first done by Lingane and Small [53], and many others have followed their lead. Polarographic measurements are simplified, and oxidation by air is avoided, by providing the working-electrode compartment of the electrolytic cell with a vertical tube through which a dropping mercury electrode can be inserted into the solution [54]. When the electrolysis is interrupted, it is easy to connect the dropping electrode and, if its resistance is low enough (commercial saturated calomel electrodes are not suitable for this purpose unless a high-input-impedance operational-amplifier polarograph is used), the reference electrode used for the electrolysis to the input terminals of a polarograph for recording the polarogram. Zuman [55] has stressed the importance of recording the entire polarogram, rather than merely monitoring the current at a fixed potential, so that new waves can be detected and so that variations of relative wave heights can be observed, and has reviewed the ways in which the data can be used to elucidate the cause of a half-reaction. For similar reasons it is of course better to obtain a complete spectrum than to monitor the absorbance at a single wavelength.

Attempts to record polarograms, or even to monitor limiting currents, in the electrolytic cell without interrupting the electrolysis are generally futile, partly because the violent motion of the solution makes it difficult to use anything simpler than a vibrating dropping electrode [56, 57], and also because the electrolysis current gives rise to an iR drop between the polarographic indicator and reference electrodes. For the detection and study of intermediates of limited stability, it is much better to circulate the solution from the working-electrode compartment through an external flow-type polarographic cell [58, 59], and the same expedient serves to prevent oxidation by air in spectrophotometric measurements [60, 61] even when the intermediates are stable. The rates of formation of gaseous products may be studied with the aid of a gas buret attached to a sealed working-electrode compartment [62], and the concentrations of free-radical intermediates may be followed by circulating the solution through a cell located in the cavity of an electron paramagnetic resonance spectrometer.

Many auxiliary techniques have been used to characterize and identify the products of controlled-potential electrolyses. Organic products have been isolated by classic techniques such as distillation and solvent extraction and by chromatographic techniques, and have often been examined by infrared, visible, ultraviolet, nuclear magnetic resonance, and electron paramagnetic resonance spectroscopy, as well as by melting-point measurements, elemental analysis, vapor-phase chromatography, mass spectrometry, and other techniques too numerous and too familiar to list. Electron paramagnetic resonance spectroscopy has assumed special importance in connection with half-reactions (especially in aprotic solvents) yielding free radicals, and the reviews by Adams [63] and Geske [64] should be consulted for information about its use in conjunction with controlled-potential electrolysis. Trapping agents, including carbon dioxide [65–67] and alkyl halides [66, 68], have been used to aid in the detection and identification of radicals formed as products.

General Technique

Controlled-potential electrolyses are generally carried out in the following fashion. The directions refer explicitly to the use of the double-diaphragm cell in Fig. 9.10; the modifications necessary for work with other cells are obvious.

One first prepares about 200 ml of the supporting electrolyte solution and dissolves an appropriate amount of the organic compound in 50 to 100 ml of this solution. The optimum concentration of the organic compound is usually between 10^{-4} and 10^{-2} M. If the solution is much more dilute than this, it may be difficult to isolate and identify the product; if it is much more concentrated, the potentiostat may be unable to maintain the desired control potential at the start of the electrolysis, as discussed above. Moreover, the limited solubility of the buffering salts may make substantial pH changes during electrolysis impossible to avoid in work with concentrated solutions unless an autotitrator is used to add acid to the working electrolyte to compensate for the consumption of hydrogen ion by the electrolytic reaction [69]. The remaining supporting electrolyte solution is used to fill the central compartment of the cell, which is then tightly stoppered, and then the auxiliary-electrode compartment is filled to a level that will be just above the final level of the solution in the working-electrode compartment. A suitable volume of the solution of the organic compound is added to the working-electrode compartment, the stirrer is started, and the solution is deaerated by passing a rapid stream of nitrogen through it for 5 to 10 min. Meanwhile the potentiostat is turned on, allowed to warm up, and adjusted so that it will provide the desired control potential. This generally must be deduced from polarograms or other voltammetric curves, which must have been obtained under conditions—temperature, pH, buffer nature, ionic strength, solvent, and so on—as nearly identical as possible to those that will prevail during the electrolysis.

When the deaeration is judged to be complete, about 30 ml of mercury is added to the working-electrode compartment. Care must be taken to ensure that the platinum lead-in wire is completely covered with mercury; to this end it is often advisable to tilt the cell slightly. The deaeration is continued to remove the air that entered the cell during the addition of the mercury, and the position and speed of the stirrer are meanwhile adjusted as described above. Finally, the potentiostat is connected to the cell and the electrolysis is then allowed to proceed without further attention.

Especially in coulometric work, such impurities as peroxides (in dioxane, Cellosolve, tetrahydrofuran, and so on) may prove troublesome. In such instances it is often beneficial to preelectrolyze the supporting electrolyte *at the same potential* that will be used for the reduction of the organic compound. This is done by the procedure just described and is continued until the electrolysis current has fallen to a negligible value. Then a concentrated stock solution of the organic compound is added, the mixture is deaerated for a few minutes, and the potentiostat is reconnected to the cell to begin the actual electrolysis.

The current during this preelectrolysis may not decrease to a negligible value. This is obviously the case if the working-electrode potential corresponds to a point on either the initial or final current rise on a polarogram of the solution. The high steady current then obtained on preelectrolysis reflects the rate of the oxidation of mercury (or platinum) or of the reduction of water, hydrogen ion, or some other major constituent of the solution. This kind of "background" current is discussed further in a later section. If it is substantial, it is likely to be fatal if any but the crudest coulometric measurements are contemplated. It can be remedied only by changing the conditions (solvent, pH, supporting electrolyte, and so on) in such a way that the wave in question becomes better separated from the initial or final current rise.

In this connection it may be pointed out that the final current rise at a large electrode appears to begin at a considerably less negative potential than it does at a dropping electrode in the same solution. This is because the comparatively large "charging" or "capacity" current, which is consumed in charging the electrical double layers surrounding successive drops up to these relatively negative potentials, masks the exponential increase of the faradaic contribution until the latter has become quite large. With the large electrode, however, the charging current decreases to zero within a very short time after the beginning of the electrolysis, and the faradaic current can therefore be detected even at relatively positive potentials.

An apparently similar, though actually distinctly different, high steady final current during preelectrolysis may be observed in single-compartment cells at working-electrode potentials quite far removed from either the initial or the final current rise. Consider a cell consisting of a mercury cathode and a silver anode immersed in a solution in which silver(I) is only moderately insoluble. As the preelectrolysis proceeds, oxygen and other impurities (peroxides, trace metal impurities in the reagents, and so on) are reduced at the cathode, and at the same time some dissolved silver(I) is formed at the anode. As this reaches the surface of the cathode, it is reduced; eventually a steady state is reached in which silver ions enter the solution at the anode just as fast as they are deposited at the cathode. This causes the flow of a steady current that obeys (9.18b), with $C_O^{b,0}$ equal to the steady-state concentration of dissolved silver(I) and that is typically 1 mA if this concentration is even 3×10^{-5} M. An internal platinum auxiliary anode gives rise to a similar cyclic process, involving the formation of oxygen at the anode and its re-reduction at the cathode. Such phenomena were described in detail by Kruse [70] and can be avoided only by using diaphragm cells.

An increase of current during an electrolysis may result from electrical heating of the solution, particularly during an interval when the current would be nearly constant if the temperature were nearly constant, as is the case when the solution is saturated with the starting material and contains an excess of the suspended solid, when the current is resistance-limited, or when the consumption of starting material is nearly counterbalanced by the formation of another electroactive substance by a homogeneous reaction [35, 36]; from the formation of an electroactive substance by a homogeneous reaction between inert ones [71] or the formation of a substance that yields its limiting current at the electrolysis potential by a homogeneous reaction between a product and a starting material that yields only a fraction of its limiting current at this potential [24]; or from the increase of the rate of reduction of water or hydrogen ion that accompanies the deposition of a noble metal such as platinum and that reflects the decrease of the overpotential for hydrogen evolution [72]. All of these conditions are fairly unusual, however, and most of them are fairly easy to recognize. By far the most common situation is that in which the current decreases continuously as the electrolysis proceeds. It may decrease to the value obtained in the pre-electrolysis at the same potential, although in many cases it decays to an appreciably higher value that remains essentially constant despite very prolonged further electrolysis. In either event the electrolysis is complete when the steady final current is attained. The mercury can then be drained out of the working-electrode compartment and the solution removed for whatever working-up may be appropriate. The other compartments of the cell should then be emptied and their contents discarded, and the cell should be washed very thoroughly, with special attention to the fritted discs. It should then be filled with water (or another solvent) or with the supporting electrolyte to be used in the next experiment: it is important that the fritted discs not be allowed to dry out. The reference electrode should be allowed to stand in a solution of potassium chloride or some other appropriate electrolyte.

4 APPLICATIONS

There are several ways in which controlled-potential electrolysis is of value in organic chemistry. One is in purifying materials and removing electroactive constituents from solutions as a preliminary to other work. It is advantageous for these purposes because it permits half-reactions to be driven to completion without contaminating the solutions with excess reagents, traces of resins, or other foreign materials. Another is in synthesizing the products of half-reactions, where it is advantageous because of its extreme selectivity and reproducibility. A third is in studying the mechanisms of half-reactions, where its peculiar advantage is that it permits the identification and investigation of chemical reactions that are coupled with the electron-transfer process and that are too slow for convenient study in other ways.

Purifications and Separations

Controlled-potential electrolysis is very much the most efficacious and convenient technique for ridding solutions of substances, such as the ions of the alkali and alkaline earth metals and of many heavy metals as well as alkynes and a number of other organic compounds, which can be converted into insoluble solids or substances soluble in mercury by electrolysis. A more specialized but related application is to the removal of minute traces of electroactive metals from mercury. In these applications the technique is employed to prepare supporting electrolytes and mercury for use in polarography, stripping analysis, and other sensitive electrochemical techniques as well as in absorption spectroscopy. For all of these purposes, one need merely subject the starting material to a sufficiently prolonged electrolysis at a potential on the plateau of a voltammetric wave known to correspond to a half-reaction yielding a suitable product.

Extreme degrees of purification are readily achieved. For a reversible process such as the deposition of copper from V_s milliliters of solution into V_a milliliters of amalgam, the numbers of moles of copper in the two phases at equilibrium are given by

$$E = E_{1/2} - \frac{RT}{nF} \ln \frac{N_a/V_a}{N_s/V_s}. \tag{9.25}$$

Suppose that an alkaline tartrate solution, in which the polarographic half-wave potential $E_{1/2}$ is -0.52 V versus SCE, is to be freed of copper(II) by an electrolysis in which $V_s = 3V_a$, which is fairly typical. At 25°C, if the electrolysis is carried out at -0.72 V versus SCE, N_a/N_s will be 1.94×10^6 at equilibrium, so that the concentration of copper(II) remaining in the solution would be only 5×10^{-7} times its initial concentration; it would of course be even smaller if the potential were more negative. An infinitely prolonged electrolysis would be required to produce this result, but one for $9.21/s$ ($= 4/0.434s$) sec would suffice to remove 99.99% of the copper initially present. An electrolysis at -0.72 V would reduce only an insignificant fraction [approximately 0.001%, according to (9.25)] of any cadmium that happened to be present, since the half-wave potential of cadmium(II), -0.856 V versus SCE in this medium, is appreciably more negative than this electrolysis potential.

The degree of completion that could be attained by a very prolonged electrolysis on the plateau of a wave exceeds that computed from (9.25) if the wave is irreversible. Here too the extent of the purification achievable in practice depends chiefly on the value of s and the patience of the operator.

The technique has been used [73] to remove zinc from sodium hydroxide, nickel and zinc together from ammoniacal ammonium citrate, iron from a weakly acidic citrate solution, and alkali and alkaline earth metals from tetramethylammonium hydroxide; to obtain mercury free from zinc by anodic stripping [74]; to remove electroactive impurities from a solution of potassium chloride before using it as a supporting electrolyte in anodic stripping voltammetry with a hanging-mercury-drop electrode [75]; and, by three successive electrolyses in different media at different potentials, to prepare a copper(II) solution entirely free from any detectable trace of any other heavy metal [76].

Separations are based on the same ideas and have usually been followed by polarographic or other electrometric analyses of the electrolyzed solutions. Although the determination of a trace of a more difficultly reducible constituent of a solution containing a large excess of a more easily reducible one is a classic source of trouble in polarography, Lingane was able to determine 10^{-4}% of zinc in cadmium and its salts after a controlled-potential electrodeposition of cadmium from an ammoniacal ammonium chloride solution [77], and as little as 10^{-5}% nickel and zinc have been determined in copper and its salts [76]. Garn and Halline [78] reduced nitrocellulose by controlled-potential electrolysis prior to determining phthalate ion polarographically; as this example indicates, physical separation is by no means essential. One might determine fumaric acid polarographically in the presence of excess maleic acid after reducing the latter at a potential and in a solution in which it is reducible while fumaric acid is not. The interference of acetaldehyde in the polarographic determination of acetone as acetone imine in a strongly ammoniacal solution [79] could be eliminated by reducing the acetaldehyde to ethanol in a weakly alkaline solution before adding ammonia.

Electrosyntheses

The first modern controlled-potential electrosynthesis was carried out by Lingane, Swain, and Fields [80]. These authors wished to prepare 9-(o-iodophenyl)dihydroacridine (II), which they were unable to secure from 9-(o-iodophenyl)acridine (I) by chemical reduction because each of the chemical reducing agents tried was either too weak to effect any reduction at all or so

$$+ \ 2H^+ \ + \ 2e \ \longrightarrow \qquad\qquad (9.26a)$$

strong as to yield 9-phenyldihydroacridine (III) in a single step. A typical polarogram of 9-(o-iodophenyl)acridine in 90% ethanol containing 0.1 F potassium hydroxide and 0.5 F potassium acetate is shown in Fig. 9.13. The

$$+ \ H^+ \ + \ 2e \ \longrightarrow \qquad\qquad + \ I^- \quad (9.26b)$$

first wave, at $E_{1/2} = -1.32$ V versus SCE, corresponds to the desired reaction [see (9.26a)], the second, at $E_{1/2} = -1.62$ V versus SCE, corresponds to reaction (9.26b). Solutions containing 0.8 g of I in 500 ml of this supporting electrolyte were electrolyzed at a working-electrode potential of -1.39 V versus SCE. This potential lies on the rising portion of the first wave [and probably, in view of (9.6), even nearer its foot on the stirred pool voltammogram than on the polarogram in Fig. 9.13]; it was chosen in preference to a more negative potential on the plateau to minimize the formation of III. Had the first wave been reversible, electrolysis on its rising portion would have led to an equilibrium mixture containing an appreciable fraction of unreduced I but, because it was actually irreversible, the electrolysis must have proceeded to completion. The electrolysis had to be performed in a diaphragm cell and in the absence of air because of the sensitivity of II to oxidation, either anodically or by dissolved oxygen. After the electrolysis had become complete, as indicated by the fall of the current to a low constant value, the product was isolated by extraction into chloroform. An average yield of 90% was obtained, the loss being attributed to incomplete extraction (there may also have been some air oxidation during the recovery process) rather than to incomplete reduction.

It is the selectivity of the technique that is especially noteworthy. The difficulty of chemical reduction depends largely on the range of potentials over which an acceptable yield of the desired product can be secured. In the reduction of p-benzoquinone under the conditions of Fig. 9.1, any potential more negative than about -0.05 V versus SCE ($+0.19$ V versus NHE) would suffice. Many reducing agents give potentials in this range; the potential varies as the reduction proceeds because the ratio [Red]/[Ox], of the concentration of the unreacted reducing agent to the concentration of its oxidized form, decreases as Red is converted into Ox by the quinone, but this variation is easy to tolerate because the range of useful potentials is so wide. In the reduction of 9-(o-iodophenyl)-acridine, however, the potential must lie between about -1.28 and -1.5 V versus SCE; if it is more positive than the first of these values, the reduction will not occur at an appreciable rate, while if it is more negative than the second the yield will decrease and the product will be more or less contaminated with **III**. A reducing agent may give a potential inside this range at the start of the reaction, but as it is consumed and its oxidized form accumulates, the potential may drift to a value so positive that the reaction virtually stops. Strictly speaking, the potentials deduced from data obtained with mercury electrodes are valid only for mercury surfaces; quite different parameters may be needed to characterize the reaction with a chemical reducing agent, but generally similar considerations must be involved.

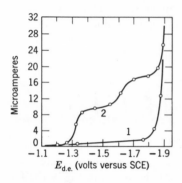

Fig. 9.13 Polarograms of (1) air-free 0.1 F potassium hydroxide–0.5 F potassium acetate in 90% ethanol and (2) the same solution after the addition of 2.22 mM 9-(o-iodo-phenyl)-acridine.

Polarography is almost always used to guide the preliminary selection of conditions for controlled-potential electrosynthesis: indeed, the most common use of controlled-potential electrosynthesis has been in preparing the products of polarographic half-reactions as an aid in the interpretation of polarographic waves. Electrolyses are customarily carried out at potentials near the midpoints of plateaus of polarographic waves and in the same supporting electrolytes in which these waves have been obtained. Citations of about 80 examples of this use were given in a recent review [36] and need not be duplicated here.

As (9.6) suggests, however, the relationship between polarography and the processes occurring at stirred electrodes is not perfectly straightforward. Intermediates that undergo further reduction in polarography because they diffuse away from the drop surface only very slowly are carried away from the electrode much more rapidly by the vigorous stirring employed in controlled-potential electrolysis and may react in the bulk of the solution to give products quite different from those obtained in polarography. There is also a difference of time scales: in an ECE mechanism

$$A \xrightarrow[1]{+n_1e} B \to C \xrightarrow{+n_2e} D \qquad (9.27)$$

for example, the chemical step is impossible to detect polarographically unless it is fast enough to proceed to an appreciable extent in the few seconds that constitute the lifetime of a single drop, whereas in a controlled-potential electrolysis the rate of formation of C must be very slow indeed if the current resulting from its reduction is to escape notice. An early example of this sort was observed by Laitinen and Kneip [81], who found that the polarographic behavior of p-N,N-dimethylaminoazobenzene in alkaline solutions corresponded to the two-electron process

$$\phi N{=}N{-}C_6H_4{-}N(CH_3)_2 + 2H^+ + 2e \to \phi NHNH{-}C_6H_4{-}N(CH_3)_2$$

while controlled-potential electrolysis yielded a mixture of aniline and p-amino-N,N-dimethylaniline. Another arises in the reduction of α-furildioxime [82], where under certain conditions eight electrons are consumed in the controlled-potential electroreduction of each molecule of the dioxime to yield 1,2-bis(2-furyl)ethylenediamine although polarograms indicate that only six electrons are involved.

Other kinds of complications may also occur and be turned to synthetic advantage. For example, the reduction of bromoacetate ion in an ammoniacal ammonium bromide solution is uninteresting if it is performed at a potential on the plateau of the wave: the half reaction is

$$BrCH_2COO^- + H^+ + 2e \to CH_3COO^- + Br^-$$

Such ways of synthesizing acetate ion are not in great demand. However, in an electrolysis at a potential very near the foot of the wave, only 0.5 faraday is consumed for each mole of bromoacetate originally present [20]. It has been suggested [36] that this reflects the formation of a radical anion intermediate

$$BrCH_2COO^- + e \to Br^- + \cdot CH_2COO^-$$

whose further reduction would be virtually instantaneous at a potential on the plateau, but is so slow near the foot of the wave that most of this product can escape into the bulk of the solution where it consumes bromoacetate ion to yield a nonreducible product. Cases similar to these might be of much interest in synthesis, and variations of experimental conditions may often yield results sufficiently unexpected to repay the small amounts of labor they entail.

5   COULOMETRY AT CONTROLLED POTENTIAL

A description of controlled-potential coulometry is inserted at this point to provide a basis for including its use in the discussion of controlled-potential electrolytic techniques for elucidating the mechanisms of electrochemical processes that follow in Section 6.

Theory

The fundamental theoretical basis of coulometry at controlled potential is described by (9.1):

$$Q_\infty = \int_0^\infty i \, dt = nFN^0, \tag{9.1}$$

which permits the calculation of either n or N^0 if the other is known. It is taken for granted that the electrolysis is carried to virtual completion, and it is also tacitly assumed that only a single substance is reduced or oxidized at the working electrode. In practical analytical applications it is also highly desirable, if not absolutely essential, that the same number of electrons be involved in the reduction or oxidation of every molecule of the starting material; this is tantamount to demanding that no significant amount of either the starting material or any intermediate be consumed by a non-electrolytic side reaction.

The first of these three requirements is easily satisfied. Because of the exponential form of the current-time curve, completion is never actually attained, but it is obviously sufficient to prolong the electrolysis until a sufficiently large fraction—for example, 99.9%—of the starting material has been oxidized or reduced.

The second requirement is less simple. Clearly one should avoid potentials so negative that a polarogram indicates that appreciable currents will be consumed by the reduction of water, hydrogen, or sodium ions, and the like, or potentials so positive that they lie on the initial current rise on a polarogram, where mercury or some constituent of the supporting electrolyte is oxidized. Nor would one ever deliberately carry out such an experiment in the presence of a foreign substance that gives a wave preceding or coinciding with the wave corresponding to the reaction of interest. Strictly speaking, however, it is impossible to avoid the consumption of electricity by such extraneous processes during a coulometric experiment, and an understanding of the sources and magnitudes of the errors thus caused is therefore vital to successful results except in the very simplest cases.

This problem was treated in detail by Meites and Moros [45], who identified a total of six components of the total quantity of electricity consumed. Those important to the present discussion are:

1. Q_{corr}, the quantity of electricity defined by (9.1), the measurement of which is the purpose of the experiment.

2. Q_c, the "charging" or "condenser" quantity of electricity, which is consumed in charging the working electrode and the electrical double layer at its surface up to the control potential. This is given by

$$Q_c = 10^{-6} \kappa A(E_{max} - E), \qquad (9.28)$$

where κ is the differential capacity of the double layer in microfarads per square centimeter and is assumed for simplicity to be constant and independent of potential, A is the electrode area in square centimeters, E is the potential of the working electrode, E_{max} is the potential of the electrocapillary maximum in the solution being studied, and Q_c is the charging quantity of electricity in coulombs. Under ordinary conditions Q_c is only of the order of a few nanofaradays, which would be entirely negligible in any experiment involving as much as a microequivalent of starting material. Meites and Moros described a procedure for the experimental evaluation of Q_c based on the fact that the accumulation of this quantity of electricity is complete within a second or less (depending on the speed of response of the potentiostat) after the start of the electrolysis.

3. $Q_{f,i}$, the faradaic impurity quantity of electricity, which results from the reduction or oxidation of impurities in the supporting electrolyte, in the sample, or in the electrode mercury. It is best eliminated by electrolytic or other purification (cf. Section 4).

4. $Q_{f,c}$, the continuous faradaic quantity of electricity, which arises from the reduction or oxidation of some major constituent of the supporting electrolyte such as water or hydrogen ion. The rate of such a process, and hence the current attributable to this cause, is described by (9.5) and is never zero. Since only a very small fraction of the hydrogen ion or water actually present would be consumed during any practical electrolysis, the continuous faradaic current remains constant throughout an electrolysis at controlled potential. This is in contrast to the current corresponding to $Q_{f,i}$, which decays to zero in the usual exponential fashion.

It is possible, in principle, to evaluate the sum of Q_c, $Q_{f,i}$, and $Q_{f,c}$ by electrolyzing the supporting electrolyte alone under exactly the same conditions and for exactly the same length of time as employed in the electrolysis of the solution of interest. This is so tedious and unsatisfactory that it is customary instead to measure the total duration (t seconds) of, and the steady final current (i_∞ amperes) attained in, the electrolysis of interest and to subtract the product $i_\infty t$ from the total number of coulombs consumed. In equating the difference to Q_{corr}, it is assumed that Q_c is negligible and that $Q_{f,i}$ has been eliminated by prior purification. It is prudent to ascertain that i_∞ is nearly the same as the current that flows in an electrolysis of the supporting electrolyte alone; if it is not, the difference may reflect a catalytic regeneration of the starting material Ox by a reaction of the form

$$\text{Red} + \text{H}^+ \rightarrow \text{Ox} + \tfrac{1}{2}\text{H}_2$$

a catalytic reduction of hydrogen ion by a scheme such as

$$\text{Red} + \text{H}^+ = \text{Red H}^+ \xrightarrow{+e} \text{Red} + \tfrac{1}{2}\text{H}_2$$

or a reaction between the product and a trace of oxygen in the stream of gas used for deaeration

$$\text{Red} + \frac{n}{2}\text{O}_2 + 2n\text{H}^+ \rightarrow \text{Ox} + n\,\text{H}_2\text{O}$$

and in any of these situations the simple correction described above may be disastrously wrong.

It is sometimes desirable to minimize the extents of competing second-order side reactions by working with extremely dilute solutions of the substance being investigated. By using sensitive apparatus for the measurement of the very small quantities of electricity involved and by exercising scrupulous experimental techniques, many investigators [49, 83–86] have succeeded in securing useful results with as little as a few micrograms of material; in one exceptionally favorable case [74], this has indeed been extended to 0.07 μg (2 nanoequivalents) of material, but the circumstances of that case have little relevance to the demands of the organic chemist. However, it is safe to say that work can be more or less routinely carried out with solutions as dilute as $2 \times 10^{-6}\ M$.

Apparatus and Techniques for Current Integration

Coulometry at controlled potential was first suggested by Hickling [29] and developed by Lingane [87, 88], who introduced the use of mercury working electrodes to facilitate the selection of operating conditions by invoking polarographic data. For many years the history of the technique consisted largely of attempts to develop more accurate, precise, sensitive, and convenient apparatus and techniques for measuring the quantity of electricity consumed, and applications to problems of chemical interest were comparatively scarce during this period. Since the problems of instrumentation were overcome several years ago, however, the emphasis has shifted, and chemical applications preponderate in the current literature.

Among the older devices, only the gas coulometer seems to deserve mention now, and this only because of its vanishingly small cost. It consists of a pair of platinum electrodes connected in series with the working electrode of an electrolysis cell and immersed in a suitable electrolyte solution in a gas buret. A dilute solution of potassium sulfate gives a mixture of hydrogen and oxygen when electrolyzed; measuring the volume of this mixture and computing the quantity of electricity from the experimental yield of 0.1739 cm^3 (at STP) per coulomb constitutes an easy way of evaluating a not too small current integral with moderate precision. Lingane's original paper [87] should be consulted for details of construction and operation. The cathodic formation and anodic reoxidation of hydrogen peroxide [89] cause the yield of gas to be smaller than the theoretical value (0.1743 cm^3/coulomb) even at moderate currents and to drop off sharply as the current decreases. Page and Lingane suggested replacing the potassium sulfate solution with $0.1\ F$ hydrazine sulfate, which on electrolysis yields a mixture of nitrogen and hydrogen. Because some ammonia is formed by the anodic oxidation of hydrazine, the yield of this mixture (0.1738 cm^3/coulomb) is also smaller than the theoretical value. Apart from the uncertainty that arises from the known error of the calibration figure and the danger that this error may be larger under the user's conditions than under those originally employed in evaluating it, the sensitivity of such a coulometer is unsuited to work much below the milliequivalent level. A Solion cell [90] is a simpler, more convenient, and only slightly more expensive alternative.

Strip-chart recording potentiometers with ball-and-disc integrators attached are nowadays reasonably common; Lingane and Jones [91] were the first to employ one in controlled-potential coulometry. If the initial current during the electrolysis is not too high, it suffices to connect the input terminals of the recorder across a suitable precision resistor in series with the cell. If the initial current and the recorder sensitivity are so high that a precision resistor small enough for the purpose would be difficult to obtain, it is easy to interpose a voltage divider between a larger resistor and the recorder. Results accurate and precise to $\pm 1\%$ or better, which are entirely sufficient for most purposes, can be obtained over a wide range of initial currents.

Another simple and inexpensive scheme [92] consists of connecting a permanent-magnet dc motor across a precision resistor in series with the working electrode. The shaft of the motor revolves at a rate that is approximately proportional to the voltage across the motor terminals, and the proportionality can be greatly improved by adding a compensation circuit [93] that counteracts the iR drop through the motor itself. A revolution counter driven by the motor shaft provides a number of counts that is correspondingly proportional to the current-time integral over the interval in question.

Two schemes for obtaining still better precision have won out over all others. One [94, 95] consists of using the output of a chopper-stabilized operational-amplifier circuit, connected across a precision resistor in series with the working electrode, to charge a precision capacitor; the capacitor is discharged before the electrolysis is begun, and the current integral at any later instant can be evaluated by measuring ths voltage across the capacitor with a precision potentiometer or, more conveniently, a digital voltmeter. The other [96–98] consists of presenting the iR drop across a precision resistor in series with the cell to the input of a dc voltage-to-frequency converter, which emits a train of pulses whose frequency at any instant is proportional to the iR drop. The pulses may be counted with an electronic counter, and the number of them accumulated during any interval is proportional to the quantity of electricity that flowed through the cell. Special provisions are needed if the current may change sign during the electrolysis.

By either of these schemes it is possible to evaluate a current integral with a precision of the order of $\pm0.02\%$, and this can be improved by employing a differential modification of the technique [99, 100]. To the analyst confronted by a half-reaction free from complications, such as side reactions that consume electroactive material and catalytic processes that regenerate it, coulometry at controlled potential is therefore an analytical technique of almost unmatched precision and one that has the advantage of yielding a datum whose ratio to the quantity of the substance being determined can be computed a priori from Faraday's law instead of having to be evaluated from measurements with standard solutions. The other side of this coin is that the technique provides an exceedingly delicate way of detecting the occurrence of a chemical complication in the overall process.

Determinations of n-Values

Controlled-potential coulometry is the standard technique for evaluating n, the number of electrons taking part in a half-reaction for each molecule or ion of material reduced or oxidized. A solution containing a known amount of this material is subjected to controlled-potential electrolysis, the value of Q_∞ is measured, and n is computed from (9.1). Citations of a great many examples may be found in the review by Perrin [101].

As an uncertainty of even $\pm1\%$ in Q_∞ is unlikely to lead to any doubt about the nature of the product, rather simple apparatus (such as a gas coulometer or Solion cell) and rough-and-ready techniques suffice for the purpose. It is sometimes advantageous to be able to evaluate Q_∞ in a fraction of the time required for the electrolysis to reach completion [102]. Four ways of doing this have been devised. One [103] involves computing Q_∞ from the slope and intercept of a plot of log i versus time, which is of course most easily obtained by means of a logarithmic recorder. Another [44] involves measuring the counting rate (which is proportional to the current) of a direct-reading integrator at two different times during the electrolysis. If R_1 is the counting rate when Q_1 microfaradays of electricity have accumulated, and if R_2 is the counting rate at some later time when Q_2 microfaradays have accumulated, then Q_∞ is given by

$$\frac{Q_\infty - Q_2}{Q_\infty - Q_1} = \frac{R_2}{R_1}. \tag{9.29}$$

Harrar et al. [39a] have employed a modification of this equation to permit on-line digital prediction of Q_∞ from data obtained as the electrolysis proceeds. The third employs integrator readings at three equally spaced times during the electrolysis; if these readings are successively Q_1, Q_2, and Q_3, then [104]

$$Q_\infty = \frac{(Q_2 - Q_1 Q_2)^2}{2Q_2 - (Q_1 + Q_3)} + Q_1, \tag{9.30}$$

from which values accurate to $\pm 0.2\%$ can generally be secured if Q_1 is at least $0.9Q_\infty$ and if $Q_2 - Q_1$ is at least $0.2(Q_\infty - Q_1)$. Values accurate to $\pm 0.5\%$ are usually obtained even if Q_1 is only $0.5Q_\infty$, so that reasonable accuracy can be secured in as short a time as that required to reach about 70% completion. A fourth technique is based on the use of a device [105] that provides a plot of i versus Q and the extrapolation of this plot to zero current. The second and third of these techniques, in which a current integrator is employed for the precise measurement of a quantity of electricity that may be a very large fraction of the whole, are superior to the others. Like the fourth technique, they are based on the tacit assumption that the current undergoes an exponential decay between the instant at which Q_1 is measured (or at which the extrapolation is begun) and the end of the electrolysis and should not be used unless this is known to be true. As they involve no assumption about the shape of the current-time curve prior to that time, they are easily able to cope with behavior similar to that shown in Fig. 9.6, as well as with many situations in which the curve is distorted by chemical complications and approaches the exponential form only when the electrolysis is nearly complete.

Several kinds of ambiguities are possible in coulometric values of n. One is that more than one half-reaction can often be written with the same n-value. For example, a one-electron anodic wave for mercaptobenzothiazole (RSH) might arise from either of the half-reactions

$$RSH + Hg \rightarrow \tfrac{1}{2}(RS)_2Hg_2 + H^+ + e$$

$$RSH \rightarrow \tfrac{1}{2}RSSR + H^+ + e$$

Another is that unless the product Red of the last electron-transfer step is electroactive so that its concentration can be followed by a current-reversal technique [106] its transformation into another substance P by hydrolysis or some other reaction

$$Ox \xrightarrow{+ne} Red \rightarrow P$$

will escape notice. The reduction of cyclohexyl nitrate [107] is one example: the nitrate ester originally formed undergoes rapid hydrolysis in an alkaline solution, yielding cyclohexanol as the product of a two-electron reduction. The oxidation of triethylamine in dimethyl sulfoxide [108], involving one electron per molecule of amine, takes the similar course

$$Et_3N \xrightarrow{-e} Et_3N^{.+}$$
$$Et_3N^{.+} + CH_3SOCH_3 \rightarrow Et_3NH^+ + \cdot CH_2SOCH_3$$

Ambiguities such as these can be resolved only by isolating and identifying the product. Another kind of ambiguity arises when the half-reaction occurs by a branched mechanism; a competing chemical step may proceed at a rate that yields an integral but misleading n-value. One such case is the reduction of picric acid from hydrochloric acid. Lingane [87] found 17.07 F of electricity to be required for the reduction of each mole of picric acid in a solution containing 0.4 mF picric acid and 0.1 F hydrochloric acid, and concluded that the primary product of the electrode reaction was bis(3,5-diamino-4-hydroxyphenyl)hydrazine:

Both Bergman and James [109] and Meites and Meites [110] were able to confirm this apparent value of n under exactly the same conditions used by Lingane, but the latter authors also showed that the results depended on the concentrations of picric and hydrochloric acids, as shown in Fig. 9.14. With very low concentrations of picric acid, n was found to be $18.00 \pm 0.01_6$, in exact agreement with the expected value for reduction to 2,4,6-triaminophenol. At higher concentrations, however, the quantity of electricity decreases, and under some conditions the apparent value of n becomes as low as 16.4. Not only is the existence of a 17-electron product difficult to credit on this evidence, but it can be inferred from the shapes of the curves in Fig. 9.14 that as few as 12 F per mole may be consumed in forming the product that competes with the triaminophenol. Similarly, although coulometric data suggest that $n = 3$ for the reduction of glyoxal [111], this is ascribed to coincidence rather than to a quantitative reduction to erythritol. Certainly, any n-value that corresponds to an unexpected or improbable product should be checked, either by product analysis or by additional measurements over the widest possible range of conditions—and preferably, of course, both. When a branched mechanism involves a competing second-order step that proceeds to an extent just sufficient to give rise to an integral value of n that lies between the sum of the n-values for the preceding electrolytic steps and

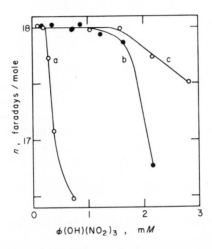

Fig. 9.14 Effect of the initial concentration of picric acid on the apparent n-value obtained on reduction at -0.40 V versus SCE in (a) 0.1 F, (b) 1 F, and (c) 3 F hydrochloric acid.

the total overall n-value, variations of the initial concentration of the substance being oxidized or reduced must have substantial effects on the coulometric results. Even if the competing step is first-order with respect to the intermediate it consumes, varying s_0, either by changing the stirring efficiency in electrolyses on the plateau or by performing electrolyses on the rising part of the wave, alters the fraction of the intermediate that undergoes this step and therefore alters the n-value obtained.

Analytical Applications

The early applications of controlled-potential coulometry to organic analysis were devoted chiefly to determinations of nitro- and halo-substituted compounds. Fukami et al. devised procedures for the determination of p,p'-DDT [112] and γ-hexachlorocyclohexane [113]. Meites and Meites developed a coulometric method for the assay of trichloroacetic acid [114] in which even relatively large amounts of di- and monochloroacetic acids do not interfere, and another for the assay of picric acid [110] in which the coulometric measurement could be combined with an alkalimetric titration to provide a value for the picric acid content of a sample contaminated with dinitrophenols. Ehlers and Sease [115] devised procedures for determining allyl bromide, chloroform, carbon tetrachloride, and several other halo-substituted compounds, as well as nitrobenzene, 1-nitropropane, and a number of isomeric substituted nitrobenzenes. Kruse [70] independently investigated the determinations of some of these compounds, and also of nitroisophthalic and nitroterephthalic acids under other conditions.

These investigations established the utility of the technique and laid the foundation for greater adventurousness. More recent applications include procedures for the determination of N-substituted phenothiazines [116]; p-phenetidine [117]; catechol [118]; phenylthiourea, thioglycolic acid, and cysteine [119]; and a number of nitroso compounds including p-nitrosophenol, 1-nitroso-2-naphthol, and cupferron [120]. The increasing interest in non-aqueous electrochemistry has led to the development of procedures for determining various isomeric nitrophenols, nitrobenzoic acids, and other nitro compounds in dimethyl sulfoxide [121] and, after removing amines by a preelectrolysis at a less positive potential, aliphatic amides in acetonitrile [122].

6 STUDIES OF REACTION MECHANISMS

The half-reaction $O + ne \rightarrow R$ that takes place at the working electrode may occur in a single step or in a series of steps of which the first involves electron transfer and is rate-determining; it may occur in a series of steps of which more than one is rate-determining; or it may involve the formation of intermediates that can undergo side reactions leading to other products. These three situations give rise to different kinds of behavior. In the first, which is the simplest:

1. R is the only product obtained.
2. The quantity Q_∞/FN_O^0 is equal to the integral value of n.
3. The value of s_O^* computed from (9.24) is constant throughout the electrolysis (so that a plot of $\log i$ versus t is truly linear).
4. The above conditions are true regardless of variations of the stirring efficiency, the initial bulk concentration of O, or (as long as the rate of the reoxidation of R remains negligible) the potential of the working electrode.

As quantitative determinations of electrolysis products are often far more difficult to execute with precisions high enough to be informative than they are to prescribe, the first of these criteria is only rather infrequently applied. Ways of applying the others have been described in the preceding pages.

The second of the above possibilities includes mechanisms such as

$$Y \underset{k_b}{\overset{k_f}{\rightleftharpoons}} O \xrightarrow{+ne} R$$

in which the current is governed by the kinetics of the prior transformation of the electroinactive precursor Y into O as well as by the factors that govern the rate of the charge-transfer step in the simple case, and the ECE mechanism mentioned several times above. Here

1. R is again the only product obtained and
2. The quantity Q_∞/FN^0 is again equal to the integral value of n, although N^0 must now be taken as the sum of the numbers of moles of O and Y present initially. These things mean that the product yield and coulometric data are alike unaffected by the occurrence of a prior or intervening chemical step. However,
3. The value of s^* computed from (9.24) is not constant throughout the electrolysis (so that a plot of log i versus t is not linear) and
4. The shape of a plot of either s^* or log i versus t is affected by variations of stirring efficiency or of working-electrode potential on the rising part of the wave if the wave is irreversible; it is also affected by changes of initial bulk concentration if a pseudo-second- or higher-order rate-determining step is involved, or by changes in pH if hydrogen ions are involved in that step.

The third of the above possibilities includes mechanisms similar to

$$O + n_1 e \rightarrow I$$
$$2I \xrightarrow{k} D$$
$$I + n_2 e \rightarrow R$$

This may be discerned in processes as diverse as the reductions of aryl ketones [123–125], picric acid [87, 109, 110], and benzyldimethylanilinium ion [126, 127] and the oxidations of hydrazine [43, 128–131] and hydroxylamine [132], although the pattern is decorated with much embroidery in each of these cases. If any competing reaction is involved:

1. The yield of R will be less than 100%
2. The quantity Q_∞/FN^0, which henceforth is called the "apparent n-value" and given the symbol n_{app}, will have a value different from that corresponding to reduction to R and will usually be nonintegral,
3. A plot of log i versus t may or may not be linear, and
4. The value of n_{app} and the shape of a nonlinear plot of log i versus t will be affected by variations of stirring efficiencies or working-electrode potential, and also by variations of initial bulk concentration if the competing

step is a pseudo-second- or -higher-order one, as well as by variations of pH if it involves hydrogen ions.

The behaviors of many specific mechanisms belonging to these general classes have been worked out in detail and have recently been reviewed at some length elsewhere [36]. The following discussion includes very abbreviated descriptions of the cases that have been investigated, with emphasis on their differential diagnosis. Frequent reference is made to procedures for the evaluation of rate constants for homogeneous reactions. As the peculiar advantage of these techniques is that they are applicable to reactions very difficult to study in other ways, there are very few situations in which rate constants thus obtained have been confirmed by other techniques. There are just enough [24, 133, 134] to show, when taken together with the nature of the theoretical foundation, that these procedures rest on a firm and trustworthy footing.

Catalytic Processes

Case Ia.   Regeneration of Electroactive Material by a Homogeneous Reaction following Electron Transfer

The mechanism

$$O + ne \rightarrow R$$

$$aR + bZ \xrightarrow{k} cO$$

where Z may be hydrogen ion, water, or some other major constituent of the solution, gives rise to plots of $\log i$ (corrected, if necessary, for the continuous faradaic current obtained with the supporting electrolyte alone) versus t that are concave upward and have a finite horizontal asymptote. This asymptote represents a steady state in which O is reduced at the surface of the working electrode at the same rate that it is produced by the homogeneous reaction. The catalytic current at the steady state is proportional to C_O, to $C_R{}^a$, and to $C_Z{}^b$, where each C is the steady-state concentration of the species denoted by the subscript, but is essentially independent of the potential of the working electrode; the order and rate constant of the homogeneous step are easily obtained. If this step is first-order with respect to R, its rate constant (or the product $kC_Z{}^b$) can also be obtained by extrapolating the final linear portion of a plot of Q versus t back to $t = 0$; the intercept is less than the expected coulometric result $nFN_O{}^0$.

References: 36, 45, 133, 135–140.

The foregoing description applies to the ordinary situation in which the product $kC_Z{}^b$ is fairly small, so that the steady-state current is much smaller than the current at the start of the electrolysis. At the other extreme, $kC_Z{}^b$ may be so large that the R produced at the electrode surface is reoxidized before it can escape from the diffusion layer into the bulk of the solution, and in this event the current is abnormally large and virtually constant. This situation arises in the reduction of molybdenum(V) in saturated hydrazine dihydrochloride [141]; other examples could be constructed in profusion, but a detailed treatment of it is neither available nor likely to be of much interest.

Case Ib. Catalytic Reduction by a Charge Transfer Step Involving the Product R

The mechanism

$$O + ne \rightarrow R$$

$$R + H^+ = RH^+ \xrightarrow{+e} R + \tfrac{1}{2}H_2$$

involving the R-catalyzed reduction of hydrogen ion (or some other major constituent of the supporting electrolyte) closely resembles case 1a but differs from it in two respects:

1. The steady-state catalytic current is potential-dependent, and
2. If the rate of protonation of R is high, as is usually true, the steady-state current will be proportional to the activity of hydrogen ion only over the range of pH-values in which most of the R remains unprotonated; it will be independent of hydrogen-ion activity in the range where most of the R is in the protonated form.

References: 24, 36, 45.

Case Ic. Catalytic Reduction by a Charge-Transfer Step Involving the Starting Material O

The related mechanism

$$O + ne \rightarrow R$$

$$O + H^+ = OH^+ \xrightarrow{+e} O + \tfrac{1}{2}H_2$$

has been referred to as an O-induced reduction of hydrogen ion because it can often be imagined to involve two alternative modes of decomposition of the product obtained when the first electron is transferred to an OH^+ ion or molecule (or to a hydrogen ion or water molecule acting as a bridge between the electrode and an ion or molecule of O).

References: 23, 24, 36, 142, 143.

Other Subsequent and Prior Reactions of R or O

Case 2a. Inactivation of R by a Homogeneous Reaction following Electron Transfer

The mechanism

$$O + ne \rightarrow R$$

$$R + Z \xrightarrow{k} P$$

in which Z is some major constituent of the solution and P is an electrolytically inert product, was discussed in another connection in Section 5. If the
electrolysis is conducted at a potential where R is not reoxidized, a plot of
log i versus t is linear and n_{app} is equal to n. The occurrence of the inactivation
step can be detected by product identification, and its pseudo-first-order rate
constant kC_Z can be evaluated by following the time dependence of the
concentration of R or P, by reversal coulometry, or by analysis of current-
time data obtained in an electrolysis on the rising portion of the polarographic
wave if the O-R couple is reversible.

References: 106, 140.

Case 2b.   Inactivation of R by Parallel Reactions

The mechanism

$$O + ne \rightarrow R$$
$$R + Z_1 \xrightarrow{k_1} P_1$$
$$R + Z_2 \xrightarrow{k_2} P_2$$

gives rise to behavior similar to that encountered in case 2a. Determin-
ing the yields of P_1 and P_2 provides a value of the ratio $k_1 C_{Z_1}/k_2 C_{Z_2}$ of the
pseudo-first-order rate constants; their sum can be evaluated in any of
the ways applicable in case 2a.

Reference: 140.

Case 2c.   Formation of O by a Homogeneous Prior Reaction

The mechanism

$$Y \underset{k_b}{\overset{k_f}{\rightleftharpoons}} O \xrightarrow{+ne} R$$

which is responsible for the kinetic current in polarography, was mentioned
above. The shape of the current-time curve depends on the relative values of
k_f, k_b, and $s_O{}^*$. At one extreme, if both k_f/k_b and $k_f/s_O{}^*$ are large, a plot of
log i versus t is so nearly linear that the occurrence of the prior reaction may
escape detection; at the other, if both of these quantities are small, a steady
state, in which O is reduced just as rapidly as it is produced by the prior reaction, is reached almost immediately after the beginning of the electrolysis. An
intermediate case of special importance is that in which k_f is considerably
smaller than $s_O{}^*$ but comparable with k_b, so that the initial equilibrium
mixture contains roughly equal concentrations of Y and O. By analogy with
case 4a, one can dissect the resulting plot of log i versus t into two linear
segments, of which the first can be ascribed to the reduction of O that was
present initially and the second to the reduction of O that is formed from Y
as the electrolysis proceeds. From the slopes and intercepts of these segments

it is possible to evaluate the rate constants (and hence the equilibrium constant of the prior reaction) and also the initial concentrations of Y and O.

References: 33, 35, 144, 145.

Case 2d.   Inactivation of O by a Homogeneous Pseudo-First-Order Reaction

The mechanism

$$O + ne \rightarrow R$$

$$O + Z \xrightarrow{k_b} Y$$

corresponds to case 2c with $k_f = 0$ but differs from it chemically: what is contemplated here is the addition of O to a supporting electrolyte containing a large excess of some substance Z that inactivates it, followed by the commencement of the electrolysis at a time, t_1 seconds later, at which the inactivation reaction is still far from complete. A plot of log i versus t is linear but $Q_\infty < nFN_O^0$. Coulometric data, which may conveniently (but need not necessarily) include information on the effect of varying t_1, permit the evaluation of k_b.

Reference: 144.

Case 2e.   Inactivation of O by Homogeneous Reactions of Other Types

If (1) the initial concentration of Z in case 2d above is not very much larger than that of O, or if (2) O is inactivated by dimerization or some other second-order process of the form

$$2O \rightarrow Y$$

Q_∞ will again be smaller than nFN_O^0 but a plot of log i versus t will not be exactly linear. No example of either case is known.

Reference: 144.

Reactions of O with a Product or Intermediate

Case 3a.   Reaction of O with R to Yield an Inert Product

The mechanism

$$O + ne \rightarrow R$$

$$O + R \xrightarrow{k} P$$

gives rise to values of n_{app} that approach n as k/s_O^* approaches zero and that approach $n/2$ as k/s_O^* increases without limit. At either of these extremes a plot of log i versus t is virtually linear. The first extreme is trivial; the second can generally be diagnosed either by product identification or from the value of n_{app}, which may be half-integral. Of more interest is the intermediate situation, in which the final mixture contains appreciable concentrations of both R and P, in which n_{app} lies between $n/2$ and n and can be used to evaluate k/s_O^*, and in which the plot of log i versus t is nonlinear. This can often be achieved by varying the initial concentration of O, since R is formed almost exclusively if $C_O{}^0k/s_O^*$ is less than about 0.03 while P is formed almost exclusively if $C_O{}^0k/s_O^*$ exceeds about 30. One may also vary s_O^* by changing the stirring efficiency or the potential of the working electrode, and it may sometimes be possible to vary k by changing the composition of the supporting electrolyte.

References: 140, 146, 147.

Case 3b.   The Effect of a Competing Reaction that Inactivates R

The mechanism

$$O + ne \rightarrow R$$

$$O + R \xrightarrow{k} P$$

$$R + Z \xrightarrow{k'} P'$$

differs from case 3a by including a pseudo-first-order inactivation of R by reaction with some species Z present in large excess and gives rise to generally similar behavior. The range of variation of n_{app} is decreased by the occurrence of the inactivation reaction, and the value of n_{app} varies with the concentration of Z.

Reference: 140.

Case 3c.   Reaction of O with an Intermediate to Yield an Inert Product

The mechanism

$$O + n_1 e \rightarrow I$$

$$O + I \xrightarrow{k} P$$

$$I + n_2 e \rightarrow R$$

is a variant of case 3a, the limits of n_{app} now being $(n_1 + n_2)$ as k approaches zero and $n_1/2$ as k increases without limit. It is worth noting that the homogeneous reaction cannot occur if the electrolysis is performed at a potential on the plateau of the wave, where I is reduced as rapidly as it is formed, for the concentration of O then cannot be appreciable at the surface of the electrode and the concentration of I cannot be appreciable anywhere. Electrolysis on the rising portion of the wave does, however, allow some of the I to survive long enough to participate in the homogeneous reaction unless it is much more easily reduced than O.

References: 20, 36.

Case 3d. Reaction of O with R to Yield an Electroactive Product

The mechanism

$$O + n_1 e \rightarrow R$$
$$O + R \rightarrow I$$
$$I + n_2 e \rightarrow R$$

can be detected only by following the time dependence of the concentration of O or I if the values of s^* for O and I are comparable. However, the overpotential for the reduction of O may so far exceed that for the reduction of I that a potential lying near the foot of the wave of O lies on or near the plateau of the wave of I. Then the current rises at the start of the electrolysis, passes through a maximum, and finally decays toward zero; the rate constant of the homogeneous reaction can be evaluated from data that include the maximum value of the current.

References: 24, 143.

Case 3e. Reaction of an Intermediate with R to Yield an Electroactive Product

The mechanism

$$O + n_1 e \rightarrow I$$
$$I + R \xrightarrow{k} J$$
$$J + n_2 e \rightarrow R$$

may, if the reaction between I and R is very slow, give rise to a current that decays nearly to zero as the reduction of O to I approaches completion and then remains small for some time while the concentration of R is slowly increased by the reactions that follow. As more and more R accumulates, the rate of the homogeneous reaction increases and the current rises in proportion to the concentration of J. After passing through a maximum when most of the I has been consumed, the current decreases again as the remaining J is reduced.

Only one example is available as yet, and its classification is rather uncertain. It arose in the reduction of 1,3-diphenyl-1,3-propanedione in dimethyl sulfoxide and was interpreted by assuming the validity of the rate equation

$$-\frac{dC_I}{dt} = kC_R,$$

according to which the homogeneous reaction is zeroth-order in I. However, the values of k thus computed were so strongly dependent on $C_O{}^0$ that the assumption must be regarded as dubious until further evidence is adduced.

Reference: 148.

Parallel Reductions

Case 4a.   Reductions of Two Chemically Unrelated Species

In the absence of any chemical interactions among the various substances involved, the controlled-potential electroreduction of a mixture of O_1 and O_2

$$O_1 + n_1 e \rightarrow R_1$$
$$O_2 + n_2 e \rightarrow R_2$$

gives rise to a plot of $\ln i$ versus t that can be dissected into two linear segments if the values of s^* for O_1 and O_2 are sufficiently different. The plot becomes linear at long times, after virtually all of the more rapidly reduced component (assumed to be O_1) has been consumed. The slope of this linear portion is $-s_2^*$, and its intercept at zero time, $\ln i_2^0$, is easily obtained by extrapolation. The quantity of electricity required to reduce all of the O_2 initially present is given by

$$Q_2 = i_2^0 / s_2^*.$$

Values of i_2 at various times during the first part of the electrolysis may be obtained from the same extrapolation. Subtracting these from the total measured currents gives values of i_1, and a plot of $\ln i_1$ versus t is linear; its slope $-s_1^*$ and intercept at zero time may be combined in the above fashion to compute Q_1, or one may write simply

$$Q_1 = Q_\infty - Q_2,$$

where Q_∞ is the total quantity of electricity consumed by the mixture, obtained with the aid of a precision current integrator. These considerations have been employed in analyzing some simple mixtures. This kind of kinetic analysis may prove useful in dealing with mixtures of substances that yield overlapping irreversible voltammetric waves. It is generally necessary (and possible) to select a working-electrode potential that lies near the plateau of one of these waves but near the foot of the other, so that very different values of s^* are obtained.

References: 31, 36.

Case 4b.   Reductions of O along Parallel Paths

The mechanism

$$O + n_1 e \rightarrow R_1$$
$$O + n_2 e \rightarrow R_2$$

gives rise to a strictly linear plot of $\log i$ versus t and a value of n_{app} intermediate between n_1 and n_2. As the heterogeneous rate constants for the two electron transfer processes are likely to have different potential-dependences, the value of n_{app} generally varies with potential.

The related mechanism

$$O + H^+ = OH^+ \qquad \text{(fast)}$$
$$O + n_1 e \rightarrow R_1$$
$$OH^+ + n_2 e \rightarrow R_2$$

behaves similarly, and in addition the value of n_{app} is pH-dependent. Some of the characteristics of case 2c may also appear as the pH is increased and the rate of protonation decreases.

Reference: 149.

Other Mechanisms

Case 5a.   Consecutive Reductions without an Intervening Chemical Step

The mechanism

$$O + n_1 e \rightarrow I$$
$$I + n_2 e \rightarrow R$$

gives $n_{app} = n_1 + n_2$ at any potential, but the shape of a plot of log i versus t depends on the heterogeneous rate constant k_I for the reduction of I. At a potential near the foot of the wave of I, where this rate constant is small, much of the I can escape into the bulk of the solution and its reduction will still continue slowly long after the last of the O has been consumed. The plot of log i versus t then has the shape described above in connection with case 4a, and can be dissected in the manner prescribed there. The ratio Q_1/Q_2 of the resulting current integrals is very nearly equal to n_1/n_2 in this region (but not at potentials nearer the plateau of the wave of I), and the potential-dependence of the slope of the more slowly decaying segment can be used to evaluate αn_a for the reduction of I. As the potential becomes more negative, the plot of log i versus t becomes more nearly linear, the slope of its first portion remaining constant while that of the second one increases, and on the plateau of the wave of I the observed behavior is that of the simple overall mechanism

$$O + (n_1 + n_2)e \rightarrow R$$

References: 13, 20, 35, 36, 147, 150.

Case 5b.   The ECE Mechanism

The mechanism

$$O + n_1 e \rightarrow I$$
$$I \xrightarrow{k} J$$
$$J + n_2 e \rightarrow R$$

gives $n_{app} = n_1 + n_2$; as explained in Section 4, this value of n_{app} exceeds the overall n-value deduced from polarographic data if k is small. The shape

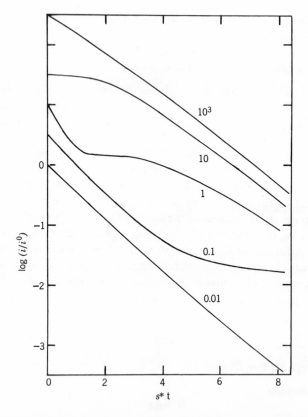

Fig. 9.15 Calculated plots of log i versus t for a controlled-potential electrolysis in which the half-reaction at the working electrode occurs by an ECE mechanism. It is assumed that $s^* = 10^{-3}\,\text{sec}^{-1}$ and that $D\ (=D_O = D_I = D_J) = 10^{-6}\,V/A\ \text{cm}^2\text{-sec}^{-1}$. The number beside each curve gives the corresponding value of k/s^*. The ordinate scale pertains to the lowest curve, for which $k/s^* = 0.01$; each successive curve above this is shifted upward 0.5 unit along the ordinate axis.

of a plot of $\log i$ versus t depends on the value of the ratio k/s^*, where $s^* = s_O^* = s_J^*$, in the fashion illustrated by Fig. 9.15, where the assumptions are equivalent to supposing that the working-electrode potential lies on the plateaus of the waves of both O and J. Since, even apart from the possibility of varying k by altering the composition of the supporting electrolyte, values of s^* ranging over about four orders of magnitude are easily obtained by adjusting the stirring efficiency and the area of the working electrode, curves corresponding to several different members of this family can be secured in successive electrolyses as an aid to diagnosis. It may further be noted that on interrupting the electrolysis for some time not too long after its beginning the transformation of I into J continues and leads to an increase in the total concentration of reducible material, so that the current which flows immediately after the electrolysis is resumed exceeds that which flowed immediately before it was interrupted.

For a long time the ECE mechanism was the only one known to give rise to curves similar to the one shown in Fig. 9.15 for $k/s^* = 1$, but at least two others are now recognized. One [151] occurs in case 4a when R_1 is a solid whose presence on the surface of the electrode retards electron transfer to the solution. The other

$$YO_3 = Y + 3O \qquad \text{(fast)}$$

$$O + n_1 e \rightarrow R$$

is one in which the electron-transfer step is preceded by a fast homogeneous equilibrium in which both Y and the compound YO_3 are electrolytically inert; this is believed to account for the current-time curves obtained in the anodic stripping of zinc from platinum-zinc amalgams [152].

Despite these complications, controlled-potential electrolysis is of great value in the diagnosis of ECE mechanisms and in evaluations of k for the chemical steps they involve. This is especially true when these steps are too slow for convenient study by faster techniques such as chronopotentiometry.

References, 20, 24, 35, 82, 126, 127, 153, 154.

Case 5c.   The ECECE Mechanism

The mechanism

$$O + n_1 e \rightarrow I$$

$$I \xrightarrow{k_1} J$$

$$J + n_2 e \rightarrow K$$

$$K \xrightarrow{k_2} L$$

$$L + n_3 e \rightarrow R$$

gives $n_{app} = n_1 + n_2 + n_3$ and plots of log i versus t more complex than those for case 5b.
Reference: 35.

Case 5d.   The ECE Mechanism with a Competing Pseudo-First-Order Inactivation of I

The mechanism

$$O + n_1 e \to I$$

$$I \xrightarrow{k_1} J$$

$$I \xrightarrow{k_2} P$$

$$J + n_2 e \to R$$

should give plots of log i versus t generally similar to those for case 5b, but values of n_{app} smaller than $n_1 + n_2$.
Reference: 126.

Case 5e.   The ECE Mechanism with a Competing Pseudo-Second-Order Inactivation of I or J

Both Bard and Mayell [126] and Meites [155] attempted to elucidate the behavior of the mechanism

$$O + n_1 e \to I$$

$$2I \xrightarrow{k} D$$

$$I + n_2 e \to R$$

by assuming that $s^* = s_O{}^* = s_I{}^*$ (i.e., that the electrolysis is performed at a potential where both O and I yield their limiting currents) and that the current at any instant is given by an equation of the form

$$i = s^* n_1 FV C_O{}^b + s^* n_2 FV C_I{}^b,$$

where C^b is the bulk concentration of the species denoted by the subscript at the instant in question. Bard and Mayell gave plots of n_{app} versus log $C_O^{b,0}$ for various values of the ratio s^*/k; Meites gave an empirical equation that could be used to compute the product $kC_O^{b,0}$ from values of n_{app} and s^* and discussed the problem of evaluating s^* from a plot of log i versus t.

Of course these attempts yielded similar results. An increase of either k or the initial bulk concentration of O tends to increase the fraction of the I consumed by the pseudo-second-order dimerization, and consequently the value of n_{app} varies from $n_1 + n_2$ when $kC_O^{b,0}$ is very small to n_1 when $kC_O^{b,0}$ is very large. Unfortunately, the two assumptions are mutually exclusive: if I yields its limiting current at the potential employed, no significant amount of it can escape from the surface of the electrode into the bulk of the solution. Hence the bulk concentration of I must always be virtually zero, so that both the extent of dimerization in the bulk of the solution and the current resulting from the reduction of I brought to the working-electrode surface from the bulk of the solution must also be virtually zero, and it is on these that the computations are based. It has been argued [36] that these authors have in fact portrayed the behavior of the related mechanism

$$O + n_1e \rightarrow I$$
$$I \xrightarrow{k_1} J$$
$$2J \xrightarrow{k_2} D$$
$$J + n_2e \rightarrow R$$

in which the pseudo-first-order transformation of I into J is just slow enough to cause most of the I to be swept away from the electrode surface into the bulk of the solution before it is converted into J, but is not so slow as to be otherwise rate-determining. The similar mechanism

$$O + n_1e \rightarrow I$$
$$I \xrightarrow{k_1} J$$
$$2I \xrightarrow{k_2} D$$
$$J + n_2e \rightarrow R$$

was discussed explicitly by Bard and Mayell and gives rise to very similar behavior.

It is very difficult to obtain an exact treatment of the case that does not involve the intervening pseudo-first-order transformation, and such a treatment may be difficult to employ because the variation of stirring efficiency from one point on the working-electrode surface to another cannot be averaged as it can when only first-order processes are involved. Rangarajan [156] has made a promising approach to the complexities of the situation, and much remains to be said in this area.

Other references: 43, 87, 110, 123–125, 127.

7 CONCLUSION

In the space available here, it has been impossible to do more than sketch the outlines of the numerous ways in which controlled-potential electrolysis can be employed in work with organic compounds. For too long there has been an artificial, and now wholly irrational, barrier between electrochemistry and organic chemistry. This is now crumbling under the assault of electrochemists who see in the organic area a rich harvest of applications for, and tests of, the techniques and theories now available. It is hoped that this review will suggest ways in which organic chemists can use these techniques and theories to advantage in work on problems confronting them, and that their success will stimulate them to join in demolishing the barrier once and for all.

References

1. S. Swann, Jr., "Electrolytic Reactions," in *Technique of Organic Chemistry*, A. Weissberger, Ed., Vol. II, Interscience, New York, 1948.
2. M. J. Allen, *Organic Electrode Processes*, Van Nostrand Reinhold, New York, 1958.
3. O. H. Müller, "Polarography," in *Physical Methods of Organic Chemistry*, Vol. I of *Technique of Organic Chemistry*, A. Weissberger and B. W. Rossiter, Eds., 4th ed., Interscience, New York, 1971.
4. L. Meites, *Polarographic Techniques*, 2nd ed., Interscience, New York, 1965.
5. The polarographic reversibility of this half-reaction has been demonstrated by O. H. Müller, *Chem. Rev.*, **24**, 95 (1939).
6. S. Glasstone, K. J. Laidler, and H. Eyring, *The Theory of Rate Processes*, McGraw-Hill, New York, 1941.
7. P. Delahay, *New Instrumental Methods in Electrochemistry*, Interscience, New York, 1954.
8. P. Zuman, *Collection Czech. Chem. Commun.*, **15**, 1107 (1950).
9. K. Schwabe, *Polarographie und Chemische Konstitution Organischer Verbindungen*, Akademie Verlag, Berlin, 1957.
10. Ref. 4, Appendix C, pp. 671–711.
11. J. Koutecký, *Collection Czech. Chem. Commun.*, **18**, 597 (1953).
12. L. Meites and Y. Israel, *J. Am. Chem. Soc.*, **83**, 4903 (1961).
13. S. Karp and L. Meites, *J. Electroanal. Chem.*, **17**, 253 (1968).
14. C. A. Streuli and W. D. Cooke, *Anal. Chem.*, **26**, 963 (1954).
15. G. Charlot, J. Badoz-Lambling, and B. Trémillon, *Electrochemical Reactions*, Elsevier, Amsterdam, 1962.
16. G. A. Rechnitz, *Controlled-Potential Analysis*, Macmillan, New York, 1963.
17. J. J. Lingane, *Electroanalytical Chemistry*, Interscience, New York, 1953, pp. 192–195.
18. L. Meites, *J. Electroanal. Chem.*, **7**, 337 (1964).
19. A. J. Bard, *Anal. Chem.*, **35**, 1125 (1963).
20. S. Karp, Ph.D. Thesis, Polytechnic Institute of Brooklyn, New York, 1967.
21. L. B. Rogers and C. Merritt, Jr., *J. Electrochem. Soc.*, **100**, 131 (1953).
22. L. Meites, unpublished experiments, 1954.
23. S. A. Moros, Ph.D. Thesis, Polytechnic Institute of Brooklyn, New York, 1961.
24. Y. Israel and L. Meites, *J. Electroanal. Chem.*, **8**, 99 (1964).

1   2   3   4   5   6   9   8   9   0

25. J. G. Jones and F. C. Anson, *Anal. Chem.*, **36**, 1137 (1964).
26. H. N. Ostensen, B.S. Thesis, Polytechnic Institute of Brooklyn, New York, 1963.
27. This is called the "working electrode." It may be either the anode or the cathode, but over the range of potentials attainable with mercury electrodes there are so many more reducible organic substances than oxidizable ones that the working electrode is generally assumed to be the cathode in this discussion.
28. This is called the "auxiliary electrode."
29. A. Hickling, *Trans. Faraday Soc.*, **38**, 27 (1942).
30. P. F. Lott, *J. Chem. Education*, **42**, A261, A361 (1965).
31. G. L. Booman and W. B. Holbrook, *Anal. Chem.*, **37**, 795 (1965); I. Shain, J. E. Harrar, and G. L. Booman, *ibid.*, **37**, 1768 (1965); D. T. Pence and G. L. Booman, *ibid.*, **38**, 1112 (1966).
32. J. E. Harrar and E. Behrin, *Anal. Chem.*, **39**, 1230 (1967).
33. H. K. Ficker and L. Meites, *Anal. Chim. Acta*, **26**, 172 (1962).
34. H. N. Ostensen and L. Meites, unpublished work, 1963.
35. R. I. Gelb and L. Meites, *J. Phys. Chem.*, **68**, 630 (1964).
36. L. Meites, *Pure Appl. Chem.*, **18**, 35 (1969).
37. A. C. Testa and W. H. Reinmuth, *J. Am. Chem. Soc.*, **83**, 784 (1961).
38. G. S. Alberts and I. Shain, *Anal. Chem.*, **35**, 1859 (1963).
39. R. S. Nicholson and I. Shain, *Anal. Chem.*, **37**, 190 (1965).
39a. F. B. Stephens, F. Jakob, L. P. Rigdon, and J. E. Harrar, *Anal. Chem.*, **42**, 764 (1970).
40. J. J. Lingane, *Anal. Chim. Acta*, **2**, 592 (1948).
41. H. Diehl, *Electrochemical Analysis with Graded Cathode Potential Control*, G. F. Smith Chemical Co., Columbus, Ohio, 1948.
42. J. J. Lingane and S. L. Jones, *Anal. Chem.*, **23**, 1804 (1951).
43. S. Karp and L. Meites, *J. Am. Chem. Soc.*, **84**, 906 (1962).
44. L. Meites, *Anal. Chem.*, **27**, 1116 (1955).
45. L. Meites and S. A. Moros, *Anal. Chem.*, **31**, 23 (1959).
46. L. Meites, unpublished results.
47. G. L. Booman and W. B. Holbrook, *Anal. Chem.*, **35**, 1793 (1963).
48. J. E. Harrar and I. Shain, *Anal. Chem.*, **38**, 1148 (1966).
49. H. C. Jones, W. D. Shults, and J. M. Dale, *Anal. Chem.*, **37**, 680 (1965).
50. E. J. Majeski, J. D. Stuart, and W. E. Ohnesorge, *J. Am. Chem. Soc.*, **90**, 633 (1968).
51. Ref. 4, p. 83.
52. J. Heyrovský and P. Zuman, *Practical Polarography*, Academic Press, New York, 1968, p. 8.
53. J. J. Lingane and L. A. Small, *J. Am. Chem. Soc.*, **71**, 973 (1949).
54. H. A. Catherino and L. Meites, *Anal. Chim. Acta*, **23**, 57 (1960).
55. P. Zuman, *The Elucidation of Organic Electrode Processes*, Vol. 2 of *Current Chemical Concepts*, L. Meites, Ed., The Polytechnic Press, Brooklyn, New York, 1969.
56. D. A. Berman, P. R. Saunders, and R. J. Winzler, *Anal. Chem.*, **23**, 1040 (1951).

57. P. W. Carr and L. Meites, *J. Electroanal. Chem.*, **12**, 373 (1966).
58. J. V. A. Novak, *Collection Czech. Chem. Commun.*, **20**, 1076, 1090 (1955).
59. W. J. Blaedel and J. H. Strohl, *Anal. Chem.*, **36**, 445 (1964).
60. R. E. Cover, Polytechnic Institute of Brooklyn, New York, unpublished experiments, 1960.
61. B. Kastening, *Collection Czech. Chem. Commun.*, **30**, 4033 (1965).
62. D. M. King and A. J. Bard, *Anal. Chem.*, **36**, 2351 (1964); *J. Am. Chem. Soc.*, **87**, 419 (1965).
63. R. N. Adams, *J. Electroanal. Chem.*, **8**, 151 (1964).
64. D. H. Geske, in *Progress in Physical Organic Chemistry*, A. Streitwieser, Jr., and R. W. Taft, Eds., Vol. 4, Interscience, New York, 1967, p. 125.
65. S. Wawzonek and D. Wearring, *J. Am. Chem. Soc.*, **81**, 2067 (1959).
66. S. Wawzonek and A. Gundersen, *J. Electrochem. Soc.*, **107**, 537 (1960).
67. S. Wawzonek and A. Gundersen, *J. Electrochem. Soc.*, **111**, 324 (1964).
68. S. Wawzonek, R. Berkey, E. W. Blaha, and M. E. Runner, *J. Electrochem. Soc.*, **103**, 456 (1956).
69. J. W. Collat and J. J. Lingane, *J. Am. Chem. Soc.*, **76**, 4214 (1954).
70. J. M. Kruse, *Anal. Chem.*, **31**, 1854 (1959).
71. R. C. Buchta and D. H. Evans, *Anal. Chem.*, **40**, 2181 (1968).
72. R. S. Rodgers and L. Meites, unpublished experiments, 1968.
73. L. Meites, *Anal. Chem.*, **27**, 416 (1955).
74. L. Meites, *Anal. Chim. Acta*, **20**, 456 (1959).
75. R. D. DeMars and I. Shain, *Anal. Chem.*, **29**, 1825 (1957).
76. L. Meites, *Anal. Chem.*, **27**, 977 (1955).
77. Ref. 17, pp. 430–432.
78. P. D. Garn and E. W. Halline, *Anal. Chem.*, **27**, 1563 (1955).
79. P. Zuman, *Collection Czech. Chem. Commun.*, **15**, 839 (1950).
80. J. J. Lingane, C. G. Swain, and M. Fields, *J. Am. Chem. Soc.*, **65**, 1348 (1943).
81. H. A. Laitinen and T. J. Kneip, *J. Am. Chem. Soc.*, **78**, 736 (1956).
82. R. I. Gelb and L. Meites, *J. Phys. Chem.*, **68**, 2599 (1964).
83. G. L. Booman, W. B. Holbrook, and J. E. Rein, *Anal. Chem.*, **29**, 219 (1957).
84. L. Meites, *Anal. Chim. Acta*, **18**, 364 (1958).
85. M. R. Lindbeck and H. Freund, *Anal. Chem.*, **37**, 1647 (1965).
86. M. T. Kelley, W. L. Belew, G. V. Pierce, W. D. Shults, H. C. Jones, and D. J. Fisher, *Microchem. J.*, **10**, 315 (1966).
87. J. J. Lingane, *J. Am. Chem. Soc.*, **67**, 1916 (1945).
88. J. J. Lingane, *Anal. Chim. Acta*, **2**, 584 (1948).
89. J. A. Page and J. J. Lingane, *Anal. Chim. Acta*, **16**, 175 (1957).
90. J. J. Banewicz, G. R. Argue, and R. F. Stewart, U.S. Patent 3,275,903, Sept. 27, 1966.
91. J. J. Lingane and S. L. Jones, *Anal. Chem.*, **22**, 1220 (1957).
92. J. J. Lingane, *Anal. Chim. Acta*, **18**, 349 (1958).
93. J. J. Lingane, *Anal. Chem. Acta*, **44**, 199 (1969).
94. G. L. Booman, *Anal. Chem.*, **29**, 213 (1957).
95. M. T. Kelley, H. C. Jones, and D. J. Fisher, *Anal. Chem.*, **31**, 488, 956 (1959); *Talanta*, **6**, 185 (1960); cf. also Ref. 49.

96. R. Ammann and J. Desbarres, *Bull. Soc. Chim. France*, 1012 (1962); *J. Electroanal. Chem.*, **4**, 121 (1962).

97. E. N. Wise, *Anal. Chem.*, **34**, 1181 (1962).

98. A. J. Bard and E. Solon, *Anal. Chem.*, **34**, 1181 (1962).

99. G. A. Rechnitz and K. Srinivasan, *Anal. Chem.*, **36**, 2417 (1964).

100. G. C. Goode and J. Herrington, *Anal. Chim. Acta*, **38**, 369 (1967).

101. C. L. Perrin, "Mechanisms of Organic Polarography," in *Progress in Physical Organic Chemistry*, S. G. Cohen, A. Streitwieser, and R. W. Taft, Eds., Vol. 3, Interscience, New York, 1965, pp. 165–316.

102. N. I. Stenina, Yu. A. Krylov, and P. K. Agasyan, *Zh. Anilit. Khim.*, **21**, 1319 (1966).

103. W. M. MacNevin and B. B. Baker, *Anal. Chem.*, **24**, 986 (1952).

104. L. Meites, *Anal. Chem.*, **31**, 1285 (1959).

105. S. Hanamura, *Talanta*, **2**, 278 (1959); **3**, 14 (1960).

106. A. J. Bard and S. V. Tatwawadi, *J. Phys. Chem.*, **68**, 2676 (1964).

107. F. Kaufman, H. Cook, and S. Davis, *J. Am. Chem. Soc.*, **74**, 4997 (1952).

108. R. F. Dapo and C. K. Mann, *Anal. Chem.*, **35**, 677 (1963).

109. I. Bergman and J. C. James, *Trans. Faraday Soc.*, **50**, 60 (1954).

110. L. Meites and T. Meites, *Anal. Chem.*, **28**, 103 (1956).

111. P. J. Elving and C. E. Bennett, *J. Am. Chem. Soc.*, **76**, 1412 (1954).

112. H. Fukami and M. Nakajima, *Botyu-Kagaku*, **18**, 6 (1953); *Chem. Abstr.*, **47**, 5815*h* (1953).

113. H. Fukami, H. Kimura, and M. Nakajima, *Botyu-Kagaku*, **18**, 51 (1953); *Chem. Abstr.*, **47**, 9180*g* (1953).

114. T. Meites and L. Meites, *Anal. Chem.*, **27**, 1531 (1955).

115. V. B. Ehlers and J. W. Sease, *Anal. Chem.*, **31**, 16 (1959).

116. F. H. Merkle and C. A. Discher, *Anal. Chem.*, **36**, 1639 (1964).

117. K. S. V. Santhanam and V. R. Krishnan, *Z. Anal. Chem.*, **206**, 33 (1964).

118. K. S. V. Santhanam and V. R. Krishnan, *Indian J. Chem.*, **4**, 104 (1966).

119. K. S. V. Santhanam and V. R. Krishnan, *Z. Anal. Chem.*, **234**, 256 (1968).

120. A. A. Podolenko, E. G. Chikrysova, and Y. S. Lyalikov, *Ukr. Khim. Zh.*, **31**, 844 (1965).

121. M. R. Lindbeck and H. Freund, *Anal. Chim. Acta*, **35**, 74 (1966).

122. J. F. O'Donnell and C. K. Mann, *J. Electroanal. Chem.*, **13**, 163 (1967).

123. P. J. Elving and J. T. Leone, *J. Am. Chem. Soc.*, **80**, 1021 (1958).

124. M. J. Allen, *J. Am. Chem. Soc.*, **72**, 3797 (1950); **73**, 3503 (1951); *J. Org. Chem.*, **15**, 435 (1950); *Proc. Intern. Comm. Electrochem. Thermodynam. Kinet.*, VI Meeting, 481 (1955).

125. M. J. Allen and A. H. Corwin, *J. Am. Chem. Soc.*, **72**, 114, 117 (1950); H. A. Levine and M. J. Allen, *J. Chem. Soc.*, **254** (1952); M. J. Allen, J. E. Fearn, and H. A. Levine, *ibid.*, 2220 (1952).

126. A. J. Bard and J. S. Mayell, *J. Phys. Chem.*, **66**, 2173 (1962).

127. J. S. Mayell and A. J. Bard, *J. Am. Chem. Soc.*, **85**, 421 (1963).

128. G. L. Booman and W. B. Holbrook, *Anal. Chem.*, **35**, 1986 (1963).

129. M. D. Morris and J. J. Lingane, *J. Electroanal. Chem.*, **8**, 85 (1964).

130. S. Szpak, P. Stonehart, and T. Katan, *Electrochim. Acta*, **10**, 563 (1965).

131. B. P. Nesterov and N. V. Korovin, *Elektrokhimiya*, **2,** 1296 (1966).
132. G. R. Rao and L. Meites, *J. Phys. Chem.*, **70,** 3620 (1966).
133. S. A. Moros and L. Meites, *J. Electroanal. Chem.*, **5,** 103 (1963).
134. J. T. Lundquist, Jr., and R. S. Nicholson, *J. Electroanal. Chem.*, **16,** 445 (1968).
135. J. Badoz-Lambling, *J. Electroanal. Chem.*, **1,** 44 (1959/60).
136. G. A. Rechnitz and H. A. Laitinen, *Anal. Chem.*, **33,** 1473 (1961).
137. G. A. Rechnitz and J. E. McClure, *Talanta*, **10,** 417 (1963); **12,** 153 (1965).
138. G. A. Rechnitz and J. E. McClure, *Anal. Chem.*, **36,** 2265 (1964).
139. J. A. Page and E. J. Zinser, *Talanta*, **12,** 1051 (1965).
140. D. H. Geske and A. J. Bard, *J. Phys. Chem.*, **63,** 1057 (1959).
141. L. Meites, unpublished results, 1955.
142. M. Spritzer and L. Meites, *Anal. Chim. Acta*, **26,** 58 (1962).
143. J. G. McCullough and L. Meites, *J. Electroanal. Chem.*, **19,** 111 (1968).
144. A. J. Bard and E. Solon, *J. Phys. Chem.*, **67,** 2326 (1963).
145. R. S. Rodgers and L. Meites, *J. Phys. Chem.*, **73,** 4348 (1969).
146. D. H. Geske, *J. Phys. Chem.*, **63,** 1062 (1959).
147. J. G. McCullough, Ph.D. Thesis, Polytechnic Institute of Brooklyn, New York, 1967.
148. R. C. Buchta and D. H. Evans, *Anal. Chem.*, **40,** 2181 (1968).
149. H. Lund, *Acta Chem. Scand.*, **13,** 249 (1959).
150. J. G. Mason, *J. Electroanal. Chem.*, **11,** 462 (1966).
151. J. E. Harrar and L. P. Rigdon, *Anal. Chem.*, **41,** 758 (1969).
152. R. S. Rodgers and L. Meites, unpublished experiments, 1969.
153. G. Costa, P. Rozzo, and P. Batti, *Ann. Chim. (Rome)*, **45,** 387 (1955).
154. G. Costa and A. Puxeddu, *Ricerca Sci.*, **27,** 894 (1957).
155. L. Meites, *J. Electroanal. Chem.*, **5,** 270 (1963).
156. S. K. Rangarajan, personal communication, 1968.

SUBJECT INDEX